Proceedings in Life Sciences

IUBS Section of Comparative Physiology and Biochemistry
1st International Congress, Liège, Belgium, August 27–31, 1984

Conference Organization

Organizing Board
R. Gilles, Chairman, Liège, Belgium
M. Gilles-Baillien and L. Bolis, Liège, Belgium/Messina, Italy.

Host Society
European Society for Comparative Physiology
and Biochemistry.

Under the Patronage of
The European Economic Community
The Fonds National de la Recherche Scientifique
The Ministére de l'Education Nationale et de la Culture Française
The Fondation Léon Fredericq
The University of Liège

The European Society for Comparative Physiology and Biochemistry
The American Society of Zoologists
The Canadian Society of Zoologists
The Japanese Society for General and Comparative Physiology

The Congress has been organized in relation with the 100th Anniversary of the School of Comparative Physiology and Biochemistry of the University of Liège.

The proceedings of the invited lectures to the different symposia of the congress have been gathered in five different volumes published by Springer-Verlag under the following titles:

Circulation, Respiration, and Metabolism
Current Comparative Approaches
Edited by R. Gilles (ISBN 3-540-15627-5)

Transport Processes, Iono- and Osmoregulation
Current Comparative Approaches
Edited by R. Gilles and M. Gilles-Baillien (ISBN 3-540-15628-3)

Neurobiology, Current Comparative Approaches
Edited by R. Gilles and J. Balthazart (ISBN 3-540-15480-9)

Respiratory Pigments in Animals, Relation Structure-Function
Edited by J. Lamy, J.-P. Truchot, and R. Gilles (ISBN 3-540-15629-1)

High Pressure Effects on Selected Biological Systems
Edited by A. J. R. Péqueux and R. Gilles (ISBN 3-540-15630-5)

Professor Dr. RAYMOND GILLES
Laboratory of Animal Physiology
University of Liège
22, Quai Van Beneden
B-4020 Liège, Belgium

Cover illustration:
The β-adrenergic pathway.
From B. Cannon, this Volume, page 502, Figure 3.

ISBN-13: 978-3-642-70612-7 e-ISBN-13: 978-3-642-70610-3
DOI: 10.1007/978-3-642-70610-3

Library of Congress Cataloging-in-Publication Data. Main entry under title: Circulation, respiration, and metabolism. (Proceedings in Life Sciences) "The proceedings of the invited lectures to the First International Congress of Comparative Physiology and Biochemistry ... at Liège (Belgium) in August 1984 under the auspices of the Section of Comparative Physiology and Biochemistry of the International Union of Biological Sciences" – P. 1. Exercise – Physiological aspects – Congresses. 2. Physiology, Comparative – Congresses. I. Gilles, R. II. International Congress of Comparative Physiology and Biochemistry (1st: 1984: Liège, Belgium). III. International Union of Biological Sciences. Section of Comparative Physiology and Biochemistry. IV. Series. QP301.C5845 1985 591.1 85-22057

2131/3130-543210

Circulation, Respiration, and Metabolism

Current Comparative Approaches

Edited by R. Gilles

With 190 Figures

Springer-Verlag
Berlin Heidelberg New York Tokyo

Foreword

This volume is one of those published from the proceedings of the invited lectures to the First International Congress of Comparative Physiology and Biochemistry I organized at Liège (Belgium) in August 1984 under the auspices of the Section of Comparative Physiology and Biochemistry of the International Union of Biological Sciences. In a general foreword to these different volumes, it seems to me appropriate to consider briefly what may be the comparative approach.

Living organisms, beyond the diversity of their morphological forms, have evolved a widespread range of basic solutions to cope with the different problems, both organismal and environmental with which they are faced. Soon after the turn of the century, some biologists realized that these solutions can be best comprehended in the framework of a comparative approach integrating results of physiological and biochemical studies done at the organismic, cellular and molecular levels. The development of this approach amongst both physiologists and biochemists remained, however, extremely slow until recently. Physiology and biochemistry have indeed long been mainly devoted to the service of medicine, finding scope enough for their activities in the study of a few species, particularly mammals. This has tended to keep many physiologists and biochemists from the comparative approach, which demands either the widest possible survey of animals forms or an integrated knowledge of the specific adaptive features of the species considered. These particular characteristics of the comparative approach have, on the other hand, been very attractive for biologists interested in the mechanisms of evolution and environmental adaptations. This diversity of requirements of the comparative approach, at the conceptual as well as at the technological level, can easily account for the fact that it emerged only slowly amongst the other new, more rapidly growing, disciplines of the biological sciences. Although a few pioneers have been working in the field since the beginning of the century, it only started effectively in the early 1960's. 1960 was the date of the organization of the periodical *Comparative Physiology and Biochemistry* by Kerkut and Scheer and of the publication of the first volumes of the comprehensive treatise *Comparative Biochemistry* edited by Florkin and Mason. These publications can be considered as milestones in the evolution of the comparative approach. They have been followed by many others which have greatly contributed to giving the field the international status it deserved. Since the 1960's, the comparative approach has been maturing and developing more and more rapidly into the independent discipline it now is, widely recognized by the international communities of physiologists, biochemists, and biologists. It is currently used as an effective tool of great help in the understanding of many research problems: biological as well as clinical, applied as well as fundamental.

The actual development of the field and the interest it arouses in a growing portion of the biological scientific community led some of us to consider the organization of an international structure, bringing together the major representative societies and groups around the world, which would aim at the general advancement and promotion of the comparative approach. This was done in 1979 with the incorporation, within the international Union of Biological Sciences, of a Section of Comparative Physiology and Biochemistry. The first International Congress of CPB I organized in Liège with the help of a few friends and colleagues, is the first activity of this newly founded Section. In 22 symposia it gathered some 146 invited lectures given by internationally renowned scientists on all major current topics and trends in the field. The proceedings of these lectures have been collected in 5 volumes produced by Springer-Verlag, a publisher long associated with the development of CPB. The organization of the CPB Section of IUBS, its first Congress and these proceedings volumes can well be considered as milestones reflecting the international status and the maturity that the comparative approach has gained, as a recognized independent discipline, in the beginning of the 1980's, some 20 years after it was effectively launched.

Finally, I would like to consider that the selection of Liège for this first International Congress has not been simply coincidental. I thus feel that this brief foreword would not be complete without noting the privileged role Liège has played in some events associated with the development of the comparative approach. Liège had a pioneer in comparative physiology already at the end of the last century with Léon Fredericq. With Marcel Florkin, Liège had its first Professor of biochemistry and one of the founding fathers of comparative biochemistry. These two major figureheads of the comparative approach founded and developed what is actually called the Liège School of Comparative Physiology and Biochemistry, which was, at the time of the Congress, celebrating its 100th anniversary. This school provided early support to the European Society for Comparative Physiology and Biochemistry organized by Marcel Florkin and myself some years ago. The society, still headquartered in Liège, was, with the CPB division of the American Society of Zoologists, at the origin of the formation of the CPB Section of IUBS under the auspices of which this first International Congress, specifically devoted to the comparative approach has been organized. An essential particularity of the Liège school of CPB is that its two founding fathers, scientists interested in general, basic aspects of the organization of living organisms, were also professors at the faculty of medicine. This largely contributed in Liège to avoiding the undesirable structuration of a so-called "zoophysiology" or "zoobiochemistry" independent of the rest of the field. The conditions were thus realized very early in Liège for CPB to play its key role in canalizing the necessary interactions between the general, pre-clinical or clinical and the environmental, ecological or evolutionary tendences of physiology and biochemistry. The possibility of stimulating such interactions has served as a major guide line in the selection of the symposia and invited lectures from which these proceedings have issued.

Liège, Belgium, June 1985 R. GILLES

Preface

Three points of view, or themes, run through this volume of the proceedings of the first congress of the Section of Comparative Physiology and Biochemistry of the International Union of Biological Sciences. On the one hand, as biochemists and physiologists, the contributors are particularly interested in principles of function (at various levels of organization, spanning the range from molecules to whole organisms) which are universally applicable to living systems. The only way to assess the universality of biochemical or physiological functions, of course, is to probe and analyze them across broad sweeps of phylogeny. Thus a second theme running through this volume explores how specific biochemical and physiological functions are put to use in different organisms, or in similar organisms living in different environmental conditions. Not only does this approach assist in identifying truly universal properties of physiology and biochemistry, it also helps to explain the immense diversity of Nature that necessarily and continuously confronts (and sometimes seduces) the comparative biologist. A third theme in this volume, as a kind of blend of the first two and perhaps best characterizing the disciplines of comparative biochemistry and physiology, is *the use of organisms as an experimental parameter per se.* The use of species-specific properties of organisms as experimental parameters in their own right for better illuminating underlying mechanisms and unifying principles is a time-honored research strategy in comparative biochemistry and physiology, going back to the origins of these disciplines. The contributions in this volume beautifully illustrate that this research strategy is as effective today as it was in August Krogh's time and in the subsequent heady days of early comparative biochemistry and physiology. The volume should therefore stand as an important milestone in the field, both in reviewing what has been done and in bringing focus on what should be done next.

P.W. HOCHACHKA

Vancouver, Canada, April 1985

Contents

List of Contributors

You will find the addresses at the beginning of the respective contributions

Symposium I

The Physiology of Exercise: Comparative Approaches

Organizer P.J. BUTLER

Respiratory, Circulatory, and Metabolic Adjustments to Exercise in Fish

C.M. WOOD[1] and S.F. PERRY[2]

1 Oxygen

The increase in muscular work and metabolic rate associated with exercise necessitates both elevated O_2 uptake by the gills ($\dot{M}O_2$) and enhanced O_2 delivery to the tissues. Both of these processes can be considered limiting factors in determining overall exercise performance. The present discussion primarily focuses on the various factors affecting the transfer of O_2 across the gill during exercise. Sections 3 and 4 deal with O_2 delivery to the tissues.

During sustained exercise $\dot{M}O_2$ can increase 12–15 times above the resting rate. Much of this increase can be attributed simply to increased bulk transfer of O_2 as a result of elevated cardiac output and gill ventilation. However, other factors, including changes in gill O_2 diffusive conductance, also contribute to the rise in $\dot{M}O_2$ and maintenance of arterial blood oxygen tensions, especially at higher swimming speeds when blood transit time through the gill vasculature is drastically reduced.

The movement of O_2 across the gill respiratory epithelium can be described by the equation:

$$\dot{M}O_2 = KO_2 \times \frac{A \times \Delta PO_2}{E} \tag{1}$$

where KO_2 = the O_2 permeation coefficient (related to the capacitance and permeability of the respiratory surface to O_2), A = the functional surface area of the gill, E = the thickness of the diffusion barrier and ΔPO_2 = the mean O_2 partial pressure gradient between blood and water ($1/2(P_IO_2 + P_EO_2) - 1/2(P_aO_2 + P_vO_2)$ is a reasonable approximation where P_IO_2 and P_EO_2 = inspired and expired O_2 tensions and P_aO_2 and P_vO_2 = arterial and venous O_2 tensions). A rearrangement of Eq. (1) yields an expression for gill O_2 diffusion conductance (GO_2):

$$GO_2 = \frac{\dot{M}O_2}{\Delta PO_2} = KO_2 \times \frac{A}{E}\,. \tag{2}$$

Thus, changes in GO_2 due to modifications of KO_2, A, and E as well as changes in ΔPO_2 will affect the overall transfer of O_2 across the gill during exercise. Three factors

1 Department of Biology, McMaster University, Hamilton, Ontario, Canada L8S 4K1
2 Department of Biology, University of Ottawa, Ottawa, Ontario, Canada K1N 6N5

Circulation, Respiration, and Metabolism
(ed. by R. Gilles)

known to affect some or all of the above variables are catecholamines, gill perfusion ($\dot{Q}$) and gill ventilation ($\dot{V}_g$).

1.1 Catecholamines

At rest, plasma catecholamines are probably close to 10^{-9} *M*, but may increase two to three orders of magnitude during and after severe exercise (Mazeaud and Mazeaud 1981; Opdyke et al. 1982; Primmett, Mazeaud, Randall, and Boutilier, pers. comm.). Recent experiments using perfused gill preparations (Pettersson and Johansen 1982; Pettersson 1983; Perry et al. 1984a) have confirmed speculation from earlier in vivo studies (Steen and Kruysse 1964; Peyraud-Waitzenegger 1979) that elevated levels of adrenaline (AD) stimulate $\dot{M}O_2$ by increasing GO_2 (Fig. 1). The observations that AD increases P_aO_2 in vivo and in perfused preparations suggests that fish gills are diffusion limited with respect to O_2 transfer. If such were the case, elevated $\dot{Q}$ would impose serious transit time limitations on gill O_2 diffusion during exercise and impair blood oxygenation in the absence of diffusive conductance changes. However P_aO_2 remains virtually unchanged over a wide range of sustained swimming speeds (Kiceniuk and Jones 1977; Boutilier et al. 1984), and may even increase after exhaustive exercise (Holeton et al. 1983; Jensen et al. 1983) even though gill transit time can be as low as 1 s during exercise compared to approx. 3 s under resting conditions (Randall and Daxboeck

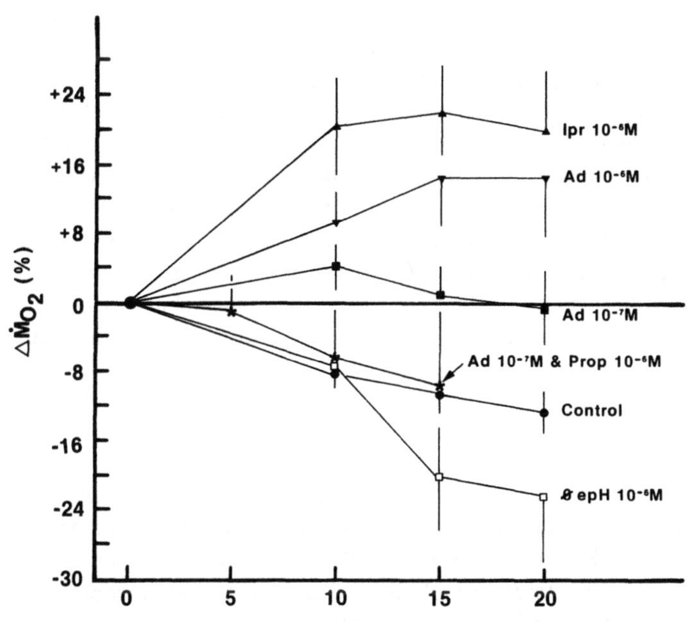

Fig. 1. The influence of various adrenergic agonists (*Ad* adrenaline; *Ipr* isoprenaline; *ϕeph* phenylephrine) and the β-adrenoceptor antagonist propranolol (*Prop*) on percentage changes in oxygen uptake ($\Delta\dot{M}_{O_2}$ %) over time in the isolated saline-perfused trout head preparation (redrawn from Perry et al. 1984a)

1982). Indeed, gill transit time limitations on $\dot{M}O_2$ have been reported by Perry et al. (1984a) by demonstrating that P_aO_2 decreased significantly as $\dot{Q}$ was elevated to exercise levels in a perfused trout head preparation. Moreover, AD elicited a greater stimulatory effect on $\dot{M}O_2$ at increased $\dot{Q}$, presumably as a result of greater initial diffusion limitations. Thus, we speculate that during exercise, gill O_2 diffusion limitations increase as a consequence of reduced gill transit time but are offset by an enhancement of O_2 diffusive conductance such that P_aO_2 remains constant. Elevation of circulating AD levels may be an important factor contributing to this response. It should be pointing out that Daxboeck et al. (1982) observed no correlation between $\dot{Q}$ and P_aO_2 in a spontaneously ventilating blood-perfused trout preparation. Lack of diffusion limitations in this preparation may have been due to abnormally high levels of circulating catecholamines caused by stress since experiments were conducted approx. 2 h following extensive surgery.

Two mechanisms have been postulated to explain the stimulatory effects of AD on GO_2. Firstly, AD is known to enhance gill functional surface area via recruitment of distal lamellae (Holbert et al. 1979; Booth 1979) as a result of β-receptor mediated dilation of afferent lamellar arterioles. Interestingly, Pettersson (1983) observed stimulation of $\dot{M}O_2$ in the perfused cod head following α-adrenergic stimulation. Since dorsal aortic flow was unaffected but afferent (input) pressure increased, he suggested the possibility of lamellar recruitment via α-receptor mediated constriction of efferent lamellar arterioles. Similar haemodynamic changes were reported in the trout head following α-adrenergic stimulation although $\dot{M}O_2$ was not stimulated (Perry et al. 1984a; Fig. 1). Additionally, α-constriction of efferent arterio-venous anastomoses (AVA's) may be important for overall O_2 transport by allowing a greater percentage of oxygenated blood to enter the dorsal aorta rather than the arterio-venous circulation. Secondly, AD may increase the permeability of the gill epithelium to O_2 (and hence KO_2) as has been proposed for small lipophilic and water soluble substances (Haywood et al. 1977). To date, no direct evidence to support this hypothesis exists. Although the gill vasculature is innervated by adrenergic fibres (Laurent and Dunel 1980), the significance of the autonomic nervous system in mediating gill surface area changes is unclear. According to Nilsson (1983), the most important control of the branchial arterio-arterial circulation is probably by circulating catecholamines released from chromaffin tissue in the head kidney.

1.2 Gill Perfusion

During exercise, cardiac output increases three- to fivefold primarily as a result of increased stroke volume (Jones and Randall 1978; Randall 1982) although cardiac frequency may also be elevated (Kiceniuk and Jones 1977). The mechanisms involved are complex and may include increased stimulation of cardiac β-adrenoreceptors by direct neural innervation or elevated levels of circulating catecholamines, decreased vagal cholinergic tone to the heart and increased venous return (Wood and Shelton 1980a; Nilsson 1983; Laurent et al. 1984; Farrell 1984; Randall and Daxboeck 1984). The increased $\dot{Q}$ at the onset of exercise is accompanied by elevated mean ventral (Pva) and dorsal (Pda) aortic pressures and pulse pressure (ΔP). Generally, pressures peak following 10–15 min of sustained swimming and then return to a constant level, only slightly above the resting value (Jones and Randall 1978).

The results of studies by Farrell et al. (1979, 1980) and Davie and Daxboeck (1982) indicate that the increase in pulsatility and Pva observed at the onset of exercise may be extremely important in enhancing gill O_2 diffusive conductance through changes in gill surface area, diffusion distance and epithelial permeability.

1.2.1 Surface Area and Diffusion Distance

At rest ~60% of lamellae are perfused (Booth 1978) but during exercise this may approach 100% due to recruitment of distal lamellae, not only as a result of active dilation of afferent lamellar arterioles (Sect. 1.1), but also as a result of passive changes caused by increased pressures and pulsatility. This is because afferent lamellar arterioles become narrower toward the distal end of filaments and therefore offer more resistance to flow (higher critical opening pressures) than proximal lamellae, a factor which can be overcome by elevated pressures due to the compliant nature of these vessels (Farrell et al. 1979). Although the increases in Pva and ΔP during exercise are transient, it is likely that the distal lamellae remain perfused throughout the exercise period since the critical opening pressure almost certainly exceeds the critical closing pressure by several torr (Randall and Daxboeck 1984). Increased Pva and ΔP not only induce lamellar recruitment but may also increase functional surface area of the gill by promoting a more even distribution of blood flow within individual lamellae as the thickness of the blood sheet increases (Farrell et al. 1980). The central and distal regions of lamellae are more compliant than the portion buried within the filament and show the largest increase in sheet thickness. Therefore $\dot{Q}$ will increase to the greatest extent in those portions of the lamellae optimized for gas exchange (i.e. exposed to water flow with minimum diffusion distance) and the overall water-to-blood diffusion barrier will be reduced (Randall and Daxboeck 1984). Furthermore, the pillar cell flanges, the epithelium, the interstitial space or a combination of all these may be thinned to accommodate the increased vascular volume associated with thickening of the blood sheet (Farrell et al. 1980).

1.2.2 Epithelial Permeability

Davie and Daxboeck (1982) have demonstrated that increased pulsatility enhances the effective permeability of the gill epithelium to ethanol. Whether or not O_2 permeability is affected in a similar manner remains unclear at this time although it would not be unexpected given that both are lipid soluble molecules. Increased fluid exchange between tissue and gill vascular compartments induced by pulsatility (Davie and Daxboeck 1982) may also be an important mechanism in preventing or reducing branchial oedema during exercise when Pva is elevated.

1.3 Gill Ventilation

Gill ventilation volume increases during exercise as a result of small increases in breathing rate and large changes in stroke volume (Jones and Randall 1978). The causes of the hyperventilation are unclear but may include feedback from body proprioceptors and central controls as in mammals, β-adrenergic effects of circulating catecholamines

(Peyraud-Waitzenegger 1979), and changes in blood acid-base status (see Sects. 3.2 and 5.2.1). While ventilation is very sensitive to decreases in arterial O_2 content (Smith and Jones 1982), there is no evidence that this occurs during exercise.

Elevated ventilation may enhance gill GO_2 by reducing unstirred or boundary layers next to lamellae. Evidence for reduced diffusion distance or alternatively, distribution dead space, is the fact that percentage utilization of O_2 from water flowing over the gills normally does not decrease at elevated $\dot{V}_g$ even though water gill residence time is drastically reduced from about 250 ms at rest to about 30 ms at high rates of exercise (Randall 1982).

1.4 ΔPO_2

During exercise ΔPO_2 increases largely as a result of lowered P_vO_2 although there is still some speculation that at very high swimming speeds, $P_IO_2-P_EO_2$ may increase as a result of reduced percentage utilization of O_2. However, direct measurements of P_IO_2 and P_EO_2 revealed no changes in rainbow trout at swimming speeds up to 100% U crit (Kiceniuk and Jones 1977). The reduction in P_vO_2 is due to increased O_2 consumption by working skeletal muscle (Randall and Daxboeck 1982) but may also involve greater O_2 utilization by cardiac muscle cells as the work of the heart increases.

2 Carbon Dioxide

Far less is known about CO_2 exchange, but most of the factors mentioned in Sect. 1 which enhance O_2 transport between water and blood during exercise should simultaneously favour CO_2 flux in the opposite direction. Only a few additional points will be addressed here.

2.1 CO_2 Excretion

Because of the difficulty of CO_2 determination in water, there have been relatively few measurements of $\dot{M}_{CO_2}$ in fish, and even less during exercise (e.g. Kutty 1968, 1972; van den Thillart et al. 1983). Unfortunately some of these are based on titration methodology and must be interpreted with caution because of possibly changing contributions of NH_3 and NH_4^+ excretion (Wright and Wood 1985) under different physiological states. To avoid these problems, many authors have simply predicted $\dot{M}_{CO_2}$ from measurements of $\dot{M}_{O_2}$ and assumptions about R.Q. This is again problematical, because CO_2 stores within the body are much larger than O_2 stores, and much more labile, especially to acid-base status (see Sect. 5). The instantaneous gas exchange ratio (R_E) varies considerably from the true metabolic R.Q. (Burggren and Cameron 1980).

In trout and goldfish, but not in tilapia, R_E was elevated at the start of sustained swimming but attenuated as exercise duration increased (Kutty 1968, 1972). This probably reflected an initial metabolic acidosis (Sect. 5.2.2) associated with excitement, but CO_2 generation by unusual metabolic pathways cannot be ruled out (Hochachka and Mommsen 1983). R_E during sustained exercise was relatively stable at ~1.0 (goldfish), ~0.85 (trout) or ~1.2 (tilapia). Recently, van den Thillart et al. (1983) have reported stable R_E's of ~0.65 during sustained swimming at 80% U crit in coho salmon. This value did not change during a 6 h period after exhaustive exercise, though the temporal resolution of the experiment would probably not have detected short-term fluctuations in $\dot{M}_{CO_2}$. R_E fell to 0.25 at unchanged $\dot{M}_{O_2}$ when the fish swam for 6 h at low pH, a result which implies profound consequences for internal acid-base status. While this was not assessed by van den Thillart et al., Graham et al. (1982) found no evidence of such internal CO_2 retention in trout exercised at an even lower environmental pH. Clearly the available information is fragmentary, and there is a need for further studies on $\dot{M}_{CO_2}$ and R_E during and after both sustained and exhaustive exercise, coupled with blood acid-base measurements.

2.2 CO_2 Movement Across the Gills

CO_2 excretion in fish undoubtedly follows the standard vertebrate scheme mediated by erythrocytic carbonic anhydrase (ĊA) and Cl^-/HCO_3^- exchange at the RBC membrane (Cameron 1979; Wood et al. 1982; Heming and Randall 1982; Perry et al. 1982). At the gill lamellae, plasma HCO_3^- is converted to molecular CO_2 within the RBC, and the diffusion of this CO_2 from the blood to the water along a P_{CO_2} gradient accounts for the bulk of CO_2 excretion. A small fraction may be reconverted to HCO_3^- by epithelial CA to participate in branchial Cl^-/HCO_3^- exchange for acid-base regulation (Perry et al. 1985; see Sect. 6.2).

During exercise, increased CO_2 production by working muscle considerably elevates Pv_{CO_2} (Stevens and Randall 1967; Wood et al. 1977) because the CO_2 dissociation curve of the blood is relatively flat in the physiological range. Theory suggests that the gills are already tremendously hyperventilated with respect to CO_2 excretion at rest (Cameron 1979), but experimental evidence shows that a fall in P_{aCO_2} may accompany increased $\dot{V}_g$ (Eddy 1974; Soivio et al. 1981). In any event, the $\dot{V}_g/\dot{Q}$ ratio increases markedly during exercise (Kiceniuk and Jones 1977). Gill transit time is reduced as $\dot{Q}$ increases, but the large amounts of CA in the blood (Heming 1984) are probably still well in excess of requirements. Randall and Daxboeck (1984) note that CO_2 excretion was perfusion-limited in the spontaneously ventilating blood-perfused trout, though the cautionary note sounded earlier about the preparation (Sect. 1.1) must be kept in mind. All these factors, together with the many others mentioned earlier which enhance branchial GO_2 (and GCO_2), should help drive increased CO_2 efflux during exercise. However this is not to say that the situation is simply the mirror image of that for O_2 flux. The theory of Haswell and Randall (1978), proposing RBC impermeability to HCO_3^-, was initially attractive as it offered a mechanism to modulate CO_2 excretion independently of O_2 uptake. Perry et al. (1982) concluded that the rate-limiting step in CO_2 excretion is HCO_3^- entry into the RBC. Factors which affect this step might independently modulate CO_2 flux; in Sect. 3.2 we argue that plasma catecholamines may be one such factor.

3 Blood Gas Transport

3.1 Haemoconcentration

During exercise, haematocrit (Hct), haemoglobin (Hb), and plasma protein levels all increase, and as a result the O_2- and CO_2-carrying, and buffer capacities of the blood are enhanced. Contributing factors include recruitment of RBC's from the spleen (Yamamato et al. 1980; Daxboeck, pers. comm.; see Fig. 2), RBC swelling, and losses of plasma water due to both diuresis (Wood and Randall 1973) and osmotic redistribution into muscle as a consequence of elevated intracellular lactate levels (Milligan and Wood, unpubl.). Elevated catecholamine levels may promote splenic contraction via α-adrenergic stimulation (see Nilsson 1983) and RBC swelling via β-adrenergic stimulation (Nikinmaa 1981, 1982). Moreover Nikinmaa (1981) has shown that injections of AD into intact trout cause plasma skimming at the gills to increase, though Olson (1984) could find no consistent effect of AD on plasma skimming in blood-perfused holobranchs.

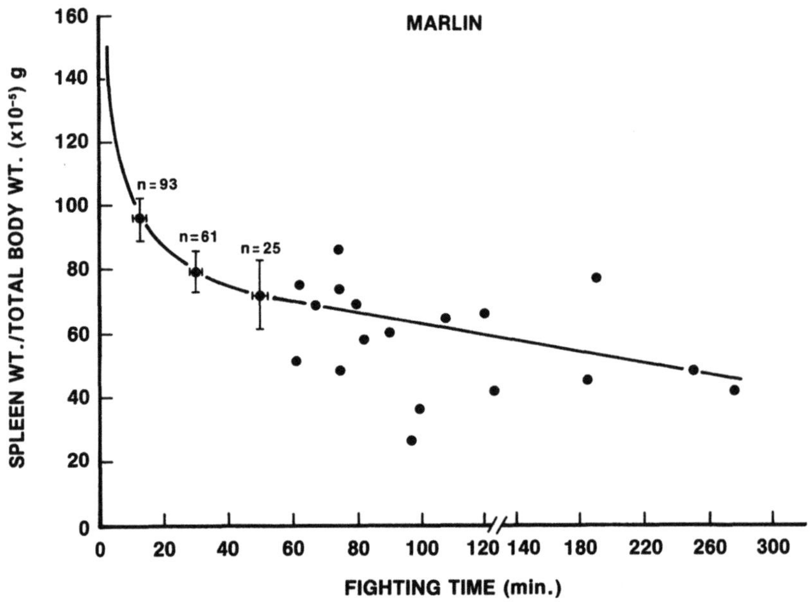

Fig. 2. Changes in relative spleen weight (as an index of splenic contraction) vs fighting time for Pacific blue marlin *(Makaira nigricans)* angled in a sport fishery (unpublished data of C. Daxboeck, by permission)

3.2 Erythrocyte Function

Until very recently, it has generally been assumed that changes in RBC intracellular pH (pHi) during exercise would parallel the decrease in plasma pH, thereby decreasing Hb O_2 affinity and augmenting O_2 delivery to muscle. However Primmett, Mazeaud,

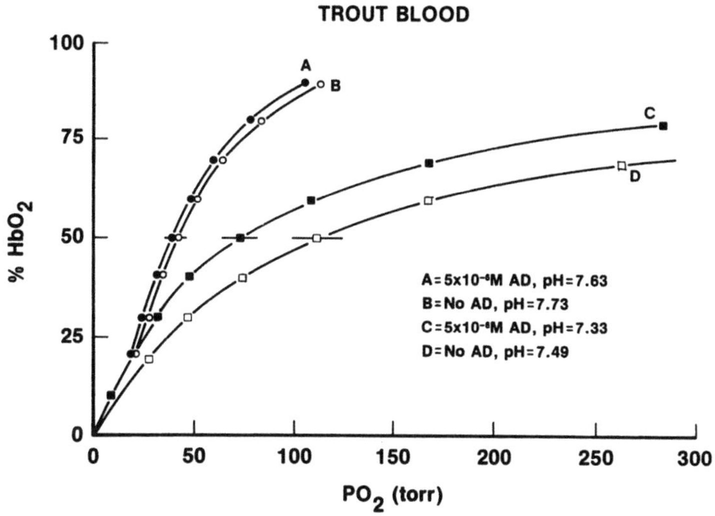

Fig. 3. The influence of adrenaline on the oxygen dissociation curves of rainbow trout erythrocyte suspensions in vitro. Note the greater effect of adrenaline at low extracellular pH (redrawn from Nikinmaa 1983)

Randall, and Boutilier (pers. comm.) have shown that after burst exercise in trout, RBC pHi is initially stable and then actually increases in the face of a marked drop in plasma pH. Catecholamine-mediated swelling of the RBC is probably responsible for this decreased proton gradient across the RBC membrane since Nikinmaa (1982) and Heming (1984) have shown that β-adrenergic stimulation raises RBC pHi in vitro. This in turn is associated with an increased Hb O_2 affinity (Fig. 3), especially at low extracellular pH, caused by the pHi change itself, the RBC swelling which dilutes the Hb and nucleoside triphosphate (NTP) levels within the RBC (Lykkeboe and Weber 1978), and possibly also a metabolic decrease in cellular NTP levels, though Jensen et al. (1983) found no change in the NTP:Hb ratio in exercised tench. The physiological significance is that O_2 loading by the blood at the gills may be maintained or actually improved during and after exercise. Interestingly, the skate, a very poor exerciser dependent mainly on anaerobic metabolism, showed no such pHi regulation and marked increases in the NTP:Hb ratio after exhaustive exercise (Wood and Graham, unpubl. data).

Catecholamines may also modulate the role of the RBC in CO_2 transport. Perry and Heming (unpubl., Fig. 4) have demonstrated that AD also inhibits HCO_3^- dehydration by trout erythrocytes in vitro in a dose-dependent fashion. This β-adrenergic effect is directly on the HCO_3^--entry step, rather than on carbonic anhydrase itself, but the consequence is a functional decrease in CA activity. Other treatments which reduce functional CA activity (e.g. acetazolamide, Haswell and Randall 1978; anaemia, Wood et al. 1982) result in substantial elevations in P_{aCO_2} and associated depressions in plasma pH – in other words, an apparent diffusive limitation of CO_2 excretion. However blood CO_2 content is only slightly elevated, so it is likely that $\dot{M}_{CO_2}$ is only transiently depressed, and returns to normal as soon as ΔP_{CO_2} across the branchial epithelium has risen sufficiently to compensate. Heming (1984) has shown that injec-

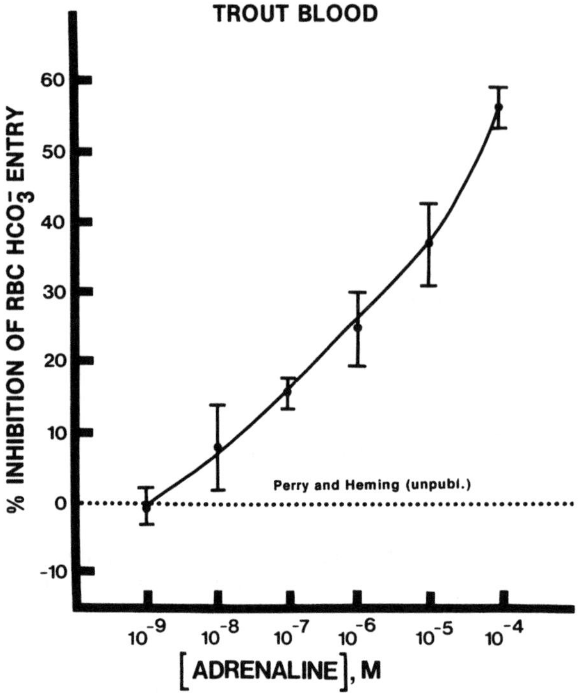

Fig. 4. The inhibitory influence of adrenaline on the rate of HCO_3^- entry into trout red blood cells in vitro determined by measurement of the rate of carbonic anhydrase catalyzed HCO_3^- dehydration (Perry and Heming, unpubl. data)

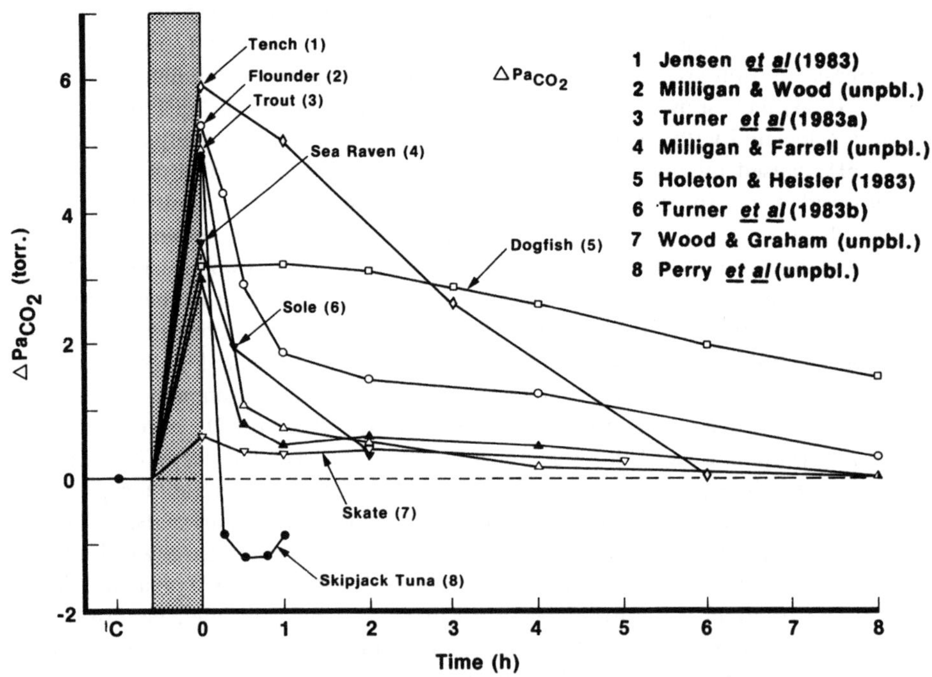

Fig. 5. Changes (ΔP_{aCO_2}) in arterial CO_2 tensions of eight fish species after exhaustive exercise (at bar)

tions of AD into intact trout causes a transient reduction in R_E at the gills. This mechanism may well operate in fish subjected to strenuous exercise. Despite the many factors which should favour CO_2 elimination (see Sect. 2.2), P_{aCO_2} increases by 1-6 torr (50%-400%) after exhaustive exercise in all species which have been examined (Fig. 5). At first glance, this appears maladaptive, but it must be remembered that this respiratory acidosis (see Sect. 5.2) may have no adverse effect on O_2 transport because of AD's simultaneous action in raising RBC pHi and O_2 affinity (Fig. 3). The blood is reset to a higher P_{CO_2} level, but Bohr and Root shifts will still occur in the standard fashion.

We speculate that this CO_2 "retention" may be adaptive for two reasons. Firstly, as suggested by Heming (1984), it may help maintain blood HCO_3^- levels in the face of metabolic acidosis; this HCO_3^- may be utilized by transmembrane exchange processes to regulate pHi in tissues of critical importance. Secondly, the rise in P_{aCO_2} and/or the associated fall in plasma pH may help maintain hyperventilation during the post-exercise period, thereby ensuring correction of the O_2 debt. Stimulation of $\dot{V}_g$ by proprioceptors responsive to muscle activity will presumably cease after exhaustion, so there is a requirement for other factors to drive hyperventilation. In the skate, experimental hypercapnia stimulates hyperventilation in resting animals under conditions where arterial O_2 content is unaffected (Graham, Wood, and Turner, unpubl. data). In contrast, Smith and Jones (1982) showed that moderate hypercapnia stimulated $\dot{V}_g$ by an indirect effect on arterial O_2 content in trout.

4 Redistribution of Blood Flow

4.1 Systemic Blood Flow

Cardiac output is elevated during exercise (Sect. 1.2) and a greater percentage of oxygenated blood may be channeled into the systemic circuit by constriction of gill AVA's (Sect. 1.1). The relative proportions flowing to "red" muscle, "white (or mosaic)" muscle, and the kidney increases, whereas flow to the liver, stomach, spleen and the rest of the body decreases. Interestingly, the flow distribution is heavily skewed towards red muscle during steady state "aerobic" exercise (Randall and Daxboeck 1982), but to white muscle after exhaustive "anaerobic" exercise (Neumann et al. 1983). Increased perfusion of working muscles is essential to meet increased metabolic requirements, while increased renal perfusion elevates urine flow rate to compensate for increased water influx across the gills (Wood and Randall 1973). The mechanisms involved in blood flow redistribution are not fully understood. Systemic vascular resistance appears to be reflexly controlled via sympathetic vasomotor tone and sensitive to passive dilation induced by elevations of Pda (Wood and Shelton 1980b). Overall, the α-constrictory effects of catecholamines dominate over their β-dilatory influence in the systemic vasculature, but total systemic resistance falls during exercise (Jones and Randall 1978). Increased flow to red muscle fibres may be due to $\beta 2$-vasodilatory receptors (Davie 1981), local metabolic influences may vasodilate both red and white muscle (Neumann et al. 1983), while muscle contraction itself and movements of the tail may also contribute (Randall and Daxboeck 1982). Adrenergic nerves and/or circulating catecholamines may induce constriction of the sympathetically innervated spleen and muscle sphincter at the base of the coeliacomesenteric artery, thereby reducing overall splanchnic flow (Randall and Daxboeck 1982).

4.2 Coronary Blood Flow

A coronary circulation, derived from efferent branchial arteries, is present in a great variety of active fish and delivers oxygenated blood to the outer myocardial layer of the heart (Farrell 1984). During rest, it is probable that the O_2 requirements of the heart can be met by venous blood O_2 content but during exercise when cardiac power output increases and venous blood O_2 content falls, the importance of the coronary circulation may increase, especially in large, active fish with thick ventricular walls. However, Daxboeck (1982) found that the swimming performance of rainbow trout was unaffected by coronary artery ablation. These experiments were complicated by the fact that delivery of arterial blood to the heart may have persisted since a collateral blood supply developed around the ablation site during the 7 day recovery from surgery. Recently, Davie and Daxboeck (1984) have shown that the perfused coronary vascular bed of Pacific blue marlin in vitro undergoes vasodilation in response to AD, an effect attributable to $\beta 1$-receptors. Certainly the role of the coronary circulation during exercise is an area that warrants further investigation.

5 Acid-Base Status

5.1 Sustained Exercise

During steady state exercise, muscle metabolism is largely aerobic, supported by the red ("slow") muscle fibres which constitute a small percentage of the total muscle mass (Bone 1978). The mainly glycolytic white ("fast") fibres, which form the bulk of the myotome, are probably involved to a minor but significant extent, especially at higher cruising speeds. While muscle lactate (La^-) levels may be elevated two- to three-fold, it seems likely that the rate of La^- oxidation eventually comes into equilibrium with the rate of La^- production (Black et al. 1962; Wokoma and Johnston 1981; Duthie 1982). There is no information on muscle acid-base status under these conditions, but blood acid-base status and La^- levels are largely unchanged from the resting state (Stevens and Randall 1967; Driedzic and Kiceniuk 1976; Kiceniuk and Jones 1977, van den Thillart et al. 1983; Boutilier et al. 1984). Thus CO_2 excretion across the gills keeps pace with elevated CO_2 generation in the tissues, and any metabolic acidosis which occurs must be confined to the intracellular compartment. Van den Thillart et al. (1983) reported a net H^+ output into the water by salmon during sustained swimming, but as ammonia excretion was not taken into account, the significance of this observation is unclear.

5.2 Exhaustive Exercise

The picture is very different after exhaustive burst swimming in which the white muscle predominates (Bone 1978). A profound acidosis (0.3–0.7 pH units) is seen in both arterial and venous blood, and blood La^- levels are elevated for up to 24 h. This acidosis can be resolved into separate "respiratory" and "metabolic" components (Wood et al. 1977).

5.2.1 Respiratory Acidosis

P_{aCO_2} increases by 1-6 torr (50%-400%) immediately after exhaustive exercise in all species (Fig. 5), and changes in venous blood are even larger (Wood et al. 1977; Turner et al. 1983b; Milligan and Wood, unpubl.). The resulting carbonic acid titrates only non-HCO_3^- buffers (Hb, plasma protein). In the absence of complicating metabolic acidosis, decreases in plasma pH and increases in plasma HCO_3^- levels are simply a function of the non-HCO_3^- buffer value (β). β values are higher in excellent swimmers such as tuna and marlin (Perry and Daxboeck, unpubl. data). It is not strictly legitimate to compare respiratory and metabolic acid levels in blood in terms of concentrations, because the latter are buffered by HCO_3^- as well as non-HCO_3^- buffers. However the relative contributions to pH change can be assessed (Wood et al. 1977) and when this is done, the respiratory component is seen to account for 45%-95% of the immediate post-exercise depression in pHa in the species shown in Fig. 5. The percentage tends to highest in relatively poor exercisers such as flatfish. The P_{aCO_2} elevation is generally corrected by 1-2 h post-exercise (Fig. 5), and from this point on metabolic acidosis predominates in all fish.

P_{aO_2} is generally unchanged or elevated by both sustained and exhaustive exercise (Sect. 1.1), but the P_{aCO_2} elevation is seen only in the latter condition. Certainly CO_2 production rates may be increased above aerobic levels due to titration of HCO_3^- stores by metabolically produced acids, but it is difficult to believe that this alone could elevate blood P_{aCO_2} levels for more than a few minutes. We speculate that the major difference between steady state and exhaustive exercise is the degree of catecholamine mobilization. In trout, plasma catecholamines do not change during mild exercise (2 BL s^{-1}; Laurent et al. 1982) but increase greatly after exhaustive exercise (Primmett, Mazeaud, Randall, and Boutilier, pers. comm.), then decline over a time course rather similar to that observed by other workers for P_{aCO_2} (Turner et al. 1983a; Holeton et al. 1983). The P_{aCO_2} elevation would therefore arise mainly from AD-mediated inhibition of HCO_3^--flux and dehydration through the RBC (Fig. 4), with possible benefits for intracellular acid-base regulation and ventilatory control, as outlined earlier (Sect. 3.2). Of course, P_{CO_2} would thereby be elevated in the tissues throughout the body, but it must be remembered that intracellular β is much greater than blood β. Indeed, pHi changes in critical tissues such as heart and brain appear to be very small after exhaustive exercise (Milligan and Farrell, pers. comm.; Milligan and Wood, unpubl. data).

5.2.2 Metabolic Acidosis

Protons produced by anaerobic metabolism in white muscle enter the bloodstream and depress plasma pH by titration of both non-HCO_3^- and HCO_3^- buffers. The latter explains the fall in blood CO_2 content which is usually seen at some point after exhaustive exercise. The blood "metabolic acid load" (ΔH^+m) peaks at 4-10 mEq. l^{-1} by the time of

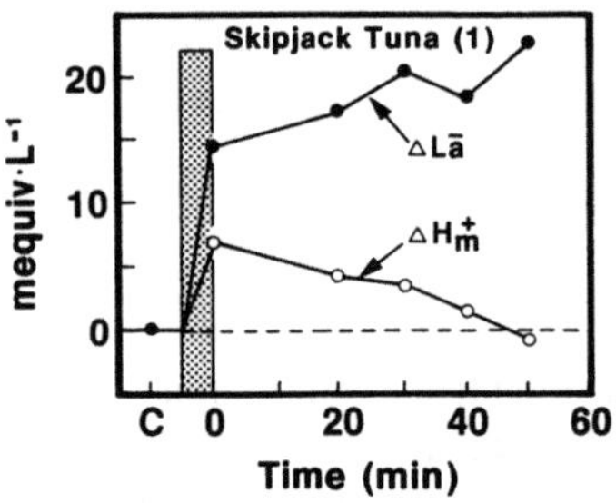

Fig. 6. Changes in the concentration of lactate anion (ΔLa^-) and "metabolic acid" (ΔH^+_m) in the arterial blood of seven fish species after exhaustive exercise (at bar). Note the different patterns in tuna, trout, and dogfish (active pelagic swimmers) vs flounder, sole, skate, and sea raven (benthic animals). Note that the time scale for tuna is in minutes

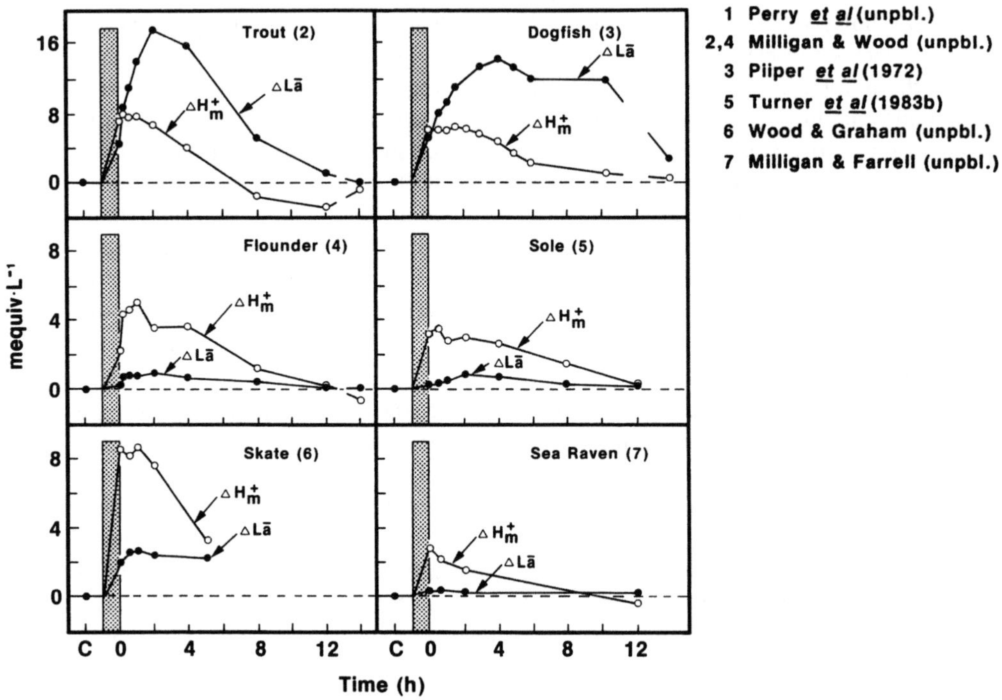

exhaustion, or shortly thereafter, and persists for up to 12 h (Fig. 6). Lactate anions also enter the blood, but ΔLa^- may follow a very different pattern from ΔH^+m and invariably peaks at a later time. There appear to be two basic patterns (Fig. 6). In relatively poor exercisers with low aerobic and anaerobic scope, such as the benthic flounder, sole, skate, and sea raven, ΔH^+m greatly exceeds ΔLa^-. Indeed the latter generally remains below 2 mEq. l^{-1}. However in good pelagic swimmers such as trout, dogfish, and tuna capable of both high aerobic metabolism and high anaerobic La^- production, ΔLa^- greatly exceeds ΔH^+m and persists after correction of the latter. The tuna is an extreme example; here ΔH^+m is corrected within 50 min, at which time ΔLa^- is 22 mEq. l^{-1} and still rising (Perry and Daxboeck, unpubl. data). The physiological explanation relates to differential transmembrane fluxes (see Sect. 6), but the adaptive significance reflects the animals' lifestyles. We speculate that the strategy of the active fish is to minimize proton loading into the blood from the high intramuscular acid load in order to minimize interference with blood O_2 transport, as the latter is necessary to support continued performance of aerobic red muscle. It is essential that a pelagic fish be able to keep swimming. The La^- released into the blood may be used as a fuel by critical organs such as heart, gills, and red muscle (Driedzic and Hochachka 1978). Conversely, the strategy of benthic sluggish fish may be to minimize La^- release from muscle so that the majority can be resynthesized to glycogen in situ (Batty and Wardle 1979). Proton release into the blood (and from there to the environment) may be tolerated so that white muscle pHi returns rapidly to a level compatible with glyconeogenesis. Here immobility and camouflage are the principal defence strategies, so any detrimental effects on blood O_2 transport would be less important.

6 Transmembrane Lactate and Proton Fluxes

6.1 The Intracellular Compartment

Muscle energy stores (creatine phosphate, ATP, glycogen) are depleted after exhaustive exercise, and intracellular La^- in white muscle may increase to over 50 mEq. l^{-1} (Black et al. 1962; Wardle 1978; Driedzic and Hochachka 1978). As Hochachka and Mommsen (1983) have elegantly pointed out, metabolic acidosis results not from the glycolytic production of lactic acid per se, but rather from the hydrolysis of ATP, which when coupled with anaerobic glycosis, results in the equimolar production of protons and La^-. A corollary is that any net degradation of ATP stores (but not creatine phosphate) will also result in equimolar H^+ production. Thus we can no longer assume, as many workers have done, that ΔLa^- is equivalent to ΔH^+m overall, let alone in any specific body compartment. A further caveat is that other unusual end products associated with H^+ production (e.g. succinate; Smith and Heath 1980) may be produced in fish tissues. While there is no direct evidence that they occur after exercise, Wood et al. (1983) have reported the appearance of an unknown anion in the plasma of exercised trout.

Clearly the safest approach is to measure ΔH^+m in the tissues. Recently, Milligan and Wood (1985) have validated the ^{14}C-DMO technique for detecting pHi transients in white muscle in vitro, and have applied the technique to the post-exercise situation in vivo in trout and flounder (unpubl. data) together with measurements in the extracellular and environmental compartments (Fig. 7). These data indicate that ΔH^+m exceeds ΔLa^- in white muscle immediately after exercise, a difference at least partially attributable to the destruction of ATP stores. The bulk of the whole body ΔH^+m and ΔLa^- load is retained within the white muscle for apparent metabolism in situ, and even in trout, blood ΔLa^- never equilibrates with muscle ΔLa^-. Intracellular acid-base status is corrected (with overshoot) prior to extracellular in the flounder, while the reverse is true in the trout, which is in accord with the arguments presented in Sect. 5.2.2.

The magnitude of ΔH^+m and ΔLa^- in the muscle is so much greater than in the ECF that it is impractical to employ intracellular data to try to explain the differential ΔH^+m vs ΔLa^- patterns seen in the blood (Sect. 5.2.2; Fig. 6). However in trout, the blood ΔH^+m vs ΔLa^- discrepancy is fully developed before transfers to the environmental compartment commence (Sect. 6.2; Fig. 7), and does not occur after the injection of lactic acid into the bloodstream of resting animals (Turner et al. 1983a). Turner and Wood (1983) have shown in a perfused trout trunk preparation that ΔLa^- release

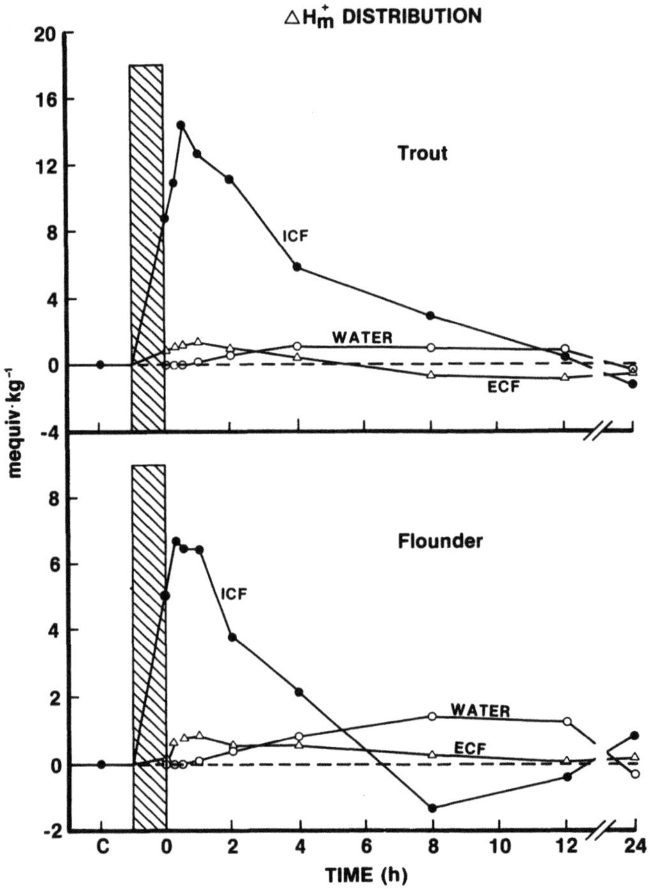

Fig. 7. The measured distribution of "metabolic acid" (ΔH^+_m) between the intracellular compartment, the extracellular compartment, and the environmental water at various times after exhaustive exercise (at bar) in the rainbow trout and the starry flounder (Milligan and Wood, unpubl. data)

equals ΔH^+m release when the perfusate has a typical resting acid-base status, but that ΔLa^- efflux is depressed relative to ΔH^+m efflux when the perfusate is acidotic. The difference in efflux rates is sufficient to explain the ΔH^+m vs ΔLa^- discrepancy seen in the blood. Thus La^- and H^+ transfer appear to be dissociated processes; H^+ efflux may be passive and subject to "equilibrium limitation" (Holeton and Heisler 1983), while La^- flux may be subject to carrier mediation (Turner and Wood 1983; Hochachka and Mommsen 1983). The measurements of Neumann et al. (1983) showing increased blood flow to white muscle after exhaustive exercise indicate that perfusion limitation is not the cause of La^- retention. Wardle (1978) has provided evidence for a β-adrenoreceptor mediated inhibition of La^- release from white muscle into blood in exercised plaice. The role of catecholamines in regulating ICF/ECF transfers clearly warrants further attention.

6.2 The Environment

Net H^+ exchange with the environment can be measured in a closed system as the sum of the titration alkalinity (or "ΔHCO_3^-") and ammonia fluxes, signs considered. The data of Milligan and Wood (Fig. 7) on trout and flounder, of Holeton et al. (1983) on trout and of Holeton and Heisler (1983) on dogfish are consistent in showing a transient "storage" of ΔH^+m in the external medium after exhaustive exercise. The actual amount transferred was a small fraction of the total body load (Fig. 7). However despite comparable post-exercise blood acid-base status, Holeton and Heisler's fish transferred about four times as much ΔH^+m, perhaps because the electric shocks used to induce exercise were more stressful than simple chasing. In all studies, the transfer occurred mainly across the gills, for the renal component was very small, and La^- movement into the environment was negligible. In the trout, the environmental "storage" appeared to be responsible for the correction and overshoot of the extracellular ΔH^+m load, while in the flounder it reflected the intracellular overshoot (Fig. 7). Later ΔH^+m uptake from the water occurred in both species to allow final acid-base homeostasis. Branchial H^+m fluxes are therefore apparently set at rates which maintain the ΔH^+m vs ΔLa^- discrepancies in the blood (Fig. 6) originally established by differential release from the myotome.

At least in freshwater fish, Na^+ vs acidic equivalent (H^+, NH_4^+) and Cl^- vs basic equivalent (HCO_3^-, OH^-) exchanges at the gills can be dynamically adjusted to correct internal acid-base status (Cameron 1976; Wood et al. 1984), and similar mechanisms are thought to occur in seawater fish. In the perfused trout head preparation, Na^+ influx is stimulated via β-adrenoceptors and Cl^- influx via α-adrenoceptors; for each, the opposite adrenoceptors may be inhibitory (Payan and Girard 1978; Perry et al. 1984b). We therefore speculate that the plasma catecholamines surge after exhaustive exercise, in which AD initially dominates over NAD (Primmett, Mazeaud, Randall, and Boutilier, pers. comm.) might initially stimulate net H^+m efflux into the water through the β-adrenergic effect, and that a later α-adrenergic effect might be responsible for the net H^+m recovery from the water which allows final metabolic homeostasis (Fig. 7).

In support of this idea, Perry (unpubl. data, Table 1) has found that infusions of AD in resting trout in vivo indeed do stimulate branchial Na^+ influx and inhibit branchial

Table 1. The effects of adrenaline (AD) infusions on unidirectional branchial sodium influx ($J^{in}_{Na^+}$) and branchial chloride influx ($J^{in}_{Cl^-}$) in resting rainbow trout in fresh water (means ± SEM; Perry, unpubl. data)

	$J^{in}_{Na^+}$ (n = 6) (μEq. kg^{-1} h^{-1})	$J^{in}_{Cl^-}$ (n = 6) (μEq. kg^{-1} h^{-1})
Pre-AD[a]	306.7 ± 16.9	261.5 ± 20.5
AD[a]	385.3 ± 23.7[b]	159.4 ± 27.8[b]
Post-AD[a]	240.4 ± 30.8	214.1 ± 25.5

[a] Fluxes were measured over 2 h periods during which l-adrenaline (saline alone in pre-AD and post-AD periods) was infused into the dorsal aorta at a rate which produced an average measured level of AD in plasma of 3×10^{-7} *M*

[b] Significantly different ($p < 0.05$) from both pre-AD and post-AD values by paired Student's t-test

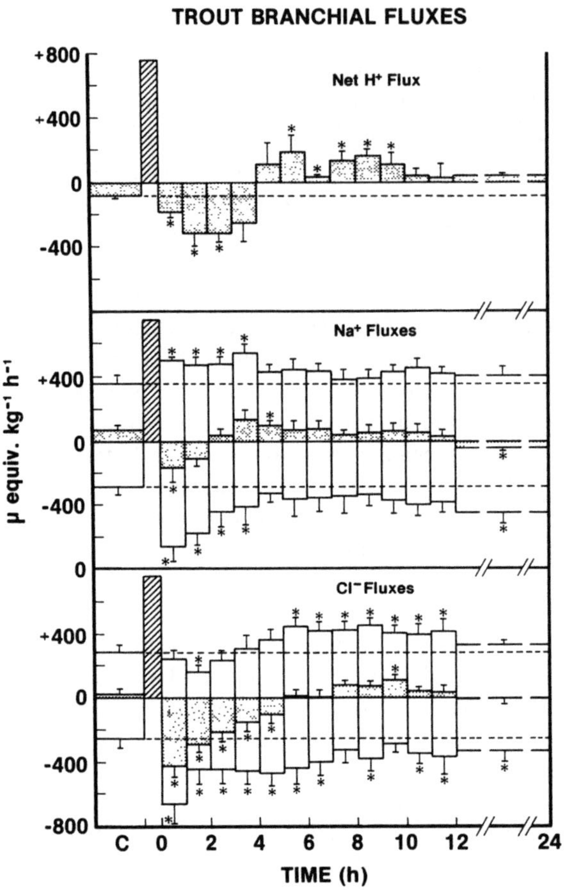

Fig. 8. Changes in branchial net H^+ flux (i.e. the sum of titratable alkalinity and ammonia flux, signs considered) and unidirectional and net Na^+ and Cl^- fluxes in rainbow trout after exhaustive exercise (at bar). Influx = *upward bars*; efflux = *downward bars*; net flux = *stippled bars*. * = Significantly different ($p < 0.05$) from appropriate control value (Wood, unpubl. data)

Cl^- influx. Furthermore Wood (unpubl. data; Fig. 8) has measured net branchial H^+m flux and unidirectional Na^+ and Cl^- fluxes in trout after exhaustive exercise. As predicted, branchial Na^+ influx was stimulated and Cl^- influx was inhibited during the first few hours post-exhaustion, during which time net H^+m excretion occurred. Thereafter, the situation was reversed, correlating with net H^+m recovery from the water. Changes in Cl^- influx were more marked than those in Na^+ influx, suggesting that modulation of $Cl^-/HCO_3^-, OH^-$ exchange is more important than modulation of Na^+/H^+, NH_4^+ exchange in dynamic acid-base regulation (see also Wood et al. 1984). However the situation was complicated by increases in the efflux components for both ions which did not always mirror the changes in influx. These may have reflected changes in the extent of simple diffusive efflux, exchange diffusion, and/or a reversal of coupled acid-base/ion exchange pumps. In any event, it is the difference between net Na^+ and Cl^- flux rates, no matter whether due to influx or efflux modulation, which should theoretically correlate with net H^+m flux (Wood et al. 1984). Both the data of Fig. 8 and those of Holeton et al. (1983) indicate that this is indeed the case.

7 Conclusion

A recurrent theme throughout this review has been the critical involvement of catecholamines in stimulating and integrating the many compensatory responses to exercise. With very few exceptions, the evidence presented here has been circumstantial rather than direct, because plasma catecholamine levels have only rarely been monitored during experimental treatments. Modern HPLC and radioenzymatic methods simplify the measurement of the low levels of AD and NAD in fish plasma. We hope that future studies on exercise in fish will incorporate plasma catecholamine measurements as an integral part of their design.

Acknowledgements. We are grateful to D.J. Randall, M. Mazeaud, B. Boutilier, D. Primmett, C. Daxboeck, P. Davie, M.S. Graham, and C.L. Milligan for permission to quote their unpublished data. Original research reported here was supported by NSERC grants to C.M. Wood and S.F. Perry.

References

Batty RS, Wardle CS (1979) Restoration of glycogen from lactic acid in the anaerobic swimming muscle of plaice, *Pleuronectes platessa* L. J Fish Biol 15:509–519

Black EC, Robertson-Connor A, Lam KC, Chiu WG (1962) Changes in glycogen, pyruvate and lactate in rainbow trout *(Salmo gairdneri)* during and following muscular activity. J Fish Res Board Can 19:409–436

Booth JH (1978) The distribution of blood flow in gills of fish: application of a new technique to rainbow trout *(Salmo gairdneri)*. J Exp Biol 73:119–129

Booth JH (1979) The effects of oxygen supply, epinephrine and acetylcholine on the distribution of blood flow in trout gills. J Exp Biol 83:31–39

Bone Q (1978) Locomotor muscle. In: Hoar WS, Randall DJ (eds) Fish physiology, vol VII. Academic, New York, pp 361–424

Boutilier RG, Aughton P, Shelton G (1984) O_2 and CO_2 transport in relation to ventilation in the Atlantic mackerel, *Scomber scombrus*. Can J Zool 62:546–554

Burggren WW, Cameron JN (1980) Anaerobic metabolism, gas exchange, and acid-base balance during hypoxic exposure in the channel catfish, *Ictalurus punctatus*. J Exp Zool 213:405–416

Cameron JN (1976) Branchial ion uptake in the Arctic grayling: resting values and effects of acid-base disturbance. J Exp Biol 65:511–515

Cameron JN (1979) Excretion of CO_2 in water-breathing animals – a short review. Mar Biol Lett 1:3–13

Davie PS (1981) Vascular resistance responses of an eel tail preparation: alpha constriction and beta dilation. J Exp Biol 90:65–84

Davie PS, Daxboeck C (1982) Effect of pulse pressure on fluid exchange between blood and tissues in trout gills. Can J Zool 60:1000–1006

Davie PS, Daxboeck C (1984) Anatomy and adrenergic pharmacology of the coronary vascular bed of Pacific blue marlin *(Makaira nigricans)*. Can J Zool 62:1886–1888

Daxboeck C (1982) Effect of coronary artery ablation on exercise performance in *Salmo gairdneri*. Can J Zool 60:375–381

Daxboeck C, Davie PS, Perry SF, Randall DJ (1982) The spontaneously ventilating blood-perfused trout preparation: oxygen transfer across the gills. J Exp Biol 101:35–45

Driedzic WR, Hochachka PW (1978) Metabolism in fish during exercise. In: Hoar WS, Randall DJ (eds) Fish physiology, vol VII. Academic, New York, pp 503–543

Driedzic WR, Kiceniuk JW (1976) Blood lactate levels in free-swimming rainbow trout *(Salmo gairdneri)* before and after strenuous exercise resulting in fatigue. J Fish Res Board Can 33: 173–176

Duthie GG (1982) The respiratory metabolism of temperature-adapted flatfish at rest and during swimming activity and the use of anaerobic metabolism at moderate swimming speeds. J Exp Biol 97:359–373

Eddy FB (1974) Blood gases of the tench *(Tinca tinca)* in well aerated and oxygen-deficient waters. J Exp Biol 60:71–83

Farrell AP (1984) A review of cardiac performance in the teleost heart: intrinsic and humoral regulation. Can J Zool 62:523–536

Farrell AP, Daxboeck C, Randall DJ (1979) The effect of input pressure and flow on the pattern and resistance to flow in the isolated perfused gill of a teleost fish. J Comp Physiol 133B:233–240

Farrell AP, Sobin SS, Randall DJ, Crosby S (1980) Intralamellar blood flow patterns in fish gills. Am J Physiol 239:428–436

Graham MS, Wood CM, Turner JD (1982) The physiological responses of the rainbow trout to strenuous exercise: interactions of water hardness and environmental acidity. Can J Zool 60: 3153–3164

Haswell MS, Randall DJ (1978) The pattern of carbon dioxide excretion in the rainbow trout *Salmo gairdneri*. J Exp Biol 72:17–24

Haywood GP, Isaia J, Maetz J (1977) Epinephrine effects on branchial water and urea flux in rainbow trout. Am J Physiol 232:110–115

Heming TA (1984) The role of fish erythrocytes in transport and excretion of carbon dioxide. Ph.D. Thesis, University of British Columbia

Heming TA, Randall DJ (1982) Fish erythrocytes are bicarbonate permeable: problems with determining carbonic anhydrase activity using the modified boat technique. J Exp Zool 219:125–128

Hochachka PW, Mommsen TP (1983) Protons and anaerobiosis. Science 219:1391–1397

Holbert PW, Boland CJ, Olson KR (1979) The effect of epinephrine and acetylcholine on the distribution of red cells within the gills of the channel catfish *(Ictalurus punctatus)*. J Exp Biol 79: 135–146

Holeton GF, Heisler N (1983) Contribution of net ion transfer mechanisms to acid-base regulation after exhausting activity in the larger spotted dogfish *(Scyliorhinus stellaris)*. J Exp Biol 103: 31–46

Holeton GF, Neumann P, Heisler N (1983) Branchial ion exchange and acid-base regulation after strenuous exercise in rainbow trout *(Salmo gairdneri)*. Respir Physiol 51:303–318

Jensen FB, Nikinmaa M, Weber RE (1983) Effects of exercise stress on acid-base and respiratory function in blood of the teleost *Tinca tinca*. Respir Physiol 51:291–301

Jones DR, Randall DJ (1978) The respiratory and circulatory systems during exercise. In: Hoar WS, Randall DJ (eds) Fish physiology, vol VIII. Academic, New York, pp 425–501

Kiceniuk JW, Jones DR (1977) The oxygen transport system in trout *(Salmo gairdneri)* during sustained exercise. J Exp Biol 69:247–260

Kutty MN (1968) Respiratory quotients in goldfish and rainbow trout. J Fish Res Board Can 25: 1689–1728

Kutty MN (1972) Respiratory quotient and ammonia excretion in *Tilapia mossambica*. Mar Biol 16:126–133

Laurent P, Dunel S (1980) Morphology of gill epithelia in fish. Am J Physiol 238:147–159

Laurent P, Ristori M, Eclancher B (1982) Blood borne catecholamines in *Salmo gairdneri*. Are they involved in physiology? Abstract Symposium. Gas exchange, gas transport, and acid-base regulation in lower vertebrates. Max-Planck Institut für Experimentelle Medizin, Göttingen, FRG, Aug 1982

Laurent P, Holmgren S, Nilsson S (1984) Nervous and humoral control of the fish heart: structure and function. Comp Biochem Physiol 76A:525–542

Lykkeboe G, Weber RE (1978) Changes in the respiratory properties of the blood in the carp, *Cyprinus carpio*, induced by diurnal variation in ambient oxygen tension. J Comp Physiol 128: 117–125

Mazeaud MM, Mazeaud F (1981) Adrenergic responses to stress in fish. In: Pickering AD (ed) Stress and fish. Academic, London, pp 49–75

Milligan CL, Wood CM (1985) Intracellular pH transients in rainbow trout tissues measured by dimethadione distribution. Am J Physiol 248:R668–R673

Neumann P, Holeton GF, Heisler N (1983) Cardiac output and regional blood flow in gills and muscles after exhaustive exercise in rainbow trout *(Salmo gairdneri)*. J Exp Biol 195:1–14

Nikinmaa M (1981) Respiratory adjustments of rainbow trout (*Salmo gairdneri* Richardson) to changes in environmental temprature and oxygen availability. Ph.D. Thesis, University of Helsinki

Nikinmaa M (1982) Effects of adrenaline on red cell volume and concentration gradients of protons across the red cell membrane in the rainbow trout, *Salmo gairdneri*. Mol Physiol 2:287–297

Nikinmaa M (1983) Adrenergic regulation of haemoglobin oxygen affinity in rainbow trout red cells. J Comp Physiol B 152:67–72

Nilsson S (1983) Autonomic nerve function in the vertebrates. Zoophysiology, vol 13. Springer, Berlin Heidelberg New York

Olson KR (1984) Distribution of flow and plasma skimming in isolated perfused gills of three teleosts. J Exp Biol 109:97–108

Opdyke DF, Carroll RG, Keller NE (1982) Catecholamine release and blood pressure changes induced by exercise in dogfish. Am J Physiol 242:306–310

Payan M, Girard JP (1978) Mise en évidence d'un échange Na^+/NH_4^+ dans la branchies de la truite adaptée à l'eau de mer: contrôle adrénergique. CR Acad Sci Paris Ser D 286:333–338

Perry SF, Davie PS, Daxboeck C, Randall DJ (1982) A comparison of CO_2 excretion in a spontaneously ventilating blood-perfused trout preparation and saline perfused gill preparations: contribution of the branchial epithelium and red blood cell. J Exp Biol 101:47–60

Perry SF, Daxboeck C, Dobson GP (1984a) The effect of perfusion flow rate and adrenergic stimulation on oxygen transfer in the isolated saline-perfused head of rainbow trout *(Salmo gairdneri)*. J Exp Biol (in press)

Perry SF, Payan P, Girard JP (1984b) Adrenergic control of branchial chloride transport in the isolated perfused head of the freshwater trout *(Salmo gairdneri)*. J Comp Physiol B 154:269–274

Perry SF, Payan P, Girard JP (1985) The effects of perfusate HCO_3^- and P_{CO_2} on chloride uptake in perfused gills of rainbow trout *(Salmo gairdneri)*. Can J Fish Aquat Sci 41:1768–1773

Pettersson K (1983) Adrenergic control of oxygen transfer in perfused gills of the cod, *Gadus morhua*. J Exp Biol 102:327–335

Pettersson K, Johansen K (1982) Hypoxic vasoconstriction and the effects of adrenaline on gas exchange efficiency in fish gills. J Exp Biol 97:263–272

Peyraud-Waitzenegger M (1979) Simultaneous modifications of ventilation and arterial PO_2 by catecholamines in the eel, *Anguilla anguilla*: Participation of alpha and beta effects. J Comp Physiol 129:343–354

Piiper J, Meyer M, Drees F (1972) Hydrogen ion balance in the elasmobranch *Scyliorhinus stellaris* after exhausting activity. Respiration Physiology 16:290–303

Randall DJ (1982) The control of respiration and circulation in fish during exercise and hypoxia. J Exp Biol 100:275–288

Randall DJ, Daxboeck C (1982) Cardiovascular changes in the rainbow trout *(Salmo gairdneri)* during exercise. Can J Zool 69:1135–1142

Randall DJ, Daxboeck C (1984) Oxygen and carbon dioxide transfer across fish gills. In: Hoar WS, Randall DJ (eds) Fish Physiology, vol X. Academic, New York (in press)

Smith FM, Jones DR (1982) The effect of changes in blood oxygen carrying capacity on ventilation volume in the rainbow trout *(Salmo gairdneri)*. J Exp Biol 97:325–334

Smith MJ, Heath AG (1980) Responses to acute anoxia and prolonged hypoxia by rainbow trout *(Salmo gairdneri)* and mirror carp *(Cyprinus carpio)* red and white muscle: use of conventional and modified metabolic pathways. Comp Biochem Physiol 66B:267–272

Soivio A, Nikinmaa M, Nyholm K, Westman K (1981) The role of the gills in the responses of *Salmo gairdneri* during moderate hypoxia. Comp Biochem Physiol 70A:133–139

Steen JB, Kruysse A (1964) The respiratory function of teleostean gills. Comp Biochem Physiol 12:127–142

Stevens ED, Randall DJ (1967) Changes of gas concentrations in blood and water during moderate swimming activity in rainbow trout. J Exp Biol 46:329–337

Turner JD, Wood CM (1983) Factors affecting lactate and proton efflux from pre-exercised, isolated-perfused rainbow trout trunks. J Exp Biol 105:395–401

Turner JD, Wood CM, Clark D (1983a) Lactate and proton dynamics in the rainbow trout *(Salmo gairdneri)*. J Exp Biol 104:247–268

Turner JD, Wood CM, Höbe H (1983b) Physiological consequences of severe exercise in the inactive benthic flathead sole *(Hippoglossoides elassodon)*: a comparison with the active pelagic rainbow trout *(Salmo gairdneri)*. J Exp Biol 104:269–288

Van den Thillart G, Randall DJ, Hoa-Ren L (1983) CO_2 and H^+ excretion by swimming coho salmon, *Oncorhynchus kisutch*. J Exp Biol 107:169–180

Wardle CS (1978) Non-release of lactic acid from anaerobic swimming muscle of plaice *Pleuronectes platessa* L.: a stress reaction. J Exp Biol 77:141–155

Wokoma A, Johnston IA (1981) Lactate production at high sustainable cruising speeds in rainbow trout (*Salmo gairdneri* Richardson). J Exp Biol 90:316–364

Wood CM, Randall DJ (1973) The influence of swimming activity on water balance in the rainbow trout *(Salmo gairdneri)*. J Comp Physiol 82:257–276

Wood CM, Shelton G (1980a) Cardiovascular dynamics and adrenergic responses of the rainbow trout in vivo. J Exp Biol 87:247–270

Wood CM, Shelton G (1980b) The reflex control of heart rate and cardiac output in the rainbow trout: interactive influences hypoxia, haemorrhage, and systemic vasomotor tone. J Exp Biol 87:271–284

Wood CM, McMahon BR, McDonald DG (1977) An analysis of changes in blood pH following exhausting activity in the starry flounder, *Platichthys stellatus*. J Exp Biol 69:173–185

Wood CM, McDonald DG, McMahon BR (1982) The influence of experimental anaemia on blood acid-base regulation in vivo and in vitro in the starry flounder *(Platichthys stellatus)* and the rainbow trout *(Salmo gairdneri)*. J Exp Biol 96:221–237

Wood CM, Turner JD, Graham MS (1983) Why do fish die after severe exercise? J Fish Biol 22: 189–201

Wood CM, Wheatly MG, Höbe H (1984) The mechanisms of acid-base and ionoregulation in the freshwater rainbow trout during environmental hyperoxia and subsequent normoxia. III. Branchial exchanges. Respir Physiol 55:175–192

Wright PA, Wood CM (1985) An analysis of branchial ammonia excretion in the freshwater rainbow trout: effects of environmental pH change and sodium uptake blockade. J Exp Biol 114: 329–353

Yamamoto K, Itazawa Y, Kobayashi H (1980) Supply of erythrocytes into the circulating blood from the spleen of exercised fish. Comp Biochem Physiol 65A:5–11

Respiratory and Cardiovascular Adjustments to Exercise in Reptiles

T.T. GLEESON [1] and A.F. BENNETT [2]

1 Introduction

An important aspect of the reptilian response to exercise is a heavy reliance on anaerobic energy production to supplement aerobic metabolism during even modest activity. The aerobic and anaerobic capacities of reptiles, and the implications of these capacities for vigorous activity, have been recently discussed (Bennett and Dawson 1976; Bennett 1978, 1980). Like other terrestrial vertebrates, reptiles have been shown to increment the rate of oxygen consumption in proportion to the intensity of locomotory exercise (Fig. 1a; Moberly 1968a; Chodrow and Taylor 1973; Prange 1976; Bennett

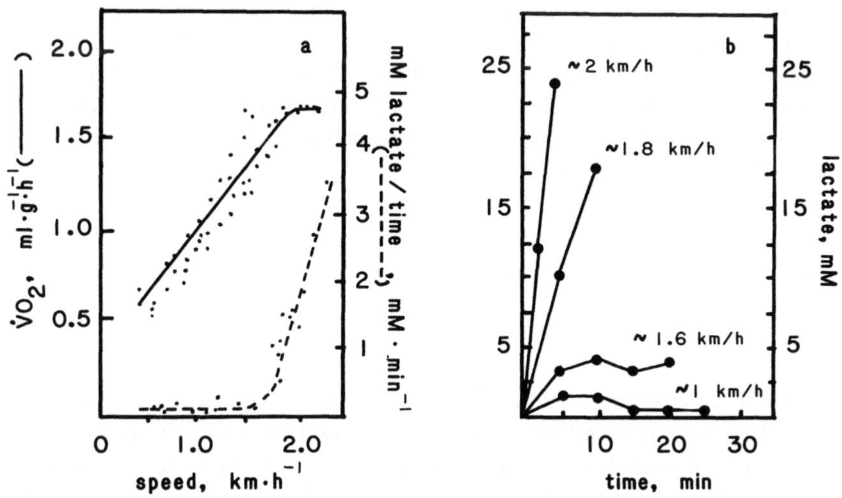

Fig. 1a,b. Oxygen consumption and lactate production in *V. exanthematicus* during exercise. **a** Rate of oxygen consumption (*solid line*) and rate of lactate production (*broken line*) as a function of treadmill running speed. **b** Change in blood lactate concentration with time at different exercise intensities. VO_{2max} in this 700 g lizard was reached at approx. 1.8 km h^{-1}. Redrawn from Seeherman et al. 1983

1 Environmental, Population, and Organismic Biology, University of Colorado, Boulder, CO 80309, USA
2 School of Biological Sciences, University of California, Irvine, CA 92717, USA

Circulation, Respiration, and Metabolism
(ed. by R. Gilles)

and Gleeson 1979; Gleeson 1979, 1981; Gleeson et al. 1980; Butler et al. 1984). Although both reptiles and mammals are capable of incrementing oxygen consumption five- to tenfold during exercise, predictive equations suggest that the maximum aerobic scope of a 100 g reptile is only 8% that of a comparable-sized mammal (Bennett and Dawson 1976; Taylor et al. 1981). The significance of this is that reptiles are capable of supporting only modest levels of activity aerobically. A 1 kg lizard will reach maximal rates of oxygen consumption ($VO_{2\,max}$) at running speeds as low as 0.5–1.0 km h^{-1} (Fig. 1a), while maximal aerobically-supported running speed for a 1 kg mammal would be 6–10 km h^{-1} (Garland 1982). This limited aerobic capacity of reptiles means that nearly all vigorous activities have an anaerobic component to their energy production. While some species may be capable of utilizing nearly their entire aerobic scope prior to supplementation with lactate production (Mitchell et al. 1981a), others experience lactate accumulations at nearly all levels of activity (Fig. 1b). The mechanism for anaerobic energy production during exercise is highly developed in this vertebrate group. Lactate accumulations of 20–30 m*M* following brief activity are common, and blood concentrations in excess of 50 m*M*, with blood pH falling as low as 6.6, have been reported to occur in crocodiles trying to escape capture (Bennett et al. 1985). Adjustments to metabolic acidosis of this magnitude are a major aspect of the reptilian response to exercise.

Physiological responses to activities having both aerobic and anaerobic components range from biochemical changes at the cellular level to changes in organismal gas exchange. We have chosen to highlight the changes that occur in respiratory and cardiovascular function during sustainable and burst activity. The data available on these subjects are mostly derived from observations on lizards and turtles. There are few data available on other groups of reptiles, and one of the goals of this review is to stimulate research of the activity physiology of these animals.

2 Ventilatory Adjustments to Exercise

2.1 Introduction

Exercise presents three challenges to the respiratory system of reptiles. Incrementing aerobic metabolism above resting levels results in increased rates of CO_2 production, and this CO_2 must be excreted by increasing ventilation in proportion to this elevated rate to maintain acid-base balance. A second challenge is that of increasing ventilation sufficiently to maintain a favorable gradient for oxygen diffusion to arterial blood. Most lizards are capable of maintaining arterial oxygen content during exercise, although they generally do not maintain the high levels of saturation typical of mammals. The third challenge is maintaining acid-base balance in the face of hydrogen ion produced as an end product of anaerobic metabolism. Buffering this hydrogen ion production is the role of the bicarbonate buffer system. Reptilian respiratory systems are surprisingly efficient at meeting these challenges despite simple lung morphologies (Tenney and Tenney 1970; Perry and Duncker 1978; Perry 1983). Readers are referred to Wood and Lenfant (1976) for a review of reptilian respiratory mechanics, ventilatory characteristics of resting animals, and the influence of temperature on ventilation. The data on ventilatory function in active animals are much less extensive. We will describe the general features of reptilian ventilatory responses to sustainable and nonsustainable exercise. A working definition of sustainable exercise is that requiring rates of VO_2 less than $VO_{2\,max}$ which does not result in continual lactate accumulation. Such exercise is illustrated by the lower curves (speeds $\leqslant 1.6$ km h^{-1}) in Fig. 1b. Rapid and continual lactic acid production has been reported at running speeds requiring 50%–95% $VO_{2\,max}$, the variation due to differences among species and experimental protocols (Bennett and Gleeson 1979; Gleeson 1980; Mitchell et al. 1981a; Seeherman et al. 1983). This division between sustainable and nonsustainable exercise allows independent focus on the maintenance of blood gases during carbonic acid loads (CO_2 production), on the one hand, and on acid-base regulation during noncarbonic acid loads (lactic acid production), on the other.

2.2 Ventilation During Sustainable Exercise

Rates of CO_2 and O_2 exchange increase five- to ninefold during sustainable exercise, and ventilation usually has an even greater factorial increment. For example, minute ventilation increases tenfold in sea turtles during terrestrial exercise, although VO_2 and VCO_2 increase only three- and sevenfold, respectively (Jackson and Prange 1979). This increase in ventilation is due entirely to increases in frequency, as tidal volume does not change. Such adjustments maintain blood gases and acid-base balance surprisingly well, given that sea turtles are unable to locomote and ventilate at the same time. This stability is even more remarkable in swimming *Chelonia*, which surface and take but one breath, whereas when walking they pause and take several. Yet, swimming turtles are also able to increase ventilation sufficiently to regulate blood gases within the range found at rest (Butler et al. 1984). Similar data on ventilatory frequency and volumes are not available for other reptiles during sustainable exercise.

An indirect assessment of ventilatory changes during exercise is provided by changes in the effective lung ventilation (V_{eff}). This calculated variable is analogous to rate of alveolar ventilation in mammals, and estimates the functional rate of lung ventilation (Mitchell et al. 1981a). Lung ventilation increases 10- to 15-fold in three lizard species during sustainable exercise (Mitchell et al. 1981a; Gleeson and Bennett 1982), but the data do not indicate how changes in frequency and tidal volume accomplish these increases. In *Varanus exanthematicus*, lung ventilation overcompensates for the metabolic demand for increased gas exchange across the lung; the factorial increase in V_{eff} exceeds the factorial increments in both VO_2 and in VCO_2 (Fig. 2a). The slight hyperventilation relative to VCO_2 results in a 5–7 torr depression in arterial and presumably also lung P_{CO_2} (Fig. 2b). Thus, sustainable exercise can cause a mild respiratory alkalosis in this species. Alkalosis occurs during such exercise in *Varanus*, but lactate production by *Iguana* early in exercise counters a respiratory alkalosis and arterial pH is maintained (Mitchell et al. 1981a).

All species of lizards studied are able to maintain Pa_{O_2} at or above resting levels throughout the period of exercise (Mitchell et al. 1981b; Gleeson and Bennett 1982). Increasing the rate of lung ventilation (V_{eff}) relative to the rate of oxygen consumption during sustainable exercise (Fig. 2a) increases lung P_{O_2} as the metabolic demand for oxygen increases (Fig. 2b). Thus, the gradient for pulmonary oxygen transport is

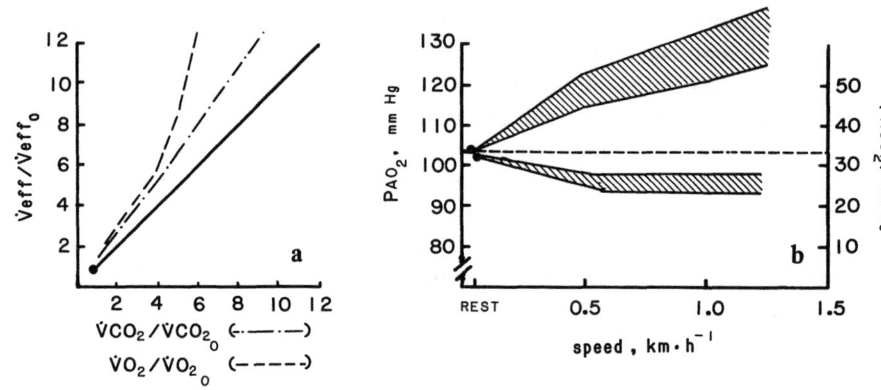

Fig. 2a,b. Changes in ventilation and lung gases during exercise in *V. exanthematicus*. **a** (left) Changes in ventilation relative to changes in gas exchange. Changes in lung ventilation from rest are expressed as a ratio (V_{eff}/V_{eff_0}), as are changes in VCO_2 and VO_2. Ventilatory change relative to VO_2, *dashed line*; relative to change in VCO_2, *broken line*. *Solid line* is line of identity, *area above line* represents hyperventilation. **b** (right) Changes in lung gas partial pressures during sustainable exercise, *upper region* reflects change in lung oxygen partial pressure; *lower region*, lung CO_2 partial pressure. Data calculated from Mitchell et al. 1981a,b

enhanced and maintains Pa_{O_2} and arterial O_2 content at or slightly above resting levels during exercise (Gleeson et al. 1980; Mitchell et al. 1981b). Any potential limitation in pulmonary oxygen transport during exercise due to insufficient or inefficient exchange surfaces is, therefore, offset by an increase in the lung-arterial P_{O_2} gradient (Mitchell et al. 1981b). An increase in lung-arterial P_{O_2} gradients is contrary to changes seen in man (Whipp and Wasserman 1969). The large gradient in resting lizards is likely due to cardiac shunts, which are known to exist in resting animals (Berger and Heisler 1977). A constant or increasing right-to-left cardiac shunt cannot explain the increasing pulmonary diffusion gradient for oxygen during exercise, however, as shunting venous blood of reduced oxygen content during exercise would result in a lower Pa_{O_2}, which does not occur (Figs. 2b and 3). The data suggest that in *Varanus* and *Iguana*, a decreasing right-left shung may occur as VO_2 increases during exercise (discussed further in Sect. 3). An alternative explanation for the increased gradient during exercise is provided by Mitchell et al. (1981b), who suggest that it might compensate for a limited lung surface area that otherwise would become limiting to oxygen uptake during exercise.

The ventilatory adjustments that occur during sustainable exercise are sufficient to maintain arterial blood gases and pH close to resting levels. Mitchell et al. (1981a) found that after prolonged exercise on a treadmill, running lizards possessed Pa_{CO_2} tensions slightly lower than at rest, hydrogen ion concentrations slightly more alkalotic than during rest, and no blood lactate accumulation. In a study which detailed the transient changes that occur during sustainable exercise, Gleeson and Bennett (1982) obtained a more complete picture of blood gas regulation during sustainable exercise. Arterial blood of the water monitor *Varanus salvator* was sampled at intermediate periods during treadmill running lasting 45 min at 85% VO_{2max}. Under these conditions, arterial oxygen tensions were maintained at a slightly elevated level relative to rest for

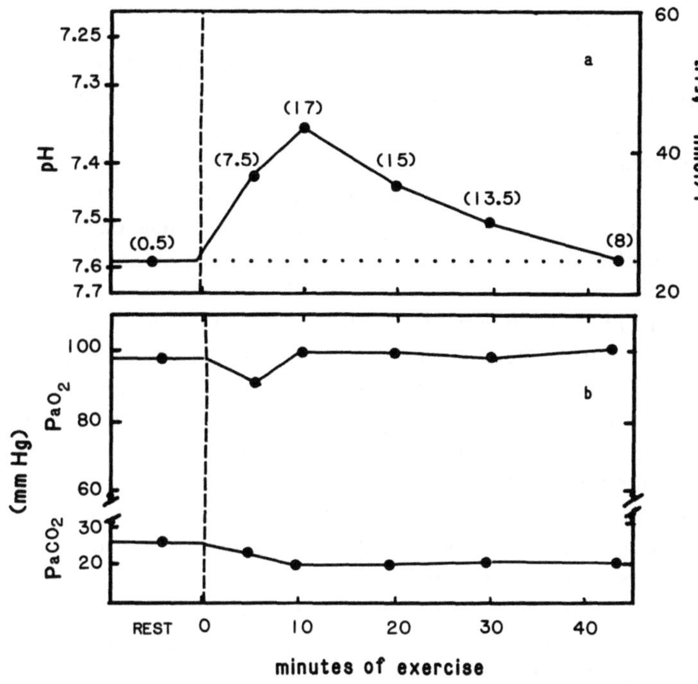

Fig. 3a,b. Blood gas and pH changes during exercise at 85% VO_{2max} in *V. salvator*. a Changes in arterial pH, with mmol l^{-1} changes in arterial lactate indicated in *parentheses*. b Arterial P_{CO_2} and P_{O_2} during exercise. Redrawn from Gleeson and Bennett 1982

all but the first few minutes of exercise (Fig. 3). The transient depression in Pa_{O_2} may indicate that the increased minute ventilation mentioned above may require a few minutes to be established and doesn't track VO_2 kinetics. Supplementary anaerobic metabolism during the initial minutes of exercise resulted in lactate accumulations of 18 m*M* and 0.2 U depression of pH after 10 min of running. Arterial P_{CO_2} declined approx. 8 torr as V_{eff}/VCO_2 increased 50%. As exercise continued, however, the metabolic acidosis was corrected by both metabolic and respiratory mechanisms. The elevated lactate levels were reduced metabolically by either oxidative or gluconeogenic mechanisms, both of which require equimolar quantities of hydrogen ions. Plasma bicarbonate levels, depressed as a result of the earlier metabolic acidosis, also begin to return towards resting concentrations. Late in the exercise period, arterial pH returns to normal (Fig. 3) and may even be slightly alkalotic, while Pa_{O_2} and Pa_{CO_2} change little relative to rest. The hyperventilation which occurs during such exercise persists during recovery in several species and results in significant alkalosis (Gleeson and Bennett 1982). The ability of *Varanus* to correct for the transient metabolic acidosis while exercising at such a high percentage of its VO_{2max} clearly indicates that ventilatory adjustments during sustainable exercise are more than adequate for excretion of the CO_2 resulting from increased rates of aerobic metabolism.

2.3 Ventilation During Exhaustive Exercise

Reptiles are noted for their ability to undergo brief bursts of exercise of very high intensity. Lizards, for example, are capable of running for a short time at speeds equal to the sprint speeds of similar-sized mammals (Garland 1985). Such exhaustive exercise is dominated by a metabolic acidosis resulting from lactate production. Large amounts of nonmetabolic CO_2 produced as the bicarbonate system buffers the fixed acid. Metabolic CO_2 production may be of secondary importance, as fatigue may occur even before maximal rates of aerobic metabolism are reached (Gleeson and Bennett 1982).

Exhaustive exercise elicits an immediate increase in lung ventilation. Minute ventilation increases five- to sixfold during such exercise in lizards and snakes (Dmi'el 1972; Bennett 1973a; Wilson 1971). Effective lung ventilation (V_{eff}) during 5 min of treadmill running increases to rates approaching 20 times standard rates in the lizard *Varanus salvator*. Increases in minute ventilation that occur during maximal exercise in lizards are due almost entirely to increases in tidal volume, as ventilatory frequency changes little if at all (Bennett 1973a). This pattern is in contrast to the response seen in sea turtles to sustainable exercise (Prange and Jackson 1976; Jackson and Prange 1979). Data on the ventilatory responses by snakes to exercise are less clear, but it appears that both frequency and tidal volume increase to some degree (Dmi'el 1972).

Both lizards and snakes hyperventilate relative to their aerobic demands for gas exchange during vigorous exercise. In the lizard *V. salvator*, ratios of V_{eff} to VO_2 indicate that this species increases ventilation threefold relative to increases in oxygen consumption. Such is also the case for other lizards and snakes (Dmi'el 1972; Wilson 1971). Ventilation is sufficient to maintain oxygenation of arterial blood, and thus, prevent an even greater reliance on anaerobic metabolism during exercise. During exhaustive exercise, saurian blood Pa_{O_2} remains constant or may rise slightly (Mitchell et al. 1981b; Gleeson and Bennett 1982) and arterial saturation may decline 10% or less (Bennett 1973b). However, ventilation in turtles is not sufficient to maintain oxygen saturation of arterial blood (Gatten 1975). During 2 min exercise at 30 °C the red earred turtle *Pseudemys scripta* experiences a fall in arterial blood saturation from 90% at rest to 34%. A similar decrement occurs in the box turtle *Terrepene ornata*. Failure to replenish arterial oxygen stores reinforces reliance of these turtles on anaerobic metabolism and probably limits their ability to sustain vigorous activity. Hypoventilation or apnea during exercise appears not to be the cause of this decreased arterial oxygen. *Pseudemys* apparently increase ventilation sufficiently to excrete the CO_2 necessary to allow the bicarbonate buffering system to maintain arterial pH precisely as blood lactate increases 6 m*M* (Gatten 1975), and it is unlikely that this regulation could be accomplished without hyperventilation. An increased shunt fraction during exhaustive exercise could explain these data.

The metabolic acidosis that occurs during exhaustive exercise makes carbon dioxide excretion of central importance for maintenance of acid-base balance. CO_2 is derived from both tissue respiration and from biocarbonate buffering of lactic acid, which frequently increases 15–30 m*M* in a few minutes. The resultant high rates of CO_2 excretion result in respiratory exchange ratios ($R_e = VCO_2/VO_2$) that peak at values of 2.0 or greater in some species (Bennett and Gleeson 1979; Gleeson 1980; Gleeson and Bennett 1982). Respiratory exchange ratios remain elevated at levels greater than 1.0

for a considerable portion of the recovery period, long after arterial lactate concentrations ([L]a) have peaked. Continued production of high levels of CO_2 is presumably due to washout of dissolved CO_2 from body tissues as arterial blood with a lower P_{CO_2} flows through it (Gleeson and Bennett 1982). The available data suggest that R_e can be used as a noninvasive indicator of the period of [L]a accumulation in lizards; decreasing values of R_e are indicative that [L]a has peaked and is declining, even though R_e may still be greater than 1.0 (Gleeson and Bennett 1982).

The magnitude of the metabolic acidosis during and after exhaustive exercise can be considerable. Lactate accumulations in snakes and lizards of 20–30 m*M* are not unusual (Bennett and Dawson 1976; Ruben 1976; Gleeson 1982). Crocodiles attempting to avoid human capture elevate [L]a more than 50 m*M*, the highest concentration ever reported in an active animal (Bennett et al. 1985). Such accumulations are very disruptive to tissue acid-base balance; blood pH in lizards may decline from around 7.5 to 7.0 (Gleeson and Bennett 1982), and crocodile blood pH may drop to 6.6 (Bennett et al. 1985; Seymour et al. 1985).

Table 1 summarizes the changes that occur in several reptiles in response to this acidosis. Most reptiles become hypocapnic during or shortly after exhaustive exercise, the result of an hyperventilation relative to VCO_2. Bicarbonate concentrations become depressed in roughly equimolar amounts to lactate accumulation (Mitchell et al. 1981a; Gleeson and Bennett 1982).

The stimulus for the hyperventilation relative to VCO_2 that accompanies both exhaustive and sustained exercise is uncertain. One possibility is that the metabolic acidosis that occurs during the onset of sustainable exercise, or throughout exhaustive exercise, provides the stimulus for hyperventilation via peripheral or central chemoreceptors. Mitchell and Gleeson (1985) recently tested this possibility by infusing lactate into resting varanid lizards. They found that respiratory compensation to the acidosis was minor and that ventilation followed changes in VCO_2 closely, regulating Pa_{CO_2} while pH fell. These results suggest that the lactacidosis attendent to exercise in reptiles may play only a minor role in stimulating the ventilatory response to exercise, leaving other factors associated with exercise as the stimulus for hyperventilation.

The recovery period following exhaustive exercise is a period of hypoventilation relative to both VO_2 and VCO_2. The result in *Varanus salvator* is that Pa_{O_2} declines relative to resting values, and both Pa_{CO_2} and $[HCO_3^-]$ slowly increase as metabolically

Table 1. Acid-base changes following exhaustive exercise in lizards and crocodile

Variable	Units	*V. salvator*[a]	*V. exanthematicus*[a]	*I. iguana*[a]	*C. porosus*[b]
R after exercise	–	1.77	1.23	1.38	–
Pa_{O_2}	mmHg	n/c	+12	+ 9	n/c
Pa_{CO_2}	mmHg	– 4	–13	– 4	n/c
pH		– 0.25	– 0.27	– 0.44	– 0.32
$[HCO_3^-]a$	m*M*	–13	–	–	–11
[L]a	m*M*	+17	+19	+14	+23

[a] Data from Gleeson and Bennett 1982
[b] Data from Seymour et al. 1985
n/c, no change

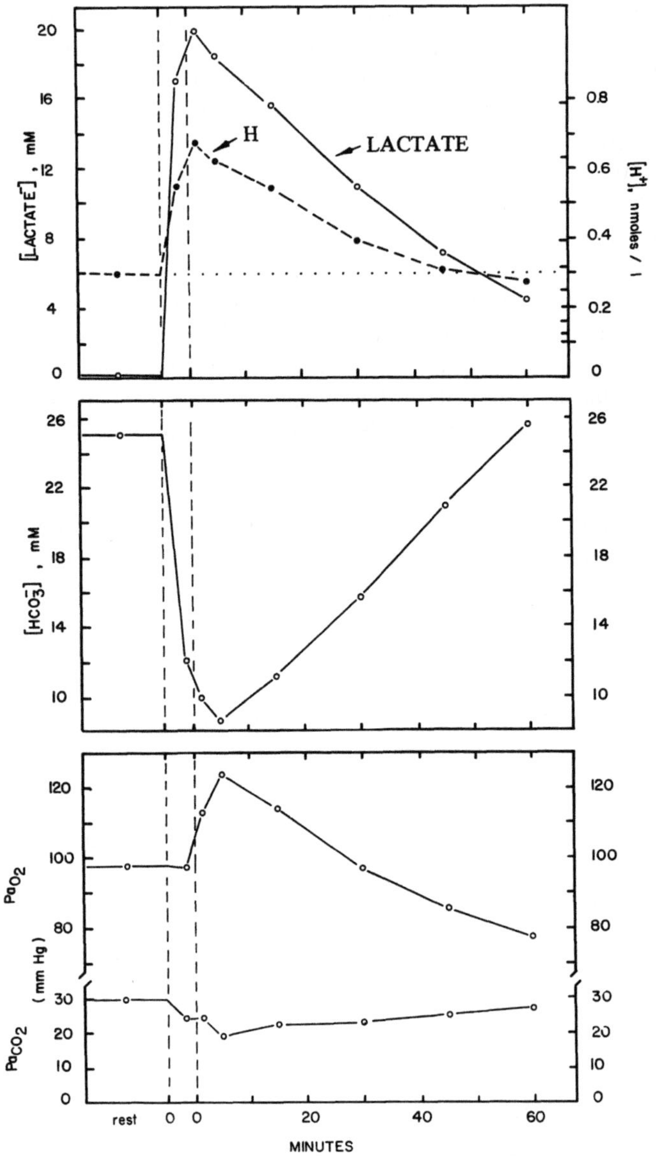

Fig. 4. Selected blood gas and acid-base variables before and after 5 min exhaustive treadmill exercise in *V. salvator* (by permission, from Gleeson and Bennett 1982)

produced CO_2 is retained and replenishes the depleted bicarbonate buffer system (Fig. 4). Values of R_e during this period may decline to 0.3–0.5. In three lizard species pH returns to preactive levels before R_e does, resulting in an alkalosis that persists for variable lengths of time following exercise. Recovery towards preexercise acid-base status is also facilitated by the metabolic removal of hydrogen ions from blood and tissues in the course of lactate catabolism. Varanid lizards become slightly alkalotic as they metabolize buffered solutions of Na^+-lactate infused into the blood (Mitchell and Gleeson 1985). Muscle gluconeogenesis in reptiles has been shown to be a possible pathway for this mechanism of metabolic recovery (Gleeson 1984).

3 Cardiovascular Adjustments to Exercise

3.1 Introduction

An increment in cardiovascular function accomodates the increased rate of gas exchange that occurs between tissue and lung during exercise. Increased blood flow facilitates the delivery of oxygen to active tissues and provides for the removal of CO_2 produced as a respiratory end product and as a consequence of lactic acid buffering. Increases in pulmonary and systemic blood flow are critically important during exercise, and towards that end changes in cardiac function play a central role. Readers are referred to White (1976) for a summary of the literature on aspects of reptilian cardiac anatomy, innervation, and function at rest.

Our understanding of how the reptilian cardiovascular system functions during exercise is rather limited. Comprehensive data on the components of cardiac output (heart rate, stroke volume) and oxygen extraction (arterial and mixed venous oxygen content. Ca_{O_2} and Cv_{O_2}, respectively) in active reptiles exist for only two lizards (Tucker 1966; Gleeson et al. 1980). Partial data exist for several other species. Confounding our understanding of reptilian cardiovascular physiology is the uncertainty over the significance of intracardiac shunts during exercise. It is apparent that right-to-left ventricular shunts occur in resting lizards, turtles, and snakes (Seymour and Webster 1975; White 1976; Berger and Heisler 1977; Seymour et al. 1981). The admixture of systemic venous blood to systemic output would reduce arterial P_{O_2} and oxygen saturation somewhat, but it is unlikely that such a shunt would seriously compromise oxygen delivery in a resting animal. Right-left shunts during running or diving, however, may be very important. Reducing pulmonary blood flow independently of systemic flow may be of some advantage to diving reptiles, and there is evidence that such a flow redistribution does occur. The sea snake *Acrochordus* responds to voluntary and forced dives by reducing pulmonary cardiac output considerably. During forced activity underwater of 2 min duration, *Acrochordus* reduces right ventricular output to 21% of its resting, eupneic rate (Seymour et al. 1981). During this period, heart rate remains unchanged and Pa_{O_2} declines, suggesting that systemic output continues at approximately preexercise rates. Seymour and Webster (1975) estimate right-left shunts in nonexercising sea snakes to be as high as 66% of total venous return. Butler et al. (1984) have shown that during periods of intermittant breathing, pulmonary blood flow in sea turtles declines between breaths, while left aortic flow remains unchanged. It is probable, therefore, that pulmonary artery and dorsal aorta blood flow may not be equal under conditions of underwater exercise in reptiles.

While right-left shunts may be advantageous during underwater exercise, shunting of venous blood during exercise on land is counterproductive to efficient gas exchange. Studies designed to evaluate the significance of and changes in cardiac shunts during exercise have not yet been performed. Data on changes in arterial and venous oxygen contents during exercise in lizards (Gleeson et al. 1980; Mitchell et al. 1981b) suggest that the shunt fraction must decrease during exercise, but these data are indirect. It would be most valuable to have direct measurements of shunt fraction, coupled with complementary data on oxygen extraction and systemic cardiac output, from exercising reptiles.

In the sections which follow we have summarized the available literature on the changes in cardiac output and oxygen extraction from arterial blood during exercise. As indicated above, these data are few, and knowledge of how they might be affected by cardiac shunts is unknown. Partly for these reasons, we have also included a discussion of the data on changes in oxygen pulse, as these changes may reflect changes in shunt fraction or aspects of cardiovascular regulation worthy of additional study.

3.2 Cardiac Output

Estimates of systemic cardiac output (Q) by the Fick equation ($Q = VO_2/Ca\text{-}\bar{v}_{O_2}$) indicate that cardiac output increases in proportion to VO_2 during exercise in the lizards *Varanus* and *Iguana* (Gleeson et al. 1980). Cardiac output itself is the product of changes in heart rate (HR) and stroke volume (SV), each of which are known to change independently during exercise.

Heart rate increases during exercise in nearly all reptiles studied, and is probably the primary mechanism for incrementing cardiac output in all groups. The single example where HR does not increase during activity is in the sea snake *Acrochordus* during forced activity while submerged (Seymour et al. 1981), suggesting that diving bradycardia may override an exercise-induced tachycardia. Factorial increments in HR range from 0.5–4 times, with twofold increments common. Gatten (1974a) has calculated from data on 14 species of reptiles at 30 °C that heart rate increments account for 19% (range 9–37%) of the total increment in oxygen delivery during forced, exhaustive exercise. During sustainable exercise on a treadmill HR increments may account for 30%–60% of the total increment in oxygen delivery (Gleeson et al. 1980). In *V. exanthematicus*, HR increments are the primary mechanism for increasing cardiac output at all levels of exercise (Fig. 5). Both the HR increment and its percentage contribution to total oxygen transport during exercise vary as a function of body temperature and may not always be optimal at preferred body temperatures (Moberly 1968b; Wilson and Lee 1970; Greenwald 1971; Bennett 1972; Dmi'el and Borut 1972; Gatten 1974a).

In *Varanus* and *Iguana*, increments in stroke volume have a secondary role in incrementing cardiac output. In *Iguana*, stroke volume decreased by 20% from rest to exercise. In this case, increments in heart rate are entirely responsible for increases in Q

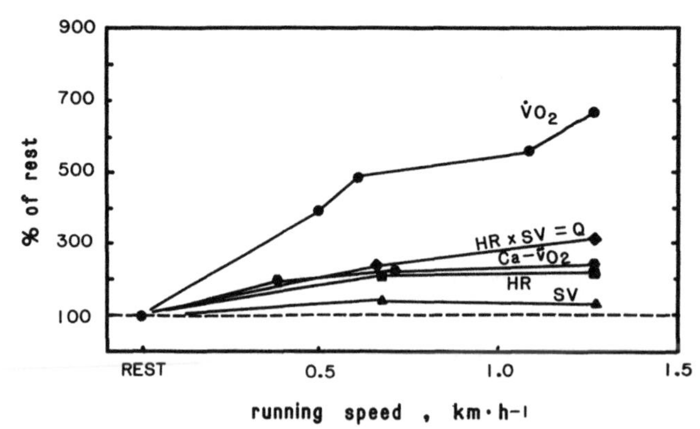

Fig. 5. Cardiovascular changes during treadmill exercise in *V. exanthematicus*. All values expressed as a percent of resting values, which were: $\dot{V}O_2$ = 0.19 ml g^{-1} h^{-1}; HR = 45 min^{-1}; SV = 0.0023 ml g^{-1} $beat^{-1}$; $Ca\text{-}\bar{v}_{O_2}$ = 2.6 vol%; Q = 6.7 ml g^{-1} h^{-1}. Data from Gleeson et al. 1980

(Tucker 1966; Gleeson et al. 1980). In *Varanus*, stroke volume increases 30% during exercise, so that stroke volume changes account for 40% of Q changes, while heart rate increments account for the remaining 60%. In this species, increments in SV contribute 10%–20% to the overall increase in oxygen transport (Fig. 5).

At preferred body temperatures exercise causes cardiac output to increase 100% to 200% over preexercise rates. Lizards appear to rely equally on increments in cardiac output and oxygen extraction to balance O_2 delivery needs, as do other vertebrates (Gleeson and Baldwin 1981). Data from *Varanus* suggest that the relationship between increments in VO_2 and Q is roughly linear throughout the range of VO_2 increments.

3.3 Oxygen Extraction

Delivery of oxygen to actively respiring tissues is restricted by the low oxygen carrying capacity reptilian blood, which typically has low hemoglobin concentrations and low hematocrits relative to the blood of endotherms. Hematocrit is usually within the 20%–30% range (Pough 1979). Blood oxygen capacities vary widely, but are low by mammalian standards, averaging 5.6 vol% in turtles, 8.4 vol% in lizards, and 10.2 vol% in snakes (Pough 1976).

Maximizing oxygen extraction from the blood during exercise requires that arterial oxygen content (Ca_{O_2}) during activity be maintained or increased, and that the volume of oxygen that passes through the capillary beds of the respiring tissue without participating in oxidative processes be minimized. Unlike the situation in mammals and birds, it is theoretically possible for reptiles to increase Ca_{O_2} as well as decrease $C\bar{v}_{O_2}$ during activity. Most arterial systemic blood in reptiles is only 70%–90% saturated with oxygen at rest, with lizards tending to the lower end and turtles the higher end of that range (Pough 1979; Stinner 1981). Optimizing pulmonary ventilation-perfusion during exercise would accomplish an increased Ca_{O_2}, as would any reduction in R–L shunting of deoxygenated blood within the pulmonary circuit or heart itself. Right-left shunting has been shown to be highly variable in *Iguana* and *Varanus* (White 1959; Tucker 1966; Baker and White 1970; Berger and Heisler 1977; Wood et al. 1977), which raises the possibility that the degree of shunting is under physiological control.

Arterial C_{O_2} is maintained during exercise at resting levels in the lizards *Iguana* and *V. exanthematicus* (Tucker 1966; Gleeson et al. 1980). Maintenance of Ca_{O_2} occurs despite increases in arterial P_{O_2} known to occur during exercise (Mitchell et al. 1981b; Gleeson and Bennett 1982). This result can occur under one of two situations; decreasing shunt fraction coupled with a oxygen dissociation curve right-shifted by decreasing pH (interested readers should read both Mitchell et al. 1981b; Wood 1982), or a constant shunt fraction coupled with an increased O_2 saturation of pulmonary venous blood (made possible by the increased lung P_{O_2}; Mitchell et al. 1981b), which would offset a lower $C\bar{v}_{O_2}$ of shunted blood. Mixed venous C_{O_2} decreases during exercise from 3–4 vol% at rest to approx. 1.8 vol% in both species. Increments in oxygen extraction account for 45%–55% of the total increment in oxygen delivery that occurs during exercise (Fig. 5). Although varanid lizards are capable of nearly twice the maximal $Ca\text{-}\bar{v}_{O_2}$ of *Iguana*, this change contributes less to incrementing total oxygen delivery than in *Iguana*.

3.4 Oxygen Pulse

Oxygen pulse has been calculated for a number of reptile species according to the following rearrangement of the Fick equation:

$$VO_2/HR = (Ca\text{-}\bar{v}_{O_2} \times \text{stroke volume}) = \text{oxygen pulse}$$

Oxygen pulse has units of volumes of oxygen delivered (or consumed) per heart beat, and in all cases has been calculated from data on oxygen consumption and HR, non-invasive measures that are relatively easy to measure in active and resting animals. Oxygen pulse (OP) is the product of the two variables most difficult to measure in active animals, arteriovenous oxygen content difference ($Ca\text{-}\bar{v}_{O_2}$) and stroke volume (SV). Because changes in oxygen pulse may be due to changes in either or both variables, knowledge of OP changes themselves are of limited value. The changes that do occur, however, point to what we feel may be a very fertile area for future research in reptilian cardiovascular physiology.

Interesting changes in OP occur as body temperature is varied. Body temperature is an important variable in reptilian biology, affecting many aspects of physiology and behavior. The changes that occur in maximal oxygen pulse (OP_{max}) with temperature suggest that its components are temperature sensitive. Oxygen pulse in resting reptiles changes little over a wide range of body temperature, i.e., resting oxygen consumption and heart rate have similar thermal dependencies. During maximal exercise, however, OP increases four to five times above resting levels, and is maximal at low body temperatures in most reptiles (Fig. 6). Maximal oxygen pulse decreases as body temperature increases in snakes (Greenwald 1971; Dmi'el and Borut 1972), lizards (Bennett 1972), and *Sphenodon* (Wilson and Lee 1970). Consequently, this aspect of oxygen delivery decreases as the metabolic demand for oxygen increases. In fact, at high body temperatures when $VO_{2\,max}$ is greatest, the factorial increment in OP during exercise may be significantly reduced (Bennett 1972; Dmi'el and Borut 1972; Wilson 1971). Decreasing maximal OP with increases in body temperature requires that either or both

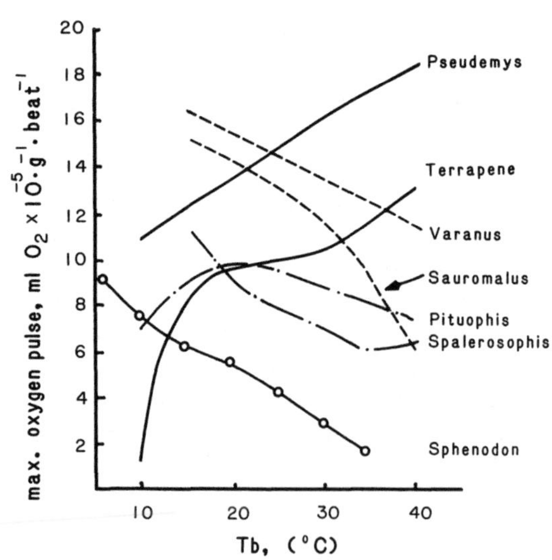

Fig. 6. Changes in maximum oxygen pulse as a function of body temperature in reptiles. Resting values of OP for all species ranged between 1 and 4 ml O_2 × 10^{-5} g^{-1} $beat^{-1}$. Data for turtles (*solid lines*) from Gatten 1974b; lizards (*dashed lines*) from Bennett 1972; snakes (*broken lines*) from Dmi'el and Borut 1972 and Greenwald 1971; *Sphenodon* from Wilson and Lee 1970

maximum oxygen extraction and stroke volume must decrease. Consequently, the three major components of cardiovascular oxygen transport, HR, SV, and Ca-$\bar{v}_{O_2}$, must change in their relative contributions to incrementing O_2 delivery during exercise at different temperatures. What these changes are or how they are regulated have not been investigated. Since OP_{max} decreases as body temperature increases, HR increments must be responsible for a greater share of oxygen delivery increments at higher temperatures, but the changes that occur in SV or Ca-$\bar{v}_{O_2}$ are now only conjectural. Known effects of temperature on cardiac muscle and on blood oxygen binding do not provide explanations for these changes in OP_{max}. High temperature may reduce SV by reducing cardiac tension generation or rate of tension generation. Both high and low temperatures decrease ventricular tension generation in a number of reptiles, however (Dawson and Bartholomew 1958; Licht et al. 1969), and this pattern of thermal sensitivity is not consistent as that for oxygen pulse. Maximum arterial-venous oxygen difference might also decrease with increased body temperature. Pough (1976) has shown that the oxygen capacity of reptilian blood decreases as temperature exceeds preferred body temperatures (PBT) in many species. Yet oxygen capacity also decreases at temperatures below PBT, a range of temperatures over which OP_{max} is increasing rather than decreasing. A third possibility is that R–L shunts during exercise increase as body temperatures increase. The increase in shunt fraction with temperature may occur as increased maximum heart rates or maximum flow rates through the ventricle disturb the functional separation of oxygenated and deoxygenated blood.

In contrast to the previous pattern, oxygen pulse in turtles has a different thermal dependence than that of other reptiles. Oxygen pulse at rest is more thermally sensitive in *Pseudemys* and *Terrapene* than in other reptiles (Gatten 1974b). Additionally, OP_{max} increases as body temperature increases in these species, contrary to the pattern found in lizards and snakes (Fig. 6). The influence of exercise on cardiovascular function in reptiles is complex, and the influence of temperature on the primary components of oxygen delivery is also complex and differs among orders of reptiles.

4 Areas of Future Research

The data on which this review is based are limited both in quantity and in their representation of the phylogenetic breadth of the class Reptilia. Varanid lizards are the only group for which we can say we have a fair knowledge of their ventilatory and cardiovascular responses to exercise. There is a clear need for more comparative studies of reptilian exercise physiology that take advantage of the wide array of anatomical and behavioral diversity within the group. For example, the evolutionary change in body morphology to the slender form of snakes necessitated the reduction of one lung in most snakes, while the other is greatly elongated. What effect might this change have had on oxygen delivery? A study of ventilatory responses of snakes to exercise could also address the effect of lung reduction on Pa_{CO_2} and Pa_{O_2} regulation. Cardiac morphology is also diverse among reptiles, ranging from forms where right-left ventricular shunts are large to forms with complete anatomical separation of ventricular flow. Study of crocodilian cardiovascular function during exercise would help us understand the consequences of incomplete ventricular flow separation in other reptilian groups.

Changes in body temperature have a major influence on the energy metabolism and exercise capacity of reptiles, yet the influence of body temperature on most aspects of activity physiology mentioned here is unknown. Body temperatures in excess of preferred body temperatures in many reptiles have a depressing effect on $VO_{2\,max}$ and consequently on maximal aerobic running speed. Does this effect have a pulmonary or, more probably, cardiovascular basis? Acid-base balance during exercise at different body temperatures is another area that has not been investigated.

Other subjects worthy of additional study include:

1. Control of ventilation during exercise. The presence of intrapulmonary chemoreceptors in lizards (Fedde et al. 1977; Scheid et al. 1977) and the relative ease of unidirectionally ventilating the reptile lung make these animals well suited for studies of mechanisms for acid-base and blood gas regulation.
2. Cardiac shunting during terrestrial exercise. Simultaneous measurements of pulmonary venous, mixed arterial, and mixed systemic venous blood O_2 content during exercise would allow evaluation of the importance of R–L shunts to gas exchange. Measurement of pulmonary versus systemic blood flow might also address this question. Do changes in shunt fraction explain the decline in arterial O_2 saturation seen during exercise in turtle blood?
3. Influence of temperature on the components of oxygen pulse. Do changes in shunt fraction, stroke volume, or oxygen extraction explain the changes in OP that occur as body temperature changes? What is the basis behind opposite temperature effects on OP seen in turtles compared to other reptiles?
4. Blood flow distribution during exercise. Although we assume that systemic blood flow increases in all reptiles during exercise, we have no knowledge of how this flow is distributed within or among organ systems. Existing techniques involving microsphere injections seen applicable to this question.

Acknowledgments. We thank Drs. G.S. Mitchell and R.S. Seymour for generously sharing unpublished data. Preparation of this manuscript was provided in part by NIH AM 03342 and by NSF PCM81-02331.

References

Baker LA, White FN (1970) Redistribution of cardiac output in response to heating in *Iguana iguana*. Comp Biochem Physiol 35:253–262

Bennett AF (1972) The effect of activity on oxygen consumption, oxygen debt, and heart rate in the lizards *Varanus gouldii* and *Sauromalus hispidus*. J Comp Physiol 79:259–280

Bennett AF (1973a) Ventilation in two species of lizards during rest and activity. Comp Biochem Physiol 46A:653–671

Bennett AF (1973b) Blood physiology and oxygen transport during activity in two lizards, *Varanus gouldii* and *Sauromalus hispidus*. Comp Biochem Physiol 46A:673–690

Bennett AF (1978) Activity metabolism of the lower vertebrates. Annu Rev Physiol 400:447–469

Bennett AF (1980) The thermal dependence of lizard behaviour. Anim Behav 28:752–762

Bennett AF, Dawson WR (1976) Metabolism. In: Gans C, Dawson WR (eds) Biology of the reptilia, vol 5. Academic, New York, pp 127–223

Bennett AF, Gleeson TT (1979) Metabolic expenditure and the cost of foraging in the lizard *Cnemidophorus murinus*. Copeia 4:573–577

Bennett AF, Seymour RS, Bradford DF, Webb GJW (1985) Mass-dependence of anaerobic metabolism and acid-base disturbance during activity in the salt-water crocodile, *Crocodylus porosus*. J Exp Biol (in press)

Berger PG, Heisler N (1977) Estimation of shunting, systemic and pulmonary output of the heart, and regional blood flow distribution in unanesthetized lizards *(Varanus exanthematicus)* by injection of radioactively labeled microspheres. J Exp Biol 71:111–121

Butler PJ, Milsom WK, Woakes AJ (1984) Respiratory, cardiovascular and metabolic adjustments during steady state swimming in the green turtle, *Chelonia* mydas. J Comp Physiol B 154:167–174

Chodrow RE, Taylor CR (1973) Energetic cost of limbless locomotion in snakes. Fed Proc 32:422

Dawson WR, Bartholomew GA (1958) Metabolic and cardiac responses in temperature in the lizard *Dipsosaurus dorsalis*. Physiol Zool 31:100–111

Dmi'el R (1972) Effect of activity and temperature on metabolism and water loss in snakes. Am J Physiol 223(3):510–516

Dmi'el R, Borut A (1972) Thermal behavior, heat exchange, and metabolism in the desert snake *Spalerosophis cliffordi*. Physiol Zool 45:78–94

Fedde MR, Kuhlmann WD, Scheid P (1977) Intrapulmonary receptors in the tequ lizard. I. Sensitivity to CO_2. Respir Physiol 29:35–48

Garland T Jr (1982) Scaling maximal running speed and maximal aerobic speed to body mass in mammals and lizards. Physiologist 25:338

Garland T Jr (1984) Physiological correlates of locomotory performance in a lizard: an allometric approach. Am J Physiol 247:R806–R817

Gatten RE Jr (1974a) Percentage contribution of increased heart rate to increased oxygen transport during activity in *Pseudemys scripta, Terrapene ornata* and other reptiles. Comp Biochem Physiol 48A:649–652

Gatten RE Jr (1974b) Effects of temperature and activity on aerobic and anaerobic metabolism and heart rate in the turtles *Pseudemys scripta* and *Terrapene ornata*. Comp Biochem Physiol 48A:619–648

Gatten RE Jr (1975) Effects of activity on blood oxygen saturation, lactate, and pH in the turtles *Pseudemys scripta* and *Terrapene ornata*. Physiol Zool 48(1):24–35

Gleeson TT (1979) Foraging and transport costs in the Galapagos Marine Iguana, *Amblyrhynchus cristatus*. Physiol Zool 52(4):549–557

Gleeson TT (1980) Metabolic recovery from exhaustive activity by a large lizard. J Appl Physiol 48(4):689–694

Gleeson TT (1981) Preferred body temperature, aerobic scope, and activity capacity in the Monitor lizard, *Varanus salvator*. Physiol Zool 54(4):423–429

Gleeson TT (1982) Lactate and glycogen metabolism during and after exercise in the lizard *Sceloporus occidentalis*. J Comp Physiol B 147:79–84

Gleeson TT (1984) Muscle gluconeogenesis as a mechanism for lactate removal in a lizard. Am Zool 24(4):109A

Gleeson TT, Baldwin KM (1981) Cardiovascular response to treadmill exercise in untrained rats. J Appl Physiol 50(6):1206–1211

Gleeson TT, Bennett AF (1982) Acid-base imbalance in lizards during activity and recovery. J Exp Biol 98:439–453

Gleeson TT, Mitchell GS, Bennett AF (1980) Cardiovascular responses to graded activity in the lizards *Varanus* and *Iguana*. Am J Physiol 239:R174–R179

Greenwald OE (1971) The effect of body temperature on oxygen consumption and heart rate in the Sonora Gopher Snake, *Pituophis catenifer affinis* Hallowell. Copeia 1971:98–106

Jackson DC, Prange HD (1979) Ventilation and gas exchange during rest and exercise in adult green sea turtles. J Comp Physiol B 134:315–319

Licht P, Dawson WR, Shoemaker VH (1969) Thermal adjustments in cardiac and skeletal muscles of lizards. Z Vgl Physiologie 65:1–14

Mitchell GS, Gleeson TT (1985) Acid-base balance during lactic acid infusion in the lizard *Varanus salvator*. Resp Physiol 60:253–266

Mitchell GS, Gleeson TT, Bennett AF (1981a) Ventilation and acid-base balance during graded activity in lizards. Am J Physiol 240:R29–R37

Mitchell GS, Gleeson TT, Bennett AF (1981b) Pulmonary oxygen transport during activity in lizards. Respir Physiol 43:365–375

Moberly WR (1968a) The metabolic responses of the common iguana, *Iguana iguana*, to walking and diving. Comp Biochem Physiol 27:21–32

Moberly WR (1968b) Metabolic responses of the common iguana, *Iguana iguana*, to activity under restraint. Comp Biochem Physiol 27:1–20

Perry SF (1983) Reptilian lungs: Functional anatomy and evolution. Adv Anat Embryol Cell Biol 79:1–81

Perry SF, Duncker HR (1978) Lung architecture, volume and static mechanics in five species of lizards. Respir Physiol 34:61–81

Pough FH (1976) The effect of temperature on oxygen capacity of reptile blood. Physiol Zool 49(2):141–151

Pough FH (1979) Summary of oxygen transport characteristics of reptilian blood. Smithson Herp Inf Service publ no 45, p 1–18

Prange HD (1976) Energetics of swimming of a sea turtle. J Exp Biol 64:1–12

Prange HD, Jackson DC (1976) Ventilation, gas exchange and metabolic scaling of a sea turtle. Respir Physiol 27:369–377

Ruben JA (1976) Aerobic and anaerobic metabolism during activity in snakes. J Comp Physiol 109:147–157

Scheid P, Kuhlmann WD, Fedde MR (1977) Intrapulmonary receptors in the tequ lizard. II. Functional characteristics and localization. Respir Physiol 29:49–62

Seeherman HJ, Dmi'el R, Gleeson TT (1983) Oxygen consumption and lactate production in varanid and iguanid lizards: A mammalian relationship. In: Knuttgen HG, Vogel JA, Poortmans J (eds) Biochemistry of exercise; Int ser Sport Sci, vol 13. Human Kinetics Publishers, Champaign, Ill, pp 421–427

Seymour RS, Webster MED (1975) Gas transport and blood acid-base balance in diving sea snakes. J Exp Zool 191:169–182

Seymour RS, Dobson GP, Baldwin J (1981) Respiratory and cardiovascular physiology of the aquatic snake, *Acrochordus arafurae*. J Comp Physiol 144:215–227

Seymour RS, Bennett AF, Bradford DF (1985) Blood gas tensions and acid-base regulation in the salt-water crocodile, *Crocodylus porosus*, at rest and after exhaustive exercise. J Exp Biol (in press)

Stinner JN (1981) Ventilation, gas exchange and blood gases in the snake, *Pituophis melanoleucus*. Respir Physiol 47:279–298

Taylor CR, Maloiy GMO, Weibel ER, Langman VA, Kamau JMZ, Seeherman HJ, Heglund NC (1981) Design of the mammalian respiratory system. III. Scaling maximum aerobic capacity to body mass: wild and domestic mammals. Respir Physiol 44(1):25–37

Tenney SM, Tenney JB (1970) Quantitative morphology of cold-blooded lungs: Amphibia and Reptilia. Respir Physiol 9:197–215

Tucker VA (1966) Oxygen transport by the circulatory system of the green iguana *(Iguana iguana)* at different body temperatures. J Exp Biol 44:77–92

Whipp BJ, Wasserman K (1969) Alveolar-arterial gas tension differences during graded exercise. J Appl Physiol 27:361–365

White FN (1959) Circulation in the reptilian heart (Squamata). Anat Rec 135:129–134

White FN (1976) Circulation. In: Gans C, Dawson WR (eds) Biology of the reptilia, vol 5. Academic, New York, pp 275–334

Wilson KJ (1971) The relationship of oxygen supply for activity to body temperature in four species of lizards. Copeia 1971(1):920–934

Wilson KJ, Lee AK (1970) Changes in oxygen consumption and heart-rate with activity and body temperature in the Tautara, *Sphenodon punctatus*. Comp Biochem Physiol 33:311–322

Wood SC (1982) Effect of O_2 affinity on arterial PO_2 in animals with central vascular shunts. J Appl Physiol 53:1360–1364

Wood SC, Lenfant CJM (1976) Respiration: Mechanics, control and gas exchange. In: Gans C, Dawson WR (eds) Biology of the reptilia, vol 5. Academic, New York, pp 225–274

Wood SC, Johansen K, Gatz RN (1977) Pulmonary blood flow, ventilation/perfusion ratio, and oxygen transport in a varanid lizard. Am J Physiol 233(3):R89–R93

Exercise in Normally Ventilating and Apnoeic Birds

P.J. BUTLER and A.J. WOAKES[1]

1 Introduction

Within the class Aves there are animals that can fly, (some for long distances and some at high altitudes) run, swim, and dive. A few, for example ducks and auks, are able to perform all of these types of locomotion. When migrating some birds may fly several thousands of kilometres without stopping. For example the blackpoll warbler, *Dendroica striata*, is one of many species that travels directly from N. to S. America across the Atlantic Ocean, a distance of 3,000–4,000 km, while black brant geese, *Branta bernicla nigricans*, fly approx. 4,800 km across the Pacific Ocean from Alaska to Mexico (Ogilvie 1978).

Although most passerines migrating at night fly below 2,000 m, some birds have been observed at extremely high altitudes. A flock of 30 swans (probably whooper swans, *Cygnus cygnus*) was located by radar off the west coast of Scotland at an altitude between 8,000–8,500 m where the temperature was −48 °C (Stewart 1978; Elkins 1979). Bar-headed geese have been observed flying at altitudes up to 9,000 m (where the partial pressure of oxygen, P_{O_2}, is 6.9 kPa) during their migration across the Himalayas (Swan 1961).

At the other extreme, it is the flightless penguins that descend to unusual depths in the sea when feeding. The deepest recorded dive performed by a bird was by an emperor penguin, *Aptenodytes forsteri*, to a depth of 265 m and the longest recorded dive was by the same species and for in excess of 15 min (Kooyman et al. 1971). When diving under water for their food, birds have to rely entirely upon the oxygen stored within their body.

These extreme examples may give some idea of the range of adjustments that must be made by the respiratory and cardiovascular systems during the various forms of exercise performed by birds.

2 Energetics of Flight

Ornithologists have been interested in the energetics of bird flight for many years and this interest relates to the non-stop migratory flights of some birds. Some of the earliest

1 Department of Zoology and Comparative Physiology, The University of Birmingham, Birmingham B15 2TT, Great Britain

Circulation, Respiration, and Metabolism
(ed. by R. Gilles)

values of metabolic rate during flight were estimated from loss of weight during a long flight with the assumption that fat constitutes by far the major part of this weight loss (e.g. Nisbet et al. 1963). In some studies the actual amount of fat that is used during flight was estimated. This was done by first of all measuring fat content of birds before any long distance flying and then taking measurements from different birds of the same species that had flown for a long distance (see LeFebvre 1964). This author also measured CO_2 production in the same birds using doubly labelled water and he obtained similar estimates of total power used during flight (power input). Unfortunately, without a knowledge of the behaviour of the birds and of the prevailing meteorological conditions, it is not always wise to apply data obtained by these techniques to flight in general.

A number of theoretical models have been presented which require minimum data to be inserted into relatively simple formulae in order to predict the mechanical power required for flight (power output) of a bird at different flight velocities (Pennycuick 1969, 1975; Rayner 1979). Useful and important though the theoretical considerations are, they do not tell us how much energy a bird actually expends when flying. The ratio of power output/power input is the efficiency of the system and is thought to be 0.2–0.3, which means that 70%–80% of power input is "wasted" as heat. It is likely that efficiency varies with flight velocity and wing beat kinematics, but as yet we do not know how (Rayner 1979). Because of these uncertainties, the most reliable way of obtaining power input is to measure it directly under controlled conditions.

The first direct measurement of power input of a bird during forward flapping flight was that of Tucker (1966), who measured oxygen uptake of budgerigars, *Melopsittacus undulatus*, trained to fly in a closed wind tunnel. In subsequent papers Tucker attached a loose fitting mask to the head of budgerigars (Tucker 1968) and laughing gulls, *Larus atricilla* (Tucker 1972), which enabled him to measure respiratory gas exchange and water loss under controlled flight conditions in open wind tunnels. A number of workers have since used basically similar techniques to measure a variety of physiological data from flying birds. Figure 1 shows power input (P_i) for a number of birds during level, flapping flight at different velocities, and only the budgerigar has the classical U-shaped curve, as predicted by the theoretical models for power output (P_o). The starling, *Sturnus vulgaris*, has a particularly flat P_i vs V curve which may result from the change in body attitude that occurs during flight (Torre-Bueno and Larochelle 1978).

The relationship between minimum power input (P_{im}) during level flapping flight of birds in wind tunnels and body mass (M) has been found to be $P_{im} = 49.9\ M^{0.72}$ W (Butler 1982a). This index is lower than that calculated by Rayner (1982). Using "most recent reliable data" he calculated that power input during flapping flight = $64.7\ M^{0.88}$ W. Although Rayner did not give the sources of his data, he did state that on theoretical grounds an index approaching 1 would be expected. The equation derived by Butler (1982a) indicates that over the size range involved (35 g to 480 g) minimum power input during flight is a constant multiple (9.1 times) of basal metabolic rate for non-passerine birds (Prinzinger and Hanssler 1980). While accepting that if data were available from smaller and from larger flying birds the index during flight might well be closer to 1, it is none the less intriguing to note from Fig. 2 that *minimum* power input during flight is 2.2 times the *maximum* power input of exercising small mammals (i.e. up to 900 g). Despite the high rate at which oxygen is used during forward flapping flight, this form of locomotion is still relatively attractive as far as long migrations are

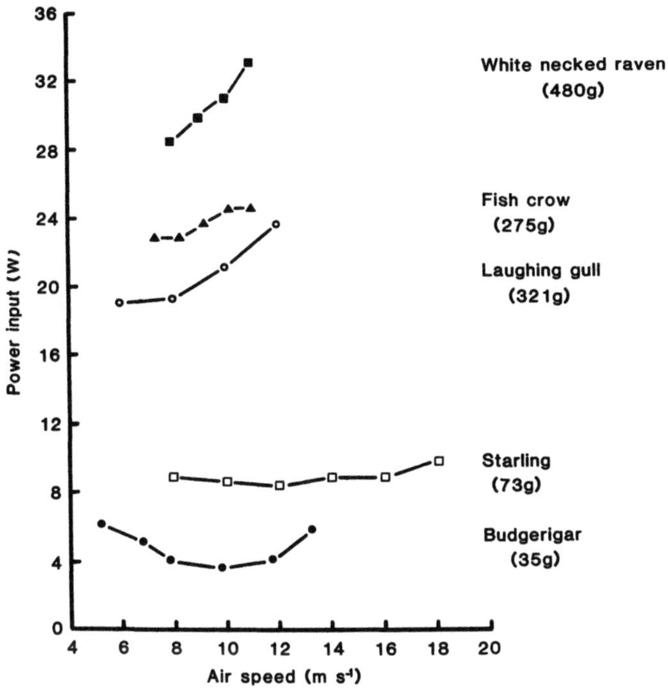

Fig. 1. Power input at different air speeds for birds of different mass during horizontal flapping flight in a wind tunnel. (After Butler 1981; Hudson and Bernstein 1983)

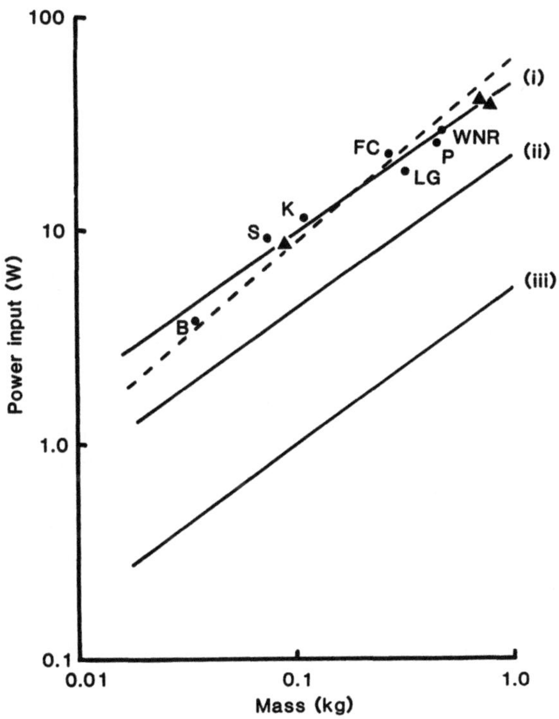

Fig. 2. The relationship between power input, Pi, and body mass, M, for: (*i*) seven species of birds during forward flapping flight in a wind tunnel; *B* budgerigar; *S* starling; *K* kestrel; *LG* laughing gull; *FC* fish crow; *P* pigeon; *WNR* white-necked raven. Minimum values of power input, Pim, were used to construct this curve, and Pim $= 50.4\ M^{0.73}$ (data from Butler 1981; Hudson and Bernstein 1983); (*ii*) small mammals during maximum exercise where $Pi = 22.6\ M^{0.73}$ (data from Pasquis et al. 1970); (*iii*) resting non-passerine birds were $Pi = 5.5\ M^{0.73}$ (data from Prinzinger and Hanssler 1980). It is interesting to note that data from three species of flying bats (▲) lie on line (*i*) (data for bats from Butler 1981). The *dashed line* is that for flying birds as calculated by Rayner (1982), where $Pi = 64.7\ M^{0.88}$

concerned. The energy cost of flight (energy required to transport a given mass a given distance) is considerably less than the energy cost of walking or running (Tucker 1970) and the higher rate of oxygen consumption is manifest as the high velocities seen during flight.

3 Physiology of Flight

The respiratory system of birds has a unique and complex structure with air flowing unidirectionally, during both phases of ventilation, through the regions of gas exchange (tertiary bronchi or parabronchi) in the lungs (Scheid 1979). Bulk air flow in the parabronchi is at right angles to bulk blood flow giving rise to the very effective cross-current system of gas exchange (Scheid and Piiper 1972). Not unreasonably, it has often been suggested that the structure of the avian respiratory system is related to the peculiarly high energy demands of flight (King and Molony 1971). However, data from bats flying in a wind tunnel indicate that their oxygen uptake (power input) is similar to that of similar-sized birds during flight (Fig. 2). Factors, other than the structure of the respiratory system, are clearly important in determining maximum oxygen uptake ($\dot{V}_{O_2 max}$) during exercise in birds and mammals, at least at sea level.

This topic has been dealt with in some detail by Weibel and Taylor (1981) working on mammals. They have proposed that the capacity of the gas transporting systems (respiratory and cardiovascular systems) is matched to the maximum demand that can be made by the muscular system. So, it could be argued that the factor which ultimately determines $\dot{V}_{O_2 max}$ in healthy animals at sea level is total mitochondrial volume (or activity of aerobic enzymes) in the exercising muscles. Certainly, increasing the partial pressure of inspired oxygen in rats has no effect on $\dot{V}_{O_2 max}$. The thought of birds flying over the top of Mount Everest is awe-inspiring, and it has been concluded that under such conditions of hypoxia the lung of birds, with its high gas exchange effectiveness, does in fact contribute significantly to their high altitude performance (Scheid 1984). Other factors, however, are also important (see later).

Unfortunately, the functioning of the cardiorespiratory system in flying birds has been studied in only a handful of papers. Direct measurements indicate that below an ambient temperature of 22°-23 °C both fish crows, *Corvus ossifragus*, and white-necked ravens, *Corvus cryptolecus*, increase ventilation volume ($\dot{V}_I$) in proportion to oxygen uptake. This indicates that oxygen extraction (O_{2Ext}) is similar to that when the birds are at rest. In the crows, tidal volume during flight is almost double the resting value, whereas in the ravens the two values are similar. So, in both instances the greater contributor to the increase in ventilation volume is respiratory frequency. Respiratory frequency and tidal volume in the fish crow (Bernstein 1976) and respiratory frequency in the barnacle goose, *Branta leucopsis* (Butler and Woakes 1980), are independent of flight velocity. This could mean that, like the starling and the fish crow, oxygen uptake of the barnacle goose is largely independent of flight velocity. It is certainly interesting to note that the budgerigar, with its distinctive U-shaped power input/velocity curve, also changes respiratory frequency with flight velocity in a similar U-shaped fashion (Tucker 1968).

Despite the apparent matching between ventilation volume and oxygen uptake at low ambient temperatures in the fish crow and white-necked raven, there does appear to be an increase in effective lung ventilation, above that required by metabolic rate, during flight in starlings at low ambient temperature; partial pressure of O_2 (P_{O_2}) in the air sacs increases and P_{CO_2} decreases (Torre-Bueno 1978). The physiological significance of this hyperventilation is unclear, but it results from the four times increase in tidal volume that occurs in this bird during flight.

There is an increase in body temperature by approx. 2.0 °C in all birds during flight at relatively low ambient temperatures (see, for example, Torre-Bueno 1976; Hudson and Bernstein 1981), and this may explain the (effective) hyperventilation in starlings. At ambient temperatures above 22°–23 °C, body temperature of the white-necked raven increases further during flight and in these birds and fish crows, ventilation volume increases progressively as ambient and body temperature rise. As oxygen uptake does not change, there is a dramatic reduction in oxygen extraction. In the fish crow, though not in the white-necked raven, the hyperventilation during flight at high ambient temperature results entirely from an increase in tidal volume. This increase in overall ventilation above metabolic demands is clearly related to temperature regulation via evaporative water loss, which will not be dealt with here. It is important to note, though, that it will tend to cause a reduction in P_{CO_2} and an increase in pH in the blood, thus making the bird alkalotic and exaggerating the process already present, in starlings at least, at low ambient temperature. It may also increase energy expenditure.

The role of the various components of the cardiovascular system in presenting oxygen to the exercising muscles can best be described by Fick's formula:

$$\dot{V}_{O_2} = \text{H.R.} \times \text{S.V.}\,(CaO_2 - C_{\bar{v}}O_2)$$

where $\dot{V}_{O_2}$ = rate of oxygen consumption
H.R. = heart rate
S.V. = cardiac stroke volume
CaO_2 = oxygen content of arterial blood
$C_{\bar{v}}O_2$ = oxygen content of mixed venous blood .

In pigeons flying at 10 m s^{-1} in a wind tunnel, steady state oxygen uptake is ten times the resting value (Fig. 3). The respiratory system maintains CaO_2 at slightly below the resting value but $C_{\bar{v}}O_2$ is halved, giving a 1.8 times increase in ($CaO_2 - C_{\bar{v}}O_2$). There is no significant change in S.V., so the major factor in transporting the extra oxygen to the muscles is the six times increase in heart rate (Butler et al. 1977). There is no significant change in arterial blood pressure during flight. Birds have larger hearts and lower resting heart rates than mammals of similar body mass (Lasiewski and Calder 1971; Grubb 1983). They also have a greater cardiac output (H.R. × S.V.) for a given oxygen consumption than similar-sized mammals (Grubb 1983). In other words ($CaO_2 - C_{\bar{v}}O_2$) is less in birds than in similar-sized mammals, because $C_{\bar{v}}O_2$ is not reduced to such a low level. The higher cardiac output in birds may be an important factor in their attaining a higher $\dot{V}_{O_2\,max}$ during flight than mammals when running. It is certainly interesting to note that bats have larger hearts and a higher blood oxygen carrying capacity than other mammals of similar body mass (Jurgens et al. 1981).

There have been no physiological studies on birds flying at high altitudes (real or simulated) but work on inactive animals has indicated that adaptations of the cardiovascular system may be very important in the altitude performance of birds. Unlike Pekin ducks, bar-headed geese, *Anser indicus*, do not increase their haematocrit (packed

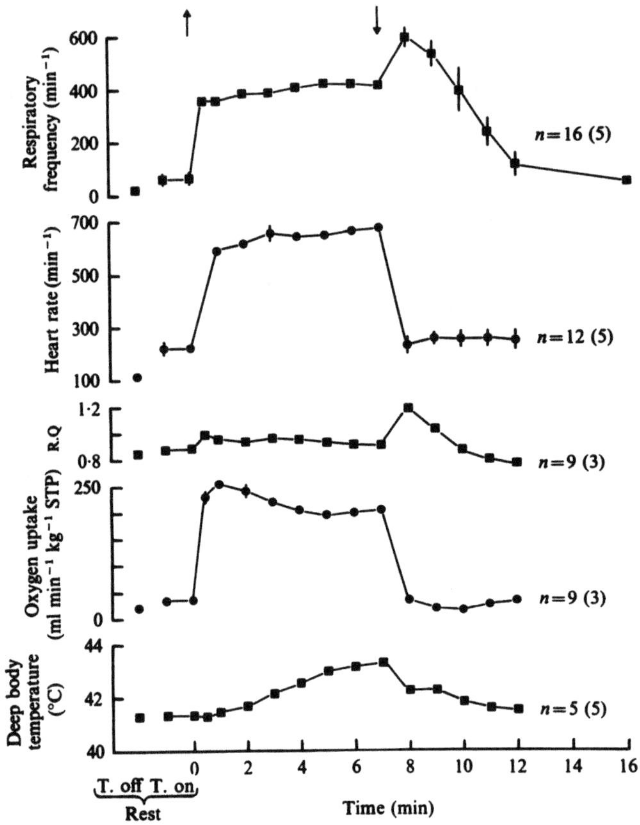

Fig. 3. Changes in respiratory frequency, heart rate, respiratory quotient (*R.Q.*) oxygen uptake and body temperature in pigeons before, during and after flying at 10 m s^{-1} for 7 min. Each *point* is the mean value of a number of observations (*n*) which is given close to each line. In *parentheses* associated with each value of *n* is the number of animals from which the observations were obtained. *Vertical lines* associated with each point are ± SE of mean. Where *vertical lines* are absent, the SE of mean is within the limits of the symbol except for *R.Q.* which was computed from the mean values of O_2 uptake and CO_2 production and thus lacks a standard error. *Arrows* indicate points of take-off and landing. Resting values were taken when the tunnel was turned off (*T. off*) and when it was turned on (*T. on*). (Butler et al. 1977)

cell volume of the blood) and haemoglobin concentration when exposed to simulated high altitudes (Black and Tenney 1980). This means that there is no increase in blood viscosity, thus preventing a possible increase in the energy cost of circulating the blood. It also means that there is no increase in the oxygen-carrying capacity of the blood. This is more than counter-balanced by the higher affinity for oxygen (low P_{50}) of the Hb in the goose (approx. 5 kPa at pH 7.5 compared with 7.5 kPa in the duck) which enables the maintenance of a higher CaO_2 (and hence $CaO_2 - C_{\bar{v}}O_2$) at high altitudes than in the duck. It is also apparent from the data of Black and Tenney (1980) that the respiratory system of the bar-headed goose is able to maintain a very small difference between the partial pressures of oxygen in inspired air (P_IO_2) and arterial blood

(PaO_2) when at high altitude. At sea level ($P_IO_2 - PaO_2$) is 7 kPa, whereas at a simulated altitude of 10,668 m it is a mere 0.5 kPa, giving proof of the effectiveness of the cross-current exchange system in birds.

Such low values of ($P_IO_2 - PaO_2$) are possible not only because of the anatomical arrangement of the lung, but also because of a large increase in ventilation. This hyperventilation in response to hypoxia initially causes a fall in $PaCO_2$ and an increase in pHa; the bird becomes hypocapnic and alkalotic, as well as hypoxic. In a number of mammals (dog, monkey, rat, man) hypocapnia causes a reduction in cerebral blood flow; this is not so in ducks (Grubb et al. 1977). Also, hypoxia causes a greater increase in cerebral blood flow in ducks than in dogs, rats, and man (Grubb et al. 1978). These two factors together, if present in other birds, will obviously be of great importance at high altitude.

It is clear from this brief account of the physiology of flight that most of the useful physiological data have been obtained from a few studies using wind tunnels. Unfortunately, wind tunnels are not ideal because the bird has to fly in a restricted space, must tolerante the noise of the fan motor and is in an optically motionless environment (Rothe and Nachtigall 1980). Also, it has been noted for the pigeon that the flight pattern is different compared with free-range flight (Butler et al. 1977) and, unless very large tunnels are used, only the smaller birds can be studied. The early use of radiotelemetry did not demonstrate this technique to be a useful alternative (see Butler 1984 for discussion). Because of the limited range of the transmitter, only very short flights (10-15 s) were possible. However, by imprinting barnacle geese, *Branta leucopsis*, on a human and training them to fly behind a truck containing the foster parent, it has been possible to record a few physiological variables, using an implantable telemetry system, from birds flying freely for several minutes (Butler and Woakes 1980). With the rapid advance of electronic technology, it may soon be possible to record a number of important respiratory and cardiovascular variables from birds freely flying under these conditions.

4 Walking, Running, Swimming, Diving

Although flight is the predominant form of exercise in birds, it is by no means the only form. In the exclusively cursorial forms, such as the cockerel, rhea, and emu (see Grubb et al. 1983), $\dot{V}_{O_2\,max}$ is close to the extrapolated line of $\dot{V}_{O_2\,min}$ vs body mass for birds flying in a wind tunnel. However, from studies on birds that have been trained to run on a treadmill or swim on a water channel, it seems that $\dot{V}_{O_2\,max}$ is very similar in those that are not primarily cursorial e.g. tufted duck, marabou-stork, emperor penguin, to that in running mammals of similar body mass (Butler 1982a). As Prange and Schmidt-Nielsen (1970) point out, this means that a bird that can fly will have a higher $\dot{V}_{O_2\,max}$ during flapping flight than when running, but that in the mallard duck at least the ratio of $\dot{V}_{O_2\,max}$ during each form of exercise is similar to the ratio of mass of the muscles involved (pectoral and leg muscles). Thus, as suggested earlier, the total volume of mitochondria in the active muscles may limit $\dot{V}_{O_2\,max}$ at sea level rather than any inadequacy of the cardiorespiratory system. The implication is that the mass of the leg muscles (or at least total mitochondrial volume in the muscles) of the cursorial birds is similar to that of the pectoral muscles of similarly-sized birds that fly. This is somewhat theoretical, however, as the largest birds are exclusively cursorial.

Although only the larger cursorial birds are likely to run for any long distances, a number of studies on the cardiorespiratory systems have been performed on running domesticated ducks and chickens, while there is only one study each on the turkey, *Meleagris gallopavo*, and emu, *Dromiceius novaehollandiae*. As might be expected the results are similar to those obtained from flying birds. An increase in heart rate is more important than a rise in cardiac stroke volume in increasing cardiac output (Grubb 1982; Grubb et al. 1983), and ventilation increases by a greater proportion than the rise in oxygen uptake (i.e. there is hyperventilation), as indicated by the animal becoming hypocapnic and alkalotic (Kiley et al. 1979; Brackenbury et al. 1981). In hens, running at a moderate ambient temperature, a seven times increase in minute ventilation volume results largely from a four times increase in respiratory frequency (Brackenbury and Gleeson 1983), whereas in running Pekin ducks tidal volume decreases to one-third of its resting value as ventilation rises to approximately five times resting (Kiley et al. 1979).

The hyperventilation may be the result of the increase in body temperature that occurs during exercise. However, if ducks run in a cold environment and body temperature is prevented from rising, there are still signs of hyperventilation (Kiley et al. 1982), although it must be pointed out that there was a metabolic acidosis (increased blood lactate and decreased pH) in these birds during exercise. This could well have caused the hyperventilation, for in chickens ventilation increases in direct proportion to gas exchange if body temperature is prevented from rising during moderate exercise (Brackenbury and Gleeson 1983). Although hyperthermia may be the cause of hyperventilation *during* running, it cannot be responsible for the rapid rise in ventilation associated with the *onset* of exercise in birds. Ventilation can still increase in running ducks in the absence of input from the CO_2 receptors in the lungs and if changes in Pa_{CO_2} and H^+ are kept to a minimum. This suggests that non-humoral factors are involved, such as muscle afferents and perhaps to a limited extent joint receptors (P.J. Butler and D.R. Jones, unpublished). There may be a "cardiodynamic hyperpnoea" in birds, as has been suggested for mammals (Whipp and Ward 1982), with the increased venous return at the onset of exercise stimulating mechanoreceptors, possibly in the heart, which causes an increase in ventilation.

The increase in heart rate in turkeys results partly from the stimulation of β-adrenergic receptors either by the sympathetic nervous system itself or by way of an increase in circulating catecholamines. However, some cardioacceleration occurs at the highest running speeds (> 3.5 km h^{-1}) in turkeys free of parasympathetic and β-adrenergic influences. The cause of this is unknown (Baudinette et al. 1982), although it could result from the increase in venous return caused by the contracting leg muscles. As might be expected, blood flow to the exercising muscles, via the sciatic artery, increases during running in the Pekin duck, but so also does blood flow to the head, via the carotid artery, by 3.7 times and 2.3 times respectively (Bech and Nomoto 1982). The latter is probably concerned with keeping the brain cool.

When walking or running, all birds that have been studied show a linear increase in oxygen uptake with increased velocity. Thus, the minimum cost of transport is at the

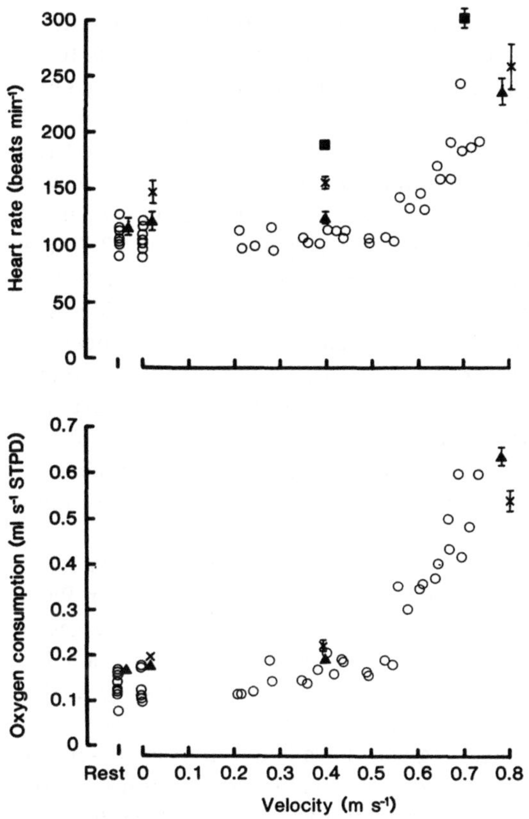

Fig. 4. Heart rate and oxygen consumption (○) of unencumbered tufted duck of mean mass 503 g at different swimming velocities. *Rest* is with motor of water channel turned off and at 0 m s^{-1} motor is turned on. Mean values (± SE of mean) from six unencumbered ducks are given at rest, 0 m s^{-1}, 0.4 m s^{-1} and at 0.78 m s^{-1} (▲). Mean values (± SE of mean) from ten ducks with a mask for measuring respiratory variables are given at 0 m s^{-1}, 0.4 m s^{-1}, and 0.8 m s^{-1} (*X*). Mean values (± SE of mean) of heart rate only are given for five ducks with a mask and with a cannulated wing artery at 0.4 m s^{-1} and at 0.7 m s^{-1} (■). (Data for unencumbered ducks, Woakes and Butler 1983)

highest sustainable speed (because when at rest the bird still consumes oxygen). However, for swimming mallard and tufted ducks, *Aythya fuligula*, oxygen uptake is relatively constant from speeds of 0.2–0.5 m s^{-1} but then increases very sharply as speed increases (Fig. 4). This gives rise to a V-shaped cost of transport/swimming speed curve with the minimum cost at the transition speed (0.5 m s^{-1}). Interestingly, Prange and Schmidt-Nielsen (1970) found that mallards freely swimming on a pond choose to swim at close to 0.5 m s^{-1}. The change in heart rate with increased swimming speed is similar to that for oxygen uptake (Woakes and Butler 1983), so there is a linear relationship between steady state oxygen uptake and heart rate.

As a duck moves through the water surface it creates waves at its bow and stern and the wave length of the waves is related to the speed at which it moves. When the bird reaches the speed where the wave length of the bow wave equals the length of the body at the water line (hull speed), the body becomes trapped in the trough between the bow and stern waves. As hull speed is approached, resistance to further increase in speed (drag) increases almost asymptotically, hence the rapid rise in oxygen uptake of the swimming ducks at a certain speed. The mallards used by Prange and Schmidt-Nielsen (1970) had a water line length of 0.33 m and a calculated hull speed of 0.71 m s^{-1}. Maximum sustainable swimming speed of these birds was 0.7 m s^{-1}.

At maximum sustainable swimming speed (0.78 m s^{-1}) oxygen uptake of unencumbered tufted ducks is 0.64 ml s^{-1} STPD (3.8 times resting) and heart rate is 235 beats

min^{-1} (2 times resting). The latter is less than half of the maximum heart rate recorded from this species (Butler and Woakes 1979). The addition of a face mask to the birds for the measurement of tidal volume, respiratory frequency and end tidal gas concentrations, and the subsequent cannulation of a branchial artery had little effect on resting oxygen uptake, whereas each procedure had a substantial effect on heart rate (Fig. 4). It is possible, therefore, that the respiratory variables mentioned above were also slightly above resting values. Bearing this in mind, at a swimming speed of 0.8 m s^{-1} respiratory minute volume is 3.1 times the resting value, largely as a result of a 2.7 times increase in respiratory frequency. Oxygen uptake is 2.8 times resting, so there is slight hyperventilation. Although this is apparent in terms of reduced Pa_{CO_2}, pHa did not increase. This is somewhat at variance to the results obtained from running ducks and chickens (see earlier), as deep body temperature at maximum sustainable swimming speed was 1 °C above resting in the masked tufted ducks and 0.5 °C above resting in the masked and cannulated birds. These data are, however, consistent with those obtained from fish crows and white-necked ravens when flying at ambient temperatures below 22°–23 °C. Blood lactate concentration in swimming tufted ducks is 0.83 m*M* at 0.3 m s^{-1} and 1.9 m*M* at 0.7 m s^{-1}.

Some birds spend extended periods of time swimming under water. An example is the tufted duck which obtains its food from the bottom of bodies of fresh water. Based largely on work performed on domestic ducks that were involuntarily submerged, it was thought for almost 40 years that when birds (and mammals) dive there is a reduction in blood flow to all tissues except the brain and heart, which continue to metabolize aerobically. The underperfused tissues metabolize anaerobically, producing lactic acid, thus saving oxygen for the oxygen-dependent regions of the body. Associated with the reduced blood flow to major regions of the body is a reduction in cardiac output resulting mainly from a large fall in heart rate (bradycardia). This bradycardia is the typical element of the response to involuntary submersion and is often taken as an indicator of the other adjustments taking place. A puzzling aspect of this oxygen conserving response is that it takes 20–40 s to reach its maximum level (τ for a 60 s "dive" is approx. 27 s) and yet, with the exception of the larger penguins, birds do not usually remain under water, when diving voluntarily, for much longer than 60 s (Butler and Jones 1982). The slowness of the cardiac response in mallard ducks, and their domesticated varieties, can be explained by the fact that 85% of the bradycardia results from the progressive stimulation of the carotid body chemoreceptors by hypercapnic and hypoxic blood (Jones et al. 1982).

By recording heart rate (and sometimes respiratory frequency) by way of a small radiotransmitter, from naturally behaving pochard, *A. ferina*, tufted ducks, and Humboldt penguins, *Spheniscus humboldti*, it was discovered that there is not an obvious, maintained reduction in heart rate during voluntary dives (Butler and Woakes 1976, 1979, 1984). Ducks and penguins dive repeatedly for extended periods without becoming exhausted. Preceding the first dive of a series there is an elevation in heart rate (Fig. 5) and an increase in respiratory frequency in the ducks. Upon submersion, which appears to require much effort as the bird arches itself into the water, there is a transient bradycardia, but heart rate increases somewhat during the first few seconds of the dive and then remains at a more or less steady rate. The ducks have to continue paddling when under water and on occasions there is a 1:1 correspondence between

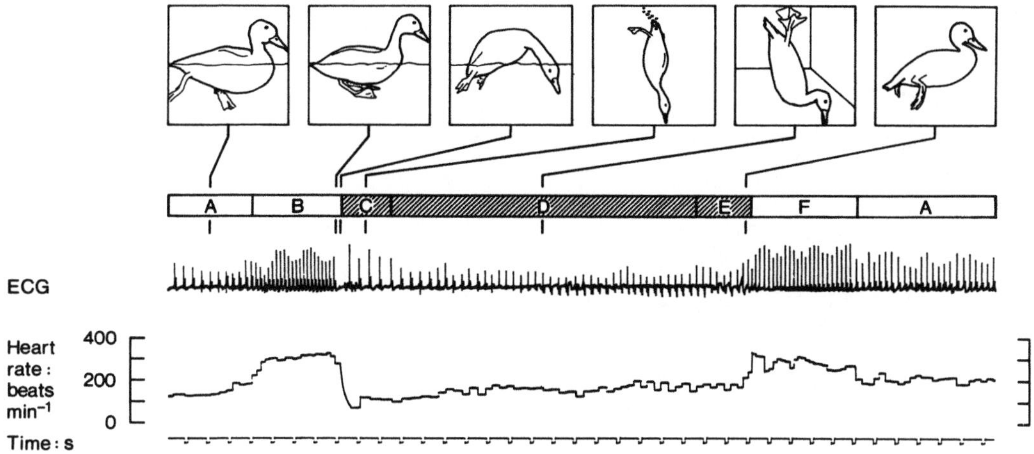

Fig. 5. Relationship between the behaviour of a tufted duck and changes in heart rate during a spontaneous dive on a glass sided tank, 1.55 m deep. From *above, downwards*: tracings of duck from cine film showing; *from left to right*, swimming, preparing to dive, moment of submersion, descending, feeding on bottom, surfacing (10 cm from surface); *A* time periods of swimming on surface; *B* cardiac acceleration before submersion; *C* descent; *D* feeding on bottom; *E* surfacing; *F* cardiac acceleration following surfacing; *ECG* heart rate, time marker(s). The *lines* between the pictures of the ducks and the time *boxes* join coincident points in time. (Butler 1982a)

heart rate and leg beat frequency (Butler and Woakes 1982b). Surfacing occurs passively once leg beating has ceased. By contrast, penguins glide under water and do not exhibit an obvious increase in heart rate before the first dive of a series (Fig. 6), neither is it necessary for them to perform perceptible locomotor activity to remain submerged.

Oxygen uptake has also been estimated for tufted ducks and Humboldt penguins during voluntary diving (Woakes and Butler 1983; Butler and Woakes 1984). For the ducks, oxygen consumption at mean dive duration is 3.5 times resting and not significantly different from that measured at maximum sustainable swimming speed. On the basis of this estimated rate of oxygen usage and the size of the oxygen storage compartments in tufted ducks (Keijer and Butler 1982), aerobic metabolism could continue for 44 s with the duck actively swimming under water. The extra oxygen taken on board during the anticipatory period before the first dive of a series increases aerobic dive duration by at least 7 s to a total of 51 s. This means that even ducks diving to a depth of 6 m in the wild, which takes an average of 28 s (Draulans 1982) do so aerobically and have oxygen to spare at the end. Heart rate increases immediately upon surfacing, largely as a result of renewed activity of inspiratory neurons in the brain stem and activation of the CO_2-sensitive receptors in the lung, both of which occur with the onset of ventilation (Butler and Taylor 1983). This ensures rapid replacement of the oxygen used during the preceding dive and enables the next dive to occur in quick succession. If mean heart rate and oxygen consumption during diving are plotted on the same graph as heart rate and oxygen consumption during surface swimming, then heart rate is lower during diving than during swimming at the same level of oxygen usage (Fig. 7). Diving heart rate is, on average, 1.5 times the resting value and 2.7 times the value after a similar period of involuntary head submersion.

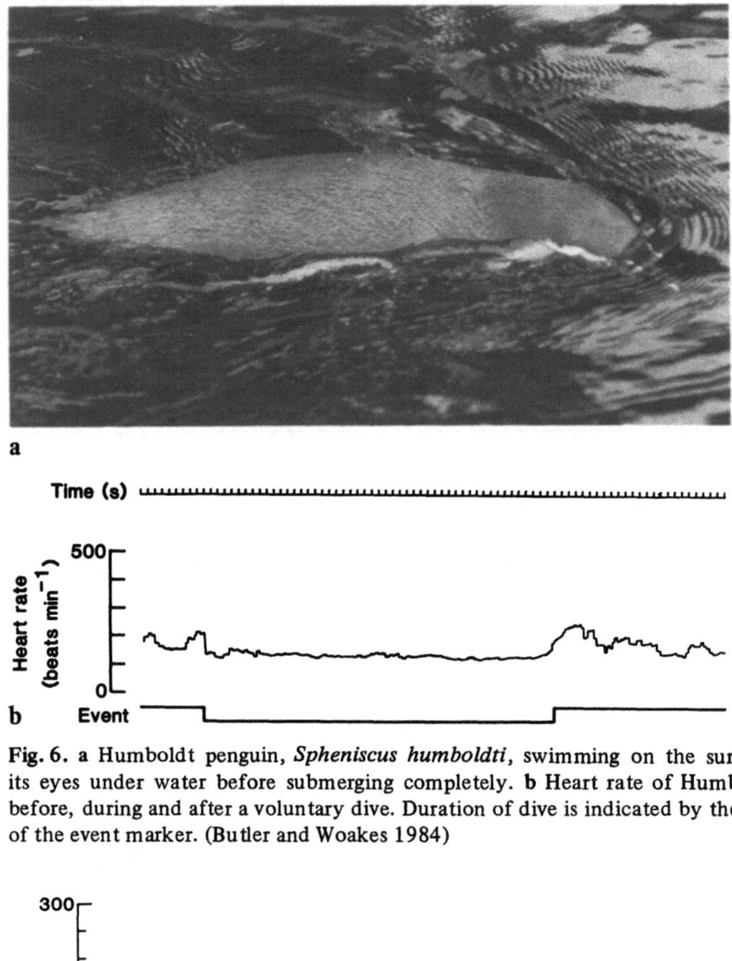

Fig. 6. a Humboldt penguin, *Spheniscus humboldti*, swimming on the surface of the pond with its eyes under water before submerging completely. **b** Heart rate of Humboldt penguin (4.5 kg) before, during and after a voluntary dive. Duration of dive is indicated by the downward deflection of the event marker. (Butler and Woakes 1984)

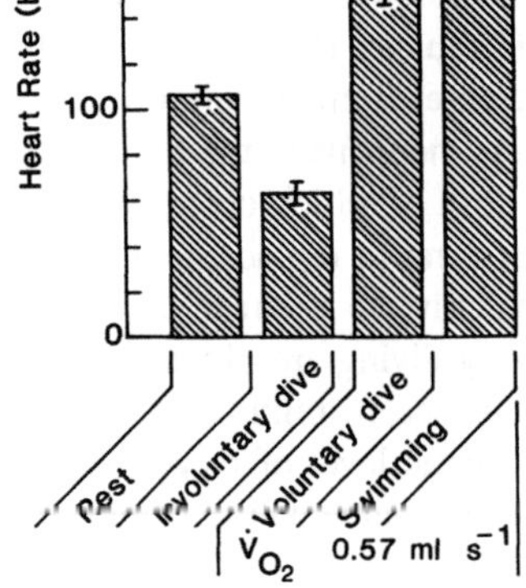

Fig. 7. Mean heart rate (± SE of mean) for six tufted ducks, *Aythya fuligula* (except during involuntary dives when ten ducks were used) at rest on water, 15 s after involuntary submersion of the head, voluntary dives of 14.4 s duration and while swimming. Oxygen consumption ($\dot{V}_{O_2}$) at mean dive duration and while swimming was the same, 0.57 ml s^{-1} STPD. (Woakes and Butler 1983)

On the basis of these results, it has been suggested that the circulatory adjustments during voluntary diving in ducks are similar to those during exercise in air in as much as the locomotory muscles as well as the heart and CNS receive an enhanced blood supply and sufficient oxygen for aerobic metabolism. The inactive muscles, viscera, kidneys, and skin may well receive a reduced supply. The lower heart rate during diving could indicate, however, that selective vasoconstriction is more intense and that oxygen extraction by the active muscles is greater than when the birds exercise in air (Butler 1982b). During voluntary dives there would appear to be, in tufted ducks at least, a balance between the cardiovascular responses to involuntary submersion and to exercise in air with the bias towards the latter (Fig. 7).

The balance can be tipped in the opposite direction if, for any reason, the duck is briefly unable to surface from a voluntary dive. As soon as the duck is aware that immediate access to air is not possible, there is a progressive bradycardia which, within 10–15 s, is similar to that seen during involuntary submersion (i.e. approx. 25 beats min^{-1}). It would seem that the duck switches to the oxygen conserving response. However, if ducks have to swim varying distances under cover in order to obtain their food, there is no initial lowering of heart rate when long distances have to be travelled, although heart rate does fall progressively during the longer dives after the first 5 s. At 10 s after submersion it is significantly lower during the longer than during the shorter dives (Fig. 8). These observations must be pertinent to ducks feeding under ice in the winter. It would seem that only if caught unawares does the animal immediately invoke its oxygen conserving response. When diving under ice, it is likely that ducks make a number of short exploratory dives initially in order to reduce their chances of becoming disoriented. If they have to swim long distances under water, they may shift progressively to the oxygen conserving response. Unlike the situation in mallard ducks, the carotid body chemoreceptors play very little part in the bradycardia seen during involuntary submersion in tufted ducks and, interestingly, the onset of the bradycardia is much more rapid in the latter case (τ for a 60 s "dive" is approx. 8 s). The carotid bodies do, however, exert an inhibitory influence on heart rate during longer voluntary dives in tufted ducks (Butler and Woakes 1982a). We are in the process of determining their role in the modified cardiac responses to voluntary diving in tufted ducks, as outlined above.

For the Humboldt penguins, neither heart rate nor oxygen uptake during diving are significantly different from the resting values. Calculations indicate that they can remain aerobic under water for 2.3 min. Observations on chinstrap penguins, *Pygoscelis antarctica*, reveal that they dive to less than 45 m and for an average of 1.6 min (Lishman and Croxall 1983). It certainly seems that penguins are able to dive and remain aerobic for longer periods than ducks. This is probably because they are more efficient at under water locomotion, partly as a result of their being almost neutrally buoyant (Butler and Woakes 1984).

In general, the behaviour pattern of diving birds consists of using stored oxygen to metabolize aerobically during relatively short dives and then quickly replacing the oxygen at the surface before the next dive. This is more economical in terms of percentage of time spent feeding during a group of dives than if longer dives are performed but proportionately longer is spent at the surface metabolizing the lactate (Kooyman et al. 1980). Also, animals in such an exhausted state would be more vulnerable to predators.

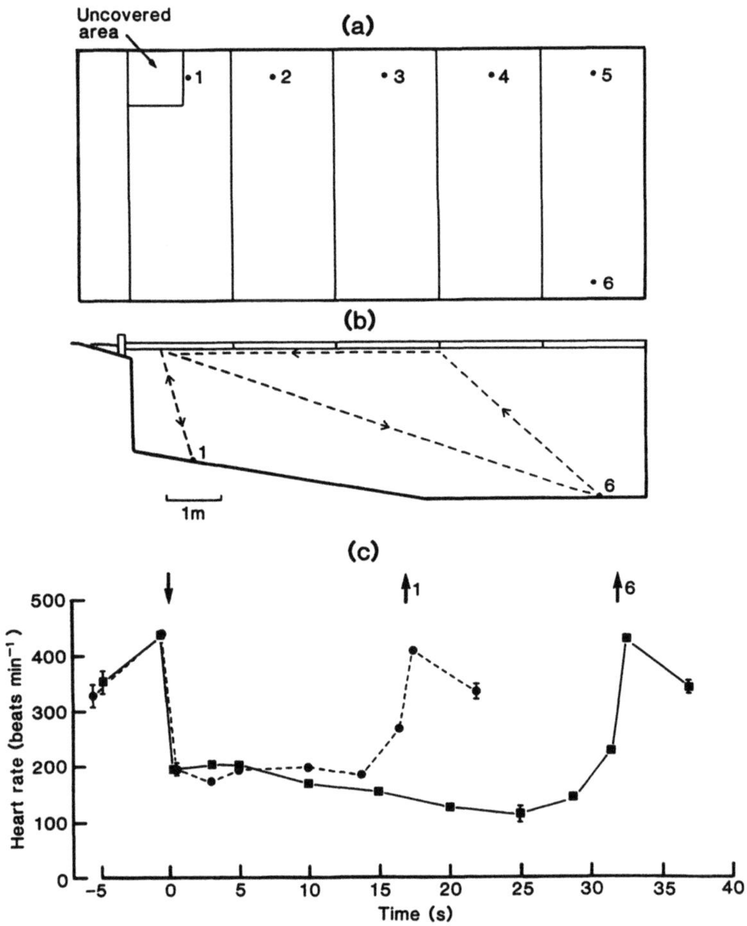

Fig. 8a–c. Plan (**a**) and side view section (**b**) of an outside pond. The surface is almost totally covered except for one small area. Food is placed on the bottom of the pool at six different distances from the uncovered area (1–6 in **a**). **b** The route taken by the ducks to and from positions *1* and *6*, c the mean dive duration and mean (± SE of mean) heart rate for six tufted ducks while making these journeys. It should be noted that for the longer diagonal journeys, the ducks have to swim actively during the return as well as during the outward leg. (R. Stephenson and P.J. Butler, unpublished data)

During exercise in air birds appear predominantly to increase the frequency of their respiratory and cardiovascular pumps to a far greater extent than the tidal or stroke volumes. In tufted ducks heart rate can vary by a factor of 20 or more depending on whether they are flying or trapped under water. Heart rate is also very sensitive to experimental manipulation. "Resting" heart rate is 25% higher in tufted ducks with a face mask compared with unencumbered birds, and after cannulation of a blood vessel it is even higher. It is important, therefore, that during future studies into the fascinating subject of exercise in birds, due care is taken to reduce the effects of measuring techniques upon the variables being measured. Such effects could be an explanation for the variable occurrence of hyperventilation, and subsequent hypocapnia and alkalosis, in birds exercising at moderate ambient temperatures.

References

Baudinette RV, Tonkin AL, Orbach J, Seymour RS, Wheldrake JF (1982) Cardiovascular function during treadmill exercise in the turkey. Comp Biochem Physiol 72A:327–332

Bech C, Nomoto S (1982) Cardiovascular changes associated with treadmill running in the Pekin duck. J Exp Biol 97:345–358

Bernstein MH (1976) Ventilation and respiratory evaporation in the flying crow, *Corvus ossifragus*. Respir Physiol 26:371–382

Black CP, Tenney SM (1980) Oxygen transport during progressive hypoxia in high-altitude and sea-level water fowl. Respir Physiol 39:217–239

Brackenbury JH, Gleeson M (1983) Effects of P_{CO_2} on respiratory pattern during thermal and exercise hyperventilation in domestic fowl. Respir Physiol 54:109–119

Brackenbury JH, Gleeson M, Avery P (1981) Effects of sustained running exercise on lung air-sac gas composition and respiratory pattern in domestic fowl. Comp Biochem Physiol 69A:449–453

Butler PJ (1981) Respiration during flight. In: Hutás I, Debreczeni LA (eds) Advances in physiological sciences, vol 10. Pergamon, Oxford, pp 155–164

Butler PJ (1982a) Respiration during flight and diving in birds. In: Addink ADF, Spronk N (eds) Exogenous and endogenous influences on metabolic and neural control. Pergamon, Oxford, pp 103–114

Butler PJ (1982b) Respiratory and cardiovascular control during diving in birds and mammals. J Exp Biol 100:195–221

Butler PJ (1984) New techniques for studying respiration in free flying birds. Proc XVIII Int Ornithol Congr (in press)

Butler PJ, Jones DR (1982) The comparative physiology of diving in vertebrates. In: Lowenstein OE (ed) Advances in comparative physiology and biochemistry, vol 8. Academic, New York, pp 179–364

Butler PJ, Taylor EW (1983) Factors affecting the respiratory and cardiovascular responses to hypercapnic hypoxia in mallard ducks. Respir Physiol 53:109–127

Butler PJ, Woakes AJ (1976) Changes in heart rate and respiratory frequency associated with natural submersion in ducks. J Physiol 256:73–74P

Butler PJ, Woakes AJ (1979) Changes in heart rate and respiratory frequency during natural behaviour of ducks, with particular reference to diving. J Exp Biol 79:283–300

Butler PJ, Woakes AJ (1980) Heart rate, respiratory frequency and wing beat frequency of free flying barnacle geese, *Branta leucopsis*. J Exp Biol 85:213–226

Butler PJ, Woakes AJ (1982a) Control of heart rate by carotid body chemoreceptors during diving in tufted ducks. J Appl Physiol 53:1405–1410

Butler PJ, Woakes AJ (1982b) Telemetry of physiological variables from diving and flying birds. Symp Zool Soc Lond 49:107–128

Butler PJ, Woakes AJ (1984) Heart rate and aerobic metabolism in Humboldt penguins, *Spheniscus humboldti*, during voluntary dives. J Exp Biol 108:419–428

Butler PJ, West NH, Jones DR (1977) Respiratory and cardiovascular responses of the pigeon to sustained, level flight in a wind tunnel. J Exp Biol 71:7–26

Draulans D (1982) Foraging and size selection of mussels by the tufted duck, *Aythya fuligula*. J Anim Ecol 51:943–956

Elkins N (1979) High altitude flight by swans. Br Birds 72:238–239

Grubb BR (1982) Cardiac output and stroke volume in exercising ducks and pigeons. J Appl Physiol 53:207–211

Grubb BR (1983) Allometric relations of cardiovascular function in birds. Am J Physiol 245:H567–H572

Grubb BR, Mills CD, Colacino JM, Schmidt-Nielsen K (1977) Effect of arterial carbon dioxide on cerebral blood flow in ducks. Am J Physiol 232:H596–601

Grubb BR, Colacino JM, Schmidt-Nielsen K (1978) Cerebral blood flow in birds: effect of hypoxia. Am J Physiol 234:H230–H234

Grubb BR, Jorgensen DD, Conner M (1983) Cardiovascular changes in the exercising emu. J Exp Biol 104:193–201

Hudson DM, Bernstein MH (1981) Temperature regulation and heat balance in flying white-necked ravens, *Corvus cryptoleucus*. J Exp Biol 90:267–281

Hudson DM, Bernstein MH (1983) Gas exchange and energy cost of flight in the white-necked raven, *Corvus cryptoleucus*. J Exp Biol 103:121–130

Jones DR, Milsom WK, Gabbott GRJ (1982) Role of central and peripheral chemoreceptors in diving responses of ducks *(Anas platyrhynchos)*. Am J Physiol 243:R537–R545

Jurgens KD, Bartels H, Bartels R (1981) Blood oxygen transport and organ weights of small bats and small non-flying mammals. Respir Physiol 45:243–260

Keijer E, Butler PJ (1982) Volumes of the respiratory and circulatory systems in tufted and mallard ducks. J Exp Biol 101:213–220

Kiley JP, Kuhlmann WD, Fedde MR (1979) Respiratory and cardiovascular responses to exercise in the duck. J Appl Physiol 47:827–833

Kiley JP, Kuhlmann WD, Fedde MR (1982) Ventilatory and blood gas adjustments in exercising isothermic ducks. J Comp Physiol 147:107–112

King AS, Molony V (1971) The anatomy of respiration. In: Bell DK, Freeman BM (eds) Physiology and biochemistry of the domestic fowl. Academic, London, pp 93–169

Kooyman GL, Drabek CM, Elsner R, Campbell WB (1971) Diving behaviour of the emperor penguin, *Aptenodytes forsteri*. Auk 88:775–795

Kooyman GL, Wahrenbrock EA, Castellini MA, Davis RW, Sinnett EE (1980) Aerobic and anaerobic metabolism during voluntary diving in Weddell seals: evidence of preferred pathways from blood chemistry and behaviour. J Comp Physiol 138B:335–346

Lasiewski RC, Calder WA (1971) A preliminary allometric analysis of respiratory variables in resting birds. Respir Physiol 11:152–166

LeFebvre EA (1964) The use of D_2O^{18} for measuring energy metabolism in *Columba livia* at rest and in flight. Auk 81:403–416

Lishmann GS, Croxall JP (1983) Diving depths of the chinstrap penguin, *Pygoscelis antarctica*. Br Antarct Surv Bull 61:21–25

Nisbet ICT, Drury WH, Baird J (1963) Weight-loss during migration. Part I: Deposition and consumption of fat by the blackpoll warbler, *Dendroica striata*. Bird Banding 34:107–159

Ogilvie MA (1978) Wild geese. T & AD Poyser, Berkhamsted

Pasquis P, Lacaisse A, Dejours P (1970) Maximal oxygen uptake in four species of small mammal. Respir Physiol 9:298–309

Pennycuick CJ (1969) The mechanics of bird migration. Ibis 111:525–556

Pennycuick CJ (1975) Mechanics of flight. In: Farner DS, King JR (eds) Avian biology, vol V. Academic, New York, pp 1–75

Prange HD, Schmidt-Nielsen K (1970) The metabolic cost of swimming in ducks. J Exp Biol 53: 763–777

Prinzinger R, Hanssler I (1980) Metabolism-weight relationship in some small non-passerine birds. Experientia 37:1299–1300

Rayner JMV (1979) A new approach to animal flight mechanics. J Exp Biol 80:17–54

Rayner JMV (1982) Avian flight energetics. Annu Rev Physiol 44:109–119

Rothe H-J, Nachtigall W (1980) Physiological and energetic adaptations of flying birds, measured by the wind tunnel technique. A survey. Proc XVII Int Ornithol Congr 400–405

Scheid P (1979) Mechanisms of gas exchange in bird lungs. Rev Physiol Biochem Pharmacol 86: 137–184

Scheid P (1984) Significance of lung structure for performance at high altitude. Proc XVIII Int Ornithol Congr (in press)

Scheid P, Piiper J (1972) Cross current gas exchange in avian lungs: effects of reversed parabronchial airflow in ducks. Respir Physiol 16:304–312

Stewart AG (1978) Swans flying at 8,000 metres. Br Birds 71:459–460
Swan LW (1961) The ecology of the high Himalayas. Sci Am 205:68–78
Torre-Bueno JR (1976) Temperature regulation and heat dissipation during flight in birds. J Exp Biol 65:471–482
Torre-Bueno JR (1978) Respiration during flight in birds. In: Piiper J (ed) Respiratory function in birds, adult and embryonic. Springer, Berlin Heidelberg New York, pp 89–94
Torre-Bueno JR, Larochelle J (1978) The metabolic cost of flight in unrestrained birds. J Exp Biol 75:223–229
Tucker VA (1966) Oxygen consumption of a flying bird. Science 154:150–151
Tucker VA (1968) Respiratory exchange and evaporative water loss in the flying budgerigar. J Exp Biol 48:67–87
Tucker VA (1970) Energetic cost of locomotion in animals. Comp Biochem Physiol 34:841–846
Tucker VA (1972) Metabolism during flight in the laughing gull, *Larus atricilla*. Am J Physiol 222: 237–245
Weibel ER, Taylor CR (eds) (1981) Design of the mammalian respiratory system. Respir Physiol 44:1–164
Whipp BJ, Ward SA (1982) Cardiopulmonary coupling during exercise. J Exp Biol 100:175–193
Woakes AJ, Butler PJ (1983) Swimming and diving in tufted ducks, *Aythya fuligula*, with particular reference to heart rate and gas exchange. J Exp Biol 107:311–329

Muscle Function During Locomotion in Mammals

R.B. ARMSTRONG and M.H. LAUGHLIN [1]

1 Introduction

Mammalian muscles are composed of fibers with different contractile and metabolic properties. Although various systems of classification have been proposed for the different fiber types (Close 1972; Saltin and Gollnick 1983), one that has been used extensively identifies the fibers as either slow-twitch-oxidative (SO), fast-twitch-oxidative-glycolytic (FOG), or fast-twitch-glycolytic (FG) (Peter et al. 1972). The most common method used for identifying the fibers is to histochemically analyze a mitochondrial oxidative enzyme and myofibrillar adenosine triphosphatase activities in muscle cross-sections. This terminology is convenient because the primary physiological and metabolic properties of the fibers are incorporated in the nomenclature. Although the two systems are not completely interconvertible (Nemeth et al. 1979), SO, FOG, and FG fibers are generally the same as those classified as type I, type IIa, and type IIb, respectively.

These three types of fibers can be identified in most mammalian species (e.g., Ariano et al. 1973). However, there are marked differences in absolute magnitude for both the physiological and metabolic characteristics of a given fiber type among different animals. For example, the FOG fibers in flight muscle of the bat have a markedly higher oxidative enzyme potential than FOG fibers in the locomotory muscles of the laboratory rat (Armstrong et al. 1977; Ianuzzo and Armstrong 1976). Similarly, fast-twitch muscle in the mouse shortens about twice as fast as fast-twitch muscle in cat (Close 1972). Thus, fiber type classification is useful for comparing muscle tissue types within a species, but care must be observed in making comparisons across species.

The three fiber types are distributed within and among the locomotory muscles of terrestrial mammals in fairly stereotyped patterns. In this paper we will refer extensively to data from Sprague-Dawley albino rats *(Rattus norwegicus)*; the distribution of the fiber types in the ankle extensor muscles for this animal is illustrated in Fig. 1. However, these same general distribution patterns are apparent in many terrestrial mammals (Ariano et al. 1973) including man (Johnson et al. 1973). The deepest extensor muscles that play an anti-gravity role are primarily composed of SO fibers (e.g., soleus muscle; Fig. 1). Deeper portions of more peripheral extensor muscles are primarily of the SO and FOG types (e.g., red gastrocnemius muscle; Fig. 1), whereas the most superficial

1 Department of Physiology, Oral Roberts University, School of Medicine, Tulsa, OK 74171, USA

Circulation, Respiration, and Metabolism
(ed. by R. Gilles)

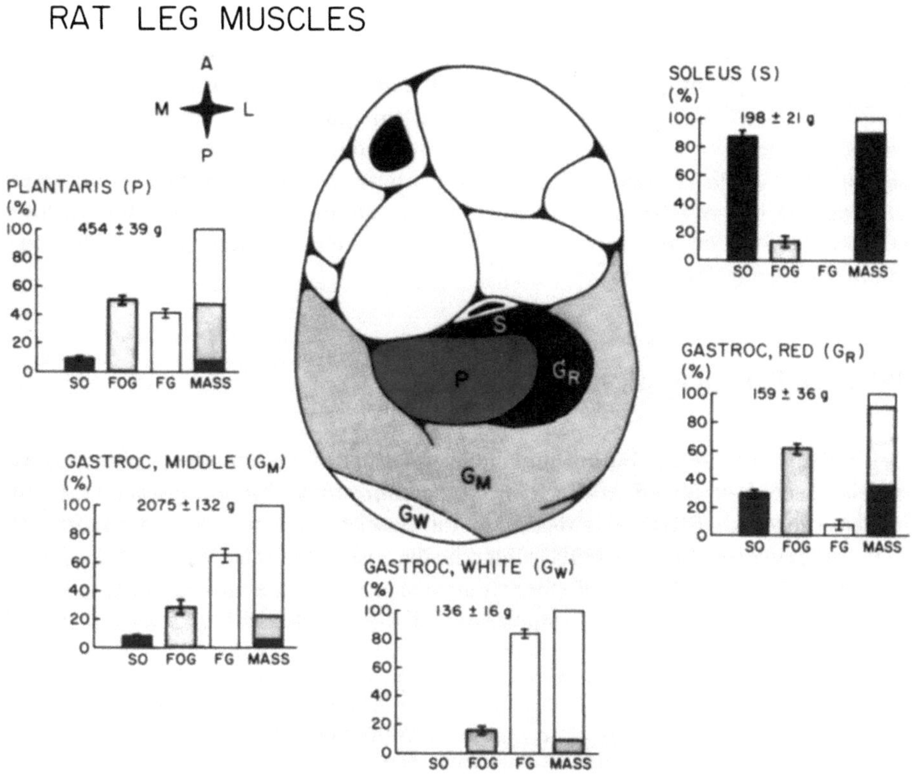

Fig. 1. Cross-section of the rat leg illustrating the distribution patterns of the slow-twitch-oxidative (SO), fast-twitch-oxidative-glycolytic (FOG), and fast-twitch-glycolytic (FG) fibers in the ankle extensor muscles. The intensity of *shading* in the cross-section is proportional to the population of high oxidative (SO + FOG) fibers. Histograms show the fiber type populations in the muscles and the relative estimated mass of the muscles composed of each of the fiber types. Data are from Armstrong and Phelps (1984)

parts of the peripheral extensor muscles have relatively high populations of FG fibers (e.g., white gastrocnemius muscle; Fig. 1). Thus, the distribution of the three fiber types within and among extensor muscles follows the general pattern of SO → FOG → FG from deep to superficial regions. This sterotyped pattern of fiber type distribution is less obvious in flexor muscle groups (Ariano et al. 1973; Armstrong and Laughlin 1983). Also, the distribution patterns are different in mammals that depend on different locomotory or postural modes (Ariano et al. 1973; Armstrong et al. 1977).

The changes in the patterns of recruitment of the fiber types as the animal progresses from postural standing through high speed running follow a similar deep to superficial spatial scheme (Burke 1981), as determined primarily from studies of electromyographic (EMG) (e.g., Walmsley et al. 1978; Gardiner et al. 1982) and glycogen loss patterns (e.g., Sullivan and Armstrong 1978) among and within the muscles. During standing the muscular force is primarily produced by the SO motor units in the extensor muscles, particularly in the deep slow muscle (e.g., soleus muscle) for the respective groups. In low-speed locomotion the SO units continue to be recruited, but the additional required muscular forces are provided by FOG fibers. With increasing locomotory speeds, progressive additive recruitment of FOG, then FG, motor units occurs (Burke 1981).

These spatial relationships among and within the muscles, i.e., the fiber type distributions and the fiber recruitment patterns, have made it convenient for studying the metabolism of the fibers during locomotory exercise. However, as may be noted in Fig. 1, none of the muscles are type-pure; all are composed of at least two types of fibers intermixed in some combination. This point should be kept in mind in interpreting the following discussion.

2 Metabolic Potentials of the Fibers

As suggested by the nomenclature, metabolic potential of the muscles within a given mammalian species is related to the fiber type composition. Figure 2 presents the succinate dehydrogenase (SDH) activities of the rat muscles as a function of the estimated mass of the oxidative (SO and FOG) fibers in the respective muscles. SDH is a Krebs cycle enzyme that is indicative of the oxidative capacity of the muscle. As is apparent from a comparison of the fiber compositions of the muscles in Fig. 1 with the SDH

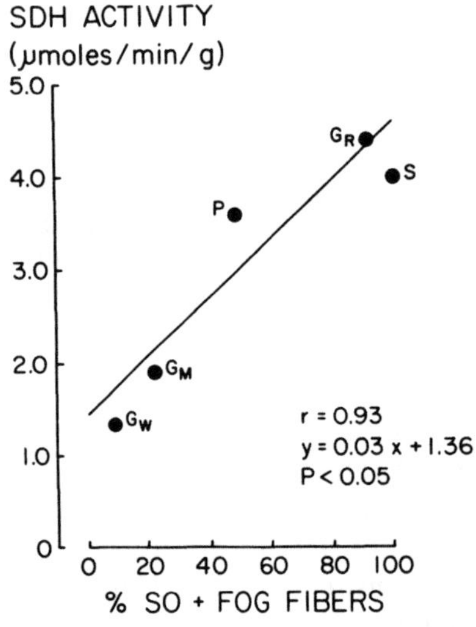

Fig. 2

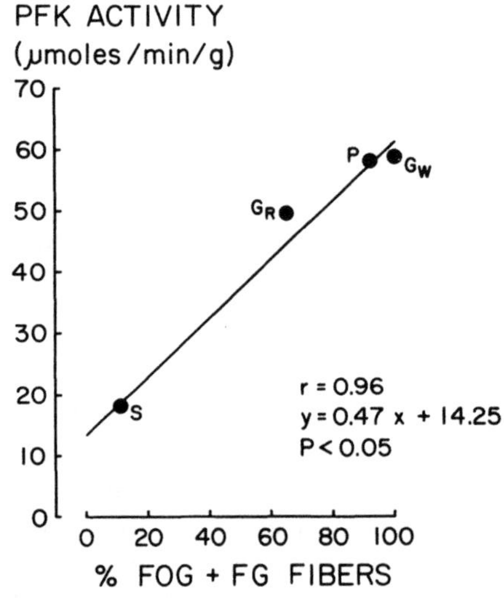

Fig. 3

Fig. 2. Succinate dehydrogenase (SDH) activities, which are indicative of the oxidative potentials of the rat ankle extensor muscles, plotted as a function of the estimated mass of oxidative (SO + FOG) fibers in the muscles. SDH data are from Armstrong and Laughlin (1984). Fiber type mass data are from Armstrong and Phelps (1984)

Fig. 3. Phosphofructokinase (PFK) activities, which are indicative of the glycolytic capacities of the rat ankle extensor muscles, plotted as a function of the estimated mass of glycolytic (FOG + FG) fibers in the muscles. PFK data are from Ianuzzo and Armstrong (1976). Fiber type mass data are from Armstrong and Phelps (1984)

data in Fig. 2, FOG fibers (red gastrocnemius muscle) have the highest oxidative potentials in rats and FG fibers (white gastrocnemius muscle) have the lowest. This relationship among the fiber types varies in different species of mammals. In many animals, including man (Saltin and Gollnick 1983), SO fibers have the highest oxidative enzyme potential.

In Fig. 3, phosphofructokinase (PFK) activity of the ankle extensor muscles is plotted against the estimated mass of the glycolytic (FOG and FG) fibers in the respective muscles. As a rate-limiting enzyme in glycolysis, PFK is indicative of the anaerobic potential of the muscles. FG fibers (white gastrocnemius muscle) have the highest PFK activity, and SO fibers (soleus muslce), the lowest (Fig. 3).

3 Spatial Metabolic Patterns During Locomotion

Metabolic events within a skeletal muscle at any given time during locomotion are dictated in large part by two major factors: (1) activity level, or recruitment of the fibers, and (2) the metabolic potentials of the fibers, including the capacity of the central circulation to deliver O_2 and substrate. The interactions of these two factors are illustrated by the alterations in glycogen loss rate that occur in the muscles as a function of increasing locomotory speed (Fig. 4). At slow speeds muscles with high populations of SO fibers (soleus muscle) lose more glycogen than those made up of FG fibers (white gastrocnemius muscle) since FG fibers are not recruited at these speeds (Sullivan and Armstrong 1978). Even though SO fibers have a relatively low glycolytic capacity (Fig. 3), they do lose glycogen when active; at higher speeds the rate of glycogen loss in the slow muscle doesn't increase (Fig. 4). The differences in glycolytic capacities in slow and fast fibers are reflected in the rates of glycogen loss during high speed exercise when all three of the fiber types are recruited to produce force. In fast running, red and white gastrocnemius muscle glycogen loss rates are several-fold higher

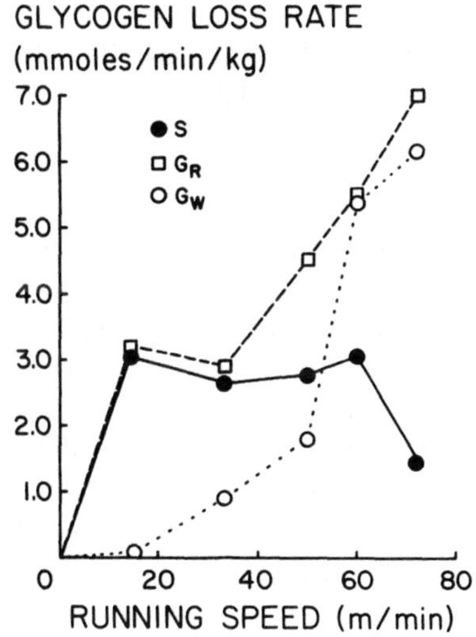

Fig. 4. Glycogen loss rates in rat ankle extensor muscles during short (1.5–8 min) bouts of treadmill running at various speeds. Data are from Sullivan and Armstrong (1978) with the exception of the 15 m min^{-1} data, which are previously unpublished

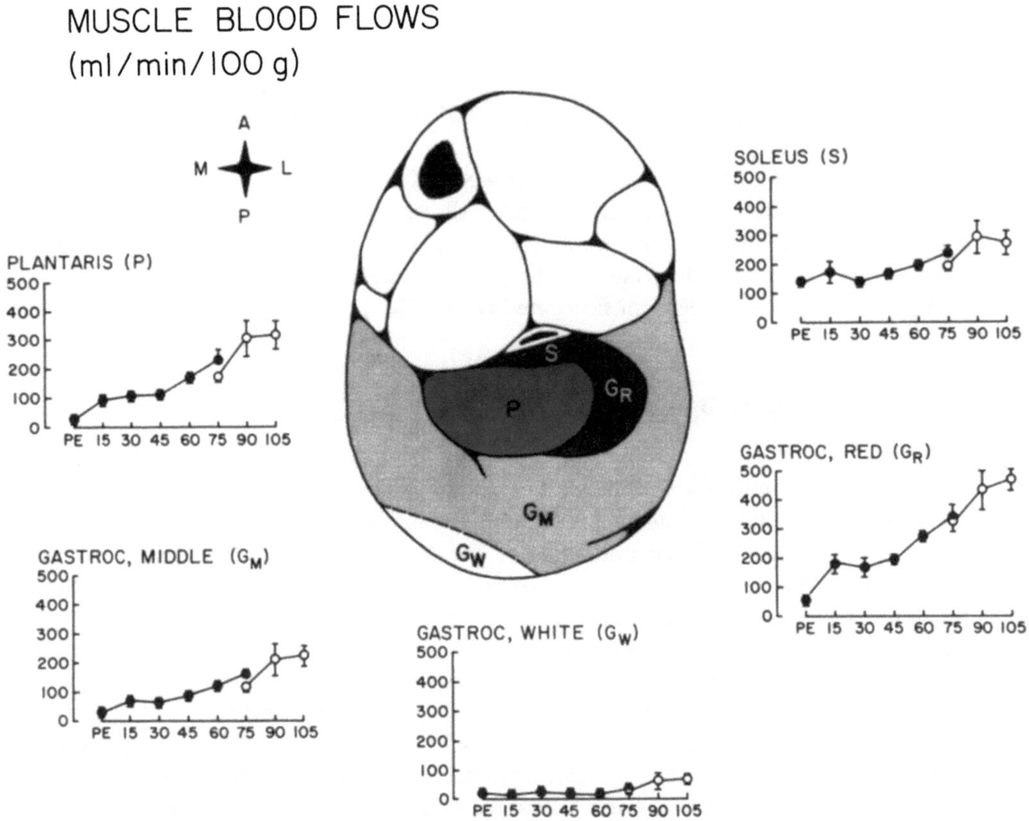

Fig. 5. Blood flows in rat ankle extensor muscles during preexercise (PE) and during 1 min of running at various speeds. Values are means ± SEM. Data for PE and 15, 30, 45, 60, and 75 ml min^{-1} are from Laughlin and Armstrong (1982). 90 and 105 m min^{-1} data are previously unpublished

than that in soleus muscle. At the high running speed the rates of glycogen loss are similar in the red and white portions of gastrocnemius muscle, even though the oxidative capacity of the red muscle is several-fold higher than the white muscle (Fig. 2). However, at higher running speeds (e.g., 105 m min^{-1}, Fig. 5) it is probable that the rate of glycogen use would increase to a greater extent in the white muscle.

Blood flow distribution within and among the muscles, which presumably is related to rates of oxidative metabolism in the muscles, also appears to reflect the interaction between activity level and metabolic potential of the fibers under a variety of postural and locomotory conditions (Fig. 5). During preexercise conditions when the animal is standing on the treadmill, the blood flow is primarily directed to the SO motor units in the extensor muscles (Laughlin and Armstrong 1982; Armstrong and Laughlin 1983). The highest blood flow during preexercise occurs in soleus muscle in the ankle extensor group. As illustrated in Fig. 6, the preexercise blood flow in the muscles is proportional to their respective populations of SO fibers. The SO fibers are responsible for providing postural support during standing (Burke 1981), so the high flows to these fibers are

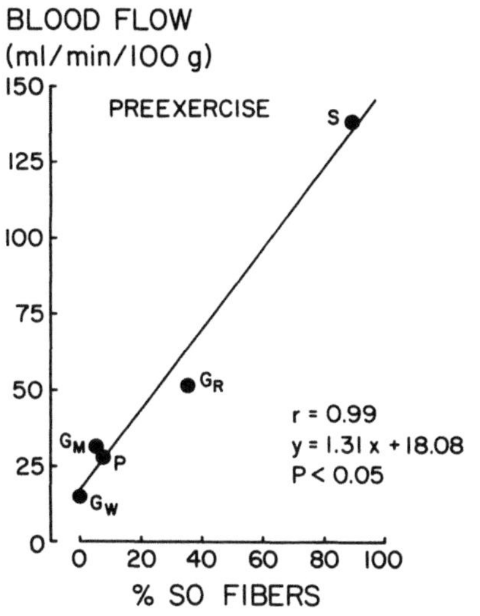

Fig. 6

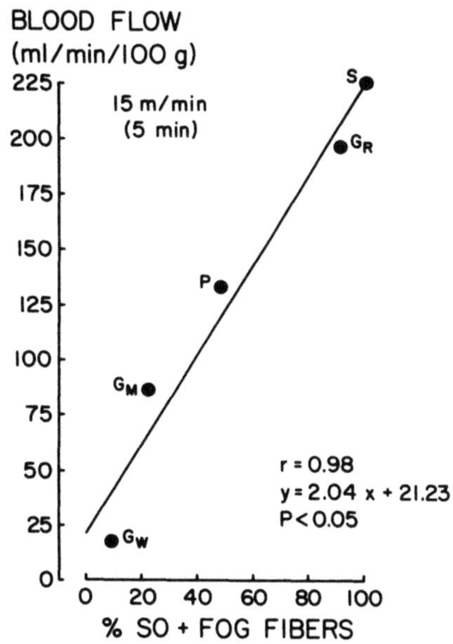

Fig. 7

Fig. 6. Preexercise blood flows in ankle extensor muscles as a function of the mass of SO fibers in the muscles. Blood flow data are from Laughlin and Armstrong (1982). Fiber mass data are from Armstrong and Phelps (1984)

Fig. 7. Blood flows in ankle extensor muscles after 5 min of treadmill walking at 15 m min^{-1} plotted as a function of SO + FOG fiber mass. Blood flow data are from Laughlin and Armstrong (1983). Fiber mass data are from Armstrong and Phelps (1984)

predictable. During walking on the treadmill, blood flow is proportional to the sum of the SO and FOG fibers in the extensor muscles (Fig. 7). This distribution also relates to the recruitment of the fibers in the muscles since the SO fibers continue to be active during walking with the additional recruitment of some FOG fibers (Burke 1981). With increasing locomotory speeds, the blood flow to FOG fibers predominates, so that during fast galloping flow becomes a function of the mass of FOG fibers in the muscles (Fig. 8). This is interesting in light of the fact that FG fibers, which are recruited at high running speeds (Burke 1981), make up about 76% of the hindlimb muscle mass in rats (Armstrong and Phelps 1984). The results reflect the lower oxidative potential of the FG fibers and the smaller vascular bed associated with these fibers (Mai et al. 1970; Armstrong et al. 1975; Saltin and Gollnick 1983).

The magnitude of blood flow to the FOG muscles in rats during high speed locomotion is impressive. For example, at 105 m min^{-1}, flow to red gastrocnemius muscle is about 475 m min^{-1} 100 g^{-1} (Fig. 5). Extrapolation of the line in Fig. 6 indicates blood flow in a hypothetical pure FOG muscle at 105 m min^{-1} would exceed 700 ml min^{-1} 100 g^{-1}. Total hindlimb muscle blood flow at this speed is about 210 ml min^{-1} 100 g^{-1}, which is similar to the muscle blood flows at $\dot{V}_{O_2\,max}$ in running dogs (Musch et al. 1985) and maximal quadriceps muscle flows in human subjects performing one leg exercise (Anderson 1982).

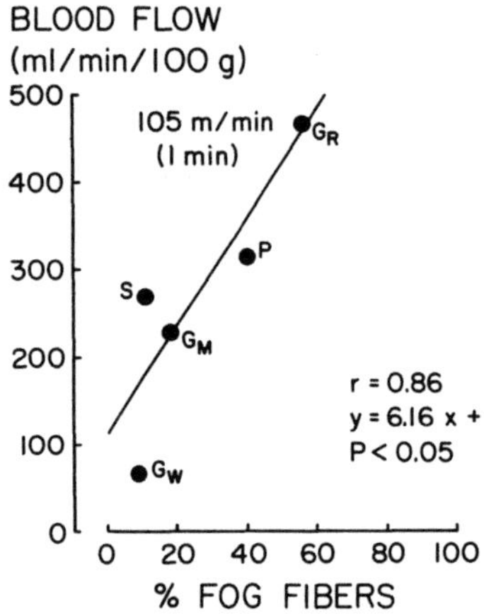

Fig. 8. Blood flows in ankle extensor muscles after 1 min of running at 105 m min^{-1} plotted as a function of FOG fiber mass. Blood flow data are previously unpublished. Fiber type mass data are from Armstrong and Phelps (1984)

These findings demonstrate that there are important functional correlates to the distribution of the fiber types in skeletal muscles, and that the spatial patterns of metabolic activity are closely related to the patterns of muscle fiber activity. However, the metabolic activities of active muscle fibers are limited by the metabolic potentials of the fibers.

Acknowledgements. The authors express appreciation to their research staff, including Ron Phelps, Ken Rouk, Robyn Stroup, Carol Vandenakker, and Judy White, for their essential contributions to this work. Special thanks to Mrs. Judy White for the artwork and Ruth Christian and Wanda Cuzick for their help in preparation of the manuscript. This research was supported by NIH Grants AM24572 and HL29428, AHA Tulsa Chapter Funds, and ORU Research Funds.

References

Anderson P (1982) Maximal blood flow and oxygen uptake of an isolated exercising muscle group in man. Acta Physiol Scand 114:37A

Ariano MA, Armstrong RB, Edgerton VR (1973) Hindlimb muscle fiber populations of five mammals. J Histochem Cytochem 21:51–55

Armstrong RB, Laughlin MH (1983) Blood flows within and among rat muscles as a function of time during high speed treadmill exercise. J Physiol (Lond) 344:189–208

Armstrong RB, Laughlin MH (1984) Influence of training on rat muscle blood flow distribution during exercise. Am J Physiol 246:H59–H68

Armstrong RB, Phelps RO (1984) Muscle fiber type composition of rat hindlimb. Am J Anat 171: 259–272

Armstrong RB, Gollnick PD, Ianuzzo CD (1975) Histochemical properties of skeletal muscle fibers in streptozoticin-diabetic rats. Cell Tissue Res 162:387–394

Armstrong RB, Ianuzzo CD, Kunz TH (1977) Histochemical and biochemical properties of flight muscle fibers in the little brown bat, *Myotis lucifugus*. J Comp Physiol 119:141–154

Burke RD (1981) Motor units: anatomy, physiology, and functional organization. Handbook of Physiology. The Nervous System. Bethesda, MD: Am Physiol Soc Sect 1, Chapt 10, pp 345–422

Burke RD, Edgerton VR (1975) Motor unit properties and selective involvement in movement. Exercise Sport Sci Rev 3:31–81

Close RI (1972) Dynamic properties of mammalian skeletal muscles. Physiol Rev 52:129–197

Gardiner KR, Gardiner PF, Edgerton VR (1982) Guinea pig soleus and gastrocnemius electromyograms at varying speeds, grades and loads. J Appl Physiol 52:451–457
Ianuzzo CD, Armstrong RB (1976) Phosphofructokinase and succinate dehydrogenase activities of normal and diabetic rat skeletal muscle. Horm Metab Res 8:244–245
Johnson MA, Polgar J, Weightman D, Appleton D (1973) Data on the distribution of fiber types in thirty-six human muscles. J Neurol Sci 18:111–129
Laughlin MH, Armstrong RB (1982) Muscular blood flow distribution patterns as a function of running speed in rats. Am J Physiol 243:H296–H306
Laughlin MH, Armstrong RB (1983) Rat muscle blood flows as a function of time during prolonged treadmill exercise. Am J Physiol H814–H824
Mai J, Edgerton VR, Barnard RJ (1970) Capillarity of red, white and intermediate muscle fibers in trained and untrained guinea pigs. Experientia 26:1222–1223
Musch TI, Haidet GC, Ordway GA, Longhurst JC, Mitchell JH (1985) Dynamic exercise training in foxhounds: distribution of regional blood flow during maximal exercise. Circ Res (in press)
Nemeth P, Hoffer HW, Pette D (1979) Metabolic heterogeneity of muscle fibers classified by myosin ATPase. Histochemistry 63:191–201
Peter JB, Barnard RJ, Edgerton VR, Gillespie CA, Stempel KE (1972) Metabolic profiles of three fiber types of skeletal muscle in guinea pigs and rabbits. Biochemistry 11:2627–2634
Saltin B, Gollnick PD (1983) Skeletal muscle adaptability: significance for metabolism and performance. Handbook of Physiology. Skeletal Muscle. Bethesda, MD: Am Physiol Soc, Sect 10, Chapt 19, pp 55–631
Sullivan TE, Armstrong RB (1978) Rat locomotory muscle fiber activity during trotting and galloping. J Appl Physiol 44:358–363
Walmsley B, Hodgson JA, Burke RE (1978) Forces produced by medial gastrocnemius and soleus muscles furing locomotion in freely moving cats. J Neurophysiol 41:1203–1216

Cardiopulmonary System Responses to Muscular Exercise in Man

B.J. WHIPP[1] and S.A. WARD[2]

1 Introduction

Muscular exercise can only be performed at the expense of energy stores which are readily accessible to the contractile mechanisms of skeletal muscle; specifically, through the utilization of the free energy of hydrolysis of ATP. Intramuscular ATP concentrations are themselves maintained at the expense of creatine phosphate breakdown and through increased rates of ATP production resulting from aerobic or anaerobic metabolism.

The cardiovascular and pulmonary systems subserve a supportive role during exercise both by providing O_2 to sustain oxidative phosphorylation rates and by clearing metabolically produced CO_2 and – at high work rates – the further CO_2 released from the (largely intramuscular) buffering of lactic and other metabolic acids. This functions to maintain the acid-base milieu of the contractile mechanism compatible with efficient energy transfer. Hence, the cooperative interaction between these two systems serves to minimize changes of PO_2 and PCO_2 of the blood perfusing the exercising tissue.

We shall, therefore, consider ventilatory control during dynamic muscular exercise not simply with respect to the handling of altered pulmonary arterial gas contents, but also with respect to the blood flow changes themselves; the combination determining the net load of gas presented to the lungs for exchange.

2 Sub-Anaerobic Threshold: Early Dynamic Phase

Cardiopulmonary interaction during exercise is perhaps best considered with respect to the early ventilatory ($\dot{V}_E$) changes which occur following the onset of work. Krogh and Lindhard (1913) were the first to document systematically that a rapid hyperpnea occurred at exercise onset. They reasoned that this had to be associated with an increase in pulmonary blood flow ($\dot{Q}$), as the O_2 uptake at the lung ($\dot{V}O_2$) also rose rapidly, despite the mixed venous blood gas tensions not having time to be influenced by metabolites from the exercising limbs. More recently, these observations have been sup-

1 Division of Respiratory Physiology and Medicine, UCLA School of Medicine, Harbor-UCLA Medical Center, Torrance, CA 90509, USA

2 Department of Anesthesiology, UCLA School of Medicine, UCLA, Los Angeles, CA 90024, USA

Circulation, Respiration, and Metabolism
(ed. by R. Gilles)

ported by several investigators who have detailed the time courses of $\dot{V}_E$, pulmonary gas exchange ($\dot{V}O_2$, $\dot{V}CO_2$), and $\dot{Q}$ response in this early phase of work (Davies et al. 1972; Wasserman et al. 1974; Linnarsson 1974; Whipp et al. 1980). Thus, both $\dot{Q}$ and $\dot{V}_E$ increase rapidly and in relatively precise proportion at exercise onset, with alveolar "end-tidal" PO_2 ($P_{ET}O_2$) and PCO_2 ($P_{ET}CO_2$) and the gas exchange ratio (R) not changing appreciably as a consequence (Fig. 1).

These rapid $\dot{V}_E$ changes have traditionally been considered to be neurogenic, originating in the exercising limbs (Kao 1963; Dejours 1967; Tibes 1977), the cerebral cortex (Krogh and Lindhard 1913) or, more recently, from the hypothalamus (Eldridge et al. 1981). The magnitude of the rapid initial hyperpnea is relatively constant in the face of progressively greater imposed work loads (Dejours 1964; Jensen 1972), and hence the change in $\dot{V}_E$ is not simply proportional to the recruitment of motor units or to the imposed work load.

Further, and perhaps more telling, evidence of there being a proportionality between the initial $\dot{V}_E$ and $\dot{Q}$ responses is that when the work is imposed from a background of light exercise (e.g., unloaded pedalling of a cycle ergometer), there is little or no rapid

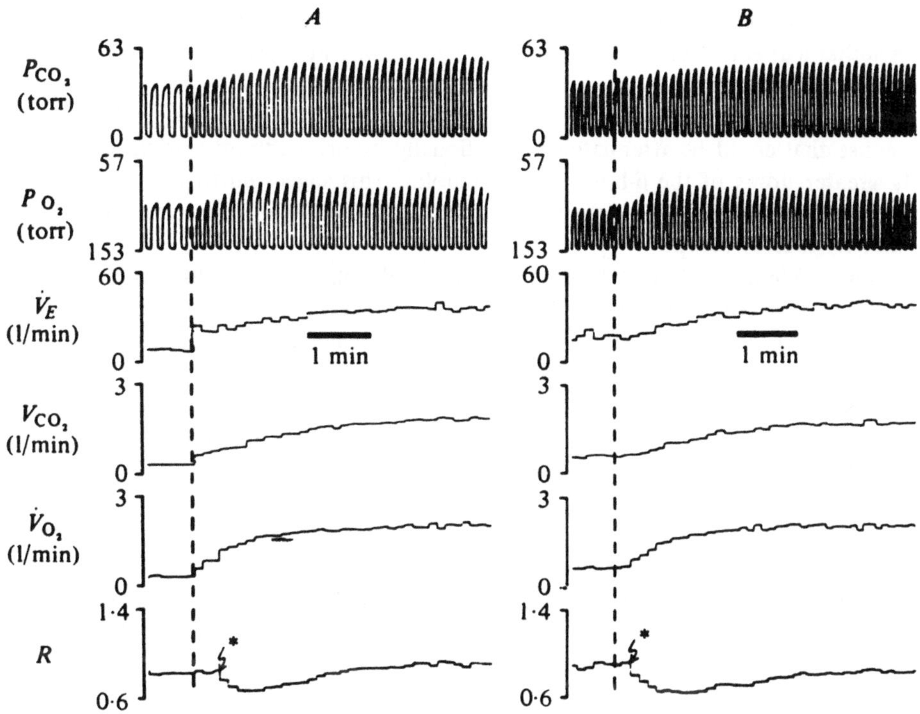

Fig. 1A,B. Ventilatory and pulmonary gas exchange responses to 125 W cycle ergometer exercise beginning from either rest (panel **A**) or from a baseline of unloaded cycling (panel **B**). *Arrow* indicates end of early phase. PCO_2 and PO_2 are CO_2 and O_2 partial pressures in respired air; $\dot{V}_E$, $\dot{V}CO_2$, $\dot{V}O_2$, and R are ventilation, CO_2 output, O_2 uptake, and the gas exchange ratio. Note that a large, abrupt hyperpnea at exercise onset is evident from prior rest, but not from prior mild work, and that in both cases neither did end-tidal PCO_2 fall nor R rise early in the work. (Reprinted from Whipp and Ward 1982)

hyperpnea. Rather, a slower pattern of $\dot{V}_E$ response is apparent (Linnarsson 1974; Whipp et al. 1980; Fig. 1). And, again, as evident from the relative stability of the end-tidal gas tensions and R, the $\dot{V}_E$ change still appears proportional to that of $\dot{Q}$. A correspondingly slow hyperpnea has also been demonstrated by Karlsson et al. (1975) and Weiler-Ravell et al. (1982) when rest-to-work transitions are imposed in the supine position.

Thus, in summary, under conditions in which a rapid increase in $\dot{Q}$ occurs at exercise onset (cf. Fig. 2) consequent to an abrupt rise of stroke volume (i.e., for rest-to-work transitions in the upright posture), a rapid hyperpnea of relatively constant magnitude is seen; whether breathing air, hypoxic, hyperoxic, or hypercapnic gas mixtures (e.g., Cunningham 1974). However, when the work transition is associated with a slower, more gradual increase in $\dot{Q}$ owing to the stroke volume already being elevated to or near the exercising value (i.e., for work-to-work transitions, or for supine rest-to-work transitions), $\dot{V}_E$ does not respond rapidly at exercise onset.

These response patterns have suggested to some investigators either that a direct link from the cardiovascular change triggers the hyperpnea (i.e., a "cardiodynamic" hyperpnea: Wasserman et al. 1974) or that "parallel activation" of cardiovascular and ventilatory control mechanisms occurs with proportional increases in both systems. To test whether a direct cardiodynamic linkage might be operative, Wasserman et al. (1974) infused small boluses of isoproterenol, a β-adrenergic stimulator, into the venous return of dogs and noted a rapid hyperpnea which began only following the increase in $\dot{Q}$; a response that could be attenuated proportionally by prior administration of progressively greater doses of the β-blocker, propranolol. This suggested to the authors that the hyperpnea was secondary to the $\dot{Q}$ change, and not merely a nonspecific stimulation of other ventilatory receptors by the drug.

However, Winn et al. (1979) and Eldridge and Gill-Kumar (1980) subsequently showed that isoproterenol does directly stimulate the carotid chemoreceptors, and hence questioned the validity of the cardiodynamic mechanisms which Wasserman et al. (1974) proposed for the response. This issue was largely resolved by the studies of Juratsch et al. (1981) who demonstrated that when the carotid bodies were temporally isolated from the thorax by interposing long delay loops into both common carotid and both vertebral arteries (care being taken to maintain cerebral perfusion by supportive pumping through the vessels) with delays to the chemoreceptors of some 2 min, a biphasic $\dot{V}_E$ response to intravenous isoproterenol was evident: an early rapid hyperpnea, followed – only after the transit delay to the carotid bodies – by a second hyperpnea. The first component of the response (taken to reflect the cardiodynamic mechanism) was associated with a relative arterial hypotension; and the second (ascribed to direct

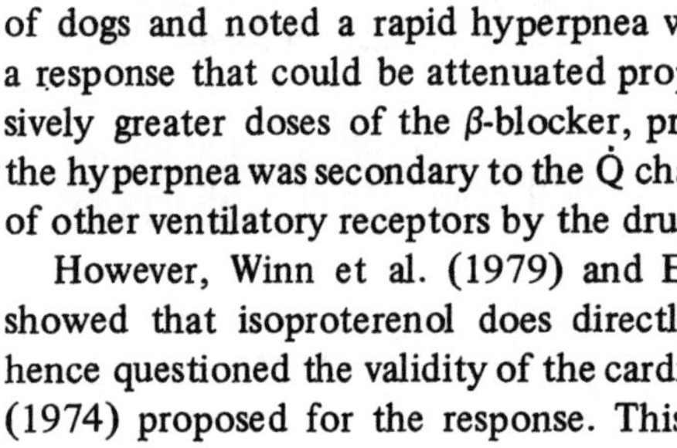

Fig. 2. Instantaneous increase of cardiac output in an anesthetized dog, caused by sudden intermittent electrical stimulation ($2\ s^{-1}$) of the lower half of the spinal cord. (Modified from Guyton et al. 1973)

carotid body stimulation) was associated with a slight hypertension. In contrast, isoproterenol injection into the proximal end of the carotid loop only induced a $\dot{V}_E$ response of the second form, thereby confirming the observations of Eldridge and Gill-Kumar (1980) and Winn et al. (1979) of there being a direct stimulatory effect of the drug on the carotid chemoreceptors; but demonstrating additionally that an earlier – presumably cardiodynamic – mechanism was still operative. Normally, however, the two mechanisms are difficult to dissociate owing to the temporal proximity of the carotid bodies to the thorax.

Jones et al. (1982) and Huszczuk et al. (1981) have attempted to determine a mechanism of the cardiodynamic hyperpnea by altering the pressure in the right side of the heart, for example, by partial occlusion of the pulmonary artery outflow tract with a balloon. They noted an excellent correlation between the magnitude of the right ventricular (RV) moving-average pressure and that of the $\dot{V}_E$ response, and also great similarities between the temporal pattern of the $\dot{V}_E$ and RV pressure changes. These authors suggested, therefore, that the right ventricle may play an important role in coupling $\dot{V}_E$ to cardiovascular changes under conditions such as exercise.

Kan et al. (1979) had also demonstrated that increasing the pressure in a closed pulmonary arterial (PA) sac of dog increased breathing frequency in proportion to the intra-sac pressure over a range of about 30–90 torr. But, whereas vagotomy abolished the frequency response to the PA pressure increases, it did not abolish the hyperpnea induced in the experiments of Jones et al. (1982). Further confusing this issue are the observations of Kostreva et al. (1979) in dog that the hyperpnea (arising from breathing frequency changes) in response to a rise of RV pressure, caused by inflating a balloon in the right ventricle, were abolished by vagotomy.

More recently, Lloyd (1984) has developed an isolated subsystem in dog, utilizing both right heart and major pulmonary arteries within which he was able to alter the pressure (i.e., attempting to couple the RV and PA mechanisms to determine whether they would yield greater $\dot{V}_E$ responses with combined stimulation). Increasing the system pressure (P_{PA} range: ~30–70 torr) led to modest breathing frequency increases that were abolished by bilateral vagotomy. Thus, Lloyd concluded that the sensitivity of this mechanism was likely to be too small to account for the results of Jones et al. (1982), but was compatible with those of Kan et al. (1979). However, directly reducing blood flow to the heart and lungs using partial venous-to-arterial bypass preparations and extracorporeal gas exchange have consistently resulted in reductions of $\dot{V}_E$ (Stremel et al. 1979; Huszczuk et al. 1983; Green and Sheldon 1983).

In summary, therefore, proportional ventilatory and cardiovascular changes suggest a possible mechanistic link between the early cardiovascular and $\dot{V}_E$ changes associated with exercise; mechanisms originating in the heart and pulmonary vasculature have been documented to lead to ventilatory stimulation, but the few direct studies on such a mechanism have produced not entirely consistent results. Clearly, much more work must be carried out for direct cardiopulmonary coupling as a major determinant of the early hyperpnea of exercise to be confidently conceded.

Evidence suggestive of parallel activation of cardiovascular and $\dot{V}_E$ responses is also available. Stimulation of small myelinated and nonmyelinated muscle afferents leads to both cardiovascular and respiratory responses, which are abolished by blockade of these afferents. In contrast, blockade of the large myelinated fibers appeared to have little or no effect (McCloskey and Mitchell 1972; Waldrop et al. 1984).

Descending pathways from the cerebral cortex or involving hypothalamic or limbic mediation can lead to rapid hyperpnea. However, when these mechanisms are directly invoked, a pattern of response is observed that is unlike the normal exercise transition in man; in that a rapid and marked hypocapnia ensues evidencing hyperventilation (Wasserman et al. 1974; Eldridge et al. 1981).

There is, however, a danger of considering these as "either/or" mechanisms. The documented presence of one of the mechanisms does not rule out the potential for one or more of the others to be operative. And, there is no reason at this stage to suggest that all might not be involved to a greater or lesser extent in certain circumstances.

3 Sub-Anaerobic Threshold: Later Dynamic Phase

As the mixed venous gas tensions begin to alter consequent to the influence of altered metabolism, a second component of $\dot{V}_E$ and gas exchange responses can be discerned. This slower, delayed hyperpnea has been thought traditionally to involve humoral mediation. However, as Eldridge (1977) has properly pointed out, the slowness of the response does not preclude neural involvement in its mediation. Thus, he has documented that the medullary respiratory neuronal complex can show a long time constant in response to an abruptly altered input; a process that he has termed "reverberation". This mechanism has been clearly and thoroughly documented for the cessation of afferent information to the respiratory center. It does not, however, appear as prominent for an on-transient of stimulation. Even so, these slow neural dynamics of the respiratory neuronal complex warrant inclusion in the conceptual control model of the exercise hyperpnea.

It is clear, however, that for work rates which do not evoke sustained increases of arterial lactate, the dynamic and steady state responses of $\dot{V}_E$ are highly correlated with those of CO_2 exchange at the lung. And, as shown in Fig. 3, a large body of confirming evidence is available on this issue. Wasserman et al. (1967) have demonstrated that for steady state work increments, $\dot{V}_E$ changes more linearly with respect to $\dot{V}CO_2$ than to $\dot{V}O_2$ in a given subject and that the spread of values for $\dot{V}_E$ at a specific $\dot{V}CO_2$ is relatively narrow for a group of subjects (panel A), compared to the relatively wide range observed when $\dot{V}_E$ is expressed as a function of $\dot{V}O_2$. A further and important piece of evidence which supports the correlation between $\dot{V}_E$ and $\dot{V}CO_2$ has been provided by N.L. Jones (1975) who caused subjects to undergo endurance training, such that the post-training $\dot{V}CO_2$ at the given work load would be less than that prior to training. He observed that $\dot{V}_E$ decreased in rather precise proportion to the CO_2 that was *not* produced at these work loads following training (panel B); a highly irregular response being noted with respect to $\dot{V}O_2$.

These observations can be extended to the nonsteady state. Thus, for both step increases and decreases of work rate, $\dot{V}_E$ responds linearly with respect to $\dot{V}CO_2$ (Herx-

Fig. 3. Representative relationships between ventilation ($\dot{V}_E$) and CO_2 output ($\dot{V}CO_2$) during exercise, to illustrate the close correlation between these two variables. See text for further details. (Reprinted from Wasserman et al. 1967; Jones 1975; Herxheimer and Kost 1932; Wasserman and Whipp 1983; Casaburi et al. 1978; Miyamoto et al. 1983)

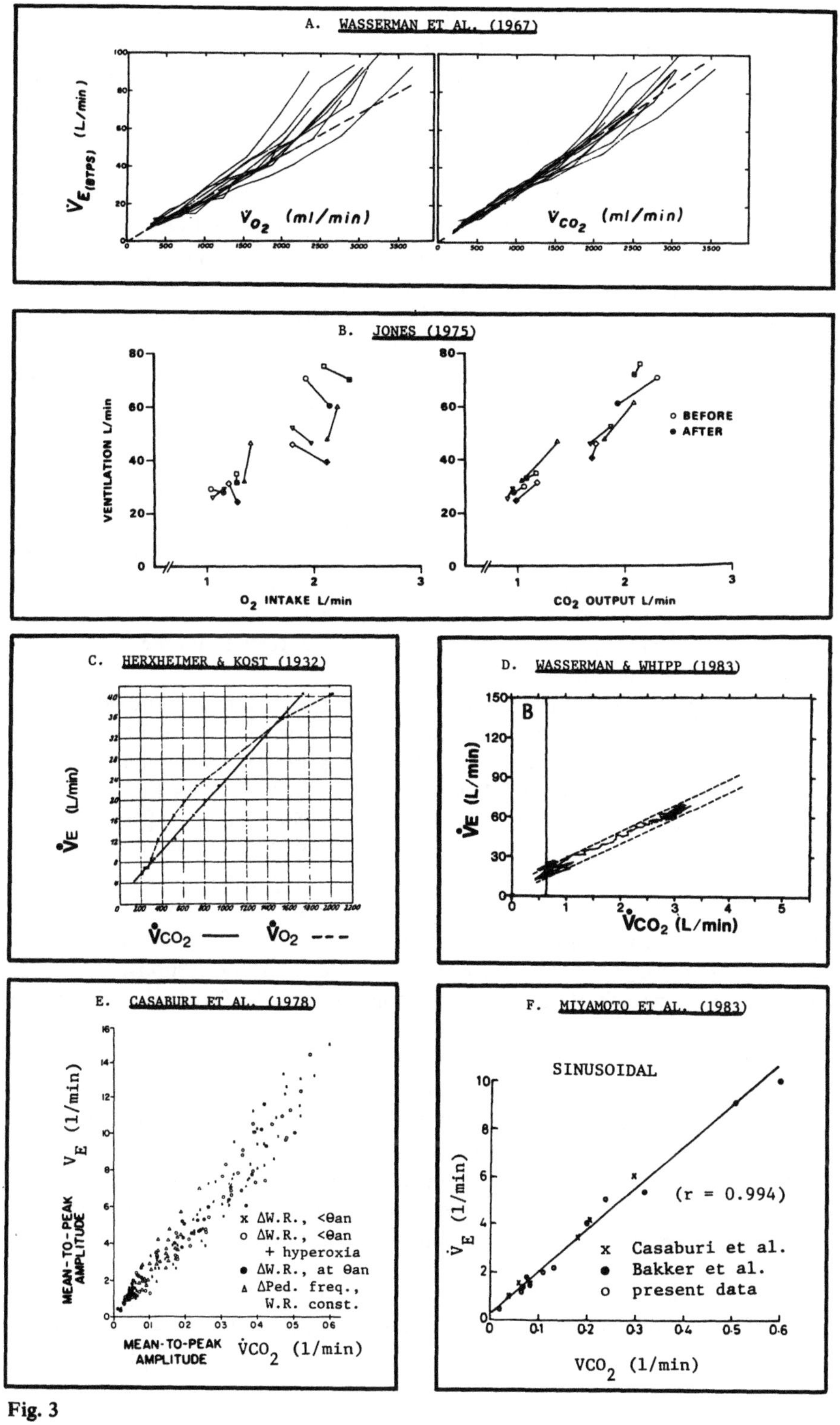
A. WASSERMAN ET AL. (1967)
$\dot{V}_{E(BTPS)}$ (L/min)
$\dot{V}_{O_2}$ (ml/min)
$\dot{V}_{CO_2}$ (ml/min)
B. JONES (1975)
VENTILATION L/min
O_2 INTAKE L/min
CO_2 OUTPUT L/min
o BEFORE
• AFTER
C. HERXHEIMER & KOST (1932)
$\dot{V}_E$ (L/min)
$\dot{V}CO_2$ —— $\dot{V}O_2$ - - -
D. WASSERMAN & WHIPP (1983)
B
$\dot{V}_E$ (L/min)
$\dot{V}CO_2$ (L/min)
E. CASABURI ET AL. (1978)
MEAN-TO-PEAK AMPLITUDE V_E (1/min)
x ΔW.R., <θan
o ΔW.R., <θan + hyperoxia
• ΔW.R., at θan
▲ ΔPed. freq., W.R. const.
MEAN-TO-PEAK AMPLITUDE $\dot{V}CO_2$ (1/min)
F. MIYAMOTO ET AL. (1983)
SINUSOIDAL
$\dot{V}_E$ (1/min)
(r = 0.994)
x Casaburi et al.
• Bakker et al.
o present data
VCO_2 (1/min)

Fig. 3

heimer and Kost 1932; Wasserman and Whipp 1983: Fig. 3, panels C and D). Others have preferred to utilize sinusoidal work-rate forcings to determine the dynamics of the exercise hyperpnea, as increasing the frequency of the work-rate sinusoid (while maintaining its amplitude constant) engenders a progressive reduction in the amplitude of the response (e.g., $\dot{V}_E$ or $\dot{V}CO_2$) and an increase in its phase lag relative to the input (i.e., the workload). Thus, Casaburi et al. (1978), Bakker et al. (1980), and Miyamoto et al. (1983) have all shown that the dynamic properties of the ventilatory and CO_2-exchanging systems lead to the progressive reduction of response amplitude as the input frequency gets progressively higher. Furthermore, $\dot{V}_E$ decreases in precise proportion to $\dot{V}CO_2$, with the $\dot{V}_E-\dot{V}CO_2$ relationship extrapolating at or close to the origin for the highest input frequencies (Fig. 3, panels E and F). The linearity of the $\dot{V}_E-\dot{V}CO_2$ response suggests a close dynamic coupling of $\dot{V}_E$ with $\dot{V}CO_2$, and the fact that the relationship extrapolated through the origin provides evidence (within this range of work rate fluctuations) that a rapid component of the $\dot{V}_E$ response which is not coupled with CO_2 exchange does not appear to be operative.

Consequent to the close matching between $\dot{V}_E$ and $\dot{V}CO_2$ during steady state exercise, $PaCO_2$ is thus regulated at or close to its control level; and during the transient differs from control by only a torr or so (Whipp 1981). This suggests, in general, that by whatever mechanism, the exercise hyperpnea is likely to be linked to a mechanism proportional to pulmonary CO_2 exchange. Because the time course of tissue CO_2 production is appreciably faster than that of lung exchange (the latter being delayed by tissue CO_2 storage during the transient), it has been suggested that the link to CO_2 exchange may be some function of CO_2 flow to the lung.

However, attempts to document such a mechanism by loading or unloading CO_2 into the venous system of several laboratory species have provided contradictory results. For example, a change of $\dot{V}_E$ in proportion to the change in CO_2 load under these conditions with consequent regulation of downstream (i.e., arterial) PCO_2 has been documented in the anesthetized (Wasserman et al. 1974) and the awake dog (Stremel et al. 1978) and also in the awake, resting, and exercising sheep (Phillipson 1981). In contrast, quite similar experiments in anesthetized (Shors et al. 1983) and awake dogs (Bennett et al. 1984) and baboon (Lewis 1975) have resulted in qualitatively similar $\dot{V}_E$ changes which, however, were quantitatively not different from those expected from the associated changes in $PaCO_2$; and thus, were characteristic of the $\dot{V}_E$ response to inhaled CO_2. Furthermore, intermediate patterns of ventilatory and $PaCO_2$ response have also been found.

In summary, there appears no debate that changes in pulmonary CO_2 load induced by experimentally-altered mixed venous PCO_2 result in consequent $\dot{V}_E$ changes. However, whether these responses are intrinsically isocapnic (i.e., operating with apparently infinite gain with respect to mean $PaCO_2$) remains the subject of debate. The resolution of this issue is likely not to result from simply adding numbers of confirming instances to one or another response pattern but, more importantly, from elucidating why isocapnia obtains in some experimental conditions while, in others, it clearly does not.

A large factor leading to the search for pulmonary mechanisms of CO_2 exchange which may control the exercise hyperpnea has resulted from the observation that $\dot{V}_E$ changes appreciably more slowly than tissue CO_2 production, with a time course similar to that of $\dot{V}CO_2$. It is possible, however, that an important potential peripheral mechanism for the coupling of $\dot{V}_E$ to CO_2 exchange may have been overlooked; that is, a CO_2-linked mechanism that could operate much more slowly than muscle CO_2 production, and that would continue to change as long as does the pulmonary CO_2 exchange.

This is contained in the notion that (e.g., for a step-increment of work rate) the amount of CO_2 stored in the tissues is represented by the difference between the exponential uprise of tissue CO_2 production (considered to be appreciably similar to that of the O_2 consumption) and that of the pulmonary CO_2 exchange. The total difference between these two exponential processes, thus, reflects the total CO_2 stored. Likewise, the time course of this CO_2 storage corresponds to the accumulating difference between these exponentials. Hence, the time course of intramuscular and venous PCO_2 will continue to change until the difference between the two exponential functions becomes zero (i.e., when $\dot{V}CO_2$ becomes constant). Intramuscular PCO_2 is thus likely to change with the time course of this accumulated CO_2 storage (transformed via the effective tissue CO_2 dissociation curve), providing a slower mechanism which could, in part, bridge the gap between those investigators who believe intramuscular chemoreception to be important, and those who assert that any coupling of $\dot{V}_E$ is likely to be proportional to pulmonary CO_2 exchange.

It is important to point out here that other investigators do not find that evidence such as that presented in Fig. 3 to be sufficiently convincing for obligatory CO_2 linkage in the dynamic and steady state $\dot{V}_E$ responses to exercise. Fordyce and Bennett (1984), utilizing mathematical modelling, have questioned whether the demonstrated regulation of $PaCO_2$ during exercise necessarily implies the existence of a precise coupling between $\dot{V}_E$ and metabolic rate. Furthermore, Dempsey et al. (1984) have actually challenged the extent to which $PaCO_2$ is a regulated variable during exercise. However, the variability of their blood gas data appears markedly greater than widely reported (1 SD for $PaCO_2$ being about 5 torr). One wonders to what extent mixing different forms and intensities of exercise might influence the results during moderate exercise, and also whether the large variability reflects individual subjects' responses being erratic or different subjects responding differently. Until these points are clarified, it is not possible to give further credence to this challenge, especially in the face of the large body of literature which demonstrates that in the steady state (a crucial requirement) of moderate exercise in a quiet, nonthreatening environment, $PaCO_2$ is regulated at or close to the resting control value (see Whipp 1981, for discussion).

It is clear that the peripheral chemoreceptors subserve a role in the hyperpneic response to steady state exercise. This role is easily missed if one simply relies on comparison of the steady state $\dot{V}_E$ response to the work for air- and O_2-breathing (the latter suppressing carotid body afferent drive); these responses differ little. However, the abrupt and surreptitious replacement of air as the inspirate with a humidified, hyperoxic gas evokes a clear transient fall of $\dot{V}_E$ whose nadir is some 15%–20% of the steady state exercise $\dot{V}_E$ (Dejours 1964; Stockley 1978; Whipp and Davis 1979; Fig. 4). Thus, $\dot{V}_E$ appears to be driven by peripheral chemoreceptor mechanisms. However, with sustained O_2-breathing, the suppression of carotid body drive appears to be compensated for by a CSF stimulus to breathe, operating via an O_2-induced reduction in cerebral $\dot{Q}$ which leads to an increase in CSF PCO_2 and $[H^+]$.

Importantly, if peripheral chemoreceptor gain is increased (e.g., by inhaling an hypoxic gas), the time constant (τ) of the delayed component of the exercise hyper-

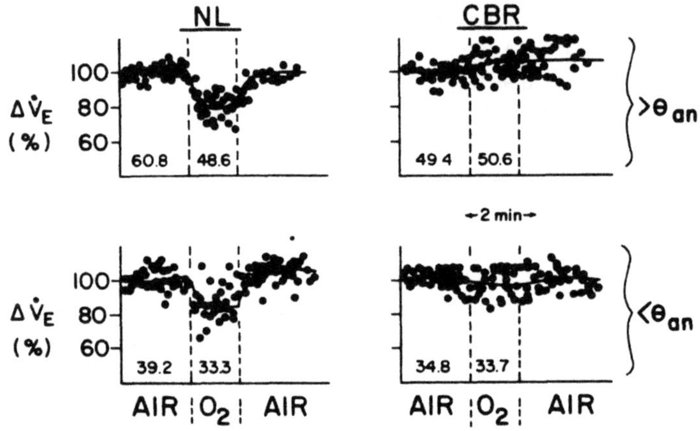

Fig. 4. Breath-by-breath changes in ventilation ($\dot{V}_E$) induced by a short bout of 100% O_2 breathing (approx. 1 min) in a group of otherwise air-breathing subjects exercising at 80% and 150% of the anaerobic threshold. *NL, CBR* represent the data for five control subjects and five carotid body resected subjects, respectively. (Reprinted from Whipp and Davis 1979)

pnea is appreciably speeded; while hyperoxic mixtures evoke a slowing. For inspirates whose O_2 fractions range from 12% to 100% (resulting in carotid body contributions to the total $\dot{V}_E$ drive ranging from some 50% with 12% O_2 to zero with hyperoxia), the $\dot{V}_E$ τ can change fourfold from less than 30 s to some 2 min (Griffiths 1980), suggesting an important role for the peripheral chemoreceptors in dictating the dynamics of the slower component of the exercise hyperpnea. Dynamic $\dot{V}_E$ responses in the face of sustained metabolic acidosis provide further support for this concept (Oren et al. 1982). That the response maintains its exponentially over this range of carotid body drive suggests less a proportional peripheral chemoreceptor component added to some underlying $\dot{V}_E$ drive, than the peripheral afferent drive itself modifying some inherently (presumably central) exponential mechanism of ventilatory control. Subjects who had both carotid bodies resected, but who (at the time of the study) had normal pulmonary mechanics and blood gases, also responded to a square-wave exercise bout with a $\dot{V}_E$ time course that was appreciably slower than seen for control subjects (Wasserman and Whipp 1976).

Several investigators have concerned themselves with the possible mechanism for the link of the purported CO_2-linkage of the exercise hyperpnea. Proposed mechanisms have included: mixed venous chemoreception, CO_2 flux across the lung (i.e., from pulmonary vasculature to the alveolar gas phase), and a systematic alteration in blood gas composition consequent to CO_2 exchange disequilibrium in the lung owing to the absence of carbonic anhydrase in the plasma. However, further detailed consideration of these mechanisms has largely ruled out their direct involvement in the normal control of the exercise hyperpnea; but with a signal proportional to CO_2 flux through the lung (i.e., pulmonary artery to pulmonary vein) still the topic of investigation.

A CO_2-linked mechanism which relates both to pulmonary blood flow and to mixed venous CO_2 content, and which currently is considered likely to be important for the exercise hyperpnea, is incorporated in what may be termed the "oscillations theories".

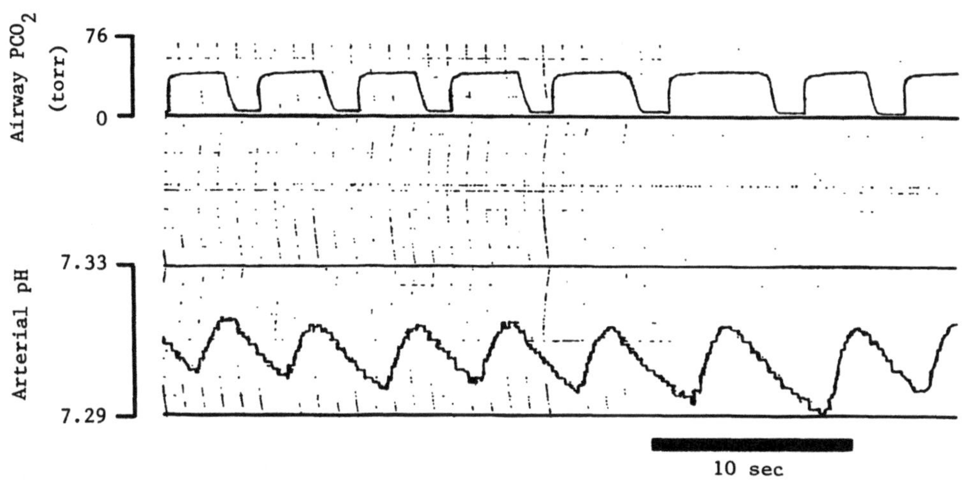

Fig. 5. On-line record of intrabreath fluctuations of arterial pH in an anesthetized dog

Owing to the phasic lung ventilation, pulmonary venous PCO_2 and $[H^+]$ will evidence an oscillating pattern; being increased during expiration, and decreased during inspiration (Fig. 5). Yamamoto (1960) originally suggested that some characteristic of this oscillating PCO_2 pattern may be sensed by elements in the respiratory control loop to provide a CO_2-linked $\dot{V}_E$ drive, even should the mean $PaCO_2$ remain unchanged. Six mechanisms have been variously proposed (Fig. 6):

1. An increase in the amplitude of the oscillation consequent to an elevated mixed venous PCO_2 during exercise, coupled with increased tidal volume, might be sensed by a rapid responding chemoreceptor; and at the frequencies of the breathing cycle, this can be demonstrated likely to be the peripheral and not the central chemoreceptors.
2. The amplitude of the oscillation above some prior set-point level may provide stimulation, while values below may provide some suppressive input. And in this scheme, the position of the set-point may not be at the midpoint of the oscillation, hence providing a rectification of the signal with consequent potential stimulation.
3. A third mechanism, and one which could operate despite no change in the amplitude of the PCO_2 signal, proposes that the system responds to the differential of the $[H^+]$ signal (Saunders 1980; Cross et al. 1982); such as occurs when the breathing frequency increases during exercise. Thus, the maximum rate of change would increase without the necessity for an amplitude increase. However, as such a signal can be asymmetrical about the set-point, especially if there is asymmetry of the overshoot of carotid body neural discharge (Black et al. 1971) as the direction of the signal changes, rectification or unidirectional rate control may be required for this mechanism.
4. It has also been suggested that the number of peaks per unit time which course pass the carotid bodies might be counted (i.e., digital counting device). One intuitive attraction of this mechanism is that as the cardiac output speeds up, a CO_2-linked signal could operate at the carotid bodies without any transit delay from the lungs

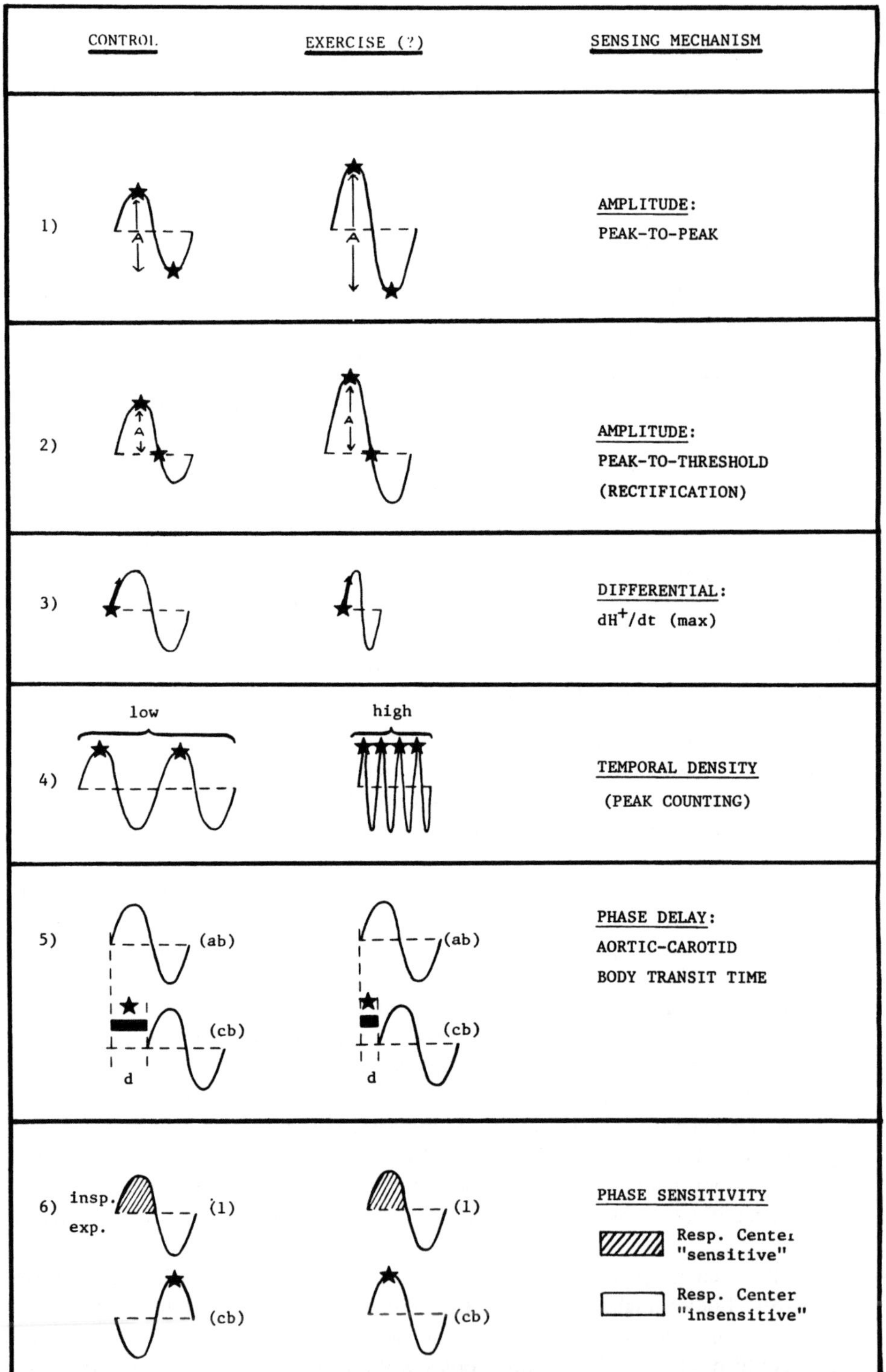

Fig. 6. Postulated mechanisms by which intrabreath oscillations of arterial PCO_2 and pH might be sensed at the carotid bodies. See text for further details

to the carotid body (i.e., the peaks would be pushed passed the sensor more rapidly: Saunders 1980).

5. A further suggestion is that the phase delay between the oscillating signal reaching the aortic bodies and the carotid bodies might be sensed, and the phase difference supply the drive to breathe (Biscoe 1977).
6. And, finally, it has been demonstrated that the sensitivity of the respiratory control loop to an afferent signal is higher in inspiration than in expiration (Black et al. 1973; Eldridge 1972). This phasic change of sensitivity, as evidenced via the inspiratory phase at the lung being higher than the expiratory phase at the lung, means that if an oscillation arrives such that the peak or the maximum rate of change occurs in the expiratory phase of the current respiratory cycle, then the system response is relatively insensitive. Were there to be a systematic shift during exercise such that the stimulus arrives progressively earlier in the sensitive inspiratory phase, then a greater stimulation of V_E could be achieved (Torrance 1974; Cunningham 1975).

It should be pointed out that the evidence generally invoked to support an appreciable role for the oscillation mechanism in the exercise hyperpnea is largely correlative. But similarly the mechanism has not been explicitly shown not to operate. Clearly, the carotid bodies have the dynamic response characteristics capable of transducing oscillatory signals into oscillating bursts of afferent dischange. And the role of these oscillations in the control of $\dot{V}_E$ is currently the subject of considerable experimental effort.

4 Supra-Anaerobic Threshold (θ an)

Above the threshold $\dot{V}O_2$ at which a sustained lactic acidosis ensues, the dynamic characteristics of the $\dot{V}_E$ response become more complex; being highly nonlinear and with steady states commonly unattainable. What has been shown for such high intensity exercise is that the lactic acidosis provides additional $\dot{V}_E$ drive which can lead to respiratory compensation for the acidosis. Thus, there is a systematic reduction of $PaCO_2$ which constrains the fall of pH. In subjects who have had both carotid bodies resected, the increase in $\dot{V}_E$ drive at these work rates is appreciably reduced or nonexistent (Wasserman et al. 1974; Whipp et al. 1980; Fig. 7). Hence, pHa falls more for a given $[HCO_3^-]$ decrease, and $PaCO_2$ can actually be higher than the subthreshold exercise level; owing to the lack of, the reduced, or the slowed response to the H^+ in the phase of the increased CO_2 load from the acid buffering.

This is not to say that the lactic acidosis is the exclusive mechanism of the additional $\dot{V}_E$ drive above θ an; only that normally it subserves the dominant component. It is well known, for example, that circulating catecholamines levels only increase at work rates which apparently exceed θ an; they may, therefore, provide some $\dot{V}_E$ stimulation. Likewise, a sufficiently high body temperature or an increased blood osmolarity may also contribute to the hyperventilation seen at these high work rates, as can anxienty regarding the high work load. In addition, conditions which are likely to lead to localized muscle pain can also evoke hyperventilation as is seen, for example, in subjects constrained to perform exercise on a cycle ergometer having the seat deliberately positioned

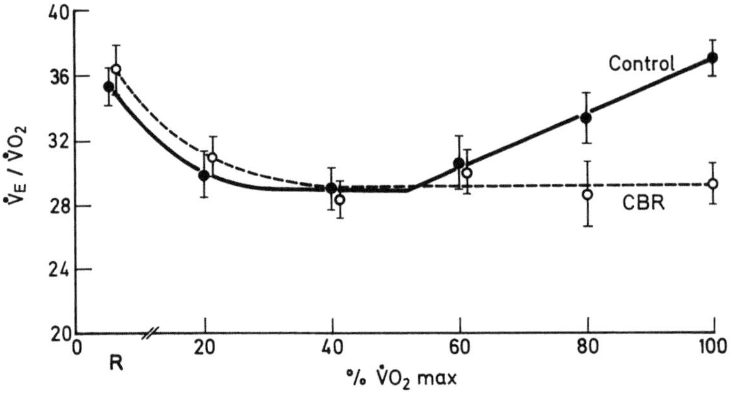

Fig. 7. Comparison of the response to incremental exercise of the ventilatory equivalent for O_2 ($\dot{V}_E/\dot{V}O_2$) between a *control group* of subjects and a group that previously had both carotid bodies surgically resected (*CBR*). Note that the *control group* increased $\dot{V}_E/\dot{V}O_2$ above 50%–60% of maximal ($\dot{V}O_2$ max), while this response was absent in the *CBR* subjects. (Reprinted from Whipp and Wasserman 1980)

too low for efficient mechanical energy transfer or in patients with peripheral vascular occlusive disease. Also patients with McArdle's syndrome, who are unable to produce lactic acid consequent to muscle phosphorylase B deficiency, actually develop a respiratory alkalosis at relatively low work rates (equivalent to 50% or more of their reduced maximum capacity: Hagberg et al. 1982).

Finally, it is important to consider the mechanical system that provides the air flow during exercise as a potential source of constraint or limitation during the exercise. The physical structure of the thorax, airways, and lung parenchyma (i.e., establishing the pulmonary resistance and recoil characteristics) cannot continually be stressed to higher and higher levels without potential air flow limitation (i.e., consequent to progressively greater metabolic rate and $\dot{V}_E$ drive as athletes reach greater performance capabilities).

Highly trained athletes have already been demonstrated to exhibit volume-specific air flow limitation (Grimby et al. 1971), from a consideration of expiratory flow-volume curves during high intensity exercise. Also, maximum $\dot{V}_E$'s of about 90% of the maximum voluntary $\dot{V}_E$ have been achieved in highly trained athletes (Follinsbee et al. 1983), compared with the 60%–70% normally achieved in unfit or moderately fit subjects; the latter are unable to extend the range of metabolic rates to sufficiently high levels. Consequently, the muscles of respiration, which provide the forces to move the thorax at these high levels of exercise, are likely themselves to be operating above their own anaerobic thresholds. At these high thoracic work loads, therefore, respiratory muscle fatigue is likely, which could further limit power output.

5 Conclusion

As shown in Fig. 8, numerous afferent drives to the respiratory neural network in the brain stem have been proposed and, to a greater or lesser extent, documented experi-

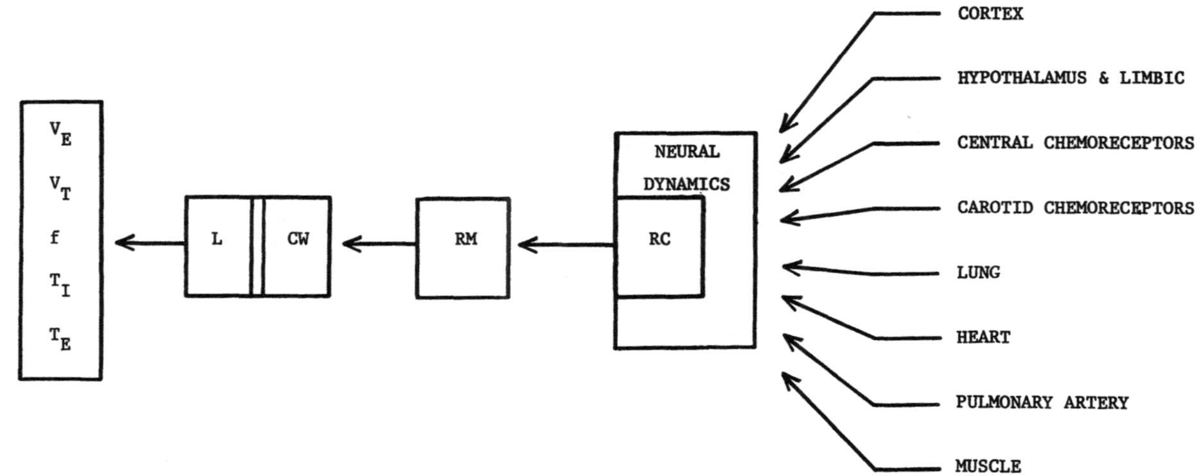

Fig. 8. Schematic depicting the various sources of afferent information postulated to be involved in the control of the exercise hyperpnea. *RC* represents the respiratory "center" complex, with its associated neural dynamics; *RM*, the respiratory muscles; *L, CW*, the lungs and chest wall. The output of the system is represented by ventilation (V_E) and its pattern components: tidal volume (V_T), frequency (f), and inspiratory and expiratory durations (T_I, T_E). See text for further details

mentally. And, therefore, the appropriate question in consideration of the exercise hyperpnea may not be why does ventilation increase, but rather why for moderate exercise (in the face of the spectrum of documented inputs) does ventilation only increase to levels commensurate with the level of CO_2 exchange.

References

Bakker HK, Struikenkamp RS, De Vries GA (1980) Dynamics of ventilation, heart rate, and gas exchange: sinusoidal and impulse work loads in man. J Appl Physiol 48:289–301

Bennett FM, Tallman RD, Grodins FS (1984) Role of VCO_2 in control of breathing of awake dogs. J Appl Physiol 56:1335–1337

Biscoe TJ (1977) The carotid body. What next? Am Rev Respir Dis 115:189–191

Black AMS, McCloskey DI, Torrance RW (1971) The response of carotid chemoreceptors in the cat to sudden changes of hypercapnic and hypoxic stimuli. Respir Physiol 13:36–49

Black AMS, Goodman NW, Nail BS, Rao PS, Torrance RW (1973) The significance of the timing of chemoreceptor impulses for their effect upon respiration. Acta Neurobiol Exp 33:139–147

Casaburi R, Whipp BJ, Wasserman K, Stremel RW (1978) Ventilatory control characteristics of the exercise hyperpnea as discerned from dynamic forcing techniques. Chest 73S:280S–283S

Cross BA, Davey A, Guz A, Katona PG, Maclean M, Murphy K, Semple SJG, Stidwell R (1982) The pH oscillations in arterial blood during exercise; a potential signal for the ventilatory response in the dog. J Physiol (Lond) 329:57–73

Cunningham DJC (1974) The control system regulating breathing in man. Q Rev Biophys 6:433–483

Cunningham DJC (1975) A model illustrating the importance of timing in the regulation of breathing. Nature 253:440–442

Davies CTM, Di Prampero PE, Cerretelli P (1972) Kinetics of cardiac output and respiratory gas exchange during exercise and recovery. J Appl Physiol 32:618–625

Dejours P (1964) Control of respiration in muscular exercise. In: Rahn H, Fenn WO (eds) Respiration. Handbook of Physiology, vol I. Am Physiol Soc, Washington DC, pp 631–648

Dejours P (1967) Neurogenic factors in the control of ventilation during exercise. Circ Res 10–21 (suppl) 1:I146–I153

Dempsey JA, Mitchell GS, Smith CA (1984) Exercise and chemoreception. Am Rev Respir Dis 129 (Suppl):S31–S34

Eldridge FL (1972) The importance of timing on the respiratory effects of intermittent carotid body chemoreceptor stimulation. J Physiol (Lond) 222:319–333

Eldridge FL (1977) Maintenance of respiration by central neural feedback mechanisms. Fed Proc 36:2400–2404

Eldridge FL, Gill-Kumar P (1980) Mechanisms of hyperpnea induced by isoproterenol. Respir Physiol 40:349–363

Eldridge FL, Millhorn DE, Waldrop TG (1981) Exercise hyperpnea and locomotion: parallel activation from the hypothalamus. Science 211:844–846

Folinsbee LJ, Wallace ES, Bedi JF, Horvath SM (1983) Respiratory pattern in trained athletes. In: Whipp BJ, Wiberg DM (eds) Elseiver, New York, pp 205–212

Fordyce WE, Bennett FM (1984) Some characteristics of a steady state model of exercise hyperpnea. Physiologist 27(4):217

Green JF, Sheldon MI (1983) Ventilatory changes associated with changes in pulmonary blood flow in dogs. J Appl Physiol 54:997–1002

Griffiths TG, Henson LC, Huntsman D, Wasserman K, Whipp BJ (1980) The influence of inspired O_2 partial pressure on ventilatory and gas exchange kinetics during exercise. J Physiol (Lond) 306:34P

Grimby G, Saltin B, Wilhelmsen L (1971) Pulmonary flow-volume and pressure-volume relationship during submaximal and maximal exercise in young well-trained men. Bull Eur Physiopathol Respir 7:157–167

Guyton AC, Jones CE, Coleman TG (1973) Circulatory physiology: Cardiac output and its regulation. Saunders, Philadelphia, chapt 25

Hagberg JM, Coyle EF, Carroll JE, Miller JM, Martin WH, Brooke MH (1982) Exercise hyperventilation in patients with McArdle's disease. J Appl Physiol 52:991–994

Herxheimer H, Kost R (1932) Das Verhältnis von Sauerstoffaufnahme und Kohlensäureausscheidung zur Ventilation bei harter Muskelarbeit. Z Klin Med 108:240–247

Huszczuk A, Jones PW, Wasserman K (1981) Pressure information from the right ventricle as a reflex coupler of ventilation and cardiac output. Fed Proc 40:568

Huszczuk A, Jones PW, Oren A, Shors EC, Nery LE, Whipp BJ, Wasserman K (1983) Venous return and ventilatory control. In: Whipp BJ, Wiberg DM (eds) Modelling and control of breathing. Elsevier, New York, pp 78–85

Jensen JI (1972) Neural ventilatory drive during arm and leg exercise. Scand J Clin Lab Invest 29: 177–184

Jones NL (1975) Exercise testing in pulmonary evaluation: rationale, methods, and the normal respiratory response to exercise. N Engl J Med 293:541–544

Jones PW, Huszczuk A, Wasserman K (1982) Cardiac output as a controller of ventilation through changes in right ventricular load. J Appl Physiol 53:218–244

Juratsch CE, Huszczuk A, Gianotta S, Whipp BJ (1981) Evidence for a 'cardiodynamic' component of the isoproterenol induced hyperpnea in the dog. Fed Proc 40:567

Kan WO, Ledsome JR, Bolter CP (1979) Pulmonary arterial distension and activity in phrenic nerve of anesthetized dogs. J Appl Physiol 46:625–631

Kao FF (1963) An experimental study of the pathways involved in exercise hyperpnea employing cross-circulation techniques. In: Cunningham DJC, Lloyd BB (eds) The regulation of human respiration. Blackwell, Oxford, pp 461–502

Karlsson H, Lindborg B, Linnarsson D (1975) Time courses of pulmonary gas exchange and heart rate changes in supine exercise. Acta Physiol Scand 95:329–340

Kostreva DR, Hopp FA, Zuperku EJ, Kampine JP (1979) Apnea, tachycardia and hypertension elicited by cardiac vagal afferents. J Appl Physiol 47:312–318

Krogh A, Lindhard J (1913) The regulation of respiration and circulation during the initial stages of muscular work. J Physiol (Lond) 47:112–136

Lewis SM (1975) Awake baboon's ventilatory response to venous and inhaled CO_2 loading. J Appl Physiol 39:417–422

Linnarsson D (1974) Dynamics of pulmonary gas exchange and heart rate changes at start and end of exercise. Acta Physiol Scand (Suppl) 415:1–68

Lloyd TC Jr (1984) Effect on breathing of acute pressure rise in pulmonary artery and right ventricle. J Appl Physiol 57:110–116

McCloskey DI, Mitchell JH (1972) Reflex cardiovascular and respiratory responses originating in exercising muscle. J Physiol (Lond) 224:173–186

Miyamoto Y, Nakazono Y, Hiura T, Abe Y (1983) Cardiorespiratory dynamics during sinusoidal and impulse exercise in man. Jpn J Physiol 33:971–986

Oren A, Whipp BJ, Wasserman K (1982) Effect of acid-base status on the kinetics of the ventilatory response to moderate exercise. J Appl Physiol 52:1013–1017

Phillipson EA, Bowes G, Townsend ER, Duffin J, Cooper JD (1981) Role of metabolic CO_2 production in ventilatory response to steady-state exercise. J Clin Invest 68:768–774

Saunders KB (1980) Oscillations of arterial CO_2 tension in a respiratory model: some implications for the control of breathing in exercise. J Theor Biol 84:163–181

Shors EC, Huszczuk A, Wasserman K, Whipp BJ (1983) Effects of spinal-cord section and altered lung CO_2 flow on the exercise hyperpnea in the dog. In: Whipp BJ, Wiberg DM (eds) Modelling and control of breathing. Elsevier, New York, pp 274–281

Stockley RA (1978) The contribution of the reflex hypoxic drive to the hyperpnea of exercise. Respir Physiol 35:79–87

Stremel RW, Huntsman DJ, Casaburi R, Whipp BJ, Wasserman K (1978) Control of ventilation during intraveous CO_2 loading in the awake dog. J Appl Physiol 44:311–316

Stremel RW, Whipp BJ, Casaburi R, Huntsman DJ, Wasserman K (1979) Hypopnea consequent to diminished blood flow in the dog. J Appl Physiol 46:1171–1177

Tibes U (1977) Reflex inputs to the cardiovascular and respiratory centers from dynamically working canine muscles. Circ Res 41:332–341

Torrance RW (1974) Arterial chemoreceptors. In: Widdicombe JG (ed) Respiration, MTP Int Rev Sci, Ser one, Physiol vol 2. Butterworths, London, pp 247–271

Waldrop TG, Rybicki KJ, Kaufman MP (1984) Chemical activation of group I and group II muscle afferents has no cardiorespiratory effects. J Appl Physiol 56:1223–1228

Wasserman DH, Whipp BJ (1983) Coupling of ventilation to pulmonary gas exchange during non-steady-state work in men. J Appl Physiol 54:587–593

Wasserman K, Whipp BJ (1976) The carotid bodies and respiratory control in man. In: Paintal AS (ed) Morphology and mechanisms of chemoreceptors. Navchetan, Delhi, pp 156–174

Wasserman K, Van Kessel AL, Burton GG (1967) Interaction of physiological mechanisms during exercise. J Appl Physiol 22:71–85

Wasserman K, Whipp BJ, Castagna J (1974) Cardiodynamic hyperpnea: Hyperpnea secondary to cardiac output increase. J Appl Physiol 36:457–464

Wasserman K, Whipp BJ, Koyal SN, Clearly MG (1975) Effect of carotid body resection on ventilatory and acid-base control during exercise. J Appl Physiol 39:354–358

Wasserman K, Whipp BJ, Casaburi R, Beaver WL, Brown HV (1977) CO_2 flow to the lungs and ventilatory control. In: Dempsey JA, Reed CE (eds) Muscular exercise and the lung. University of Wisconsin Press, Madison, pp 105–135

Weiler-Ravell D, Cooper DM, Whipp BJ, Wasserman K (1982) Effect of posture on the ventilatory response at the start of exercise. Fed Proc 41:1102

Whipp BJ (1981) The control of the exercise hyperpnea. In: Hornbein T (ed) The regulation of breathing. Dekker, New York, pp 1069–1139

Whipp BJ, Davis JA (1979) Peripheral chemoreceptors and exercise hyperpnea. Med Sci Sports 11: 204–212

Whipp BJ, Ward SA (1982) Cardiopulmonary coupling during exercise. J Exp Biol 100:175–193

Whipp BJ, Wasserman K, Davis JA, Lamarra N, Ward SA (1980) Determinants of O_2 and CO_2 kinetics during exercise in man. In: Ceretelli P, Whipp BJ (eds) Exercise bioenergetics and gas exchange. Elsevier, Amsterdam, pp 175–185

Winn R, Hildebrant JR, Hildebrant J (1979) Cardiorespiratory responses following isoproterenol injection in rabbits. J Appl Physiol 47:352–359

Yamamoto WS (1960) Mathematical analysis of the time course of alveolar CO_2. J Appl Physiol 15:215–219

Symposium II

Comparative Physiology of Gas Exchange and Transport

Organizers J. PIIPER and P. SCHEID

Gas Transport Properties of Fish Blood

C. ALBERS[1]

1 Introduction

One important aspect of respiration is the exchange of the respiratory gases between blood and respiratory media. As respiratory gases I shall consider only the two classical ones, oxygen and carbon dioxide. In spite of the evident physical differences between the respiratory media, air and water, all vertebrates are using hemoglobin as a common though chemically modified respiratory pigment. In consequence the blood gas transport in all vertebrates shares some basic physicochemical properties. On the other hand, water as breathing medium imposes some peculiarities on blood gas transport in fish, as for instance, the low PCO_2 which results stringently from the Fenn-Rahn gas exchange diagram for water. In addition, water offers, frequently the problem of poor oxygen availability which has lead during evolution to special features of the hemoglobin molecule. To illustrate some of these points I have chosen the blood gas transport in carp. Carp hemoglobin has been very thoroughly studied biochemically [for ref. see Chien and Mayo (1980a, b) and Perutz and Brunori (1982), and there are also extensive studies as to the adaptation to hypoxia (for ref. see Weber and Lykkeboe 1978)].

2 Oxygen Transport

2.1 The Root Effect

The Bohr effect and the Root (Fig. 1) effect both describe effects of pH on the O_2 dissociation curve. Whereas blood having only a Bohr effect shows a horizontal displacement of the O_2 dissociation curve, but attains full oxygenation at low pH, blood characterized by a Root effect does not attain full oxygenation at low pH even if PO_2 is raised to the highest values in the physiological range. According to Perutz and Brunori (1982) the Root effect is an extreme stabilization of the T structure of hemoglobin due to the replacement of some residues in the β chain, particularly the replacement of cysteine in F9 by serine which forms hydrogen bonds to histidine HC3. Most interestingly in this position and in two other ones characterizing the carp hemoglobin, completely different replacements are found in hemoglobin I of trout which lacks the Root effect as well as the Bohr effect. It has been reported that in extreme cases of the Root effect

1 Institut für Physiologie, Universität Regensburg, Universitätsstraße 31, 8400 Regensburg, FRG

Circulation, Respiration, and Metabolism
(ed. by R. Gilles)

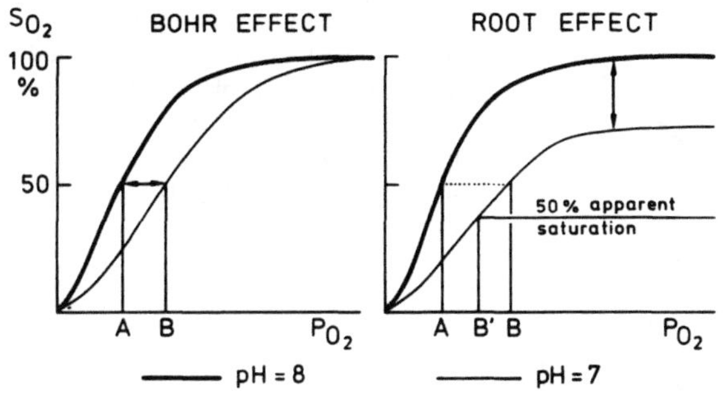

Fig. 1. Bohr and Root effect schematically presented. Points *A* and *B*: P_{50} refered to total hemoglobin as oxygen capacity. *B'*: P_{50} refered to an apparent loss of oxygen capacity

only half of the hemes in the hemoglobin can undergo a reversible oxygenation (Antonini and Brunori 1971), the other half remaining unliganded even at high pressure of oxygen. Such a point will be difficult to demonstrate, indeed, since at low pH the hemoglobin of carp, for instance, is extremely susceptible to oxidation into the ferric form. Furthermore, the allosteric model proposed by Perutz and Brunori still treats the oxygenation of hemoglobin as an equilibrium which implies that full oxygenation should be attained at a sufficiently high partial pressure of oxygen, whereas some conflicting experimental data are reported in the literature. As the O_2-binding curve describes an equilibrium, the Root effect should not be described as loss of oxygen capacity. For physicochemical reasons it seems more appropriate to state that the binding of oxygen does not exceed, e.g., 65% at a low pH rather than to call it a 35% loss of oxygen capacity. Otherwise the half-saturation point would be hard to define. In addition, the interaction of hydrogen ions and O_2 binding would be obscured as can be seen by the points B and B' (Fig. 1).

2.2 The Bohr Effect

Only the alkaline Bohr effect will be discussed since it is only the alkaline Bohr effect which operates in the physiological pH range. The ratio $\Delta \log P_{50}/\Delta pH$ is called the Bohr coefficient. If ΔpH results from the addition of acids other than carbonic acid, it is called fixed-acid Bohr coefficient; if it comes from carbonic acid, it is called CO_2 Bohr coefficient. In carp the fixed-acid Bohr coefficients is larger than in mammals and comes close to unity. There are no problems, if the Bohr coefficient is estimated in hemoglobin solutions. In experiments with whole blood, however, ΔpH should refer to the intraerythrocytic pH rather than to the plasma pH. In general, the red cell pH changes less than the pH of the plasma surrounding the cells as shown in Fig. 2 (Albers et al. 1983b). Like in the sheatfish, *Silurus glanis* (Albers et al. 1981), the slope of the lines relating red cell pH to plasma pH in carp is clearly smaller than that of the identity line. As a corollary, the Bohr coefficient based on plasma pH will be underestimated. Actually the CO_2 Bohr coefficient is about -1.3 when related to the red cell pH, whereas it is only -1.0 if related to the plasma pH. The Bohr coefficient obtained in solutions of carp hemoglobin is in the order of -0.67 to -1.0. These values, however, were fixed-acid Bohr coefficients. Thus, the CO_2 Bohr coefficient tends to be higher than the fixed-acid Bohr coefficient. This is a first hint to the existence of carbamate which will be discussed below.

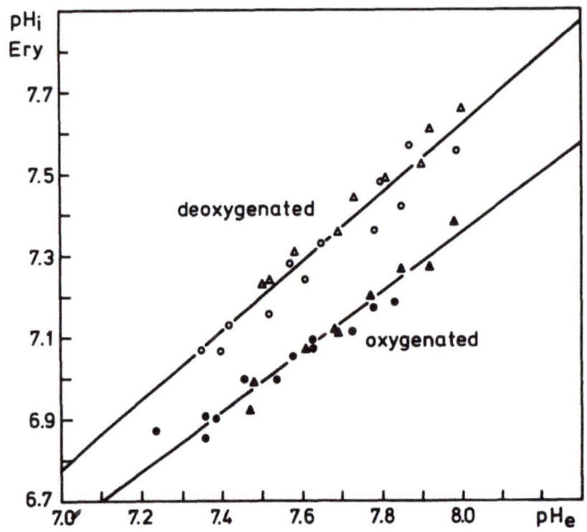

Fig. 2. Intraerythrocytic pH (pH_i) vs pH of true plasma in carp acclimated at 20° and 10 °C. $(pH_i - 6.10) = (0.853 - 0.159\ s) \cdot (pH_e - 6.21)$. From Albers et al. (1983b), with permission of Elsevier Science Publishers B.V.

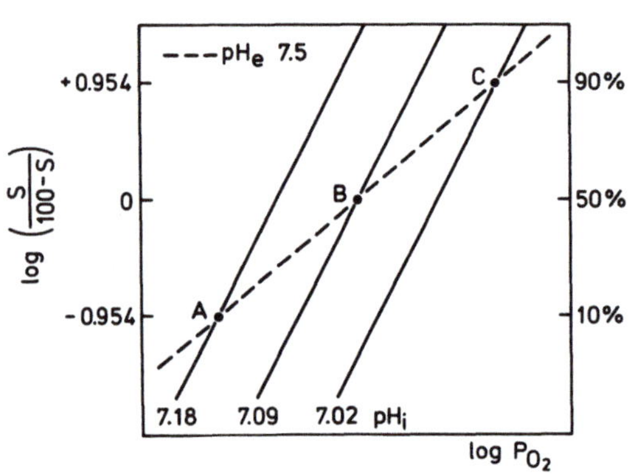

Fig. 3. Hill plot assuming parallel straight lines for each pH_i. *Dotted line:* Hill plot for a constant plasma pH

Not only the Bohr coefficient, but also the exponent n in the Hill equation is affected by the site of pH measurement. Misleading results are obtained if O_2 binding curves of whole blood at a constant plasma pH are used for the Hill plot. As shown in Fig. 2, the relationship between plasma pH and red cell pH strongly depends on the oxygenation of the hemoglobin. Thus, if the plasma pH is kept at a constant value, the red cell pH will decrease with increasing oxygen saturation. Figure 3 shows the Hill plots at three values of red cell pH corresponding to a constant plasma pH of 7.5. The points A, B, and C refer to saturation of 10%, 50%, and 90%, respectively. The dotted line drawn through these points represents the Hill plot for a constant plasma pH. Obviously, this line yields a much lower n value. Sometimes a fictitious discrepancy seems to exist between the results of O_2-binding studies carried out either in whole blood or in hemoglobin solutions. Since such studies aim at reconciling the biochemical facts obtained in purified and controlled solutions with the physiological facts prevailing in the highly complex reality, a careful distinction between the pH in blood (pH in true plasma) and red cell pH is required.

2.3 Problems of Heterogeneity

There is, however, another problem which complicates the O_2 binding by the blood in fish. In contrast to mammals, fish hemoglobin displays a marked heterogeneity. Up to 11 fractions of hemoglobin in one species have been obtained electrophoretically (Riggs 1970). According to Weber (1982), the different fractions may all be sensitive to temperature and pH; they may be only sensitive to pH, but not to temperature; or neither sensitive to pH nor to temperature. The hemoglobin pattern may depend on age (Wilkins and Iles 1966; as quoted by Riggs 1970) or on season (Fourie and van Wuren 1976). Figure 4 shows the major fractions of hemoglobin from carp, sheatfish, and eel as compared to human hemoglobin A. The physiological significance of the multiplicity is not yet evident. However, the interpretation of O_2-binding curves of whole blood must be intricate or even impossible in the case of functionally different hemoglobin fractions.

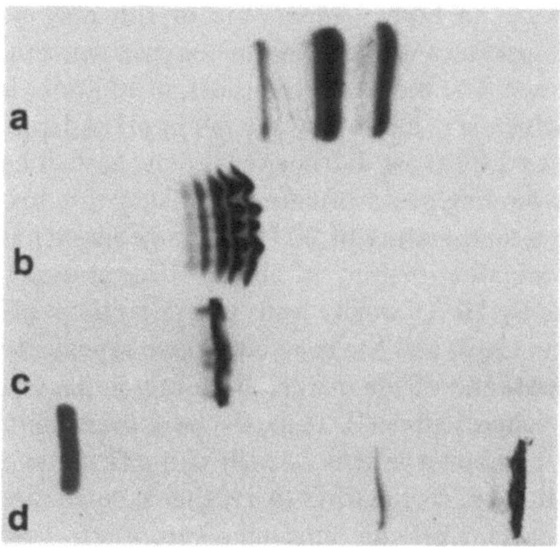

Fig. 4. Isoelectric focusing of major hemoglobin fractions. From *top to bottom*: *a* carp, *Cyprinus carpio*; *b* sheatfish, *Silurus glanis*; *c* human Hb A; *d* eel, *Anguilla anguilla*

Heterogeneity does not only exist among hemoglobins, but also among red cells. It has been claimed that activation of the erythropoiesis is of minor importance for adaptative processes in fish (Weber 1982). However, recent observations of Schindler (personal communication) point to the existence of a stress-induced enhancement of red cell proliferation. In carp usually 1% of the red cells are precursor cells with a low hemoglobin content. Two weeks after a severe blood loss (50% of the total blood volume), abundant precursor cells emerge within a couple of days. Figure 5 gives an example. The precursor cells are round and have a low optical density. Microphotometric analysis at the Soret band of hemoglobin yields at that time two morphologically different populations of red cells, the precursor cells amounting to nearly 50%. Within 4 weeks the two populations gradually fuse until finally the precursor cells are back at the level of about 1%. Due to the low hemoglobin content, the precursor cells presumably differ in Donnan equilibration and pH from the adult cells. It would not be surprising to observe difference in the O_2-binding curve. This aspect certainly disserves further research.

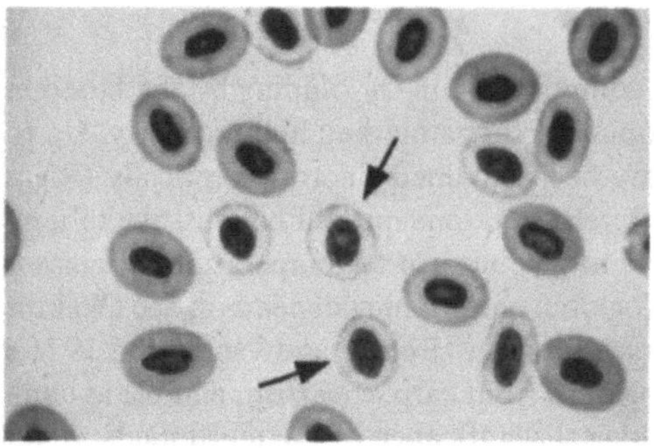

Fig. 5. Erythrocytes of carp. Pappenheim staining. *Arrows* mark precursor cells with low hemoglobin content. J. Schindler, unpublished, with permission of the author

2.4 Effect of Temperature

An increase in water temperature and, hence, in body temperature of fish may be hazardous for several reasons. The high temperature will increase the oxygen demand, but decrease the oxygen available in the water. The blood gas transport, in addition, is hampered by a decrease in oxygen affinity which is enhanced by the fall in pH. Adaptation to a higher temperature reduces at least partly these detrimental effects as shall be demonstrated in Fig. 6. Curve A is the O_2-binding curve of carp acclimated to a low temperature (10 °C) and at a pH of 8.0. At a temperature of 20 °C the same blood has a lower O_2 affinity corresponding to an overall enthalphy of about − 10 kcal mol^{-1} (curve B). An increase in body temperature by 10 °C usually leads to a decrease in pH of about 0.17 units as was shown by Rahn (1967) and has been confirmed repeatedly since. As a consequence of the lowering of pH and of the marked Bohr effect, the O_2-binding curve is shifted to even higher PO_2 values (curve C). Thus, the poor availability of O_2 in warm water is aggravated by the combined effects of high temperature and low pH on the O_2 affinity of hemoglobin. In carp, acclimation to a higher temperature increases the O_2 affinity. Though for a constant pH the pure temperature effect on P_{50} is nearly abolished, the marked effect of the pH change still keeps the O_2-binding

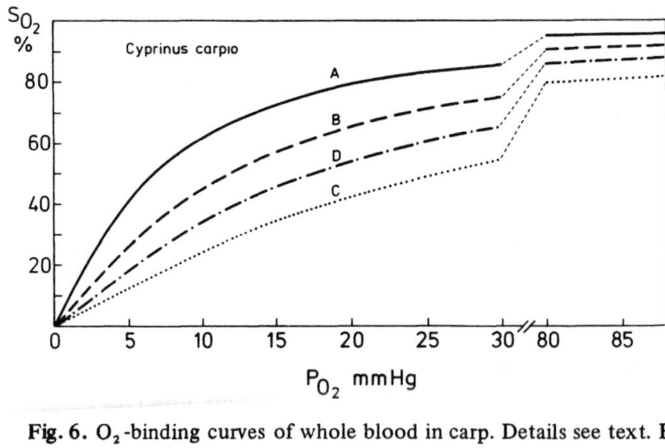

Fig. 6. O_2-binding curves of whole blood in carp. Details see text. From Albers et al. (1983a), with permission of Elsevier Science Publishers B.V.

curve far away from the region of an optimal O_2 uptake as curve D demonstrates. Grigg (1969) found the same type of temperature adaptation in the brown bullhead. On the other hand, Weber et al. (1976) did not see any adaptational change in the trout and Dobson and Baldwin (1982) observed an enhanced right shift of the O_2-binding curve in the Australian blackfish during acclimation to a high temperature. Only an intra-erythrocytic mechanism can account for short-term adaptations in O_2 affinity (Albers et al. 1983a). Since the most potent effectors of O_2 binding by hemoglobin in fish are ATP and GTP, the metabolic control of these nucleoside triphosphates could play a key role in regulation of O_2 affinity as has been repeatedly stressed by Weber and his group (Weber 1982). In carp acclimated at 20 °C we found a 30% decrease of GTP and also a decrease in the GTP/ATP ratio (Albers et al. 1983b). Further experiments are needed to firmly establish the involvement of the nucleoside triphosphates in temperature acclimation.

3 CO_2 Transport and pH

Like in mammalian blood, CO_2 is transported in fish blood in three fractions: physically dissolved CO_2, bicarbonate, and carbamate. In contrast to the CO_2 transport in mammals, the partial pressure of CO_2 in fish is fairly low, depending on temperature, as has been stressed by Rahn (1967). Nevertheless, the partial pressure difference for CO_2 between blood and water is sufficiently high to allow a sufficient CO_2 elimination. We obtained in the dorsal aorta of free swimming carp at a temperature of 20 °C a PCO_2 of 5.6 mmHg and a pH of 7.8. This pH value is well in line with the data of Rahn and Baumgardner (1971) and Weber and Lykkeboe (1978).

3.1 Nonbicarbonate Buffer Value

How does pH change, if carbonic acid and/or strong acids enter the blood? To answer this question we equilibrated carp blood in vitro at three PCO_2 levels in either oxygen or nitrogen. The resulting plot of bicarbonate vs pH is shown in Fig. 7. The concentra-

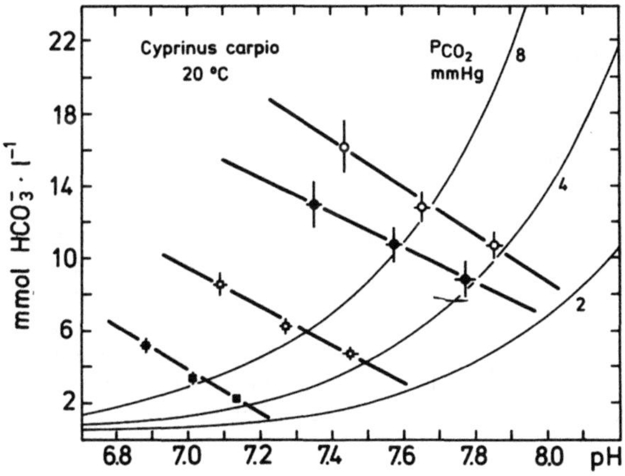

Fig. 7. Davenport diagram for true plasma and red cells in carp at 20 °C. Circles = plasma, squares = red cell, open symbols = deoxygenated blood, closed symbols – oxygenated blood. Modified from Albers et al. (1983a), with permission of Elsevier Science Publishers B.V.

tion of bicarbonate was calculated from pH and PCO_2 using appropriate constants for the solubility and pK′. What can be called appropriate is a question not yet fully settled since we lack data describing the dependence of pK′ on temperature and pH as far as plasma and notably as red cell water in fish is concerned. However, the calculated bicarbonate concentrations are well in line with data obtained by Baumgardner (1971) by means of direct CO_2 determinations. Likewise the nonbicarbonate buffer values obtained from these data agree with those obtained in hemoglobin solutions by Chien and Mayo (1980b).

3.2 Effect of Oxygenation

At a given PCO_2 pH as well as bicarbonate concentrations are strongly influenced by the degree of oxygenation. If the oxygen saturation changes from 0% to 100% the plasma pH decreases in the average by 0.13 and the red cell pH by 0.20 or even more. The latter effect was negatively correlated with pH. Based on 56 paired observations of pH_i in the presence and absence of O_2, the following regression equation was obtained: $\Delta pH_i/\Delta S = 3.115-0.4815\ pH_i$. According to this equation the effect of oxygenation on $\Delta pH_i/\Delta S$ would vanish at a pH of about 6.4. This pH corresponds to the value at which in the presence of high concentrations of organic phosphate, such as inositol hexaphosphate, the O_2 affinity of carp hemoglobin becomes insensitive to pH (Chien and Mayo 1980a). Not only bicarbonate concentration and pH are sensitive to pH, also the apparent buffer values of the red cell depend on oxygenation. In fully deoxygenated blood the nonbicarbonate buffer of carp erythrocytes were −8.4 as compared to −4.4 in fully oxygenated blood. This points to the existence of carbamate.

3.3 Carbamate

The question whether or not carp hemoglobin forms carbamate was investigated by Gros (1983), using the stop-flow technique of Crandall and Forster. The hemoglobin was prepared from the same carp acclimated at 10 °C and 20 °C which we had used for the experiments on oxygen equilibrium and pH. Figure 8 demonstrates that carp hemoglobin forms less carbamate than human hemoglobin. In addition, the carbamate formation in carp hemoglobin is highly sensitive to temperature in contrast to human

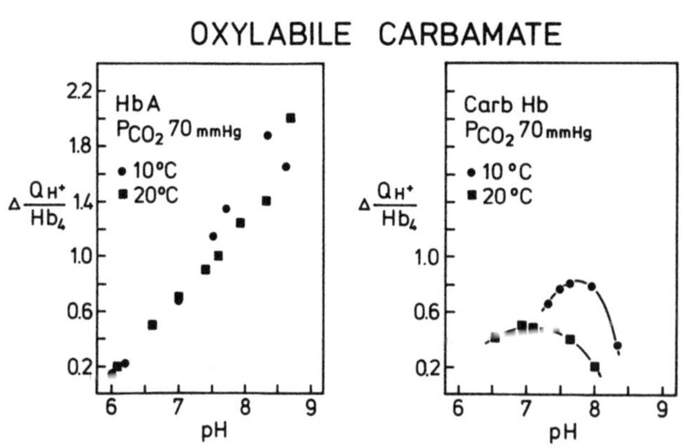

Fig. 8. Total carbamate of human deoxy-hemoglobin and of carp deoxy-hemoglobin measured at 10 °C (*filled circles*) and at 20 °C (*filled squares*). Unpublished figure of G. Gros (1983); with permission of the author

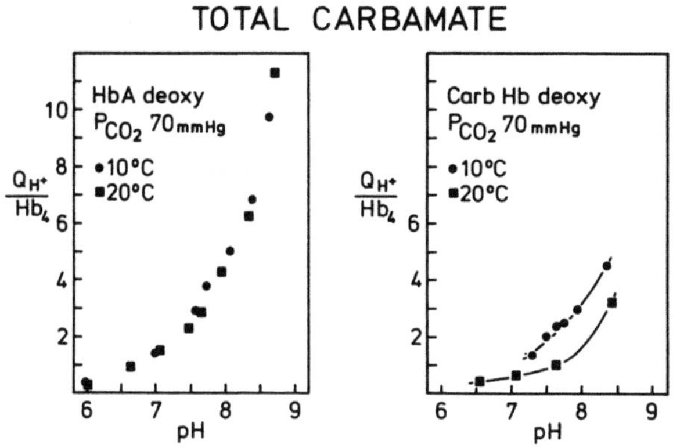

Fig. 9. Oxylabile carbamate of human hemoglobin and carp hemoglobin. Same symbols as in Fig. 8. Unpublished figure of the author

hemoglobin. The same holds for the oxylabile fraction of carbamate, as shown in Fig. 9. Obviously total as well as oxylabile carbamate decrease appreciably if the temperature increases. Such decrease would help to counteract the decrease of the intrinsic oxygen affinity which results from an increase in temperature. This feature of the carbamate formation in carp blood may, thus, have some adaptational value, though the carbamate formation remains unaltered during temperature acclimation. The amount of carbamate at a physiological PCO_2 presumably is smaller than shown here. But also the bicarbonate concentration in the red cells is low. From our data we obtained about 0.6 mmol bicarbonate per mole hemoglobin tetramer. This value has the same order of magnitude as the carbamate. Thus, carbamate may well be of importance to the blood gas transport as well as the acid-base regulation.

3.4 Donnan Equilibrium

The low bicarbonate concentration in the red cells is not only due to the low PCO_2 prevailing in fish blood, but also to the Donnan distribution. As already shown by Steen and Turitzin (1968), the Donnan ratio r is smaller in fish than in mammals. This was ascribed to the lower electrophoretic mobility of fish hemoglobin. It was also described that the Donnan ratio for chloride did not correspond quantitatively to the Donnan ratio of hydrogen ions. We obtained the Donnan ratio for hydrogen ions in carp and in

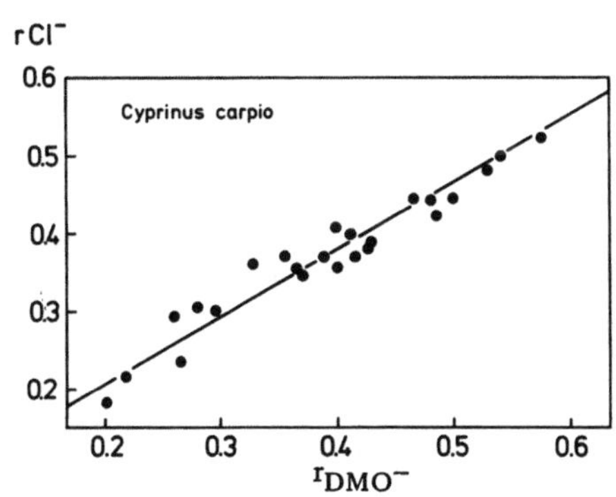

Fig. 10. Donnan distribution of monovalent anions in whole blood of carp in vitro at 20 °C. r_{Cl^-} vs r_{DMO^-} (unpublished data)

the European sheatfish from the intraerythrocytic and plasma pH values, already shown. When plotted against pH, the values were close to the values obtained in the eel by Steen and Turitzin. However, in carp we did not find an extreme discrepancy between the distribution of hydrogen ions and chloride and, even more convincingly, between chloride and DMO (Fig. 10). Since the distribution of DMO closely reflects the distribution of hydrogen ions, we conclude that in carp monovalent anions are distributed according to a Donnan equilibrium just as is the distribution in mammalian blood.

References

Albers C, Götz KH, Welbers P (1981) Oxygen transport and acid-base balance in the blood of the sheatfish, *Silurus glanis*. Respir Physiol 46:223–236

Albers C, Manz R, Muster D, Hughes GM (1983a) Effect of acclimation temperature on oxygen transport in the blood of the carp, *Cyprinus carpio*. Respir Physiol 52:165–179

Albers C, Götz KH, Hughes GM (1983b) Effect of acclimation temperature on intraerythrocytes acid-base balance and nucleoside triphosphates in the carp, *Cyprinus carpio*. Respir Physiol 54: 145–159

Antonini E, Brunori M (1971) Hemoglobin and myoglobin in their reactions with ligands. In: Neuberger A, Tatum EL (eds) Frontiers of biology, vol 21.

Baumgardner FW (1971) Acid-base balance in vertebrates. Thesis. State University of New York at Buffalo, N.Y.

Chien JCW, Mayo KH (1980a) Carp hemoglobin I. Precise oxygen equilibrium and analysis according to the models of Adair and of Monod, Wyman and Changeux. J Biol Chem 255:9790–9799

Chien JCW, Mayo KH (1980b) Carp hemoglobin II. The alkaline Bohr effect. J Biol Chem 255: 9800–9806

Dobson GP, Baldwin J (1982) Regulation of blood oxygen affinity in the Australian blackfish, Gadopsis marmoratus. J Exp Biol 99:245–254

Fourie F le R, van Vuren JHJ (1976) A seasonal study on the hemoglobins of Carp, *Cyprinus carpio*, and yellowfish, Barbus holub, in South Africa. Comp Biochem Physiol 55B:523–525

Grigg GC (1969) Temperature induced changes in the oxygen equilibrium curve of the blood of the brown bullhead, *Ictalurus nebulosus*. Comp Biochem Physiol 28:1203–1223

Gros G (1983) Biophysics of carbon dioxide binding to the hemoglobin molecule. Satellite Symp of the 29th Congr. of the Int. Unio of Physiol. Sci. on "Physiology and Biochemistry of Blood Gas Transport", Canberra

Perutz MF, Brunori M (1982) Stereochemistry of cooperative effects in fish and amphibian hemoglobins. Nature 299, p 421

Rahn H (1967) Gas transport from the external environment to the cell. In: Ciba Foundation Symposium on Development of the Lung. J.A. Churchill, London, pp 3–23

Rahn H, Baumgardner FW (1971) Temperature and acid-base regulation in fish. Respir Physiol 14: 171–182

Riggs A (1970) Properties of fish hemoglobins. In: Hear WS, Randall DJ (eds) Fish physiology, vol IV, pp 209–252

Steen JB, Turitzin SN (1968) The nature and biological significance of the pH difference across red cell membranes. Respir Physiol 5:234–242

Weber RE (1982) Intraspecific adaptation of hemoglobin function in fish to oxygen availability. In: Addink ADF, Spronk N (eds) Exogenous and endogenous influences on metabolic and neural control. Invited lectures. Pergamon, Oxford

Weber RE, Lykkeboe G (1978) Respiratory adaptations in carp blood. Influences of hypoxia, red cell organic phosphates, divalent cations and CO_2 on hemoglobin-oxygen affinity. J Comp Physiol 128:127–137

Weber RW, Wood SC, Lomholdt JP (1976) Temperature acclimation and oxygen-binding properties of blood and multiple hemoglobins of rainbow trout. J Exp Biol 65:333–345

Ammonia Transfer Across Fish Gills: A Review

J.N. CAMERON[1] and N. HEISLER[2]

1 Introduction

In almost all aquatic animals, the primary end product of nitrogen metabolism is ammonia, which must be excreted to the external medium. Due to its rather complex chemistry, the exact mechanism(s) by which ammonia is transferred from internal to external media in various organisms and under various conditions is still not a matter of complete agreement. The purpose of this review is to describe the chemistry of ammonia in water, briefly, and to review some of the major pieces of evidence for transfer in one form or another.

2 Aqueous Chemistry of Ammonia

2.1 Chemical Dissociation

Ammonia exists in water in two chemical forms, the nonionized NH_3, and the protonated ion NH_4^+. Since these two forms exist in equilibrium, according to:

$$NH_3 + H^+ <==> NH_4^+ \tag{1}$$

ammonia can be seen to act as a weak base, and the equilibrium can be described by the Henderson-Hasselbalch equation:

$$pH = pK + \log\{[NH_3]/[NH_4^+]\}. \tag{2}$$

The pK for this reaction is quite sensitive to both temperature and ionic strength, and may range between 9.00 and 10.00 over the physiological range (Cameron and Heisler 1983). For the purposes of our present discussion of fish, the value for 20 °C and the ionic strength typical for teleost plasma would be 9.44. Since a typical blood pH value at this temperature would be 7.85, the ratio of NH_3 to NH_4^+ in the blood would be about 0.026, or nearly 40 to 1 in favor of NH_4^+.

1 The University of Texas at Austin, Department of Zoology and Marine Science Institute, Port Aransas, TX 78373, USA

2 Abteilung Physiologie, Max-Planck-Institut für experimentelle Medizin, Hermann-Rein-Straße 3, 3400 Göttingen, FRG

Circulation, Respiration, and Metabolism
(ed. by R. Gilles)

2.2 Ammonia as a Dissolved Gas

In addition to its behavior as a weak base and electrolyte, the NH_3 portion must also be considered as a dissolved gas. The solubility of NH_3 in aqueous solution is very high, with a value of 43.7 $mmol\,l^{-1}\,torr^{-1}$ in plasma at 20 °C, about 27,000-fold the solubility of oxygen (Cameron and Heisler 1983). The partial pressure of NH_3 gas, therefore, calculated as:

$$P_{NH_3} = [NH_3]/\alpha \tag{3}$$

where α is the solubility coefficient, is very low, generally in the microtorr range. For a normal, resting rainbow trout in water with low ammonia content, the average arterial NH_3 partial pressure is about 40 μtorr (Cameron and Heisler 1983).

2.3 Diffusion and Permeability

As a dissolved gas, NH_3 diffuses according to the Fick equation:

$$dNH_3/dt = -\, A\, D\, \alpha\, (P_{NH_3}/x) \tag{4}$$

where A is the total surface area (of gills, in this case), D the diffusion coefficient, α the solubility coefficient, and x the thickness of the diffusion barrier. There is considerable inconsistency in the presentation of permeability information, since the values for D, the true diffusion coefficient, may be given; values of Krogh's permeation coefficient, equal to $\alpha \cdot D$, or D', which incorporate the solubility into the coefficient; or finally, permeability coefficients are often given when the thickness, x, is not known, these values being equal to $\alpha \cdot D/x$.

In addition to diffusive movement of the nonionic form, there is also some movement of the charged NH_4^+ form across biological membranes (Jacobs 1940; Pitts 1973; Evans 1977; Goldstein et al. 1982; Schwartz and Tripolone 1983; Arruda et al. 1984). It is generally agreed that since the NH_4^+ ion is large and carries a charge, its permeability, like other ions of similar size and charge, will not be high. Comparisons of actual values for the permeability suffer the same kinds of difficulties with nonstandard units as those for NH_3, but most estimates are one to two orders of magnitude lower than NH_3, but of similar magnitude to other cations.

There do seem to be considerable tissue and species differences, however: the ratio of NH_4^+ to sodium permeability given for the bullfrog gallbladder is 2.48 and for the rabbit gallbladder 3.23 (Diamond 1975), but a value given recently for the turtle urinary bladder is 0.08 (Arruda et al. 1984), about 30-fold smaller. The gallbladder is known to be a "leaky" epithelium, and the main route of NH_4^+ movement is thought to be paracellular, thus, affected importantly by the tightness of cell junctions. It is unfortunately not possible to study a geometrically complex tissue like the fish gill in such detail.

What is more important than the absolute rates of permeation of NH_3 and NH_4^+, however, is the ratio of permeabilities of these two species. Some values from the literature are summarized in Table 1, but here again considerable variability is encountered, even between two studies of the same tissue (urinary bladder of the turtle *Pseudemys scripta*). The estimates of the permeability ratio given by Schwartz and Tripolone (1983) are already three to four times higher, and they considered their

Table 1. Permeability and permeability ratios for NH_3 and NH_4^+ in various tissues

Tissue	PM_{NH_3} [a]	$PM_{NH_4^+}$ [a]	PM_{NH_3}/NH_4^+ [a]	Source [d]
Turtle bladder	7.5×10^{-5}	4.1×10^{-6}	18.4	1
Turtle bladder	2.6×10^{-4}	4.9×10^{-6}	53	2
	3.0×10^{-4}	4.3×10^{-6}	70[b]	2
Dogfish gills	7.6×10^{-4}	2.6×10^{-6}	292	3
Crab gills	–	–	53[c]	4

[a] The symbol PM is used for permeability to avoid confusion with partial pressure notation
[b] Estimate two times higher due to tissue damage
[c] Estimated from Na^+ permeability
[d] 1, Arruda et al. 1984; 2, Schwartz and Tripolone 1983; 3, Evans 1985, this volume; 4, Kormanik and Cameron 1981b

values to be low by a factor of two due to probable edge damage in their mounted tissues. The correlation between "tightness" of an epithelium and the relative NH_4^+ permeability is suggestive, and may relate to fresh- vs seawater differences in the fish.

3 Experimental Evidence from Fish Gills

In the remainder of this paper, I will discuss some of the different experiments which have been performed in an attempt to unravel the principal route of ammonia movement across the gills of fish (and other aquatic animals). The pattern for many of these experiments was developed by Krogh (1938), but some new approaches have also been explored in more recent years. Krogh's experiments showed that in freshwater, where ions must be actively transported from the external environment to replace passive losses, the uptake of Na^+ and Cl^- were demonstrably electroneutral, and he proposed that there were linked ionic exchanges of Cl^- for HCO_3^-, and of Na^+ for either H^+ or NH_4^+.

3.1 Ammonium Salt Injection

Given the existence of a Na^+/NH_4^+ exchange, an injection of an ammonium salt should stimulate Na^+ uptake. So reasoning, Maetz and Garcia-Romeu (1964) performed a series of such experiments with the goldfish, and various others have obtained similar results since (e.g., Evans 1977). What was not fully appreciated in these experiments was that the stimulation of Na^+ uptake could also be accounted for by indirect effects on the alternate exchange proposed by Krogh, i.e., Na^+/H^+ exchange. The injected ammonium salt remains in equilibrium in the body fluids with NH_3, and diffusive loss of the NH_3 will lead to an internal acidosis. The effects of infusion of a 2 mmol kg^{-1} load of NH_4Cl into catheterized channel catfish *(Ictalurus punctatus)* are shown in Fig. 1 (Cameron and Kormanik 1982). There is clearly a rapid and severe acidosis, which can only be caused by excretion of at least part of the ammonia load as NH_3. Recently we repeated

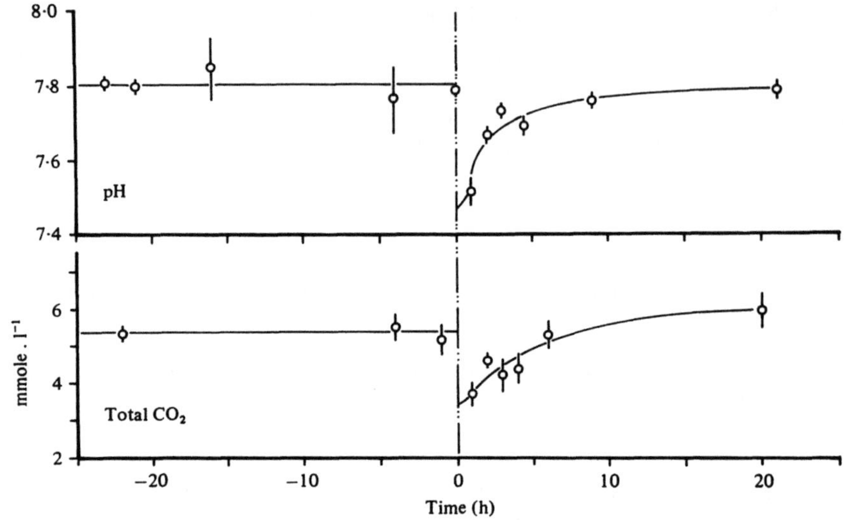

Fig. 1. The effects of infusion of a 2 mmol kg^{-1} load of NH_4Cl into channel catfish, *Ictalurus punctatus*. The *upper panel* shows the arterial pH, and the *lower panel* the arterial plasma total CO_2 during a control period (−24 to 0 h), and after infusion of the NH_4Cl load at time 0

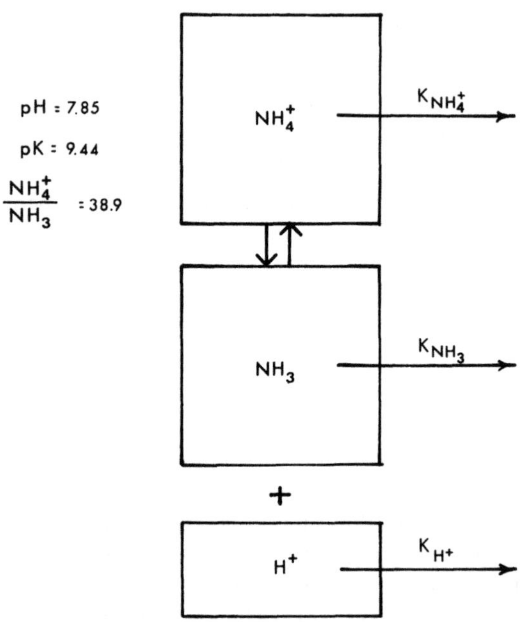

Fig. 2. A simple model of ammonia excretion, incorporating direct losses from NH_3 and NH_4^+ compartments, as well as an excretion rate for the H^+ generated by diffusive loss of NH_3. The model is only semiquantitative, but realistic rate constants may be employed for any particular case. The equilibrium between NH_3 and NH_4^+ is assumed to be instantaneous

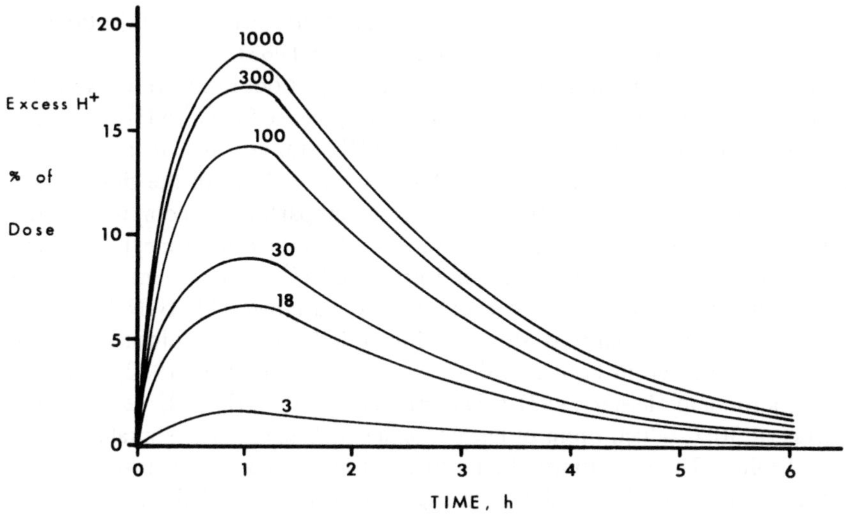

Fig. 3. The time course and severity of acidosis generated by NH_4Cl infusion, as a function of the ratio of NH_3 to NH_4^+ permeability employed in the model shown in Fig. 2. The observed acidosis in both channel catfish (Cameron and Kormanik 1982) and in rainbow trout (Cameron and Heisler 1983) indicated ratios of 100 or more

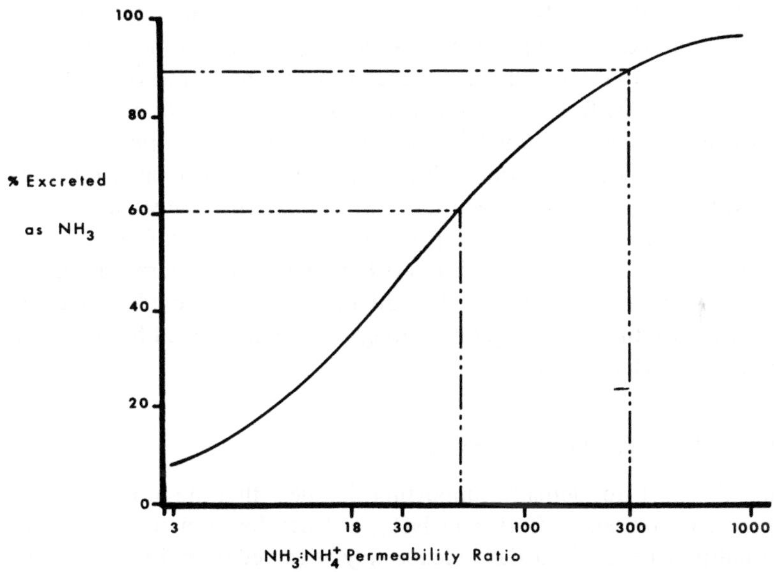

Fig. 4. The percentage of an infused ammonium salt load excreted as nonionized NH_3, as a function of the permeability ratio. The area demarcated by *dashed lines* shows the most likely range, based on published permeability ratios (see text and Table 1)

Maetz and Garcia-Romeu's experiments with equivalent intraperitoneal injections of $(NH_4)_2SO_4$, with very similar results (Cameron and Heisler 1983).

Because of the acid-base complications of ammonia injections or infusions, stimulation of Na^+ uptake by this treatment cannot be taken as evidence for Na^+/NH_4^+ exchange. The exact kinetics of excretion of the ammonia and H^+ loads are difficult to measure, but at least an approximate estimate of the relative permeabilities of the gills to NH_3 and NH_4^+ may be made. To illustrate this, a simplified compartmental computer model was made, shown schematically in Fig. 2. By varying the ratios of permeabilities (export rates, actually) from the NH_3, NH_4^+, and generated H^+ compartments, the time course and severity of acidosis could be estimated over a range of NH_3/NH_4^+ permeability ratios. The model calculations, shown in Fig. 3, indicate that acidosis of the approximate severity observed in both catfish (Cameron and Kormanik 1982) and rainbow trout (Cameron and Heisler 1983) is only predicted with NH_3/NH_4^+ permeability ratios in excess of 30:1, which more or less coincides with the actual estimates of this ratio given in Table 1. The percentage of an infused or injected ammonia load that will be excreted as nonionized NH_3 is shown in Fig. 4 as a function of the permeability ratio, with the most probable range indicated, based on literature values. Thus, it appears from ammonium salt infusion studies, and from the literature values on permeability ratios that between 60% and 90% of an infused load will be excreted as NH_3.

3.2 Effects of High External Ammonia on Na^+ Uptake

A corollary prediction of a Na^+/NH_4^+ exchange hypothesis is that elevated external $[NH_4^+]$ will reduce Na^+ influx. This experiment was also performed on the goldfish by Maetz and Garcia-Romeu (1964), and has been repeated with other species (e.g., Evans 1977). It too, however, suffers from the ambiguity imposed by indirect effects from the proposed alternate exchange, Na^+ for H^+. Concentrated external ammonium solutions, since they are in chemical equilibrium [Eq. (2)] with NH_3, will cause diffusive movement of NH_3 into the animal, which will result in an internal alkalosis, a frequent maneuver employed in single cell studies of pH regulation (Roos and Boron 1981). This nonrespiratory alkalosis has been observed in fish with elevated external ammonia experiments (Cameron and Heisler 1983), and might be expected to inhibit a Na^+/H^+ exchange, particularly since these exchanges are thought to be modulated for acid-base regulation (Cameron 1976; Heisler 1982).

3.3 Estimates of NH_3 Flux Based on Permeability

The evidence cited above, if not actually supporting the view that the principal route of ammonia excretion is nonionic diffusion of NH_3, at least does not contradict that hypothesis. Some additional circumstantial evidence may be gained from the calculations of partial pressure gradients, based upon careful measurements of internal ammonia concentration, pH, and appropriate physicochemical parameters. Such calculations for the rainbow trout (Cameron and Heisler 1983) show that the measured total ammonia excretion of the animal can easily be accounted for on the basis of nonionic diffusion. Similar calculations for the aquatic blue crab *(Callinectes sapidus)* gave similar results (Kormanik and Cameron 1981a). Maetz (1973) reached different conclusions, but was using an erroneously high solubility coefficient for the blood, which gave partial pressure values for blood NH_3 that were also consequently too low.

3.4 Amiloride Inhibition of Ammonia Excretion

The commercial drug Amiloride (Merck) is generally thought to block Na^+ entry at the apical cell surface, and if ammonia were excreted significantly via a Na^+/NH_4^+ exchange, one would predict a reduction of ammonia excretion upon external administration of amiloride. The literature on this subject is contradictory: Kirschner et al. (1973) reported a 30% reduction in ammonia efflux in the trout, and 54% in the crayfish, in spite of 79% and 90% reductions in Na^+ uptake, respectively. In the blue crab, Kormanik and Cameron (1981a) found no effect of amiloride on ammonia excretion, but Pressley et al. (1981) reported a 65% reduction. In various other animal tissues, highly variable reductions in Na^+ uptake and ammonia excretion are reported for amiloride, so generalizations are difficult to support. It should be noted, however, that these effects may in some cases be explained by an indirect pH mechanism, much like the external ammonia and ammonia injection experiments. That is, if amiloride blocks a Na^+/H^+ exchange mechanism important in maintaining internal pH, then the internal pH may drop, reducing the partial pressure of NH_3, and consequently its excretion rate.

3.5 Prolonged High External Ammonia and Ammonia Excretion

The elevated external ammonia experiments described above have most often been performed for relatively short periods, up to 1 or 2 h. As we have recently discovered (Cameron and Heisler 1983), there is a transient period of at least that duration during which events are quite different than what occurs over longer periods. The results of one such experiment are shown in Fig. 5. In the first hour or so, the gradients for both NH_3 partial pressure and NH_4^+ concentration are reversed, and the net movement of total ammonia is inward (Fig. 6, upper panel). After some time, however, ammonia excretion resumes, but against both the partial pressure gradient for NH_3 and the electrochemical gradient for NH_4^+ (Fig. 6, lower panel). A similar experiment on the channel catfish (Kormanik and Cameron, unpublished data) shows the same pattern: an initial period of reversed gradients and reversed flux, and a later period of renewed excretion against the gradients. This may be the most unequivocal evidence to date of the presence of a Na^+/NH_4^+ exchange, or at least of an energy-requiring active transport of NH_4^+ in the fish gill. Abundant evidence for such carrier-mediated transport may be found in the amphibian, reptile, and mammal literature (e.g., Schwartz and Tripolone 1983). There are very few other studies of fish in which the blood acid-base status has been simultaneously measured, and so the possibility of indirect pH effects cannot be ruled out. A relatively small change in the blood pH may make a significant change in the NH_3/NH_4^+ distribution ratio, and in the resulting gradients.

3.6 Other Evidence for Na^+/NH_4^+ Exchange

Regardless of its quantitative importance in normal ammonia excretion, there are several other lines of evidence indicating at least the existence of, or potential for this pathway. In several studies it has been demonstrated that ouabain inhibits both Na^+

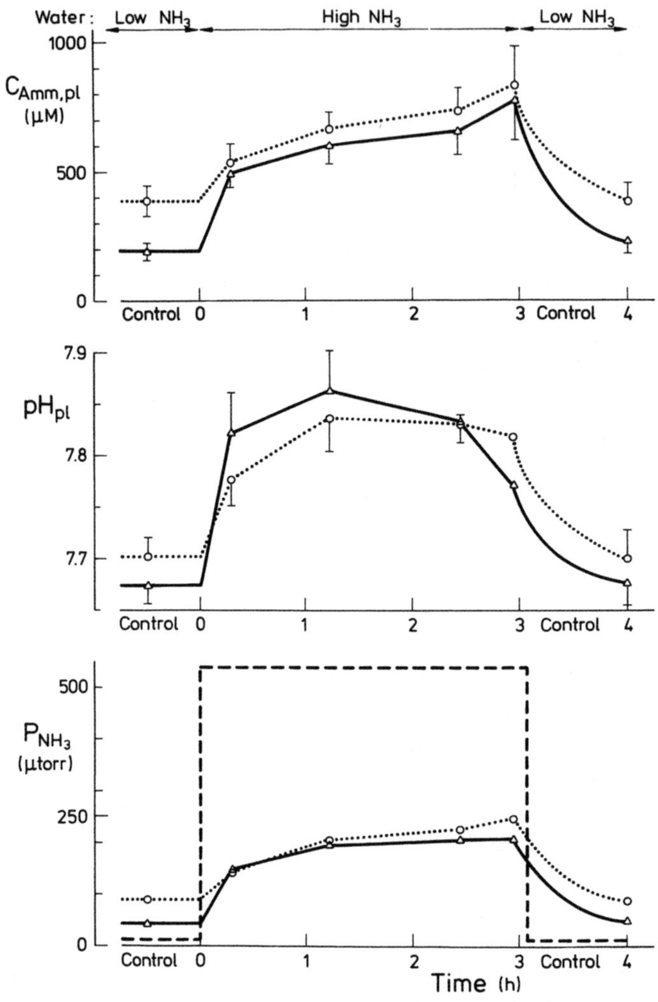

Fig. 5. Data from an experiment on rainbow trout, in which the external ammonia was increased to 1 mmol l^{-1} after an initial control period. The control data are shown at the left, with the increase in ammonia indicated at the top from time 0 to just past 3 h. Total ammonia concentrations and pH of arterial (*solid line*) and venous blood (*dotted line*) were measured, and the partial pressures calculated from solubility and pK data gathered from the same fish. The data appear to show attainment of a steady state after about 1–2 h, with reversed ammonia gradients (*bottom panel*)

and ammonia fluxes (Claiborne et al. 1982; Payan 1978), and a ouabain-sensitive NH_4^+ stimulation of a Na/K-ATPase isolated from fish gills has also been demonstrated (Mallery 1979). There are many problems in the interpretation of isolated gill studies, however, and extension of the findings to intact fish is somewhat uncertain. There is always the possibility that the observed effects may also be due to indirect pH effects in the intracellular fluids. For example, ouabain treatment may cause an intracellular alkalosis, inhibiting Na^+/H^+ exchange.

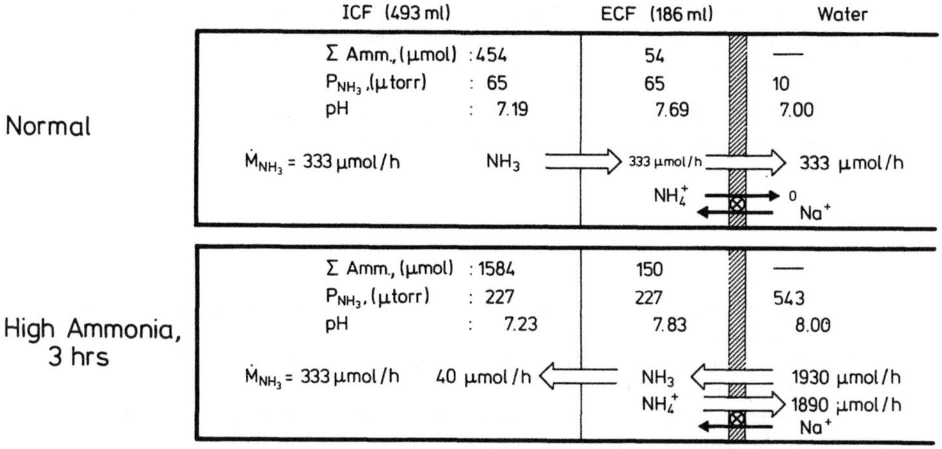

Fig. 6. A hypothetical model of ammonia distribution and movement based on data from Fig. 5. The ICF and ECF compartments are drawn roughly to scale of their relative volumes for a 1 kg fish, with the gill tissue (*hatched, right*) separating the ECF from a water compartment of infinite volume. The principal assumptions are that ammonia production in tissues is constant at 333 μmol h^{-1}, that ammonia movement across the gills is proportional to the NH_3 gradient, and that only diffusion of NH_3 takes place during the control period. (From Cameron and Heisler 1983)

4 Conclusions

The normal condition for aquatic animals is to have a significant ammonia gradient from inside to outside, both a chemical gradient of the ionic NH_4^+ and a partial pressure gradient of NH_3. The ammonia concentration in natural waters is almost always very low, in large part due to the efficient uptake of ammonia by microorganisms as a nutrient. In the normal situation, then, it appears likely from most of the evidence that the majority, 60% to 90% or perhaps even more, of the total ammonia excretion occurs via nonionic diffusion of NH_3. Arteriovenous differences in NH_3 partial pressure of the correct magnitude have been directly observed in the trout (Cameron and Heisler 1983).

Many experiments reported in the literature as verification of the existence of Na^+/NH_4^+ exchange may be given alternate interpretations, based upon the acid-base consequences of the ammonia manipulations employed, and on the probable existence of a Na^+/H^+ exchange in the gills. For most of these experiments, the appropriate data on blood acid-base status is missing. In some further cases, however, notably prolonged elevated external ammonia experiments, and some recent perfused gill studies (Evans 1985, this conference; Goldstein et al. 1982), there is also evidence that ammonia can be excreted against both partial pressure and chemical gradients, or that ammonia excretion responds to NH_4^+ gradients in the absence of change in the NH_3 partial pressure gradient. These experiments support the existence of a Na^+/NH_4^+ exchange in the gills.

Since the normal aquatic situation does not seem to require such an exchange, one is tempted to speculate that the NH_4^+ may only incidentally substitute for H^+, in artificial experimental situation, and that the normal functions of the exchange are for ion replacement and acid-base regulation.

Note added in proof. In recent studies of prolonged high external ammonia in channel catfish, Na^+ uptake was unchanged. It appears that internal NH_4^+ is exchanged for external H^+, rather than Na^+ (cf. Sect. 3.5, above). (JN Cameron, unpublished data)

Acknowledgements. These studies were supported by NSF Grants PCM80-20982 and PCM83-15833 to J.N.C.; and by funds from the Deutsche Forschungsgemeinschaft and the Max-Planck-Gesellschaft.

References

Arruda JAL, Dytko G, Withers L (1984) Ammonia transport by the turtle urinary bladder. Am J Physiol 246:F635–F647

Cameron JN (1976) Branchial ion uptake in Arctic grayling: resting values and effects of acid-base disturbance. J Exp Biol 64:711–725

Cameron JN, Heisler N (1983) Studies of ammonia in the rainbow trout: physico-chemical parameters, acid-base behaviour and respiratory clearance. J Exp Biol 105:107–125

Cameron JN, Kormanik GA (1982) The acid-base responses of gills and kidneys to infused acid and base loads in the channel catfish, *Ictalurus punctatus*. J Exp Biol 99:143–160

Claiborne JB, Evans DH, Goldstein L (1982) Fish branchial Na^+/NH_4^+ exchange is via basolateral Na^+-K^+-activated ATPase. J Exp Biol 96:431–434

Diamond JA (1975) How do biological systems discriminate among physically similar ions? J Exp Zool 194:227–240

Evans DH (1977) Further evidence for Na/NH_4 exchange in marine teleost fish. J Exp Biol 70: 213–220

Evans DH (1985) Modes of ammonia transport across fish gills (this volume)

Goldstein L, Claiborne JB, Evans DH (1982) Ammonia excretion by the gills of two marine teleost fish: the importance of NH_4 permeance. J Exp Zool 219:395–397

Heisler N (1982) Transepithelial ion transfer processes as mechanisms for fish acid-base regulation in hypercapnia and lactacidosis. Can J Zool 60:1108–1122

Jacobs MH (1940) Some aspects of cell permeability to weak electrolytes. Cold Spring Harbor Symp Quant Biol 8:30–39

Kirschner LB, Greenwald L, Kerstetter TH (1973) Effect of amiloride on sodium transfer across body surfaces of fresh water animals. Am J Physiol 224:832–837

Kormanik GA, Cameron JN (1981a) Ammonia excretion in the seawater blue crab *(Callinectes sapidus)* occurs by diffusion, and not Na^+/NH_4^+ exchange. J Comp Physiol B 141:457–462

Kormanik GA, Cameron JN (1981b) Ammonia excretion in animals that breathe water: a review. Mar Biol Lett 2:11–23

Krogh A (1938) The active absorption of ions in some freshwater animals. Z Vgl Physiol 25:335–350

Maetz J (1973) Na^+/NH_4^+, Na^+/H^+ exchanges and NH_3 movement across the gills of *Carassius auratus*. J Exp Biol 58:255–275

Maetz J, Garcia-Romeu F (1964) The mechanisms of sodium and chloride uptake by the gills of a fresh water fish, *Carassius auratus*. II. Evidence for NH_4^+/Na^+ and HCO_3^-/Cl^- exchanges. J Gen Physiol 47:1209–1227

Mallery CH (1979) Ammonium stimulated properties of K-dependent ATPase in *Opsanus beta*, a teleost with an NH_4^+/Na exchange pump. Am Zool 19:944 (Abstract)

Payan P (1978) A study of the Na^+/NH_4^+ exchange across the gill of the perfused head of the trout *(Salmo gairdneri)*. J Comp Physiol 124:181–188

Pitts RF (1973) Production and excretion of ammonia in relation to acid-base regulation. In: Orloff J, Berliner RW (eds) Renal Physiology. Am Physiol Soc, Washington, pp 445–496 (Handbook of Physiology, 1, Section 8)

Pressley TA, Graves JS, Krall AR (1981) Amiloride sensitive ammonium and sodium ion transport in the Blue Crab. Am J Physiol 241:R370–R378

Roos A, Boron W (1981) Intracellular pH. Physiol Rev 61:296–435

Schwartz JH, Tripolone M (1983) Characteristics of NH_4^+ and NH_3 transport across the isolated turtle urinary bladder. Am J Physiol 245:F210–F216

The Regulation of Cutaneous Gas Exchange in Vertebrates

M.E. FEDER[1] and W.W. BURGGREN[2]

1 Introduction

Vertebrates that breathe with lungs or gills can use a variety of mechanisms to regulate the magnitude of respiratory gas exchange. These mechanisms include variation in blood flow to the respiratory organs, variation in ventilation, acclimatory and evolutionary changes in respiratory surface area or the diffusion barrier, and changes in the properties of the blood itself. In concert, these mechanisms maintain relatively constant partial pressures of gases in the blood yet enable overall gas transport to or from the tissues to increase or decrease.

At first consideration, it would seem that vertebrates breathing with their skin cannot avail themselves of these potential regulatory mechanisms. Because the skin is external, it must resist mechanical abrasion and consequently cannot be as thin and as gas-permeable as the respiratory surfaces of internal gills or lungs. The skin is large and anatomically diffuse, which seemingly must complicate control of its perfusion and render its ventilation difficult or impossible. Moreover, many studies (see Sect. 3) suggest that the diffusive resistance of skin is so large that only gross changes in the levels of blood gases can increase or decrease cutaneous gas exchange to meet changing respiratory needs. Thus, cutaneous gas exchange in vertebrates has come to be regarded as "limited", "inefficient", "uncontrolled", "passive", and "poorly regulated" (e.g., see Gottlieb and Jackson 1976; Mackenzie and Jackson 1978; Jackson and Braun 1979).

These characteristics of cutaneous gas exchange clearly are problems for vertebrates that breathe with their skin. Nonetheless, skin-breathing vertebrates have a large, but often unrecognized repertoire of mechanisms with which they may achieve some regulation of cutaneous gas exchange (Feder and Burggren 1985). Here we discuss these mechanisms through a consideration of theory and results of current research, and provide a prospectus for future research.

Potential regulatory mechanisms of cutaneous gas exchange are of two types, the first of which includes acclimatory and evolutionary changes in the skin, particularly in its anatomy. Such responses occur over relatively long times and, while effective in that time frame, are ineffective in dealing with acute variation in the gas exchange requirement. The second type of potential regulatory response includes changes in skin

1 Department of Anatomy and The Committee on Evolutionary Biology, The University of Chicago, 1025 East 57th Street, Chicago, IL 60637, USA

2 Department of Zoology, University of Massachusetts, Amherst, MA 01003, USA

Circulation, Respiration, and Metabolism
(ed. by R. Gilles)

ventilation, skin perfusion, functional surface area, and the source of the blood that perfuses the skin. These responses, which occur rapidly, may regulate cutaneous gas exchange from minute to minute.

2 Long-Term Regulatory Mechanisms

The quantity of O_2 and CO_2 that diffuses through the skin depends on both morphological and physiological variables. Fick's equation summarizes these relationships (Dejours 1981):

$$\dot{M}_x = K_x \cdot A \cdot (Pext_x - Pint_x) / L$$

where $\dot{M}_x$ is the mass of gas x transferred per unit time, K_x (Krogh's diffusion constant) is the product of the diffusivity of gas x (D_x) and the skin capacitance for gas x (β_x), A is the surface area for diffusion, ($Pext_x$ - $Pint_x$) is the partial pressure gradient of gas x across the skin, and L is skin thickness. This equation expresses all of the factors that a vertebrate potentially can vary to increase or decrease the diffusion of gas through the skin. For example, increases in K_x, A, and ($Pext_x$ - $Pint_x$), or decreases in L should in theory promote cutaneous gas exchange.

Anatomical surface area [A] and diffusion distance [L] can undergo adaptive modification during the lifetime of an individual organism, although not on an acute basis. Such variation in A and L is typically associated with changes in the total metabolic requirement or the availability of environmental O_2. For example, metabolic rates increase greatly during courtship in the males of many amphibians (Taigen and Wells 1985), in which the skin often supports a significant fraction of total gas exchange (see Feder and Burggren 1985). During the courtship season, the males of many species develop folds or papillae that increase the skin surface area. These include the dermal papillae of the "hairy frogs" (Noble 1925) and the heightened tail fin of newts (Czopek 1965). Such changes do not occur in the relatively quiescent females. The skin also becomes thinner in some species (Czopek 1965), which should promote gas exchange. Inasmuch as these changes may function in aspects of courtship other than gas exchange and have not been shown to increase skin respiration, direct analyses of the respiratory consequences of this cutaneous hypertrophy would be welcome.

Similar changes occur in larval amphibians, although in a very different context. During chronic exposure to aquatic hypoxia (70–80 torr), bullfrog *(Rana catesbeiana)* larvae double the capillary mesh density in the skin and halve the distance between the water and the respiratory capillaries in the skin (see Fig. 1 and Burggren and Mwalukoma 1983). Both changes should promote O_2 uptake even in hypoxic water, given the high O_2 affinity of tadpole blood. By contrast, the development of the lungs during metamorphosis parallels a decline in the height of the tail fin, a large skin surface in tadpoles (Burggren and West 1982). Thus, in each case the variables governing gas exchange through the skin are adjusted in accordance with the overall gas exchange requirement of the organism.

In some species that rely heavily upon cutaneous gas exchange, hypertrophy of the skin surface area and "thinning" of the skin have become fixed during evolution. One

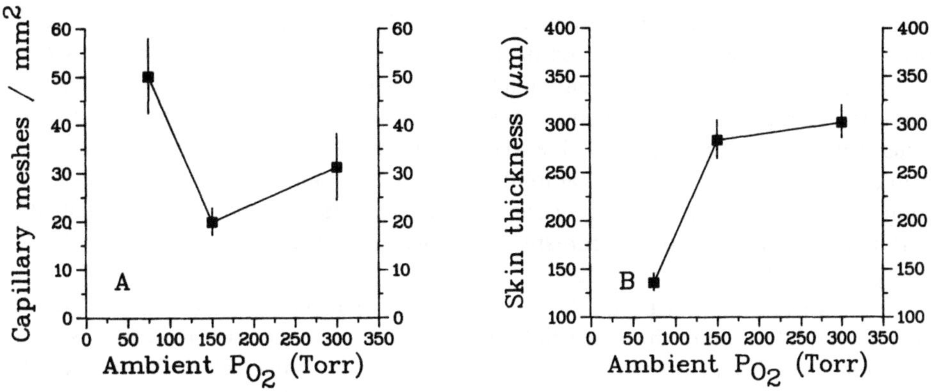

Fig. 1A,B. Effect of ambient P_{O_2} on the skin morphometrics of frog *(Rana catesbeiana)* larvae. Tadpoles were maintained at the indicated P_{O_2} for 4 weeks (see Burggren and Mwalukoma 1983). Mean values ± 1 SE are given. **A** Thickness of the flank skin. **B** Capillary mesh density in the flank skin

noteworthy example of the former response concerns several genera of frogs (*Batrachophrynus* and *Telmatobius*) in which large flaps of skin hang from the body (Noble 1925; Hutchison et al. 1976; W. Duellman, personal communication). Although less dramatic, this response occurs in other forms. The skin of the large aquatic salamander *Cryptobranchus alleganiensis* is in numerous minute, but well-vascularized folds (Guimond and Hutchison 1973). The flattened tail fin of amphibian larvae has long been recognized as a specialized respiratory organ (Medvedev 1937). In other species, the skin itself is not especially folded; however, the body is elongate, with a large increase in surface area: volume ratio. For example, whereas many fish have a relatively globose body form, some species that rely heavily upon cutaneous respiration [e.g., the eel, the reedfish *Calamoichthys*, the mudfish *Neochanna* (reviewed by Feder and Burggren 1985)] are particularly elongate.

In determining the skin's significance as a gas exchanger for the rest of the organism, the absolute area of the skin should be less important than the total surface area of the capillaries in the skin. However, interspecific differences in skin capillary density bear little relationship to dependence upon cutaneous gas exchange. For example, in the lungless plethodontid salamanders, which rely almost exclusively upon cutaneous gas exchange, skin capillary networks are no more dense than in anurans or salamanders with lungs (Czopek 1965). Indeed, the proportion of cutaneous capillaries that are ventilated and perfused at any given time seems a more important determinant of cutaneous gas exchange than the absolute number of cutaneous capillaries (Sect. 4.2).

Evolutionary *decreases* in skin thickness seem problematic. Water loss and susceptibility to injury might well result from very thin skin. Thus, whereas the gas diffusion barrier in lungs and internal gills can often be as little as 0.1 μm (Dejours 1981), the epidermis of amphibians is seldom less than 20 μm thick (Czopek 1965). Some vertebrates, however, circumvent this problem in that their skin is alternately thick and thin. In snakes, lizards, and many teleosts, relatively thick scales (which protect the skin) alternate with patches of capillarized skin (reviewed by Feder and Burggren 1985). An extreme example of this is evident in the eleotrid fish *Dormitator*, in which a highly vascularized epithelial surface on top of the head is specialized for gas exchange, while the remainder of the skin is unremarkable (discussed by Feder and Burggren 1985). Of course, *increases* in skin thickness are less problematic than decreases (provided alternative gas exchangers are available to compensate for any diminution in cutaneous gas exchange). Some fishes and amphibians increase "skin thickness" facultatively during estivation by constructing cocoons of shed epidermis, mucus, and dried soil about the skin. Such cocoons decrease cutaneous gas exchange, but increase reliance on lungs (Loveridge and Withers 1981).

The thickness of the skin may be less important than the placement of cutaneous capillaries in determining the magnitude of cutaneous gas exchange. Relatively superficial capillaries should promote cutaneous gas exchange, while relatively deep capillaries should retard it. Moreover, even if portions of the skin (e.g., scales) are relatively impermeable to the diffusion of gases, an organism may still achieve high rates of cutaneous gas exchange if the cutaneous capillaries are located above these barriers to diffusion. Such an arrangement is evident in fishes (e.g., Mittal and Datta Munshi 1971). In reptilian species that rely heavily upon cutaneous gas exchange, the blood vessels may underlie the scale hinges or even penetrate the scales, and are relatively superficial (reviewed by Feder and Burggren 1985). Conversely, in at least one species of xeric frog, the cutaneous blood vessels are relatively deep, which presumably curtails water loss (Drewes et al. 1977).

Both acclimatory and evolutionary changes in the skin clearly must result in some "regulation" of cutaneous gas exchange. Yet, because the time course of these changes is so great, skin area and thickness are fixed on a short-term basis. The following section will consider potential options for acute regulation of cutaneous gas exchange.

3 Diffusion Limitation in Cutaneous Gas Exchange

Changes in ventilation and perfusion of the skin are two potential means by which vertebrates could regulate cutaneous gas exchange on a short-term basis. However, the consensus of experimental studies during the past 20 years is that variation in skin ventilation is likely to be irrelevant and that variation in skin perfusion is likely to be ineffective in regulating cutaneous gas exchange. Although this consensus requires some modification (Sect. 4), the experimental studies from which it stems clearly establish that diffusion limitation constrains the options for regulation of cutaneous gas exchange.

In an extensive analysis of cutaneous gas exchange in an exclusively skin-breathing salamander *(Desmognathus fuscus)*, Gatz et al. (1975), Piiper and Scheid (1975), and Piiper et al. (1976) assessed the relative roles of diffusion and perfusion in setting the magnitude of cutaneous gas exchange. Their calculations were based on a model with a single blood compartment experiencing variable blood flow, and yielded indices of diffusion limitation (L_{diff}) and perfusion limitation (L_{perf}). For mass transfer of both O_2 and CO_2, L_{perf} was less than 0.2, while L_{diff} was greater than 0.7. Essentially, factors regulating diffusion (changes in the partial pressure gradient across the skin, K_x, L, or A) had a much greater effect on cutaneous gas exchange in *Desmognathus* than even quite gross changes in the rate of skin perfusion. Thus, if Dx, L, and A are assumed to be fixed at any one time, the only means for regulating cutaneous gas exchange lies in changing the P_{O_2} and P_{CO_2} gradient by altering internal P_{O_2} and P_{CO_2}. Similarly, according to a one blood compartment model of cutaneous gas exchange in adult bullfrogs *(Rana catesbeiana)*, diffusion is also more likely to limit cutaneous gas exchange than perfusion (Moalli et al. 1980; Moalli 1981; Burggren and Moalli 1984; Feder and Burggren 1985).

Although these largely theoretical studies have certain intrinsic difficulties (Piiper et al. 1976; Piiper 1982), their conclusions are remarkably consistent with many experimental studies of cutaneous gas exchange. These in vivo studies, mainly with amphibians, emphasize the problems posed by diffusion limitation of cutaneous gas exchange. Presumably because of diffusion limitation, the blood flowing out of the skin of *Desmognathus* never approaches atmospheric P_{O_2}, and its P_{O_2} remains at approx. 100 torr less than the P_{O_2} of air (Piiper 1982). In bullfrogs, the skin is capable of only modest increases in O_2 uptake and CO_2 excretion as the metabolic rate increases; the lungs, by contrast, can achieve much larger increases in gas transfer (Gottlieb and Jackson 1976; Mackenzie and Jackson 1978; Jackson and Braun 1979; Moalli et al. 1981). In accordance with the predicted importance of the partial pressure gradient in determining cutaneous gas exchange, skin-breathing amphibians in hypoxic water lose O_2 to the environment and skin-breathing amphibians in hypercapnic environments gain CO_2 from the environment (West and Burggren 1982; Heisler et al. 1982; Feder 1984; Stiffler et al. 1983). These effects by and large appear undesirable, and it is generally assumed that they represent an *inability* of skin-breathing vertebrates to regulate cutaneous gas exchange satisfactorily.

Because the skin is in immediate proximity with the respiratory medium, the above studies generally assume that the skin is in contact with an "infinite pool" (Piiper and Scheid 1975) of respiratory medium with a typical atmospheric P_{O_2} and P_{CO_2}. Inasmuch as one of the major functions of ventilation is the transport of respiratory medium to the gas exchanger, ventilation seems unnecessary for cutaneous gas exchange. Indeed, in models of cutaneous gas exchange, ventilation limitation (L_{vent}) is assumed to equal zero (Piiper and Scheid 1975). According to this reasoning, ventilation should be even less effective than perfusion in regulating cutaneous gas exchange.

The theoretical studies that suggest that vertebrates cannot regulate skin breathing by varying skin perfusion or skin ventilation rest upon several assumptions. Although these assumptions were necessary if the theoretical models were to be formulated and verified, the assumptions themselves do not always hold. [This has been recognized from the onset by Gatz et al. (1975), Piiper et al. (1976), and Piiper (1982), and is clearly stated in their discussions of the models]. Some of the significant assumptions are: (1) skin surface area (A) is constant; (2) even immediately next to the skin, the respiratory medium has a single, uniform composition; (3) blood flows through the skin in a fixed number of capillaries and is at uniform gas tensions as it does so. These assumptions have, to a large extent, become embodied in the experimental studies that have been used to support the conclusions of the theoretical studies. For example, in their studies of cutaneous gas exchange in bullfrogs, Jackson and his colleagues used a respirometer in which the medium (water) was well-stirred; this technique assured that the respiratory medium had a uniform composition, even immediately next to the skin. As we discuss below, each of these assumptions is violated by real organisms in natural situations; as these assumptions are violated, variation in skin ventilation and skin perfusion may become profoundly important in determining the level of cutaneous gas exchange.

4 Regulation of Cutaneous Gas Exchange by Ventilation and Perfusion

4.1 Ventilation

One function of ventilation, the transport of respiratory medium to the gas exchanger is probably irrelevant in cutaneous gas exchange because the skin contacts the respiratory medium directly. A second major function of ventilation is the dissipation of stratified respiratory medium about the respiratory surface. Inadequate mixing of the medium next to the skin can lead to the formation of a "diffusion boundary layer" with a low P_{O_2} and a high P_{CO_2}. If the diffusion boundary layer is large, skin ventilation can profoundly affect cutaneous gas exchange by dissipating this stagnant layer about the skin (reviewed by Feder and Burggren 1985).

Theoretical considerations suggest that a diffusion boundary layer can be a significant barrier to cutaneous gas exchange. The thickness of the diffusion boundary layer is related to the velocity of the ventilatory flow across the skin by the following equation (see Feder and Burggren 1985 and references cited therein):

$$T = K^{0.33}\ \nu^{0.66}\ (X/V)^{0.5}$$

where T is the thickness of the diffusion boundary layer, V and ν are the velocity and kinematic viscosity of the respiratory medium about the skin, K is Krogh's diffusion constant for the gas in question, and X is the linear dimension of the animal across which respiratory flow occurs. For the boundary layer not to limit cutaneous gas exchange, the following must hold:

$$T \ll K_{water} \cdot T_{skin}/K_{skin}$$

where K_{water} and K_{skin} are the diffusivity of a gas in water and skin, respectively, and T_{skin} is skin thickness. Inserting reasonable values for the variables and solving for V′, the ventilatory flow velocity at which T does not limit cutaneous gas exchange, yields the following (see Feder and Burggren 1985):

$$V'\ (\text{cm s}^{-1}) > 1.743 \times 10^{-5}\ X\ (\text{cm}) \cdot [T_{skin}\ (\text{cm})]^{-2}\ .$$

For typical values of these variables, $V' = 4$ cm s^{-1} (see Feder and Burggren 1985). Interestingly, these equations suggest that as skin diffusion limitation increases (i.e., T_{skin}/K_{skin} becomes large), V′ decreases; i.e., ventilation becomes increasingly ineffective. This corresponds to the assumptions of Piiper and Scheid (1975) regarding the relative limits of diffusion and ventilation to cutaneous gas exchange. Moreover, the equations suggest that in still air (with a large K) or in rapidly moving water, diffusion boundary layers (and, therefore, their dissipation by skin ventilation) should have relatively little impact on cutaneous gas exchange. These are exactly the experimental conditions and the findings of many previous studies of cutaneous gas exchange (see above).

Given the predicted importance of skin ventilation to the dissipation of diffusion boundary layers, it is not surprising that several amphibians and fishes undertake what are evidently skin ventilatory behaviors. The large aquatic salamander *Cryptobranchus alleganiensis* rocks its body in the water (Guimond and Hutchison 1973; Boutilier and Toews 1981) and the aquatic frog *Telmatobius culeus* flexes its legs to wave their large skin flaps in the water (Hutchison et al. 1976). These movements are more frequent in hypoxia or hypercapnia. The larval stages of several fishes have ciliated epithelia capable of generating a flow of water over the skin surface (reviewed by Feder and Burggren 1985). In the larvae of an air-breathing teleost fish, *Monopterus*, movement of the pectoral fins generates a discrete flow of water in a direction countercurrent to cutaneous blood flow (Liem 1981). When these larvae are placed in a concurrent (instead of countercurrent) flow regime, the rate of O_2 consumption decreases (Liem 1981). If the respiratory medium in contact with the skin were actually an infinite and uniform pool, none of these behaviors would be of value.

A fundamental question is whether ventilation of the skin actually leads to an increase in cutaneous gas exchange. To examine this question, we exposed bullfrogs alternately to stirred or unstirred water, thereby either ventilating or not ventilating the skin (Burggren and Feder 1986). Each frog was placed in a screen container beforehand so that its entire surface area was in contact with the water except for the nares (which were in air). Each time stirring was ceased, these frogs underwent a decline in cutaneous O_2 uptake that averaged 30% of the "stirred" O_2 consumption (Fig. 2). The transition from stirred to unstirred water had little or no effect on heart rate, lung ventilatory rate, systemic arterial P_{O_2}, and pulmocutaneous arterial P_{O_2}, suggesting

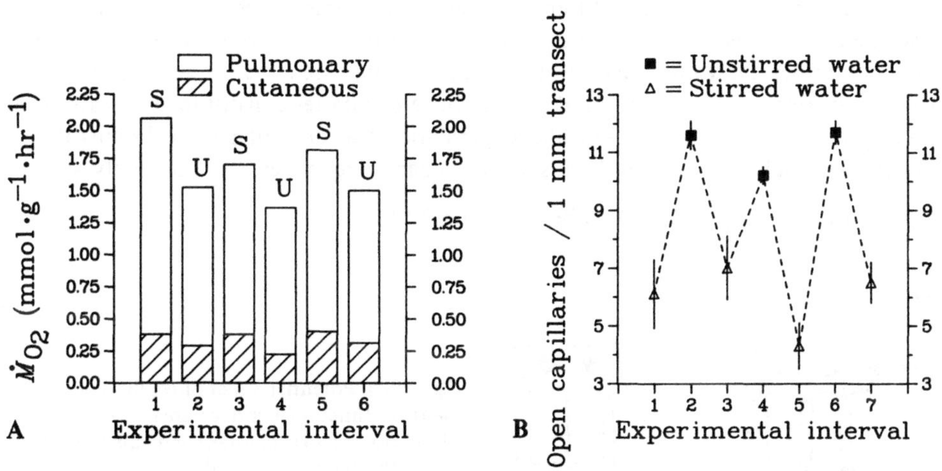

Fig. 2A,B. Effect of experimental ventilation of the skin on gas exchange and cutaneous capillary recruitment in bullfrogs, ***Rana catesbeiana***. Animals were exposed alternately to stirred and unstirred (i.e., still) water. **A** Pulmonary and cutaneous O_2 consumption ($\dot{M}_{O_2}$) of a typical frog. The height of each bar corresponds to the total $\dot{M}_{O_2}$. **B** Capillary recruitment in the hindlimb web of another frog. See Sect. 4.1 and Burggren and Feder (1986) for details

that none of these factors were responsible for the decline in cutaneous O_2 consumption. However, each time skin ventilation was ceased, additional capillaries were perfused in the hindlimb webs of frogs (Fig. 2). This increase averaged 29%, and should have led to an increase (rather than the observed decrease) in cutaneous gas exchange. Thus, the changes in cutaneous O_2 consumption associated with experimental skin ventilation appear to result from the dissipation of the diffusion boundary layer in the water adjacent to the skin and not from changes in internal physiological variables.

4.2 Capillary Recruitment and Functional Surface Area

The characterization of cutaneous gas exchange as "diffusion limited" is based in part upon studies that model the skin as a single compartment through which blood flows; i.e., essentially as a single large capillary. According to this model, increases in blood flow through this single capillary are unlikely to augment respiratory gas transfer across the skin (Gatz et al. 1975; Piiper et al. 1976; Moalli 1981; Piiper 1982). We infer from these studies that any increase in blood flow through already perfused ("open") capillaries will have few consequences for cutaneous gas exchange. Alternatively, if a change in blood flow results in the perfusion of formerly nonperfused capillaries (capillary recruitment) or in the derecruitment of formerly perfused capillaries, cutaneous gas exchange may vary markedly even though gas exchange along any given open capillary is diffusion limited. Here we consider (1) the theoretical consequences of capillary recruitment (or derecruitment); (2) the capacity of skin-breathing vertebrates to alter capillary recruitment; and (3) the actual consequences of changes in capillary recruitment for cutaneous gas exchange.

Respiratory gas transport from any given patch of skin to the rest of an organism will occur only if blood perfuses the capillaries in that patch of skin. Yet all capillaries in the skin of vertebrates are seldom perfused all the time. Accordingly, at any given time the surface area (A) in Fick's equation represents not the *total* surface area of the skin, but rather the *functional* surface area; i.e., that area that is actually perfused (Fig. 3). In this sense, patches of skin not above capillaries constitute an "anatomical dead space" and patches of skin above capillaries that are not perfused constitute a "functional dead space". Thus, Fick's equation can be rewritten more precisely as:

$$\dot{M}_x = (n_{open}/n_{total}) \cdot K_x \cdot A \cdot (Pext_x - Pint_x) / L$$

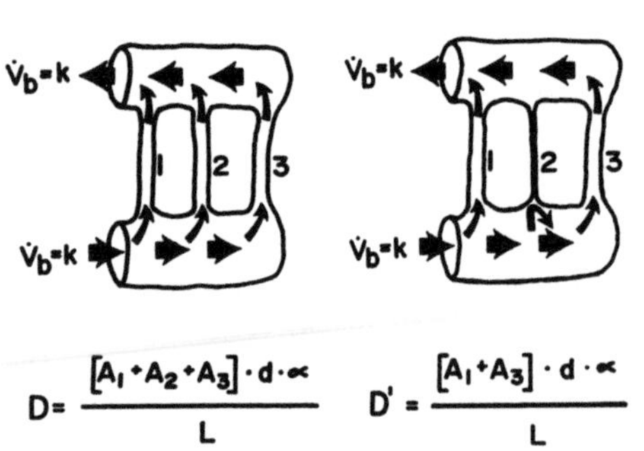

Fig. 3A,B. Potential effect of capillary recruitment on functional surface area of the skin. Blood flow ($\dot{V}_b$) is constant and equal into and out of the capillary network in both **A** and **B**. In **A**, all three capillaries are perfused, and the functional surface area, A, is the sum of A1 + A2 + A3. In **B**, only two capillaries are perfused, and the functional surface area (A1 + A3) is a third smaller. See Burggren and Moalli (1984) and Sect. 4.2

where n_{open} is the number of perfused capillaries and n_{total} is the total number of capillaries, both perfused and nonperfused (Burggren and Moalli 1984). This relationship suggests that skin-breathing vertebrates can regulate cutaneous gas exchange by altering the proportion of cutaneous capillaries that are perfused. Burggren and Moalli (1984) have shown that this multiple compartment model of the skin reasonably predicts in vivo rates of CO_2 elimination in bullfrogs with changing degrees of cutaneous capillary recruitment associated with alternate exposure to water and air (Fig. 4). Similarly, Klocke et al. (1963) have documented increases in the CO_2 diffusing capacity of skin in humans undergoing peripheral vasodilation at high temperatures.

The concept of functional dead space can be extended still further. Except when an animal is entirely off the substrate, some portion of its skin will contact the substrate and, thus, will be unavailable for cutaneous gas exchange, even if this skin is perfused. In frogs, for example, the ventral surface is the most heavily capillarized region of the skin (Czopek 1965), but is typically on the ground rather than exposed to the air. Postures in which the skin is folded against itself will similarly reduce the functional surface area. Thus, many skin-breathing vertebrates may undergo large changes in functional surface area each time they move from place to place or even change posture.

In many skin-breathing vertebrates, the proportion of capillaries that are perfused and, hence, the functional surface area is under active control. Often capillary recruitment can be related to respiratory considerations; i.e., capillaries are recruited when gas exchange becomes problematic and are derecruited in more favorable circumstances (see Feder and Burggren 1985). For example, the number of skin capillaries perfused increases by 30% when frogs *(Rana esculenta)* breathe a 5% CO_2-air mixture or when pulmonary respiration is prevented by blocking the glottis (Poczopko 1957). Capillary recruitment increases when bullfrogs *(Rana catesbeiana)* dive and decreases by nearly 50% when they emerge from water (Moalli et al. 1980; Burggren and Moalli 1984). As

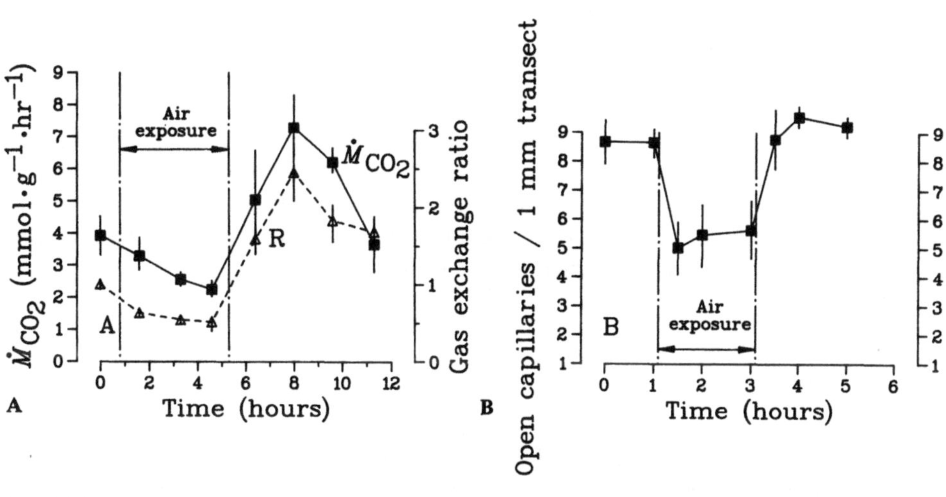

Fig. 4A,B. Gas exchange and capillary recruitment before, during, and after air exposure in bullfrogs, *Rana catesbeiana*. **A** CO_2 elimination (*squares*) and the gas exchange ratio (*triangles*). **B** Capillary recruitment in a second sample of frogs. Each *symbol* represents the mean of five frogs; *bars* indicate the SE. See Burggren and Moalli (1984) and Sect. 4.2

discussed above, initiation or cessation of skin ventilation results in a 30% change in capillary recruitment in bullfrogs (Burggren and Feder 1986). Although recruitment has not been quantified, skin of frog larvae visibly "blushes" during sustained activity (see Feder and Burggren 1985). Only meager data are available on the control of cutaneous capillary recruitment. Burggren and Moalli (1984) demonstrated α-adrenergic control of cutaneous capillary recruitment by administering phenoxybenzamine, an α-antagonist, to bullfrogs. Thereafter exposure to air, which previously resulted in capillary derecruitment, had no effect. Burggren and Feder (1986) showed that experimental changes in the ventilation of the head and body of a bullfrog in water significantly affected capillary recruitment in the hindlimb web, which was in air. Obviously, a comprehensive portrait of these responses is yet to emerge and requires further study.

Interestingly, equally large changes in capillary recruitment may come about for reasons that have little to do with gas exchange (e.g., thermoregulation, hydroregulation, camouflage). Connoly (1926), for example, described adrenergically-mediated vasodilation of the skin and fin rays of the teleost *Fundulus*. Alterations in skin perfusion and subsequent changes in the depth of skin color in this fish were stimulated by changes in substrate color! Many reptiles undergo large increases in total skin perfusion during basking (Bartholomew 1982), and skin-breathing sea snakes may literally turn pink during exposure to cold (Heatwole and Seymour 1978). Brown (1972) has shown that lungless salamanders *(Aneides)* derecruit cutaneous capillaries as these animals become progressively dehydrated, probably to curtail cutaneous water loss. Although these and similar changes are clearly not in response to respiratory demands, they nonetheless should cause large changes in the functional surface area for cutaneous gas exchange.

Unfortunately, there has been little or no systematic experimentation to determine if recruitment or derecruitment of cutaneous capillaries, whether in response to respiratory concerns or for some other reason, actually have the anticipated consequences for cutaneous gas exchange. Burggren and Moalli (1984) reported a 50% decrease in the gas exchange ratio that paralleled a 50% decrease in capillary recruitment during air exposure in bullfrogs *(Rana catesbeiana)*, which excrete CO_2 predominantly through the skin (Fig. 4). Although this result is consistent with expectations, it is insufficient to establish a general correlation between the diverse changes seen in capillary recruitment and cutaneous gas exchange. This area should have priority in future research.

Yet another regulatory mechanism may be the preferential distribution of systemic vs pulmocutaneous arterial blood to the skin (Moalli et al. 1980). Malvin and Boutilier discuss this process at length elsewhere in this volume.

5 Conclusion

We have presented two contrasting views of the potential for regulation of cutaneous gas exchange. According to the first, cutaneous gas exchange is predominantly diffusion limited and, thus, appears difficult or impossible to regulate short of gross changes in internal gas tensions. According to the second view, cutaneous gas exchange is subject to regulation by changes in ventilation, capillary recruitment, skin thickness, and other determinants of functional surface area. Perhaps the best way to reconcile these contrasting views is to regard diffusion limitation as setting a "ceiling" to cutaneous gas exchange at any one time. Skin ventilation and other factors that determine the size of the diffusion boundary layer (e.g., the nature of the respiratory medium) will, thus, determine the rate of cutaneous gas exchange below this ceiling. Capillary recruitment and other changes in functional surface area will alter the level of the ceiling itself.

Although many hypotheses regarding the regulation of cutaneous gas exchange are yet to be tested, cutaneous gas exchange is clearly *not* "passive" and incapable of being regulated. Indeed, the potential complexity of cutaneous gas exchange is astounding! Consider the variables at work from minute to minute. Each patch of skin may receive blood from two different sources, the pulmocutaneous and systemic circulations, each of which may vary in flow rate and partial pressures of the respiratory gases. The consequences of this variation will differ if flow through already perfused capillaries simply increases or if capillaries are recruited. Animals can ventilate their skin themselves, but skin ventilation will also vary any time water or air currents change next to an animal or if an animal begins locomotor movements. When peripheral vasodilation or vasoconstriction occurs for any reason, be it respiratory or otherwise (e.g., thermoregulation, hydroregulation), cutaneous capillary recruitment may affect gas exchange. Any change in posture or other movement may change the surface area for cutaneous gas exchange. Furthermore, each portion of the skin may differ in each of these respects. This complex of effects potentially poses serious "functional conflicts" for skin-breathing vertebrates. What occurs, for example, if curtailment of cutaneous water loss dictates capillary derecruitment at the same time that physical activity or metabolic acidosis dictate capillary recruitment? Thus, the real problem for skin-breathing vertebrates may not be "Can cutaneous gas exchange be varied?" but rather "How can this enormous complex of variables be controlled and conflicts reconciled?". Understanding the resolution of this problem remains a challenge for future research.

Acknowledgements. Research was supported by NSF Grants BSR 83-07089 to MEF and PCM 83-09404 to WWB, and NIH Biomedical Research Support Grant RR-05367-22 to The University of Chicago. Travel to the International Congress of Comparative Physiology and Biochemistry was supported by the NSF. We thank Juan Markin for continuing inspiration and technical advice.

References

Bartholomew GA (1982) Physiological control of body temperature. In: Gans C, Pough FH (eds) Biology of the Reptilia, vol 12. Academic, New York, pp 167–212

Boutilier RG, Toews DP (1981) Respiratory, circulatory and acid-base adjustments to hypercapnia in a strictly aquatic and predominantly skin-breathing urodele, *Cryptobranchus alleganiensis*. Respir Physiol 46:177–192

Brown AG (1972) Responses to problems of water and electrolyte balance by salamanders (Genus *Aneides*) from different habitats. PhD dissertation, University of California, Berkeley

Burggren WW, Feder ME (1986) Effect of experimental ventilation of the skin on cutaneous gas exchange in amphibians. J Exp Biol (in press)

Burggren WW, Moalli R (1984) "Active" regulation of cutaneous gas exchange by capillary recruitment in amphibians: experimental evidence and a revised model for skin respiration. Respir Physiol 55:379–392

Burggren WW, Mwalukoma A (1983) Respiration during chronic hypoxia and hyperoxia in larval and adult bullfrogs. I. Morphological responses of lungs, gills, and skin. J Exp Biol 105:191–203

Burggren WW, West NH (1982) Changing respiratory importance of gills, lungs and skin during metamorphosis in the bullfrog, *Rana catesbeiana*. Respir Physiol 47:151–164

Connolly CJ (1926) Vasodilation in *Fundulus* due to a color stimulus. Biol Bull 50:207–209

Czopek J (1965) Quantitative studies on the morphology of respiratory surfaces in amphibians. Acta Anat 62:296–323

Dejours P (1981) Principles of comparative respiratory physiology, 2nd edn. North-Holland, Amsterdam

Drewes RC, Hillman SS, Putnam RW, Sokol OM (1977) Water, nitrogen and ion balance in the African treefrog *Chiromantis petersi* Boulenger (Anura: Rhacophoridae) with comments on the structure of the integument. J Comp Physiol 116:257–268

Feder ME (1984) Consequences of aerial respiration for anuran larvae. In: Seymour RS (ed) Respiration and metabolism of embryonic vertebrates. Junk, The Hague, pp 71–86

Feder ME, Burggren WW (1985) Cutaneous gas exchange in vertebrates: design, patterns, control, and implications. Biol Rev 60:1–45

Gatz RN, Crawford EC, Piiper J (1975) Kinetics of inert gas equilibration in an exclusively skin-breathing salamander, *Desmognathus fuscus*. Respir Physiol 24:15–29

Gottlieb G, Jackson DC (1976) Importance of pulmonary ventilation in respiratory control in the bullfrog. Am J Physiol 230:608–613

Guimond RW, Hutchison VH (1973) Aquatic respiration: an unusual strategy in the Hellblender *Cryptobranchus alleganiensis alleganiensis* (Daudin). Science (NY) 182:1263–1265

Heatwole H, Seymour RS (1978) Cutaneous O_2 uptake in three groups of aquatic snakes. Aust J Zool 26:481–486

Heisler N, Forcht G, Ultsch GR, Anderson JF (1982) Acid-base regulation in response to environmental hypercapnia in two aquatic salamanders, *Siren lacertina* and *Amphiuma means*. Respir Physiol 49:141–158

Hutchison VH, Haines HB, Engbretson G (1976) Aquatic life at high altitude: respiratory adaptations in the Lake Titicaca frog, *Telmatobius culeus*. Respir Physiol 27:115–129

Jackson DC, Braun BA (1979) Respiratory control in bullfrogs: cutaneous versus pulmonary response to selective CO_2 exposure. J Comp Physiol 129:339–342

Klocke RA, Gurtner GH, Farhi LE (1963) Gas transfer across the skin in man. J Appl Physiol 18: 311–316

Liem K (1981) Larvae of air-breathing fishes as countercurrent flow devices in hypoxic environments. Science (NY) 211:1177–1179

Loveridge JP, Withers PC (1981) Metabolism and water balance of active and cocooned African bullfrogs *Pixicephalus adspersus*. Physiol Zool 54:203–214

Mackenzie JA, Jackson DC (1978) The effect of temperature on cutaneous CO_2 loss and conductance in the bullfrog. Respir Physiol 32:313–323

Medvedev L (1937) The vessels of the caudal fin in amphibian larvae and their respiratory function. Zool Zh 16:393–403

Mittal AK, Datta Munshi JS (1971) A comparative study of the structure of the skin of certain air-breathing fresh-water teleosts. J Zool (Lond) 163:515–532

Moalli R (1981) The effect of temperature on skin blood flow, gas exchange and acid-base balance in the bullfrog. PhD dissertation, Brown University, Providence

Moalli R, Meyers RS, Jackson DC, Millard RW (1980) Skin circulation of the frog, *Rana catesbeiana*, distribution and dynamics. Respir Physiol 40:137–148

Moalli R, Meyers RS, Ultsch GR, Jackson DC (1981) Acid-base balance and temperature in a predominantly skin-breathing salamander, *Cryptobranchus alleganiensis*. Respir Physiol 43:1–11

Noble GK (1925) The integumentary, pulmonary and cardiac modifications correlated with increased cutaneous respiration in the Amphibia: a solution of the "hairy frog" problem. J Morphol 40:341–416

Piiper J (1982) A model for evaluating diffusion limitation in gas-exchange organs of vertebrates. In: Taylor CR, Johansen K, Bolis L (eds) A companion to animal physiology. Cambridge University Press, Cambridge, pp 49–64

Piiper J, Scheid P (1975) Gas transport efficacy of gills, lungs and skin: theory and experimental data. Respir Physiol 23:209–221

Piiper J, Gatz RN, Crawford EC (1976) Gas transport characteristics in an exclusively skin-breathing salamander, *Desmognathus fuscus* (Plethodontidae). In: Hughes GM (ed) Respiration of amphibious vertebrates. Academic, New York, pp 339–356

Poczopko P (1957) Further investigations on the cutaneous vasomotor reflexes in the edible frog in connexion with the problem of regulation of the cutaneous respiration in frogs. Zool Pol 8: 161–175

Stiffler DF, Tufts BL, Toews DP (1983) Acid-base and ionic balance in *Ambystoma tigrinum* and *Necturus maculosus* during hypercapnia. Am J Physiol 245:R689–R694

Taigen TL, Wells KD (1985) Energetics of vocalization by an anuran amphibian *(Hyla versicolor)*. J Comp Physiol 155:163–170

West NH, Burggren WW (1982) Gill and lung ventilatory responses to steady-state aquatic hypoxia and hyperoxia in the bullfrog tadpole *(Rana catesbeiana)*. Respir Physiol 47:165–176

Ventilation-Perfusion Relationships in Amphibia

G.M. MALVIN[1] and R.G. BOUTILIER[2]

1 Introduction

Amphibians respire through a number of respiratory organs, the most common being skin and lungs. Patterns of gas exchange across these organs vary to meet different metabolic and environmental challenges. Alterations in gas exchange patterns generally are facilitated by adjustments in ventilation and/or blood flow. Initially, this paper will examine pulmonary gas exchange and focus on the matching of perfusion to ventilation in the lung. This will be followed by a discussion of skin perfusion and its relation to cutaneous gas exchange. The final sections of this paper will investigate possible control mechanisms acting to adjust respiratory organ perfusion.

2 Pulmonary Ventilatory Patterns

Almost all amphibians ventilate their lungs in an intermittant fashion with periods of breath-holding separating the individual ventilations. In diving anurans, such as the exclusively aquatic *Xenopus laevis*, these respiratory pauses may constitute short breath-holds within a burst of ventilations at the surface or represent long apneic periods associated with diving (Boutilier 1984). Although amphibian ventilatory patterns are variable, they are regulated. For example, decreases in water PO_2 will elicit marked increases in pulmonary ventilation of swimming amphibians. In the neotenic tiger salamander, *Ambystoma tigrinum*, a decrease in water PO_2 from about 150 torr to 25 torr causes over a 500% increase in pulmonary ventilatory frequency (Malvin and Heisler unpublished).

3 Pulmonary Perfusion Patterns

Perfusion of amphibian lungs is also intermittant. Blood flow in the pulmocutaneous arch of *Xenopus* increases during the ventilations of a breathing burst and progressively declines in the interim between breaths (Shelton 1970, 1976). Direct measurements of

1 Department of Physiology and Biophysics, SJ-40, School of Medicine, University of Washington, Seattle, WA 98195, USA
2 Department of Biology, Dalhousie University, Halifax, Nova Scotia B3H 4J1, Canada

Circulation, Respiration, and Metabolism
(ed. by R. Gilles)

pulmonary blood flow during periods of ventilation and breath-holding in the semiterrestrial toad, *Bufo marinus* (West and Burggren 1984) also show such blood flow variations. In the neotenic tiger salamander, severe water hypoxia (water $PO_2 < 10$ torr) causes approximately a threefold increase in the fraction of cardiac output flowing to the lungs over normoxic levels (Malvin and Heisler 1984). Thus, variations in pulmonary ventilation are associated with corresponding changes in lung perfusion.

Changes in pulmonary blood flow act to match lung perfusion to pulmonary oxygen availability, which may be important during a dive. Experiments on *Xenopus* have shown that the rate of oxygen depletion from the lung, as well as the PO_2 gradients between lung and blood may vary considerably as breath-holding continues (Boutilier and Shelton, unpublished). The variation in pulmonary perfusion was recently examined in studies on *Xenopus* in which the alveolar gas tensions were measured continuously in an extracorporeal loop (Boutilier, Butler, and Evans, unpublished). The first section of Fig. 1 shows the lung gas PO_2 and PCO_2 over the last few minutes of a voluntary

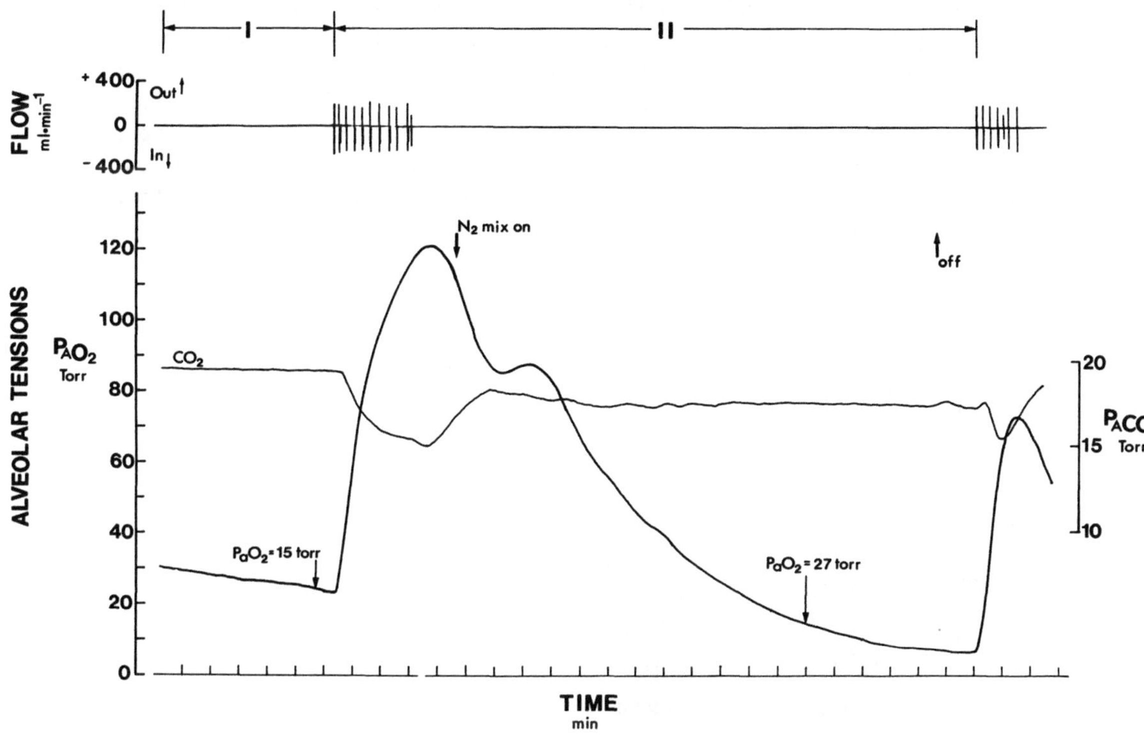

Fig. 1. Ventilatory air flow and alveolar gas tensions of a single *Xenopus laevis* voluntarily diving and surfacing at a blowhole fitted with a pneumotachograph. Lung gas tensions were measured continuously in an extracorporeal loop fitted with electrodes through which alveolar gases were circulated using a perstaltic pump. A second loop connected in series with the first allowed various gas mixtures to be introduced into the circulation. In the experiment shown, a nitrogen mixture was introduced soon after a voluntary dive so as to purge the lung of its oxygen supplies, while maintaining PCO_2 at physiological levels. Blood samples taken towards the end of a normal dive (*section I*) and after purging the lungs to low oxygen tensions (*section II*) were analyzed for arterial PO_2 (PaO_2)

dive. Towards the end of this dive, a blood sample was taken from an indwelling cannula for measurement of arterial PO_2. As was observed on numerous occasions previously, the arterial blood PO_2 was approx. 5 to 10 torr lower than that of the lung. In the dive which followed (Fig. 1, II), a nitrogen/carbon dioxide gas mixture was used to purge the lungs of their oxygen stores. This was achieved without changing lung volume by connecting the input and output lines of the nitrogen mixture in series with the extracorporeal loop. After an initial period of oxygen decline, alveolar oxygen tensions actually increased for a brief time, indicating loss of oxygen from blood to lung. The subsequent decline of alveolar PO_2 to values approaching zero occurred much more rapidly than did the decline in arterial PO_2 which towards the end of the dives was usually 10 to 20 torr higher than alveolar tensions. The data suggest that as the lung was rendered progressively less useful as an oxygen store during the dive, pulmonary perfusion became markedly reduced, ameliorating further loss of oxygen from the blood to the lung. Thus, it appears that through regulation of pulmonary blood flow, amphibians can economize the oxygen contained in blood.

Recently completed microsphere studies on *Rana catesbiana* (Boutilier, Glass, and Heisler, unpublished) are consistent with the foregoing data on *Xenopus*. In these studies, the distribution of pulmocutaneous blood during a dive was examined under various conditions of hypoxia delivered selectively to either the lungs or skin. Under all conditions of oxygenation, the lungs rather than skin received the greatest proportion of pulmocutaneous blood flow (Fig. 2). In comparison to normoxic conditions, however, the relative flow of pulmocutaneous blood to the skin increased during those dives which followed periods of breathing hypoxic gas mixtures. Aquatic hypoxia, on

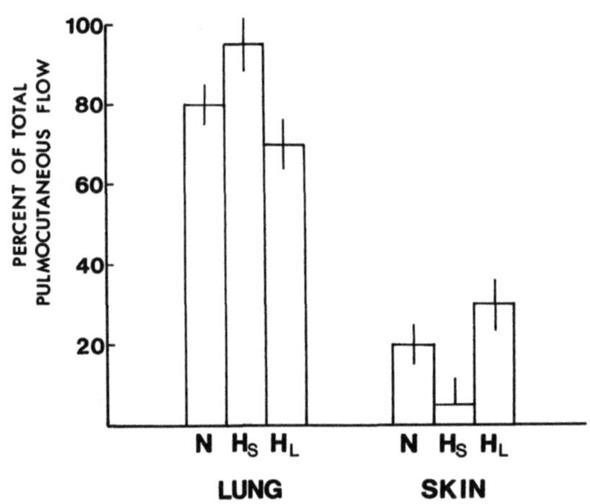

Fig. 2. Histogram showing the relative flow of pulmocutaneous blood in bullfrogs subjected to normoxia (*N*), aquatic hypoxia (P_{O_2} ≅ 40 torr) delivered only at the skin (H_s) and aerial hypoxia (P_{O_2} ≅ 40 torr) delivered only at the lungs (*HL*). Radioactive microspheres were injected approx. 10 min into a voluntary dive. Relative flows were estimated by counting the labels in the left lung and entire skin of the animal. Retrograde flow upon injection was later judged to be quite small owing to the low levels of microsphere labels foung in the right lung of the animals. The data are the means (± 1 SEM) of seven experiments at 20 °C (Boutilier RG, Glass ML, Heisler N, unpublished). Significant changes throughout ($p < 0.05$, Student's paired t-test)

the other hand, reduced relative blood flow to the skin during a dive (Fig. 2), favoring increased perfusion of the lung. Again, respiratory organ blood flow was adjusted to match oxygen availability.

The skin of amphibians also constitutes an important site for respiratory gas exchange. It is the major organ for carbon dioxide elimination and substantial cutaneous oxygen uptake occurs, particularly at low temperatures (Whitford and Hutchinson 1965). Since most amphibians do not ventilate their skin, alterations in cutaneous gas exchange can occur only by changes in the partial pressure gradients across the skin or by changes in perfusion patterns. Altered partial pressure gradients have marked effects on cutaneous gas exchange. For example, below a PO_2 of about 125 torr, cutaneous oxygen uptake is directly proportional to environmental PO_2 for a number of amphibians (Shield and Bentley 1973). The role of skin perfusion in affecting cutaneous gas exchange, however, is not clear. An analysis of gas transport characteristics in the lungless salamander, *Desmognathus fuscus*, indicate that cutaneous gas exchange is primarily a diffusion-limited process (Piiper et al. 1976). Thus, changes solely in the magnitude of cutaneous blood flow should have little influence on gas transfer across the skin. However, in a recent study on bullfrogs, Burggren and Moalli (1984) argue that despite a pronounced diffusion-limitation, changes in skin perfusion that affect the number of open skin capilaries will significantly alter cutaneous gas exchange by changing the effective surface area for gas exchange. Regardless of the precise role of perfusion in cutaneous gas exchange, skin blood flow patterns do change, and such changes are associated with corresponding alterations in gas exchange patterns. As previously mentioned (Boutilier, Glass, and Heisler, unpublished), aquatic hypoxia acts to direct a greater fraction of pulmocutaneous blood flow away from the skin, whereas intrapulmonary hypoxia directs a greater fraction of pulmocutaneous flow to the skin. Other microsphere studies (Moalli et al. 1980) have shown that diving elicits a dramatic redistribution of cutaneous perfusion in the bullfrog. After 10 min of submergence, the absolute blood flow reaching the skin via the cutaneous artery fell by half, while cutaneous flow supplied by the systemic circulation increased threefold. In yet another study on bullfrogs (Burggren and Moalli 1984), exposure to air also had marked effects on the skin: total cutaneous blood flow, the number of open skin capillaries, and carbon dioxide elimination all were significantly reduced.

4 Control of Respiratory Organ Blood Flow

What controls these changes in respiratory organ blood flow? This next section will address this question using: (1) studies performed on the neotenic tiger salamander to investigate the control of lung perfusion, and (2) work done on anurans which suggest regulatory mechanisms for skin perfusion. Studies performed on the neotenic tiger salamander, will be discussed in regard to the control of pulmonary perfusion during aquatic hypoxia. As previously mentioned, water hypoxia augments both pulmonary ventilatory frequency and pulmonary blood flow. The cardiac anatomy, vascular organization, and quality of pulmonary arterial blood will be presented first followed by three possible mechanisms serving to match perfusion to ventilation in the lung.

4.1 Control of Lung Perfusion

4.1.1 Anatomical Considerations

In the neotenic salamander, as in most amphibians, systemic venous blood enters the heart through the sinus venosus which empties into the right atrium. Pulmonary venous blood flows directly into the left atrium. Both atria empty simultaneously into a common ventricle. From the ventricle, blood is pumped through the conus arteriosus and then into the bulbous arteriosus which gives rise to four pairs of aortic arches. The first three are afferent branchial arteries and the fourth is the pulmonary artery (Fig. 3).

Within the gill, there are two major circulatory pathways: (1) a respiratory pathway from the afferent branchial artery through the respiratory lamellae and into the efferent branchial artery; and (2) a shunt directly connecting the afferent and efferent arteries at the base of the gill. Both the external and internal carotid arteries emerge from the first branchial arch. Thus, the head receives most of its blood from the first branchial arch. The efferent arteries from each of the gills unite to form the dorsal aorta. The ductus aerteriosus connects the pulmonary artery with the efferent artery of the third gill. This places the lung in series with the third gill and in parallel with both the first two gills and also the systemic circulation. Proximal to the ductus arteriosus, the pulmonary artery forms a vascular plexus. The vascular resistance of this plexus, however, is quite high (Malvin 1985a), and microsphere experiments have indicated that only about 20% of pulmonary blood flow passes through the proximal part of the pulmonary artery (Malvin and Heisler 1984). Most of the blood which reaches the lung first flows to the third gill, then through the ductus arteriosus and into the pulmonary artery.

4.1.2 Nature of Pulmonary Arterial Blood

Despite only one ventricle, pulmonary and systemic venous blood are selectively distributed into the different aortic arches. Simultaneous infusions of differently labeled microsphere into the systemic and pulmonary venous systems indicated that the fraction of left atrial output flowing into the first branchial arch was two times the fraction of right atrial output entering that arch. In contrast, the fraction of right atrial output flowing into the third branchial arch was over three times the fraction of left atrial output perfusing that arch (Malvin 1985b). Since the lung derives most of its blood from the efferent artery of the third gill, pulmonary arterial blood must be largely systemic venous blood. Similar intracardiac shunting patterns have been demonstrated in other amphibian species (Johansen 1963, Tazawa et al. 1979). Whether this selective distribution is regulated to meet different environmental or metabolic challenges is not known.

4.1.3 Possible Mechanisms Regulating Blood Flow to the Lung

Vagal Control of Pulmonary Vascular Resistance. Since flow varies inversely with resistance, a change in pulmonary vascular resistance will cause a reciprocal alteration in pulmonary perfusion. To evaluate possible autonomic control over pulmonary vascular resistance, the isolated-perfused pulmonary circuit (lung plus distal half of the pulmonary artery) was tested for responsiveness to the common autonomic transmit-

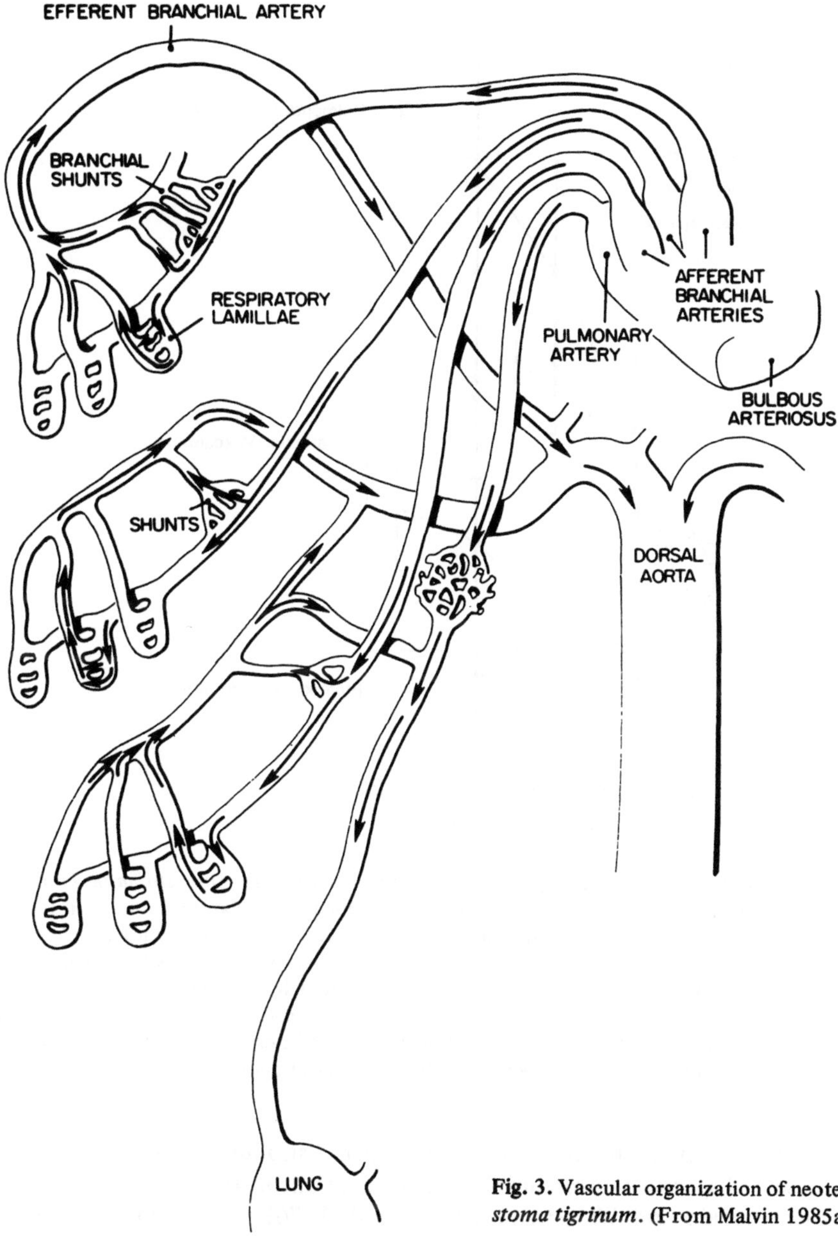

Fig. 3. Vascular organization of neotenic *Ambystoma tigrinum*. (From Malvin 1985a)

ters, acetylcholine, epinephrine, and norepinephrine (Malvin 1985b). Acetylcholine was an extremely strong pressor agent, while the catecholamines had no effect on the resistance of the pulmonary circuit (Fig. 4). This suggests that pulmonary vascular resistance is under cholinergic control, but not direct adrenergic control. Although a morphological examination of the pulmonary circuit has not been performed, the cholinergic response is probably exercised via a vagally-innervated sphincter located in

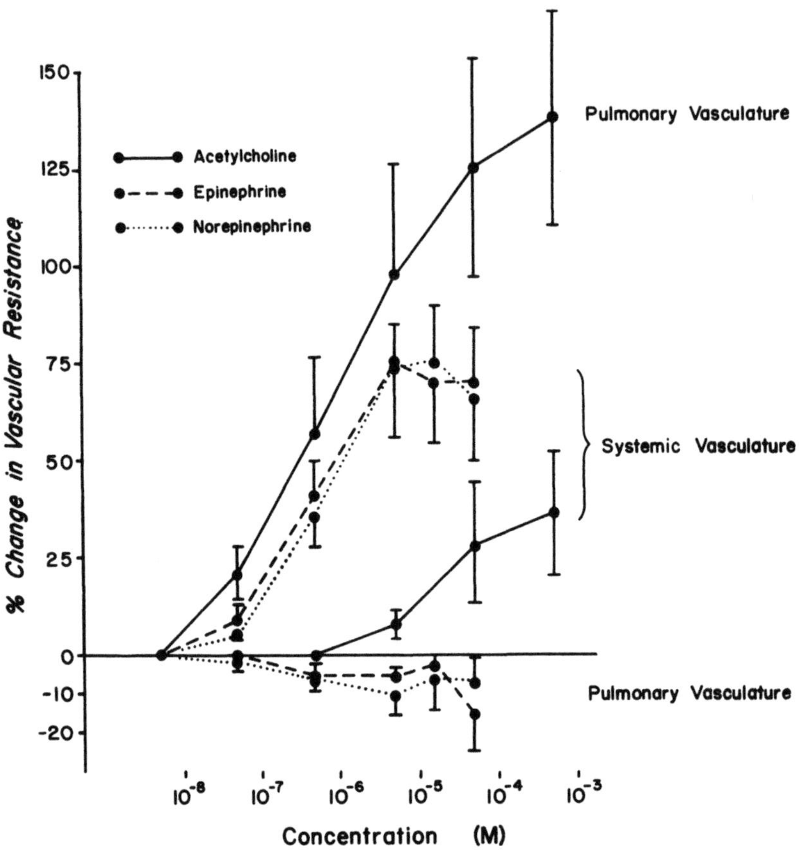

Fig. 4. Responses of the pulmonary and systemic vasculatures of neotenic *Ambystoma tigrinum* to acetylcholine, epinephrine, and norepinephrine. Either the pulmonary vasculature via the pulmonary artery or the systemic vasculature via the dorsal aorta were selectively perfused with Ringer's solution under constant, pulsatile flow while perfusion pressure was measured. Drugs were added cumulatively to the perfusate. Dose-response curves were generated by calculating the changes in vascular resistance from alterations in perfusion pressure that were caused by drug administration. Values are means + SEM; n = number of animals. Epinephrine and norepinephrine caused no significant change in pulmonary vascular resistance ($p > 0.05$; Student's paired t-test)

the distal segment of the pulmonary artery, since such a structure has been described previously in several species of Amphibia (Luckhart and Carlson 1921; de Saint-Aubain 1982; de Saint-Aubain and Wingstrand 1979; Smith 1976). Vagal stimulation and acetylcholine exert a strong pressor influence on the sphincter. In situ, the sphincter relaxes with lung inflation and contracts as lung volume diminishes. This structure may be influenced also by the composition of intrapulmonary gas, since lung inflation with gas high in oxygen increases pulmocutaneous blood flow more than lung inflation with oxygen-poor gas (Emilio and Shelton 1972). A central mediation of the oxygen effect, presumably via the vagus nerve, is indicated by tissue-perfusion experiments that show no local effect of oxygen on the resistance of the pulmonary circuit (Malvin and Heisler, unpublished). Such coupling of pulmonary arterial resistance to lung volume and/or intrapulmonary PO_2 may facilitate the matching of pulmonary perfusion to ventilation. For example, if during aquatic hypoxia when pulmonary ventilatory frequency is high, mean lung volume is also high, then the mean tension in the pulmonary arterial sphincter will be reduced. This should lower the mean resistance of the pulmonary circuit and augment pulmonary perfusion to match ventilation. In addition, it is possible that the marked increase in pulmonary ventilation during aquatic hypoxia leads to a rise in mean intrapulmonary PO_2. This would also act to decrease pulmonary vascular resistance. However, the effect of alterations in ventilatory patterns on lung volume and intrapulmonary PO_2 has not been studied directly making it difficult to assess the importance of such mechanisms in the control of pulmonary vascular resistance.

Adrenergic Control over Systemic Vascular Resistance. The position of the ductus arteriosus and pulmonary artery places the lungs in parallel with the systemic circulation. Thus, changes in systemic vascular resistance will directly affect pulmonary blood flow. For example, an increase in systemic vascular resistance will increase the fraction of cardiac output perfusing the lung by shunting blood away from the systemic circulation and into the lung.

In contrast to the pulmonary circuit, acetylcholine has little effect on the resistance of the systemic circulation (Fig. 2), indicating that cholinergic control over total systemic vascular resistance is not of primary importance. However, significant adrenergic control of systemic vascular resistance is likely. Both epinephrine and norepinephrine are effective pressor agents in the isolated-perfused systemic circulation (Fig. 2), and adrenergic innervation is present in the systemic arterial system. This contrasts with the pulmonary circuit which is unresponsive to catecholamines.

The difference in the action of catecholamines on the pulmonary and systemic vasculatures suggests a role for the adrenergic nervous system in regulating blood flow distribution between the lung and the systemic circulation. For example, an increase in either circulating catecholamines or general adrenergic outflow should constrict systemic vascular smooth muscle and, thus, increase systemic vascular resistance. Resistance in the pulmonary circuit should be unaffected. This will act to shunt blood away from the systemic circulation and into the lungs. Both microsphere experiments and blood gas measurements are consistent with such a regulatory mechanism (Malvin and Heisler, unpublished). Intravenous injection of norepinephrine into unanesthetized animals caused a twofold increase in the fraction of systemic venous blood reaching the lung. Norepinephrine administration during the first few minutes of a breath-hold caused a 20% increase in arterial PO_2, which again suggests a catecholamine-induced increase in lung perfusion. However, the precise response of the adrenergic nervous system to aquatic hypoxia, as yet, is not known.

Shunting Between the Gills and Lung. Since the ductus arteriosus places the lung in series with the third branchial arch, the first two gill arches lie in parallel with the lung. Thus, changes in the ratio of the vascular resistance of the first two gill arches and the third gill arch will affect the amount of blood flowing toward the pulmonary circuit. Microsphere experiments indicate that this ratio changes during aquatic hypoxia (Malvin and Heisler 1984). When water PO_2 was lowered below 10 torr, the fraction of cardiac output perfusing the respiratory section of the first gill decreased by approx. 50%, and the fraction reaching the head (which receives most of its blood from the first arch) fell approx. 30%. In contrast, there was no change in the fraction of cardiac output perfusing the respiratory sections of the other two gills. This suggests that resistance in the first branchial arch had increased with respect to the other arches. Such a change will shunt blood away from the first gill arch and into the other two arches. An increase in flow through the third branchial arch will make more blood available for lung perfusion via the ductus arteriosus. This redistribution of blood flow may have contributed to an estimated threefold increase in the fraction of cardiac output flowing to the lung under these hypoxic conditions. The mechanism mediating this apparent increase in the resistance of the first gill is not known. However, the presense of dense adrenergic innervation and catecholamine-containing cell bodies in the shunt vasculature of the first gill suggest possible adrenergic involvement (Malvin 1983).

4.2 Control of Skin Perfusion

In anurans, the pulmocutaneous arch branches into a pulmonary and a cutaneous artery. During nondiving conditions, a little more than one-half of the total blood flow to the skin is supplied by the cutaneous branch of the pulmocutaneous arch, the remainder coming from the systemic arterial circulation (Moalli et al. 1980). This circulatory arrangement makes cutaneous blood flow a function of the vascular resistance of both cutaneous circuits, as well as the resistance of the pulmonary circuit. As previously mentioned, pulmonary vascular resistance is probably under the influence of a vagally-innervated sphincter located in the pulmonary artery. The cutaneous artery emerges from the pulmocutaneous branch just proximal to this sphincter. Tension in the sphincter is inversely related to lung volume. Thus, a decrease in lung volume may act to increase skin perfusion by constricting the sphincter causing blood to be shunted away from the lung and into the cutaneous artery. Additional factors influencing the contractile state of this sphincter, such as intrapulmonary PO_2, are likely.

Vasomotor control of the cutaneous vessels probably involves a myriad of effector systems. For example, a sphincter located in the proximal segment of the cutaneous branch of the pulmocutaneous arch receives a rich innervation of adrenergic constrictory fibers (Smith 1976). Thus, blood flow through the cutaneous artery can be altered by changes in adrenergic outflow to this vascular segment. Active regulation of the peripheral cutaneous vessels is also likely. Poczopko (1958, 1959) has documented that diving and increases in PO_2 and PCO_2 all act to increase the number of open skin capillaries in the frog, *Rana esculenta*. In addition, Burggren and Moalli (1984) recently showed that exposure to air decreased the number of perfused skin capillaries in *Rana catesbiana*. Phenoxybenzamine, an α-adrenergic antagonist, abolished this response, indicating that the adrenergic nervous system can influence cutaneous perfusion patterns. Finally, amphibian skin is rich in vasoactive peptides (Montecucci et al. 1981; Nakajima 1981). Perhaps local hormonal control of skin perfusion exists through these substances.

5 Summary

Gas exchange, ventilation, and blood flow across amphibian gas exchange organs are variable. Generally, these parameters all vary in concert. Intermittant pulmonary ventilation is matched by changes in lung perfusion, and increases in pulmonary ventilatory frequency are concomitant with increases in pulmonary blood flow. Blood flow to the lung also appears to vary directly with intrapulmonary oxygen content. Such matching of perfusion to ventilation acts to control gas transfer across the lung in a dynamic fashion. Blood flow to the skin also changes in response to different environmental and physiological conditions. Increases in cutaneous blood flow occur during diving and intrapulmonary hypoxia when pulmonary gas exchange is markedly diminished. Exposure to air elicits simultaneous decreases in both cutaneous blood flow and carbon dioxide excretion.

The matching of pulmonary perfusion to ventilation may be realized by: (1) cholinergic control of pulmonary vascular resistance; (2) adrenergic control of systemic vascular resistance; and (3) redistribution of cardiac output into the different aortic arches. Active control of skin perfusion may involve: (1) cholinergically-mediated changes in pulmonary vascular resistance; (2) adrenergically-mediated changes in the resistance of the proximal segment of the cutaneous artery and the peripheral skin vessels; and (3) local control of skin blood flow by oxygen, carbon, dioxide, and vasoactive skin peptides.

References

Boutilier RG (1984) Characterization of the intermittent breathing pattern in *Xenopus laevis*. J Exp Biol 110:291–309

Burggren WW, Moalli R (1984) 'Active' regulation of cutaneous gas exchange by capillary recruitment in amphibians: experimental evidence and a revised model for skin respiration. Respir Physiol 55:379–392

Emilio MG, Shelton G (1972) Factors affecting blood flow to the lungs in the amphibian, *Xenopus laevis*. J Exp Biol 56:67–77

Heath AG (1975) Respiratory responses to hypoxia by *Ambystoma tigrinum* larvae, paedomorphs, and metamorphosed adults. Comp Biochem Physiol 55A:45–49

Johansen J (1963) Cardiovascular dynamics in the amphibian *Amphiuma tridactylum*. Acta Physiol Scand 60 (Suppl):1–82

Luckhart AB, Carlson AJ (1921) Studies on the visceral sensory system. VIII: On the presense of vasomotor fibers in the vagus nerve to the pulmonary vessles of the amphibian and reptilian lung. Am J Physiol 56:72–112

Malvin GM (1983) Control of respiratory organ blood flow in a tri-modal breather, *Ambystoma tigrinum*. Ph.D. dissertation, University of New Mexico

Malvin GM (1985a) Vascular resistance and vasoactivity of gills and pulmonary artery in the salamander, *Ambystoma tigrinum*. J Comp Physiol 155:241–249

Malvin GM (1985b) Cardiovascular shunting during amphibian metamorphosis. In: Johansen K, Burggren WW (eds) Cardiovascular shunts: phylogenetic, ontogenetic and clinical aspects. Munksgaard, Copenhagen, pp 163–178

Malvin GM, Heisler N (1984) Central vascular shunting of pulmonary and systemic venous blood in the amphibian, *Ambystoma tigrinum*. Fed Proc 43:639

Moalli R, Meyers RS, Jackson DC, Millard RW (1980) Skin circulation of the frog, *Rana catesbiana*: distribution and dynamics. Respir Physiol 40:137–148

Montecuchi PC, de Castiglione R, Piani S, Gozzini L, Erspamer V (1981) Amino acid composition and sequence of Derorphin, a novel opiate-like peptide from the skin of *Phyllomedusa sauvagi*. J Peptide Res 17:275–283

Nakajima T (1981) Active peptides in amphibian skin. Trends Pharmacol Sci 2:202–205

Piiper J, Gatz RN, Crawford EC (1976) Gas transport characteristics in an exclusively skin-breathing salamander, *Desmognathus fuscus* (plethodontidae). In: Hughes GM (ed) Respiration of the Amphibious vertebrates. Academic, New York, pp 339–356

Poczopko P (1958) The influence of the atmosphere enriched in CO_2 and in O_2 on skin capillaries of the edible frog (*Rana esculenta* L.). Zool Pol 9:115–129

Poczopko P (1959) Changes in blood circulation in Rana esculenta L. while driving. Zool Pol 10: 29–43

de Saint-Aubain ML (1982) Vagal control of pulmonary blood flow in *Ambystoma mexicanum*. J Exp Zool 221:155–158

de Saint-Aubain ML, Wingstrand K (1979) A sphincter in the pulmonary artery of the frog *Rana temporaria* and its influence on blood flow in skin and lungs. Acta Zool (Stockh.) 60:163–172

Shelton G (1970) The effect of lung ventilation on blood flow to the lungs and body of an amphibian *Xenopus laevis*. Respir Physiol 9:183–196

Shelton G (1976) Gas exchange, pulmonary blood supply, and the partially divided amphibian heart. In: Spencer Davies P (ed) Perspectives in experimental biology. Pergamon, Oxford, pp 247–283

Shield JW, Bentley PJ (1973) Respiration of some urodele and anuran amphibia – I. In water, role of the skin and gills. Comp Biochem Physiol 46A:17–28

Smith DG (1976) The innervation of the cutaneous artery in the toad, *Bufo marinus*. Gen Pharmacol 7:405–409

Tazawa H, Mochizuki M, Piiper J (1979) Respiratory gas transport by the incompletely seperated double circulation in the bullfrog, *Rana catesbiana*. Respir Physiol 35:77–95

West NH, Burggren WW (1984) Control of pulmonary and cutaneous blood flow in the total, *Bufo marinus* (submitted)

Whitford WG, Hutchison VH (1965) Gas exchange in salamanders. Physiol Zool 38:228–242

Mechanisms of Intracardiac Shunting in Reptiles

N. HEISLER [1]

1 Introduction

Central vascular shunting (i.e., mixing of oxygenated blood returning from the lungs and of deoxygenated blood returning from the systemic tissues) is in higher vertebrates (birds, mammals) generally a feature of a pathological status and considered as a factor of inefficiency, whereas in amphibians and noncrocodilian reptiles the incompletely divided ventricle system implies intracardiac shunting also during the physiological status. The intriguing questions evolving have stimulated numerous studies already during the last century, which were primarily concerned with anatomical description and functional analysis of reptilian hearts and central vessel system, and resulted often in extremely contradictory conclusions (e.g., Brücke 1852; Goodrich 1916, 1919, 1930; Greil 1903; Benninghof 1933; Hopkinson and Pancoast 1937; Mahandra 1942; Mathur 1944, 1946; Mertens 1942; O'Donoghue 1918; Rathke 1857; Thapar 1924; Vorstman 1933). More recent studies utilizing physiological techniques (see below) resulted in similarly conflicting results between the two extremes of no shunting at all, and shunting to a large extent in both directions (L → R, R → L, or both simultaneously).

This extreme diversity has been attributed to anatomical species differences, differences in experimental conditions, or to physiological modulation of shunting. The functional anatomy has actually been described to vary considerably between different orders of reptiles (Chelonia and Squamata), between suborders (Lacertilia and Ophidia), and even within suborders (Varanidae and nonvaranid Lacertilia) (cf. Webb et al. 1971; Webb 1979; Mathur 1944). Also some physiological measurements related to intracardiac hemodynamics indicated species differences. Similar blood peak pressures between aortic arches and pulmonary artery have been reported for the turtle (*Pseudemys scripta*; Chelonia) (White and Ross 1966; White 1968) and the grass snake (*Tripodonotus natrix*; Ophidia) (Johansen 1959). In contrast, the aortic peak pressure is about 50% higher in *Iguana iguana* (nonvaranid Lacertilia) (White 1968), and in varanids the aortic peak blood pressure is about threefold higher than in the pulmonary artery (Harrison 1965; Millard and Johansen 1974; Burggren and Johansen 1982; Heisler et al. 1983). These considerable interspecies differences in hemodynamic parameters, however, provide no explanation for the considerable intraspecies variability of reported shunting patterns (e.g., *Iguana iguana*: Tucker 1966; Baker and White 1970; *Varanus*

1 Abteilung Physiologie, Max-Planck-Institut für Experimentelle Medizin, 3400 Göttingen, FRG

Circulation, Respiration, and Metabolism
(ed. by R. Gilles)

exanthematicus: Berger and Heisler 1977). Accordingly, methodological problems cannot be excluded from being at least partially responsible for the observed scatter.

Injection of radioopague substances and X-ray observation of the flow path is a rather unreliable approach. This method can a priori be applied at any one time to only one of the venous return systems and, thus, is incapable to demonstrate simultaneous shunting in both directions (R → L or L → R). Moreover, even qualitative evaluation is difficult because of the complicated three-dimensional arrangement of the heart chamber system (see later) with overprojection of pulmonary and systemic intracardiac blood flow beds. Quantitative evaluation is, furthermore, complicated by nonlinearities of the concentration/radiodensity relationship and the variable background radiodensity. Accordingly relatively small shunts (10%–20%) may not be detected at all even with most advanced X-ray/computer technology (Bürsch, Heintzen, Neumann, and Heisler, unpublished data).

Application of the Fick principle requires simultaneous anaerobic blood sampling from at least five different sites (right and left atrium, pulmonary artery, and right and left aortic arches). All five samples have to be analyzed for O_2 content immediately and simultaneously because of the considerable metabolism of nucleated reptile erythrocytes. Besides the rather disturbing effect of indwelling catheters in all heart output vessels, additional uncertainties are provided by the required assumptions on flow distribution to both aortic arches.

The last alternative, recording of flow in at least three, or – because of the anatomical situation four to five – vessels requires the same number of electromagnetic flow probes perfectly matched to the vessel diameter and kept in close to 90° angle to the respective vessel during the experiment. This task, combined with the considerable problem of in vitro periarterial flow probe calibration (cf. publications of Statham Instruments Inc.) renders this method inapplicable at least in fully recovered, concise and moving animals.

A much more attractive alternative to methodological imperfectness as explanation for controversial results on intracardiac shunting in reptiles has been provided by White and Ross (1965). They reported a correlation of net shunting direction with the type of respiratory period (i.e., ventilation or apnea). Since ventilation has not been monitored in other studies, modulation of shunting for the purpose of gas exchange optimization or for energy conservation could be an explanation for the divers shunting pattern reported for reptiles.

This hypothesis was recently tested in two species of different orders of reptiles by continuous monitoring of pulmonary ventilation, and simultaneous determination of intracardiac shunting by application of the unconventional (though most direct and appropriate) microspheric method.

2 Application of the Microsphere Method for Determination of Intracardiac Shunting

2.1 Principle

Microspheres (MS) manufactured from plastic material and labeled with various different γ-emitting isotopes, are injected into the blood stream. The rigid MS, which are chosen similar to, or only slightly larger than the erythrocytes are distributed very similar to the erythrocytes – as long as blood and MS are well mixed – and lodge in the tissue vessels, whereas the erythrocytes are sludged through the capillaries. The distribution of flow is then proportional to the amount of lodged MS determined from radioactivity in the respective tissue.

Intracardiac shunting is studied by injection of differently labeled MS into pulmonary and systemic venous return and determination of the relative quantity of MS "spilled" over into the respective inadequate circulatory bed. Systemic and pulmonary heart output can be quantified by application of the "artificial organ" or "reference sample" technique. Activity of MS in blood withdrawn from the arterial pool at known rate, correlated with the injected activity, yields cardiac output (Berger and Heisler 1977).

Essential prerequisites for proper application of this method are good mixing of the MS with the blood (best achieved by jet injection into various directions via indwelling catheters, cf. Heisler et al. 1983) and appropriate size of the injected MS in relation to the erythrocytes. MS should be chosen as small as possible, but checks should be provided to detect MS tissue passage (pulmonary or systemic) and recirculation, and taken into account for shunt calculations. Such corrections should not be extended to experiments with recirculation of more than 5%, since methodological uncertainties may become unacceptable (cf. Heisler et al. 1983).

2.2 Surgery

The surgical preparations required for application of the MS method are relatively extensive, though, after recovery less disturbing to nonanesthetized animals than any of the other methods for comparable measurements.

Halothane anesthetized varanus lizards *(Varanus exanthematicus)* were fitted with atrial PE50 catheters, mounted with 5-hole injection tips (pointing into 5 by 90° different directions by application of microsurgery (Fig. 1). A pneumotachograph sensor was implanted into the trachea and an arterial catheter fitted occlusively into the right common carotid artery. All five catheters were fed under the skin to a skin incision in the shoulder range and led out of the terrarium armed with a stainless steel spring sewn to the back of the animal (Fig. 2) (cf. Heisler et al. 1983).

Freshwater turtles *(Chrysemys picta bellii)* were equipped with atrial and branchial artery catheters via a plastron hole cut with a circular saw with similar techniques. The excised plastron disk was replaced and fixed with dental acrylics. Ventilation was monitored by application of the breath-hole technique (Glass et al. 1983).

The animals recovered from anesthesia after surgery within 30 min and were allowed to readapt to their environment for at least 2 days. The animals appeared completely healthy moving around in their relatively large terrarium or aquarium, respectively. The status of well-recovered animals was well documented by the typical aggressiveness of the varanids and resumption of the normal, highly intermittent breathing pattern of ventilatory periods and apnea (cf. Heisler et al. 1983; Milsom and Jones 1980; Glass et al. 1983).

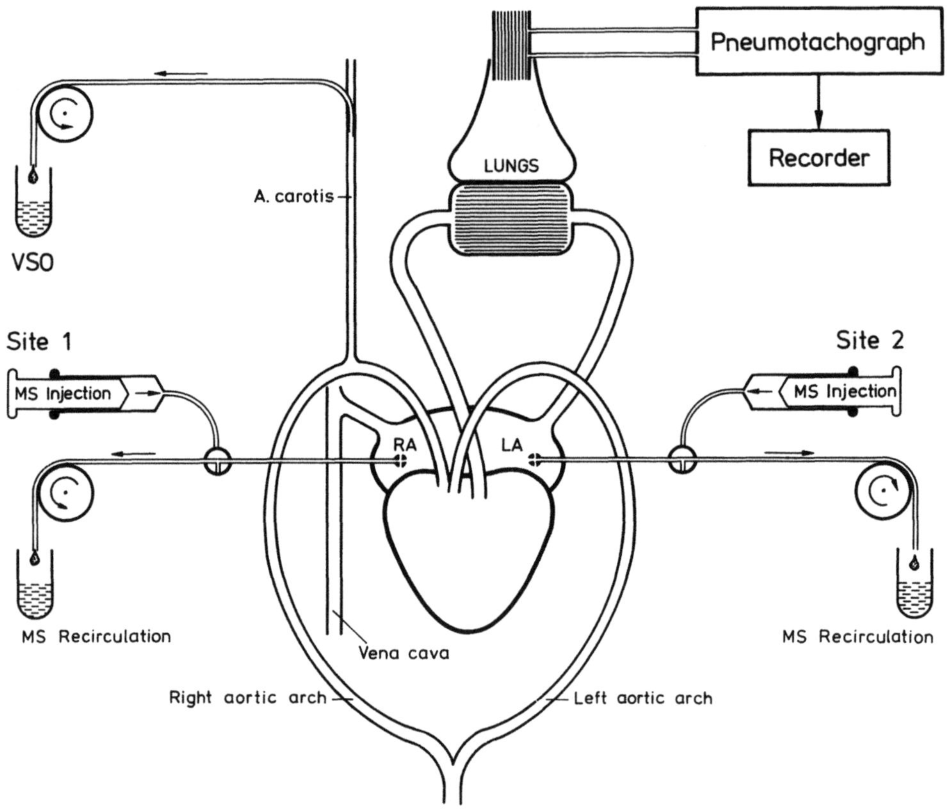

Fig. 1. Application of the microspheric (MS) method for determination of central vascular shunting in *Varanus*. Indwelling catheters in both atria (*RA* = right atrium, *LA* = left atrium) serve for injection of microspheres and for withdrawal of blood for determination of MS recirculation. Withdrawal of blood from the carotid artery yields ventricular systemic output (*VSO*). Ventilation is monitored by intratracheal pneumotachography (see text for details) (from Heisler et al. 1983)

2.3 Procedure

After complete recovery the animals were injected with differently labeled MS into both atria simultaneously and the catheters flushed with dextran solution within 20 s. Immediately after injection blood was withdrawn from the atria in order to detect eventual recirculation of MS (cf. Fig. 1). In varanids recirculation of 15 μ MS was insignificant (cf. Heisler et al. 1983), in turtles 7%–15% of 15 μ MS passed the capillary bed. Therefore, turtle experiments were continued with 25 μ MS, which better matched the erythrocyte diameter of the animals. Withdrawal of blood for determination of cardiac output from carotid or brachial artery in varanids or turtles, respectively, was started simultaneously with the MS injection (cf Heisler et al. 1983; Neumann et al. 1983; Berger and Heisler 1977).

The injection procedure was conducted during periods of ventilation and, with a different set of MS, during apnea which had lasted for at least 2 min in 30 °C experiments

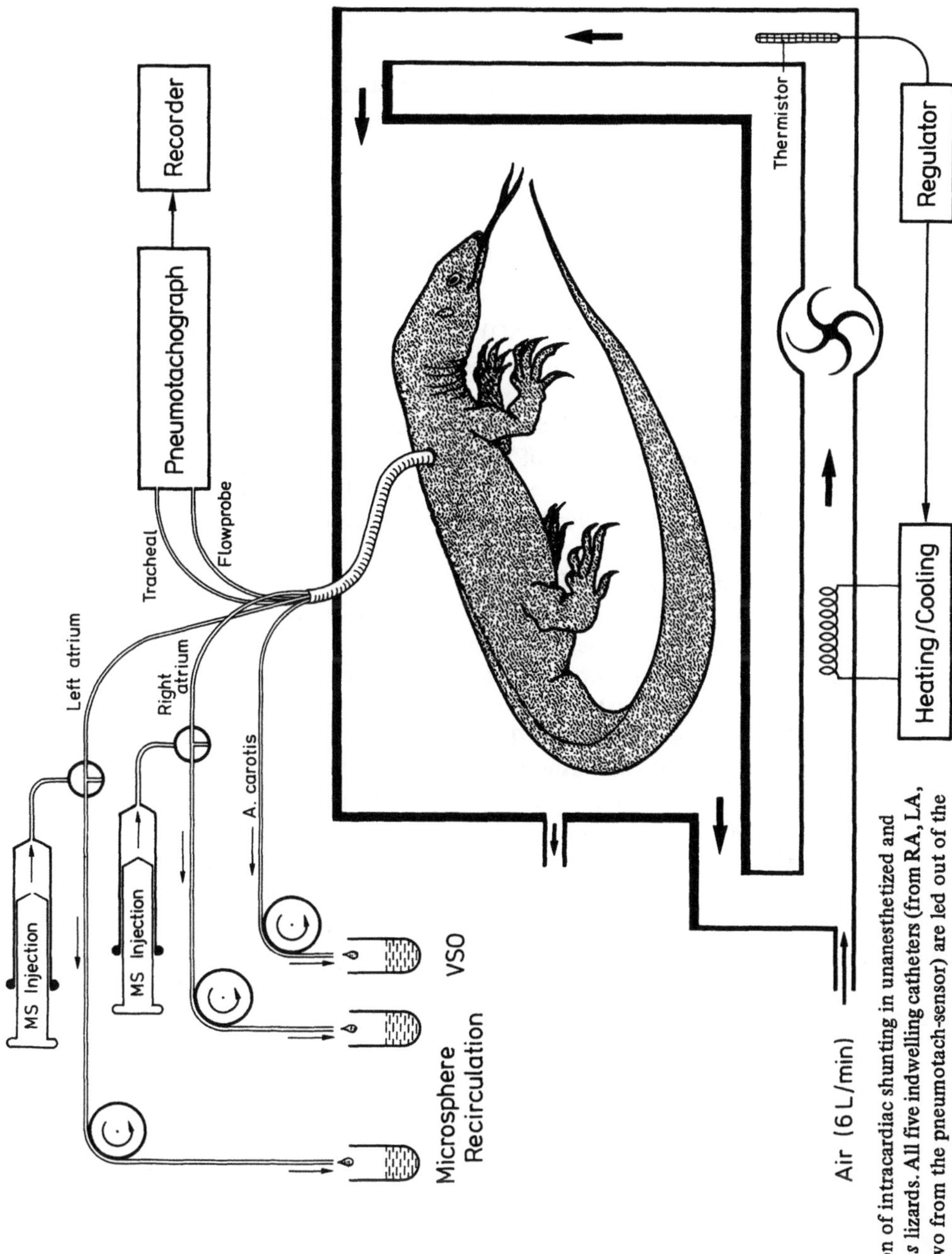

Fig. 2. Determination of intracardiac shunting in unanesthetized and unrestrained *Varanus* lizards. All five indwelling catheters (from RA, LA, carotid artery and two from the pneumotach-sensor) are led out of the body protected against destruction by a light stainless steel spring via a skin hole at the back of the animal, and fed out of the top of the large thermostated terrarium (see text for details) (from Heisler et al. 1981)

in varanids and turtles, and in turtles at 15 °C after 10 min of apnea (with reference to longer apneic periods of up to 1 h).

Details of radioactivity determination, discrimination of various radioactive labels and calculations have been reported previously (cf. Berger and Heisler 1977; Neumann et al. 1983; Heisler et al. 1983).

3 Ventilatory Periods and Cardiac Hemodynamics

In varanids, ventricular systemic output (VSO) and ventricular pulmonary output (VPO) remained unaffected by the type of ventilatory period (VSO = 35.3 ± 10.5 ml min^{-1} kg^{-1} and VPO = 26.9 ± 7.2 ml min^{-1} kg^{-1}. In turtles at 30 °C, only a relatively small reduction of VSO from 85.7 ± 13.3 during breathing to 65 ± 5.6 ml min^{-1} kg^{-1} occurred during apnea, whereas the effect of different ventilatory periods was rather pronounced at 15 °C. VSO was largely reduced by more than threefold from 41.6 ± 8.4 to 13.2 ± 2.6 ml min^{-1} kg^{-1} with similar behavior of VPO (30.0 ± 3.6 during breathing to 6.3 ± 3.2 ml min^{-1} kg^{-1} in apnea).

The differences between VSO and VPO reflect intracardiac shunting, which was little affected by the type of respiratory period (Table 1). Differences could only be noted among species and acclimation temperatures. In *Varanus* the R → L shunt was much larger than the L → R shunt, but this pattern was completely unaffected by the type of respiratory state. Similar relationships were found in turtles at 15 °C, though the magnitude of shunting was much larger. The R → L shunt in turtles at 30 °C was still larger than the L → R shunt, the ratio, however, was reduced to about 1.2. The possible mechanisms for these changes in shunting pattern between 15° and 30 °C turtles are discussed below.

A tendency (though only marginally significant, $2p < 0.1$) for an influence of the ventilatory state on the magnitude of intracardiac shunting, which could be interpreted as an expression of mechanisms for optimization of gas exchange and energy consumption during diving, could only be found in turtles at 15 °C. The reduction in cardiac output induced by apnea had a much more pronounced effect on lung perfusion than any change in shunting pattern. With constant shunting, VPO would fall from 30 to

Table 1. Central vascular shunting in *Varanus exanthematicus* and *Chrysemys picta bellii*

Species	Temperature (°C)	R → L Shunt (% of VSO) Ventilation	Apnea	L → R Shunt (% of VSO) Ventilation	Apnea	Number of animals
Varanus exanthematicus	30	28.7 ± 5.4	30.9 ± 5.7	10.9 ± 1.1	11.2 ± 2.7	9
Chrysemys picta bellii	15	41.0 ± 8.4	59.5 ± 6.0	18.3 ± 5.3	14.8 ± 3.6	5
	30	31.8 ± 7.5	27.0 ± 5.8	27.7 ± 5.8	22.4 ± 4.6	6

$\bar{x}$ ± SE; VSO = ventricular systemic output; VPO = ventricular pulmonary output

9 ml min^{-1} kg^{-1}, whereas the change in shunting contributes not more than an additional reduction to 6.3 ml min^{-1} kg^{-1}.

There are also doubts that R → L shunt-induced reduction of lung perfusion during apnea and increased lung perfusion during periods of lung ventilation would actually be advantageous to the animal. The majority of incorporeal oxygen is stored during diving in the lungs, a compartment with negligible oxygen consumption. Thus, increased R → L shunting would lead, after short apnea, to further desaturation and lowered P_{O_2} in the blood with the deleterious effect of incomplete O_2 supply to sensitive tissues. Since reptiles are considered oxygen regulators, conservation of oxygen at lowered arterial P_{O_2} cannot be expected. In contrast, reduction of peripheral circulation (i.e., to muscle tissues) would contribute significantly to conservation of O_2 for supply of more sensitive tissues.

The negative energetical aspect of recirculating blood through the lungs during breathing cycles by increased L → R shunting (which would not improve the already complete oxygen saturation of pulmonary return blood), as well as the energy conservation by blood bypassing the lungs during apnea can quantitatively be neglected. The contribution of the heart to the whole animal energy consumption amounts to only a few percent. Less than one-third of this is attributable to pulmonary circulation, which in turn could be reduced by only a fraction by an increase in R → L shunting.

4 Functional Anatomy of Reptile Hearts

The complex pattern of cardiac shunting in reptiles can only be understood after careful analysis of the functional heart anatomy.

Transverse sections of *Varanus* hearts slightly above and in one-third distance between valve plane and apex indicate the unique configuration of the valves in correlation to the U-shaped arrangement of ventricular cava (Fig. 3). Unfolding of the U-shaped chamber arrangement reduces the complex three-dimensional configuration to a more easily comprehensible two-dimensional representation.

The cavum arteriosum (CA) is characterized by connection to the left atrium (LA) via the left atrioventricular valve and is a relatively deep, largely trabecularized chamber. The vertical septum between CA and cavum venosum (CV), which is usally in line with the atrial septum (Fig. 3) may also be shifted (one-third of the cases) in the direction of the CV to be in line with the midpoint of the right atrioventricular valve. The CA is connected to the CV (which is smaller than the CA by 60%-70%) via the relatively narrow interventricular canal (IC). The CV widens again close to the „Muskelleiste“ (muscular ridge, MR), which separates CV with the ostia of the aortic arches, and the cavum pulmonale (CP) with the ostium of the pulmonary artery (PA). The atrioventricular valves are usually capable of covering the opening of the IC during diastole on both sides and to provide very good separation between the CA and CV during diastole. It is evident from this description and from Fig. 3 that a functional crossover of oxygenated and deoxygenated blood in the CV is the unavoidable consequence of valve and chamber arrangement in *Varanus* hearts (cf. Fig. 3; for further details of the functional anatomy see Heisler et al. 1983).

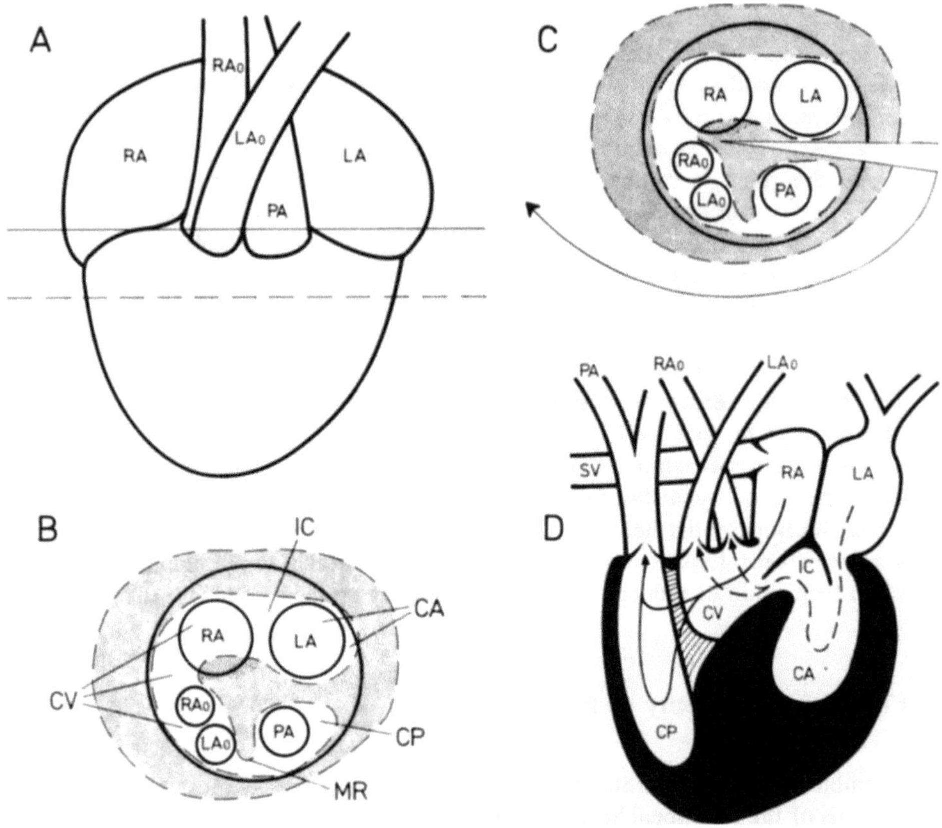

Fig. 3A–D. **A** Ventral aspect of the heart of *Varanus*. **B** Transverse sections slightly above the valve plane (*solid line*) and at about one-quarter of the distance between valve plane and apex (cf. **A**). **C** Unfolding of the three-dimensional chamber arrangement displayed in **B**. **D** Schematic two-dimensional presentation of the *Varanus* heart after unfolding according to **C**. Note the functional crossover of pulmonary and systemic blood return. *Striped area* and *MR* = „Muskelleiste" (muscular ridge); *CP* cavum pulmonale; *CV* cavum venosum; *CA* cavum arteriosum; *PA* pulmonary artery; *IC* interventricular canal; *RAo* right aortic arch; *LAo* left aortic arch; *SV* sinus venosus (from Heisler et al. 1983)

In *Chrysemys* the general anatomical arrangement of heart chambers, valves, and „Muskelleiste" (muscular ridge, MR) is very similar to *Varanus*, except that the relative volume of the CV is larger. The significance of this detail is discussed later.

5 Blood Pressure Separation in the Heart of Varanus

As first pointed out by Harrison (1965) and later by Millard and Johansen (1974) the systemic blood pressures in *Varanus* are much higher than the pulmonary artery blood pressure. These observations were confirmed by Burggren and Johansen (1982). Their

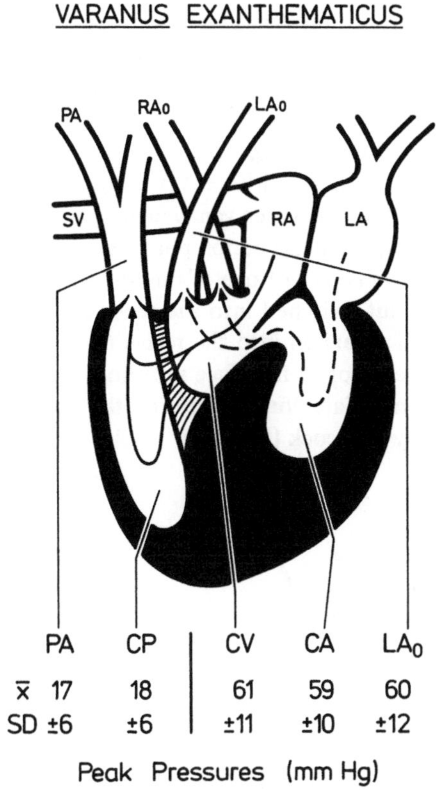

Fig. 4. Average peak blood pressures in cava and in the large effluent vessels of the heart of *Varanus exanthematicus*

conclusion that pressure separation between the two circulatory stream beds occurred between CA and the major portion of CV and CP by pressing the vertical septum against the "aorticopulmonary septum bringing the cavum arteriocum in direct contact with the right and left systemic arteries" was, however, invalidated by recent pressure measurements in effluent vessels, CA, CP, and in the lower extension of the CV close to the „Muskelleiste" (MR) (Heisler et al. 1983). As indicated by the average peak blood pressures presented in Fig. 4 pressure separation in *Varanus* hearts is achieved by pressure-tight contact between muscular ridge (MR) and external heart wall during systole. Though this mechanism was first suggested by Greil (1903) and discussed various times (e.g., Thapar 1924; White 1959; Webb et al. 1971) physiological evidence for this phenomenon has not been provided before.

6 Principles of Intracardiac Shunting

In mammals and birds, central vascular shunting is generally an expression of a pathological status characterized by anatomical defects of the separating structures between pulmonary and systemic circulation (atrial or ventricular "septum defect"), or by remainders of ontogenetic development (e.g., ductus Botalli). Since a hydrostatic blood pressure difference is also required, this type of shunt may be called "pressure difference shunt" (Fig. 5A). Only with extensive septum defects or complete lack of septa, like in amphibia, oxygenated and deoxygenated blood may mix as a result of blood turbulence induced by different blood flow velocities at the open interface, or by extensive ventricular trabeculation ("open interface shunt") (Fig. 5B).

In reptiles, however, the functional heart anatomy is rather different from the above mentioned configurations in mammals, birds, and amphibia. Three ventricular chambers are connected to inflow and outflow systems in a complex, heart action phase-dependent way (Fig. 5C). The intermediate of the three chambers is part of one of the functionally crossing blood stream beds during heart phase I, but switched into the second blood stream bed during heart phase II (Fig. 5C). Blood diversion into the respective other stream beds does not occur by pressure difference or open interface shunting, but by washout of the blood that remained in the intermediate chamber during the switch comparable to the dead space volume of a four-way stopcock (Fig. 5C).

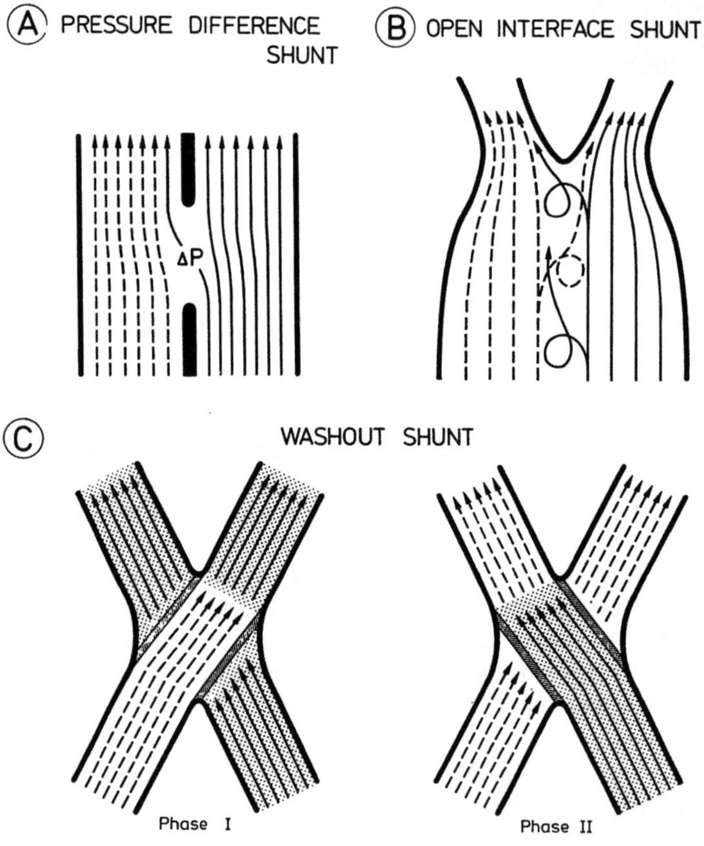

Fig. 5A–C. Schematic presentation of shunting mechanisms. **A** Pressure difference shunt. Characteristics: pressure difference at direct anatomical connection between pulmonary and systemic circulation. **B** Open interface shunt. Characteristics: no anatomical separation between pulmonary and systemic blood. No significant pressure difference. Mixing by turbulance. **C** Washout shunt. Characteristics: a blood compartment common to both pulmonary and systemic circulation during different circulatory phases

7 Shunt Mechanisms in Reptiles

The in vivo conditions of *Varanus* are very well resembled by the simple model of Fig. 5C. Separation between the two blood streams (oxygenated and deoxygenated) is extremely well performed by the functional anatomical features of *Varanus* hearts. During diastole separation is provided by the large atrioventricular valves occluding the interventricular canal, and during systole the „Muskelleiste" (MR) efficiently separates CV and CP (cf. Figs. 3 and 4). However, the site of separation is shifted from the range of the interventricular canal during diastole to the MR, such that the CV is, depending on the heart action phase, part of either pulmonary or systemic circulation. Blood remaining in the CV from the filling phase during diastole (deoxygenated) is washed out by the oxygenated blood ejected from CA into the aortic arches (Fig. 6, systole). In turn, the oxygenated blood remaining in the cavum venosum at the end of systole is washed during diastolic filling into CP by deoxygenated blood and ejected into the pulmonary artery during the following heart contraction (Fig. 6, diastole). The relative amount of blood transferred into the respective "inappropriate" circulation is, thus, dependent on the ratio of the respective "stroke" volume (systemic or pulmonary) to the volume of the CV at the end of the preceding half cycle of heart action: small volume of oxygenated blood at the end of systole (= small L → R shunt) and large volume of deoxy-

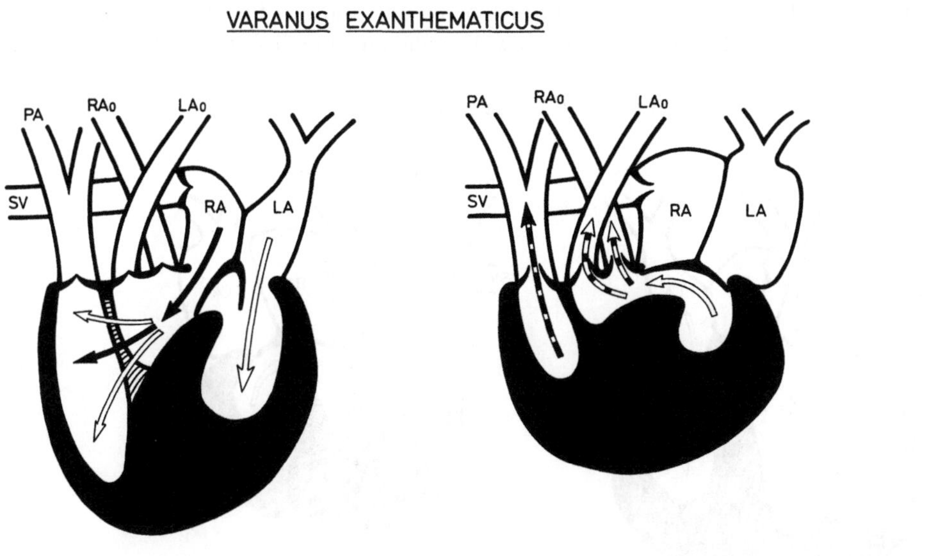

Fig. 6. Intracardiac shunting due to washout of the cavum venosum. *Diastole:* oxygenated blood (*open arrows*) remaining in the CV from the preceding systole is washed into the CP by deoxygenated blood (*black arrows*). CA is filled with oxygenated blood. Separation between CV and CA by atrioventricular valve occluding interventricular canal (IC). *Systole:* deoxygenated blood remaining in CV from the preceding diastole is flushed by oxygenated blood from CA into the systemic circulation. Deoxygenated blood with admixture of oxygenated blood in the CP is expelled into the pulmonary circulation. Separation between CV and CP by the muscular ridge (*striped area*) pressed against the outer heart wall (from Heisler et al. 1983)

genated blood at the end of diastole (= large R → L shunt). This pattern conforms very well with the results actually obtained in *Varanus*, and also the variability of shunting between individuals is well explained on the basis of the different anatomical sizes of the CV (cf. Heisler et al. 1983).

According to the similar cardiac configuration, also in *Chrysemys*, the "washout" shunt is an unavoidable feature of heart action. It appears to be the dominating factor at least in animals at 15 °C. This is indicated by the much larger R → L than L → R shunt, the ratio of which is very similar to *Varanus*. In absolute terms the shunts are larger in turtles, which is expected on the basis of the larger CV. The data suggest that blood flow separation by valves and interventricular canal during diastole and by the „Muskelleiste" during systole is functional also in turtles at low temperature.

At 30 °C, however, additional factors may be involved. The increase in L → R shunt and the reduction in R → L shunt (as compared to 15 °C) may be an expression of a pressure difference shunt at the „Muskelleiste" (muscular ridge). A systolic shift of deoxygenated blood remaining in the cavum venosum and of oxygenated blood from cavum arteriosum into cavum pulmonale at the beginning of contraction (Fig. 7) could well produce the pattern deviating from that expected on the basis of a pure washout shunt (Figs. 5 and 6). Also a diastolic pressure difference shunt from cavum arteriosum into cavum venosum via interventricular canal during early diastolic filling of the cavum pulmonale could be claimed to contribute to the rise in L → R shunt. It is, however, unlikely that the required large quantities of blood would be transferred through this narrow pathway by the prevailing very low diastolic pressure differences. The reduction

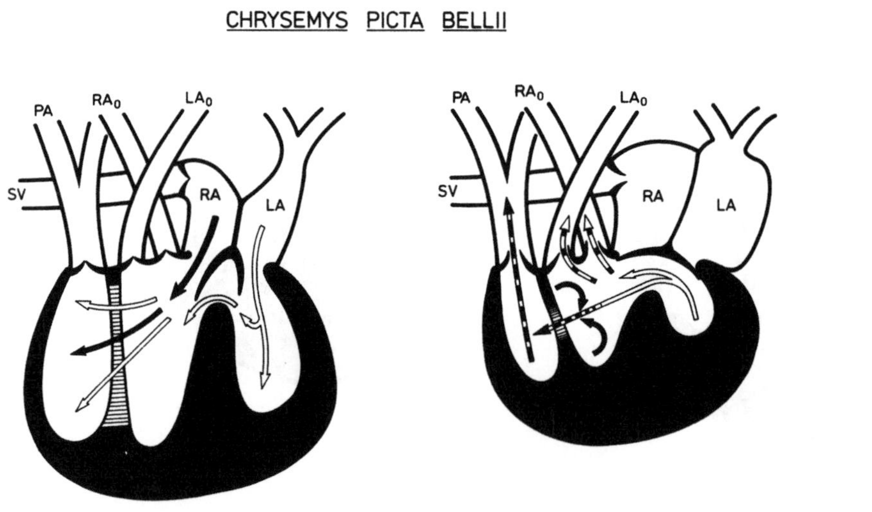

Fig. 7. Intracardiac shunting in *Chrysemys*. Anatomical differences reside mainly in the larger CV of *Chrysemys*. The basic mechanism (washout of CV) is the same as in *Varanus*. At high temperature (30 °C), however, shunting may also be influenced by the indicated mechanisms. *Diastole:* incomplete separation between CA and CV by atrioventricular valves with oxygenated admixture to systemic venous blood in CV. *Systole:* incomplete separation between CV and CP resulting in transfer of deoxygenated CV-blood and oxygenated CA-blood to CP

in R → L and the increase in L → R shunt could, however, also be explained exclusively on the basis of the washout model combined with incomplete washout of the much larger and more recessed cavum venosum of *Chrysemys* combined with temperature-induced changes in stroke volume. Further experiments are required to shed more light on the role of these additional mechanisms.

Considerable shunting in both directions on the basis of cavum venosum washout is evidently an inherent feature of the anatomical configuration of reptile hearts with their functional crossover of pulmonary and systemic circulation. Shunting in only one of the two possible directions (R → L, L → R) can occur only with complete shutdown of either pulmonary or systemic circulation, which leads to cessation of the respective venous return, and implies 100% R → L or L → R shunting, respectively. Apart from these very special conditions, shunting is – at least in its basic component – determined by the respective stroke volume in relation to the volume of blood remaining in the preceding half cycle of heart action.

In principle this ratio of stroke volume/washout volume could be modulated by various factors, such as diastolic filling, systolic expulsion, or changes in the ratio of pulmonary to systemic resistance (e.g., Shelton and Burggren 1976; Burggren et al. 1977; White 1976; Burggren 1977a,b). In our experiments, however, no ventilatory period-related changes in shunting pattern could be detected. Accordingly, we conclude on the basis of the above considerations, experimental data, and analysis of functional anatomy that the extent of intracardiac shunting in reptiles is primarily determined by the ratio of stroke volume to the volume of the cavum venosum, is essentially unaffected by the type of ventilatory period, and is generally an unavoidable consequence of the reptilian heart anatomy.

References

Baker LA, White FN (1970) Redistribution of cardiac output in response to heating in *Iguana iguana*. Comp Biochem Physiol 35:253–262

Benninghof A (1933) Das Herz (d) Reptilien. In: Handbuch der Vergleichenden Anatomie der Wirbeltiere, vol VI, pp 502–556

Berger PJ, Heisler N (1977) Estimation of shunting, systemic and pulmonary output of the heart, and regional blood flow distribution in unanaethetized lizards *(Varanus exanthematicus)* by injection of radioactively labelled microspheres. J Exp Biol 71:111–121

Brücke E (1852) Beiträge zur vergleichenden Anatomie und Physiologie des Gefäß-Systemes. Denkschr Akad Wiss Wien 3:335–367

Burggren WW (1977a) The pulmonary circulation of the chelonian reptile: morphology, haemodynamics and pharmacology. J Comp Physiol 116:303–323

Burggren WW (1977b) Circulation during intermittent lung ventilation in the garter snake *Thamnophis*. Can J Zool 55:1720–1725

Burggren WW, Johansen K (1982) Ventricular haemodynamics in the monitor lizard *Varanus exanthematicus*: pulmonary and systemic pressure separation. J Exp Biol 96:343–354

Burggren WW, Glass ML, Johansen K (1977) Pulmonary ventilation/perfusion relationship in terrestrial and aquatic chelonian reptiles. Can J Zool 55:2024–2034

Glass ML, Boutilier RG, Heisler N (1983) Ventilatory control of arterial P_{O_2} in the turtle *Chrysemys picta bellii*: effects of temperature and hypoxia. J Comp Physiol 151:145–153

Goodrich ES (1916) On the classification of the reptilia. Proc R Soc Ser B 89:261–276

Goodrich ES (1919) Note on the reptilian heart. J Anat 53:298–304

Goodrich ES (1930) Studies on the structure and development of vertebrates. Macmillan, London, pp 553–561

Gordon AS, Elazar S, Austin S (1971) Practical aspects of blood flow measurements. Statham Instruments, Oxnard, CA, USA, p 78

Greil A (1903) Beiträge zur vergleichenden Anatomie und Entwicklungsgeschichte des Herzens und des *Truncus arteriosus* der Wirbeltiere. Morphol Jahrb 31:123–310

Harrison JM (1965) The cardiovascular system in reptiles with special reference to the goanna, *Varanus varius*. Unpubl BSc (Med) Thesis, University of Sidney

Heisler N, Neumann P, Maloiy GMO (1983) Mechanism of intracardiac shunting in the lizard *Varanus exanthematicus*. J Exp Biol 105:15–31

Hopkinson JP, Pancoast J (1837) On the visceral anatomy of the python (Cuvier) described by Daudin as the *Boa reticulata*. Trans Am Philos Soc 5:121–136

Johansen K (1959) Circulation in the three-chambered snake heart. Circ Res 7:828–832

Mahendra BC (1942) Contribution to the bionomics, anatomy, reproduction and development of the Indian house-gecko, *Hemidactylus flaviviridis* Rüppel. III. The heart and the venous system. Proc Indian Acad Sci 15B:231–252

Mathur PN (1944) The anatomy of the reptilian heart. I. *Varanus monitor*. Proc Indian Acad Sci Sect B 20:1–29

Mathur PN (1946) The anatomy of the reptilian heart. Part II. Serpentes, testudinata and loricata. Proc Ind Acad Sect B 23:129–152

Mertens R (1942) Die Familie der Warane *(Varanidae)*. I. Senckenb Naturforsch Gesellsch 462:1–116

Millard RW, Johansen K (1974) Ventricular outflow dynamics in the lizard, *Varanus niloticus*: responses to hypoxia, hypercarbia and diving. J Exp Biol 60:871–880

Milsom WK, Jones DR (1980) The role of vagal afferent information and hypercapnia in the control of the breathing pattern in chelonia. J Exp Biol 87:52–63

Neumann P, Holeton GF, Heisler N (1983) Cardiac output and regional blood flow in gills and muscles after strenuous exercise in rainbow trout *(Salmo gairdneri)*. J Exp Biol 105:1–14

O'Donoghue CH (1918) The heart of the leathery turtle, *Dermochelys (Spargis) coriacea*. With a note on the septum ventriculorum in the reptiles. J Anat 3rd series 52:467–480

Rathke H (1857) Untersuchungen über die Aortenwurzeln und die von ihnen ausgehenden Arterien der Saurier. Denkschr Acad Wiss Wien 13:51–142

Shelton G, Burggren W (1976) Cardiovascular dynamics of the Chelonia during apnoea and lung ventilation. J Exp Biol 64:323–343

Thapar GS (1924) On the arterial system of the lizard *Varanus bengalensis* (Daud.), with notes on *Uromastix* and *Hemidactylus*. J Asiatic Soc Bengal New Ser 19:1–13

Tucker VA (1966) Oxygen transport by the circulatory system of the green Iguana *(Iguana iguana)* at different body temperatures. J Exp Biol 44:77–92

Vorstman AG (1933) The septa in the ventricle of the heart of *Varanus komodoensis*. Proc R Acad Amsterdam 36:911–913

Webb GJW (1979) Comparative cardiac anatomy of the reptilia. III. The heart of crocodilians and an hypothesis on the completion of the interventricular septum of crocodilians and birds. J Morphol 161:221–240

Webb G, Heatwolfe H, DeBavay J (1971) Compartive cardiac anatomy of the reptilia. I. The chambers and septa of the varanid ventricle. J Morphol 134:335–350

White FN (1959) Circulation in the reptilian heart (Squamata). Anat Rec 135:129–134

White FN (1968) Functional anatomy of the heart of reptiles. Am Zool 8:211–219

White FN (1976) Circulation. In: Gans C (ed) Biology of the Reptilia, Physiology A, vol 5. Academic, New York, pp 275–334

White FN, Ross G (1965) Blood flow in turtles. Nature 208:759–760

White FN, Ross G (1966) Circulatory changes during experimental diving in the turtle. Am J Physiol 211:15–18

Gas Exchange in Intermittently Breathing Turtles

F.N. WHITE, P. BICKLER, and M. YACOE [1]

1 Introduction

The differing behaviors of the respiratory gases in media (air, water, blood) and, the special anatomical and functional characteristics of gas exchangers (gills, avian lung, alveolar lung, skin) have introduced anatomical and conceptual difficulties in arriving at a unitary approach to comparative vertebrate gas exchange. A useful step toward the resolution of these problems is the model analysis of Piiper and Scheid (1972, 1975, 1977). They have related their model to selected sets of data for dogfish, bird, dog, and amphibian skin, each analysis emphasizing the peculiarities of these distinct gas exchange systems.

The present study, based on the model of Piiper and Scheid for the alveolar lung (ventilated pool), is designated to extend our knowledge of gas exchangers to a reptile, *Pseudemys scripta*. In addition, the effects of body temperature (BT) are included in order to better appreciate the influence of this natural perturbation on interactions among metabolism, blood transport properties, and ventilation. Correlates of blood transport properties at different BT's will be related to observed alterations in ventilation levels and air convection requirements (ACR = liters of gas ventilated per mmol O_2 consumed or, CO_2 produced: $1 \cdot mmol^{-1}$). It will be shown that specific blood gas transport properties are predictive of both absolute levels of ventilation and ACR at the BT's studied and that these blood properties may be utilized in the construction of a model of central neural receptor control of the alterations in ventilation which accompany changes in body temperature.

2 Elements of the Model Analysis

The ventilated pool model perceives the gas exchanger to be composed of two convective processes (ventilation, perfusion) and an intervening diffusion barrier (Fig. 1). Each convective process is analyzed through an expanded form of the Fick equation in which the more conventional relationship $[\dot{M}_g = \dot{V}_g \cdot (C_i \cdot C_e)]$ is replaced by: $\dot{M}_g = \dot{V}_g \cdot \beta \cdot (P_i - P_e)$ where β is the capacitance coefficient for the gas under consideration. The introduction of β allows meaningful comparative analyses of gases which differ in

1 The Physiological Research Laboratory, Scripps Institution of Oceanography, University of California, San Diego, La Jolla, CA 92014, USA

Circulation, Respiration, and Metabolism
(ed. by R. Gilles)

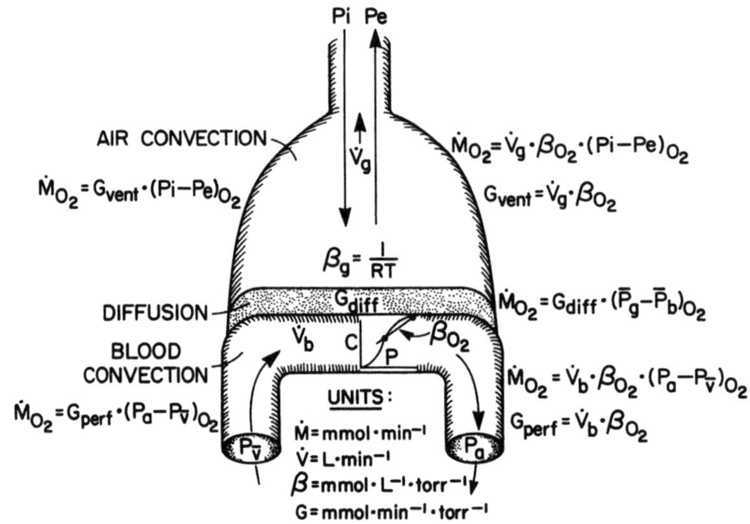

Fig. 1. Ventilated pool model of cardiopulmonary gas exchange

their chemical-binding characteristics, such as O_2 and CO_2 in blood. Knowledge of the O_2 and CO_2 dissociation curves is required. For present purposes β is estimated from the slopes of lines connecting pulmonary arterial and venous points after due consideration of the Bohr and Haldane effects. For the gas phase β is derived from the general gas law and is equivalent for both O_2 and CO_2 ($\beta = 1/RT$).

The product of convective flow and β represents conductance: $G = V \cdot \beta$. This allows, for both O_2 and CO_2, the use of ventilation to perfusion conductance ratios (G_{vent}/G_{perf}) and replaces the ventilation to perfusion ratio, so widely used for O_2 transport in mammals, with terms which incorperate the differing chemical binding characteristics of O_2 and CO_2 in blood.

Reciprocals of G's have the meaning of resistances, e.g., $R_{vent} = 1/G_{vent}$. When the two convective conductances and the total conductance of the system [$G_{tot} = \dot{M}/(Pi-Pv)$] are known, R_{diff} and G_{diff} may be calculated. It is important to recognize that these calculated values may incorporate components other than those which are strictly diffusive in nature, e.g., stratification, ventilation-perfusion inequalites, or shunt.

Partial pressure differences may be utilized for the derivation of *relative* resistances (Δ p values) attributable to each of the processes. Thus, $\Delta p_{vent} = Pi-Pe/Pi-Pv$; $\Delta p_{perf} = Pa-Pv/Pi-Pv$; and, $\Delta p_{diff} = Pe-Pa/Pi-Pv$. The relative resistance add up to unity. The Δp values may be expressed as functions of two conductance ratios: G_{vent}/G_{perf} and, G_{diff}/G_{perf}. Piiper and Scheid (1977) provide useful equations which allow calculation of Δp's over a range of G_{vent}/G_{perf} and G_{diff}/G_{perf} values.

In addition to the utility the model provides for appreciating the influence of conductance ratios on relative resistances of the processes of gas exchange, so-called limitation indices (L values) may be derived. Limitation indices take on the general form: $L_x = 1 - G_{tot}(\text{limited by x})/G_{tot}(\text{not limited by x})$, where x may be ventilation, perfusion, or diffusion, e.g., $L_{vent} = 1 - G_{tot}(\text{actual})/G_{tot}(G_{vent} = \infty)$ and, so on for the other processes. $L_{vent} = 0.85$ signifies that G_{tot} is reduced by 85% from the value which would exist in the absence of any ventilation limitation. Theoretical limitation indices may be calculated from a range of G_{vent}/G_{perf} and G_{diff}/G_{perf} values by the use of convenient equations developed by Piiper and Scheid (1977).

3 Assumptions and Sources of Data

The analysis utilizes metabolic, pulmonary blood flow, and blood gas data of Kinney et al. (1977) where pulmonary arterial and venous blood samples were collected over the period of the commencement of one ventilatory cycle to the next. We have assumed, for the purposes of the model analysis, that both blood and air convection are equivalent, in net effect, to steady nonpulsatile processes. Mean values for Pe were derived at the intersection of an RQ 0.75 line with the $P_{O_2}-P_{CO_2}$ lung gas diagrams for each BT (White and Somero 1982).

Pulmonary arterial (pa) and venous (pv) points for CO_2 were established in the following manner. We assumed a partial pressure difference between Pe and pv of 1 torr, a value which is in close agreement with Burggren and Shelton (1979) and the values found by Kinney et al. (1977) for pv blood when related to the estimates of Pe (see above). This allows the establishment of a pv point on the CO_2 dissociation curve (Weinstein et al., in press) after correction for the Haldane effect. This correction utilizes the O_2 saturation values for pv of Kinney et al. (1977). From the Fick equation $[\dot{M}_{CO_2} = \dot{V}_b \cdot (Cpa-Cpv)CO_2]$ the Cpa–Cpv values were calculated. Knowledge of the O_2 saturation values for pa blood and Cpa–Cpv allows the establishment of a pa point at the intersection of the pa CO_2 content line and the CO_2 dissociation curve of blood at the appropriate O_2 saturation level. Thus, the Haldane effect is incorporated. β_{CO_2} is estimated from the slopes of lines connecting pa and pv points at the three BT's utilized (15°, 25°, 35 °C).

The CO_2 partial pressures for pa and pv were used to calculate the pH's of these bloods utilizing the CO_2 contents (see above) and the Henderson-Hasselbach relationship. This allowed correction of the O_2 dissociation curves of Maginniss et al. (1980) for the Bohr effect. The O_2 saturation data of Kinney et al. (1977) for pa and pv blood were applied to the O_2 dissociation curves to establish Ppa and Ppv values for O_2 at each temperature. β_{O_2} was estimated from the slopes of lines connecting pa and pv points. Figure 2 displays the disposition of the arterial and venous points for O_2 and CO_2 in regard to the dissociation curves.

The analysis of the relationship of CO_2 perfusive characteristics to the behavior of a putative proton activated central neural receptor utilized acid-base characteristics of cerebrospinal fluid (csf) which was considered, for heuristic reasons, to be a protein-free ultrafiltrate of plasma. The analysis follows the principles of Stewart (1981). CO_2 dissociation curves and the pH–C_{CO_2} relationship, as functions of temperature, were generated by a Basic language computer program. The putative proton receptor is conceptualized as existing in an environment similar to csf. It is further assumed that the ventilatory level is governed by the time required for the receptor to oscillate from a state of protonization below threshold (Zsth) to a threshold state (Zth) which activates

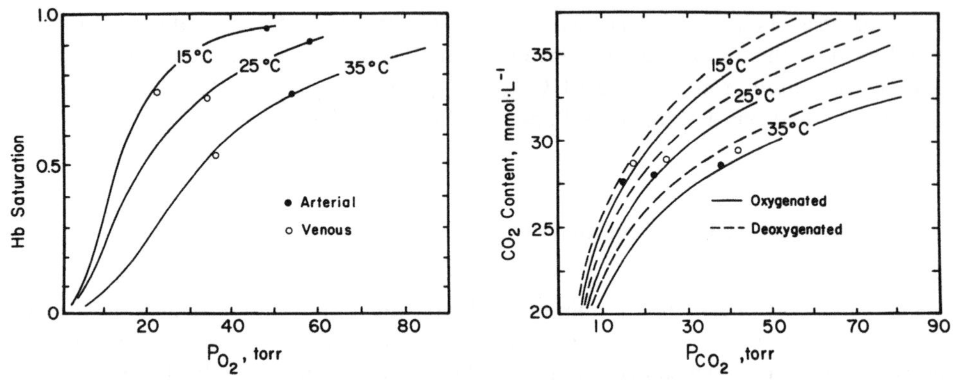

Fig. 2. O_2 and CO_2 dissociation curves for *P. scripta*. *Data points* are those found in the analysis. *Venous points* which fall below the O_2 saturation curves incorporate the Bohr effect. Haldane effect is utilized in establishing arterial and venous points on the CO_2 dissociation curves. O_2 dissociation curves based on Maginniss et al. (1980); CO_2 dissociation curves are from Weinstein et al. (in press)

ventilation. There being no data on blood flow characteristics influencing the environment of such a receptor, the time to effect a standard perturbation of 0.1 pH unit in 1 liter of csf was calculated by applying the CO_2 perfusive values observed for a 1 kg turtle at the three BT's utilized for the Piiper and Scheid analysis, i.e., $\dot{M}_{CO_2} = \dot{V}_b \cdot \beta \cdot (Pv-Pa)$. Thus, the CO_2 transport properties of blood were related to the acid-base characteristics of csf in order to calculate a "fictitious" time required to produce a stated alteration if csf pH. A csf pH$-C_{CO_2}$ position around which the pH alteration was calculated was fixed at the mean Pa_{CO_2}'s observed at each temperature. This allows the calculation of ΔC_{CO_2} values associated with the standard pH alteration from the pH$-C_{CO_2}$ curves for csf. Since more protracted times are expected to be associated with lower levels of ventilation, the reciprocals of the fictitious times were related to measured levels of ventilation at the three BT's. A constant slope of such a relationship is considered as a preliminary test of the utility of the model. This is a more stringent test than might at first be imagined since the pH$-C_{CO_2}$ curves for csf are nonlinear, are not proportionately displaced with BT alterations, and require significantly different ΔC_{CO}'s per 0.1 pH alteration at each of the temperatures studied.

4 Results

Table 1 exhibits observed and calculated values essential to the model analysis. Of special interest are: (1) the inverse relationship between ACR and BT; (2) relative constancy of β_{O_2} and the inverse relationship of β_{CO_2} to BT; (3) the consistently high G_{perf} values for CO_2 relative to O_2; and (4) low G_{diff} (high R) for O_2.

In Fig. 3 Δp values are related to G_{vent}/G_{perf} at stated G_{diff}/G_{perf}. The latter values varied over a relatively small range for both O_2 and CO_2 and exhibited no trend with BT. For heuristic reasons the G_{diff}/G_{perf} values utilized represent means for the three tem-

Table 1. Temperature and gas exchange parameters: turtle

°C:	15			25			35		
Gas:	O_2		CO_2	O_2		CO_2	O_2		CO_2
$\dot{M}_g$	0.0087		0.0065	0.0257		0.0193	0.0757		0.0568
$\dot{V}_E$		0.0104			0.0203			0.0396	
ACR	1.20		1.60	0.79		1.05	0.52		0.70
$\dot{V}_b$		0.0085			0.0267			0.0845	
β_b	0.035		0.440	0.034		0.300	0.045		0.200
β_g		0.0556			0.0538			0.0520	
G_{vent}[a]	5.68^{-4}		5.68^{-4}	1.06^{-3}		1.06^{-3}	2.01^{-3}		2.01^{-3}
G_{perf}[a]	2.96^{-4}		3.72^{-3}	9.08^{-4}		8.01^{-3}	3.80^{-3}		1.69^{-2}
G_{diff}[a]	7.81^{-5}		4.98^{-3}	3.18^{-4}		2.03^{-2}	1.20^{-3}		2.97^{-2}
Ppa–Pva	−25.5		0.75	−23.5		0.91	−20.5		2.86
RQ		0.75			0.75			0.75	

[a] $\times 10^{-x}$; $\dot{M}$ = mmol • min^{-1}; $\dot{V}_E$, $\dot{V}_b$ = L • min^{-1}; ACR = L • mmol^{-1}; β = mmol • L^{-1} • torr^{-1}; G = mmol • min^{-1} • torr^{-1}; Ppa, Ppv = partial pressure, pulmonary artery, vein; g = gas; b = blood

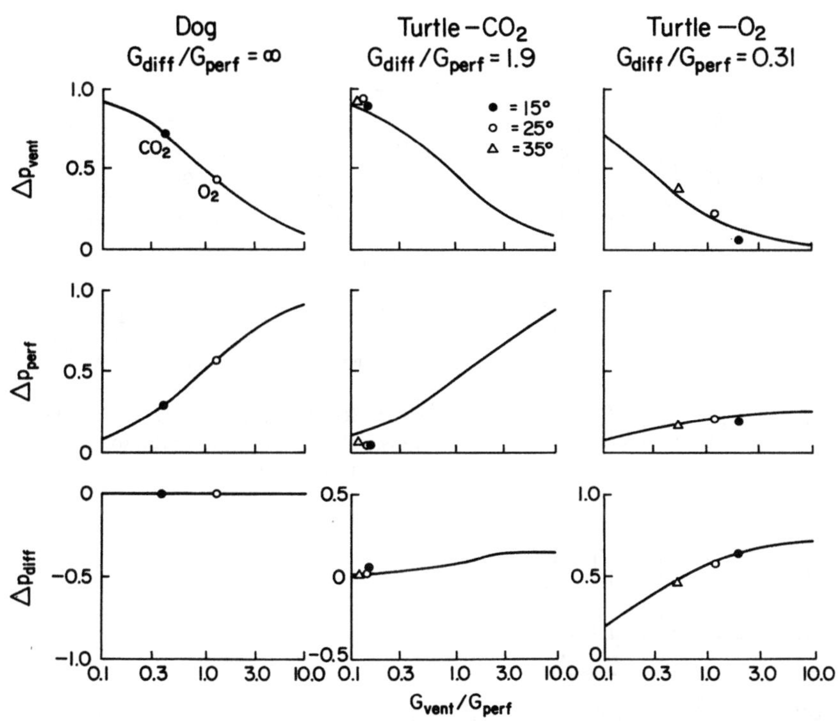

Fig. 3. Relative partial pressure differences as functions of G_{vent}/G_{perf} and G_{diff}/G_{perf}. In the case of $G_{diff}/G_{perf} = \infty$, *data points* for the dog are shown. The two *right-hand frames* are for the turtle

peratures. For comparative purposes, a case in which G_{diff}/G_{perf} is infinitely large is presented along with data points for the dog (Piiper and Scheid 1972). In the case of the turtle it is noteworthy that G_{vent}/G_{perf} for O_2 varies inversely with BT over the range of 0.53 to 1.92, while Δp values for ventilation and diffusion are unstable with temperature. In contrast, both G_{vent}/G_{perf} and Δp values for CO_2 are constrained to very small variations.

Limitation indices for the turtle are exhibited along with those for other species in Fig. 4. These data emphasize a high degree of similarity for CO_2 among lung breathers. Distinct differences exist for O_2. These are especially evident for L_{vent} and L_{diff}.

Figure 5 displays correlations between blood CO_2 transport characteristics and ventilatory parameters. Ventilation varies directly, in a linear fashion, with G_{perf} CO_2 (panel A), while ACR is directly related to β_{CO_2} (panel B). No such correlations were found for O_2 transport parameters.

The CO_2 dissociation curves for blood and csf are shown in Fig. 6. The difference in slopes of these curves is reflected in high blood: csf ratios for β_{CO_2}. The relationship between csf pH and C_{CO_2} is shown in Fig. 7A along with C_{CO_2}'s associated with a 0.1 pH alteration around the points corresponding to observed mean Pa_{CO_2}'s. Using values for blood transport of CO_2 representative of a 1 kg animal, the C_{CO_2} values for csf were utilized to calculate fictitious times required to perturb 1 liter of csf by 0.1 pH units (see assumptions and sources of data). The reciprocals of these times are related to measured ventilation values in Fig. 7B. The $\Delta pH/^{\circ}C$ was found to be −0.012.

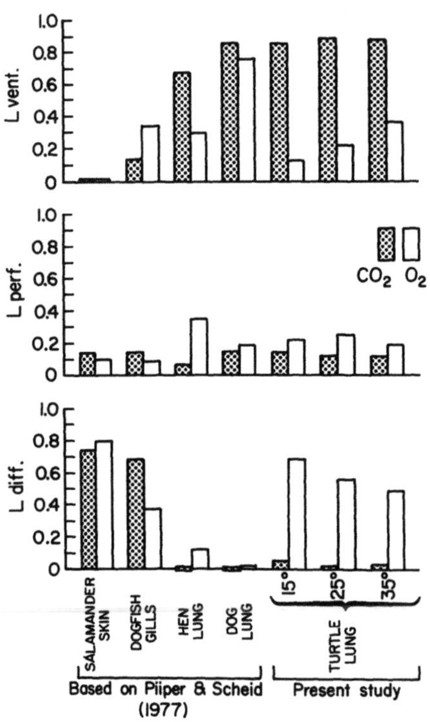

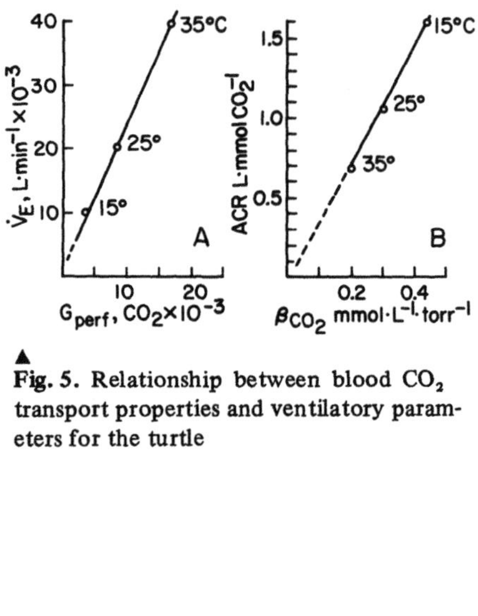

▲ **Fig. 5.** Relationship between blood CO_2 transport properties and ventilatory parameters for the turtle

◀ **Fig. 4.** Limitation indices for the processes of ventilation, perfusion, and diffusion. Note similarities among lung breathers

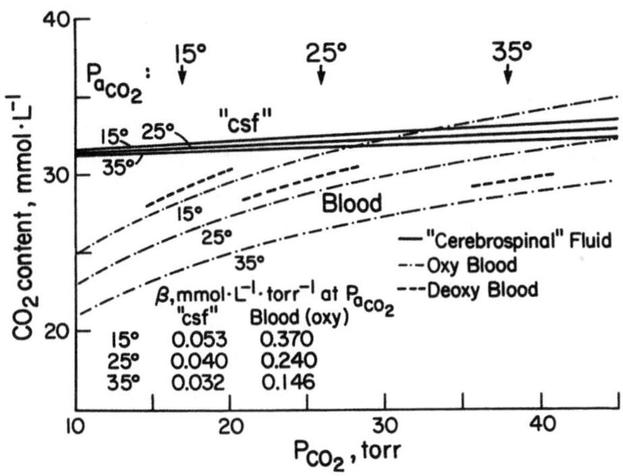

Fig. 6. CO_2 dissociation curves for turtle blood and csf (cerebrospinal fluid)

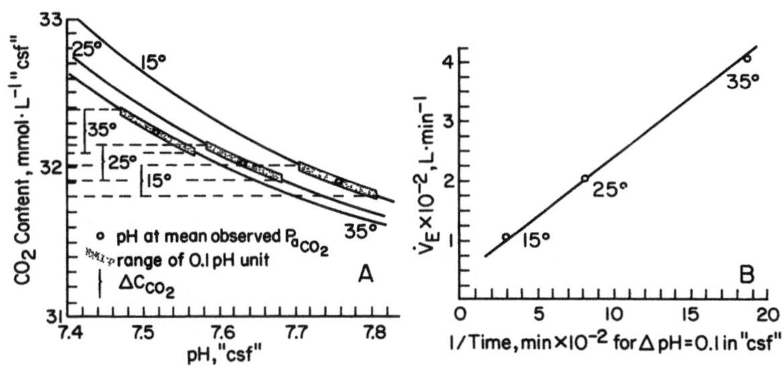

Fig. 7. A Relationship between csf pH and CO_2 content of csf. ΔC_{CO_2}'s associated with a 0.1 pH change are indicated by *brackets*. **B** Ventilation as a function of the reciprocals of ficitious times required to perturb csf pH by 0.1 U

5 Discussion

Incorporation of measurements into the model of Piiper and Scheid resulted in close agreement between experimentally and theoretically derived values. This is especially evident for the relationships between G_{vent}/G_{perf} and Δp values for the primary processes of cardiopulmonary gas exchange. For O_2, no specific G_{vent}/G_{perf} prevails over a range of BT's and the greatest relative resistance resides in the diffusive process. This is in agreement with the finding of low O_2 diffusion capacity in *P. scripta* (Crawford et al. 1976). G_{vent}/G_{perf} is low for CO_2 and stable with BT. Unlike the situation for O_2, the highest resistance for CO_2 is associated with ventilation, the lowest with diffusion. High Δp_{vent} for CO_2 also characterizes the dog as does G_{vent}/G_{perf} at values below those obtaining for O_2. So far as limitation indices for CO_2 are concerned, the turtle resembles the bird and is hardly distinguishable from the dog. Differences between fish gills and amphibian skin are largely explicable on the basis of characteristics of the diffusion barrier and the medium breathed (Piiper and Scheid 1977).

This analysis of gas exchange in the turtle demonstrates that the processes determining O_2 transport are unstable between 15° and 35 °C. Blood O_2 saturation values are inversely related to BT, a trend which is accentuated at higher temperatures. The inverse relationship between BT and G_{vent}/G_{perf} for O_2 and the temperature-dependent shift in the O_2 dissociation curve, in part, explain the decline in blood O_2 saturation.

Carbon dioxide transport properties are remarkably stable over the temperature range studied. The relative constancy of G_{vent}/G_{perf} and Δp values are essential to the regulation of the CO_2 stores of the body at a constant level, a prerequisite to effective acid-base regulation in the face of alterations in BT and metabolism. The physiochemical factors producing the inverse relationship between β_{CO_2} and BT are important determinants of G_{perf} and, thus, the overall stability of G_{vent}/G_{perf}.

The linear relationship between G_{perf} CO_2 and ventilation and that between β_{CO_2} and ACR are possibly of regulatory significance. The decline in ACR, with increasing BT, results in the primary alteration responsible for the stabilization of CO_2 content through its effect on P_{CO_2}. This establishes conditions which are prerequisite to α-stat regulation (Reeves 1972).

Central neural receptors, associated with the cerebral ventricles of intact *P. scripta*, have been verified by Hitzig and Jackson (1978). The receptors appear to be exceedingly sensitive to acid-base alterations of csf as judged by ventilatory responses. Based on response times, the receptors appear to be superficial and exposed to the immediate environment of csf. The large alterations in $\dot{V}_E/\dot{V}_{O_2}$ produced in association with a decline in csf pH of 0.33 U was similar at 20° and 30 °C, indicating that sensitivity of the receptor is temperature independent. This is in agreement with the demonstration of similar temperature-independent ventilatory responses to inspired CO_2 (Jackson et al. 1974).

When CO_2 transport properties of blood were related to those of csf at each BT, a linear relationship between the reciprocal of fictitious times to alter csf pH by at standard amount and the observed ventilation was found. It should be emphasized that the inverse relationship between β_{CO_2} and BT exerts a strong influence on perfusive transport of CO_2 in these calculations. This, and the fact that the pH–C_{CO_2} curves for csf are nonlinear and of differing slope at each BT and require differing changes in ΔC_{CO_2} for an equivalent pH shift, argues that the relationship between fictitious times and ventilation are unlikely to be fortuitous.

Figure 8 presents a hypothetical model of ventilatory control by a putative proton sensitive receptor, the immediate environment of which resembles csf. The alterations in csf acid-base status are here supposed to be influenced by blood CO_2 transport properties similar to those observed for cardiorespiratory gas exchange and by the physicochemical properties which are specific to csf. The rate of alteration between a subthreshold level of protonization of the receptor (Zsth) and its threshold (Zth) is governed by dZ/dt. G_{perf} is here viewed as the primary blood variable which governs dZ/dt, while physicochemical characteristics of csf, which vary with BT, are also interactive. Examination of the components of G_{perf} reveals that $\dot{V}_b$ is directly proportional to metabolism (Kinney et al. 1977). Yet, G_{perf} for CO_2 increases only 4.5 times in the face of an 8.7 times increase in $\dot{M}_{CO_2}$ over the range of 15°–35 °C. This disproportion-

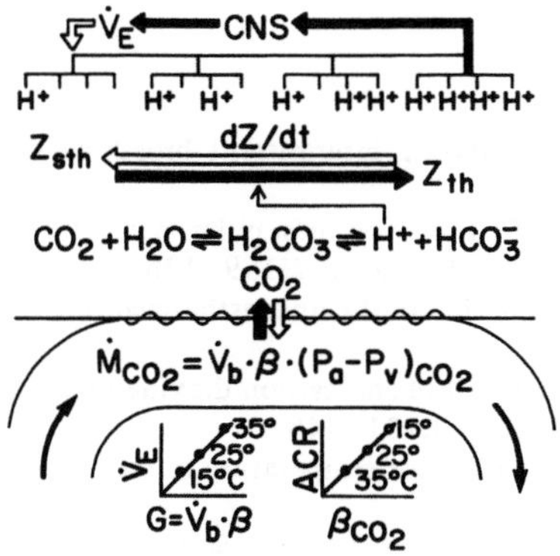

Fig. 8. Hypothetical model in which blood CO_2 transport properties are related to the behavior of proton-activated receptors which are influenced by the acid-base status of csf

ality is directly attributable to the observed alterations in β_{CO_2}. It, thus, appears that β_{CO_2} is the primary blood transport property which can account for the observed inverse relationship between ACR and BT. This interpretation provides a tentative explanation of the interactions of temperature, blood CO_2 transport properties, and central receptor control of ventilation.

Reeves (1972) proposed that the regulation, through ventilation, of constant net protein charge of the principal protein buffer, imidazole of histidine, was reason to suppose that the receptor driving ventilation may well be imidazole. The present analysis produces a $\Delta pH/^\circ C = -0.012$ for csf which is not far removed from the change in pK of imidazole required for constant net charge (α-stat regulation). This charge in pH is also similar to that observed in arterial blood of *P. scripta* ($\Delta pH/^\circ C = -0.013$) by Kinney et al. (1977).

6 Conclusion

The theoretical model of gas exchange provided by Piiper and Scheid provides a coherent reference to which experimental data may be usefully related. The present analysis for *P. scripta* finds the overall relationships among the primary processes of pulmonary gas exchange to resemble those of mammals so far as CO_2 exchange is concerned. In the turtle these processes are stabilized for CO_2 over a range of BT's, unlike the case for O_2.

While the present analysis has treated blood and air convections as equivalent, in net effect, to steady, nonpulsatile processes, derivations of the model provide tentative insights into the control of nonsteady state processes, such as ventilation. The incorporation of temperature effects in the analysis of gas exchange is useful toward the goal of understanding interactions among metabolism, ventilation, perfusion, shifting blood gas transport properties, and the control of ventilation.

Acknowledgement. Supported by National Science Foundation Grant PCM 82-04545 to F.N.W.

References

Burggren WW, Shelton G (1979) Gas exchange and transport during intermittent breathing in chelonian reptiles. J Exp Biol 82:75–92

Crawford EC, Gatz RN, Magnussen H, Perry SF, Piiper J (1976) Lung volumes, pulmonary blood flow and carbon monoxide diffusing capacity of turtles. J Comp Physiol 107:169–178

Hitzig BM, Jackson DC (1978) Central chemical control of ventilation in the unanesthetized turtle. Am J Physiol 235:R257–R264

Jackson DC, Palmer SE, Meadow WL (1974) The effects of temperature and carbon dioxide breathing on ventilation and acid-base status of turtles. Respir Physiol 20:131–146

Kinney JL, Matsuura DT, White FN (1977) Cardiorespiratory effects of temperature in the turtle. Respir Physiol 31:309–325

Maginniss LA, Sang YK, Reeves RB (1980) Oxygen equilibria of ectotherm blood containing multiple hemoglobins. Respir Physiol 42:329–343

Piiper J, Scheid P (1972) Maximum gas transfer efficacy of models for fish gills, avian lungs and mammalian lungs. Respir Physiol 14:115–124

Piiper J, Scheid P (1975) Gas transport efficacy of gills, lungs and skin: theory and experimental data. Respir Physiol 23:209–221

Piiper J, Scheid P (1977) Comparative physiology of respiration: functional analysis of gas exchange organs in vertebrates. In: Widdicombe JG (ed) International review of physiology, respiratory physiology II, Vol 14. University Park Press, Baltimore, p 219

Reeves RB (1972) An imidazole alphastat hypothesis for vertebrate acid-base regulation: tissue carbon dioxide and body temperature in bullfrogs. Respir Physiol 14:219–236

Stewart PA (1981) How to understand Acid-base, 1st edn. Elsevier, New York

Weinstein Y, Ackerman RA, White FN (in press) Influence of temperature on the CO_2 dissociation curve of the turtle, *Pseudemys scipta*. Respir Physiol

White FN, Somero G (1982) Acid-base regulation and phospholipid adaptations to temperature. Time courses and physiological significance of modifying the milieu for protein function. Physiol Rev 62:40–90

Cardiopulmonary Adaptations in Birds for Exercise at High Altitude

M.R. FEDDE[1], F.M. FARACI[2], D.L. KILGORE, Jr.[3], G.H. CARDINET, III[4], and A. CHATTERJEE[1]

1 Introduction

In general, birds are more tolerant of hypoxia than mammals. At an altitude of 6,100 m, unacclimatized house sparrows *(Passer domesticus)* are as active and alert as at sea level and can readily fly (Tucker 1968a). On the other hand, unacclimatized mice in a hypobaric chamber at the equivalent of 6,100 m are comatose, have difficulty moving, and show a 10 °C drop in their body temperature. Even at 3,700 m, mice are lethargic, walk slowly, or do not move at all around the chamber. Resting and unacclimatized men taken to 6,100 m in a hypobaric chamber are in a state of hypoxic collapse after only 10 min and even after acclimatization to this altitude can only do heavy work for 5 min (Tucker 1972).

Certain species of birds are exceptionally resistant to hypoxia, having the ability to exercise at extremely high altitude. It is not uncommon for birds to fly at altitudes above 6,000 m (Lack 1960). Several species that have been identified flying at these elevations are listed in Table 1.

Table 1. Avian species reported to fly at extreme altitude (> 6,000 m)

Species	Common name	Altitude (m)	Reference
Anas platyrhynchos	Mallard	6,401	Manville 1963
Anser indicus	Bar-headed goose	>8,848	Swan 1961, 1970
Aquila rapax	Steppe eagle	7,500	Brown 1976
Grus grus	Common crane	6,096, >7,620	Meinertzhagen 1955; Walkinshaw 1973
Grus leucogeranus	Siberian white crane	>6,096	Walkinshaw 1973
Gypaetus barbatus	Lammergeier	7,620	Meinertzhagen 1955
Gyps rueppellii	Ruppell's griffon	11,278	Laybourne 1974
Pyrrhocorax graculus	Cough	8,230	Swan 1961
Tichodroma muraria	Wall creeper	6,400	Meinertzhagen 1955
Vultur gryphus	Andes condor	6,035	Meinertzhagen 1955

1 Department of Anatomy and Physiology, Kansas State University, Manhattan, KS 66506, USA
2 Department of Internal Medicine and Cardiovascular Center, 200 Medical Laboratories, University of Iowa College of Medicine, Iowa City, IA 52242, USA
3 Department of Zoology, University of Montana, Missoula, MT 59812, USA
4 Dean's Office, College of Veterinary Medicine, University of California, Davis, CA 95616, USA

Circulation, Respiration, and Metabolism
(ed. by R. Gilles)

The species that has attracted recent attention of physiologists is the bar-headed goose *(Anser indicus)*. These birds fly over the Himalayan mountains during their normal migration between Tibet and India and have been sighted flying directly over the summit of Mount Everest (8,848 m) (Swan 1961, 1970). Beginning their northward flight from near sea level in India, bar-headed geese may reach altitudes of 9,000 m in less than 1 day, allowing essentially no time for acclimatization. Bar-headed geese remain conscious and are able to stand in a hypobaric chamber at simulated altitudes of 12,190 m (Black and Tenney 1980a).

Migratory flights often are prolonged over many hours or days without rest, food, or water (Tucker 1968b, 1974). Because of the great length of many migrations, the birds cannot incur accumulative oxygen debts; they must maintain aerobic metabolism. This means that an adequate O_2 delivery system must be present to supply the working muscles and other tissues with sufficient O_2 to produce the required ATP from mitochondrial sources. Such a delivery system must have unique characteristics, especially in those birds that can fly at high altitudes, because the oxygen consumption in nonpasserine birds increases during flapping flight by at least elevenfold over rest (Butler 1982). That value is over twice that incurred during maximal running exercise in mammals of similar mass.

The specific components of the O_2 delivery system that must be examined in analyzing the ability of birds to fly at extreme altitude are illustrated in Fig. 1. Move-

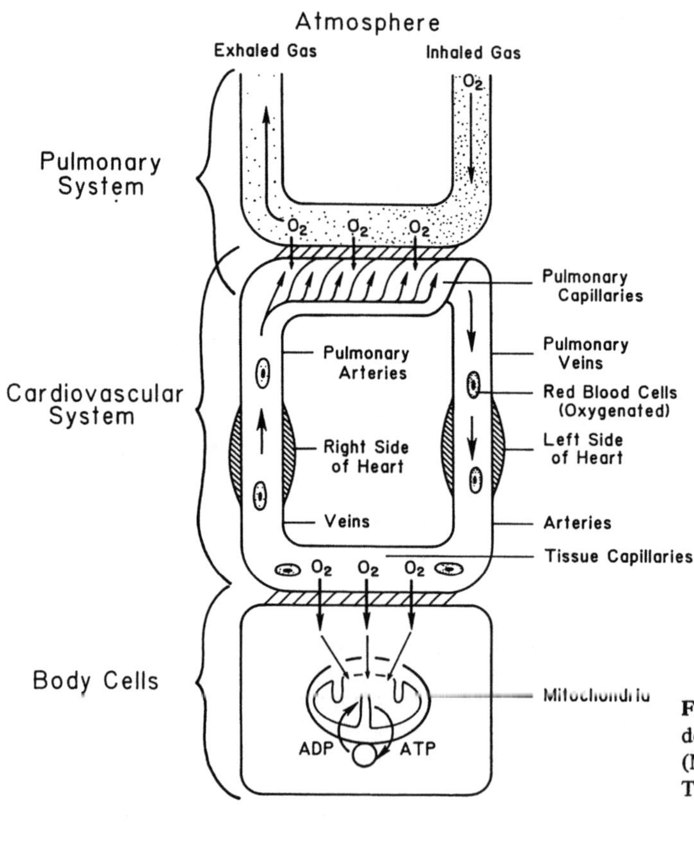

Fig. 1. Model of the oxygen delivery system in birds. (Modified from Fig. 1 of Taylor and Weibel 1981)

ment of O_2 from the atmosphere to the gas exchange surfaces of the lung constitute the first step in the O_2 transport chain. Oxygen then must move across the barrier in the lung and combine with hemoglobin in the pulmonary capillary blood. Sufficient cardiac output and flow of blood to all organs, especially skeletal muscle, brain, and heart, must occur to adequately distribute the O_2 required for metabolic purposes. Finally, transport of O_2 from the tissue capillaries to the mitochondria in proportion to the use by specific cells is necessary to allow the production of ATP required for the activity of the cells.

In the following discussion, each of the major components of the O_2 delivery system will be considered, with emphasis on the specializations that allow the bar-headed goose to achieve its amazing ability to fly at high altitude.

2 Pulmonary Specialization

2.1 Cross-Current Gas Exchange System

As the bird breathes, gas moves by convective flow *through* the tertiary bronchi (parabronchi) of the lungs on its way to and from the air sacs. Extending radially from the lumen of each parabronchus are evaginations, which ultimately lead to cylindrical, tortuous, thin-walled air capillaries ranging from 3 to 10 μm in diameter in various birds. Oxygen moves by diffusion into the air capillaries from the parabronchial lumen (carbon dioxide moves in the opposite direction). Blood capillaries, emanating from intraparabronchial arterioles, intertwine with the air capillaries forming a large surface area for gas exchange. Because mixed venous blood, returning to the lung, enters a parabronchus all along its length at a right angle to the convective flow of gas through the parabronchus, a cross-current gas exchange system exists. Such a gas exchange system has been modeled and extensively studied both theoretically and experimentally and has been found to have a higher efficacy for gas exchange than that occurring in the mammalian lung (Piiper and Scheid 1975). The avian cross-current gas exchanger permits the partial pressure of CO_2 in blood leaving the lung ($PaCO_2$) to be lower, and the partial pressure of O_2 (PaO_2) to be higher, than the respective end-expired partial pressures. This efficient gas exchange system is characteristic of all birds and is not unique to those capable of high altitude flight. However, it may constitute one important reason for the general tolerance of birds to hypoxia compared to mammals.

2.2 Extreme Hyperventilation and Alkalosis in Response to Hypoxia

Birds and mammals exposed to extreme hypoxia markedly increase their ventilation. When at rest on the summit of Mount Everest [moist inspired PO_2 (P_IO_2) = 43 torr], the severe hyperpnea in man causes an alveolar CO_2 partial pressure of 7.5 torr (West et al. 1983). When Pekin ducks and bar-headed geese are exposed to hypoxia even more extreme than that found at the summit of Mount Everest, they also hyperventilate, with $PaCO_2$ becoming as low as 6.2 torr (Table 2). As shown later, such increased ventilation may be extremely beneficial for both gas exchange and blood gas transport.

Table 2. Hyperventilation and alkalosis during hypoxia[a]

Variable	Pekin duck		Bar-headed goose		
P_IO_2 (torr)	141	34	142	34	28.5
$PaCO_2$ (torr)	31.3	6.4	21.6	7.5	6.2
pHa	7.466	7.543	7.548	7.696	7.654

[a] P_IO_2, partial pressure of oxygen in the inspired gas when completely humidified at body temperature; $PaCO_2$, partial pressure of carbon dioxide in arterial blood; pHa, pH of arterial blood. Data from Faraci et al. (1984a)

2.3 Transformation to Nearly an Ideal Lung During Hypoxia

Like mammals, resting birds exhibit a PaO_2 that is substantially lower than the P_IO_2. The difference is attributed to lowered intrapulmonary PO_2 because of addition of CO_2 to the intrapulmonary gas, ventilation-perfusion inequality in the lung, possible diffusion limitation by the gas-blood barrier, and possible limitation of the reaction of hemoglobin with oxygen in the erythrocyte. As illustrated in Fig. 2, the difference between the P_IO_2 and the PaO_2 becomes very small in bar-headed geese when they breathe a hypoxic gas mixture (the most hypoxic point on the graph represents a fractional inspired O_2 concentration of 0.04). At a P_IO_2 of 28.5 torr, the PaO_2 is 25.4 torr (a difference of only 3.1 torr). Similar results have been reported by Black and Tenney (1980a). This finding indicates that with the hyperventilation accompanying this degree of hypoxia, the gas exchange surfaces in the lung are almost exposed to the moist inspired gas. Furthermore, there appears to be essentially no diffusion limitation at the gas-blood barrier, no limitation of the chemical reaction rate of hemoglobin with oxygen, and almost a perfect balance of ventilation to perfusion in the lung.

The mechanisms responsible for the improvement in gas exchange efficiency with hypoxia remain to be explained. Furthermore, it remains to be determined if this degree of efficiency occurs during exercise under hypoxic conditions. However, such improvement would prevent the pulmonary system from being the limiting factor in determining the upper threshold for high altitude flight in these birds.

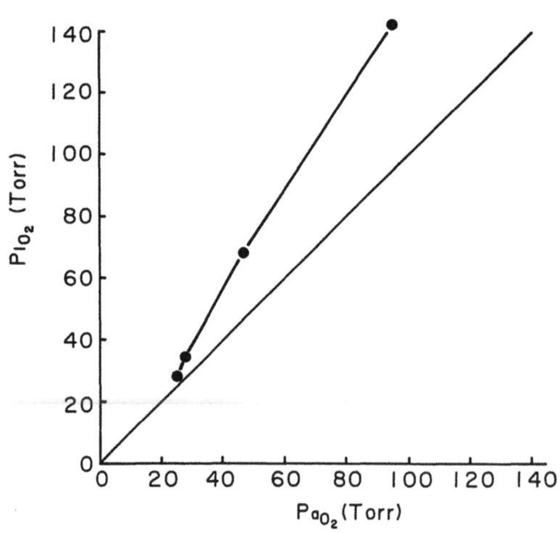

Fig. 2. Relationship between moist inspired PO_2 (P_IO_2) and arterial PO_2 (PaO_2) as bar-headed geese become progressively more hypoxic. *Solid line* is the line of identity. (Data from Faraci et al. 1984a)

3 Cardiovascular Specialization

3.1 Heart Rate and Cardiac Output Reserve During Hypoxia

Resting bar-headed geese have the potential to increase their cardiac output sevenfold when exposed to a P_IO_2 of 27 torr (Black and Tenney 1980a). However, at a P_IO_2 of 37 torr, cardiac output is not increased over that at a P_IO_2 of 146 torr. By contrast, the cardiac output in Pekin ducks is increased more than twofold at a P_IO_2 of 37 torr, but only fourfold at a P_IO_2 of 27 torr. The increase in cardiac output is probably due principally to the increases in heart rate (B.R. Grubb 1982) that occur with hypoxia. As seen in Table 3, heart rate in Pekin ducks substantially increases at a P_IO_2 of 34 torr, but heart rate in bar-headed geese is no higher than when these birds were exposed to a P_IO_2 of 142 torr. Lowering the P_IO_2 further, however, causes a large rise in heart rate in the geese, consistent with the increase in cardiac output found by Black and Tenney (1980a). Bar-headed geese apparently have a much greater cardiac output reserve that can be used during exercise under hypoxic conditions than Pekin ducks.

Mean systemic arterial blood pressure is not greatly altered by hypoxia at these levels in either species, indicating that the vascular pressure control system is not disrupted.

Table 3. Heart rate and blood pressure during hypoxia[a]

Variable	Pekin duck		Bar-headed goose		
P_IO_2	141	34	142	34	28.5
Heart rate (beats min^{-1})	181	268	213	216	297
Mean arterial blood pressure (torr)	148	159	153	161	142

[a] Data from Faraci et al. (1984a)

3.2 Organ Blood Flow During Hypocapnic Hypoxia

Large increases in blood flow to the heart, brain, and flight muscles are necessary to sustain aerobic metabolism during flying, especially under hypoxic conditions. Grubb et al. (1978) demonstrated that blood flow to the brain in Pekin ducks increased about six times when arterial PO_2 was reduced from about 91 to 30 torr. Similar results have been obtained by Faraci et al. (1984a). However, during similar degrees of hypoxia, cerebral blood flow increases only about threefold in bar-headed geese. Coronary blood flow during hypoxia is also much higher in the ducks than in the geese at comparable degrees of hypoxia. Thus, the geese apparently have a large reserve in cerebral and coronary blood flow that could be used during exercise.

The large blood flow increase to the brain in these birds during hypoxia is even more remarkable in the presence of the severe hyperventilation and alkalosis. In mammals, the increase in cerebral blood flow due to hypoxia may be abolished if the degree of hypocapnia is sufficient (Davies et al. 1983; Heistad and Kontos 1983). However, in all avian species thus far tested (ducks, chickens, bar-headed geese); decreasing the $PaCO_2$ from 30 to as low as 7 torr in normoxic birds causes no change in cerebral blood flow (Grubb et al. 1977; Wolfenson et al. 1982; Faraci 1984). Furthermore, severe hypocapnia does not affect coronary or skeletal muscle blood flow in bar-headed geese (Faraci 1984). The lack of a blood flow reduction during hypocapnia may represent another adaptation enhancing the ability of these birds to fly at high altitude.

3.3 Minimization of Pulmonary Hypertension During Hypocapnic Hypoxia

Another feature potentially enhancing the performance of the cardiovascular system during hypoxia in bar-headed geese is the attenuated pulmonary arterial hypertension (Black and Tenney 1980b; Faraci et al. 1984b). In geese, mean pulmonary arterial blood pressure remains unchanged when PaO_2 is reduced from 95 to 46 torr and rises only 3 torr when PaO_2 is reduced further to 28 torr. A PaO_2 decrease to 29 torr in Pekin ducks results in an 11 torr rise in pulmonary arterial pressure. Thus, in the geese, right ventricular work is not increased as a result of pulmonary arterial hypertension and thereby the oxygen consumption of the right ventricular myocardium is likely to be lower than in those animals with a strong pulmonary pressor response to hypoxia.

3.4 Distribution of Organ Blood Flow During Hypocapnic Hypoxia

Although birds that fly at high altitude probably do not remain there for prolonged periods, it is likely that they can sustain activity at these elevations for several hours without interruption. To prevent cellular damage, there must be an adequate oxygen supply to all organs, even those not directly involved with flying. Thus, marked reduction or cessation of blood flow to some organs for several hours duration with redistribution to more essential organs could be detrimental. Faraci (1984) measured regional blood flow to a variety of organs in resting, unanesthetized Pekin ducks and bar-headed geese during normoxia and during moderate and severe hypoxia (PaO_2 of 28–29 torr). Severe hypocapnic hypoxia produced a change in the pattern of blood flow in the ducks with blood flow to adrenal glands and eyes increasing and flow to the liver, spleen, small intestine, and shell gland decreasing below normoxia values. Blood flow to these organs was either unchanged from normoxia or was increased during hypoxia in bar-headed geese. Thus, in addition to increasing blood flow to the brain, heart, and flight muscles during hypoxia, flow to other organs in the geese is maintained at levels necessary to provide at least minimal O_2 requirements.

3.5 Hemoconcentration During Hypoxia

Most animals produce more red blood cells, and hence more hemoglobin, when chronically exposed to hypoxia. Although this results in an increase in the O_2-carrying capacity of the blood, it also increases the viscosity of the blood. The blood is, therefore, more difficult for the heart to pump and the work of the heart is increased. It is not clear if the advantages of the higher O_2-carrying capacity outweigh the disadvantages to the heart. Bar-headed geese do not exhibit polycythemia when chronically exposed to an altitude of 5,640 m for 4 weeks; on the other hand, the packed cell volume in Pekin ducks increases over 20% (Table 4). Hence, at this degree of hypoxia, at least, the heart is not burdened by the additional work of pumping more viscous blood in geese.

Table 4. Changes in red blood cell numbers following high altitude acclimation (5,640 m)[a]

Variable	Pekin duck		Bar-headed goose	
	Sea level	High altitude	Sea level	High altitude
Packed-cell vol. (%)	45.4	55.9	47.8	43.9
Hemoglobin (gm%)	15.5	24.0	13.9	12.9

[a] Data from Black and Tenney (1980a)

3.6 Oxygen Transport by Hemoglobin

One of the principal features of birds adapted to high altitude appears to be an increased affinity of their hemoglobin for oxygen. Hemoglobin of bar-headed geese is one-half saturated with O_2 when the PO_2 is only 29 torr. Erythrocytes in these birds have the same concentration of organic phosphates as other birds normally residing near sea level, but the binding of the principal organic phosphate, myoinositol pentophosphate, to hemoglobin appears to be decreased, resulting in the high O_2 affinity (Petschow et al. 1977; Rollema and Bauer 1979). At a P_IO_2 of 34 torr, blood of resting bar-headed geese (hematocrit of 41.6%) has an O_2 content of 10.4 vol%, while that of Pekin ducks (hematocrit of 39.2%) has an O_2 content of only 4.1 vol% (Faraci et al. 1984a). Much more O_2 can, thus, be delivered to the tissues per milliliter of blood in the geese than is possible in the ducks. The geese, especially, are aided in this regard by the alkalosis that results from hyperventilation. The high pH in these birds further increases the affinity of hemoglobin for O_2 and at a PaO_2 of 28 torr may result in a hemoglobin saturation of some 12% higher than if the alkalosis was not present. Whether the same degree of alkalosis also occurs during exercise at high altitude remains to be determined.

3.7 Oxygen Delivery to the Tissues During Hypocapnic Hypoxia

The net result of the pulmonary and cardiovascular specialization of both Pekin ducks and bar-headed geese is an actual increase in O_2 delivery to the brain and heart during

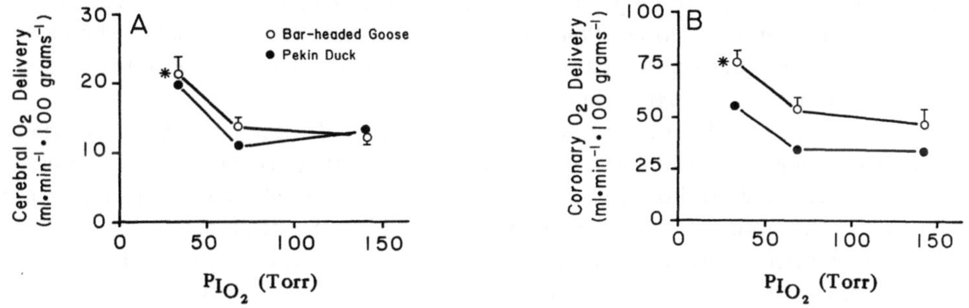

Fig. 3A,B. Cerebral and coronary oxygen delivery during increased degrees of hypoxia in Pekin ducks and bar-headed geese. Oxygen delivery was calculated as arterial O_2 content × organ blood flow. Means ± SE are shown. * Significant differences from normoxia (142 torr P_IO_2). (From Faraci et al. 1984a)

severe hypoxia as compared to normoxia (Fig. 3). The increase occurs because of different mechanisms in the two species. It is accomplished by a greatly increased blood flow at lower blood O_2 content in the ducks and by much lower blood flow at higher blood O_2 content in the geese. Hence, the geese would appear to have the advantage by possessing a substantial reserve for even higher O_2 delivery during exercise by further increases in tissue blood flow.

4 Skeletal Muscle Specialization

4.1 Capillary Density

The high rate of delivery of O_2 to the mitochondria in exercising skeletal muscle requires a large surface area for diffusion and this implies a high capillary density. It is well established that more capillaries surround those muscle fibers with highest oxidative enzyme activities (Gray and Renkin 1978; Gray et al. 1983) and that at rest, only a small fraction of the capillaries are open at any one time (29% in anterior latissimus dorsi of the chicken, a slow-tonic muscle with most fibers high in oxidative enzyme activity, and 11% in the posterior latissimus dorsi, a fast-twitch muscle with most fibers low in oxidative enzyme activity).

Although quantitative estimates of the number of capillaries surrounding fibers of various histochemical types or of the capillary density per mm^2 for the flight muscles of bar-headed geese are not available, Fig. 4 illustrates that the small fibers in the pectoralis muscle of both Pekin ducks and bar-headed geese are in contact with a substantial number of capillaries. As will be shown later, the small fibers possess high oxidative enzyme activity. It would appear that these fibers possess more capillaries per fiber than found in the anterior latissimus dorsi of the chicken (1.34 capillaries/fiber, Gray et al. 1983) and may approach or exceed the number found for the soleus of the rabbit (Gray and Renkin 1978) or guinea pig (Mai et al. 1970).

4.2 Oxidative Enzyme Activity

Most avian skeletal muscles possess fibers with multiple structural and functional characteristics (George and Berger 1966; Fedde and Cardinet 1977; Talesara and Goldspink 1978; Rosser and George 1984). The ability of these fibers to produce abundant quantities of adenosine triphosphate depends on the size and number of mitochondria and on the activity of the oxidative enzymes present therein.

The pectoralis muscle of both Pekin ducks and bar-headed geese contains two types of fibers with regard to size and oxidative enzyme activity (Fig. 5). The nicotinamide adenine dinucleotide diaphorase (NADDase) activity in the small fibers is much greater than in the large ones. In addition, there is a considerable difference in the level of activity of this oxidative enzyme in the small fibers between the two species; the activity is much greater in bar-headed geese with especially dense staining reactions in the subsarcolemmal regions of many fibers. It would appear, as expected, that this principle flight muscle of the strong flying geese should be able to remain aerobic during muscular contraction at a much higher work rate than would be the case with the nonflying Pekin ducks.

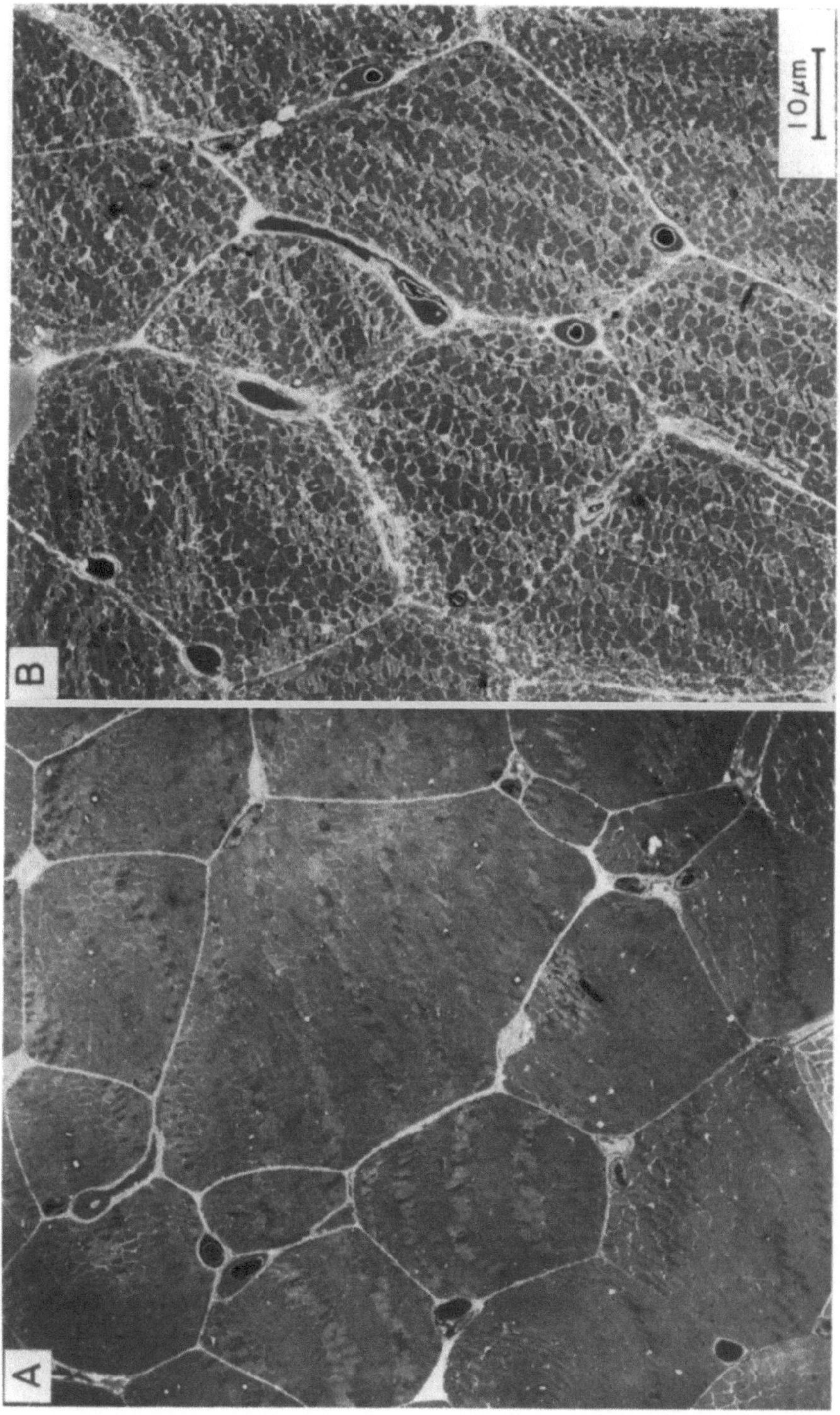

Fig. 4A,B. Low power electron micrographs illustrating the abundance of capillaries in the pectoralis muscle of both the Pekin duck (A) and the bar-headed goose (B)

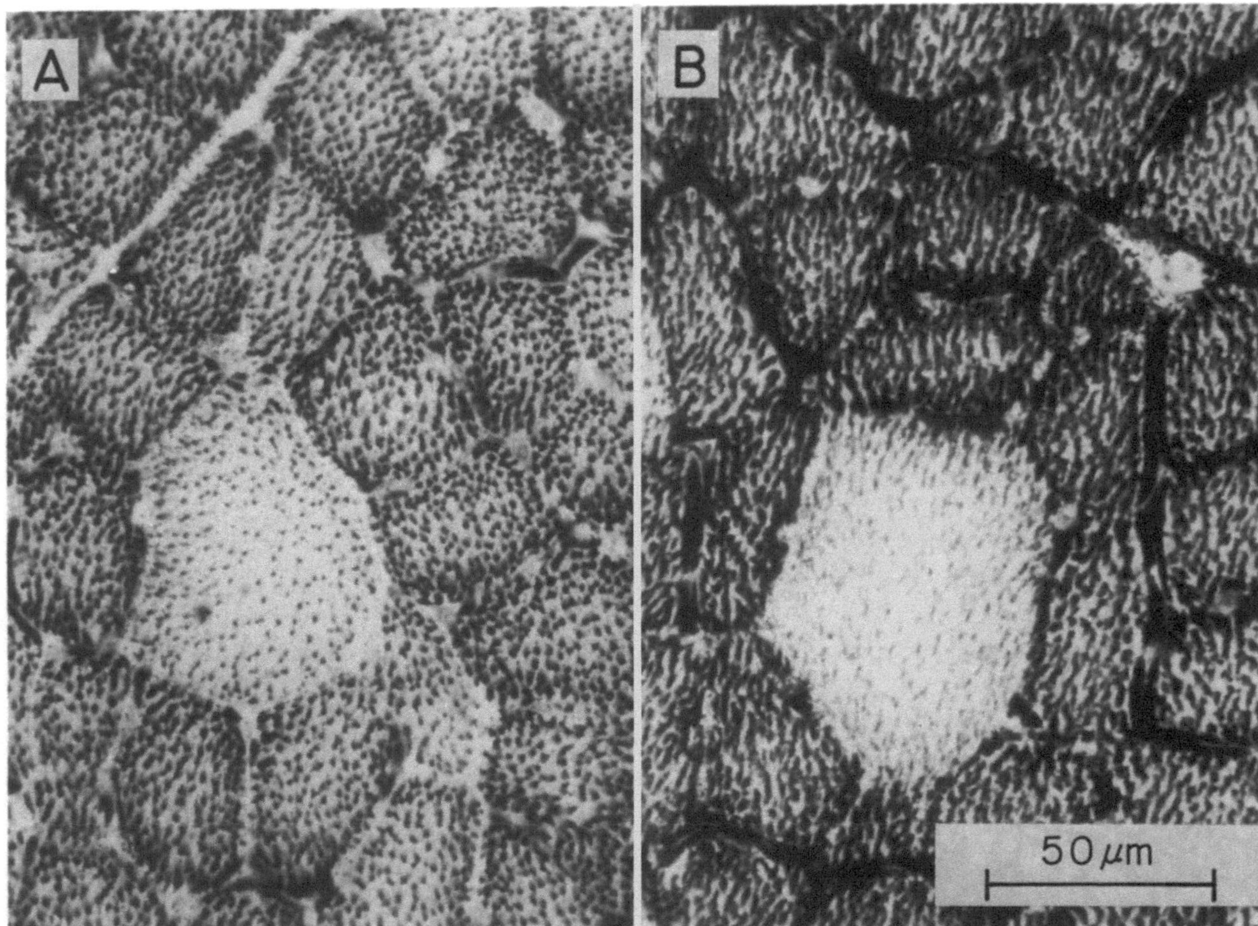

Fig. 5A,B. Histochemical demonstration of the activity of the oxidative enzyme NADDase in pectoralis muscle fibers of the Pekin duck (**A**) and bar-headed goose (**B**)

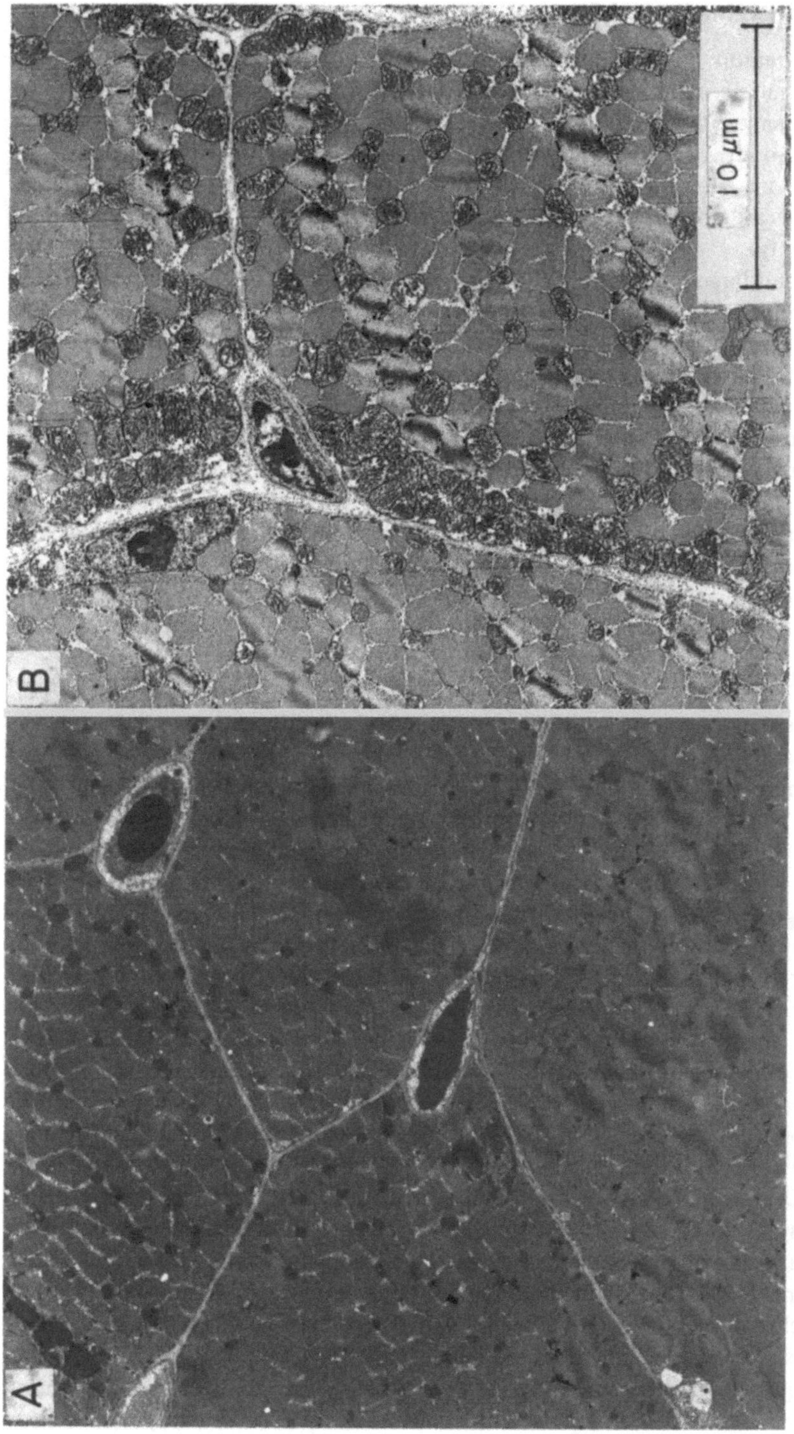

Fig. 6 A,B. Electron micrographs showing differences in number and size of mitochondria in small muscle fibers of the pectoralis muscle of the Pekin duck (A) and bar-headed goose (B)

4.3 Mitochondrial Number and Size

The definitive reason for the higher NADDase activity in the small muscle fibers in the pectoralis muscle of bar-headed geese compared to Pekin ducks can be seen from the great differences in number and size of mitochondria (Fig. 6). High concentrations of large mitochondria are present in the subsarcolemmal regions in the muscle from geese, as well as a greater number of much larger mitochondria between the myofibrils. Such differences are not present between these two species in the larger fibers with low NADDase activity.

Numerous and large mitochondria also have been reported in pectoralis muscles of the little penguin, a bird with a high wing beat frequency during swimming (Mill and Baldwin 1983). Measurements of the mitochondrial volume fractions in high (small fibers) and low (large fibers) oxidative fibers in the pectoralis muscle of pigeons are approx. 0.30 and 0.04, respectively (James and Meek 1979). The mitochondria: myofibril volume ratio for the oxidative fibers of the pectoralis muscle of quails and sparrows has been estimated to be 0.18 and 0.45, respectively (Pennycuick and Rezende 1984). Thus, great differences exist between fiber types within a species and among species and may be related to the type of flight activity. Whether chronic hypoxia and exercise increase the mitochondrial numbers in the flight muscles of bar-headed geese is currently not known.

4.4 Myoglobin Content

It is now quite certain that myoglobin in muscle fibers facilitates intracellular oxygen diffusion from the region nearest capillaries to the mitochondria (J.B. Wittenberg 1970; Wittenberg and Wittenberg 1975). Myoglobin recently has been shown to be highly mobile with an intracellular diffusion rate nearly half that in dilute solution (Livingston et al. 1983). Myoglobin is only partially oxygenated in muscle, implying a steep gradient within the fiber. Such a gradient is required for operation of any carrier-mediated oxygen transport system. Indeed, at low oxygen partial pressures, this transport system appears to contribute a major fraction of the total oxygen uptake, with the remainder being the simple diffusive flow of dissolved oxygen (Wittenberg and Wittenberg 1975).

The rate at which O_2 could move from the muscle fiber surface to the mitochondria by facilitated diffusion as oxymyoglobin would depend on the concentration of myoglobin in the specific fiber type. Myoglobin concentrations have been measured in various muscles of a variety of birds and have been summarized by Pages and Planas (1983). These concentrations vary for the pectoralis muscle from very low values for the chicken (0.3 mg g^{-1} wet weight) to extremely high values for Gentoo, Royal, and King Penguins (43 to 47 mg g^{-1} wet weight, Weber et al. 1974; Baldwin et al. 1984). The concentration of myoglobin decreases with reduced muscle use (Catlett et al. 1978; Pages and Planas 1983) and is low in young penguins before they are fully fledged and use their muscles for swimming (Weber et al. 1974). Muscles containing high myoglobin concentrations are very dark in color (and may even look almost black).

Although quantitative measurements of the myoglobin in the pectoralis muscle of bar-headed geese have not been reported, the muscle is very dark in color, even in birds that have not flown. It is likely that these muscles would have a high myoglobin concentration, especially in geese that have exercised their muscles by flying.

Myoglobin often has been considered as functioning as an oxygen storehouse to provide a continuous flow of O_2 to mitochondria during bursts of muscle activity. However, evidence has been put forth that oxymyoglobin in oxidative fibers of the pigeon pectoralis muscle is capable of storing O_2 for no longer than 120 ms of activity and, therefore, does not have an important storage function (James 1972).

5 Summary

Specialization at all levels from the pulmonary system, to the cardiovascular system, to the flight muscles appear to enable some birds to fly at extremely high altitudes. Details about the exact physiological mechanisms operating in birds at each site in the transfer of O_2 from the atmosphere to the mitochondria are only beginning to be understood. Comparison of these mechanisms with those existing in mammals should provide insight into improved ways to extend mammalian capabilities to perform work during severe hypoxia.

Acknowledgments. The authors thank Susan Dolezal for expert technical assistance. This study was supported, in part, by a grant-in-aid from the American Heart Association, Kansas Affiliate and by NIH Grant NS-10264 to Jane A. Westfall. Contribution No. 85-122-A, Department of Anatomy and Physiology, College of Veterinary Medicine, Kansas Agricultural Experiment Station, Manhattan, Kansas.

References

Baldwin J, Jardel J-P, Montague T, Tomkin R (1984) Energy metabolism in penguin swimming muscles. Mol Physiol 6:33–42

Black CP, Tenney SM (1980a) Oxygen transport during progressive hypoxia in high-altitude and sea-level waterfowl. Respir Physiol 39:217–239

Black CP, Tenney SM (1980b) Pulmonary hemodynamic responses to acute and chronic hypoxia in two waterfowl species. Comp Biochem Physiol 67A:291–293

Brown L (1976) Birds of prey, their biology and ecology. Hamlyn, London, p 63

Butler PJ (1982) Respiration during flight and diving in birds. In: Addink ADF, Sprank N (eds) Exogenous and endogenous influences on metabolic and neural control, vol I. Pergamon, Oxford, pp 103–114

Catlett RH, Walters TW, Dutro PA (1978) The effect of flying and not flying on myoglobin content of heart muscle of the pigeon *Columbia livia domestica*. Comp Biochem Physiol 59A:401–402

Davies DG, Nolan WF, Sexton J (1983) Medullary blood flow during hypocapnic hypoxia. Physiologist 26:40

Faraci FM (1984) Control of the circulation in high and low altitude adapted birds. PhD Thesis, Kansas State University, Manhattan, Kansas, USA

Faraci FM, Kilgore DL Jr, Fedde MR (1984a) Oxygen delivery to the heart and brain during hypoxia: Pekin duck vs bar-headed goose. Am J Physiol 247 (Regulatory Integrative Comp Physiol 16): R69–R75

Faraci FM, Kilgore DL Jr, Fedde MR (1984b) Attenuated pulmonary pressor response to hypoxia in bar-headed geese. Am J Physiol 247 (Regulatory Integrative Comp Physiol 16):R402–R403

Fedde MR, Cardinet GH III (1977) Histochemical studies of respiratory muscles of chicken. Am J Vet Res 38:585–589

George JC, Berger AJ (1966) Avian myology. Academic, New York

Gray SD, Renkin EM (1978) Microvascular supply in relation to fiber metabolic type in mixed skeletal muscles of rabbits. Microvasc Res 16:406–425

Gray SD, McDonagh PF, Gore RW (1983) Comparison of functional and total capillary densities in fast and slow muscles of the chicken. Pflügers Arch 397:209–213

Grubb BR (1982) Cardiac output and stroke volume in exercising ducks and pigeons. J Appl Physiol: Respir Environ Exercise Physiol 53:207–211

Grubb B, Mills CD, Colacino JM, Schmidt-Nielsen K (1977) Effect of arterial carbon dioxide on cerebral blood flow in ducks. Am J Physiol: Heart Circ Physiol 1:H596–H601

Grubb B, Colacino JM, Schmidt-Nielsen K (1978) Cerebral blood flow in birds: effect of hypoxia. Am J Physiol 234 (Heart Circ Physiol 3: H230–H234

Heistad DD, Kontos HA (1983) Cerebral circulation. In: Shepherd JT, Abboud FM (eds) The cardiovascular system. Peripheral circulation and organ blood flow. American Physiological Society, Bethesda, Maryland, pp 137–182 (Handbook of physiology, sect 2, vol III, part 1, chapt 5)

James NT (1972) A study of the concentration and function of mammalian and avian myoglobin in Type I skeletal muscle fibres. Comp Biochem Physiol 41B:457–460

James NT, Meek GA (1979) Stereological analysis of the structure of mitochondria in pigeon skeletal muscle. Cell Tissue Res 202:493–503

Lack D (1960) The height of bird migration. Br Birds 53:5–10

Laybourne RC (1974) Collision between a vulture and an aircraft at an altitude of 37,000 feet. Wilson Bull 86:461–462

Livingston DL, La Mar GN, Brown WD (1983) Myoglobin diffusion in bovine heart muscle. Science 220:71–73

Mai JV, Edgerton VR, Barnard RJ (1970) Capillarity of red, white and intermediate muscle fibers in trained and untrained guinea-pigs. Experientia 26:1222–1223

Manville RH (1963) Altitude record for mallard. Wilson Bull 75:92

Meinertzhagen R (1955) The speed and altitude of bird flight. Ibis 97:81–117

Mill GK, Baldwin J (1983) Biochemical correlates of swimming and diving behavior in the little penguin *Eudyptula minor*. Physiol Zool 56:242–254

Pages T, Planas J (1983) Muscle myoglobin and flying habits in birds. Comp Biochem Physiol 74A: 289–294

Pennycuick CJ, Rezende MA (1984) The specific power output of aerobic muscle, related to the power density of mitochondria. J Exp Biol 108:377–392

Petschow D, Würdinger I, Baumann R, Duhm J, Braunitzer G, Bauer C (1977) Causes of high blood O_2 affinity of animals living at high altitude. J Appl Physiol:Respir Environ Exercise Physiol 42:139–143

Piiper J, Scheid P (1975) Gas transport efficacy of gills, lungs and skin: Theory and experimental data. Respir Physiol 23:209–221

Rollema HS, Bauer C (1979) The interaction of inositol pentaphosphate with the hemoglobins of highland and lowland geese. J Biol Chem 254:12038–12043

Rosser BWC, George JC (1984) Some histochemical properties of the fiber types in the pectoralis muscle of an emu *(Dromaius novaehollandiae)*. Anat Rec 209:301–305

Swan LW (1961) The ecology of the high Himalayas. Sci Am 205:68–78

Swan LW (1970) Goose of the Himalayas. Nat Hist 79:68–75

Talesara GL, Goldspink G (1978) A combined histochemical and biochemical study of myofibrillar ATPase in pectoral, leg and cardiac muscle of several species of bird. Histochem J 10:695–710

Taylor CR, Weibel ER (1981) Design of the mammalian respiratory system. I. Problem and strategy. Respir Physiol 44:1–10

Tucker VA (1968a) Respiratory physiology of house sparrows in relation to high-altitude flight. J Exp Biol 48:55–66

Tucker VA (1968b) Respiratory exchange and evaporative water loss in the flying budgerigar. J Exp Biol 48:67–87

Tucker VA (1972) Respiration during flight in birds. Respir Physiol 14:75–82
Tucker VA (1974) Energetics of natural avian flight. In: Paynter RA Jr (ed) Avian energetics. Nuttall Ornithological Club, Cambridge, Mass., pp 298–328
Walkinshaw L (1973) Cranes of the world. Winchester, New York, p 4, 47
Weber RE, Hemmingsen EA, Johansen K (1974) Functional and biochemical studies of penguin myoglobin. Comp Biochem Physiol 49B:197–214
West JB, Hackett PH, Maret KH, Milledge JS, Peters RM JR, Pizzo CJ, Winslow RM (1983) Pulmonary gas exchange on the summit of Mount Everest. J Appl Physiol:Respir Environ Exercise Physiol 55:678–687
Wittenberg BA, Wittenberg JB (1975) Role of myoglobin in the oxygen supply to red skeletal muscle. J Biol Chem 250:9038–9043
Wittenberg JB (1970) Myoglobin-facilitated oxygen diffusion: Role of myoglobin in oxygen entry into muscle. Physiol Rev 50:559–636
Wolfenson D, Frei YF, Berman A (1982) Blood flow distribution during artificially induced respiratory hypocapnic alkalosis in the fowl. Respir Physiol 50:87–92

Symposium III

The Biochemistry of Exercise: Insights from Comparative Studies

Organizer P.W. HOCHACHKA

Low and High Power Output Modes of Anaerobic Metabolism: Invertebrate and Vertebrate Strategies

A. DE ZWAAN[1] and G. v.d. THILLART[2]

1 Introduction

During muscular contraction ATP is hydrolysed. The cell does not store large amounts of ATP and this compound has, therefore, to be almost simultaneously resynthesised at about the same rate. Depending on the type and function of the muscle, the driving force is mainly delivered by either aerobic or anaerobic processes. In vertebrates three primary types of muscle are found: cardiac muscle, smooth muscle and skeletal muscle. Smooth muscle which lack a regular pattern of banding contract slowly. Vertebrate skeletal and heart muscle show cross-striation due to the orderly arrangement of the myofibrils within the fibers. The same muscle types are also common in invertebrates. Based on differences in the pattern of innervation muscle fibres are classified as tonic (slow) fibres or twitch fibres. Tonic fibres receive multiple innervation and their contractile response is confined to the immediate region of the nerve-muscle junction. Twitch fibres receive only a single motor nerve terminal on their surface and its impulse causes contraction throughout the whole length. On the basis of the contraction time two types of muscles can be distinguished: the fast twitch and the slow twitch muscles. The latter are red, while the fast twitch muscles are often pale in colour. Red muscles have a high myoglobin and mitochondria content and are particularly suited for sustained aerobic activity. Red muscle can be either of the slow and of the fast twitch type. The slow type is almost exclusively aerobic, while the fast red muscle has also a high anaerobic capacity and is, therefore, called intermediate type. Fast twitch white fibres are highly anaerobic and have together with the intermediate fibre type a high glycolytic and ATPase activity and considerable amounts of phosphagen and glycogen. The fast twitch fibres are primarily suited for burst activity which can be maintained only for a short time due to endogenous substrate depletion and the accumulation of lactate (and/or other acids). A specific muscle may contain different fibre types depending on the work they perform. Uniform fibres within one muscle are often located within distinct layers or bundles. Examples of invertebrates are squid mantle muscle (Bone et al. 1981), lobster claw closer muscles (Costello and Govind 1983) and adductor muscle of bivalves. In scallops the large central adductor muscle is divided by

1 Laboratory of Chemical Animal Physiology, State University of Utrecht, Padualaan 8, 3508 TB Utrecht, The Netherlands

2 Department of Animal Physiology, Gorlaeus Laboratories, University of Leiden, P.O. Box 9502, 2300 RA Leiden, The Netherlands

Circulation, Respiration, and Metabolism
(ed. by R. Gilles)

a connective tissue sheet into two different parts, one of which is composed of smooth muscle fibres (catch part) and one of striated muscle fibres (phasic part). The smooth part is responsible for keeping the valves closed and the phasic part for rapid closures of the shell. Since the beginning of this century studies on energetic aspects of muscular exercise of vertebrates have been carried out, but the past few years a number of studies on muscle metabolism concerned also invertebrates during strenuous exercise. Metabolic events during burrowing, swimming, escape or catching prey movements or by subjecting the whole animal to electrical shocks have been followed in a number of marine gastropods, bivalves and cephalopods, in marine and freshwater annelids and in crustaceans. A problem arising from invertebrate studies is the question whether burst activity was reached or not since experimental conditions are not always well defined both with respect to work force displayed and the total time of muscular activity. During burst activity the energy needs exceed the maximal capacity of aerobic ATP yielding processes and are therefore also covered by depletion of phosphagen stores and the production of organic acids (e.g. lactate). We will call anaerobiosis caused by locomotory activity of the animal exercise anaerobiosis (in recent invertebrate literature often called functional anaerobiosis; see Zebe et al. 1981; Gäde 1983a). During burst activity the total energy output can be raised one to two orders of magnitude by anaerobic metabolism (high power output), albeit for only a few minutes.

Whole animals may be forced to survive anaerobically due to environmental change. Especially water-breathing animals may be confronted with low oxygen levels due to the low oxygen solubility of water. Small changes in temperature, total biomass or wind force can induce large changes in oxygen availability. A special case is the littoral zone where species live in burrows (annelids, crustaceans) or are attached to substratum (sea anemones, sessile bivalves). The mode of gas exchange in many intertidal animals precludes making use of environmental oxygen, whereas the interstitial water of burrows may contain very low oxygen partial pressures. For these reasons many invertebrates have to sustain regularly hypoxic or anoxic conditions during low tide. Also diving vertebrates are deprived from oxygen uptake during underwater excursion (turtles, Weddell seals, whales). Some species of Cyprinidae, including goldfish, demonstrate a remarkable tolerance to long-term anoxia. In Mid-European ponds crucian carps hibernate for several months in an environment that becomes anoxic as a consequence of ice and snow covering. The majority of vertebrate species, however, are strict aerobes and not able to survive low oxygen tensions for extended periods. Environmental hypoxia or anoxia is mostly coupled to a reduction of metabolic rate (low power output) and may be imposed for periods up to several months. We will discuss especially the power output rates during both exercise and environmental anaerobiosis as well as the sources which contribute to the energy needs. Concerning the substrates and routes of anaerobic metabolism there are a number of differences between invertebrates and vertebrates. We, therefore, have divided further discussion into two main parts, one dealing with invertebrates and one with vertebrates. Within each part exercise anaerobiosis will be discussed separately.

2 Anaerobiosis in Invertebrates

2.1 Anaerobiosis During Exercise

Anaerobic functioning excludes oxidation of NADH by oxygen. Therefore, oxidative and reductive steps have to be balanced. Anaerobic utilisation of carbohydrate can meet this condition while still generating ATP by substrate level phosphorylation. In the classical glycolysis glycogen is converted into lactate and the redox balance is maintained by a functional 1:1 coupling of the steps catalysed by GAPDH (oxidation) and LDH (reduction). All invertebrate phyla possess the lactate pathway, although in the lower phyla LDH may co-exist or be replaced by pyruvate oxidoreductases which need besides pyruvate an amino acid as co-substrate. The iminodicarboxylic acid compounds formed by reductive imination of pyruvate are collectively known as opines. To date three opines have been identified in relation to anaerobic metabolism; octopine, strombine, and alanopine. The amino acid co-substrates of pyruvate are arginine, glycine, and alanine, respectively. Opine formation is restricted to marine invertebrates. Like the vertebrates, the higher invertebrate phyla (Arthropoda, Echinodermata) do not possess opine dehydrogenase activities, whereas the annelids are marked by the virtual absence of the octopine pathway (Livingstone et al. 1983). The meaning of the presence of multiple pyruvate oxidoreductases within one tissue is not yet understood. Suggestions in the literature have recently been discussed (Fields 1983; De Zwaan and Dando 1984). Octopine is the favourite end product during escape burst activity in the pedal retractor muscle of cockles and whelks, in the adductor muscle of swimming bivalves (file shells and scallops), in the mantle muscle of cephalopods and in the body wall muscle of *Sipunculus nudus*. These muscles contain high resting levels of arginine phosphate (20–35 m*M*) and exercise anaerobiosis is for 50% to 90% driven by hydrolysis of this compound (Table 1). This leads to a large increase of arginine levels which in turn selectively stimulates octopine dehydrogenase. Phosphoarginine breakdown, therefore, generally preceeds octopine formation and once glycolysis becomes the main driving force the total power output is reduced. This is reflected by a lower work force; excape movements are, for instance, slower and less powerful. This is well demonstrated in experiments with *Argopecten irradians* (Chih and Ellington 1983) and *Nassa mutabilis* (Gäde et al. 1984) in which phosphoarginine pools were (partly) depleted before eliciting vigorous escape reactions by subjecting the animals previously to a hypoxic atmosphere. In crustacean tail muscle, which lacks the possibility of octopine formation, it was also noticed that lactate did not begin to accumulate until phosphoarginine reserves were essentially exhausted (England and Baldwin 1983). Again resting arginine phosphate levels are very high (above 20 m*M*; Table 1) and there appears to be a clear correlation between the total power output and the resting phosphagen level. In both molluscs and crustaceans displaying burst activity during escape or catching prey ATP turnover rates of 15 to 30 μmol g^{-1} wet wt^{-1} min^{-1} are reached which are high and comparable to those found in lower aquatic vertebrates and even mammalian muscle during exercise (see Tables 1 and 4). It appears that during burst activity the total energy output can be raised about 20-fold in molluscs and about 60-fold in crustaceans as compared to the aerobic resting state (Table 3). At ATP turnover rates of about 1 (many molluscs and annelids) glycolysis is often more important than hydro-

Table 1. Exercise anaerobiosis in invertebrates

Species	Tissue[e]	Conc (μmol g⁻¹)			$\dot{M}$ ATP[a]	%[b]			°C	Condition	Ref[f]
		ATP	GP[c]	ASP (MAL*)		ATP	GP	Cat[d]			
Mollusca											
Nassarius coronatus	F	3.1	6.8	–	1.4	–	36	64	23	8 s, 125 lashing mov.	1
Nassa mutabilis	F	3.2	9.3	5.7	10.6	1.5	51	47.5	16	73 s, 37 escape mov.	2
					5.7	2	26	72		132 s, 35 escape mov., previously 4 h anoxia	2
Buccinum undatum	F	2.1	8.5	–	0.76	4	32	64	18	15 min, 30 escape mov.	3
Busycon contrarium	RRM	–	10.0	15.4	0.134	–	23	77	25	KCL contractures, in vitro	4
Strombus luhwanus	F	–	11.5	–	2.93	–	41	59	23	2 min, 31 leaps	5
Cardium tuberculatum	F	3.3	24.1	5.8	5.6	0	72	28	20	53 s, 7–11 leaps	6
						1	88	11		85 s, 19–31 leaps	6
Cardium edule	F	–	7.4	–	0.77	–	78	22	13	10 min, electr. stim.	7
Lima hians	A	8.5	22.6	–	11.6	2	59	39	20	3 min, 65 valve snaps	8
Chlamys opercularis	A	5.2	20.4	–	15.42	22	78	0	18	90 s, 45–55 snaps	9
Pecten jacobaeus	A	7.5	12.6	–	2.38	13	57	30	–	5 min, 23 snaps in SW	10
					6.50	7	38	55		4 min, 42 snaps in air	
Argopecten irradians	A	6.7	20.20	1.48	9.66	7	50	43	24	3.5 min, 54 snaps	11
Hapalochlaena maculosa	M	4.8	14.9	–	6.39	14	31	56	23	2.5 min, 75 swimming mov.	12
	TN	5.1	26.8	–	15.97	9	44	47		2.5 min, 75 swimming mov.	
Annelida											
Nereis diversicolor	WA	1.95	–	3.78	0.63	–	–	<100	12	15 min, electr. stim.	13
Nereis pelagica	WA	2.25	–	1.54	0.62	–	–	<100	12	15 min, electr. stim.	13
Nereis vireus	WA	2.03	–	3.36	0.43	–	–	<100	12	14 min, electr. stim.	13
Hirudo medicinalis	BWM	0.92	–	7.86*	0.31	3	–	< 97	20	20 min, swimming	14
Arthropoda											
Cherax destructor	T	7.43	24.9	–	25.53	6	48	46	25	2 min, tail flips	15
					32.70	9	72	18		1 min, tail flips	15
					62.55	8	92	0		20 s, tail flips	15

Table 1 (continued)

Species	Tissue[e]	Conc (µmol g^{-1})			$\dot{M}$ ATP[a]	%[b]			°C	Condition	Ref[f]
		ATP	GP[c]	ASP (MAL*)		ATP	GP	Cat[d]			
Crangon crangon	T	4.08	20.7	0.86	62.28	2	98	0	8	10 s, electr. stim.	16
	T				24.31	7	88	5	8	50 s, escape swimming	
	T				10.77	9	86	5	8	120 s, escape swimming	17
Orconectes limosus	T	4.09	20.8	–	16.90	0.5	90	9.5	14	59 s, tail flips	

a $\dot{M}$ATP, ATP turnover rate in µmol ATP g^{-1} wet wt^{-1} min^{-1}
b Percentual contribution to ATP turnover rate
c GP, guanidino ~P (arginine phosphate)
d Cat (anaerobic) catabolism
e Tissue abbreviations: F, foot; RRM, radula retractor muscle; A, adductor muscle; M, mantle; TN tentacle; WA, whole animal; BWM, body wall muscle; T, tail muscle
f References: 1. Baldwin et al. 1981; 2. Gäde et al. 1984; 3. Koormann and Grieshaber 1980; 4. Ellington 1982b; 5. Baldwin and England 1982; 6. Gäde 1980. 7. Meinardus and Gäde 1981; 8. Gäde 1981; 9. Grieshaber 1978; 10. Grieshaber and Gäde 1977; 11. Chih and Ellington 1983; 12. Baldwin and England 1980; 13. Schöttler 1979; 14. Zebe et al. 1981; 15. England and Baldwin 1983; 16. Onnen and Zebe 1983; 17. Gäde 1984

Table 2. Environmental anaerobiosis in invertebrates[a]

Species	Tissue	Conc (µmol g^{-1})			$\dot{M}$ ATP	%			°C	Condition	Ref[b]
		ATP	GP	ASP (MAL*)		ATP	GP	Cat			
Mollusca											
Nassa mutabilis	F	3.2	9.3	5.7	0.036	0	48	52	16	4 h anoxia	2
Busycon contrarium	RRM	–	10.0	15.4	0.048	–	0	100	25	3 h anoxia, in vitro	3
Cardium tuberculatum	F	3.3	24.1	5.8	0.030	0	67	23	20	10 h anoxia	4
Cardium edule	F	3.3	9.9	3.0	0.035	–	46	54	13	4 h anoxia	5
					0.005		6	94		24–48 h anoxia	

Lima hians	A	6.6	21.9	2.9	0.121	4		65	31	2.5 h anoxia	6
					0.021	14	24	52		2.5–15 h anoxia	
Argopecten irradians	A	5.7	20.20	0.95	0.045	4	80	7	24	4 h anoxia	7
Crassostrea virginica	V	–	–	6.7	0.073	–	–	<100	22.5	90' anoxia, in vitro	8
Marisa cornaurietis	F	–	–	0.3	0.017	–	–	<100	20	17 h anoxia	9
Annelida											
Nereis diversicolor	WA	1.95	–	3.78	0.042	1	–	< 99	12	24 h anoxia	10
Nereis pelagica	WA	2.25	–	1.54	0.020	3	–	< 97	12	24 h anoxia	10
Nereis vireus	WA	2.03	–	3.36	0.028	1	–	< 99	12	24 h anoxia	10
Hirudo medicinalis	BWM	0.75	–	7.76*	0.030	0	–	<100	20	48 h anoxia	11
Arthropoda											
Artemia	WA	–	–	0.11	0.150	–	–	<100	25	2 h anoxia	12
Callianassa californiensis	WA	–	–	0.87	0.020	–	–	<100	12	24 h anoxia	13
Upogebia pugettensis	WA	–	–	0.49	0.040	–	–	<100	12	24 h anoxia	13
Orconectes limosus	T	4.1	20.8	–	0.036	0	27	73	14	16 h anoxia	14
Chaoborus crystallinus	WA	1.65	3.68	15.54*	0.035	1	4	95	14	24 h anoxia	15

[a] For notes and abbreviations see Table 1

[b] References: 2. Gäde et al. 1984; 3. Ellington 1982b; 4. Gäde 1980 and 1983; 5. Meinardus and Gäde 1981; 6. Gäde 1983b; 7. Chih and Ellington 1983; 8. Foreman and Ellington 1983; 9. De Zwaan et al. quoted in Livingstone and De Zwaan 1983; 10. Schöttler 1979; 11. Zebe et al. 1981; 12. Bernaerts 1982; 13. Zebe 1982; 14. Gäde 1984; 15. Englisch et al. 1982

Table 3. Anaerobiosis in invertebrates; comparison of metabolic rates with resting rates[a]

Species	Tissue	Conc (μmol g^{-1})			$\dot{M}$ ATP (μmol min^{-1} g^{-1})			%			°C	Condition	Ref[c]
		ATP	GP	ASP(MAL)*	Anaerobic	Resting[b]	A/R	ATP	GP	Cat			
A. EXERCISE													
Mollusca													
Placopecten magellanicus	PA	8.8	22.3	0.5	12.65	0.430[(a)]	29.42	12.5	69	18.5	5	2 min, 30 snaps	1
Limaria fragilis	CA	3.4	5.1	1.5	5.07	0.430	11.79	10	30	60	5	2 min, 30 snaps	2
Sipunculida													
Sipunculus nudus	BWM	2.42	30.5	1.35	0.68	0.084	8.09	0	43	57	15	45 s, 15 digging cycl	3
	IRM	2.29	12.2	1.66	0.24	0.076[(b)]	3.16	0	46	54	15	45 s, 15 digging cycl	3
Annelida													
Arenicola marina	BWM	3.0	8.86	14.44	2.93	0.276[(c)]	10.62	5	20	75	12–16	10 min, 4 digging cycl	4
Lumbriculus variegatus	WA				3.43	0.55[(d)]	6.24	6	0	94	20	2.5 min (anoxic) electr. stim.	5
B. ENVIRONMENTAL													
Cnidaria													
Bunodosoma cavernata		0.7	–	1.4	0.03	0.071	0.04	–	–	<100	22	96 h anoxia	6
Mollsuca													
Mytilus edulis	A	27	20–8.8	4–14	0.100	0.339[(c)]	0.29	2	27	71	12	2 h anoxia, in vitro	7
					0.072	0.339	0.21	–	–	–	12	6 h forced value closure	8
					0.013	0.339	0.038	–	–	–	12	0–12 h anoxia	8
					0.008	0.339	0.023	0	0	100	12	20–24 h anoxia	9

Placopecten magellanicus	CA	3.4	5.1	1.5	0.115	0.430[(a)]	0.26	8	25	67	5	90 min, hypoxia	2
Lymnaea stagnalis		–	–	0.1	0.011	1.07[(c)]	0.010	–	–	–	20	15 h anoxia	10
Sipunculida													
Sipunculus nudus	BWM	2.42	30.5	1.35	0.028	0.084	0.33	0.0	32	68	15	24 h anoxia	3
Annelida													
Arenicola marina	BWM	2.32	8.8	7.4	0.040	0.276[(c)]	0.145	1	12	87	12–16	24 h anoxia	4
Nephtys hombergii	WA	–	–	5.6	0.023	0.530[(e)]	0.043	–	–	–	12	3 h anoxia	11
					0.032	0.530	0.060	–	–	–	12	30 h anoxia	11
Lumbriculus variegatus	WA	2.62	2.03	8.85*	0.144	0.55[(d)]	0.26	–	–	–	20	0–4 h anoxia	5
					0.058	0.55	0.11	0	0	100	20	12–48 h anoxia	5
Tubifex sp.	WA	1.56	1.74	–	0.047	0.322	0.145	1	4	95	10–13	0–12 h anoxia	12
	WA				0.039	0.322	0.12	0.5	0	99.5	10–13	12–48 h anoxia	12
Arthropoda													
Chironomus plumosus larvae		–	–	–	0.093	21.2[(c)]	0.004	–	–	–	10	24 h anoxia	13
Cirolana borealis	WA	–	–	1.06	0.054	0.071	0.76	–	–	–	13	18 anoxia	14

[a] For notes and abbreviations see Table 1. PA; phasic adductor, CA; catch adductor

[b] Resting M ATP is calculated from respiration at rest, assuming 1 mol O_2 ~6 mol ATP. $\dot{V}O_2$ taken from: (a) Thompson et al. 1980; (b) Pörtner 1982; (c) Prosser 1973; (d) Kaufmann 1984; (e) Banse et al. 1971

[c] References: 1. De Zwaan et al. 1980; 2. Baldwin and Morris 1983; 3. Pörtner et al. 1984; 4. Schöttler et al. 1983; Surholt 1977; Siegmund and Grieshaber 1984; 5. Putzer 1984; 6. Ellington 1981, 1982a; 7. De Zwaan et al. 9182; 8. De Zwaan et al. 1983; 9. Ebberink et al. 1979; 10. De Zwaań et al. 1976; 11. Schöttler 1982; 12. Seuss et al. 1983; Hoffman 1981; 13. Frank 1983; 14. De Zwaan and Skjoldal 1979

lysis of phosphoarginine. The larger the contribution of glycolysis to the total ATP needs of the cell during activity the lower the metabolic rate. In molluscs there is at relatively low metabolic rates no longer preference for conversion of pyruvate into octopine. In the pedal retractor muscle of Strombidae (Baldwin and England 1982) and the introvert retractor muscle of *Sipunculus nudus* strombine accumulates in equal or even larger amounts than octopine. Body wall muscle of *Arenicola marina* accumulates alanopine during digging movements (Siegmund and Grieshaber 1984). In the sea mussel *Mytilus edulis* pyruvate derivatives are hardly formed with shell closure during aerial expose. When, however, the animals are stressed by forcing them to shell closure with a rubber band the ATP turnover rate and the glycolytic flux appear to increase by about one order of magnitude ($\dot{M}$ ATP increases from 0.013 to 0.072, Table 3) and concomitantly pyruvate derivative formation is induced. Especially strombine accumulates, but also octopine and lactate (De Zwaan et al. 1983).

In scallops (De Zwaan et al. 1980; Chih and Ellington 1983) annelids (Siegmund and Grieshaber 1984) and crustaceans (Zebe 1982; Gäde 1984) formation of succinate also occurs during exercise anaerobiosis, but in low quantities in comparison to pyruvate derivatives (see De Zwaan and Putzer 1985). In most studies it is assumed that exercise anaerobiosis is driven by oxygen-independent sources of ATP generation. However, in a few cases oxygen consumption rates were also established. The file shell *Limaria fragilis* increased its $\dot{V}O_2$ by about eight-fold at sustained swimming (Baldwin and Lee 1979). The extra oxygen consumption represents a $\dot{M}$ ATP of 10.5, whereas the anaerobic compounds (phosphagen hydrolysis and anaerobic glycolysis) account for only 3.9 or 27% of the total ATP turnover rate (Baldwin and Morris 1983). When the freshwater annelid *Lumbriculus variegates* is stimulated electrically, exhaustion is reached within 90 s. When the same experiment is carried out in oxygen-free water exhaustion already occurs within 60 s (Putzer 1984). An increase of $\dot{V}O_2$ during aerobic stimulation of about six times has been established by Kaufmann (1984) which can account for about the same ATP output as anaerobic sources. These few examples may indicate that most calculated $\dot{M}$ ATP values in Tables 1 and 3 will be underestimated, because a possible aerobic contribution has not been taken into consideration in the other cited studies.

2.2 Anaerobiosis During Environmental Hypoxia/Anoxia

In the annelids, the molluscs and lower invertebrate phyla there is a large variability and flexibility in the way carbohydrate is utilised. End products identified in relation to environmental anaerobiosis are lactate, octopine, strombine, alanopine, alanine, ethanol, CO_2 and the volatile fatty acids acetate and propionate. The various metabolic pathways involved have been the subject of many reviews (Schöttler 1980; Livingstone 1983; Fields 1983; De Zwaan 1983).

Arthropods and probably echinoderms possess a well-developed classical glycolysis for the conversion of carbohydrate, but in the other invertebrate phyla, the succinate pathway is relatively more important during environmental anaerobiosis. The carbon for succinate can be derived from glycogen, but in marine invertebrates also from aspartate and in freshwater species from malate (Zebe et al. 1981; Englisch et al. 1982; Putzer 1985; De Zwaan et al. 1984). In marine invertebrates the conversion of aspartate into succinate is coupled to the conversion of glycogen into alanine (Felbeck 1980;

Foreman and Ellington 1983; De Zwaan et al. 1982, 1983) and in muscle tissue also to the conversion of aspartate into alanine in a reaction sequence including one decarboxylation (aspartate → alanine + CO_2). For this reason aspartate utilisation exceeds the formation of succinate in muscle tissue. It has been assumed that the latter transformation of aspartate is linked to the transport of hydrogen from the cytosol through the inner mitochondrial membrane (De Zwaan et al. 1983). The connection of the two linear pathways (glycogen → alanine and aspartate → succinate) by transaminase reactions and the involvement of the aspartate conversion in hydrogen transport is depicted in Fig. 1. It has been discussed that the affinity of the "aspartate coupled hydrogen transfer system" for NADH, generated in the cytosolic GAPDH reaction, is much higher than that of the pyruvate oxidoreductases, but that its capacity to deal with NADH is much lower. It, therefore, can only out compete the pyruvate oxidoreductases when the metabolic rate is low (see for a detailed discussion De Zwaan and Putzer 1985). Formation of succinate from glycogen is initiated by carboxylation of the glycolytically-formed phosphoenolpyruvate into oxaloacetate. Oxaloacetate is subsequently converted via malate and fumarate into succinate. This route contains two reductive steps. The first one, catalysed by malate dehydrogenase, is compensated for by cytosolic NADH; the second one, catalysed by fumarate reductase, needs generation of reduced co-enzyme in the mitochondrion. Therefore, a minor part of malate is converted into acetate or even succinate, mainly by exploiting a part of the citric acid cycle operating in "aerobic" (clockwise) direction (Schöttler 1977; De Zwaan et al. 1981). The reduction of fumarate is coupled to the electron transfer linked phosphorylation of ADP (Schöttler 1977; Holwerda and De Zwaan 1979). Succinate is end product until a certain level is reached and from then on is a mainstream intermediate in the route terminating into propionate. Again one extra ATP is gained with the decarboxylation of D-methylmalonyl-CoA (Schroff and Zebe 1980; Schulz and Kluytmans 1983). In energetic terms the advantage of the formation of succinate and volatile acids above lactate is obvious. When glycogen is converted into lactate or an opine 3 ATP equivalents are formed per glycosyl, in the case of propionate 6.4 and when acetate and propionate accumulate in a ratio of one to two 5.7.

There is a strong reduction of energy demand during environmental anaerobiosis. In annelids and crustaceans the ATP turnover rate may drop to about 10% of aerobic resting rates, whereas in molluscs the reduction may be even larger (Table 3). Soon after the onset of anaerobiosis the ATP turnover rate appears to be below the aerobic resting rate (Ebberink et al. 1979; Seuss et al. 1983). It takes several hours before the ATP turnover rate reaches its lower limit. Initially, when aspartate is still precursor of succinate, the rate is three to five times higher as compared to the anaerobic steady state. In this period glycolysis leads to both alanine and opine formation (for example, strombine in *Sipunculus nudus*, Pörtner et al. 1984; strombine in *Mytilus edulis*, Zurburg et al. 1981; strombine in *Arenicola marina*, Siegmund and Grieshaber 1984; strombine and alanopine in *Crassostrea virginica*, Eberlee et al. 1983) and is phosphagen hydrolysed, whereas ATP levels somewhat decrease. When the steady state is reached, the glycolytic flux is reduced and channelled towards malate, whereas the phosphagen pool is depleted.

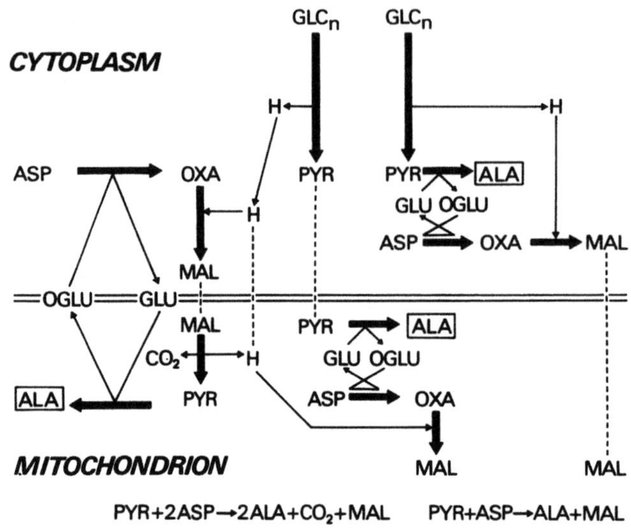

Fig. 1. Metabolic maps showing two extreme possibilities for the redox coupling of the glycogen-alanine pathway with the conversion of aspartate into malate in marine invertebrates. Alanine accumulates, but malate is further metabolised within the mitochondrion. The *left* part is assumed to occur in (muscle) tissue which exploits the "classical glycolysis" during exercise anaerobiosis (opine formation), but couples conversion of pyruvate into alanine to the aspartate-succinate pathway during environmental anaerobiosis. In the latter case the transport of hydrogen through the inner mitochondrial membrane is assumed to depend on the transformation of aspartate into CO_2 and alanine (analogous to the aspartate-malate shuttle when oxygen is available). The *right* part depicts the metabolic situation in tissues in which pyruvate reductases do not compete with glutamate pyruvate transaminase for cytosolic pyruvate. The overall reaction at the *bottom* illustrates the stoichiometric differences with respect to the simultaneous mobilisation of glycogen and aspartate (in muscle tissue aspartate utilisation exceeds the formation of succinate)

3 Anaerobiosis in Vertebrates

3.1 Anaerobiosis During Exercise

Energy metabolism during anaerobic exercise in vertebrate muscle is mainly studied on lactate accumulation and phosphagen hydrolysis; in only a few studies are other metabolites measured simultaneously. Of course, this is mainly due to the fact that energy balance studies with isolated anaerobic vertebrate muscle (Homsher and Kean 1978) indicated the quantitative importance of anaerobic glycolysis and phosphagen hydrolysis. In Table 4 representative data are given for fish, amphibia, reptiles, and mammals of metabolites during strenuous exercise or tetanic stimulation of muscle tissue. As can be seen, lactate and creatine phosphate levels change rapidly during the first 2 min and exhaustion is reached between 5 and 15 min. When comparing initial lactate fluxes with those at exhaustion, we find for fish muscle values of about 7 and 1 $\mu \cdot g^{-1}$. Initial lactate fluxes of rat muscle were also around 7 $\mu \cdot g^{-1}$, although when measured over the first 10 s of isometric contraction in rat quadriceps muscle, the lactate fluxes are about 40 $\mu \cdot g^{-1}$ (Westra et al. 1982), under these conditions the creatine phosphate levels were exhausted in 5 s. Whole animal lactate production rates of most investigated species are of a similar magnitude (Table 4).

Stimulation of isolated muscle causes depletion of creatine phosphate levels at a slightly faster rate than the lactate increases (Driedzic et al. 1981; Westra et al. 1982), while during recovery creatine phosphate increase is simultaneous with lactate increase at a coupling rate of about 1.5 (Ambrosoli and Cerretelli 1973), again indicating glycolysis as the main energy generating process in the anaerobic working muscle. The anaerobic energy production rate of whole animals, measured by lactate accumulation, varied between 10% and 90% of the maximal energy output (Bennett and Licht 1972, 1973; Ruben and Battalia 1979). The energy equivalent of creatine phosphate depletion is in most muscle tissues similar to that of lactate production. We, therefore, assume that the anaerobic component of the energy production during strenuous exercise, as mentioned above, may be significantly higher, due to substantial contribution of the creatine phosphate reserves.

During exercise two different muscle types may be used: (1) red muscle for long-term aerobic exercise and (2) white muscle used for short-term high power anaerobic exercise. In fish it is established that white muscle is also used at intermediate and higher sustained swimming speeds (Johnston and Goldspink 1973a; Hudson 1973). Since oxidative capacity of white muscle is about ten times as low as that of red muscle (Van den Thillart and Smit 1984) and white muscle is poorly perfused (Randall and Daxboeck 1982), we assume that the energy in active white muscle is produced mainly by anaerobic glycolysis in which case lactate may either accumulate within the tissue or diffuse into the blood and metabolise in other parts of the body. There is ample evidence that lactate is a primary oxidative fuel, especially for red muscle tissues, both in fishes and mammals (Bilinski 1974; Pearce and Connett 1980). The lactate oxidation rate in exercising trout (Van den Thillart 1984) and rats (Brooks et al. 1984) is also very high; therefore, it is likely that any lactate diffusing into the circulation will be oxidized at a high rate. In case lactate does not diffuse into the circulation, the accumulation of lactate in the active white muscle must be double the depleted amount of glycogen on a molar base. As can be seen in Table 4, this relation was found with *Salmo gairdneri* and *Pleuronectes platessa*, but not with *Carassius carassius* and *Macrozoarces americanus*. In the first two experiments the animals were chased until exhaustion, while *Carassius carassius* was swum in a swim tunnel and *Macrozoarces* was electrically stimulated under anaesthesia. It was demonstrated by Wardle (1978) that non-release of lactate from exercised flounder white muscle is a "stress" reaction and also that lactate release is apparently under hormonal control. So it seems that lactate can diffuse out of the white muscle as long as the animal is not highly stressed.

Vertebrate white muscle is shown to be active at sustained activity and also glycogen is depleted without comparable increase of lactate (Johnston and Goldspink 1973b; Green et al. 1983). This mechanism enables local anaerobiosis, while the animal as a whole remains aerobic, mainly due to the high lactate oxidation rates of active red muscle. When work rate is progressively increased, a stage is reached whereafter blood lactate levels increase very fast, which is called lactate anaerobic threshold (LAT), indicating the inability of the animal to metabolise more lactate. The LAT has been demonstrated for mammals (Green et al. 1983) and for reptiles (Mitchell et al. 1981). High blood levels could not be found in trout when swimming at the maximal sustained

Table 4. Exercise anaerobiosis in vertebrates

Species	Tissue	Δ μmol g⁻¹				ṀATP (μmol min⁻¹ g⁻¹)			Condition		Ref[d]
		ATP	CrP	lac	glyc	Anaerobic[a]	Resting[b]	A/R	Temp	Other[c]	
Pisces											
Carassius auratus	RM	−0.3	–	3.1	–	0.50 (T)	0.19(a)	2.6	20°	10′ electr. stim. a.v.	1
	WM	−1.0	–	12.4	–	1.96 (T)	0.19(a)	10		electr. stim. a.v.	
Carassius carassius	RM	–	–	– 2.5	−29	5.8 (G)	0.10	58	10°	15′ stren. exercise	2
	WM	–	–	5.7	−42	8.4 (G)	0.10	84		stren. exercise	
Salmo gairdneri	WM	–	–	17.6	− 8.6	12.9 (G)	0.26(b)	50	12°	2′ stren. exercise	3
	WM	–	–	25.4	−12.9	2.6 (G)	0.26(b)	10		15′ stren. exercise	
Salmo salar	WM	–	–	39.3	–	5.9 (L)			10°	10′ stren. exercise	4
Pleuronectes platessa	WM	–	–	40	−20	4.0 (L)	0.09(c)	44	9°	15′ stren. exercise	5
Macrozoarces americanus	WM	0	−12	7	−10	32.0 (T)			13°	1′ electr. stim. a.v.	6
	WM	−0.5	−20	12	−12	11.2 (T)			13°	5′ electr. stim. a.v.	6
Cyprinus carpio	WM	−2.2	–	– 8.9	–	1.6 (T)	0.16(b)	10	12°	10′ stren. exercise	7
Amphibia											
Batrachoseps attenuatus	WA	–	–	12.0	–	9.8 (L)	0.27	36	20°	2′ stren. exercise	8
Hyla regilla	WA	–	–	6.8	–	5.1 (L)	0.27	19	20°	2′ stren. exercise	8
Bufo boreas	WA	–	–	2.6	–	2.0 (L)			20°	2′ stren. exercise	8
Reptilia											
Diposaurus dorsalis	WA	–	–	13.9	–	10.4 (L)	0.27	38.5	30°	2′ electr. stim.	9
	WA	–	–	20.4	–	15.3 (L)	0.67	22.8	40°	2′ electr. stim.	9
Anolis carolinensis	WA	–	–	7.9	–	23.7 (L)			30°	30″ stren. exercise	10
	WA	–	–	10.9	–	16.4 (L)			30°	90″ stren. exercise	10
	WA	–	–	12.3	–	3.7 (L)			30°	5′ stren. exercise	10
Mammalia											
Microtus montanus	WA	–	–	8.4	–	2.5 (L)	13.7	0.2	RT	5′ stren. exercise	11
Dipodomys muriana	WA	–	–	5.7	–	1.7 (L)	7.8	0.2	RT	5′ stren. exercise	11
Rattus	WM	−0.5	−16.9	7.3	–	28.4 (T)	3.9(d)	7.3	37°	1′ electr. stim.	12
	WM	−0.7	−18.9	17.1	–	15.1 (T)	3.9(d)	3.8	37°	3′ electr. stim.	

Footnotes to Table 4: page 179

speed (Driedzic and Kiceniuk 1976); however, they do rise during recovery following a burst swim (Black et al. 1962).

The fast decline of the anaerobic energy production and simultaneous force development of vertebrate muscle appears to be correlated to the decreasing intracellular pH rather than to complete exhaustion of energy sources (Sahlin et al. 1981). Low pH is known to depress muscle activity in invertebrates and is caused by the accumulation of lactic acid (Sahlin et al. 1981; Seo et al. 1983). So it is evident that the anaerobic capacity will be increased by perfusion of the anaerobic muscle, since in that way lactic acid may leave the tissue. Therefore, when glucose is available in the perfusate, the glycolytic flux may stay high for long periods, such as with the perfused anaerobic rat heart (Schertzer and Cascarano 1978).

3.2 Anaerobiosis During Environmental Hypoxia/Anoxia

While during high power output anaerobiosis is limited to a few minutes, low power output anaerobiosis may last for several months in some vertebrates. Metabolic research on this topic is mainly done with fish, sparsely with amphibia and reptiles, while with mammals only in vitro experiments were carried out. Representative data are presented in Table 5. At first glance, the most remarkable difference with the data in Table 4 is the fact that the metabolic rates in Table 5 are between 10 and 10^4 times lower. The lower the rates, the longer the animal survives the imposed conditions, which means suppressed metabolic rate. However, absolute anaerobic conditions were not met with all animals, mainly because they would die immediately. So, the anoxic trout (ref. 2, Table 5), was in a death struggle which lasted about the same period as the fatiguing time during strenuous exercise (ref. 3, Table 4). In both experiments the lactate production rates were of the same magnitude.

Under hypoxic conditions, glycolytic rates are lower, animals survive longer, but energy production is mainly aerobic (ref. 1, 4, 10, in Table 5). All hypoxia and anoxia resistent animals have very well-developed oxygen extraction capacities; not only fish, but also amphibians (G. Ultsch 1974; Leivestad 1960) and even reptiles (Ultsch and Jackson 1982) are able to extract oxygen dissolved in water. So, we left out most of the hypoxia experiments, and data summarised in Table 5 refer mainly to anoxic conditions, in order to obtain a better insight in the anaerobic capacity.

Anaerobiosis during environmental hypoxia/anoxia is mainly characterised by glycolytic lactate production and, as for white muscle tissue, depletion of creatine phos-

---

Footnotes from Table 4:

[a] $\dot{M}$ATP is calculated from lactate (L), glycogen (G) or from all measured metabolites (T), assuming 1 mol lactate ~1.5 mol ATP and 1 mol glycosyl units ~3 mol ATP

[b] Resting $\dot{M}$ATP is calculated from whole animal respiration at rest, assuming 1 mol O_2 ~6 mol ATP. Metabolic rates taken from: (a) Beamish and Mookherjii (1964); (b) Ott et al. (1980); (c) Jϕrgensen and Mustafa (1980); (d) Prosser (1973)

[c] a.v.: Animal was artificially ventilated; WM-RM: white-red muscle; WA: whole animal

[d] *References:* 1. Van Waarde and Kesbeke 1983; 2. Johnston and Goldspink (1973a); 3. Black et al. 1962; 4. Börjeson and Fellenius 1976; 5. Wardle 1978; 6. Driedzic et al. 1981; Driedzic and Hochachka 1976; 8. Bennett and Licht 1973; 9. Bennett and Dawson 1972; 10. Bennett and Licht 1972; 11. Ruben and Batalia 1979; 12. Sahlin et al. 1981

Table 5. Environmental Anaerobiosis in Vertebrates[a]

Species	Tissue	Δμmol g^{-1}							$\dot{M}ATP$ (μmol min^{-1} g^{-1})				°C	Condition	Ref[b]
		ATP	CrP	lac	glyc	ala	Other		Anaerobic		Resting	A/R			
Pisces															
Salmo	RM	–	–	4.9	-14.6	1.3	–		0.73	(G)	0.17(a)	4	15	60 min at 50 T	1
gairdneri	WM	–	–	15.6	-9.6	0.7	–		0.48	(G)	0.17(a)	3			
	RM	–	–	8.0	–	3.8	0.4	(S)	1.48	(L+A)	0.17(a)	9	15	12 min anoxia	2
	WM	–	–	10.7	–	3.0	0.1	(S)	1.71	(L+A)	0.17(a)	10			
Cyprinus	RM	–	–	15.9	–	4.9	−0.3	(S)	0.65	(L+A)	0.26(a)	2.5	25	46 min anoxia	3
carpio	WM	–	–	16.9	–	−1.3	−0.2	(S)	0.51	(L+A)	0.26(a)	2.0			
Carassius carassius	RM	–	–	16	-62	2	–		2.07	(G)	0.10	21.0	15	90 min at about 10 T	4
	WM	–	–	50	-99	0.9	–		3.30	(G)	0.10	33.0			
Carassius	WA	–	–	2.7	-4.9	–	–		0.020	(G)	0.15(b)	0.13	20	12 h anoxia	5
auratus	WA	–	–	5.6	–	–	11.2	(E)	0.035	(E+L)	0.02(b)	1.8	5	12 h anoxia	6
	WA	–	–	–	–	–	23.3	(E)	0.048	(E)	0.15(b)	0.32	20	12 h anoxia	
	WA	–	–	–	–	–	60.6	(E)	0.018	(E)	0.05(b)	0.36	10	86 h anoxia	
	WA	–	–	–	–	–	13.7	(E)	0.010	(E)	0.02(b)	0.50	5	36 h anoxia	
	RM	−0.8	-2.4	0.2	-18.6	1.2	0.5	(S)	0.089	(G+P)	0.15(b)	0.59	20	11 h anoxia	7/8
	WM	−1.3	-6.0	2.0	-1.7	0.7	0.6	(S)	0.010	(G+P)	0.15(b)	0.07			
Anguilla	H	−0.4	–	3.8	-4.9	–	–		0.043	(G)	0.39(c)	0.11	15	5.7 h anoxia	9
anguilla	WM	−0.7	−14.2	6.4	-1.3	−0.2	1.0	(S)	0.070	(L+P)	0.39(c)	0.18			
Platichthys	H	−0.5	0.0	7.0	-11.3	0.5	0.0	(S)	0.014	(G)	0.09	1.6	10	4 h at 15 T	10
flesus	WM	−0.7	−12.2	0.8	-7.3	0.4	0.0	(S)	0.15	(G+P)	0.09	1.7			
Amphibia															
Bufo cognatus	WA	–	–	4.5*	–	–	–		0.084	(L)	0.30(c)	0.28	20	80 min anoxia	11
Rana pipiens	WA	–	–	15.7	-12.3	–	–		0.005	(G)	–	–	5	120 h anoxia	12
Reptiles															
Pseudemys	WA	–	–	24*	–	–	–		0.025	(L)	0.21(d)	0.12	22	24 h anoxia	13
scripta elegans	H	–	–	75	−310	–	–		0.65	(G)	0.21(d)	3.1			

Chrysemys picta bellii	WA	–	–	133*	–	–	–		0.001	(L)	–	–	3	126 days anoxia	14
Mammalia															
Dog	H	–1.6	– 8	11.3	–	–	–		6.64	(L+P)	1.6[(f)]	4.1	37	4 min systemic anoxia	15
Human	WM	–	–	24	–	–	–		0.30	(L)	0.93[(c)]	0.3	37	2 h anoxia	16
Rat	H	–2.8	– 3.0	136[+]	– 3.4	–	0.6	(S)	10.5	(L+P)	3.93[(c)]	2.7	37	20 min anoxia	17

[a] Abbreviations: H, heart; WM-RM, white-red muscle; WA, whole animal. Metabolites: S, succinate; E, ethanol. Anaerobic $\dot{M}$ATP: ATP equivalents for glycogen (G), lactate (L), or alanine (A) and ATP plus CrP (P) are, respectively, 3, 1.5, and 1 on mole basis. Resting $\dot{M}$ATP is based on whole animal respiration assuming 6 ATP equivalent per mole O_2. Metabolic rates taken from: (a) Ott et al. 1980; (b) Beamish and Mookherjii 1964; (c) Prosser 1973; (d) Jackson and Silverblatt 1974. *: lactate concentration estimated from 2/3 of blood level. +: lactate concentration includes: lactate efflux

[b] *References:*

1. Johnston 1975a; 2. Smith and Heath 1980; 3. Johnston 1975b; 4. Van den Thillart and Kesbeke 1978; 5. Shoubridge and Hochachka 1980; 6. Van den Thillart et al. 1983; 7. Van den Thillart et al. 1980; 8. Van Waarde et al. 1982; 9. Van Waarde et al. 1983; 10. Jørgensen and Mustafa 1980a, b; 11. Armentrout and Rose 1971; 12. Christiansen and Penney 1973; 13. Penney 1974; 14. Ultsch and Jackson 1982; 15. Fellenius and Samuelson 1973; 16. Petterson et al. 1982; 17. Penney and Cascarano 1970

phate stores. Still, there are individual differences, both with respect to survival as to metabolic changes, which are not always corroborative. Goldfish, for example, has high glycogen stores and survives 12 h anoxia at 20 °C (Van den Thillart and Kesbeke 1978; Van den Thillart et al. 1980). Eel, however, has very low glycogen stores (Van Waarde et al. 1983), much lower than trout (Johnston 1975a), but still survives 6 h anoxia at 15 °C, while trout dies after about 12 min. There are no indications yet that other metabolic processes are involved, which suggests that metabolic suppression in eel is the only means to survive extended anoxic periods. This corresponds to the observed activity levels; eel becomes quiescent, while trout starts to panic at low PO_2. Similarly, when goldfish were transferred suddenly from normoxic into anoxic water, they die within 2 h (Van den Thillart, unpublished), which contrasts to the 12 h survival of unstressed fish.

The occurrence of metabolic depression due to anoxia can be measured by direct calorimetry. A reduction to 20% of the aerobic resting rate and even lower was observed with goldfish (Anderson 1975), toad (Leivestad 1960), and turtle (Jackson 1968). Heat flux measurements with anoxic vertebrates demonstrated a similar metabolic depression (Gnaiger 1983). Reduction of metabolic rate is demonstrated by the low glycolytic fluxes shown in Table 5; the longer the anoxia period, the lower the fluxes. Based on glycogen depletion or lactate accumulation, the glycolytic flux in fish ranged from about $1{,}000 \times 10^{-3}\ \mu \cdot g^{-1}$ for periods less than 1 h to about $10 \times 10^{-3}\ \mu \cdot g^{-1}$ for periods longer than 80 h. Low glycolytic fluxes are also observed with submerged amphibia and turtles (Table 5): at room temperature the estimated whole body rate of toad (ref. 11) was about $60 \times 10^{-3}\ \mu \cdot g^{-1}$; at 5 °C the whole body rate of frog (ref. 12) was found to be $2 \times 10^{-3}\ \mu \cdot g^{-1}$, while for the turtle at 22 °C and 3 °C rates were found of $18 \times 10^{-3}\ \mu \cdot g^{-1}$ (ref. 13) and $0.7 \times 10^{-3}\ \mu \cdot g^{-1}$ (ref. 14). Remarkable is the high glycolytic rate of the heart in anoxic frog and turtle, which remains about ten times higher than the whole body rate.

Apart from a depression of metabolic rate, animals may use other strategies to survive anoxia. One way would be the use of a modified metabolic pathway, resulting in more ATP production. As a consequence of the finding of exotic metabolic pathways in vertebrates, succinate and volatile fatty acids were measured in several studies regarding vertebrate anaerobiosis (Smith and Heath 1980; Burton and Spehar 1971; Van Waarde et al. 1983; Driedzic and Hochachka 1975; Van Waarde et al. 1982; Jørgensen and Mustafa 1980; Penney 1974). Although succinate levels are indeed higher under anaerobic conditions, the increase is not of quantitative importance in the muscle tissues. An appreciable increase of succinate concentration was observed in the liver of eel (Van Waarde et al. 1983), flounder (Jørgensen and Mustafa 1980a,b), and turtle (Penney 1974). Also by mammalian tissues, some succinate is produced under anaerobic stress; blood levels of seal, sea lion and porpoise (Hochachka et al. 1974) were significantly increased albeit to a maximum level of 0.3 m*M*. However, succinate production by perfused anaerobic rat hearts (Penney and Cascarano 1970) and papillary muscle (Taegtmeyer 1978) can be more substantial, especially when TCA-cycle intermediates are added to the perfusate. It has been demonstrated that the fumarate reductase reaction is coupled to Ca^{2+} uptake by the heart mitochondria (Schertzer and Cascarano 1978) indicating that succinate production can be coupled to energy transfer. Indeed, addition of fumarate to perfused rat hearts not only increases succinate production, but also increases the energy expenditure (Penney and Cascarano 1970). In invertebrates the pathway to succinate starts with the formation of oxaloacetate by phosphoenolpyruvate carboxykinase. This enzyme, although present in the liver, is almost inactive in the muscle tissues of vertebrates (Crabtree et al. 1972), which explains the succinate formation in the liver and low succinate levels in muscle tissues. In muscle tissues aspartate seems to be the best substrate for succinate formation. In the anoxic goldfish muscle aspartate depletion may indeed be partially responsible for succinate increase (Van Waarde et al. 1982). Also, aminooxyacetate, a potent transaminase inhibitor, suppresses the formation of succinate in anoxic rabbit heart muscle (Taegtmeyer 1978), which indicates that transaminases are involved in the formation of succinate. Since the conversion of aspartate to succinate has two reductive steps, this pathway may drive, in addition, the oxidation of glutamate to succinate, which has the advantage of substrate level phosphorylation. Although glutamate levels do not fall during anoxia (Van Waarde et al. 1982, 1983; Driedzic and Hochachka 1975), it has been found that anoxic rabbit heart muscle (Sanborn et al. 1979) and goldfish muscle (Mourik 1982) decarboxylate $1-C^{14}$ glutamate at the same rate as under normoxic conditions, indicating that glutamate to succinate conversion does proceed. When aspartate and glutamate are used, some other amino acid or ammonia must accumulate. In all anoxia/hypoxia experiments, alanine is found to accumulate to levels exceeding succinate, suggesting that alanine is not only formed by coupling to the succinate pathway. One good explanation may be the continuous amino acid supply from proteolysis, a pathway which has not had much attention in studies on anaerobic metabolism. In carp white muscle (Driedzic and Hochachka 1975) the amino acid pool was significantly increased after 4 h of severe hypoxia, indicating that proteolysis is not suppressed under anoxia; a similar conclusion was drawn from metabolic changes in anoxic goldfish (Van Waarde et al. 1982; Van den Thillart et al. 1983).

While typical invertebrate anaerobic end-products were not found in vertebrates, a novel exotic end-product was found in goldfish: ethanol (Shoubridge and Hochachka 1980). There are indications that ethanol is produced in the cytoplasm of the red muscle tissue via acetaldehyde (Mourik 1982). The conversion of pyruvate to ethanol is not coupled to energy conservation (Mourik 1982) and since ethanol and CO_2 diffuse into the surrounding water, the pathway functions as a means to prevent a high metabolic acidosis. Because pyruvate to ethanol conversion is a downhill reaction coupled to NADH oxidation, the pathway also stabilised the cytoplasmic redox state. The total ethanol production capacity varies and appears to be temperature dependent (Van den Thillart et al. 1983). When expressed as ethanol produced per gram of whole goldfish, the amount ranges between 3 and 60 μmol. The highest values were observed with 10 °C acclimated goldfish after 86 h anoxia. There is remarkably little knowledge on the occurrence of ethanol as an anaerobic end-product in animals. Thus far ethanol is found in anoxic *Chironomus* larvae (Wilps and Schöttler 1980).

High lactate levels associated with glycogen depletion after long-term anoxia were observed in amphibia and turtles (ref. 11, 12, 13, 14 in Table 5). An extremely high blood lactate concentration of 200 m*M* was observed in *Chrysemys picta bellii* after 126 days of submersion in anoxic water at 3 °C. Since blood pH in these animals was only reduced from 8.0 to 7.0, it is evident that there was a very efficient buffer. The buffer is probably $CaCO_3$, since there was an extremely high blood Ca^{2+} level. A similar buffer is available in frogs (Simkiss 1968). These animals possess $CaCO_3$ deposits, which dissolve during acidosis. Most remarkable is the location of the deposits: the crystals are found in the endolymphatic sacs, which overlie the brain and the spinal ganglia. So it seems that the $CaCO_3$ deposits are in the first place a protection for the central nervous system. High buffer capacities are also found in the pericardial fluids of the turtle. Although these are not used during submersion at room temperature (Jackson and Silverblatt 1974), it is, however, very likely that the buffers will be used during long-term anaerobiosis.

In mammals environmental anaerobiosis is hardly studied on whole animals. Some indications can be obtained from oxygen debt measurements after diving. Metabolism during a dive is strongly depressed; it is even observed that resting metabolism of seals is higher than the swimming metabolism under water (Kooyman et al. 1980). The animals reduce heart frequency and have peripheral vasoconstriction, which may be the major reason for metabolic depression, since metabolic rate seems to depend on the blood flow (Wahlen et al. 1973). During anoxia, the mammalian heart function depends completely on the availability of blood glucose (Penney and Cascarano 1970). Despite the high glycolytic fluxes, the anaerobic energy generation seems unsufficient, since the energy expenditure falls to below 20% of the aerobic rates (Penney and Cascarano 1970; Fellenius and Samuelsson 1973). In contrast, very low glycolytic fluxes are found in anoxic smooth muscle preparations (Petterson et al. 1982), about 50 times lower than that of perfused anaerobic rat heart (Table 3). The importance of glycolysis in anaerobiosis is also expressed by the observation that mammals have relatively high glycogen levels built up in muscle and liver before birth, while during partus a substantial depletion of glycogen reserves is observed (Shelley 1961).

From the above discussion and data it is obvious that vertebrates rely mainly on the classical glycolysis for their anaerobic energy production. Animals with extreme anoxia resistence are characterised by strongly reduced metabolic rates and enhanced buffering capacities to prevent metabolic acidosis. These adaptations are specifically suited when the animals become lethargic during hibernation. At higher metabolic rates the anaerobic capacity is restricted and, therefore, also the anaerobic tolerance of the animal.

4 pH Effects of Anaerobiosis

During anaerobiosis in all animals a metabolic acidosis is building up, due to the production of acidic end-products, such as lactic, pyruvic, succinic, and propionic acid (see above). The effect of these products on intracellular pH (pHi) depends on both the amount of metabolic acid and the available buffer capacity. During environmental anaerobiosis, a significant decrease in blood pH can be seen: in trout pH 7.96–7.35 (Thomas 1983), in diving seal pH 7.4–6.8 (Kooyman et al. 1980) and in diving turtle pH 7.6–6.8 (Jackson and Silverblatt 1974); also, the extrapallial fluid of *Mytilus* decreases from 7.6 to 6.6 during shell closure (Wijsman 1975). Less measurements are made on intracellular pH; most of them are carried out by a recently developed, nondestructive technique: high field strength nmr. In this way a decrease of pHi was measured in anoxic welk heart pHi 7.1–6.9 (Ellington 1983), in *Artemia* embryos the pHi dropped from 8.0 to 6.9 (Busa et al. 1982) and in anoxic frog muscle pHi fell from 7.1 to 6.5 (Seo et al. 1983). In the case of frog muscle, it was shown that the pHi decrease was caused by lactic acid production, both during rest and electrical stimulation. Also in rat skeletal muscle anaerobic electrical stimulation shows a close coupling between decrease of pHi 7.1–6.76 and increase of lactate concentration (Sahlin et al. 1981). Stimulation of IAA poisoned muscle does not cause a pHi change. Also in human muscle with a phosphorylase deficiency pHi does not fall during stimulation (Bore et al. 1982), which points to lactic acid as the major acidifying agent.

A high intracellular pH seems important for many cellular functions. Not only is there a strong correlation between muscle performance and high pH (Sahlin et al. 1981; Jacobus et al. 1982), but there is also a correlation between increase of pHi and glycolytic activity, cell motility, development, membrane transport, etc. (Nuccitelli and Heiple 1982). So it appears that acid production is a potential threat to most cellular activities. Obviously, animals have developed several protection mechanisms against acidosis. In muscle, especially white muscle, there are three H^+ absorbing mechanisms: (1) creatine phosphate hydrolysis consumes H^+ ions. Although ATP hydrolysis generates H^+ ions (Gevers 1977), the creatine phosphate concentration is several times higher than the ATP concentration, which could even increase pHi. Indeed, this has been observed with human ischemic muscle at rest (Bore et al. 1982). Arginine phosphate which is the vertebrate equivalent of creatine phosphate, also consumes H^+ ions upon hydrolysis (Hochachka and Mommsen 1983). (2) ATP-IMP conversion is found in fish (Driedzic and Hochachka 1976; Van den Thillart et al. 1980), as well as in rat muscle (Meyer and Terjung 1980), due to anaerobic activity. Since AMP-deaminase converts AMP into IMP + NH_3, this reaction may compensate partially for H^+ ions generated by ATP hydrolysis. (3) The muscle tissue has a variable buffer capacity, which appears to be related to the anaerobic capacity, both in vertebrates (Castellini and Somero 1981) and in vertebrates (Morris and Baldwin 1984) and may consume up to 80 μmol H^+ g^{-1} tissue for a pH drop of 1 U. Apart from the above mentioned H^+ absorbing mechanisms in muscle tissues, there are others which function on the organismal level. In amphibia (Simkiss 1968), as well as in turtles (Ultsch and Jackson 1982), $CaCO_3$ deposits play a major role in the neutralisation of lactic acid formed during anaerobiosis. Similarly, bivalves use their shell to neutralise succinic acid, formed during anoxia (Crenshaw and Neff 1969). An indirect way to protect the animal from severe metabolic acidosis is the confinement of the acid to the muscle by way of vasoconstriction in diving mammals (see Kooyman et al. 1980). Restricted blood flow is also found in highly active trout white muscle (Randall and Daxboeck 1982); the non-release of lactactic acid of fish muscle seems to be under hormonal control; when the hormonal inhibition was relieved by propanolol, most animals died from metabolic acidosis (Wardle 1978). From a metabolic point of view, the most efficient way of preventing metabolic acidosis, is to produce either neutral or volatile end products. In the latter case, products like CO_2 (coupled to f.e. ethanol) and propionic acid, will leave the animal by way of diffusion.

5 Summary

There are two conditions in which anaerobic energy metabolism is involved in providing ATP: (1) when the energy demand exceeds the capacity of the aerobic energy production and (2) when environmental change blocks or reduces oxygen supply (environmental anaerobiosis). The first condition applies to short-term vigorous activity (exercise anaerobiosis) and the initial phase of restoration of normal physiological functions following periods of exercise or environmental anaerobiosis (recovery anaerobiosis; see De Zwaan and Putzer 1985).

In arthropods, echinoderms and vertebrates, anaerobiosis is almost exclusively driven by depletion of phosphagen and production of lactate, both in high and low output modes. The metabolic diversity, such as occurring in most invertebrates, is not observed in the many vertebrate species studied so far. Anaerobic glycolysis occurs always in its simplest form, namely, the conversion of glycogen into lactate. In the other invertebrates, the pattern of end product accumulation depends on the metabolic rate. A low rate can be covered by the simultaneously operating glycogen-alanine and aspartate-succinate pathway, for example, during environmental anaerobiosis when metabolic rate is strongly reduced compared to the aerobic resting rate. During burst activity the ATP demand exceeds by far the ATP output capacity of the aspartate-succinate pathway and simultaneously pyruvate derivatives accumulate in amounts high relative to succinate. Recovery anaerobiosis leads to accumulation of pyruvate derivatives (De Zwaan and Putzer 1985).

During burst activity ATP generation may almost entirely rely on phosphagen depletion, whereas in the first hours of environmental anaerobiosis phosphagen also contributes considerably to the ATP formation. In lower marine invertebrates, up to the arthropods, four different pyruvate oxidoreductases may be present for the reduction of pyruvate leading to lactate or iminodicarboxylic acid (opine) formation. Three different opines have been identified as end product being formed in relation to exercise, environmental or recovery anaerobiosis (see De Zwaan and Putzer 1985). In some anoxia tolerant fish lactate formed in glycolytic tissue is transported to red muscle and there further converted into ethanol.

In marine invertebrates aspartate can be converted into CO_2 and alanine in a route which involves the malate dehydrogenase/malic enzyme redox couple. This conversion may serve to transport cytosolic hydrogen (from the GAPDH step) to the mitochondrion and thus competes with the pyruvate oxidoreductases. Its higher affinity for cytosolic NADH explains why pyruvate derivatives only accumulate when the ATP demand exceeds the ATP generating capacity of the aspartate-succinate pathway. In some freshwater invertebrates high malate pools have been found which serve as an energy source during the first hour(s) of environmental anaerobiosis.

References

Ambrosoli G, Cerretelli P (1973) The anaerobic recovery of frog muscle. Pflügers Arch 345:131–143

Anderson JR (1975) Anaerobic resistance of *Carassius auratus* (L). Ph.D. Thesis, Australian National University

Armentrout D, Rose FL (1971) Some physiological responses to anoxia in the great plains toad, *Bufo cognatus*. Comp Biochem Physiol 39A:447–455

Baldwin J, England WR (1980) A comparison of anaerobic energy metabolism in mantle and tentacle muscles of the blue ringed octopus *Hapalochlaena maculosa* during swimming. Aust J Zool 28: 407–412

Baldwin J, England WR (1982) The properties and functions of alanopine dehydrogenase and octopine dehydrogenase from the pedal retractor muscle of Strombidae (Class Gastropoda). Pac Sci 36:381–394

Baldwin J, Lee AK (1979) Contribution of aerobic and anaerobic energy production during swimming in the bivalve mollusc *Limaria fragilis* (Family Limidae). J Comp Physiol 129:361–364

Baldwin J, Morris GM (1983) Re-examination of the contributions of aerobic and anaerobic energy production during swimming in the bivalve mollusc *Limaria fragilis* (family Limidae). Aust J Mar Freshwater Res 34:909–914

Baldwin J, Lee AK, England WR (1981): The functions of octopine dehydrogenase and D-lactate dehydrogenase in the pedal retractor muscle of the dog whelk, *Nassarius coronatus* (Gastropoda: Nassariidae). Mar Biol 62:235–238

Banse K, Nichols FN, May DR (1971) Oxygen consumption by the reached III on the role of mavrofauna at three stations. Vie Mulien, Suppl 22, 31–52

Beamish FW, Mookherjii PS (1964) Respiration of fishes with special emphasis on standard oxygen consumption. Can J Zool 42:355–366

Bennett AF, Dawson WR (1972) Aerobic and anaerobic metabolism during activity in the lizard *Dipsosaurus dorsalis*. J Comp Physiol 81:289–299

Bennett AF, Licht P (1972) Anaerobic metabolism during activity in lizards. J Comp Physiol 81: 277–288

Bennett AF, Licht P (1973) Relative contributions of anaerobic and aerobic energy production during activity in amphibia. J Comp Physiol 87:351–360

Bernaerts F (1982) Het aerobe en anaerobe metabolisme van *Artemia*. Ph.D. Thesis, University of Antwerp, Belgium

Bilinski E (1974) Biochemical aspects of fish swimming. In: Malins DC, Sargent JR (eds) Biochemical and biophysical perspectives in marine biology, vol 1. Academic, New York, pp 239–288

Black EC, Connor AR, Lam K, Chiu W (1962) Changes in glycogen, pyruvate and lactate in rainbow trout *(Salmo gairdneri)* during and following muscular activity. J Fish Res Board Can 19: 409–436

Bone Q, Pulsford A, Chubb AD (1981) Squid mantle muscle. J Mar Biol Assoc UK 61:327–342

Bore PJ, Chan L, Gadian DG, Radda GK, Ross BD, Styles P, Taylor DJ (1982) Noninvasive pH measurements of human tissue using ^{31}P-nmr. In: Intracellular pH: its measurement, regulation and utilization in cellular functions. Kroc Foundation series 15. A.R. Liss, New York, pp 527–535

Börjeson H, Fellenius E (1976) Towards a valid technique of sampling fish muscle to determine redox states. Acta Physiol Scand 96:202–206

Brooks GA, Donovan CM, White TP (1984) Estimation of anaerobic energy production and efficiency in rats during exercise. J Appl Physiol 56:520–525

Burton DT, Spehar AM (1971) A re-evaluation of the anaerobic endproducts of fresh-water fish exposed to environmental hypoxia. Comp Biochem Physiol 40A:945–954

Busa WB, Crowe JH, Matson GB (1982) Intracellular pH and the metabolic status of dormant and developing Artemia embryos. Arch Biochem Biphys 216:711–718

Castellini MA, Somero GN (1981) Buffering capacity of vertebrate muscle: Correlations with potentials for anaerobic function. J Comp Physiol 143:191–198

Chih CP, Ellington WR (1983) Energy metabolism during contractile activity and environmental hypoxia in the phasic adductor muscle of the baby scallop *Argopecten irradians* concentricus. Physiol Zool 56:623–631

Christiansen J, Penney D (1973) Anaerobic glycolysis and lactic acid accumulation in cold submerged *Rana pipiens*. J Comp Physiol 87:235–245

Costello WJ, Govind CK (1983) Contractile responses of single fibers in lobster claw closer muscles: Correlation with structure, histochemistry and innervation. J Exp Zool 227:381–393

Crabtree B, Higgins SJ, Newsholme EA (1972) Activities of pyruvate carboxylase, phosphoenolpyruvate carboxylase, and fructose diphosphatase in muscles from vertebrates and invertebrates. Biochem J 130:391–396

Crenshaw MA, Neff JM (1969) Decalcification of the mantle-shell interface in molluscs. Am Zool 881–885

Driedzic WR, Hochachka PW (1975) The unanswered question of high anaerobic capabilities of carp white muscle. Can J Zool 53:706–712

Driedzic WR, Hochachka PW (1976) Control of energy metabolism in fish white muscle. Am J Physiol 230:579–582

Driedzic WR, Kiceniuk JW (1976) Blood lactate levels in free swimming rainbow trout *(Salmo gairdneri)* before and after strenuous exercise resulting in fatigue. J Fish Res Board Can 33: 173–176

Driedzic WR, McGuire G, Hatheway M (1981) Metabolic alterations associated with increased energy demand in fish white muscle. J Comp Physiol 141:425–432

Ebberink RHM, Zurburg W, Zandee DI (1979) The energy demand of the posterior adductor muscle of *Mytilus edulis* in catch during exposure to air. Mar Biol Lett 1:23–31

Eberlee JC, Storey JM, Storey KB (1983) Anaerobiosis, recovery from anoxia and the role of strombine and alanopine in the oyster Crassostrea virginica. Can J Zool 61:2682–2687

Ellington WR (1981) Effect of anoxia on the adenylates and the energy charge in the sea anemone, *Bunodosoma cavernata* (Bosc). Physiol Zool 54:415–422

Ellington WR (1982a) Metabolic responses of the sea anemone, *Bunodosoma cavernata* (Bosc) to declining oxygen tensions and anoxia. Physiol Zool 55:240–249

Ellington WR (1982b) Metabolism at the pyruvate branch point in the radula retractor muscle of the whelk, *Busycon contrarium*. Can J Zool 60:2973–2977

Ellington WR (1983a) The recovery from anaerobic metabolism in invertebrates. J Exp Zool 228: 431–444

Ellington WR (1983b) Phosphorus nuclear magnetic resonance studies of energy metabolism in molluscan tissue: Effect of anoxia and ischemia on intracellular pH and high energy phosphates in the ventricle of the whelk, *Busycon contrarium*. J Comp Physiol 153:159–166

England WR, Baldwin J (1983) Anaerobic energy metabolism in the tail musculature of the australian yabby, *Cherax destructor* (Crustacea, Decapoda, Parastacidae): Role of phosphagens and anaerobic glycolysis during escape behavior. Physiol Zoll 56:614–622

Englisch H, Opalka B, Zebe E (1982) The anaerobic metabolism of the larvae of the midge, *Chaoborus crystallinus*. Insect Biochem 12:149–155

Felbeck H (1980) Investigations on the role of amino acids in anaerobic metabolism of the lugworm *Arenicola marina* L. J Comp Physiol 137:183–192

Fellenius E, Samuelsson R (1973) Effects of severe systemic hypoxia on myocardial energy metabolism. Acta Physiol Scand 88:256–266

Fields JHA (1983) Alternatives to lactic acid: Possible advantages. J Exp Zool 228:445–457

Foreman RA, Ellington WR (1983) Effects of inhibitors and substrate supplementation on anaerobic energy metabolism in the ventricle of the oyster, *Crassostrea virginica*. Comp Biochem Physiol 74B:543–547

Frank C (1983) Ecology, production and anaerobic metabolism of Chironomus plumosus L. larvae in a shallow lake II. Anaerobic metabolism. Arch Hydrobiol 96:345–362

Gäde G (1980) The energy metabolism of the foot muscle of the jumping cockle, *Cardium tuberculatum*: Sustained anoxia versus muscular activity. J Comp Physiol 137:177–182

Gäde G (1981) Energy production during swimming in the adductor muscle of the bivalve, *Lima hians*: comparison with the data from other bivalve molluscs. Physiol Zool 54:400–406

Gäde G (1983a) Energy metabolism of arthropods and molluscs during environmental and functional anaerobiosis. J Exp Zool 228:415–429

Gäde G (1983b) Energy production during anoxia and recovery in the adductor muscle of the file shell, *Lima hians*. Comp Biochem Physiol 76B:73–78

Gäde G (1984) Effects of oxygen deprivation during anoxia and muscular work on the energy metabolism of the crayfish, *Orconectes limosus*. Comp Biochem Physiol 77A:495–502

Gäde G, Carlsson KH, Meinardus G (1984) Energy metabolism in the foot of the marine gastropod *Nassa mutabilis*, during environmental and functional anaerobiosis. Mar Biol 80:49–56

Gevers W (1977) Generation of protons by metabolic processes in heart cells. J Mol Cell Cardiol 9: 867–872

Gnaiger E (1983) Heat dissipation and energetic efficiency in animal anaerobiosis: economy contra power. J Exp Zool 228:471–490

Green HJ, Hughson RL, Orr GW, Ronney DA (1983) Anaerobic threshold, blood lactate and muscle metabolites in progressive exercise. J Appl Physiol 54:1032–1038

Grieshaber M (1978) Breakdown and formation of high-energy phosphates and octopine in the adductor muscle of the scallop, *Chlamys opercularis* (L.), during escape swimming and recovery. J Comp Physiol 126:269–276

Grieshaber M, Gäde G (1977) Energy supply and the formation of octopine in the adductor muscle of the scallop, Pecten jacobaeus (Lamarck). Comp Biochem Physiol 58B:249–252

Hochachka PW, Mommsen TP (1983) Protons and anaerobiosis. Science 219:1391–1397

Hochachka PW, Owen TG, Allen JF, Whitlow GC (1974) Multiple endproducts of anaerobiosis in diving vertebrates. Comp Biochem Physiol 50B:17–22

Hoffman KH (1981) Phosphagens and phosphokinases in Tubifex sp. J Comp Physiol 143:237–243

Holwerda DA, Zwaan A de (1979) Fumarate reductase of *Mytilus edulis* L. Mar Biol Lett 1:33–40

Homsher E, Kean CJ (1978) Skeletal muscle energetics and metabolism. Annu Rev Physiol 40:90–131

Hudson RCL (1973) On the function of the white muscle in teleosts at intermediate swimming speeds. J Exp Biol 58:509–522

Jackson DC (1968) Metabolic depression and oxygen depletion in the diving turtle. J App Physiol 24:503–509

Jackson DC, Silverblatt (1974) Respiration and acid-base status of turtles following experimental dives. Am J Physiol 226:903–909

Jacobus WE, Pores IH, Lucas SK, Kallman CH, Weisfeldt ML, Flaherty JT (1982) The role of intracellular pH in the control of normal and ischemic myocardial contractility: A ^{31}P nuclear magnetic resonance and mass spectrometry study. In: Intracellular pH: its measurement, regulation and utilization in cellular functions. Kroc foundation series, vol 15. A.R. Liss, New York, pp 537–565

Johnston IA (1975a) Studies on the swimming musculature of the rainbow trout II. Muscle metabolism during severe hypoxia. J Fish Biol 7:459–467

Johnston IA (1975b) Anaerobic metabolism in the carp (*Carassius carassius* L.). Comp Biochem Physiol 51B:235–241

Johnston IA, Goldspink G (1973a) A study of the swimming performance of the Crucian carp *Carassius carassius* in relation to the effects of exercise and recovery on biochemical changes in the myotomal muscles and liver. J Fish Biol 5:249–260

Johnston IA, Goldspink G (1973b) A study of glycogen and lactate in the myotomal muscles and liver of the coalfish (Gadus virens L.) during sustained swimming. J Mar Biol Assoc UK 53:17–26

Jφrgensen JB, Mustafa T (1980a) The effect of hypoxia on carbohydrate metabolism in flounder (*Platichthys flesus* L.) I. Utilization of glycogen and accumulation of glycolytic end products in various tissues. Comp Biochem Physiol 67A:243–248

Jφrgensen JB, Mustafa T (1980b) The effect of hypoxia on carbohydrate metabolism in flounder (*Platichthys flesus* L.) II. High energy phosphate compounds and the role of glycolytic and glucoenogenetic enzymes. Comp Biochem Physiol 67A:249–256

Kaufmann R (1984) Relationship of activity and respiration of *Lumbriculus variegatus* (Oligochaeta) at various temperatures and under declining oxygen conditions. Abstract First International Congress C.P.B., Liege, Belgium

Koormann R, Grieshaber M (1980) Investigations on the energy metabolism and on octopine formation and recovery of the common welk, *Buccinum undulatum* L., during escape. Comp Biochem Physiol 65B:543–547

Kooyman GL, Wahrenbrock EA, Castellini MA, Davies RW, Sinett EE (1980) Aerobic and anaerobic metabolism during voluntary diving in weddell seals: evidence of preferred pathway from blood chemistry and behaviour. J Comp Physiol 138:335–346

Leivestad H (1960) The effect of prolonged submersion on the metabolism and the heart rate of the toad *(Bufo bufo)*. Arboh University Bergen Mat Naturv 5:1–15

Livingstone DR (1983) Invertebrate and vertebrate pathways of anaerobic metabolism: evolutionary considerations. J Geol Soc 140:27–37

Livingstone DR, Zwaan A de (1983) Carbohydrate metabolism in gastropods. In: Hochachka PW (ed) The mollusca, vol 1. Metabolic biochemistry and molecular biomechanics. Academic, New York, pp 177–242

Livingstone DR, Zwaan A de, Leopold M, Marteijn E (1983) Studies on the phylogenetic distribution of pyruvate oxido reductase. Biochem Syst Ecol 11:415–425

Meinardus G, Gäde G (1981) Anaerobic metabolism of the common cockle, *Cardium edule* IV. Time dependent changes of metabolites in the foot and gill tissue induced by anoxia and electrical stimulation. Comp Biochem Physiol 70B:271–277

Meyer RA, Terjung RL (1980) AMP deamination and IMP reamination in working skeletal muscle. Am J Physiol 239:C32–C38

Mitchell GS, Gleeson TT, Bennett AF (1981) Ventilation and acidbase balance during graded activity in lizards. Am J Physiol 240:R29–R37

Morris GM, Baldwin J (1984) pH buffering capacity of invertebrate muscle: correlations with anaerobic muscle work. Physiol 5:61–70

Mourik J (1982) Anaerobic metabolism in red skeletal muscle of goldfish (*Carassius auratus* L.). Ph.D. Thesis, University of Leiden, The Netherlands

Nuccitelli R, Heiple JM (1982) Summary of the evidence and discussion concerning the involvement of pHi in the control of cellular functions. In: Intracellular pH: its measurement, regulation and utilization in cellular functions. Kroc Foundation series, vol 15. A.R. Liss, New York, pp 567–586

Onnen T, Zebe E (1983) Energy metabolism in the tail muscles of the shrimp *Crangon crangon* during work and subsequent recovery. Comp Biochem Physiol 74A:833–838

Ott ME, Heisler N, Ultsch GR (1980) A revaluation of the relationship between temperature and the critical oxygen tension in fresh water fishes. Comp Biochem Physiol 67A:337–340

Pearce FJ, Connett RJ (1980) Effect of lactate and palmitate on substrate utilization of isolated rat soleus. Am J Physiol 238:C149–C159

Penney DG (1974) Effects of prolonged diving anoxia on the turtle, *Pseudemys scripta elegans*. Comp Biochem Physiol 37A:933–941

Penney DG, Cascarano J (1970) Anaerobic rat heart. Effects of glucose and tricarboxylic acid-cycle intermediates on metabolism and physiological performance. Biochem J 118:231–227

Petterson G, Arnqvist HJ, Varenhorst E (1982) Pasteur effect in human arterial tissue. Acta Physiol Scand 114:639–640

Pörtner HO (1982) Biochemische und physiologische Anpassungen an das Leben im marinen Sediment: Untersuchung am Spritzwurm *Sipunculus nudus* L. Dissertation, Universität Düsseldorf

Pörtner HO, Kreutzer U, Siegmund B, Heisler N, Grieshaber MR (1984) Metabolic adaptation of the intertidal worm *Sipunculus nudus* to functional and environmental hypoxia. Mar Biol 75: in press

Prosser CL (1973) Comparative animal physiology. W Saunders, Philadelphia

Putzer V (1984) Energy production and glycolytic flux during functional and environmental anoxia in *Lumbriculus variegatus*. Abstract First Int Congr C.P.B., Liege, Belgium

Putzer V (1985) Biochemische Anpassung von *Lumbriculus variegatus* an ökologische und physiologische Sauerstofflimitierung. Dissertation, Universität Innsbruck, Austria, in press

Randall DJ, Daxboeck C (1982) Cardiovascular changes in rainbowtrout (*Salmo gairdneri* R.) during exercise. Can J Zool 60:1135–1140

Ruben JA, Battalia DE (1979) Aerobic and anaerobic metabolism during activity in small rodents. J Exp Zool 208:73–76

Sahlin K, Edström L, Sjöholm H, Hultman E (1981) Effects of lactic acid accumulation and ATP decrease on muscle tension and relaxation. Am J Physiol 240:C121–C126

Sanborn T, Gavin W, Berkowitz S, Perille T, Lesch M (1979) Augmented conversion of aspartate and glutamate to succinate during anoxia in rabbit heart. Am J Physiol 237:H535–H541

Schertzer HG, Cascarano J (1978) Anaerobic rat heart: mitochondrial role in calcium uptake and contractility. J Exp Zool 207:337–350

Schöttler U (1977) The energy-yielding oxidation of NADH by fumarate in anaerobic mitochondria of *Tubifex sp.* Comp Biochem Physiol 58B:151–156

Schöttler U (1979) On the anaerobic metabolism of three species of Nereis (annelida). Mar Ecol Prog Ser 1:249–254

Schöttler U (1980) The energy metabolism during facultative anaerobiosis: investigations on Annelids. Verh Dtsch Zool Ges: 228–240

Schöttler U (1982) An investigations on the anaerobic metabolism of *Nephtys hombergii* (Annelida: Polychaeta). Mar Biol 71:265–269

Schöttler U, Wienhausen G, Zebe E (1983) The mode of energy production in the lugworm *Arenicola marina* at different oxygen concentrations. J Comp Physiol 149:547–555

Schroff G, Zebe E (1980) The anaerobic formation of propionic acid in the mitochondria of the lugworm *Arenicola marina*. J Comp Physiol 183:35–41

Schulz TKF, Kluytmans JH (1983) Pathway of propionate synthesis in the sea mussel *Mytilus edulis* L. Comp Biochem Physiol 75B:365–372

Seo Y, Yosizaki K, Morimoto T (1983) A ^{1}H-nuclear magnetic resonance study on lactate and intracellular pH in frog muscle. Jpn J Physiol 33:721–731

Seuss J, Hipp E, Hoffmann KH (1983) Oxygen consumption, glycogen content and the accumulation of metabolites in Tubifex during aerobic-anaerobic shift and under progressing anoxia. Comp Biochem Physiol 75A:557–562

Shelley HJ (1961) Glycogen reserves and their changes at birth and in anoxia. Br Med Bull 17: 137–143

Shoubridge E, Hochachka PW (1980) Ethanol: novel end product of vertebrate anaerobic metabolism. Science 209:308–309

Siegmund B, Grieshaber MK (1984) Opine metabolism of *Arenicola marina* L. Abstract First Int Congr C.P.B., Liege, Belgium

Simkiss K (1968) Calcium and carbonate metabolism in the frog *(Rana temporaria)* during respiratory acidosis. Am J Physiol 214:627–634

Smith MJ, Heath AG (1980) Responses to acute anoxia and prolonged hypoxia by rainbow trout *(Salmo gairdneri)* and mirror carp *(Cyprinus carpio)* red and white muscle: use of conventional and modified metabolic pathways. Comp Biochem Physiol 66B:267–272

Sürholt B (1977) Production of volatile fatty acids in anaerobic carbohydrate catabolism of *Arenicola marina*. Comp Biochem Physiol 58B:147–150

Taegtmeyer H (1978) Metabolic responses to cardiac hypoxia. Increased production of succinate by rabbit papillary muscles. Circ Res 43:808–815

Thomas S (1983) Changes in blood acid-base balance in trout (*Salmo gairdneri* R.) following exposure to combined hypoxia and hypercapnia. J Comp Physiol 152:52–57

Thompson RJ, Livingstone DR, Zwaan A de (1980) Physiological and biochemical aspects of the valve snap and valve closure responses in the giant scallop *Placopecten magellanicus*. J Comp Physiol 137:97–104

Ultsch G (1974) Gas exchange and metabolism in the sirenidae (Amphibia: candala)-I. Oxygen consumption of submerged sirenids as a function of body size and respiratory surface area. Comp Biochem Physiol 47A:485–498

Ultsch GR, Jacuson DC (1982) Long-term submergence at 3 °C of the turtle *Chrysemys picta bellii*, in normoxic and severely hypoxic water. I-Survival, gas exchange and acid-base status. J Exp Biol 96:11–28

Van den Thillart G (1982) Adaptations of fish energy metabolism to hypoxia and anoxia. Mol Physiol 2:49–61

Van den Thillart G (1984) Energy metabolism of swimming trout (*Salmo gairdneri)*. Oxidation rates of intra-arterially injected glucose, palmitate, glutamate, alanine and leucine. J Comp Physiol (in press)

Van den Thillart G, Kesbeke F (1978) Anaerobic production of carbon dioxide and ammonia by goldfish, *Carassius auratus* (L.). Comp Biochem Physiol 59A:393–400

Van den Thillart G, Smit H (1984) Carbohydrate metabolism of goldfish (Carassius auratus L.). Effects of long term hypoxia acclimation on enzyme patterns of red muscle, white muscle and liver. J Comp Physiol 154:477–486

Van den Thillart G, Kesbeke F, Van Waarde A (1980) Anaerobic energy metabolism of goldfish, *Carassius auratus* (L.). Influence of hypoxia and anoxia on phosphorylated compounds and glycogen. J Comp Physiol 136:45–52

Van den Thillart G, Van Berge Henegouwen M, Kesbeke F (1983) Anaerobic metabolism of goldfish, *Carassius auratus* (L.): Ethanol and CO_2 excretion rates and anoxia tolerance at 20, 10 and 5 °C. Comp Biochem Physiol 76A:295–300

Van Waarde A, Kesbeke F (1983) Goldfish muscle energy metabolism during electrical stimulation. Comp Biochem Physiol 75B:635–639

Van Waarde A, Van den Thillart G, Dobbe F (1982) Anaerobic energy metabolism of goldfish, *Carassius auratus* (L.). Influence of anoxia on mass-action ratios of transaminase reactions and levels of ammonia and succinate. J Comp Physiol 147:53–59

Van Waarde A, Van den Thillart G, Kesbeke F (1983) Anaerobic energy metabolism of the european eel, Anguilla anguilla L. J Comp Physiol 149:469–475

Wardle CS (1978) Non-release of lactic acid from anaerobic swimming muscle of plaice *Pleuronectes platessa* L: a stress reaction. J Exp Biol 77:141–155

Whalen WJ, Buerk D, Thuning CA (1973) Blood flow limited oxygen consumption in resting cat skeletal muscle. Am J Physiol 224:763–768

Westra HG, Haan A de, van Doorn H, Haan EJ de (1982) Short term persistent metabolic changes as induced by exercise. In: Addink, Spronk (eds) Exogenous and endogenous influences on metabolic and nerval control. Pergamon, Oxford, pp 285–295

Wilps H, Schöttler U (1980) In vitro studies on the anaerobic formation of ethanol by the larvae of *Chironomus thummi thummi* (diptera). Comp Biochem Physiol 67B:239–242

Wijsman TC (1975) pH fluctuations in *Mytilus edulis* L. in relation to shell movements under aerobic and anaerobic conditions. In: Barnes H (ed) Proc 9th Europ Mar Biol Symp. Aberdeen University Press, pp 139–149

Zebe E (1982) Anaerobic metabolism in *Upogebia pugettensis* and *Callianassa californiensis* (Crustacea, Thalassinidae). Comp Biochem Physiol 72B:613–617

Zebe E, Salge U, Wiemann C, Wilps H (1981) The energy metabolism of the leech *Hirudo medicinalis* in anoxia and muscular work. J Exp Zool 218:157–163

Zurburg W, Bont AMT de, Zwaan A de (1981) Recovery from exposure to air and the occurrence of strombine in different organs of the sea mussel *Mytilus edulis*. Mol Physiol 2:135–147

Zwaan A de (1983) Carbohydrate catabolism in bivalves. In: Hochachka PW (ed) The Mollusca, vol 1. Metabolic biochemistry and molecular biomechanics. Academic, New York, pp. 13–175

Zwaan A de, Dando PR (1984) Phosphoenolpyruvate-pyruvate metabolism in bivalve molluscs. Mol Physiol 5:285–310

Zwaan A de, Putzer V (1985) Metabolic adaptations of intertidal invertebrates to environmental hypoxia – Comparison of environmental anoxia exercise anoxia. In: Laverack MS (ed) Physiological adaptations of marine animals. SEB Symp No 39 (in press)

Zwaan A de, Skjoldal HR (1979) Anaerobic energy metabolism of the scavenging isopod *Cirolana borealis* (Lilljeborg). J Comp Physiol 129:327–331

Zwaan A de, Mohamed AM, Geraerts WPM (1976) Glycogen degradation and the accumulation of compounds during anaerobiosis in the fresh water snail *Lymnaea stagnalis*. Neth J Zool 26: 549–557

Zwaan A de, Thompson RJ, Livingstone DR (1980) Physiological and biochemical aspects of the valve snap and valve closure responses in the giant scallop *Placopecten magellanicus*. J Comp Physiol 137:105–114

Zwaan A de, Holwerda JA, Veenhof PR (1981) Anaerobic malate metabolism in mitochondria of the sea mussel *Mytilus edulis* L. Mar Biol Let 2:131–140

Zwaan A de, Bont AMt de, Verhoeven A (1982) Anaerobic energy metabolism in isolated adductor muscle of the sea mussel *Mytilus edulis* L. J Comp Physiol 149:137–143

Zwaan A de, Bont AMT de, Hemelraad J (1983) The role of phosphoenolpyruvate carboxykinase in the anaerobic metabolism of the sea mussel Mytilus edulis L. J Comp Physiol 153:267–274

Zwaan A de, Bont AMT de, Nilsson P (1984) Anaerobic energy metabolism in two organs of the fresh water mussel Anodonta cygnea L. Abstract. First Int Congr C.P.B., Liège, Belgium

Metabolic Biochemistry of Insect Flight

K.B. STOREY[1]

1 Introduction

An overview of the biochemistry of insect flight muscle will be presented incorporating information on the intermediary metabolism of fuels, enzyme regulation, mitochondrial function and neuronal and hormonal control of metabolism.

Insect species can be placed into one of three groups (transcending phylogenetic relationships) depending upon the metabolic fuels used for flight. Group 1 contains insects which oxidize carbohydrate (glycogen, blood sugars) plus proline providing the C2 and C4 inputs for the tricarboxylic acid cycle, respectively. Either input can be maximized to suit the individual species producing insects which utilize primarily carbohydrate or primarily proline. Group 2 contains species which fuel flight by lipid oxidation utilizing either carbohydrate or proline as an anapleurotic source of C4. Group 3 contains those species which can switch from the use of carbohydrate/proline during short-term flight to lipid oxidation for long-term flight. These metabolic groupings result from the requirements of an obligately aerobic flight muscle metabolism and the need for essentially instantaneous activation of metabolism and for maximal functioning of energy producing pathways. The factors important in activating and sustaining muscle metabolism during flight will be considered for each group.

Hormonal control of flight will be considered with respect to the activation of flight muscle metabolism and the long-term mobilization of fuels for sustained flight. The relative effects of individual hormones appear to be tuned to the type of fuel utilization (group 1, 2, or 3) such that anomalies arise when the effects of hormones are tested between species without regard for the metabolic make-up of the species.

Enzymatic control of flight muscle metabolism will consider regulatory enzymes such as glycogen phosphorylase, hexokinase, phosphofructokinase, glutamate dehydrogenase, and lipase. Regulatory mechanisms promoting enzyme activation during the initiation of flight will be discussed and new areas of metabolic control (covalent modification by phosphorylation-dephosphorylation, the allosteric effector fructose-2,6-biphosphate) will be evaluated.

Insect flight muscle appears to be the ultimate aerobic catabolic machine adapted at all levels of biological organization (anatomical, physiological, biochemical) for maximal energy output.

1 Institute of Biochemistry and Department of Biology, Carleton University, Ottawa, Ontario, Canada K1S 5B6

Circulation, Respiration, and Metabolism
(ed. by R. Gilles)

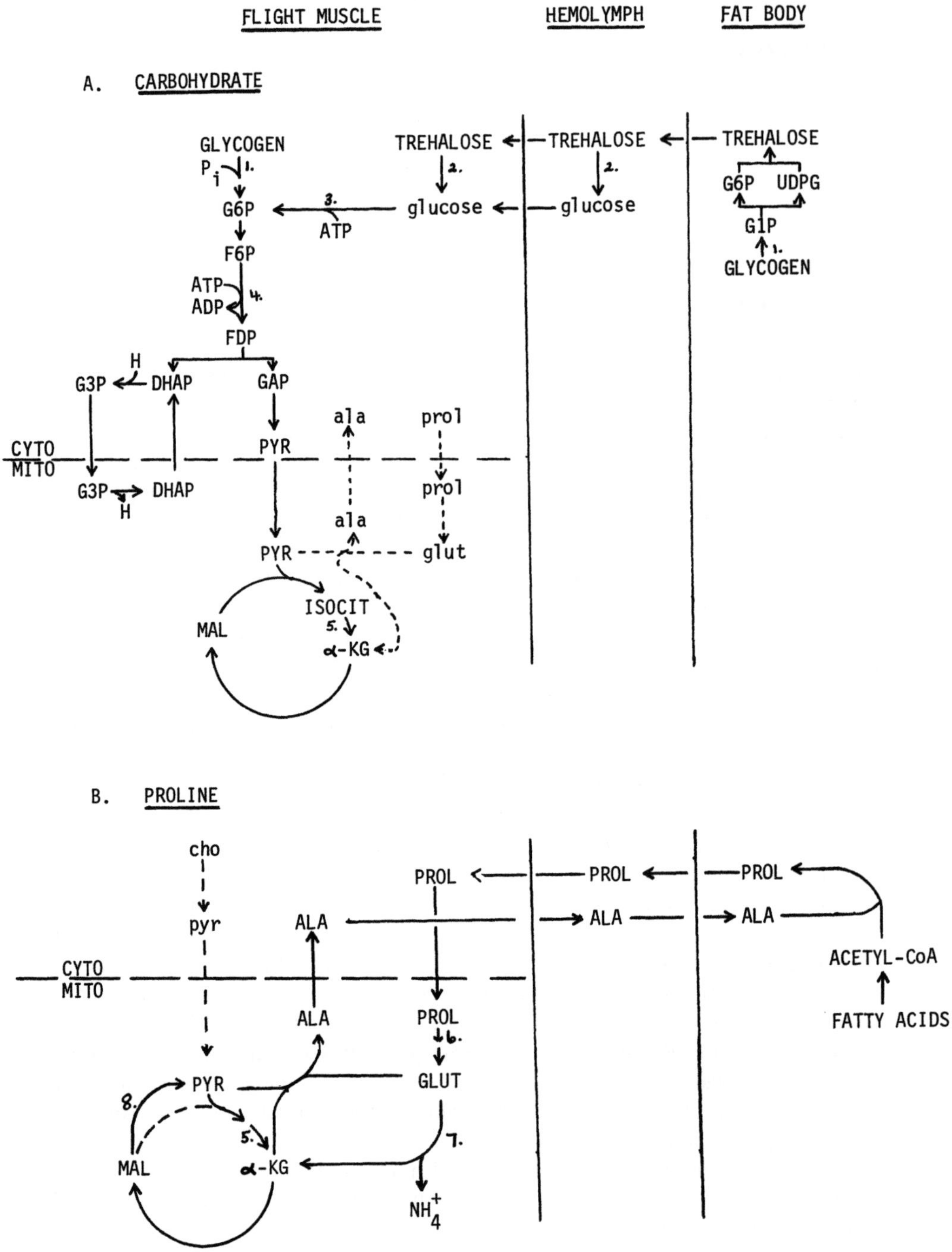
FLIGHT MUSCLE
HEMOLYMPH
FAT BODY
A. CARBOHYDRATE
GLYCOGEN
TREHALOSE
TREHALOSE
TREHALOSE
G6P
UDPG
G1P
GLYCOGEN
G6P
ATP
glucose
glucose
F6P
ATP
ADP
FDP
G3P
DHAP
GAP
ala
prol
CYTO
MITO
PYR
G3P
DHAP
prol
ala
PYR
glut
ISOCIT
MAL
α-KG
B. PROLINE
cho
pyr
PROL
PROL
PROL
ALA
ALA
ALA
ACETYL-CoA
FATTY ACIDS
CYTO
MITO
ALA
PROL
PYR
GLUT
MAL
α-KG
NH4+

Fig. 1A,D

Numerous reviews of the biochemistry of insect flight can be found including those by Sacktor (1970, 1975), Crabtree and Newsholme (1975), Steele (1981), Bursell (1981), and Beenakkers et al. (1981a,b, 1984a,b). My intention here is to concentrate on principles of metabolic control including some of the new developments in enzyme regulation, the role of hormones in flight and a brief examination of the parallels in muscle exercise between insects and mammals.

2 Flight Fuels

Flight in insects is powered by the aerobic oxidation of fuels. Figure 1 outlines the options for fuel utilization.

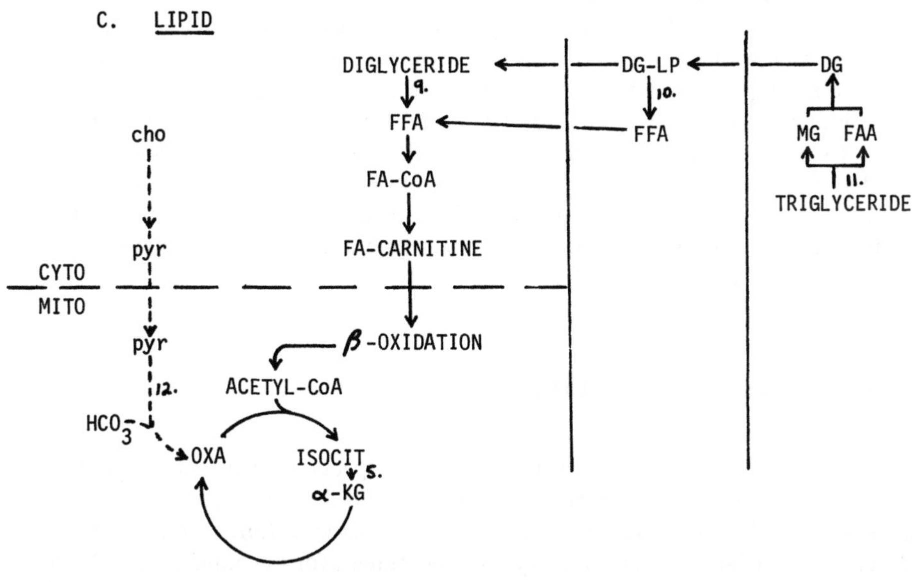

Fig. 1A–C. Pathways of fuel utilization in insect flight muscle. **A** Routes of carbohydrate catabolism showing the use of muscle glycogen, hemolymph sugars, trehalose output from fat body and the coupled oxidation of proline. Carbohydrate as the major flight fuel typically is used by many species of Diptera and Hymenoptera. **B** Routes of proline oxidation. A partial oxidation, using transamination for production of α-ketoglutarate, leads to the synthesis of alanine as the end product and the potential for resynthesis of proline in the fat body. This route is typical of species which use proline as the major flight fuel (with minor carbohydrate utilization) such as the tsetse fly, *G. morsitans*. Complete oxidation of proline to CO_2, H_2O, and NH_4^+ requires the action of glutamate dehydrogenase and may occur in species with a more balanced oxidation of proline and carbohydrate such as the Japanese beetle, *P. japonica*. **C** Routes of lipid catabolism showing the oxidation of diglyceride and replenishment from hemolymph lipoprotein and fat body reserves. A coupled minor catabolism of carbohydrate provides 4-carbon units for the TCA cycle. Lipid as the major flight fuel is typical of many Lepidoptera and Orthoptera. Enzymes: *1* Glycogen phosphorylase; *2* trehalase; *3* hexokinase; *4* phosphofructokinase; *5* isocitrate dehydrogenase; *6* proline dehydrogenase; *7* glutamate dehydrogenase; *8* malic enyzme; *9* diglyceride lipase; *10* lipoprotein lipase; *11* triglyceride lipase; *12* pyruvate carboxylase

Insects which power flight primarily from the oxidation of carbohydrate reserves can draw upon two sources, glycogen in the muscle and the hemolymph disaccharide, trehalose. Both are oxidized by the classical pathways of glycolysis and the tricarboxylic acid (TCA) cycle with the α-glycerophosphate cycle used for redox balance. Glycogen is often used to power take-off or preflight warm-up only with sustained flight drawing upon hemolymph trehalose which can be replenished by the mobilization of fat body glycogen reserves.

Fatty acids are oxidized by the familiar pathways of β-oxidation. Lipid as a flight fuel predominates amongst migratory species and those which are non-feeding as adults. Most often lipid is supplied as diglyceride attached to hemolymph lipoproteins and originating from fat body triglyceride reserves. Various Hemipterans, however, maintain large triglyceride stores within the flight muscle (Ward et al. 1982).

Proline is used as a supplimentary fuel by many species and as the major flight fuel by selected species such as the tsetse fly, *Glossina morsitans* and the Colorado potato beetle, *Leptinotarsa decemlineata*. Two oxidative reactions convert proline to glutamate which then enters the TCA cycle at α-ketoglutarate after transamination or deamination. When used as the major flight fuel the predominant fate of proline is conversion to alanine, an incomplete oxidation but one which nets an energy yield, 0.52 mol ATP g^{-1}, comparable to that of lipid oxidation (0.65 mol ATP g^{-1} vs 0.18 for carbohydrate) (Bursell 1981). Alanine is returned to the fat body where in combination with acetyl-CoA (derived from lipids) proline is resynthesized. Two other fates of proline can occur: (1) incorporation of C_4 to augment TCA cycle intermediates, or (2) complete oxidation to CO_2 and H_2O requiring glutamate dehydrogenase (deamination) for the entry of carbon into the TCA cycle.

3 Three Patterns of Fuel Utilization

Probably 500,000 species of flying insects occur. Aspects of flight muscle metabolism have been examined in perhaps 50 while only a handful have received detailed attention. However, although some details differ, three basic patterns of fuel utilization emerge. These are largely determined by the requirement for the coupling of a 2-carbon and a 4-carbon input for TCA cycle function. This coupling most often takes the form of a major (generally C_2 acetyl-CoA) and a minor substrate input. Complete in vivo studies invariably reveal these couplings although in many instances the minor substrate input has been neglected. Studies on isolated mitochondria often yield confusion as, in vitro, mitochondria can respire on one substrate only. Metabolism in many species needs to be re-examined with these substrate pairings in mind; experiments employing injected radiolabelled substrates added individually would be very useful in this regard.

3.1 Group 1 Insects: Carbohydrate/Proline Powered Flight

The coupled oxidation of carbohydrate and proline powers flight in many species. Both fuels can be rapidly mobilized and are particularly suited to the instantaneous energy generation required during take-off and the high speed flight of many Diptera and Hymenoptera. The proportions of the two fuels used can vary enormously from species which burn predominantly carbohydrate (e.g. the blowfly, *Phormia regina*) to those which burn predominantly proline *(G. morsitans, L. decemlineata)* (Fig. 1). Even in *G. morsitans*, where significant changes in carbohydrate cannot be demonstrated during flight, changes in the levels of glycolytic intermediates during flight clearly show an activation of glycolysis and a role for carbohydrate as the coupled substrate for proline oxidation (Olembo and Pearson 1982).

The flexibility for choosing either carbohydrate or proline as the major flight fuel is due to the special status of proline. Via transamination from aspartate mammalian systems directly feed oxaloacetate into the TCA cycle. Provision of oxaloacetate from proline, however, requires five oxidative reactions and provides instantaneous aerobic energy production while priming the TCA cycle. Parenthetically the rapid ATP production from proline oxidation may be the reason that only very minor changes in ATP and arginine phosphate are seen in insect muscle during flight initiation and the activation of glycolysis. In addition proline has a high energy yield on a per gram basis making the amino acid an effective fuel for long-term flight in species which have very high proline reserves or which can regenerate proline from alanine during flight (e.g. *L. decemlineata*) (Beenakkers et al. 1984b).

3.2 Group 2 Insects: Lipid Powered Flight

Oxidation of diglycerides provides the high flux C_2 substrate for muscle energy production in many species. The light weight of lipid reserves coupled with the high energy yield per gram makes this fuel particularly suited for long-term flight. The minor, C_4, TCA cycle input appears to be carbohydrate in some species, pyruvate carboxylase catalyzing the conversion of pyruvate to oxaloacetate. It is also possible that proline could serve in this role although such a pattern of flight muscle metabolism has not yet been shown. Mitochondria from group 2 insects oxidize both fatty acyl carnitines and pyruvate (Hansford and Johnson 1976). Carbohydrate is likely the major fuel utilized during the period of preflight warm up in these species although this has largely gone untested.

3.3 Group 3 Insects: Short-Term Carbohydrate, Long-Term Lipid

Carbohydrate/proline oxidation offers the instantaneous energy production needed for take-off and short bursts of intense flight. Lipid, because of the longer time involved in mobilizing and transporting reserves, is a poor substrate for short-term flight but a highly efficient fuel for the long-term. Various species of insects, typified by the locust, have lifestyles requiring both short-term burst flights and long-term migratory flights. Flight muscle metabolism in these species makes use of carbohydrate, proline, and lipid. Take-off in the locust is powered by the instantaneous oxidation of muscle glycogen and proline (Rowan and Newsholme 1979). Muscle glycogen is quickly exhausted and short-term flights of 20–30 min are largely fueled by the oxidation of hemolymph trehalose (Beenakkers et al. 1981b). During this time lipid mobilization under the control of adipokinetic hormone begins and lipid takes over as the major flight fuel (with continued low levels of trehalose used) after flight times exceeding about 30 min. Group 3 insects have the most generalized options for fueling flight and this is perhaps the basic pattern from which the more specialized metabolism of group 1 and 2 insects has developed. Indeed, group 2 insects which use a carbohydrate-based thermogenesis during preflight are actually following a group 3 pattern although passing through the carbohydrate phase before take-off.

4 Control of Metabolism During Flight

The initiation of flight results in a 20- to 100-fold increase in oxygen consumption and metabolic rate compared to only a 7- to 10-fold increase for aerobic work in mammalian muscles (Sacktor 1975; Beenakkers et al. 1984a). The metabolic demands of flight therefore require an enormous increase in energy production as well as highly coordinated and extremely rapid activation of aerobic metabolism. In addition a comparison of metabolic rate during flight (C_6 utilization rate) with the maximal activities of key regulatory enzymes (phosphorylase, hexokinase, phosphofructokinase) demonstrates that flight requires full expression of the enzymatic potential of the muscle (Crabtree and Newsholme 1975). This requires a much stronger control of enzymes by activating and deinhibiting modulators than is required in mammalian systems where the enzymatic potential of the cell typically exceeds maximal flux rates by 5- to 10-fold.

Control of metabolism during flight can be broken down into two areas. The first is the activation of muscle energy production during take-off. This largely involves allosteric control of key regulatory enzymes within the muscle. The second area of control is the regulation of sustained fuel supplies for long-term flight. In many instances this control occurs outside of the muscle and is based upon hormonal regulation of fuel production and release from the fat body.

4.1 Regulation of Carbohydrate Metabolism

The activation of glycolysis at the initiation of flight is controlled at three sites. Glycogen phosphorylase and hexokinase control carbon input from glycogen or blood sugar, respectively, while phosphofructokinase controls the overall rate of C_6 utilization. At the TCA cycle, NAD-isocitrate dehydrogenase provides the primary regulation of the rate of acetyl-CoA utilization. During long-term flight sustained fuel supply is dependent upon the activation of glycogen phosphorylase in the fat body to supply glucosyl units for trehalose synthesis. Thus take-off and the early minutes of flight rely upon endogenous glycogen reserves in muscle and existing trehalose (glucose) supplies in hemolymph. Some species, such as the cockroach (Elliot et al. 1984), appear to be largely limited in their flight capabilities by these supplies. Most species, however, begin a hormone mediated production of trehalose in the fat body within the first few minutes of flight and long-term flight is characterized by steady state levels of trehalose in hemolymph (generally lower than resting levels) determined by the balanced rates of trehalose oxidation by muscle and production by fat body.

4.1.1 Kinetic Control of Carbon Input into Glycolysis

Glycogen phosphorylase *a* and *b* from blowfly flight muscle have been purified and characterized and the control of glycogen breakdown has been extensively reviewed (Sacktor 1975). The primary control of phosphorylase is covalent modification via phosphorylation/dephosphorylation. The transition from rest to flight in the blowfly increases the percentage of phosphorylase in the active *a* form from 18% to 70% within 15 s of flight. Ca^{2+} activation of phosphorylase *b* kinase mediates this conversion.

The use of hemolymph sugars is regulated by hexokinase. The primary regulatory feature of the locust muscle enzyme is potent product inhibition by glucose-6-P (Storey 1980a). Activation of the enzyme depends on the reversal of glucose-6-P inhibition and this occurs via deinhibition by P_i, alanine, and glycerol-3-P. Thus although take-off may elevate glucose-6-P in locust muscle (Rowan and Newsholme 1979) increased levels of alanine (from the primary catabolism of glycogen to alanine) and P_i deinhibit hexokinase and promote the second stage of flight, the utilization of hemolymph sugars as fuels.

Use of hemolymph trehalose may also be regulated by trehalase which occurs in both the muscle and hemolymph. Kinetic studies of the insect enzyme are lacking but the enzyme in yeast is now known to be regulated by protein kinase-mediated phosphorylation (Uno et al. 1983). There is conflicting evidence as to whether muscle trehalase activity can be modified during flight or in response to hormones (Beenakkers et al. 1984a); obviously the enzyme must be characterized and examined for covalent modification, particularly in group 1 insects.

4.1.2 Kinetic Control of Glycolytic Rate

Phosphofructokinase (PFK), which controls the rate of hexose phosphate utilization by glycolysis, typically has complex properties including sigmoidal fructose-6-P kinetics, substrate inhibition by high levels of ATP, activation by AMP, NH_4^+, P_i and the products ADP and fructose-1,6-P_2 and inhibition by citrate and glycerol-3-P. Since 1980 regulatory control of PFK has been extensively re-evaluated due to the discovery of fructose-2,6-P_2 (Hers and van Schaftingen 1982). This compound is an extremely potent activator of PFK acting at levels of 1 μM or less and provides a mechanism for hormonal control of PFK since hormone stimulated phosphorylation of 6-phosphofructo-2-kinase regulates fructose-2,6-P_2 production.

PFK from insect flight muscle lacks activation by ADP and fructose-1,6-P_2 and inhibition by citrate. The enzyme from cockroach flight muscle has an extremely high affinity constant for fructose-6-P ($S_{0.5}$ = 16 mM) compared to levels of the compound in flight muscle (0.04 to 0.11 μmol g^{-1}) and is also strongly inhibited by physiological (5 μmol g^{-1}) levels of ATP (0.2 mM is the enzyme optimum) (Storey 1983, 1984). Expression of enzyme activity in vivo is absolutely dependent, therefore, on the actions of allosteric activators. Activation constants for NH_4^+, P_i, AMP, and fructose-2,6-P_2 (1 mM, 0.7 mM, 8 μM, and 0.2 μM, respectively) are all well within physiological levels of these compounds. The effects of activators are also additive with synergistic interactions between AMP and fructose-2,6-P_2 such that near physiological levels (3 mM NH_4^+, 10 mM P_i, 0.4 mM AMP, and 1 μM fructose-2,6-P_2) lower $S_{0.5}$ for fructose-6-P by 640-fold to 0.025 mM and raise the I_{50} for ATP to 18 mM (Storey 1984). Although

P_i content of cockroach muscle remains constant during flight initiation, levels of NH_4^+, AMP, and fructose-2,6-P_2 increase by 2.8-, 2.7-, and 2.2-fold (Storey 1983) to achieve a maximal activation of PFK during flight.

4.1.3 Inhibitory Control of Glycolysis During Lipid Oxidation

A common pattern of fuel utilization in both vertebrates and invertebrates is the use of carbohydrates for short-term (often burst) work followed by, if work becomes sustained, the mobilization of lipid reserves for long-term energy production. The switch from carbohydrate to lipid oxidation is determined partly by the relative availability of the two substrates but also has a kinetic basis in inhibitory effects of lipid metabolites on glycolytic rate. In vertebrates, citrate inhibition of PFK 'spares' carbohydrate oxidation during lipid catabolism. However insect PFK is not citrate inhibited and perfusion studies of locust flight muscle suggested that the inhibitory effect of lipid catabolism was localized at the aldolase reaction (Ford and Candy 1972). Studies of purified aldolase from locust flight muscle (Storey 1980b) have now shown that inhibition of this enzyme by citrate (K_i = 1.9 m*M*) and long chain acyl carnitines (K_i palmitoyl carnitine = 18 μ*M*), which accumulate during lipid oxidation (Worm et al. 1980), provides the mechanism for inhibitory control of glycolysis during lipid catabolism.

4.2 Regulation of the TCA Cycle

Primary control of flux through the TCA cycle in insect flight muscle is at the NAD-isocitrate dehydrogenase reaction. This has been convincingly shown in isolated mitochondria studies with insects utilizing carbohydrate *(P. regina)*, lipid *(M. sexta)* or proline *(Popillia japonica)* as major flight fuels (Johnson and Hansford 1975; Hansford and Johnson 1975, 1976) and is also supported by metabolite changes in vivo during flight (Olembo and Pearson 1982). NAD-isocitrate dehydrogenase is ADP activated and, at least in asynchronous flight muscle, is Ca^{2+} inhibited. The metabolic changes at the initiation of flight, hydrolysis of ATP to ADP and an efflux of mitochondrial Ca^{2+}, result in the activation of the enzyme and the increase in TCA cycle flux.

4.3 Regulation of Lipid Metabolism

The activation of lipid catabolism in muscle is initiated by lipases, either diglyceride lipase in the muscle or lipoprotein lipase in the hemolymph or both. Control properties of these enzymes have not been studied; they may be sensitive to hormone-mediated activation via phosphorylation. The rate of β-oxidation of fatty acids is likely passively controlled by substrate availability while the rate of oxidation of acetyl-CoA by the TCA cycle is for lipid, as it is for carbohydrate, controlled by a NAD-isocitrate dehydrogenase.

During long-term flight sustained lipid supply from the fat body is required and this is controlled by hormone-sensitive lipases which break down fat body triglycerides to monoglycerides and fatty acids. 1,2-Diglycerides are then synthesized and released into the hemolymph. Both octopamine and adipokinetic hormone stimulate the production and release of diglycerides from fat body.

4.4 Regulation of Proline Metabolism

The rate determining step for proline oxidation in flight muscle is again NAD-isocitrate dehydrogenase in the TCA cycle (Hansford and Johnson 1975). This appears to be true even in species which primarily convert proline to alanine although α-ketoglutarate dehydrogenase may also have a regulatory role in these species (Olembo and Pearson 1982). Lesser points of control may include proline dehydrogenase, glutamate dehydrogenase and malic enzyme. Proline dehydrogenase is ADP activated and glutamate inhibited so that the initiation of muscle work, with a rise in ADP and a fall in glutamate (as the amino acid is utilized by the TCA cycle), would activate proline oxidation (Bursell 1981). Flight muscle glutamate dehydrogenase (as opposed to the fat body isozyme) is tightly controlled by nucleotides, activation by ADP and inhibition by GTP and ATP (Male and Storey 1983) and as such is activated to allow the oxidative deamination of glutamate at the initiation of flight. Flight muscle malic enzyme is affected by adenylates and various organic acids (pyruvate, succinate) (Hansford and Johnson 1975).

4.5 Co-Ordinated Control of Flight Muscle Metabolism by Ca^{2+}

Ca^{2+} forms the basis for the integrated activation of muscle metabolism at the initiation of flight (Sacktor 1975). At rest the ion is sequestered in the sarcoplasmic reticulum or the mitochondria. Nervous stimulation releases Ca^{2+} into the cytoplasm activating actinomysin ATPase to initiate contraction. Ca^{2+} also activates glycerol-3-P oxidase to stimulate the NADH shuttle while the movement of Ca^{2+} out of the mitochondria releases inhibition of NAD-isocitrate dehydrogenase. Ca^{2+} influences the activity of key enzymes through modulation of phosphorylation state. Thus phosphorylase *b* kinase is Ca^{2+} activated as is the protein phosphatase responsible for dephosphorylation of pyruvate dehydrogenase, producing the active enzyme form. Other potential sites for Ca^{2+} mediated phosphorylation control include trehalase, 6-phosphofructo-2-kinase and diglyceride lipase; controls of these enzymes by both allosteric effectors and covalent modification require investigation.

5 Hormonal Control of Flight

Hormonal control of flight muscle metabolism has been reviewed by a number of authors including Candy (1981), Goldsworthy and Gade (1983), Steele (1983), and Beenakkers et al. (1984a). Two types of hormones participate during flight. The catecholamine, octopamine, increases in concentration in hemolymph within the first minutes of flight and mediates the general arousal (stimulation of muscle contraction, release of fuels from storage organs) known as 'fight or flight'. Peptide hormones from the glandular lobe of the corpus cardiacum (CC) begin to increase in the hemolymph after several minutes of flight. These hormones, either hypertrehalosemic hormone (HTH) (also called trehalagon) or adipokinetic hormone (AKH) mediate the sustained release of either trehalose or diglyceride from the fat body to support long-term flight.

The functions of CC hormones are only now becoming clarified and have suffered from years of confusing data brought about by unfortunate experimental design including: (1) the use of whole CC extracts containing a mixture of hormonal factors; (2) the mixing of species (from groups 1, 2, and 3) in chosing donor and recipient for CC extracts; and (3) the use of pharmacological doses. With the purification of AKH, studies are now sorting out the physiological functions of this hormone, particularly in locusts. However after surveying the literature, I would like to offer a few guidelines for investigation. (1) *The major fuel for flight will determine the fat body response to CC extracts.* Hormone stimulation increases cAMP and/or Ca^{2+} levels. However the response of fat body to these messengers differs for the three metabolic groups. The primary hormone-mediated response of fat body in group 1 insects is carbohydrate catabolism, phosphorylase activation and trehalose output. Group 2 and 3 insects are primed to activate lipid output under hormone stimulation. Thus anomalous results occur when donor and recipient of CC extracts are from different metabolic groups; for example, locust CC extract or even AKH itself raises hemolymph lipid in locusts and hemolymph trehalose in cockroaches (Beenakkers et al. 1984a). The primary experiment with any new species must therefore be to determine the major substrate mobilized from fat body during long-term flight; this knowledge will indicate the enzyme machinery most sensitive to hormone stimulation and will suggest the hormone factor, AKH or HTH, to be sought. (2) *Purified species specific hormones must be used whenever possible.* Whole CC extracts contain several hormone factors and physiological doses cannot be determined. Evidence suggests that AKH varies in structure between species and that AKH and HTH likely have similar structures. Indeed the so-called AKH-II of locusts actually has strong hypertrehalosemic effects and is likely the HTH (Orchard and Lange 1983a). Thus physiological functions can only be reliably determined using hormones purified from the species under examination. (3) *An integrated approach to studying the effects of hormones must be taken.* Hormones can have a variety of molecular effects, some physiological, some pharmacological. For example, physiological effects of adrenaline on mammalian muscle are mediated by Ca^{2+} but pharmacological levels of the catecholamine will also elevate cAMP. A preselected parameter (e.g. cAMP levels) should not, therefore, be assumed as the sole index of hormone action when dealing with new systems. To pinpoint physiological effects of hormone action, hormones should be administered at varying doses and effects monitored over a time course with the following parameters measured: (i) *Ca^{2+}, cAMP (cGMP)*: hormones may act through either Ca^{2+} or cyclic nucleotides or, as growing evidence indicates, these compounds may be synarchic messengers (Rasmussen 1981). Levels of both must be measured to determine the primary physiological response. Ca^{2+} can now be quantitated intracellularily using QUIN-II (Tsien et al. 1984) while cAMP and cGMP are determined by radioimmune assay. (ii) *Protein kinase and adenyl cyclase*: activities of both increase under cAMP-mediated hormone stimulation and clearly signal when cAMP is the physiological messenger. (iii) *Fructose-2,6-P_2 and 6-phosphofructo-2-kinase*: levels of fructose-2,6-P_2 rise under hormone-mediated stimulation of glycolytic rate. Hormone-mediated phosphorylation also increases the activity of the kinase (Hers and van Schaftingen 1982). Determination of both clearly defines the state of glycolytic flux; this may be most important in tissues such as fat body where an activation of glycogen phosphorylase does not necessarily indicate an increase in glycolytic flux. In fat body the hormone mechanisms stimulating glycogen breakdown and leading to trehalose output should be quite distinct from those which activate glycolysis. (iv) *Enzyme activities*: hormone-mediated phosphorylation alters measureable kinetic parameters of enzymes. For phosphorylase, the activity of the active *a* (AMP-independent) form is increased; for pyruvate kinase the $S_{0.5}$ for phosphoenolpyruvate is increased. These effects on enzyme kinetics can be used to identify hormone action and may be the primary means by which hormone effects on enzymes such as trehalase and lipase can be attacked. (v) *Two-dimensional gel electrophoresis*: many cellular responses to hormone action involve protein phosphorylation. The effects of flight or hormone injection can be readily surveyed using ^{32}P. $^{32}P_i$ is injected at rest and allowed to equilibrate with the cellular ATP pool. After stimulation by flight or hormones, phosphorylated enzymes and proteins are extracted in buffers containing EDTA/EGTA and NaF and are subjected to non-denaturing 2-D electrophoresis (isoelectric focusing in the first direction, a gradient gel in the second). Staining for enzyme activity coupled with autoradiography identifies phosphorylated enzymes.

5.1 Octopamine

Roles for octopamine as a neurotransmitter, neurohormone, and neuromodulator in insects have been identified (Orchard 1982). As a neurotransmitter octopamine release from the nervous corpus-cardiacum II stimulates the release from the CC of AKH in locusts and HTH in cockroaches (Orchard 1982; Downer et al. 1984); this is apparently the mechanism for AKH/HTH release during flight. As a neurohormone, circulating levels of octopamine rise in hemolymph during the early minutes of flight. Octopamine increases the strength of flight muscle contractions and stimulates the oxidation of exogenous fuels in perfused muscle preparations (Candy 1981). At the fat body octopamine stimulates trehalose production in cockroaches and lipid release in locusts. Orchard and Lange (1983b) have identified two phases of lipid elevation in hemolymph during flight, the early elevation (10 min) apparently due to octopamine and the later (after 20 min) being due to AKH.

Octopamine elevates cAMP levels in the CC and fat body and octopamine sensitive adenylate cyclase occurs (Orchard 1982; Downer et al. 1984). However α-antagonists block octopamine effects while β-antagonists do not. In mammals catecholamine effects mediated through α-receptors act by elevating intracellular Ca^{2+} while actions mediated through β-receptors are cAMP dependent. The apparent dichotomy in the insect situation will be resolved by experiments which measure both Ca^{2+} and cAMP levels during octopamine stimulation. It is likely that these will point to Ca^{2+} as the primary messenger or that Ca^{2+} and cAMP will be found to be synarchic messengers (Rasmussen 1981). Indeed in blowfly salivary glands 5-hydroxytryptamine stimulation increases both intracellular Ca^{2+} and cAMP with Ca^{2+} increasing adenylate cyclase activity (Litosch et al. 1982).

The functions of octopamine during the early minutes of flight parallel those of adrenaline in mammalian exercise. Thus adrenaline action on skeletal muscle or heart increases the force of contraction, increases glucose uptake and activates glycolysis (at PFK) via α-receptor mechanisms; β-receptor mechanisms stimulate muscle glycogenolysis (Clark et al. 1983). Adrenaline effects on liver act through α-receptors to stimulate glycogenolysis and gluconeogenesis.

5.2 Adipokinetic and Hypertrehalosemic Hormones

AKH titre in hemolymph of locusts is elevated after 15 min of flight and peaks after about 30 min; this correlates well with the pattern of lipid release from the fat body (Orchard and Lange 1983b). Levels of AKH-II increased more slowly but peaked at 30 min of flight. The hypertrehalosemia stimulated by AKH-II suggests that this is actually the HTH (Orchard and Lange 1983a); if so then peak levels of the hormone coincide with the establishment of steady state trehalose levels in hemolymph (Beenakkers et al. 1981b) brought about as flight muscle oxidation rate and fat body production rate come equal. Direct measurements of HTH titre in hemolymph of carbohydrate utilizing (group 1) insects during flight have not yet been made although the blowfly is clearly dependent upon CC hormones for fuel supply during flight (Goldsorthy and Gade 1983).

AKH stimulates the release of lipid from the fat body to elevate hemolymph diglyceride content. The molecular site of AKH actions is likely to be a hormone sensitive triglyceride lipase although studies to date have not demonstrated an AKH activation of the enzyme. Pines et al. (1981) were able, however, to activate fat body lipase in vitro with dibutyryl cAMP or the red pigment concentrating hormone from crustaceans (a structural analogue to AKH). AKH also promotes the association of hemolymph lipoproteins to form the high molecular weight A^+ and 0 proteins which take up the increased diglyceride load during flight (Beenakkers et al. 1984a).

CC extracts containing hypertrehalosemic factors increase the percentage of phosphorylase *a* in fat body, stimulate fat body glycogenolysis and increase the level of trehalose in hemolymph (Goldsworthy and Gade 1983). A highly purified HTH from cockroach CC acts similarily (McClure and Steele 1981).

CC extracts as well as purified AKH raise the content of cAMP in fat body but do not affect cAMP in flight muscle (Candy 1981). CC extracts also raise the activity of cAMP-dependent protein kinase in fat body. Dibutyryl cAMP or theophylline applied to fat body mimics the lipid releasing action of AKH or the phosphorylase activation of HTH (Beenakkers et al. 1984a; Goldsworthy and Gade 1983). The actions of both hormones therefore appear to be mediated through cAMP-dependent phosphorylation of enzymes. Effects of covalent modification of fat body enzymes during flight should include activation of glycogen phosphorylase and triglyceride lipase, inhibition of key glycolytic enzymes, phosphofructokinase, pyruvate kinase, and pyruvate dehydrogenase and activation of key enzyme(s) in the conversion of alanine to proline. In addition McClure and Steele (1981) have also demonstrated an important role for Ca^{2+} in the action of HTH; activation of fat body phosphorylase could be accomplished by Ca^{2+} alone (probably via the calmodulin subunit of phosphorylase *b* kinase) and HTH, theophylline or cAMP stimulation of the enzyme required the presence of Ca^{2+}. Synergistic effects of Ca^{2+} and cAMP are probable.

AKH and HTH effects in insects parallel those of glucagon in mammals (Newsholme and Leech 1983). Insects appear to have two hormones (structurally similar) which accomplish the release of either lipid or carbohydrate from the fat body during flight while mammals have one hormone which acts on two tissues releasing carbohydrate from liver and lipid from adipose tissue during exercise. Glucagon levels in blood rise over the course of long-term aerobic exercise and elevate blood lipids at a time coincident with the depletion of muscle glycogen reserves. Glucagon raises cAMP levels in both liver and adipose. In liver this leads to increased glycogenolysis, gluconeogenesis, and glucose output into blood and decreased glycolytic rate (mediated via lowered fructose-2,6-P_2 levels). In adipose glucagon stimulates hormone sensitive lipase to increase the export of lipid. Growing evidence implicates a role for Ca^{2+} in the hormone action; liver has high and low affinity sites for glucagon, the low affinity sites apparently mediate hormone action through Ca^{2+}. There are no receptor sites for glucagon on skeletal muscle.

5.3 Parallels Between Insect Flight and Mammalian Exercise

Sustained aerobic work in insects and mammals is remarkably similar at the biochemical level. Long-term flight in group 3 insects (e.g. locusts) can be compared to sustained running in humans on several levels.

5.3.1 Fuel Utilization

In both instances muscle glycogen supports the initiation of work and reserves are rapidly depleted over the initial minutes of work. Blood or hemolymph sugars are then utilized and these are replenished from the breakdown of liver or fat body glycogen. Lipid levels in blood or hemolymph begin to rise as exercise becomes sustained, mobilized from triglyceride reserves in adipose or fat body. In the long-term lipid becomes the primary fuel for sustained work with low level oxidation of sugars also retained.

5.3.2 Hormone Involvement

In both instances the activation of muscle metabolism is stimulated by nervous input and mediated by Ca^{2+} influx into the cytoplasm. Circulating levels of biogenic amines rise during the first few minutes of exercise. These initiate a general arousal stimulating muscle contractility, muscle glycolysis and the release of fuels from storage organs. Effects of octopamine and adrenaline on muscle appear to be mediated largely through Ca^{2+}. Catecholamines also stimulate the release of peptide hormones, AKH or HTH in insects and glucagon in mammals and in mammals also inhibit insulin release.

AKH, HTH, and glucagon control long-term fuel supply for sustained exercise. Circulating levels of the hormones begin to increase after several minutes of exercise and stimulate carbohydrate and/or glyceride output from storage organs. These peptide hormones have no direct effects on muscle. Hormone action appears to be largely cAMP mediated controlling enzyme activation through protein phosphorylation.

Acknowledgements. My thanks to Dr. L. Hue for many excellent discussions on hormone function and to Drs. R. Downer, J. Steele, and A. Beenakkers for helpful talks and preprints of new reviews.

References

Beenakkers AMT, van der Horst DJ, van Marrewijk WJA (1981a) Role of lipids in energy metabolism. In: Downer RGH (ed) Energy metabolism in insects. Plenum, New York, pp 53–100

Beenakkers AMT, van der Horst DJ, van Marrewijk WJA (1981b) Metabolism during locust flight. Comp Biochem Physiol 69B:315–321

Beenakkers AMT, van der Horst DJ, van Marrewijk WJA (1984a) Biochemical processes directed to flight muscle metabolism. In: Kerkut GA, Gilbert LI (eds) Comprehensive insect physiology, biochemistry and pharmacology, vol 10. Pergamon, Oxford

Beenakkers AMT, van der Horst DJ, van Marrewijk WJA (1984b) Insect flight muscle metabolism. Insect Biochem 14:243–260

Bursell E (1981) The role of proline in energy metabolism. In: Downer RGH (ed) Energy metabolism in insects. Plenum, New York, pp 135–154

Candy DJ (1981) Hormonal regulation of substrate transport and metabolism. In: Downer RGH (ed) Energy metabolism in insects. Plenum, New York, pp 19–52

Clark MG, Patten GS, Filsell OH, Rattigan S (1983) Co-ordinated regulation of muscle glycolysis and hepatic glucose output in exercise by catecholamines acting via α-receptors. FEBS Lett 158:1–6

Crabtree B, Newsholme EA (1975) Comparative aspects of fuel utilization and metabolism by muscle. In: Usherwood PNR (ed) Insect muscle. Academic, New York, pp 405–500

Downer RGH, Orr GL, Gole JWD, Orchard I (1984) The role of octopamine and cyclic AMP in regulating hormone release from corpora cardiaca of the American cockroach. J Insect Physiol 30:457–462

Elliot J, Hill L, Bailey E (1984) Changes in tissue carbohydrate content during flight of the fed and starved cockroach, *Periplaneta americana* L. Comp Biochem Physiol 78A:163–165

Ford WCL, Candy DJ (1972) The regulation of glycolysis in perfused locust flight muscle. Biochem J 130:1101–1112

Goldsworthy GJ, Gade G (1983) The chemistry of hypertrehalosemic factors. In: Downer RGH, Laufer H (eds) Endocrinology of insects. Alan R. Liss, New York, pp 109–119

Hansford RG, Johnson RN (1975) The nature and control of the tricarboxylate cycle in beetle flight muscle. Biochem J 148:389–401

Hansford RG, Johnson RN (1976) Some aspects of the oxidation of pyruvate and palmitoylcarnitine by moth *(Manduca sexta)* flight muscle mitochondria. Comp Biochem Physiol 55B:543–551

Hers H-G, van Schaftingen E (1982) Fructose-2,6-bisphosphate 2 years after its discovery. Biochem J 206:1–12

Johnson RN, Hansford RG (1975) The control of tricarboxylate-cycle oxidations in blowfly flight muscle. The steady-state concentrations of citrate, isocitrate, 2-oxoglutarate and malate in flight muscle and isolated mitochondria. Biochem J 146:527–535

Litosch I, Fradin M, Kasaian M, Lee HS, Fain JN (1982) Regulation of adenylate cyclase and cyclic AMP phosphodiesterase by 5-hydroxytryptamine and calcium ions in blowfly salivary gland homogenates. Biochem J 204:153–159

Male KB, Storey KB (1983) Tissue specific isozymes of glutamate dehydrogenase from the Japanese beetle, *Popillia japonica*: Catablic vs. anabolic GDH's. J Comp Physiol 151:199–205

McClure JB, Steele JE (1981) The role of extracellular calcium in hormonal activation of glycogen phosphorylase in cockroach fat body. Insect Biochem 11:605–613

Newsholme EA, Leech AR (1983) Biochemistry for the medical sciences. John Wiley, New York

Olembo NK, Pearson DJ (1982) Changes in the contents of intermediates of proline and carbohydrate metabolism in flight muscle of the tsetse fly *Glossina morsitans* and the fleshfly *Sarcophaga tibialis*. Insect Biochem 12:657–662

Orchard I (1982) Octopamine in insects: neurotransmitter, neurohormone, and neuromodulator. Can J Zool 60:659–669

Orchard I, Lange AB (1983a) Release of identified adipokinetic hormones during flight and following neural stimulation in *Locusta migratoria*. J Insect Physiol 29:425–429

Orchard I, Lange AB (1983b) The hormonal control of haemolymph lipid during flight in *Locusta migratoria*. J Insect Physiol 29:639–642

Pines M, Tietz A, Weintraub H, Applebaum SW, Josefsson L (1981) Hormonal activation of protein kinase and lipid mobilization in the locust fat body in vitro. Gen Comp Endocrinol 43:427–431

Rasmussen H (1981) Calcium and cAMP as synarchic messengers. John Wiley, New York

Rowan AN, Newsholme EA (1979) Changes in the contents of adenine nucleotides and intermediates of glycolysis and the citric acid cycle in flight muscle of the locust upon flight and their relationship to the control of the cycle. Biochem J 178:209–216

Sacktor B (1970) Regulation of intermediary metabolism with special reference to the control mechanisms in insect flight muscle. Adv Insect Physiol 7:267–347

Sacktor B (1975) Biochemistry of insect flight. I. Utilization of fuels by muscle. In: Candy DJ, Kilby BA (eds) Insect biochemistry and function. Chapman and Hall, London, pp 3–88

Steele JE (1981) The role of carbohydrate metabolism in physiological function. In: Downer RGH (ed) Energy metabolism in insects. Plenum, New York, pp 101–133

Steele JE (1983) Endocrine control of carbohydrate metabolism in insects. In: Downer RGH, Laufer H (eds) Endocrinology of insects. Alan R. Liss, New York, pp 427–439

Storey KB (1980a) Regulatory properties of hexokinase from flight muscle of *Schistocerca americana gregaria*. Role of the enzyme in control of glycolysis during the rest-to-flight transition. Insect Biochem 10:637–645

Storey KB (1980b) Kinetic properties of purified aldolase from flight muscle of *Schistocerca americana gregaria*. Role of the enzyme in the transition from carbohydrate to lipid-fueled flight. Insect Biochem 10:647–655

Storey KB (1983) Regulation of cockroach flight muscle phosphofructokinase by fructose 2,6-bisphosphate. FEBS Lett 161:265–268

Storey KB (1984) Phosphofructokinase from flight muscle of the cockroach, *Periplaneta americana*: Control of enzyme activation during flight. Insect Biochem 15:663–666

Uno I, Matsumoto K, Adachi K, Ishikawa T (1983) Genetic and biochemical evidence that trehalase is a substrate of cAMP-dependent protein kinase in yeast. J Biol Chem 258:10867–10872

Tsien RY, Pozzan T, Rink TJ (1984) Measuring and manipulating cytosolic Ca^{2+} with trapped indicators. Trends Biochem Sci 9:263–266

Ward JP, Candy DJ, Smith SN (1982) Lipid storage and changes during flight by triatomine bugs (*Rhodnius prolixus* and *Triatoma infestans*). J Insect Physiol 28:527–534

Worm RAA, Luytjes W, Beenakkers AMT (1980) Regulatory properties of changes in the contents of coenzyme A, carnitine and their acyl derivatives in flight muscle metabolism of *Locusta migratoria*. Insect Biochem 10:403–408

Lactate: Glycolytic End Product and Oxidative Substrate During Sustained Exercise in Mammals – The "Lactate Shuttle"

G.A. BROOKS [1]

1 Introduction

Results of kinetic tracer studies on several mammalian species during rest and prolonged sustained exercise indicate that lactate is a metabolic intermediate which is very active in supplying carbon for a number of important physiological processes. Lactate is a product of glycolysis and glycogenolysis, and it is a precursor for glucose and glycogen resynthesis. Additionally, lactate produced at some sites during exercise is also a substrate for oxidative energy transduction at other sites. This "shuttling of lactate" through the interstitium and vasculature can be quantitatively as important as the release of glucose from the liver for supplying oxidizable substrate. Because much of the glycolytic and gluconeogenic flux passes through the lactate pool, the metabolism of lactate emerges as a critically important component in the overall integration and regulation of intermediary metabolism. This dynamic role of lactate in mammals, as described here, is different from the stagnant role of lactate portrayed from measurements of lactate concentration in the blood and muscle of mammals and other species. Rather than a dead-end metabolite which accumulates as the result of muscle anoxia during exercise and waits until the recovery ("O_2 debt") period to be returned to glucose and glycogen, lactate is a dynamic metabolite which turns over rapidly and which can participate in a number of processes during rest, exercise, and recovery from exercise.

2 Lactate Metabolism

2.1 Lactate Turnover During Rest

Under resting conditions, the circulating level of lactate remains low and constant because the rate of appearance of lactate in the blood (Ra) equals the rate of disappearance (Rd) from the blood. The condition of a constant blood concentration where Ra=Rd is termed a dynamic steady state (DSS). Under conditions of the DSS, the Ra is alternatively termed the rate of production; together the Ra and Rd in the DSS are often termed the turnover rate (Rt). The term "turnover" denotes a renewal process in which lactate (and other metabolites) are continuously formed and utilized. In some

1 Exercise Physiology Laboratory, Department of Physical Education, University of California, Berkeley, CA 94270, USA

Circulation, Respiration, and Metabolism
(ed. by R. Gilles)

situations, for reasons of methodology, the Rd is termed the irreversible disposal rate (Ri) (Hetenyi et al. 1983).

The procedures to estimate turnover, appearance, and disappearance rates involve the principle of isotope dilution. Basically, the greater or more rapid the dilution of tracer by tracee (endogenous metabolite of interest), the more rapid the turnover of that metabolite. Technically, estimation of turnover rate involves the injection or infusion of a minute amount of carbon or hydrogen labeled material into a large vein and the sampling of blood from an arterial site. These procedures insure the complete mixing of tracer with tracee in the blood, and the presentation of a known input function (specific activity) of lactate tracer at all sites of metabolism.

Blood lactate turnover has been measured in several mammalian species; among these are dogs (Depocas et al. 1969; Eldridge 1975; Issekutz et al. 1976), rats (Freminet et al. 1972; Katz et al. 1981; Donovan and Brooks 1983), guinea pigs (Freminet and Le Clerc 1980), sheep (Reilly and Chandrasena 1977), and humans (Kreisberg et al. 1970; Searle and Cavalieri 1972; Brooks et al. 1984; Stanley et al. 1984). A comparison of blood lactate turnover rates as estimated using carbon labeled tracers (i.e., ^{13}C and ^{14}C) for several species during rest is given in Table 1A. Use of ^{3}H-labeled lactate to

Table 1. Lactate and glucose turnover (Rt) and oxidation (Rox) rates in several mammalian species during rest and exercise[a]

Lactate	Rt ($mg\ kg^{-1}\ min^{-1}$)	Rox ($mg\ kg^{-1}\ min^{-1}$)	%Ox	References
A. Rest				
Rats	19.3	8.8	46	Donovan and Brooks (1983)
	5.6[b]	–	–	Freminet and LeClerk (1980)
Dogs	2.0	1.0	50	Issekutz et al. (1976)
Humans	1.4	0.2	14	Kreisberg et al. (1970)
	1.6	1.4	88	Searle and Cavalieri (1972)
	1.1	–	–	Stanley et al. (1984)
Glucose				
Rats	5.9	2.6	44	Brooks and Donovan (1983)
	9.5	–	–	Katz et al. (1981)
Dogs	3.6	–	–	Issekutz (1970
Humans	2.1	0.3	14	Kreisberg et al. (1970)
	2.4	–	–	Stanley et al. (1984)
B. Exercise				
Rats	46.6	33.4	71.7	Donovan and Brooks (1983)
Dogs	8.0	5.2	65	Issekutz et al. (1976)
Humans	5.9	4.6	78	Brooks et al. (1984)
	12.8	–	–	Stanley et al. (1984)
Glucose				
Rats	12.7	10.7	84	Brooks and Donovan (1983)
Dogs	6.0	–	–	Issekutz (1978)
Humans	6.0	–	–	Stanley et al. (1984)

[a] Data selected to compare results reported by the same investigators
[b] Anaesthetized animals

estimate blood lactate turnover results in even greater (approx. twofold) turnover estimates (Katz et al. 1981; Donovan and Brooks 1983), but the ^3H-derived values are probably too high to represent net lactate production (Donovan and Brooks 1983).

In Table 1A values for glucose turnover obtained on the same species are also presented. These values have been obtained with ^3H-labeled tracer. A comparison of the glucose and lactate turnover values in resting mammals reveals that the turnover rate of glucose is significantly greater than that of lactate, a possible exception being the rat which has a high resting metabolic rate.

Carbon-labeled tracers allow estimation of the rate of oxidation and fractional conversion of turnover removed through oxidation. The procedures involve the collection of expired air and the assay for labeled CO_2. On the basis of results compiled in Table 1A it is apparent that oxidation represents a major avenue of lactate disposal during rest.

In the resting mammal the low and invariate levels of lactate in blood and other tissues belie the very high rates of lactate turnover and oxidation as measured by carbon tracers.

2.2 Lactate Turnover During Sustained Exercise

During exercise, the level of circulating lactate depends on a number of factors. In the exercising human Bang (1936) determined that exercise intensity and duration are primarily important. If exercise intensity is maximal, blood lactate concentration will increase continuously until cessation of work. If exercise is less intense, such that it can be maintained for a finite period, blood lactate concentration will follow a distinct pattern: blood lactate increases at the beginning of exercise, peaks after several minutes, and declines to a stable level. During submaximal, sustained exercise by a normal human, blood lactate level will remain relatively constant from approx. 15 min to the cessation of exercise. In mammalian species other than the human, such as the rat, the pattern of blood lactate response is similar, but the time course of change (increase, decrease, and establishment of a steady blood lactate level) is shorter (Brooks and Donovan 1983).

During sustained exercise, the maintenance of a steady blood lactate level results in a condition analogous to the DSS observed during rest. Under these conditions, lactate Rt can be estimated by isotope dilution techniques. Ra differs from Rd, and equations exist to obtain quantitative estimates of Ra and Rd (Steele 1959).

Lactate turnover and oxidation have been found to have a direct, positive relationship to metabolic rate (VO_2) during prolonged, steady state exercise by dogs (Depocas et al. 1969; Issekutz et al. 1976), rats (Donovan and Brooks 1983), and humans (Mazzeo et al. 1982; Brooks et al. 1984). In rats (Donovan and Brooks 1983), lactate oxidation was a linear function of VO_2 up to an intensity of 75% of maximal O_2 consumption ($VO_{2\,max}$) (Fig. 1).

A comparison of glucose and lactate turnover rates in several species during exercise is presented in Table 1B. In the transition from rest to exercise, both glucose and lactate Rt increase, but the increase in lactate Rt is far greater. The difference between the Rd of glucose and Ra of lactate reflects the amplified role of glycogen in supplying glucosyl units for glycolysis and oxidative catabolism during exercise.

A comparison of the rates of lactate and glucose oxidation (Table 1B) during exercise reveals that the oxidation of both substances increases in the transition from rest

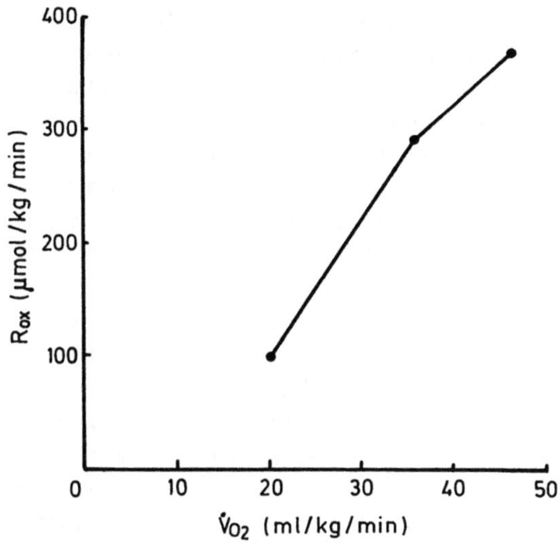

Fig. 1. Rate of lactate oxidation as a function of metabolic rate (VO_2) in laboratory rats during rest and two intensities of exercise; each *point* represents the mean of 20 animals. Modified from Donovan and Brooks (1983)

to exercise. However, the oxidation of lactate increases relatively more during exercise than does glucose.

When viewed from the context of glucose and glycogen oxidation, the apparently large oxidation of lactate (Table 1) deserves comment. Although the simultaneous oxidations of glucose, lactate, and glycogen have not been simultaneously studied, based on the results of studies on either glucose or lactate oxidation during rest (Table 1A), or exercise (Table 1B), it appears that glucose, glycogen, and lactate oxidations are not additive functions. Rather, lactate formation occurs due to glycogenolysis at one site (e.g., fast glycolytic, type IIb skeletal muscle fibers), whereas lactate oxidation occurs at other sites (e.g., heart and types I and IIa muscle fibers). Therefore, lactate circulating in the blood represents a source of substrate for oxidation. In this sense, lactate released into the circulation is analogous to glucose released from the liver. The distinction during exercise is that the amount of lactate shuttled from one site to another is quantitatively greater than that of glucose. Under exercise conditions the rate of glycogen catabolism may be close to the difference between the Ra of lactate and the Rd of glucose (Donovan and Brooks 1983).

In the exercise condition, blood lactate concentrations several times higher than those observed during rest can be maintained for extended periods (30-120 min) (Bang 1936; Mazzeo et al. 1982; Donovan and Brooks 1983). As in the resting condition, the constancy of blood lactate level during sustained exercise belies the active turnover and oxidation of this metabolite.

3 Why Lactate is Formed and Removed

3.1 Local Anoxia as a Cause of Lactate Production

Classically, lactate was thought to be produced as the result of anoxia. With unicellular organisms studied in vitro, removal of O_2 accelerates the rate of lactate formation. The question arises, however, whether the appearance of lactate in the circulation of a resting or exercising mammal indicates the presence of anoxic areas. This issue is unresolved.

Several lines of evidence suggest that local anoxia may not always be the cause of lactate production. Consider that significant amounts of lactate are produced in resting mammals (Table 1). Consider also that skeletal muscle produces lactate during sustained exercise at intensities significantly below maximal oxygen consumption ($VO_{2\,max}$), when significant reserves exist in cardiac output, muscle blood flow, capillary dilitation, and arterial-venous O_2 difference [$(a-v)O_2$]. Further, consider that the healthy heart produces lactate at rest.

The heart represents the epitome of an oxidative muscle. Recently, Gertz et al. (1981) using a continuous infusion of carbon-labeled lactate, along with arterial and venous (coronary sinus) catheterization procedures, found that the myocardium in a healthy, resting human produces lactate during net lactate extraction. These investigators confirmed that the heart is a net consumer of lactate [as measured by $(a-v)$] and they determined that over 85% of the lactate removed was oxidized. However, even duing this period of net lactate extraction they found that the heart produces lactate. Across the heart, the isotopic enrichment of tracer lactate decreased by approx. 50%, whereas, the net chemical decrease in lactate across the heart approximated 25%. These findings indicate that at rest approx. 25% of the lactate contained in the coronary sinus blood draining the heart was derived from lactate produced in the heart.

The critical mitochondrial O_2 tension is the partial pressure of O_2 below which the maximal mitochondrial respiratory rate (state 3) cannot be supported. Estimates by Chance and Quintorff (1978) of the critical mitochondrial O_2 tension place it between 0.1 and 0.5 torr. During maximal leg exercise by humans, Pirnay et al. (1972) obtained blood samples from the deep femoral vein. Results indicated that the O_2 tension in venous blood did not decline below 12 torr during maximal leg exercise. During submaximal exercise the PvO_2 ranged between 40 and 20 torr. These values are many times greater than the critical mitochondrial O_2 tension. Therefore, it is unlikely that the critical muscle mitochondrial O_2 partial pressure is achieved in the muscles of mammals during submaximal, sustained exercise at sea level altitudes.

In experiments on dog muscle contracting in situ, Jöbsis and Stainsby (1968) utilized fluorometric techniques to study the mitochondrial redox state ($NADH/NAD^+$). When those muscle preparations were stimulated to contract at intensities sufficient to produce maximal O_2 consumption and a significant net efflux of lactate, the mitochondrial redox state was more oxidized than during rest. These data on dog muscle preparations stimulated to contract at high intensities in situ indicate that the critical mitochondrial tension was not achieved.

Recently, Connett et al. (1984) concluded that lactate accumulation occurs in dog gracilis muscle contracting in situ for reasons other than a simple limit on mitochondrial ATP production. Dog gracilis is a pure red muscle containing only types I and IIa fibers. In their experiments, Connett et al. observed lactate accumulation in dog gracilis during mild (10% $VO_{2\,max}$) exercise. Further, up to a contraction rate which elicited 70% of $VO_{2\,max}$ (where in tissue ATP and CP levels were maintained), lactate accumulation was linearly related to twitch (work) rate. This result suggests that lactate accumulation is related to increments in work and metabolic rate. Additionally, Connett et al. observed that lactate accumulation was not reduced by increasing blood flow or inducing capillary dilitation. Further, these investigators utilized myoglobin cryomicroscopic techniques to determine the O_2 tension throughout the muscle tissue. Anoxic areas (i.e., areas in which the local PO_2 approached the critical mitochondrial O_2 tension) were found neither during submaximal exercise nor during the transition from rest to exercise.

On exercising humans, Green et al. (1983) have utilized muscle biopsy and other procedures to study the interrelationships among muscle lactate level, blood lactate level, and pulmonary ventilation during progressive exercise. Green et al. observed that muscle lactate accumulation occurs in mild exercise intensities which do not elicit an increase in the circulating blood lactate level. Thus, the results of Green et al. on human leg muscle during exercise are consistent with those of Connett et al. on dog gracilis contracting in situ; muscle lactate accumulation occurs during submaximal exercise when mitochondrial function is not limited by the presence of oxygen.

3.2 Enzyme Kinetics and Thermodynamic Equilibrium as Causes of Lactate Production

The view has been expressed (Brooks and Fahey 1984) that the production of lactate occurs because of a competition between cytoplasmic lactate dehydrogenase (LDH) and the mitochondrial capacity for pyruvate oxidation (pyruvate oxidase). Of the glycolytic enzymes, LDH possesses by far the greatest catalytic activity (V_{max}) (Newsholme and Leech 1983). In comparison with the activities of enzymes which compete for pyruvate removal, the V_{max} of LDH is so great it is surprising that lactate is not the exclusive product of glycolysis. On rat skeletal muscle, Molé et al. (1973) determined LDH activity to be 677 μmol g^{-1} wet wt. min^{-1}. In the same laboratory, Fitts et al. (1975) determined pyruvate oxidase capacity to approximate 0.65 μmol g^{-1} min^{-1}. This is a 1,000-fold difference. Further, if the activities of pyruvate oxidase are combined with those of malic enzyme and glutamate-pyruvate transaminase [GPT] (Fitts et al. 1975), the maximal catalytic capacity of LDH exceeds those of all the competitors by approx. 30-fold (Brooks 1985).

Not only is the catalytic capacity of LDH much higher than that of other glycolytic enzymes and enzymes which compete for glycolytic carbon flow, but also the thermodynamic characteristics of the LDH reaction are such that it is likely to result in lactate accumulation under physiological conditions. The equilibrium constant (K'_{eq}) is a measure of the reactant to product conversion under standard conditions. For the conversion of pyruvate to lactate, the K'_{eq} exceeds 1,000, and the resulting free energy change (G') exceeds -6 kcal mol^{-1}, which is the highest for any step in the glycolytic pathway (Lehninger 1975).

To a certain extent, the activity of all enzymes depends upon the mechanisms of their interactions with substrate. The Michaelis-Menten constant (k_M) is a measure of the interaction between substrate concentration and activity of the enzyme. Most simply defined, the k_M is the substrate concentration which yields $V_{max}/2$. In the human (Karlsson et al. 1974) and other mammalian species (Everse and Kaplan 1973), the k_M for conversion of pyruvate to lactate is on the order of 0.08 m*M*. In resting mammalian skeletal muscle, pyruvate concentration approximates 0.08 m*M*, while during sustained, submaximal exercise pyruvate concentration easily exceeds 0.15 m*M*. During difficult exercise by humans, blood pyruvate can exceed 0.25 m*M* (Karagiorgos et al. 1979). Because under exercise conditions, the pyruvate concentration exceeds the

k_M of LDH for pyruvate by several-fold, LDH is saturated with its substrate. Therefore, for reasons of maximal catalytic activity (V_{max}), thermodynamic equilibrium (K'_{eq}), and enzyme-substrate interaction (k_M), lactate production in skeletal muscle is an inevitable result of glycolysis. This production of lactate appears to be related to the overall metabolic rate, and not necessarily to the unavailability of O_2 in active muscle (Fig. 1).

3.3 Enzyme Kinetics and the Removal of Lactate

The removal of lactate begins with its conversion to pyruvate. This reverse reaction is mediated by LDH, and several factors appear to be important. These are the concentration of lactate, the k_M for the reversal of LDH, the glycolytic carbon flux, and the mitochondrial activity. These factors are related to the muscle fiber type.

The isoenzyme pattern of LDH varies among tissues of the same individual, ranging from tissues which contain a high proportion of heart (H) type, to those which contain a high proportion of muscle (M) type (Nisselbaum et al. 1964). In more oxidative muscles, the prepondrance of H-type isoenzymes of LDH (York et al. 1974), with their relatively lower k_M for lactate and higher k_M for pyruvate, means that the conversion of lactate to pyruvate is more likely to occur than with M-type isoenzymes.

In highly oxidative muscle, such as in the heart and red skeletal muscle fibers of endurance trained rats (Davies et al. 1981), the conversion of lactate to pyruvate is influenced by two factors: a greater reliance on lipid as an oxidizable fuel source, and high mitochondrial activity (Fig. 2). In these tissues use of lipid as a fuel retards the rate of glycolysis, and thus, the rate of pyruvate formation. The lesser pyruvate flux puts less of a "pressure" on LDH for the formation of lactate. In concert, the draw of mitochondria for oxidizable substrate will effectively remove pyruvate and lactate which diffuse or are transported into the oxidative fibers. For this reason, mitochondria in heart (Gertz et al. 1981) and red skeletal muscle fibers can effect a net uptake of lactate from the circulation.

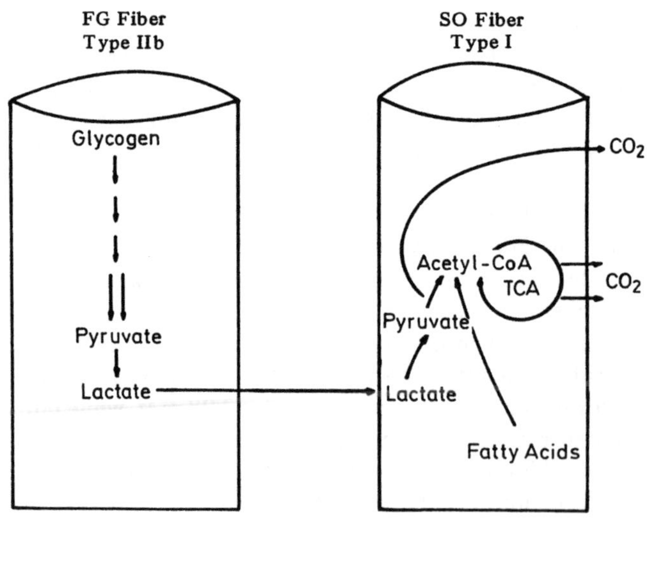

Fig. 2. The postulated "lactate shuttle". Lactate is formed in pale, type IIb, fast-glycolytic (*FG*) fibers when they are recruited. Lactate released by FG fibers is consumed and combusted by pigmented, type I, slow-oxidative (*SO*) fibers

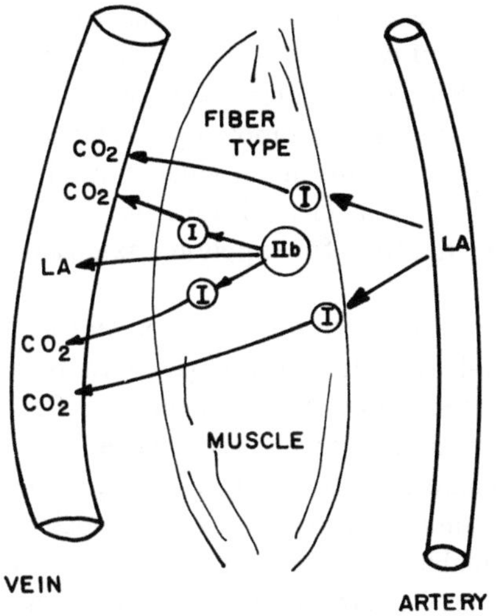

Fig. 3. How a contracting muscle could simultaneously release, consume, and oxidize lactate as part of the postulated "lactate shuttle". Some lactate molecules released from type IIb fibers could be combusted in type I fibers without leaving the tissue. Other lactate molecules which escape from the tissue could be consumed on recirculation to the tissue

Very recently, Stanley et al. (1984) have utilized the continuous infusion of tracer, with arterial and femoral vein catheterization techniques, to study lactate metabolism across the legs of humans during cycle ergometer exercise. Arterial-venous difference measurements for lactate indicate that over 90% of the body's total lactate production is derived from the exercising muscle. However, (a–v) measurements of tracer lactate indicate significant uptake of lactate tracer and conversion to labeled CO_2, which appeared in the venous circulation. How the exercising legs might simultaneously produce, consume, and oxidize lactate is illustrated in Fig. 3. Recruitment of type IIb fibers will result in glycogenolysis and lactate production. Some of this lactate can diffuse to type I fibers and be combusted without leaving the muscle tissue. Additionally, lactate released into the venous blood can recirculate and be taken up by type I fibers. These data reinforce the conclusion that blood concentration measurements, even (a–v) measurements, cannot be utilized to estimate rates of lactate production.

3.4 Matching Glycogenolysis to Pyruvate Oxidation: The "Lactate Shuttle"

Key control points exist along the metabolic pathway leading from glycogen to CO_2. These are at the levels of glycogen, glucose 6-phosphate, and pyruvate. On occasion, physiological functions depend upon coordinated function of the pathway segments between each key site. On other occasions, physiological function depends on various sections of the pathway operating at different rates. One example is the release of glucose into the circulation from hepatic glycogenolysis.

The release of lactate, pyruvate, and alanine from muscle during exercise is another example of an instance when sections of the pathway from glycogen to CO_2 function at different rates. These 3-carbon substances are gluconeogenic precursors which function as part of the Cori and Glucose-Alanine Cycles to maintain glucose homeostasis during sustained exercise. In postabsorptive rats during prolonged (2 h) exercise, it has been estimated (Donovan and Brooks 1983) that a majority of the glucose produced comes from lactate as part of Cori Cycle activity. And, as described above (Sect. 3.2), lactate produced in one locus (e.g., type IIb muscle fibers) during exercise can serve as a source of oxidizable substrate at other sites (heart, and types I and IIa muscle fibers). This shuttling of lactate from the site of production through the interstitium or vasculature to the sites of oxidation represents an important means of supplying substrate to active areas during exercise. Based upon quantitative comparisons of glucose and lactate turnover rates during sustained exercise (Table 1B), lactate release from muscle compares in importance with glucose release from the liver for supplying oxidizable substrate.

In active skeletal muscle glucose uptake proceeds by both insulin-dependent and insulin-independent mechanisms (Richter et al. 1984). In active muscle, glycogenolysis is stimulated by both catecholamine (C-AMP mediated) and C-AMP independent (Ca^{++} mediated) mechanisms. Because the production of lactate during sustained exercise comes from muscle glycogen (Table 1B), and because the respiratory exchange ratio during sustained exercise is below unity (indicating some lipid oxidation), it is apparent that an ample supply of oxidizable substrate is available in active muscle. Therefore, it is possible to conclude that glycogenolysis and glycolysis proceed faster than the immediate need for pyruvate as an oxidizable substrate in active muscle. The "overstimulation" in glycogenolysis during exercise may be associated with an activation of the autonomic nervous system and the secretion of catecholamines (Lehmann et al. 1982). It is known that part of the adaption to endurance training is a dampening of catecholamines secretion for given exercise intensities (Winder et al. 1978). This damped catecholamine response may be responsible for the lower levels of circulating lactate in trained individuals during given absolute or relative intensities of submaximal exercise. Thus, we may conclude that lactate production occurs in muscle during sustained exercise because various segments of the glycogenolytic, glycolytic, and oxidative pathways can function at separate rates. This production of lactate is, therefore, not necessarily indicative of muscle anoxia.

4 Conclusion

It is postulated that during sustained exercise, lactate represents an important source of oxidizable substrate. The recent realization that oxidation is the major fate of lactate formed during sustained exercise, together with the well-established finding that glycogen is the major precursor to lactate formation during exercise, leads to the conclusion that glycogen reserves in some active muscles can provide oxidizable substrate to other active muscles. The route of this shuttling of oxidizable substrate is the release of lactate from tissue sites of glycogenolysis, transport through the interstitium and vasculature, and consumption by highly oxygenated and active tissues. The quantitative significance of this shuttling of oxidizable substrate through lactate can, under some exercise circumstances, compare with the release of glucose from liver.

Acknowledgment. Supported by NIH Grant AM19577.

References

Bang O (1936) The lactate content of the blood during and after muscular exercise in man. Scand Arch Physiol 74 (Suppl 10):49–82
Brooks GA (1985) Anaerobic threshold: Review of the concept and directions for future research. Med Sci Sports Exercise (in press)
Brooks GA, Donovan CM (1983) Effect of endurance training on glucose kinetics during exercise. Am J Physiol 244 (Endocrinol Metab 7):E505–E512
Brooks GA, Fahey TD (1984) Exercise physiology: Human biogenergetics and its applications. John Wiley, New York, p 79
Brooks GA, Mazzeo RS, Schoeller DA, Budinger TF (1984) Lactate turnover in man during rest and greated exercise. Med Sci Sports Exercise 16:121
Chance B, Quinstorff B (1978) Study of tissue oxygen gradients by single and multiple indicators. Adv Exp Med Biol 94:331–338
Connett RJ, Gaueski TEJ, Honig CR (1984) Lactate accumulation in fully aerobic, working dog gracilis muscle. Am J Physiol 246 (Heart Circ Physiol 15):H120–H128
Davies KJA, Packer L, Brooks GA (1981) Biochemical adaptation of mitochondria, muscle and whole animal respiration to endurance traning. Arch Biochem Biophys 209:539–554
Depocas F, Minaire Y, Chatonnet J (1969) Rates of formation and oxidation of lactic acid in dogs at rest and during moderate exercise. Can J Physiol Pharmacol 47:603–610
Donovan CM, Brooks GA (1983) Training affects lactate clearance, not lactate production. Am J Physiol 244 (Endocrinol Metab 7):E83–E92
Drury DR, Wick AN (1956) Metabolism of lactic acid in the intact rabbit. Am J Physiol 184: 304–308
Eldridge FL (1975) Relationship between turnover and blood concentration in exercising dogs. J Appl Physiol 39:231–234
Everse J, Kaplan NO (1973) Lactate dehydrogenases: structure and function. Adv Enzymol 37: 61–134
Fitts RH, Booth FW, Winder WW, Holloszy JO (1975) Skeletal muscle respiratory capacity, endurance, and glycogen utilization. Am J Physiol 228:1029–1033
Freminet A, Le Clerc L (1980) Effect of fasting on glucose lactate and alanin turnover in rats and guinea pigs. Comp Biochem Physiol 65:363–367
Freminet A, Bursax E, Poyart CB (1972) Mesure de la vitesse de renouvellement du lactate chez le rat per perfusion de 14C-lactate. Pflügers Arch 334:292–297
Gaesser GA, Brooks GA (1984) Metabolic bases of excess post-exercise oxygen consumption: a review. Med Sci Sports Exercise 16:29–43
Gertz EW, Wisneski JA, Nesse R, Bristow JD, Searle GL, Hanlon JT (1981) Myocardial lactate metabolism: evidence of lactate release during net chemical extraction in man. Circulation 1273–1279
Green HJ, Hughson RL, Orr GW, Ranney DA (1983) Anaerobic threshold, blood lactate, and muscle metabolites in progressive exercise. J Appl Physiol:Respir Environ Exercise Physiol 54:1032–1038
Hetenyi G, Perez G, Vranic M (1983) Turnover and precursor-product relationships of nonlipid metabolites. Physiol Rev 63:606–667
Issekutz B Jr (1970) Studies on hepatic glucose cycles in normal and methylprednisolone treated dogs. Metabolism 26:157–170
Issekutz B Jr (1978) Role of beta-adrenergic receptors in mobilization of energy sources in exercising dogs. J Appl Physiol:Respir Environ Exercise Physiol 44:869–876
Issekutz B Jr, Paul P, Miller HI (1967) Metabolism in normal and pancreatomized dogs during steady-state exercise. Am J Physiol 213:857–860
Issekutz B Jr, Shaw WAS, Issekutz AC (1976) Lactate metabolism in resting and exercising dogs. J Appl Physiol 40:312–319
Jöbsis FF, Stainsby WN (1968) Oxidation of NADH during contractions of circulated mammalian skeletal muscle. Respir Physiol 4:292–300

Karagiorogos A, Garcia JF, Brooks GA (1979) Growth hormone response to continuous and intermittent exercise. Med Sci Sport 11:302–307

Karlsson J, Hulten B, Sjodin B (1974) Substrate activation and product inhibition of LDH activity in human skeletal muscle. Acta Physiol Scand 92:21–26

Katz JF, Okajima F, Chenoweth M, Dunn A (1981) The determination of lactate turnover in vivo with ^{3}H and ^{14}C-labelled lactate. Biochem J 194:513–524

Kreisberg RA, Pennington LF, Boshell BR (1970) Lactate turnover and gluconeogenesis in normal and obese humans. Diabetes 19:53–63

Lehmann M, Wybitul K, Spori U, Keul J (1982) Catecholamines, cardiocirculatory and metabolic response during graduated and continuously increasing exercise. Int Arch Occup Environ Health 50:261–271

Lehninger AL (1975) Biochemistry: The molecular basis of cell structure and function. Worth, New York, p 431

Lusk G (1924) Analysis of the oxidation of mixtures of carbohydrate and fat. J Biol Chem 59: 41–42

Mazzeo RS, Brooks GA, Budinger TF, Scholler DA (1982) Pulse injection, ^{13}C tracer studies of lactate metabolism in humans during rest ad two levels of exercise. Biomed Mass Spectrom 9: 310–314

Mole PA, Baldwin KM, Terjung RL, Holloszy JO (1973) Enzymatic pathways of pyruvate metabolism in skeletal muscle: adaptations to exercise. Am J Physiol 224:50–54

Newsholme EA, Leech AR (1983) Biochemistry for the Medical Sciences. John Wiley, Chichester

Nisselbaum JS, Packer DE, Bodansky O (1964) Comparison of the actions of human, brain, liver, and heart lactic dehydrogenase variants on nucleotide analogs on the substrate analogues in the absence and in the presence of oxalate and oxanmate. J Biol Chem 239:2830–2834

Pirnay F, Lamy M, Dujardin J, Deroanne R, Petit JM (1972) Analysis of femoral venous blood during maximum muscular exercise. J Appl Physiol 33:289–292

Reilly PKR, Chandrasena LG (1977) Sheep lactate entry-rate measurements: error due to sampling jugular blood. Am J Physiol (Endocrinol Metab Gasterointest Physiol 2):E138–E140

Richter EA, Ploug T, Galbo H (1984) Increased muscle glucose uptake during contractions: no need for insulin. Med Sci Sports Exercise 16:173

Searle GL, Cavalieri RR (1972) Determination of lactate kinetics in the human analysis of data from single injection vs. continuous infusion methods. Proc Soc Exp Biol Med 139:1002–1006

Stanley WC, Gertz EW, Wisneski JA, Neese RA, Brooks GA (1984) Glucose and lactate turnover in man during rest and exercise studied with simultaneous infusion of 14C-glucose and 13C-lactate. Med Sci Sports Exercise 16:136

Stanley WC, Gertz EW, Wisneski JA, Morris DL, Neese R, Brooks GA (unpublished) Systemic lactate turnover: lactate extraction and release by skeletal muscle during graded exercise in man

Steele R (1959) Influence of glucose loading and of injected insulin on hepatic glucose output. Ann NY Acad Sci 82:420–430

Winder WW, Hagberg JM, Hickson RC, Ehsani AA, McLane JA (1978) Time course of sympathoadrenev adaptation to endurance exercise training in man. J Appl Physiol 45:370–374

York J, Oscai LB, Penny DG (1974) Alterations in skeletal muscle lactic dehydrogenase isozymes following exercise traning. Biochem Biophys Res Commun 61:1387–1393

Closed Systems: Resolving Potentially Conflicting Demands of Diving and Exercise in Marine Mammals

M.A. CASTELLINI[1]

1 Introduction

Since terrestrial mammals usually have continuous access to air while exercising, most investigations on the oxygen demands of exercise have looked at the physiological or biochemical limits to oxygen transport and utilization. Recently there has also been a substantial interest in exercise at altitudes where oxygen supplies start to become limiting (Sutton et al. 1982). However, there have been only a few studies of mammalian exercise under the extreme conditions of complete inaccessibility to oxygen: marine mammals swim, hunt, and work underwater without access to air – they must carry their oxygen supplies with them while diving. It was quickly realized that diving mammals possess two major adaptations that allow such underwater work: increased oxygen storage capacity relative to terrestrial mammals and the ability to utilize a "diving reflex" – a physiological response that shunts the oxygen-enriched blood away from the peripheral organs towards the primary aerobic tissues such as the heart, lung, and brain (Irving 1939; Scholander 1940). In recent years it has been found that this reflex is actually a response of variable nature and tends to be maximized under laboratory diving conditions and minimal during natural diving within the animals aerobic dive limits (Hill et al. 1984; Kooyman et al. 1980, 1983). However, as more information on natural diving becomes available another interesting controversy has arisen: the requirements for diving, which include bradycardia, reduced metabolism and some peripheral vasoconstriction, conflict with the standard physiological adjustments seen in aerobically exercising terrestrial mammals. Consequently, the control of heart rate, for example, which in a quiet, laboratory dive is presumably regulated for diving conditions, would have to be modified to fit the needs of exercise. These conflicts arise, however, by assuming that exercise while diving is just as energetically demanding as land exercise is for terrestrial mammals. We know that surface exercise for both types of mammals is similar in terms of increasing metabolic rate (Castellini et al. 1985). However, the only values we have for freely diving seals suggest that while swimming and working underwater the whole animal metabolic rate is actually less than during resting conditions (Kooyman et al. 1973). Is the seal then swimming and diving at no extra metabolic cost? To examine these problems more closely, we must separate diving into its components of oxygen conservation and exercise requirements.

1 Zoology Department, University of British Columbia, Vancouver, British Columbia, Canada, V6T 2A9

Circulation, Respiration, and Metabolism
(ed. by R. Gilles)

2 Requirements of Diving

The major physiological adaptations for diving in mammals have been well described and reviewed (Butler and Jones 1982; Kooyman et al. 1980). These responses include a strong and immediate bradycardia upon immersion, peripheral vasoconstriction and a pulsing of lactate into the circulation from hypoperfused tissues after the dive is finished. Combined with a higher oxygen carrying capacity in the blood and both an elevated buffering and oxygen capacity in the muscles (Castellini et al. 1981; Lenfant et al. 1970), these responses allow maximal diving times of over an hour in some species (Kooyman et al. 1980). There is also some evidence that tissues of marine mammals may have enzymatic adaptations that allow them to better function during periods of hypoxia and during recovery after the dive (Murphy et al. 1980), although these changes are relatively minor compared to the physiological adjustments (Castellini et al. 1981). The physiological responses were first thoroughly documented under laboratory diving conditions (Irving 1939; Scholander 1940) and assumed to take place under natural diving situations (Blix 1976; Hochachka 1981). However, experiments by Kooyman et al. on the Weddell seal *(Leptonychotes weddelli)*, first documented that freely diving animals usually stay underwater for times less than their aerobic diving limit. In these seals, a species capable of at least 73 min dives, over 97% of their natural dives are within 20–25 min and little lactic acid washes out into the circulation after such dives (Kooyman et al. 1980). Coupled with experiments showing that diving heart rates in the field were not as low as those seen in the laboratory (Hill et al. 1983; Kooyman and Campbell 1972), the evidence is very convincing that the dive response is a graded reaction based on the animals individual diving requirements. A short dive can be carried out using stored oxygen supplies without an apparent maximum diving response, with reduced bradycardia, minimal lactic acid accumulation and apparent normal liver and kidney function (Davis et al. 1983a). A long dive, however, or one in the laboratory, requires a selective partitioning of the oxygen stores and thus a stronger bradycardia, lactate accumulation and apparent reduction of liver and kidney function (Davis et al. 1983a; Hill et al. 1984; Kooyman et al. 1980; Zapol et al. 1979). Overall, field dives show less physiological modifications than do laboratory dives and this has led to some suggestions that laboratory dives are more similar to a fright response than to natural diving (Kanishwer et al. 1981). However, another major difference between laboratory and field diving is that in the field, the animals are exercising while laboratory dives are usually quiet. The standard responses of mammals to exercise include elevated heart rates, increased cardiac output, increased oxygen consumption and peripheral vasodilation. These responses conflict directly with those seen during a full dive response in marine mammals. How does the animal balance these two needs – one to conserve and the other to maximally utilize oxygen? Perhaps, the true "diving" response in a free animal is just as strong as seen in the laboratory, but modified by the requirements of exercise. Therefore, a short hardworking dive alters the dive response a great deal while a quiet laboratory dive might not put such an exercise demand on the seal.

It is important to realize, however, that regardless of the mechanisms involved, most dives by a seal are probably aerobic and can be continued in a long series that may last for hours (Kooyman et al. 1980). The controversy is assessing whether free diving is more like surface aerobic exercise or if it has a strong dive component. That is, does the animal merely hold its breath on a short dive with no major physiological or metabolic adjustments other than those required by exercise or does it use a diving response that is modified by the level of exercise?

3 The Energetics of Swimming

Very little information is available on the metabolic cost of swimming at the surface for marine mammals and even less on the cost of swimming while diving. Recently, exercise chambers (swimming mills) have become available that allow the investigator to swim seals at different speeds while monitoring a variety of physiological and metabolic responses. At the University of Guelph, Canada, grey seals *(Halicherous grypus)* have been used for such experiments (Castellini et al. 1985). In these studies, the seals were monitored for heart rate, respiratory frequency and quotient, metabolic rate and blood metabolites while swimming at speeds which increased their metabolic rate twofold over resting. Under these conditions, changes in the above parameters were indistinguishable from those seen in submaximally exercising terrestrial mammals. Since the seals had continual access to oxygen while swimming, it is reasonable to expect that exercise should not have appeared different from any other working mammal since diving responses would not have been necessary. Attempts to force both diving conditions and exercise onto the seals were difficult and the heart rate values were indicative of standard forced dive conditions (Fedak et al. 1983). Similar experiments at the Scripps Institution of Oceanography on harbor seals *(Phoca vitulina)* have shown basically the same results in regards to metabolic rate increasing with swimming speed (Davis et al. 1983b). The only information on the exercise metabolic requirements for freely diving animals comes from studies on Weddell seals (Kooyman et al. 1973). It was shown that the average oxygen consumption over a 5-6 h period while the seals were freely diving was less than required during periods of rest! This means either the diving response dominated the exercise requirements in terms of metabolic demand or that the rate taken at rest was abnormally high. The authors could find no evidence that the resting rate was not correct and noted that this meant the diving conditions were certainly not energetically demanding. At this point, it is very tempting to suggest that natural diving involves a substantial "diving response" so that organs such as the kidney, liver, and gut, which in resting mammals can utilize up to 50% of the resting metabolic rate (Friedman and Selkurt 1966), could be "turned down" during such dives. In 1981 Weddell seals again were used to test this hypothesis and the results indicated that liver and kidney, and perhaps even the intestinal organs, appear to function normally during most of the diving period (Davis et al. 1983a). During 5-6 h of periodic diving then, the metabolic rate appears to be lower than at rest and simultaneously, liver and kidney function are also normal. This apparent conflict has lead some to suggest that "normal" conditions for a seal are when it is underwater and that the recovery periods on the surface represent times of "tachymetabolism" (Lin et al. 1972). Obviously, attempts to compare surface resting or exercise conditions with diving exercise become difficult since we do not even know the speed at which the seals swim while underwater, let alone metabolic rate, organ blood flow or metabolite turnover. The continual monitoring of heart rate during free diving is possible and the results indicate, as mentioned earlier, that the bradycardia is not as intense as seen during extremely long dives or during forced dives (Hill et al. 1983). But heart rate alone does not answer the exercise/diving conflict since it can be adjusted by both requirements. Metabolite tracer experiments, however, might be able to separate underwater work into its diving and exercise factors.

4 Metabolite Turnover While Diving and Swimming

Standard methods for measuring metabolic rate (oxygen consumption, CO_2 production) cannot be used for diving seals since they do not have access to air. Methods must be used that can measure the consumption of internal oxygen stores. The first attempts under laboratory conditions to do this noted that the seal incurred an "oxygen debt" after the dive that was less than what should have occurred had the seal continued to use oxygen at the pre-dive rate (Scholander 1940). This hypometabolism is probably the primary adaptation for long diving in seals and for surviving periods of hypoxia in many other animals (Hochachka and Dunn 1983). When these experiments were attempted on freely diving animals, the results again indicated hypometabolism (Kooyman et al. 1973). However, these experiments could not distinguish between the possibility that diving was "normal" and that between dive periods was above normal in terms of metabolic rate.

As mentioned in the previous section, it is now possible to conduct experiments that examine the turnover and oxidation of various blood-borne metabolites while a seal is swimming in an exercise chamber. Turnover values in grey seals for glucose and free fatty acids (FFA) are similar during exercise to values found in submaximally exercising terrestrial mammals (Castellini et al. 1985). If exercise without major physiological adjustments necessary for diving is the primary component of aerobic diving, then it stands to reason that turnover values for carbohydrates and FFA should be similar to those seen during surface exercise conditions. When a bolus of radiolabelled tracer is injected into a surface swimming seal, the specific activity (SA) curve falls off with time at a rate faster than normal and similar to such a curve in an exercising land mammal (Castellini et al. 1985). Under the extreme conditions of a forced dive, however, the same bolus of radiotracer follows a radically different curve (Fig. 1).

There are several aspects of this diving curve that are important. First, the tracer is injected in the aortic bulb so that the bolus is directed downstream away from the quickly circulating blood pools in the seal. Thus, after some initial turbulent mixing, the specific activity gradually rises as the bolus of label first moves away from the sampling site and then spreads into the total plasma pool. Second, the rate of turnover is obviously slower than during rest or exercise although that rate cannot be quantified since the dive is not a steady state condition. Third, as the animal surfaces, the turnover rapidly increases to a rate very similar to that seen during rest or exercise.

A critical point to this description is whether the bolus of tracer truly equilibrates in the total plasma pool or whether it stays in some smaller central pool throughout the dive. The same shape of curve could be generated if the label slowly leaked from a small central pool into the periphery until the dive ended and then washed out into the larger pool. Two simple calculations show that the label is mixed into the entire

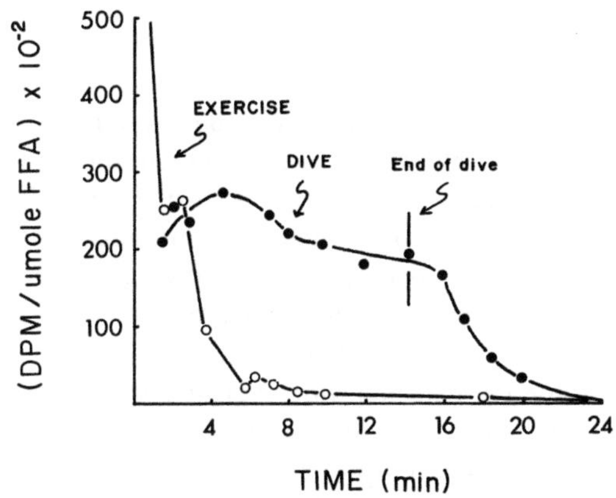

Fig. 1. Specific activity of ^{14}C–U-palmitate in the plasma of a grey seal after a bolus injection of approx. 100 μCi at T = 0. In the exercise experiment, the seal was in steady state surface swimming (approx. 1.5 $M\ s^{-1}$) at T = 0. In the diving experiment, the seal was underwater for at least 1.0 min by T = 0

pool while the animal is still diving. First, values for radioactivity per milliliter of plasma indicate that the bolus was injected into a large pool approximately the size of the blood volume. It if was injected into a smaller central pool, radioactivity values would be six to seven times higher. As the SA curve slowly moves upward during the dive, the values finally peak at the equilibrated value that they would have reached had the bolus simply been injected into a controlled volume the size of the blood pool. Second, calculations of the metabolite mass (Ms) show that the radiotracer was injected into a FFA or glucose pool that approximates the mass of the total plasma volume (Castellini et al. 1985). Similar experiments in our laboratory on other diving species suggest that this "tracer equilibration curve" is an excellent indicator of the diving response. Essentially, the bolus circulates so slowly after injection relative to the normal or exercise condition that the rising phase of the SA curve becomes visible. In a normal condition, the circulation is so fast that samples cannot be taken quickly enough to catch this part of the curve. If the animal will dive for a long enough time, the bolus will equilibrate and start to be metabolized but at a rate much slower than normal. Unfortunately, the apparent turnover rate cannot be quantified during such forced dives because metabolite concentrations usually change (Davis 1983; Murphy et al. 1980) and the animal is therefore not in steady state.

Applying such experiments to field conditions, however, could be very informative. Since natural diving does not appear to greatly alter metabolite levels, perhaps a true turnover could be calculated. If blood samples could also be analyzed for ^{14}C activity in the CO_2 pool, oxidation rates could be obtained. Thus with both turnover and oxidation values, estimates of metabolic rate compared to resting and exercise could be obtained. However, such experiments require the ability to inject and withdraw samples anaerobically from a freely diving animal. This has only recently been attempted on Weddell seals by Zapol and co-workers. Future experiments on this species may provide us with a reasonable estimate of metabolite turnover while the seal is swimming underwater.

Regardless of whether a true turnover and oxidation value could be obtained, the simple shape of the tracer equilibration curve should answer the more basic question of whether the seal is diving with exercise or merely exercising while diving in terms of physiological responses. In the former case, the curve should appear as the complex dive curve seen in Fig. 1 and in the exercise instance, it should appear as a simple hyperbola.

5 Aerobic Diving

It is important to stress that all evidence to date confirms that diving species routinely work within their aerobic dive limits. How they achieve this generates our present investigations into metabolic rate, metabolite turnover, blood flow patterns, etc. This is a quite different concern than existed 10-15 years ago when most investigations focused on the anaerobic abilities of seals with the assumption that diving was an anaerobic feat (Blix 1976; Hochachka and Storey 1975). Calculations of the aerobic dive limit for any given species show values very close to dive limits seen in nature (Kooyman 1982). Some sort of signal, or some learned behavior, indicates to the seal when to surface. This is an interesting signal for the longer dives of the Weddell seal in that the animal must start returning to its breathing hole up to 30 min before the dive is actually finished. Perhaps it is a critical level of oxygen in the blood, or a critical concentration of lactate that signals the half-way point for such long dives. But what about the shorter aerobic dives? What is the signal to finish the hunt and return to the surface? Could it also be low oxygen in the blood? The seal has a finite store of oxygen available - it can use that supply slowly as in a long dive or quickly as in a short dive, but the end result is the same: when the animal surfaces, little available oxygen remains. This is the central point to the calculation of the aerobic dive limit: oxygen stores vs metabolic rate. Metabolic rate of course is extremely difficult to assess for a swimming and diving seal. Most calculations have used approximately two times resting as the estimate of diving metabolic rate (Kooyman et al. 1980, 1983) but this could be off by a significant amount. Estimates for whole animal and muscle metabolic rate in resting, surface exercising and freely diving seals vary by such a wide range that realistic estimates of metabolic rate while the seal is working underwater become extremely difficult (Hochachka 1981; Kooyman 1981). Metabolite turnover studies, however, should give us a direct measurement of diving metabolic requirements.

6 Conclusion

Diving and exercise in marine mammals calls for potentially conflicting metabolic and physiological responses. Diving requires hypometabolism in order to conserve oxygen supplies while exercise, at least in terrestrial mammals and surface exercising seals, calls for an elevated metabolism. Quiet laboratory diving experiments on seals demonstrate the maximum diving modification but call for no exercise requirement. In nature, however, seals both dive and swim: how are these two activities coordinated in terms of physiological responses? Does the seal utilize a dive response only to have it attenuated by exercise requirements? Does the diving seal exercise underwater without any major diving adaptation? Unfortunately, most past studies could only measure heart rate or lactate levels in the blood - these two indices are affected by both diving and exercise, however, and cannot answer the diving/exercise dilemma. Experiments must be designed that can separate natural diving into its exercise and diving components. Metabolite turnover experiments can be used to distinguish exercise from diving since the tracer equilibration curve is radically different from one condition to the other, at least as seen in laboratory conditions. If such techniques could be used in the field, they might help settle controversies such as: whether the diving state is normal and time at the surface represents tachymetabolism; whether diving is more like surface exercise or more like laboratory diving in terms of blood flow patterns; and whether metabolic rate goes up or down as compared to rest conditions. These controversies are specific to the diving mammal since terrestrial mammals have continuous access to oxygen while exercising. They represent some of the unique problems that face the diving seal in adapting to its underwater world.

References

Blix AS (1976) Metabolic consequences of submersion asphyxia in mammals and birds. Biochem Soc Symp 41:169–178

Butler PJ, Jones DR (1982) Respiratory and cardiovascular control during diving in birds and mammals. J Exp Biol 100:195–221

Castellini MA, Somero GN (1981) Buffering capacity of vertebrate muscle: correlations with potentials for anaerobic function. J Comp Physiol 143:191–198

Castellini MA, Somero GN, Kooyman GL (1981) Glycolytic enzyme activities in tissues of marine and terrestrial mammals. Physiol Zool 54(2):242–252

Castellini MA, Murphy BJ, Fedak M, Ronald K, Gofton N, Hochachka PW (1985) Potentially conflicting metabolic demands of diving and exercise in seals. J App Physiol 58(2):392–399

Davis RW (1983) Lactate and glucose metabolism in the resting and diving harbor seal *(Phoca vitulina)*. J Comp Physiol 153:275–288

Davis RW, Castellini MA, Kooyman GL, Maue R (1983a) Renal glomerular filtration rate and hepatic blood flow during voluntary diving in Weddell seals. Am J Physiol 245:R743–R748

Davis RW, Williams TW, Kooyman GL (1983b) Swimming metabolism in yearling and adult harbor seals *(Phoca vitulina)*. Abstract. Fifth biennial conference on the biology of marine mammals. Boston

Fedak MA, Kanishwer J, Lapennas G (1983) Patterns of breathing and heart rate in relation to energy requirements of swimming and diving seals. Abstract. Fifth biennial conference on the biology of marine mammals. Boston

Friedman JJ, Selkurt EE (1966) Circulation in specific organs. In: Selkurt EE (ed) Physiology. Little, Brown, Boston

Hill RD, Schneider RC, Liggins GC, Hochachka PW, Schuette AH, Zapol WM (1983) Microprocessor controlled recording of bradycardia during free diving in the antarctic Weddell seal. Abstract. Fed Proc 42(3):470

Hill RD, Zapol WM, Liggins GC, Schneider RC, Schuette AH, Hochachka PW (1984) Metabolite biochemistry of the Weddell seal diving at sea. Am J Physiol (Regulatory) (in press)

Hochachka PW (1981) Brain, lung, and heart functions during diving and recovery. Science 212: 509–514

Hochachka PW, Dunn JF (1983) Metabolic arrest: the most effective means of protecting tissues against hypoxia. In: Hypoxia, exercise, and altitude: Proceedings of the Third Banff International Hypoxia Symposium. Alan Liss, New York, p 297

Hochachka PW, Storey KB (1975) Metabolic consequences of diving in animals and man. Science 187:613–621

Irving L (1939) Respiration in diving mammals. Physiol Rev 19:112–134

Kanwisher JW, Gabrielson G, Kanwisher N (1981) Free and forced diving in birds. Science 211: 717–719

Kooyman GL (1981) Weddell seal-consummate diver, Ist edn. Cambridge, New York

Kooyman GL (1982) How marine mammals dive. In: Taylor CR, Johansen K, Bolis L (eds) A companion to animal physiology. Cambridge, London, p 151

Kooyman GL, Campbell WB (1972) Heart rates in freely diving Weddell seals, *(Leptonychotes weddelli)*. Comp Biochem Physiol 43A:31–36

Kooyman GL, Kerem DH, Campbell WB, Wright JJ (1973) Pulmonary gas exchange in freely diving Weddell seals, *(Leptonychotes weddelli)*. Respir Physiol 17:283–290

Kooyman GL, Wahrenbrock EA, Castellini MA, Davis RW, Sinnett EE (1980) Aerobic and anaerobic metabolism during voluntary diving in Weddell seals: evidence of preferred pathways from blood chemistry and behavior. J Comp Physiol 138:335–346

Kooyman GL, Castellini MA, Davis RW, Maue R (1983) Aerobic diving limits of immature Weddell seals. J Comp Physiol 151:171–174

Lenfant C, Johansen K, Torrance SD (1970) Gas transport and oxygen storage capacity in some pinnipeds and the sea otter. Respir Physiol 9:277–286

Lin YC, Matsuura DT, Whittow GC (1972) Respiratory variation of heart rate in the California sea lion. Am J Physiol 222:260–264

Murphy BJ, Zapol WM, Hochachka PW (1980) Metabolic activities of heart, lung, and brain during diving and recovery in the Weddell seal. J Appl Physiol 48(4):596–605

Scholander PF (1940) Experimental investigatory function in diving mammals and birds. Hvalradets Skr 22:1–131

Sutton JR, Jones NL, Houston CS (ed) (1982) Hypoxia: Man at altitude. Thieme-Stratton, New York

Zapol WM, Liggins GC, Schneider RC, Qvist J, Snider M, Creasy T, Hochachka PW (1979) Regional blood flow during simulated diving in the conscious Weddell seal. J Appl Physiol 47:968–973

Thoroughbreds and Greyhounds: Biochemical Adaptations in Creatures of Nature and of Man

D.H. SNOW and R.C. HARRIS [1]

1 Introduction

The purpose of this paper is to describe how adaptations, especially within the muscular system, may have bestowed advantages that may account for the athletic abilities of two species, canine and equine. Although both species may not represent the fastest of mammalian animals known nor the best for endurance, selection processes imposed during their domestication and development for specific tasks initially connected with hunting, farming, and warfare, and more recently for leisure activities, led to a wide spectrum of breeds with differing capabilities. A study of these provides an interesting comparison to elite human athletes whose training today tries to lift them above their natural limitations. It could be argued that information obtained from studies of these two athletic species could be more useful in an understanding of muscular development in man rather than the extensive studies carried out in the most commonly used model, the rat. In a recent review on skeletal muscle adaptability (Saltin and Gollnick 1983), studies in the rat were generally referred to when the required data was lacking from investigations in man.

Within the canine species the greyhound is commonly acknowledged as the fastest breed, attaining speeds of just under 1,000 m min^{-1} over 400 m. The greyhound, evolved for hunting by keeness of sight and fleetness of foot, originated in Babylon and Egypt, and reference to its participation in coursing was recorded in Ancient Rome by Ovid. Development for speed with stamina has led to the modern greyhound, a description of which is best given in the following quote, "The modern greyhound the most elegant of the canine race, the highest achievement of man's skill in manipulating the plastic nature of the dog ... as he is stripped in all his beauty of outline and wonderful development, not only of muscle but of the hidden fire which gives dash, energy and daring, stands revealed a manufactured article, the acme of perfection in beauty of outline and fitness of purpose. He is a combination of art and Nature that challenges the world unequalled in speed, spirit, and perseverance, and in elegance and beauty of form as far removed from many of his clumsy ancestors as the English Thoroughbred from a coarse drayhorse" (Dalziel 1868). At the other end of the spectrum are the sled dogs, huskies, which can race over distances of up to 1,710 km in 12–14 days. Twelve dog teams have been recorded as pulling a sled and driver over 32.5 km cross-country in sub-zero temperatures in 77 min (Van Citters and Franklin 1969).

1 Physiology Unit of the Animal Health Trust, Balaton Lodge, Snailwell Road, Newmarket, Great Britain

Circulation, Respiration, and Metabolism
(ed. by R. Gilles)

In comparison, the development of equine species for true speed has been a much more recent occurrence. Breeds have been selected from both cold-blooded and hot-blooded types, the former giving rise to pony and draught horse breeds, whilst the latter to those endowed with speed such as the Thoroughbred, Arab, and Standardbred (Trotter). The Thoroughbred, which is portrayed as the epitome of speed, has only been developed within the last 300 years. The breed was first raced over distances of 3–4 miles, but changes in fashion have seen still faster animals being developed and racing distances now vary between 1,000–3,200 m; although jump races are held over longer distances. In the USA the further pursuit of speed has given rise to the Racing Quarterhorse, with speeds of up to 1,200 m min^{-1} over 400 m.

The eliteness of these two species can be seen when their maximum oxygen consumption is studied. Although in both species the rate of oxygen consumption at rest is as for other mammalian species, proportional to their body weight raised to the power of 0.75, both deviate from the predicted scaling effect when $VO_{2\,max}$ is compared to body weight in that they have higher than expected values (Fig. 1). As the results shown in Fig. 1 were determined from non-elite breeds, an even higher $VO_{2\,max}$ could be expected in the elite.

This high aerobic capacity indicates a very high metabolic activity of working muscle, and is related to the sustained speed of these species. The ability to attain this high

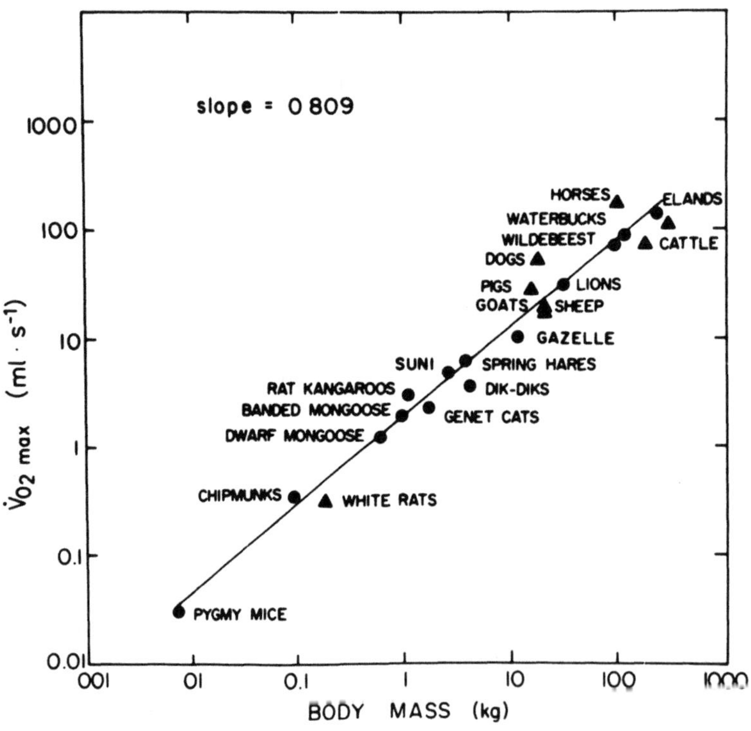

Fig. 1. Average values of $V_{O_2\,max}$ for 14 species of wild animals (*circles*) and 7 species of laboratory/domestic animals (*triangles*) are plotted as a function of body mass on logarithmic co-ordinates. (From Taylor et al. 1980)

aerobic capacity could only occur if adaptations had occurred both within the cardiovascular and muscular systems.

2 Cardiovascular Adaptations

Briefly, within the cardiovascular system the ability to transport large quantities of oxygen to muscle has been brought about in a number of ways which are summarised and compared to man in Table 1. The greatest adaptations occur in the elite athletic breeds. Cardiac output is high due to both the high heart rates attained during exercise and the large heart size relative to body weight permitting a large stroke volume. Blood volume during exercise is also increased due to the ability to markedly increase the number of circulating red blood cells by the emptying of reserves in the spleen. These changes allow the increase in blood flow to muscle to be much greater than that seen in man, with increases in ponies being in the order of 75-fold.

3 Muscular Adaptations

3.1 General Body Composition

In both adult greyhounds and other breeds, it has been found that muscle constitutes a greater proportion of live weight than the 40% found in man, or domestic species bred for meat production. In average dogs, muscle is about 44% of live weight, whilst in the greyhound it is 57%, a figure considered unique for terrestrial mammals (Gunn

Table 1. Comparison of some cardiovascular parameters in horse, dog, and man[a]

	Hot-blood breeds[b]		Dog[c]		Man[d]	
	Rest	Max Ex	Rest	Max Ex	Rest	Max Ex
Haematocrit (l/l)	0.40	0.65	0.55	0.70	0.47	0.50
Heart rate (beats min^{-1})	34	240	100	300	60	190
Cardiac output (ml min^{-1} kg^{-1})	75	545	(277)	(802)	67	333(500)
O_2 consumption (ml min^{-1} kg^{-1})	3	~130	(16)	(112)	5	40
Muscle blood flow (ml min^{-1} 100 g^{-1})	(2)	(155)	(5)	–	3	60
Heart wt:body wt (g kg^{-1})	8.6		12.0		4.0(6.3)	

a Data compiled from several sources: Astrand and Rodahl (1977); Donald et al. (1968); Keele et al. (1982); Ordway et al. (1984); Parks and Manohar (1983); Snow (unpublished data); Steel et al. (1976); Stewart and Steel (1970); Thomas and Fregin (1981)

b Figures in parenthesis: pony

c Figures in parentheses: non-greyhound breeds

d Figures in parentheses: elite atheltes

1978a). Webb and Weaver (1979) have reported the proportion of muscle to live weight in a number of ponies and horses to be 40.1%. Unfortunately, although some Thoroughbreds were studied individual figures for this breed are not given, and in any case would have been unrepresentative as they examined animals suffering from debilitating diseases.

Despite the high proportion of muscle in dogs, the muscle to bone ratio is similar to that in meat producing animals (Gunn 1978a). In both the greyhound and Thoroughbred the proportion of muscle in the femoral region is greater than in other breeds of their respective species (Gunn 1978b). As Thoroughbreds do not have longer legs relative to body weight than non-athletic breeds, Gunn (1983) suggested that this larger pelvic muscle mass would favour a high natural frequency of hindlimb movement, facilitating a high stride frequency. The increased muscle mass in the pelvic region can be explained by both increased cell numbers and greater fiber areas in these elite breeds compared to others of the same species (Gunn 1978b, 1979). In contrast in the transverse pectoral, no difference in fiber area was found between breeds.

Although the greyhound does have some muscles in which the fibers have relatively large areas, within the canine species fiber areas are generally small when compared to man and horse, and are of the same size as in the cat (Maxwell et al. 1977). The relatively small size of canine muscle fibers may be related to the high oxidative capacity of all fibers. This establishes a high surface area to volume ratio enabling a more rapid diffusion of oxygen. The fiber areas of the equine (Snow 1983) are similar or slightly less than those reported in man (Saltin et al. 1977), and considerably less than in athletes involved in power events.

3.2 Fiber Types

From studies using convential histochemical methods to study the basic fiber types, the horse has been found to have similar types to that reported in man. Using both metabolic and contractile indicators, the three types, slow twitch (type 1), and high and low oxidative fast twitch (type II), can be identified. Determination of myosin ATPase activity after acid preincubation results in the identification of type I, IIA, and IIb fibers, the first two having a high oxidative capacity, whilst the IIB may be either high or low. More recently a combination of both systems has been used (Essen-Gustavsson et al. 1984). As in man, training results in a decrease of fast twitch low oxidative or type IIB fibers (Snow 1983).

In the canine both type I and type II fibers can be described. However, sub-division of type II fibers has proved difficult. Numerous studies (Maxwell et al. 1977; Gunn 1978c; Braund et al. 1978; Guy and Snow 1981; Armstrong et al. 1982) have shown that all type II fibers have similar intermediate to high oxidative capacity. This finding corresponds to the common observation that canine muscles generally have a very red appearance and to the physiological observations on the endurance capability of the species. Braund et al. (1978) has claimed that type IIB fibers do not exist in the dog and that only IIA are present. Recently, however, using a combination of methods involving enzyme histochemistry, immunocytochemistry for myosins, and peptide mapping of myosin heavy chains, Snow et al. (1982) concluded that although the typical type IIA fiber exists, another sub-type (perhaps to be called a type IID) occurs which differs from the normal type IIB. Further studies on the identification of the myosin in this fiber type are being carried out. Therefore two sub-types of type II fibers occur both having similar oxidative-glycolytic capacity. The physiological significance of this with respect to contractile characteristics awaits elucidation, but one can only speculate that it is associated with all canine fiber being fatigue resistant as shown by Maxwell et al. (1977), bestowing speed and stamina when all fibers are recruited. As power output and speed are generally associated with fibers of large cross-sectional areas, which the dog does not possess, it is possible that the capacity to develop power is related to its ability to sustain force output by a continuous recruitment of numerous fatigue resistant fibers.

Initial studies in the horse (Gunn 1978c; Snow and Guy 1980) indicated that, as for man, fiber type distribution was relatively uniform throughout the muscle. This was in contrast to most other species. However, more recent studies in man (e.g. Henriksson-Larsen et al. 1983) and a detailed investigation of the middle gluteal in the horse (Bruce and Turek 1985) have shown that variation does occur, with a higher proportion of type I fibers being found in the deeper parts of the muscle. Thus when biopsies are taken in the course of either cross-sectional or longitudinal studies, great care must be taken to ensure a uniform site. From studies of a number of limb muscles, Snow and Guy (1980) found a lower proportion of type II fiber in the forelimb than hindlimb, although in all muscles type II fibers predominated. It was suggested that this related to locomotory style, in that the thoracic limbs act as brakes while the pelvic limbs propel the body.

In the dog typical compartmentalisation as observed in many mammals occurs, with the deeper parts of muscles containing the greatest proportion of type I fibers (Gunn 1978c; Armstrong et al. 1982; Snow, unpublished data). Interestingly Armstrong et al. (1982) found that in the antigravity (extensor) groups in the arm and thigh, the deepest muscles have the highest proportion of type I fibers, whilst in the forearm and leg the pattern was reversed.

In attempting to relate performance characteristics of horses to muscle composition, Snow and Guy (1980) studied six limb muscles (two forelimb, four hindlimb) and found that the middle gluteal gave the best relationship when different breeds were compared. This is not surprising, as this muscle is one of the largest in the body, and important in the generation of propulsive force. Webb and Weaver (1979) have shown

Table 2. Fibre composition in the middle gluteal of untrained animals from different breeds of horses (% fibre type, mean ± SEM) from Snow (1983)

	n	ST	FTH	FT
Quarterhorse	28	8.7 ± 0.8	51.0 ± 1.6	40.3 ± 1.6
Thoroughbred	50[a]	11.0 ± 0.7	57.1 ± 1.3	32.0 ± 1.3
	149[b]	14.7 ± 0.4	65.1 ± 0.5	20.2 ± 0.5
Arab	6	14.4 ± 2.5	47.8 ± 3.2	37.8 ± 2.8
Standardbred	8	18.1 ± 1.6	55.4 ± 2.2	26.6 ± 2.0
Shetland Pony	4	21.0 ± 1.2	38.8 ± 1.9	40.2 ± 2.7
Heavy Hunter	7	30.8 ± 3.1	37.1 ± 3.3	32.1 ± 3.4
Donkey	5	24.0 ± 3.0	38.2 ± 3.0	37.8 ± 2.8

[a] Elite broodmares
[b] Moderate 2 year olds

Table 3. The percentage of high myosin ATPase activity fibres in the various muscles of the greyhound, foxhound, and cross-bred dogs (mean ± SEM) (from Guy and Snow 1981)

	Greyhound	Cross-bred	Foxhound
No. of animals	6	5	4
M. deltoideus	99.8 ± 0.2	74.4 ± 3.0	56.6 ± 6.8[a,b]
M. triceps brachii caput longum	94.2 ± 5.4	77.2 ± 2.7[a]	64.9 ± 4.4[a,b]
M. vastus lateralis	96.6 ± 1.7	61.4 ± 10.2[a]	80.7 ± 8.2[a]
M. gluteus medius	97.4 ± 0.7	68.6 ± 5.6[a]	65.3 ± 4.9[a]
M. biceps femoris	88.6 ± 2.2	67.2 ± 1.3[a]	63.0 ± 1.5[a]
M. semitendinosus	98.9 ± 0.8	85.3 ± 2.6[a]	69.6 ± 6.8[a,b]

[a] $P < 0.05$ when cross-bred or foxhound compared to greyhound
[b] $P < 0.05$ when cross-bred compared to foxhound

that this muscle contributes a considerably greater proportion of muscle mass in the equine than bovine. Further studies (Table 2) have shown that the fastest breeds have the highest proportion of type II fibers, whilst the lowest are found in endurance animals (Snow and Guy 1981; Snow 1983). However, in addition to the proportion of type II fibers, individual fiber areas and oxidative capacity have to be considered in estimating performance characteristics (Snow 1983).

When comparing the greyhound to other canine breeds (Table 3) the effect of selection for speed over thousands of years can be clearly seen. Guy and Snow (1981) demonstrated that in six limb muscles examined almost all fibers are of type II, and that these are of the atypical type II previously mentioned (Rowlerson, personal communication).

3.3 Biochemistry

In this section, because of the difficulties in comparing results between different laboratories with respect to enzyme activities, work carried out initially comparing equine and canine breeds by Guy (1978), Snow and Guy (1981), and Guy and Snow (1981), and a more recent comparative study by the authors will be referred to. In this recent study the middle gluteal of well-trained Thoroughbred horses was compared to the vastus lateralis of trained humans and the semitendinosus of greyhounds of unspecified training status. Enzyme activities from the first studies are shown in Table 4, whilst those from the latter study are given in Table 5.

3.3.1 Anaerobic Metabolism

Phosphagen Utilisation. The ATP and total creatine levels are similar in all three muscle systems, being approx. 6 and 20 mmol kg^{-1} wet muscle respectively. However, there are marked differences in the activity of CPK. This is almost twice as high in greyhound as in human muscle, whilst horse lies in between. Conceivably this permits a more rapid breakdown of phosphcreatine in the dog during high burst activity and could account for the extremely rapid acceleration at the start of a race, for which this animal is well noted.

Table 4. Enzyme activities (mmol min^{-1} kg^{-1} dry wt) in middle gluteal of equine and canine breeds (mean)

	Quarterhorse	Thoroughbred	Pony	Heavy Hunter	Donkey	Greyhound	Cross-bred
No. of Animals	8	8	8	9	4	6	5
CPK	19,940[a]	20,220[a]	17,578[a]	16.930[a]	10.346	34,250	22,849[d]
LDH	6,213[a]	2,613[a,b]	2,947[a,b]	2,885[a,b]	1,548	2,201	2,330
ALD	632[a]	348[a,b]	402[a,b]	391[a,b]	203	442	245[d]
AST	486	618[a]	448	553[a]	340	1,038	915
ALT	53.7[a]	44.8[a]	41.1[a]	49.3[a]	17.5	76	48[d]
CS	8.2[a]	15.1[a,b]	10.7[a,b,c]	12.7[a,b]	2.8	23	8[d]
HK	8.7[a]	11.1[a]	7.0[a,c]	7.6[c]	5.9	5.2	7.0[d]
HAD	112	159[b]	160	187[b]	148	115	291[d]
GDH	181[a]	142[a]	149	157[c]	53	143	156
Glycogen	252[a]	329[a,b]	300[a]	321[a]	234	149	192

Assay temperature 37 °C. Adapted from Snow and Guy (1981), and Guy and Snow (1981)

[a] $P < 0.05$ compared to donkey
[b] $P < 0.05$ compared to quarterhorse
[c] $P < 0.05$ compared to thoroughbred
[d] $P < 0.05$ compared to greyhound

Table 5. Comparison of enzyme activities between horse, dog, and man (mmol min^{-1} kg^{-1} wet muscle, assay temperature (30 °C)

	Thoroughbred (6)	Greyhound (4)	Man (8)	Assay pH
CPK	4,352	5,018	3,224	7.0
Phosphorylase[a]	58		38	7.0
PFK	89	76	35	8.2
LDH	980	994	220	7.0
ALT	9.0	6.9	3.4	7.5
OGDH	4.0	1.2	0.6	7.4
HAD	36	17	7.3	7.0
GPDH	68	66	23	7.0
AST	234	257	86	7.5
AMP deaminase	52	42	69	6.8

[a] Assayed at 35 °C

Glycolysis and Lactate Production. When comparing blood lactate concentrations over similar periods of maximal exercise it can be seen (Table 6) that both the greyhound and Thoroughbred attain higher values than man. Over similar distances, lower concentrations occur in horses involved in trotting races (Krzywanek 1974). The very high blood lactates seen in Thoroughbreds is reflected in the amounts of lactate which are accumulated in working muscle. For example, lactate accumulation at the end of a 1,500 m maximal gallop can reach as high as 36 mmol kg^{-1} wet muscle and, after four consecutive 600 m gallops with 5 min rest periods, as high as 53 mmol kg^{-1} wet muscle (Snow et al. 1984). In human muscle lactate contents have rarely been observed to exceed 25–30 mmol kg^{-1} wet muscle, even under the most exacting of exercise conditions. Despite this the buffering capacity of both horse and human muscle is about the same, about 70 slykes (Harris et al. 1984) with the result that the pH in horse muscle can reach extremely low values during maximal exercise. In one horse a pH value of 6.2 was calculated following four consecutive gallops (Snow et al. 1985). A high anaerobic capacity and tolerance of low pH is thus a feature of the Thoroughbred muscle.

Table 6. Blood lactate following maximal exercise[a] (mean [SD])

Species	Number	Distance (m)	Speed (m min^{-1})	Lactate (mmol l^{-1})
Dog (Greyhound)	5[b]	300	941	15.1 (1.4)
	4[b]	480	932	22.9 (1.4)
Horse (Thoroughbred)	6[b]	600	1,002	14.2 (2.1)
	19[c]	1,200	954	26.6 (4.3)
	4[c]	2,400	924	27.3 (2.4)
Man	4[d]	400	397	12.3 (1.6)
	6[e]	800	384	12.6 (0.3)

[a] Samples collected approx. 5 min after exercise
[b] Snow, unpublished data
[c] Snow et al. (1983)
[d] Costill et al. (1983)
[e] Wilkes et al. (1983)

Enzyme studies confirm a much higher anaerobic capacity of horse muscle and also of greyhound muscle, compared to man (Table 5). Thoroughbred and greyhound muscle are rather similar to each other exhibiting two and three times as much glycogen phosphorylase, phosphofructokinase (PFK), and lactic dehydrogenase (LDH) activity per gram of muscle than human muscle.

To some extent these findings are at variance with Emmett and Hochachka (1981) showing a positive scaling between glycolytic enzymes and body mass (the larger the animal, the higher the activity of glycolytic enzymes in muscle). As discussed by these authors the positive scaling probably follows from a higher relative power cost of burst type activity in larger animals and in this respect the greyhound, in particular, stands apart from the normal trend. As in the case of oxygen uptake capacity, selective breeding for speed seems to have lifted this animal above the natural line, and this is probably also true of the Thoroughbred. The higher levels of these glycolytic enzymes are in accordance with the high proportion of type II muscle fibers in the Thoroughbred and greyhound compared to man, where the admixture of types I and II fibers in the vastus lateralis is on average close to 50:50 (as was also the case in the present study). Given a distribution of 2:1 or more of phosphorylase, PFK and LDH between type II and type I fibers (Harris et al. 1976; Essen et al. 1975; Lowry et al. 1978), then the intrinsic activity of these enzymes per fiber is probably much closer between the three species.

The two- to three-fold higher activity of alanine aminotransferase (ALT) in Thoroughbred and greyhound muscle, compared to man, points to a higher capacity to transaminate pyruvate to alanine - an alternative end product of anaerobic glycolysis to lactate, and an important substrate for gluconeogenesis in the liver. The possible importance of this enzyme in the horse may be indicated by the doubling of activity with training (Guy and Snow 1977).

Within the spectrum of horse breeds a similar gradation of anaerobic capacity with running speed also occurs (Snow and Guy 1981), the elite sprinters, the quarterhorse and Thoroughbred, showing much higher levels of LDH and ALT [and hexokinase (HK) and aldolase] than the donkey. A similar trend can be seen when the greyhound is compared to the cross-bred dog (Guy and Snow 1981).

3.3.2 Carbohydrate and Free Fatty Acid (FFA) Oxidative Capacity

Measurements of enzymes of carbohydrate and FFA oxidative capacity both within and between species show a more confused picture. Oxoglutarate dehydrogenase (OGDH), which reportedly is the limiting enzyme in the TCA cycle, is four times higher in Thoroughbred muscle compared to man (Table 5). In the greyhound OGDH activity is about twice as high as in man. The very high level of OGDH activity measured in the Thoroughbred in this study is probably reflective of the fact that all type II fibers were highly oxidative. In contrast to OGDH, citrate synthase (CS), a further marker of mitochondrial activity was found to be roughly equal to the activity in Thoroughbred and greyhound muscle (Table 4). This discrepancy of activities of OGDH and CS activities between greyhounds and Thoroughbreds may be attributed to the fact that in the earlier study when CS was measured the Thoroughbreds were in an untrained state.

Relatively few comparative data on mitochondrial capacity are available between different breeds of horses and dogs. Within the equine family, higher activities of CS are found in quarterhorse and Thoroughbred muscle compared to donkey, and in the greyhound compared to the cross-bred. Although the current picture is still somewhat confused the evidence points to a higher oxidative capacity in the muscle of the elite runners both within and between species, commisserate with the high proportion of high oxidative type II fibers found in these animals, i.e. the Thoroughbred and greyhound. It is well documented that in the horse as for other species, training results in an increase in oxidative capacity.

3-Hydroxyacyl coenzyme A dehydrogenase (HAD), which catalyses the reversible oxidation of 3-hydroxyacyl CoA to acetoacetyl CoA from which acetyl CoA is split by the action of β keto thiolase, shows a clear degradation between the Thoroughbred, greyhound and man, with muscle of the former exhibiting twice as much activity as greyhound muscle; greyhound muscle has twice the activity of human muscle. The gradation of HAD activity between the three species parallels closely with that of OGDH (Table 5). Within species, the elite sprinter – the quarterhorse – exhibits a lower activity of HAD than either the Thoroughbred or donkey, which are themselves very similar. Again, the sprinting greyhound exhibits only half the activity of HAD compared to the cross-bred mongrel. Development of sprinting ability within species thus seems to be associated with a reduction in the capacity to utilise FFA. Despite this, surprisingly, HAD activity in both the Thoroughbred and greyhound is, as earlier indicated, far higher than that in human muscle, although the proportion of type I muscle fibers in the v. lateralis of the latter is much higher than in the comparable muscles of the other two species.

3.3.3 Shuttle Enzymes

In further agreement with the higher metabolic activity of Thoroughbred and greyhound muscle are the two- to three-fold higher activities of the so-called shuttle enzymes, glycerol-3 phosphate dehydrogenase (GPDH) and aspartate aminotransferase (AST), compared to their activity in human muscle. Of the two, AST is probably the most important quantitatively in mammalian muscle. In 1977, Guy and Snow reported that AST increased markedly with training, whilst no change occurred in GPDH. Within species there is a trend towards higher AST activity in muscle of the elite runner – the Thoroughbred, quarterhorse and greyhound – compared to their slower relatives, the donkey and cross-bred.

3.3.4 Purine Nucleotide Degradation

In line with a recent investigation (Snow et al. 1985) showing a decrease in ATP concentration with maximal exercise, enzymes involved in purine nucleotide degradation have been examined. In the horse it has been found both from the study of adenosine and inosine concentrations (Snow et al. 1985), and enzymes involved in their production (Cutmore, Snow, and Newsholme, unpublished data), that the inosine pathway is more important. AMP deaminase has been found in higher concentrations than in the rat, and increases with traning. This enzyme has been reported to occur mainly in type II fibers (Fishbein et al. 1980; Meyer et al. 1980). Its physiological function is unknown, but one possibility is that it is important in preventing the build-up of ADP during burst movement, which could possibly inhibit myosin ATPase and in the extreme contribute to the development of force fatigue. Patients with skeletal muscle AMP deaminase deficiency exhibit profound muscle weakness and cramping in response to exercise (Fishbein et al. 1978).

In our recent study (Table 5) the Thoroughbred and greyhound had similar AMP deaminase activities, ranging between 30–68 units. Interestingly, in man a much broader spectrum of activity (14–133 units) was seen, with four of the seven individuals having a higher activity than the highest seen in either of the other two species. Within the human group there was no correlation between fiber distribution and AMP deaminase, but CPK activity tended to be higher in those with the highest level of AMP deaminase.

4 Conclusion

In this paper we have tried to illustrate how generations of breeding have resulted in many specific adaptations at the cardiovascular and muscular levels which are aimed at solving the problem of maintaining a high energy output without incurring excessive fatigue. In many instances parallel adaptations have occurred within the two species relative to their non-racing counterparts. It is our belief that studies both between and within breeds for each of these species constitute interesting models for investigations into exercise physiology and adaptations with training.

Acknowledgements. Much of the work reported in this paper has been supported by grants from the Horserace Betting Levy Board. Human muscle samples were generously provided by Professor E. Hultman, and taken as part of a study supported by the Swedish Medical Research Council (Project No 02647).

References

Armstrong RB, Saubert SW IV, Seeherman JJ, Taylor CR (1982) Distribution of fiber types in locomotory muscles of dogs. Am J Anat 163:87–98

Astrand RO, Rodahl K (1977) Textbook of work physiology. McGraw Hill, New York

Braund KG, Hoff EJ, Richardson KEJ (1978) Histochemical identification of fibre types in canine skeletal muscle. Am J Vet Res 39:561–565

Bruce VL, Turek RJ (1985) Muscle fibre variation in the gluteus medius of the horse. Equine vet J 17:317–321

Costill DL, Barnett A, Sharp R, Fink WJ, Katz A (1983) Leg muscle pH following sprint running. Med Sci Sports Exercise 15:325–329

Dalziel H (1868) The greyhound. L. Upcott Gill, London

Donald DE, Ferguson DA, Milburn SE (1968) The effect of β-adrenergic receptor blockade on racing performance of greyhounds with normal and with denervated hearts. Circ Res 22:127–134

Emmett B, Hochachka PW (1981) Scaling of oxidative and glycolytic enzymes in mammals. Respir Physiol 45:261–272

Essen B, Jansson E, Henriksson J, Taylor AW, Saltin B (1975) Metabolic characteristics of fibre types in human skeletal muscle. Acta Physiol Scand 95:153–165

Essen-Gustavsson B, Karlstrom K, Lindholm A (1984) Fibre types, enzyme activities and substrate utilisation in skeletal muscles of horses competing in endurance rides. Equine Vet J 16:197–202

Fishbein WN, Armbrustmacher VW, Griffin JL (1978) Myoadenylate deaminase deficiency: A new disease of muscle. Science 200:545–548

Fishbein WN, Griffin JL, Armbrustmacher VW (1980) Stain for skeletal muscle adenylate deaminase. Arch Pathol Lab Med 104:462–466

Gunn HM (1978a) The proportion of muscle, bone and fat in two different types of dog. Res Vet Sci 24:277–282

Gunn HM (1978b) The mean fibre areas of the semitendinosus, diaphragm and pectoralis transversus muscles in differing types of horse and dog. J Anat 127:403–414

Gunn HM (1978c) Differences in the histochemical properties of skeletal muscle of different breeds of horses and dogs. J Anat 127:615–634

Gunn HM (1979) Total fibre numbers in cross-sections of the semitendinosus in athletic and non-athletic horses and dogs. J Anat 128:821–828

Gunn HM (1983) Morphological attributes associated with speed of running in horses. In: Snow DH, Persson SGB, Rose RJ (eds) Equine exercise physiology. Granta Editions, Cambridge, pp 271–274

Guy PS (1978) Factors influencing muscle fibre composition in the horse. PhD Thesis, University of Glasgow

Guy PS, Snow DH (1977) The effect of training and detraining on muscle composition in the horse. J Physiol 269:33–51

Guy PS, Snow DH (1981) Skeletal muscle fibre composition in the dog and its relationship to athletic ability. Res Vet Sci 31:244–248

Harris RC, Essen B, Hultman E (1976) Glycogen phosphorylase activity in biopsy samples and single muscle fibres of musculus quadriceps femoris of man at rest. Scand J Clin Lab Invest 36:521–526

Harris RC, Katz A, Sahlin K, Snow DH (1984) Measurement of muscle pH in horse muscle and its relation to lactate content. J Physiol 357:119P

Henriksson-Larsen KB, Lexell J, Sjostrom M (1983) Distribution of different fibre types in human skeletal muscles. I. Method for the preparation and analysis of cross-sections of whole tibialis anterior. Histochem J 15:167–178

Keele LA, Neil E, Joels N (1982) Samson Wright's Applied Physiology, 13th edn. Oxford University Press, Oxford

Krzywanek H (1974) Lactic acid concentration and pH values in trotters after racing. J S Afr Vet Assoc 45:355–360

Lowry CV, Kimmey JS, Felder S, Chi M M-Y, Kaiser KK, Passonneau PN, Kirk KA, Lowry OH (1978) Enzyme patterns in single human muscle fibres. J Biol Chem 253:8269–8277

Maxwell LC, Barclay JK, Mohrmann DE, Faulkner JA (1977) Physiological characteristics of skeletal muscles of dogs and cats. Am J Physiol 133:C14–C18

Meyer RA, Gilloteaux J, Terjung RL (1980) Histochemical demonstration of differences in AMP deaminase activity in rat skeletal muscle fibers. Experientia 36:676–677

Ordway GA, Floyd DL, Longhurst JC, Mitchell JH (1984) Oxygen consumption and hemodynamic responses during graded treadmill exercise in the dog. J Appl Physiol 57:601–607

Parks CM, Manohar M (1983) Distribution of blood flow during moderate and strenuous exercise in ponies (Equus caballus). Am J Vet Res 44:1861–1866

Saltin B, Gollnick PD (1983) Skeletal muscle adaptability: significance for metabolism and performance. In: Peachey CD (ed) Handbook of physiology, sect 10. Am Physiol Soc, Maryland, pp 555–631

Saltin B, Henricksson J, Nygaard E, Andersen P, Jansson E (1977) Fiber types and metabolic potential of skeletal muscles in sedentary man and endurance runners. Ann NY Acad Sci 301:3–29

Snow DH (1983) Skeletal muscle adaptations: A review. In: Snow DH, Persson SGB, Rose RJ (eds) Equine exercise physiology. Granta Editions, Cambridge, pp 160–183

Snow DH, Guy PS (1980) Muscle fibre type composition of a number of limb muscles in different types of horse. Res Vet Sci 28:137–144

Snow DH, Guy PS (1981) Fibre type and enzyme activities of the gluteus medius in different breeds of horse. In: Poortmans J, Niset G (eds) Biochemistry of exercise IV–B. University Park Press, Baltimore, pp 275–282

Snow DH, Billeter R, Mascarello F, Carpene E, Rowlerson A, Jenny E (1982) No classical type IIB fibres in dog skeletal muscle. Histochemistry 75:53–65

Snow DH, Mason KD, Ricketts SW, Douglas TA (1983) Post-race blood biochemistry in Thoroughbreds. In: Snow DH, Persson SGB, Rose RJ (eds) Equine exercise physiology. Granta Editions, Cambridge, pp 389–399

Snow DH, Harris RC, Gash S (1985) Metabolic response of equine muscle to intermittent maximal exercise. J Appl Physiol 58:1689–1697

Steel JD, Taylor RI, Davis PE, Stewart GA, Salmons PW (1976) Relationships between heart score, heart weight and body weight in greyhound dogs. Aust Vet J 52:561–564

Stewart GA, Steel JD (1970) Electrocardiography and the heart score concept. Proc Am Assoc Equine Practitioners 16:363–381

Taylor CR, Maloiy GMO, Weibel ER, Langman VA, Kamau JMZ, Seeherman HJ, Heglund NC (1980) Design of the mammalian respiratory system. III. Scaling maximum aerobic capacity to body mass: Wild and domestic mammals. Resp Physiol 44:25–37

Thomas DP, Fregin GF (1981) Cardiorespiratory and metabolic responses to treadmill exercise in the horse. J Appl Physiol 50:864–868

Van Citters RL, Franklin DL (1969) Cardiovascular performance of Alaska sled dogs during exercise. Circ Res 24:33–42

Webb AI, Weaver RMQ (1979) Body composition of the horse. Equine Vet J 11:39–47

Wilkes D, Gledhill N, Smyth R (1983) Effect of acute induced metabolic alkalosis on 800-m racing time. Med Sci Sports and Exercise 15:277–280

Exercise Limitations at High Altitude: The Metabolic Problem and Search for Its Solution

P.W. HOCHACHKA [1]

1 Introduction

Despite the declining availability of O_2 with decreasing barometric pressure, man is able to meet the O_2 demands of basal aerobic metabolism (about 0.17 mmol O_2 kg^{-1} min^{-1}) on the planet's highest peaks. In contrast, the elevated demands of aerobic exercise metabolism may not be satisfiable at high altitudes because of a progressive drop in metabolic scope for activity. At low altitudes, the scope for activity (difference between basal and maximum metabolic rates) for most mammals is about tenfold (Taylor et al. 1980). Trained athletes may nearly double this value, yet at altitude man's maximum metabolic rate or $\dot{V}_{O_2 (max)}$ decreases on average by about 11% per 1,000 m of altitude (Frisancho 1983). At 6,000 m, $\dot{V}_{O_2 (max)}$ is decreased to about 50% of normal. At altitudes as high as the top of Mt. Everest, the aerobic scope for activity for man is so low that scaling the mountain without supplementary O_2, as first achieved by Peter Habeler and Reinhold Messner in 1978, has been viewed by high altitude physiologists as an entirely remarkable feat and reviewed in at least two scientifically reputable places (Sutton et al. 1983; West 1983). In a comparative context, such a feat is not all that impressive; a common house sparrow can readily match it and the bar-headed goose routinely *flies* over Everest. Nevertheless, the triumph of man on Everest serves to emphasize not only our order-of-magnitude limitations (compared to altitude-adapted species), but also the most essential metabolic problem faced by all high altitude animals: *how to maintain an acceptably high scope for sustained aerobic metabolism in spite of reduced availability of oxygen in the atmosphere.*

The problem at the working tissue level revolves around the way mitochondrial metabolism interacts with O_2. The paradox is that even on Everest, the concentration of O_2 in arterial plasma, about 44 μM (West 1983), seems easily adequate to maintain intracellular O_2 concentrations much higher than the apparent K_m for isolated mitochondria (about 0.3 μM). Why then should muscle metabolism behave as if it were severely O_2 limited as soon as exercise is initiated? Although this area is still controversial, recent studies with whole cell suspension (hepatocytes, myocytes, microorganisms) indicate (1) that the apparent K_m for O_2 is at least an order of magnitude higher than with isolated mitochondria (Jones and Kennedy 1982a,b; Wilson et al. 1979a,b), and (2) that the K_m increases substantially to values of about 6–10 μM as

1 Department of Zoology, University of British Columbia, Vancouver, British Columbia, V6T 2A9, Canada

Circulation, Respiration, and Metabolism
(ed. by R. Gilles)

mitochondrial respiration increases, presumably because of diffusional limitation within the cellular compartment (Jones and Kennedy 1982a). These values are high enough to easily explain O_2-limited muscle metabolism in man on Everest (West 1983), and thus to explain the drastic drop in aerobic metabolic scope. Unlike man and other lowland adapted species, high altitude animals manage to maintain mitochondrial O_2 concentrations at least at K_m levels or higher and are thus able to maintain a near-normal scope for activity. The question is how.

In addressing this problem, it is worth remembering that the difficulties of functioning in chronic hypoxia are not unique to high altitude animals; these problems are encountered in numerous environments and are successfully circumvened by numerous groups of organisms usually by using one or only a few recurring metabolic strategies (Hochachka 1980). In the absence of major anaerobic adaptations, the three most common strategies involve:

(1) redistribution of cardiac output so as to conserve available O_2 for working tissues (see Zapol et al. 1979, for an example);
(2) preferential utilization of fuels which improve the yield of ATP per mole of O_2 consumed (see Vik-Mo and Mjos 1981) so as to maximize the effectiveness of that O_2 which is available; and/or
(3) elevation of O_2 flux capacities from lung to working mitochondria, so as to compensate for reduced O_2 availability by increased rate of delivery (Hochachka et al. 1982).

In principle, any or all of these metabolic strategies for aerobic function under hypoxic conditions could be utilized by high altitude animals. The aim of this paper is to explore the theoretical basis for, and potential effectiveness of, each of the above mechanisms, and then assess current evidence on when, where, and how they are used.

2 Redistribution of Cardiac Output

Probably the two best studied examples of controlled redistribution of cardiac output under potentially O_2-limiting conditions are diving in marine animals and exercise in terrestrial ones. As emphasized elsewhere (Castellini et al. 1985), the demands of diving under natural conditions in fact may be competitive with those of exercise. During exercise under normoxic conditions in terrestrial mammals, however, there is a well-described redistribution of cardiac output away from the splanchnic region, the kidneys and the skin (Armstrong and Laughlin 1984); in man, an additional quantity of about 3 liters of blood is thus made available for working muscle (Rowell 1974). In terms of O_2 delivery, the quantitative importance of this mechanism depends upon the $\dot{V}O_{2(max)}$ of different individuals: the improvement is 9%, 15%, and 40% respectively, in trained athletes, in sedentary individuals, and in patients with mitral stenosis disease. In other words, *the lower the maximum metabolic rate, the larger the impact of releasing an additional (3 liter) quantity of blood for the support of muscle work* (Rowell 1974). This graded effectiveness implies that the strategy could be useful in high altitude animals, but to our knowledge it has never been quantified in any well-adapted species. However, even in low altitude animals, such as rats during maximal work rates under normoxic conditions, the mechanism can be activated so strongly it may fully inhibit flows to organs, such as the liver and kidneys (Armstrong and Laughlin 1984). That is why we conclude there is little room left for further adaptation along these lines in high altitude animals. At best, this strategy could improve performance at altitude by about 10%-15%, not by the several-fold improvement that separates performance capacities of high vs low altitude species.

3 Improving ATP Yield per Mole of Oxygen Consumed

Whereas redistribution of cardiac output may be viewed as a strictly physiological mechanism for extending aerobic muscle work in the face of limited atmospheric O_2, the preferential use of fuels which maximize the yield of ATP/mol of O_2 consumed is a purely biochemical adaptation. Because the relative roles of NAD^+ vs FAD^+-linked dehydrogenases differ in the oxidative catabolism of different substrates, the yield of ATP/mole of O_2 consumed can also be rather different (Hochachka and Somero 1984). The accepted stoichiometry for palmitate oxidation, for example, yields 129 mol ATP mol^{-1} of palmitate oxidized by 23 mol of O_2 or 5.6 mol ATP/mol O_2 consumed. During glucose oxidation, 6 mol ATP are formed/mol O_2 consumed, while during glycogen oxidation, 6.2 mol ATP/mol O_2 are obtained. Thus, the preferential use of carbohydrate rather than fat theoretically may net a high altitude animal at least a 10% energetic advantage. Empirically, the difference between fatty acids and carbohydrates is actually somewhat larger (25%-40%), because of an 'O_2 wasting effect' of fatty acids which is thus far not properly explained (Vik-Mo and Mjos 1981; Hochachka and Somero 1984).

To our knowledge there are no direct tests of altered fuel preferences in any high altitude adapted species. However, there is indirect evidence for this strategy from studies of enzyme profiles in the gastrocnemius of three high altitude adapted mammals – the taruca (a high altitude deer), the llama, and the alpaca (Table 1). An interesting feature of these enzyme profiles is that the ratio of pyruvate kinase (PK)/lactate dehydrogenase (LDH) is notably high compared to other species (Tables 1, 2). Whereas PK may be used in both anaerobic and aerobic glycolysis, LDH is presumed to operate mainly in the former during high intensity work; hence, a high PK/LDH ratio is taken to indicate a high capacity for aerobic glycogen catabolism (Hochachka et al. 1982), with the consequent gain in yield of ATP/mol of O_2 consumed. Interestingly, this is most notable in the taruca, which, of the three species examined, is probably the most vigorous performer. As evident in Table 2, this indicator of metabolic organization varies somewhat between species, but only in shrew muscle (which has a very low anaerobic capacity, a very high oxidative one) is the PK/LDH activity ratio similar to that in the high altitude adapted animals. Thus it is tempting to suggest that high altitude adapted animals may preferentially utilize a carbohydrate-based metabolism in their muscles, and thus gain the advantage of increased ATP yield/mol O_2 consumed. However, as with strategic redistribution of O_2-rich blood, this mechanism can only bring about a 10%-40% improvement, which by itself is inadequate to account for observed differences between a llama-like species and low altitude species like man. That leaves us with our third option.

Table 1. Enzyme activities in gastrocnemius in high altitude species[a]

Enzyme	Species		
	Taruca	Llama	Alpaca
Lactate dehydrogenase (LDH) (2 m*M* pyruvate)	1,214 ± 304	2,116 ± 123	2,225 ± 211
Lactate dehydrogenase (10 m*M* pyruvate)	1,177 ± 218	1,418 ± 96	1,514 ± 144
Pyruvate kinase (PK)	1,083 ± 187	1,374 ± 56	1,356 ± 79
Citrate synthase (CS)	48 ± 10.9	7.1 ± 0.9	9.7 ± 0.9
β-hydroxyacylCoA (HOAD) dehydrogenase	15 ± 3.2	12.4 ± 0.9	12.4 ± 0.7

[a] From Hochachka and Stanley (unpublished data) using methods given in Hochachka et al. (1982) Activity is expressed in μmol substrate converted g wet wt.$^{-1}$ min^{-1} at 37 °C. Each value represents the mean from eight individuals (five in the case of taruca) ± SE

Table 2. Enzyme activity ratios in gastrocnemius of various mammals

High altitude species[a]	PK/LDH	PK/HOAD
Taruca	0.92	72
Llama	0.97	111
Alpaca	0.89	109
Low altitude species[b]		
Ox	0.56	280
Rabbit	0.55	264
Rhesus monkey	0.75	193
Deer	0.80	98
Vole	0.67	97
Pig	0.76	74
Rat	0.70	33
Guinea pig	0.48	22
Deer mouse	0.53	5
Shrew	0.93	2

[a] From Hochachka and Stanley (unpublished data)
[b] From Emmett and Hochachka (1982)

4 Improving Oxygen and Substrate Flux Capacities

4.1 Enzyme and Metabolic Responses to Altitude Hypoxia

We have argued elsewhere that an invariable result of long-term adaptation to high altitude is an elevated capacity for fluxing O_2 from alveolar atmosphere to working mitochondria (Hochachka et al. 1982). This inevitable result appears to require 'tuning up' two parts of the system, the O_2-delivery part and the tissue metabolism part.

Historically, the former was first noticed and there are now numerous demonstrations (1) of left-shifted oxygen dissociation curves (ODCs) for blood of high altitude animals; (2) of reduced blood viscosity (reduced hematocrit, low hemoglobin levels); (3) of decreased diffusion distances; (4) of increased capillarity; and (5) of increased facilitated diffusion capacity due to increased myoglobin levels (see Lahiri 1977; Wood and Lenfant 1979; Heath et al. 1984). The implicit message of all these studies is that natural selection favors high O_2 delivery (flux) capacity in altitude adapted animals, while a large O_2 carrying capacity apparently is not advantageous. Our view (Hochachka et al. 1982) is that the system is adapted for operation at high speed but at low O_2 concentrations, which indeed fits the environment.

Tuning up mechanisms in the second (cellular metabolic) part of the system are not so easily understood. Nevertheless, recent studies of three altitude adapted mammals (Table 3) quite unequivocally indicate two kinds of adjustments in the myocardium: *firstly, an upward scaling of absolute oxidative capacity, and secondly, a downward scaling of ratios of anaerobic/aerobic metabolic capacities of the heart in all three species.* The capacity adjustments are indicated by absolute activities of CS and HOAD, two mitochondrial marker enzymes, and by the ratios of HOAD/CS, PK/LDH, and PK/HOAD. Interestingly, the upward scaling of myocardial oxidative capacity in both camelid species preferentially emphasizes fat-based metabolism (hence, the unusually high HOAD activities), despite the disadvantages of a lower ATP yield per mole of O_2 consumed.

Table 3. Heart enzyme activities in U g^{-1} of left ventrile from taruca, llama, and alpaca compared to other species[a]

	μmol Substrate converted g^{-1} min^{-1}			
a) Species	CS	HOAD	HOAD/CS	LDH/CS
Taruca	217	78	0.36	2.5
Llama	88	134	1.52	3.3
Alpaca	93	156	1.68	3.3
Ox	62	22	0.35	9.0
Weddell seal	29	16	0.55	35.8
Others	31	27	0.87	8.0
Sheep	69			
Pig	59			
Ox	64			
b) Species	PK	LDH	PK/LDH	PK/HOAD
Taruca	731	536	1.4	9.4
Llama	223	287	0.8	1.7
Alpaca	212	308	0.7	1.4
Ox	133	556	0.2	6.0
Weddell seal	217	1,032	0.2	13.6

[a] From Hochachka et al. (1982) with modification
(a) Data for CS and HOAD; (b) data for PK and LDH

Our conclusion that the myocardium in all three species displays an enhanced ratio of aerobic/anaerobic glycolysis relies on PK participation in both processes; as mentioned above, a relative indication of the two processes can be obtained from PK/LDH ratios, particularly since all three species display similar myocardial LDH/CS ratios. As indicated in Table 3, myocardial PK/LDH ratios are near unity for all three species. Such high PK/LDH ratios in cardiac musle are rare in nature; in the ten terrestrial and marine species studied by Castellini et al. (1981), in the ox, Weddell seal, and in lower vertebrates (Hochachka 1980), heart PK/LDH activity ratios are about one-fourth the values for the alpaca and llama, one-seventh the value for taruca heart (Table 2). That is why we previously concluded that in all three high altitude species heart metabolism displays a higher capacity for aerobic glycolysis than in other vertebrates (Hochachka et al. 1982). Exactly the same impression is gained from PK/LDH ratios in the skeletal muscle from the three species (Table 2), although absolute catalytic activities of oxidative enzymes here are not so obviously scaled upwards [cf. Table 1 with Emmett and Hochachka (1982)].

These data on altitude adapted camelids and deer are similar to earlier studies of oxidative enzyme activities in cardiac and skeletal muscles of dogs, guinea pigs, and rabbits resident at high altitude (Harris et al. 1970; Barrie et al. 1975). In addition, there are numerous clinical studies which provide clues pointing to this as a general response to conditions of chronic shortfall of O_2, providing no major investments in anaerobic adaptations are made (Hochachka 1980). For example, Bylund-Fellenius et al. (1981) report that in patients afflicted with peripheral vascular occlusive disease the activities of HOAD, CS, and cytochrome oxidase in calf muscle are some two-fold higher than in controls, while no adjustments in enzymes such as LDH occur. In a different approach, Young et al. (1982) studied the effect of chronic high altitude (4,300 m) exposure on aerobic and anaerobic contributions to exercise in healthy, physically fit army volunteers; although during exercise at altitude a smaller amount of muscle glycogen is utilized for a given amount of work, *a higher proportion is oxidized (much less lactate accumulates)* than during the same exercise at low altitude. Thus, for a variety of reasons it seems fair to conclude that a standard adaptational response to altitude hypoxia is to substantially increase the oxidative capacity of cardiac muscle and (perhaps to a lesser extent) of skeletal muscle as well, while anaerobic potentials are either dampened or simply unaltered. Within any related group of organisms, this may simply involve increased mitochondrial abundance (Ou and Tenney 1970), although, as indicated in Table 3, between-species specializations in fuel preferences may well occur (presumably for biological reasons unrelated to altitude). Even increased affinity for ADP by myocardial mitochondria is known for some high altitude species (Reynafarje 1971).

4.2 Meaning of Enzyme Adaptations in Altitude Adapted Animals

In trying to understand how, in high altitude animals, working muscle can turn over ATP at essentially normoxic rates, it is necessary to review the nature of low and high activity states of mitochondria from 'normoxic' systems. Although many fine-tuning control processes may be involved in activating and maintaining high rates of mitochondrial metabolism (see Hochachka and Somero 1984), the major or 'coarse control' signals apparently revolve around acceptor control of flux. According to most recent studies (for example, see Chance et al. 1981) mitochondria in nonworking muscle are in state 4: flux rates through the electron transfer system (ETS) are low because of limiting [ADP] or [P_i]. When exercise work is initiated, mitochondrial respiration and phosphorylation rates increase (the mitochondria are said to enter state 3) as [ADP] and/or [P_i] increase. The overall response in vitro is hyperbolic, but at K_m substrate levels respiration increases directly with [ADP] (Jacobus et al. 1982) or [P_i], which is why work rates vary in a linear fashion with in vivo estimates of [P_i]/[PCr] (Chance et al. 1981).

These characteristics spell out what may be the most critical advantage of increasing oxidative capacities during altitude adaptation: *for any given rate of heart or muscle work, flux through the ETS per gram will be lower, because the amount of ETS per gram is 1.5- to 2-fold higher.* Heart and muscle mitochondria from high altitude animals thus almost necessarily operate at lower [ADP]. In addition to improved responsiveness (the mitochondria are operating on the steep part of the ADP saturation curve), a secondary advantage derives from this arrangement: because in vivo function occurs at lower ADP concentrations *at the same time as the maximum catalytic capacity of state 3 flux is elevated, the overall metabolic scope for activity of the tissue is proportionately expanded.*

As in high altitude adaptation, endurance training at low altitude leads to upward scaling of citrate synthase and of other Krebs cycle enzymes, of various ETS components in a constant proportion manner, and of F_1 ATP synthase (see Davies et al. 1981, for literature). In species genetically adapted for sustained exercise, the activity levels of oxidative enzymes are also substantially higher than in homologous muscles from more sedentary relatives. Within a given species, these adjustments may be represented mainly by increased mitochondrial abundance, but between-species specializations in mitochondrial metabolism clearly also can occur (Hochachka and Somero 1984). Be that as it may, in all these cases, exactly the same advantages (improved responsiveness and increased scope) accrue as in altitude adapted species, *providing O_2 supplies do not become limiting.* If O_2 concentrations *do* fall into the range of the apparent K_m or less, then the situation at the mitochondria is altered and the above considerations with respect to limiting [ADP] now apply to ETS flux as a function of O_2 concentration: for any given muscle work rate in altitude adapted animals, flux through the ETS per gram will be lower, because the amount of enzyme machinery per gram is substantially higher. At any given work rate, therefore, muscle mitochondria from altitude adapted animals *may well operate at lower O_2 concentrations.* Because of the higher flux capacity of oxidative metabolism (more enzyme per gram muscle), the difference between flux at rest and flux during maximum work (i.e. the scope for activity) again may be expanded, so long as increased perfusion can lead to increases in O_2 concentrations.

As mentioned above, none of these conditions would have been considered likely until recently. However, it now appears that the apparent K_m values for O_2 in vivo are higher than observed with isolated mitochondria and *increase rather markedly with respiration rate (as may be expected during exercise)* to values in the 10 μM range (Jones and Kennedy 1982a,b). A part of this difference is ascribed to intracellular diffusion limitations, which may be reduced in tissues such as the heart in high altitude animals (decreased diffusion distances because of increased mitochondrial numbers). Nevertheless, the overall O_2 saturation curves for activated metabolism should be the same (because of higher apparent K_m values at high respiration rates) even though they may be left-shifted at lower respiration rates (Jones 1984). If we assume arterial plasma O_2 levels of about 45 μM and mixed venous levels of about 15–20 μM, as in man on Everest (West 1983), then it is not unreasonable to assume that intramuscular O_2 concentrations at the mitochondria are close to the apparent K_m, even at rest. In such extreme conditions, the only way high altitude adapted organisms could sustain a several-fold increase in respiration during exercise (coincident with increased K_m values for O_2) would be to increase the steady state concentration of O_2; i.e., *to increase perfusion by even a greater factor than would normally be required*. Man is clearly unable to achieve this, which is why scope for activity at these extremes is so low. That high altitude animals are able to achieve this is consistent with the high flux capacities of their physiological O_2 transport systems and may imply that at least at hypobaric extremes these animals may have to sustain *higher cardiac outputs per unit of skeletal muscle work* than in lowland species in normoxia. This apparent need may explain why heart oxidative capacities in taruca, llama, and alpaca are so outstanding when compared with other mammals (Table 3), but requisite cardiac output measurements, to our knowledge, have not been made.

5 Summary Assessment

It is well known that the scope for sustained activity (defined as the difference between maximum and resting rates of oxidative metabolism) steadily decreases as barometric pressure decreases. At extreme altitudes such as the top of Mt. Everest, the scope becomes vanishingly small for lowland mammals such as man, even if enough oxygen is available for resting metabolism. The fundamental metabolic problem for high altitude adapted organisms therefore is how to maintain an acceptably high scope for oxidative metabolism in the face of low barometric pressures.

In principle this problem could be resolved in one of three ways or by a combination of them all. Firstly, to circumvene the debilitating effects of limited oxygen availability during periods of activated metabolism, high altitude adapted organisms could increase the fractional cardiac output being distributed to working tissues, with concurrent hypoperfusion of non-working muscles and other peripheral tissues and organs. Secondly, high altitude organisms could improve the yield of ATP per mole of oxygen consumed by adjusting the fuels being preferentially used at the level of working skeletal and myocardial muscles. The third possibility is to improve oxygen flux capacities proportionately at all steps in the path of oxygen from alveolar air to mitochondria.

When the theoretical basis for these mechanisms and their potential contribution to circumvening hypoxic limitations to exercise at high altitudes are assessed, the third is found to be by far the most effective.

References

Armstrong RB, Laughlin MH (1985) Metabolic indicators of fiber recruitment in mammalian muscles during locomotion. J Exp Biol 115:201–213

Barrie E, Heath D, Arias Stella J, Harris P (1975) Enzyme activities in red and white muscles of guinea-pigs and rabbits indigenous to high altitudes. Environ Physiol Biochem 5:18–26

Bylund-Fellenius AC, Walker PM, Elander A, Holm S, Holm J, Schersten T (1981) Energy metabolism in relation to oxygen partial pressure in human sekletal muscle during exercise. Biochem J 200:247–255

Castellini MA, Somero GN, Kooyman GL (1981) Glycolytic enzyme activities in tissues of marine and terrestrial mammals. Physiol Zool 54:242–252

Castellini MA, Murphy BJ, Fedak M, Ronald K, Gofton N, Hochachka PW (1985) Potentially conflicting demands of diving and exercise: Insights from metabolite replacement rates in gray seals. J Appl Physiol 58:392–399

Chance B, Eleff S, Leigh JS Jr, Sokolow D, Sapega A (1981) Mitochondrial regulation of phosphocreatine/inorganic phosphate ratios in exercising human muscle: A gated ^{31}P NMR study. Proc Natl Acad Sci (USA) 78:6714–6718

Davies KJA, Packer L, Brooks GA (1981) Biochemical adaptation of mitochondria, muscle, and whole-animal respiration to endurance training. Arch Biochem Biophys 209:539–554

Emmett B, Hochachka PW (1982) Scaling of oxidative and glycolytic enzymes in mammals. Respir Physiol 45:261–267

Frisancho AR (1983) Perspectives on functional adaptation of the high altitude native. In: Proceeding Third Banff Int Hypoxia Symp. Alan R Liss, New York, pp 383–407

Harris P, Castillo Y, Gibson K, Heath D, Arias-Stella J (1970) Succinic and lactic dehydrogenase activity in myocardial homogenates from animals at high and low altitude. J Mol Cell Cardiol 1:189–193

Heath D, Williams D, Dicksinson J (1984) The pulmonary arteries of the yak. Cardiovasc Res 18: 133–139

Hochachka PW (1980) Living without oxygen. Harvard University Press, Cambridge, Mass, pp 1–181

Hochachka PW, Somero GN (1984) Biochemical adaptation. Princeton University Press, Princeton, New York, pp 1–537

Hochachka PW, Stanley C, Merkt J, Sumar-Kalinowski J (1982) Metabolic meaning of elevated levels of oxidative enzymes in high altitude adapted animals: An interpretative hypothesis. Respir Physiol 52:303–313

Jacobus WE, Moreadith RW, Vandegaer KM (1982) Mitochondrial respiratory control. Evidence against the regulation of respiration by extramitochondrial phosphorylation potentials or by [ATP]/[ADP] ratios. J Biol Chem 257:2397–2402

Jones DP (1984) Effect of mitochondrial clustering on O_2 supply in hepatocytes. Am J Physiol 247:C83–C89

Jones DP, Kennedy FG (1982a) Intracellular oxygen supply during hypoxia. Am J Physiol 243: C247–C253

Jones DP, Kennedy FG (1982b) Intracellular O_2 gradients in cardiac myocytes. Lack of a role for myoglobin in facilitation of intracellular O_2 diffusion. Biochem Biophys Res Commun 105: 419–424

Lahiri S (1977) Physiological responses and adaptations to high altitude. Int Rev Physiol Environ Physiol II, 15:217–251

Ou LC, Tenney SM (1970) Properties of mitochondria from hearts of cattle acclimatized to high altitude. Respir Physiol 8:151–159

Reynafarje B (1971) Effect of chronic hypoxia on the kinetics of energy transformation in heart mitochondria. Cardiology 56:206–208

Rowell LB (1974) Human cardiovascular adjustments to exercise and thermal stress. Physiol Rev 54:75–145

Sutton JR, Jones NL, Pugh LGCE (1983) Exercise at altitude. Annu Rev Physiol 45:427–437

Taylor CR, Maloiy GMO, Weibel ER, Langman VA, Kamau JMZ, Seeherman HJ, Heglund NC (1980) Design of the mammalian respiratory system. III. Scaling maximum aerobic capacity to body mass: wild and domestic mammals. Respir Physiol 44:25–37

Vik-Mo H, Mjos OD (1981) Influence of free fatty acids on myocardial oxygen consumption and ischemic injury. Am J Cardiol 48:361–365

West J (1983) Climbing Mt. Everest without oxygen: An analysis of maximal exercise during extreme hypoxia. Respir Physiol 52:265–279

Wilson DF, Erecinska M, Drown C, Silver IA (1979a) The oxygen dependence of cellular energy metabolism. Arch Biochem Biophys 195:485–493

Wilson DF, Owen CS, Erecinska M (1979b) Quantitative dependence of mitochondrial oxidative phosphorylation on oxygen concentration: a mathematical model. Arch Biochem Biophys 195: 494–504

Wood SC, Lenfant C (1979) Oxygen transport and oxygen delivery. In: Wood SC, Lenfant C (eds) Evolution of respiratory processes. A comparative approach. Marcel Dekker, New York, pp 193–223

Young AJ, Evans WJ, Cymerman A, Pandolf KB, Knapik JJ, Maher JT (1982) Sparing effect of chronic high altitude exposure on muscle glycogen utilization. J Appl Physiol 52:857–862

Zapol WM, Liggins GC, Schneider RC, Qvist J, Snider MT, Creasy RK, Hochachka PW (1979) Regional blood flow during simulated diving in the conscious Weddell seal. J Appl Physiol 47: 968–973

Scaling of Oxidative and Glycolytic Enzyme Activities in Fish Muscle

G.N. SOMERO[1] and J.J. CHILDRESS[2]

1 Introduction

We are unaware of any topic within the broad field of comparative physiology in which the ratio of "factual information" to "verified explanatory principles" is as high as it is in the case of scaling relationships. Physiologists of every bent, ranging from experimentalists who have laboriously gathered data on the relationships of different physiological, anatomical, or biochemical properties to body size, to theorists who eschew laboratory work have analyzed and debated the meaning of scaling in diverse biological systems. Peters (1983) has recently published a volume, *The Ecological Implications of Body Size*, in which he presents the most extensive listing available of phenomena in which size-related variation in populational, anatomical, physiological, and biochemical properties has been reported. In reviewing these multifarious scaling patterns, Peters emphasizes that there appears to be no single broad explanatory principle which can account for these scaling relationships (although, as Peters' analysis shows, this is not from a lack of trying on the part of many physiologists over the past century!). In fact, Peters seems to view attempts to establish an all-encompassing principle with suspicion. In another recent review, Smith (1984) offers a seminal analysis of the assumptions and, closely related to these, the statistical approaches used in the analysis of scaling relationships. His analysis, like that of Peters, sounds appropriate notes of caution concerning the pitfalls lying in the path of anyone wishing to develop all-encompassing theoretical constructs in which the allometric scaling equation is wedded to physiological, biochemical, or ecological theory.

In this survey of scaling relationships involving enzymatic activity in vertebrates, fishes in particular, we shall not attempt to develop broad explanatory schemes that are designed to explain scaling "in general." Further, we shall not attempt to determine the "real" scaling coefficient for the phenomena discussed, a goal chosen by many workers in studies of scaling functions. However, we shall not merely present an empirical story without any attempt being made to explain "why" the scaling patterns we obseve might have biological significance, and why the particular scaling relationships we have discovered, notably those linking glycolytic enzyme activity to body size in fishes, may reflect important aspects of locomotory power costs at different speeds of swim-

1 Marine Biology Research Division, Scripps Institution of Oceanography, University of California, San Diego, La Jolla, CA 92093, USA

2 Department of Biological Sciences, University of California, Santa Barbara, CA 93106, USA

Circulation, Respiration, and Metabolism
(ed. by R. Gilles)

ming. We approach this subject with a predictive model in mind, one which is based on established hydrodynamic theories and a data set gathered in studies of the swimming performances of fishes. The predictions made by this model are directly testable using measurements of enzymatic activity. Our results have revealed a heretofore unrecognized pattern of metabolic scaling, one which has serious implications for certain of the conventionally held views about the exact values of scaling exponents and the ultimate causes of metabolic scaling.

2 Basic Attributes of Scaling Relationships

One basic equation will occupy our attention for most of this essay:

$$Y = aX^b .$$

In this familiar allometric equation, some dependent variable, e.g., metabolic rate, heart rate, swimming speed, etc., is related to an independent variable (typically, body mass) by an exponential function. Much discussion has centered around the numerical value of the b exponent and its meaning. For a large number of different characteristics of organisms, the b exponent is near 0.75 when the allometric relationship is expressed in terms of total organismal mass (or near − 0.25, when the allometric equation is given in terms of unit mass, e.g., oxygen consumption rate per gram). These exponents have often been viewed as "magic" numbers in the minds of some physiologists, whereas others, notably Heusner (1982), regard these magic numbers with suspicion. The tabulation of b values made by Peters (1983) is, to our knowledge, the most comprehensive compilation of such data that exists, and his list should be consulted by those interested in the broad scope, and the fine details, of scaling parameters.

The equation given above can be restated if logarithms of the terms are taken. When this is done, the equation becomes

$$\log \cdot Y = \log a + b \cdot \log X .$$

Plots of log Y versus log X yield a Y intercept of log a, and have a slope equal to b. This is the most commonly used allometric scaling equation. However, as Smith (1980, 1984) has clearly shown, there are a number of pitfalls and questionable assumptions linked to use of this equation, including the tendency for the human eye to see a close fit of log transformed data to the allometric scaling regression line, and the assumption that this magic equation should fit the data set in the first place. Ideally, one should have a quantitative hypothesis in mind prior to gathering ones data, so that the data collected can serve to test an hypothesis, and not merely serve as an empirical resource for post hoc rationalizations.

3 Scaling of Enzymatic Activity in Fishes

3.1 Activities of Enzymes of Aerobic Pathways of Energy Metabolism

Even though the primary focus of our discussion is on the scaling of enzymatic activities that are indicative of a tissue's potential for anaerobic glycolysis, it is pertinent to begin with a brief review of the scaling of activities of enzymes of aerobically poised pathways. The discussion of aerobic scaling will reveal possible proximal causes for the scaling of aerobic metabolism, and will also establish a basis for appreciating how very differently the anaerobic and the aerobic pathways of energy metabolism scale with body size.

Prior to discussing our data, some methodological points should be made. First, we have employed assay conditions yielding maximal velocities for the reactions; thus, our data provide a quantitative estimate of the peak metabolic power of the enzyme systems. Second, we have expressed our data in terms of international units of activity per gram fresh weight of muscle (all measurements have been made at 10 °C). Our choice of this means of normalizing units of activity, instead, for example, of expressing activity in protein-specific form, is based on the fact that the muscle power of a fish is due to the metabolic potential in the total muscle mass. Thus, using fresh (wet) weight as the normalizing factor allows the most meaningful expression of the enzymatic potential of a muscle. In addition, for most of the species discussed in this paper, we have measured the total muscle mass of different sized individuals in order to allow calculation of total activity of a given enzyme in the entire locomotory musculature. Third, we

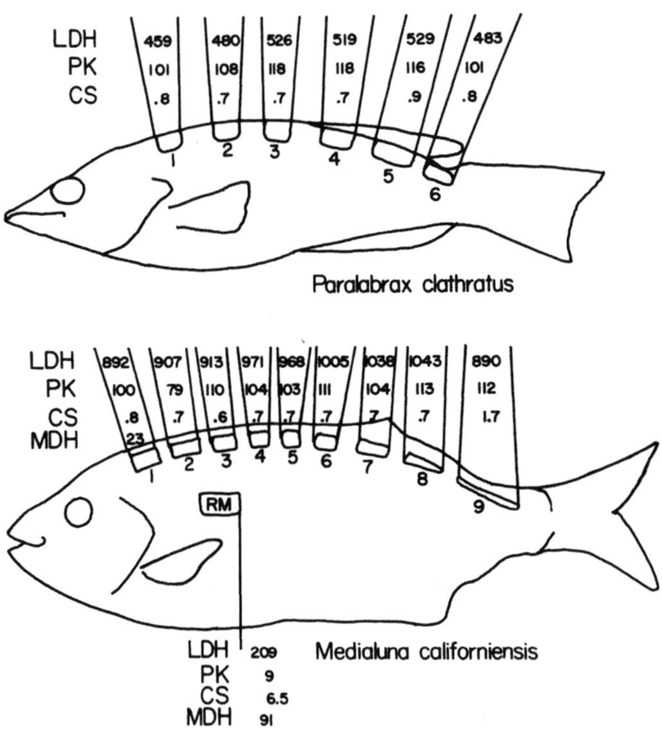

Fig. 1. The activities of lactate dehydrogenase (*LDH*), pyruvate kinase (*PK*), malate dehydrogenase (*MDH*), and citrate synthase (*CS*) as a function of body site. In *Paralabrax clathratus LDH, PK,* and *CS* activities were determined only in epaxial white muscle. In *Medialuna californiensis* activities were determined in red muscle (*RM*) as well. Modified from G.N. Somero and J.J. Childress (1980) with permission

have carefully studied the possible variation that may exist in different regions of the epaxial white musculature to ensure that our muscle samples, normally taken just posterior to the opercular region, and near the top of the body, provide a valid estimate of average muscle enzymatic activity. As shown in Fig. 1, there is, in fact, extremely little variation in enzymatic activity along the length of a fish's body compared to the variation noted among species and among different-sized individuals of a single species.

Very few investigators have systematically examined how the activities of enzymes vary as a function of body size, either among individuals of a single species or among different species. Moreover, there are extremely few data available on enzymatic activity scaling in locomotory muscle, the tissue of central importance to our analysis, and to this symposium. In our analyses (Somero and Childress 1980; Siebenaller et al. 1982), we have emphasized intraspecific scaling patterns. This emphasis has been chosen because the absolute levels of enzymatic activity in muscle vary considerably between fish species, as do metabolic rates (cf. Torres et al. 1979), confusing any analysis of broad interspecific scaling patterns. For example, we have shown in comparisons of teleost fishes having radically different locomotory habits (e.g., strong subcarangiform swimmers vs labriform swimmers or benthic "sit and wait" predators) and habitat preferences (shallow-living vs deep-sea species), that the wet-weight-specific activity of a particular enzyme in white muscle may differ by three orders of magnitude among species of roughly similar size (Fig. 2). [Details concerning the species studied can be found in Childress and Somero (1979), Sullivan and Somero (1980), and Siebenaller et al.

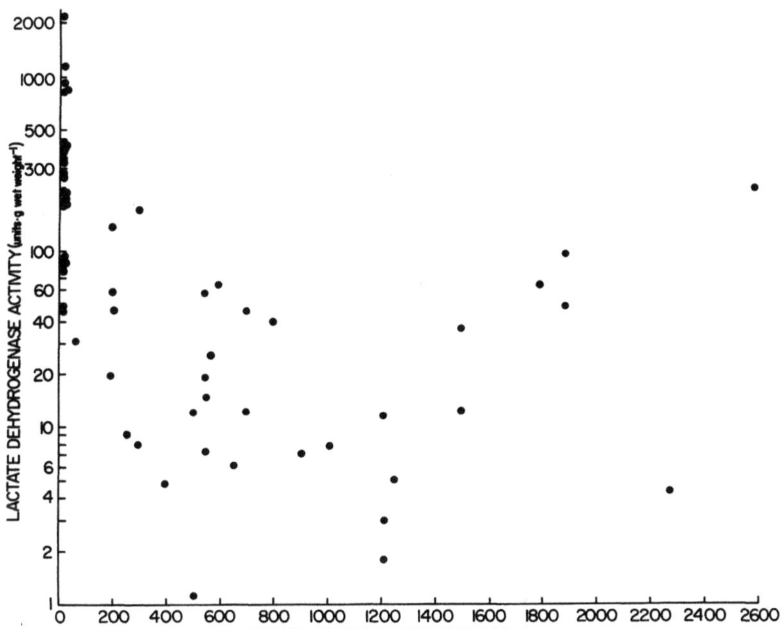

Fig. 2. Lactate dehydrogenase activity (U g^{-1} wet wt.) in epaxial white muscle of marine teleost fishes having different depths of occurrence and locomotory habits. From Hand and Somero (1983). Data on the species studied are found in Childress and Somero (1979), Sullivan and Somero (1980), Siebenaller et al. (1982), and Hand and Somero (1983)

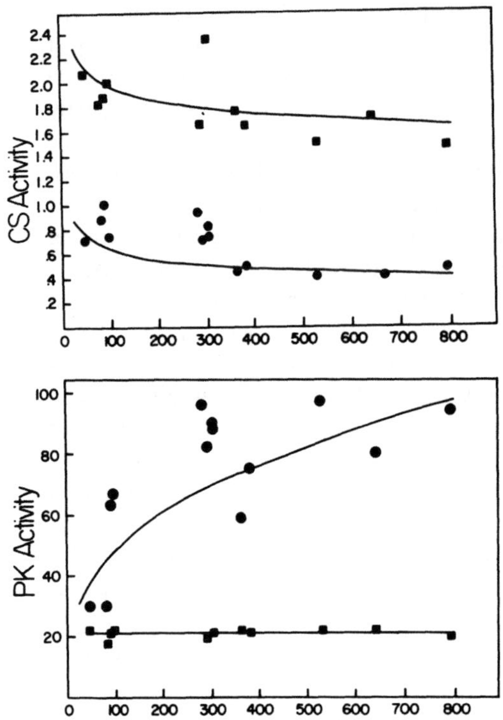

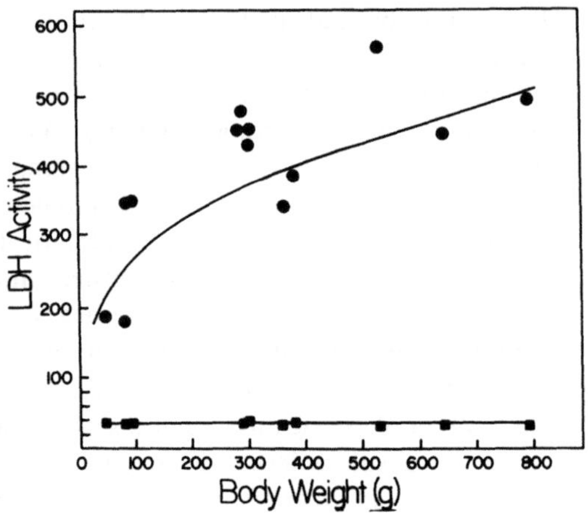

Fig. 3. The scaling of *LDH, PK*, and *CS* activities in white skeletal muscle (*circles*) and brain (*squares*) of *Paralabrax clathratus* and *Paralabrax nebulifer* (data are pooled). The *curve* has been drawn to fit the data for *P. nebulifer*. Modified after G.N. Somero and J.J. Childress (1980), with permission

(1982).] It is obvious from the data in Fig. 2 that great care must be taken to avoid making comparisons of enzymatic activities that are influenced by differences in locomotory habit and general life-style, as well as by variations in body size.

We have examined the scaling of enzymatic activity in approx. 20 species of marine teleost fishes. The most pronounced scaling relationships have been noted for actively swimming pelagic species which employ a subcarangiform mode of locomotion. We

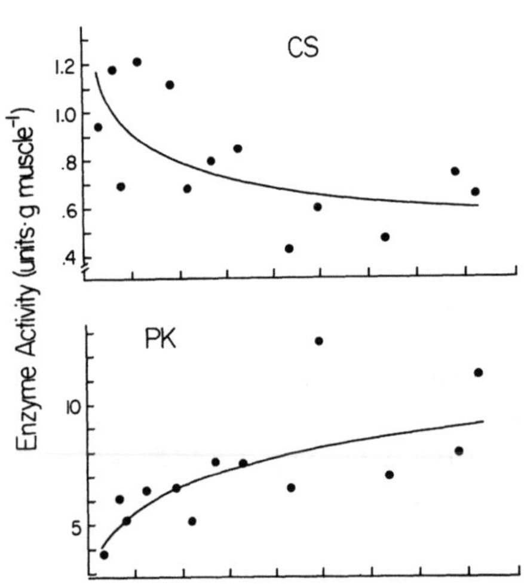

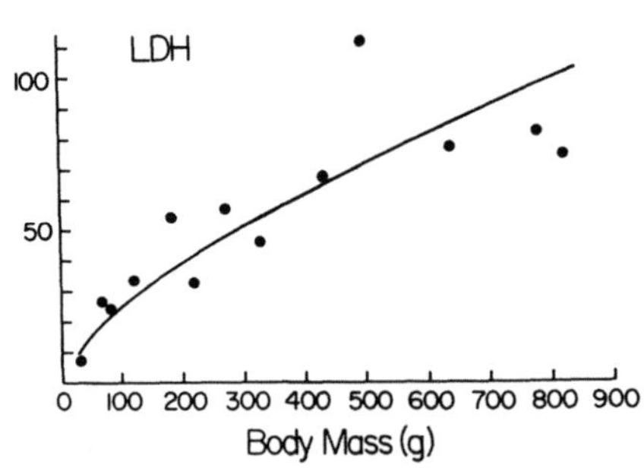

Fig. 4. The scaling of *CS, PK*, and *LDH* activity in white skeletal muscle of the deep-sea rattail fish, *Coryphaenoides armatus*. The scaling equations relating activity (A) to mass (M) in grams are as follows: CS $A = 1.0\ M^{-0.21}$, PK $A = 1.83\ M^{0.24}$, LDH $A = 1.16\ M^{0.66}$. From Siebenaller et al. (1982)

emphasize fishes of this type in our discussion. Figures 3 and 4 present data sets for three pelagic fishes, two of which are congeners inhabiting shallow, near-shore environments, and one a deep-sea species. These three species exhibit strong scaling of aerobically and anaerobically poised enzymes (Figs. 3 and 4). The two species of the genus *Paralabrax, P. clathratus* (the kelp bass) and *P. nebulifer* (the sand bass) show strong, and statistically indistinguishable, scaling patterns in white muscle and brain for the enzyme citrate synthase (CS) which is a diagnostic enzyme of the Krebs citric acid cycle (Fig. 3); data for these two species, from Somero and Childress (1980), are thus pooled in our analyses. Scaling of CS activity is also apparent in white skeletal muscle of the deep-sea rattail fish, *Coryphaenoides armatus* (Fig. 4); brain activities were not measured in this case. The b exponents for the muscle scaling relationships (log CS activity = log a + b log body weight) are - 0.23 for the *Paralabrax* species, and - 0.21 for *C. armatus*. For most of the other fishes so examined, CS activity scaled with exponents similar to those found for these three species (Table 1). The scaling of CS activity in the white muscle of these three species, thus, closely resembles the magic number found for aerobic scaling. Therefore, the scaling of aerobically-poised enzymes like CS may be a proximal cause of the scaling found in whole organism oxygen consumption rates. This correlation in b exponents does not, of course, provide an ultimate explanation of metabolic scaling. That is, we are left without an account of why CS activity, and aerobic metabolism per se scale as they do. We shall not attempt here to review the several hypotheses that attempt to explain the scaling of aerobic respiration. We refer the interested reader to Chap. 13 of Peters' book, to the recent book by McMahon (1984), and to Schmidt-Nielsen (1983) for critical reviews of these ideas.

3.2 Scaling of Activities of Enzymes of Anaerobic Glycolysis

Extremely limited attention has been paid to the scaling of anaerobic metablism. Perhaps this is because the anaerobic processes of energy metabolism are less easily measured than oxygen consumption. Or, perhaps, physiologists have concluded, a priori, that anaerobic metabolism, with at least its short-term independence of oxygen supply, is not apt to show interesting scaling relationships that, in the case of aerobic respiration, may be founded on limitations of oxygen supply to the mitochondria. For whatever reasons, when we began our studies of the scaling of glycolytic enzyme activity in fishes, no data on this subject were available in the literature.

Our reasons for examining scaling patterns in the activities of glycolytic enzymes, notably lactate dehydrogenase (LDH), which is the strongest indicator of a locomotory muscle's capacity for anaerobic glycolysis (Somero and Childress 1980), were twofold. First, in the interspecific comparisons we were engaged in making to learn more about the biochemical concomitants of interspecific differences in locomotory habits and habitat preference (see Fig. 2), we wished to control for size effects in our data. Second, we were aware of an apparent paradox discovered in studies of high-speed swimming in fishes. As discussed by Webb (1977) and Wu (1977), high-speed, "burst" swimming in fishes apparently displays a size independence when swimming speed is expressed in terms relative to body length, i.e., as body lengths swum per unit time. Both small and large individuals of active, pelagic species like salmonids appear to be capable of maximal burst swimming speeds in the range of 8-12 body lengths per second (Webb

Table 1. Scaling relationships for lactate dehydrogenase (LDH) and citrate synthase (CS) activities in different sized individuals of several species of marine teleost fishes[a]

	Mass to length[b]		Enzymatic activity (A)[c]							
			LDH				CS			
	$m = a \cdot L^B$		$A = a \cdot m^b$		$A = a \cdot L^B$		$A = a \cdot m^b$		$A = a \cdot L^B$	
	a	b	a	b	a	B	a	b	a	B
Species										
Bajacalifornia burragei	1.78	3.67	2.77	0.29	0.132	4.58	0.63	−0.23	29.12	2.86
Engraulis mordax	6.72	3.35	317	0.21	267	3.90	1.78	−0.05	28.09	3.14
Gillichthys mirabilis	38.4	2.99	269	0.08	3,548	3.26	1.24	−0.16	370.6	2.46
Lampanyctus regalis	3.01	3.58	2.11	0.57	0.59	4.89	ND	ND	ND	ND
Lampanyctus ritteri	6.80	3.45	5.00	0.61	0.01	6.21	1.12	−0.13	33 39	3.07
Paralabrax clathratus	18.19	3.20	38.9	0.42	12.29	4.33	2.74	−0.23	2 277.8	2.27
Paralabrax nebulifer	19.54	3.17	67.4	0.30	23.74	4.19	1.76	−0.22	852.5	2.38
Parvilux ingens	0.78	3.79	1.57	0.61	0.01	5.57	ND	ND	ND	ND
Poromitra crassiceps	1.36	3.86	2.40	0.30	0.08	4.87	ND	ND	ND	ND
Vinciguerria lucetia	53.6	2.82	124.9	0.66	2.82	4.83	ND	ND	ND	ND

[a] Data are from Somero and Childress (1980). Statistical information and size ranges of fishes are found in that paper. Body lengths (L) are in cm. Muscle mass (m) is in grams

[b] Mass refers to the total epaxial muscle mass, which was determined by dissecting the muscle from the specimen. Red muscle was excluded

[c] Units are international units per gram wet (fresh) weight of muscle, measured at 10 °C. The expression $A = a \cdot m^b$ relates $U\,g^{-1}$ muscle to total body mass. The expression $A = a \cdot L^B$ relates the total units in the epaxial musculature to body length. The a values for the relationship $A = a \cdot L^B$ have been multiplied by 10^{-}

1977; Wu 1977). Phrased in terms of scaling functions, burst swimming speeds, expressed in body lengths (L) per second, scale in proportion to L raised to a power of approx. 0.9–1.0. Thus, larger size individuals of a species seem capable of swimming at much higher *absolute* velocities than smaller size individuals, but all sizes of individuals can swim at similar peak *relative* velocities.

These data on burst swimming were difficult to reconcile with the aerobic metabolic capacities of fishes. While power requirements for overcoming drag increase rapidly with increases in swimming speed and, therefore, in peak burst swimming would rise very rapidly with increasing body size, fishes normally show typical scaling of aerobic metabolism. That is, b coefficients relating log oxygen consumption to body weight typically are near −0.25. And, as discussed above, this is the type of scaling noted for an aerobically-poised enzyme, CS, in white muscle. Thus, to account for the power needed to drive length-independent burst swimming, we were required to invoke a dominant role for anaerobic mechanisms of ATP generation in white muscle. The question we had to address then became how glycolytic capacity in white muscle scaled with body size, and how closely this scaling of glycolytic activity matched the scaling of power requirements with body size.

To make this analysis, we had to develop a predictive model that could provide a quantitative estimate of how the power requirements for burst swimming scaled with body size. Because we were using relative swimming speeds in our comparisons, it was necessary to express the scaling relationship in terms of body length, not body weight. Thus, we needed to develop an expression that would allow us to predict how the total glycolytic activity of white skeletal muscle would have to change with body length to provide adequate power to support length-independent burst swimming.

One method for developing a predictive relationship between burst swimming speed power costs and enzymatic activity involves the use of hydrodynamic relationships that relate the drag on a rigid body to velocity of movement through the water (Bainbridge 1961; Webb 1977; Wu 1977). The equation expressing this relationship is:

$$P_t = D \cdot V = 1/2 \cdot \rho \cdot S \cdot V^3 \cdot C_D \tag{1}$$

where P_t is the power required to drive the body through the water, D is the total drag on the body, V is the velocity with respect to the medium, ρ is the density of water, S is the wetted surface area of the body (proportional to L^2), and C_D is the drag coefficient (Bainbridge 1961). For laminar flow in the boundary layer,

$$C_D \propto 1.32 \cdot R_L^{-0.5} \tag{2}$$

and for turbulent flow over a smooth surface,

$$C_D \propto 0.078 \cdot R_L^{-0.2}\ , \tag{3}$$

where R_L is the Reynolds' number,

$$R_L = L \cdot V/\nu \tag{4}$$

where L is body length and ν is the kinematic viscosity of water. The equation relating power costs (P_t) to drag times velocity can be solved for relative velocity, i.e., velocities expressed in body lengths swum s^{-1}, by substituting

$$V = \text{constant} \cdot L^B \tag{5}$$

into the above equations.

Table 2. Metabolic power scaling predictions and scaling of enzymatic activity in white locomotory muscle of fishes[a]

B Exponents for scaling of power costs (P) with body length (L) for burst and sustained swimming		Total muscle enzymatic activity B values that are not significantly different from predicted values	
		LDH (10 Total)	CS (6 Total)
1) Predictions based on drag model (Bainbridge 1961)			
Burst velocity $\propto L^1$			
Laminar flow	4.00	8	2
Turbulent flow	4.60	9	0
Burst velocity $\propto L^{0.9}$			
Laminar flow	3.65	7	3
Turbulent flow	4.30	8	0
Sustained swimming $\propto V\ L^{0.5}$			
Laminar flow	2.75	4	6
Turbulent flow	3.20	5	4
2) Predictions from bulk momentum model (Webb 1977)			
Burst velocity $\propto L^{1.0}$	4.02	8	2
Burst velocity $\propto L^{0.9}$	3.92	8	3
Sustained $\propto L^{0.5}$	3.52	7	3

[a] All enzymatic activities are based on the total activity in the white epaxial musculature. The activity per gram wet weight was multiplied by the total weight of the muscle, determined from dissecting and weighing the entire epaxial musculature, minus any red muscle present. Data are from Somero and Childress (1980)

The power requirements for burst swimming ($V \propto L^{0.9-1.0}$) are predicted, therefore, to scale according to body length raised to the power of 4.0 for laminar flow, and to the power 4.6 for turbulent flow (Table 2). Wu (1977) did, in fact, note these high predicted B values for burst swimming, and regarded them as unexplained.

A second hydrodynamic basis for computing the predicted scaling of anaerobic energy metabolism in burst swimming can be made using the bulk momentum model of Lighthill, as developed by Webb (1977). This approach models the power exerted by the tail of the fish in moving through water. Using data from studies of sockeye salmon, Webb (1977) suggested that

$$P \propto V \cdot L^{3.02} \quad (6)$$

where V is the absolute swimming speed. Substituting into this equation the velocity relationship given in Eq. (5), i.e., to convert the relationship given in Eq. (6) to relative swimming speed, one again arrives at estimates for the length-related scaling coefficient for peak burst swimming ability near 4 (Table 2). Note that we use an upper case B to designate the scaling coefficients relating enzymatic activity to body length, L, and reserve the lower case b for coefficients expressing scaling in terms of body weight.

Against the predictions of this model we examined the scaling of both aerobically- and anaerobically-poised enzymes of white muscle (Tables 1 and 2). We included in our analysis only those species for which we measured total muscle mass of different-sized specimens, in order to allow calculation of enzymatic activity in the total white musculature. As shown by the data in these Tables and in Figs. 3 and 4, radically different scaling relationships exist for CS and LDH. For CS, all of the b values are negative and, therefore, all of the B coefficients are smaller than the scaling coefficients relating total muscle mass to body length (Table 1). In striking contrast to CS activity, LDH activity in white muscle increases dramatically with increasing body size. All of the b values are greater than zero, and the B coefficients all are larger than the scaling coefficients relating muscle mass to L.

The agreement between the theoretically developed B values for burst swimming power requirements and the empirically determined B values for LDH activity is striking (Table 2). For predictions based on the drag model (Bainbridge 1961), under laminar flow and turbulent flow conditions, eight and nine (out of a total of ten) of the empirically determined B values for LDH activity, respectively, are not statistically different from the predicted values. For predictions based on the bulk momentum model (Webb 1977) eight of ten B values for LDH are not statistically different from the theoretically predicted B coefficients. In contrast, while the scaling of CS activity is significantly different from the theoretically predicted B values for burst swimming in almost all cases, it does agree well with B values calculated assuming that sustained swimming scales with body length with a B value near 0.5, as shown by Brett (1965) for salmonids (Table 2). Thus, both the anaerobic and aerobic enzyme scaling patterns agree with theoretically derived B values, and we view this excellent agreement between theory and observation as a strong indication that the size-dependent variation in enzymatic activity in white muscle is causally related to the size relationships observed in both aerobically-powered and anaerobically-powered swimming.

The variation in the scaling coefficients among species is not surprising in view of the different swimming modes and life-styles of the species examined. The strongest scaling of LDH activity is found in strong subcarangiform swimmers like the *Paralabrax* species. Demersal species like *Gillichthys mirabilis*, which lack the capacities of the strong subcarangiform swimmers for burst locomotion, show much weaker scaling of LDH activity. Labriform swimmers, which rely strongly on sculling motions of the pectoral fins for propulsive force, also show less marked scaling than the strong subcarangiform swimmers [see Somero and Childress (1980) for data on additional species]. We conclude, therefore, that the scaling of anaerobic metabolic power is strongest in actively swimming pelagic fishes that rely on burst locomotion for predation and for predator avoidance, and we have proposed that predator-prey relationships may be a strong selective factor favoring body-length-independent burst swimming abilities [cf. Somero and Childress (1980)].

It is pertinent to note that the size of the b exponent is not determined by the pre-exponential term, a (Table 1). For example, despite the greatly lower levels (= lower a value) of LDH activity in muscle of *C. armatus* compared to the *Paralabrax* species, a difference which reflects the more sluggish locomotory habit of *C. armatus* (cf. Siebenaller et al. 1982), the b coefficients are not significantly different (Figs. 3 and 4). Likewise, the absolute level of CS activity in a species does not determine the size of the b coefficient. Therefore, the factors that establish the intraspecific scaling patterns are independent of the absolute levels of enzymatic activity, i.e., the a values, in the species.

Because LDH is the best indicator enzyme of a muscle's capacity for anaerobic glycolysis, it is not surprising that LDH activity exhibited the strongest scaling among the enzymes of white muscle examined. Pyruvate kinase (PK), another glycolytic enzyme, generally increased in activity with body size (Figs. 3 and 4), but the scaling usually was not as strong as for LDH (Somero and Childress 1980). We interpret the difference between LDH and PK to reflect the fact that LDH is employed in anaerobic glycolysis, while PK is common to both aerobic and anaerobic modes of glycolytic function.

An additional argument supporting our conjecture that the scaling of enzymatic activity in white muscle reflects a causal relationship between ATP (power) supplying systems and locomotory performance can be developed on the basis of size-related enzymatic activity in brain tissue (Somero and Childress 1980). Figure 2 illustrates the scaling of CS and LDH activity in brains of the two *Paralabrax* species. Unlike CS activity, which scales in a fashion akin to aerobic metabolic scaling, as discussed earlier, LDH activity is size-independent in brain. In fact, there is extremely little variation in LDH activity of brain with body size, suggesting that the larger amount of variation in LDH activity noted in muscles of similar sized individuals of a species (Figs. 3 and 4) may reflect natural variation, due to factors such as diet or conditioning (Somero and Childress 1980). Therefore, the size independence of LDH activity in brain, a tissue which is not involved in locomotion, is further support for the arguments developed above suggesting a key role for scaling of LDH activity in white muscle.

4 Conclusions

We began this essay with a statement that very few scaling relationships have been given a well-established causal interpretation either in the sense of explaining why the scaling pattern is necessary or why it is of benefit to the organism. In the case of anaerobic enzyme activities in fish white muscle, we believe that the close agreement between our theoretically derived B values relating locomotory power requirements to body size and the empirically determined B values for LDH activity suggests a strong causal link. The conservation of similar capacities for burst swimming, as measured in $L\ s^{-1}$, in different sized individuals of a fish species seems to demand large increases in anaerobic energy metabolism with increasing body size. Through mechanisms which remain unknown, the LDH activity of white muscle is elevated in larger size individuals, even though the concentration of contractile proteins appears to be independent of size and, within wide limits, of species' locomotory habits and habitats (Swezey and Somero 1982). That is, the power (ATP) supply system can scale strongly, while the power utilizing system (actomyosin ATPase activity) appears to remain size independent.

The only data set at all similar to the one we have gathered is that of Emmett and Hochachka (1981), who looked at enzymatic activities in locomotory muscle of terrestrial mammals differing by nearly six orders of magnitude in body weight. CS activity again decreased in larger species, while LDH activity increased. The scaling coefficients for the log transformed allometric equation were −0.106 and +0.150, for CS and LDH, respectively. Some reservations apply to this study because it involved only interspecific comparisons. As in the case of fishes, similar-sized species may differ substantially in locomotory habits, leading to a major source of variation in enzymatic activity that is unrelated to size per se. For example, in the Emmett and Hochachka study, the rabbit, a strong burst runner, had much higher glycolytic potential than "predicted" by the allometric scaling function. Despite the influences of locomotory habit, however, the trends noted by Emmett and Hochachka (1981) seem real, and provide further grounds for speculating about the scaling of aerobically and anaerobically powered locomotion in vertebrates. We refer the reader to their paper for a discussion of what the trends noted in the comparisons of different sized mammals could mean.

An obvious conclusion arising from our data set and that of Emmett and Hochachka is that restrictions on metabolic power output imposed by inherent limitations in skeletal strength may not be of key importance in the ultimate causation of scaling relationships (cf. McMahon 1984). Arguments that the decreased aerobic potential per unit weight in larger size individuals is due, in whole or part, to mechanical constraints imposed by strength limitations of skeletal elements seem to be disproven by the scaling patterns of LDH activity. In fact, power output for burst (anaerobically powered) swimming in fishes may be 10–100 times those characteristics of sustained (aerobic) swimming (Brett and Groves 1979). If skeletal strength limitations were an ultimate cause of the scaling of metabolic rates, therefore, anaerobic metabolism should display even lower scaling exponents than aerobic metabolism! This obviously is not the case. Furthermore, the scaling of enzymatic activity in brain, a tissue where mechanical constraints to metabolic power output do not pertain, argues against such mechanical explanations of aerobic scaling patterns.

In conclusion, the strikingly different scaling patterns noted for aerobically- and anaerobically-poised enzymes of ATP generation in white muscle of actively swimming pelagic fishes correlate well with the size dependence of aerobic swimming, and the size independence ($L\ s^{-1}$) of burst swimming powered by anaerobic metabolism. This newly discovered scaling of anaerobic metabolism requires a reevaluation of certain hypotheses that attempt to account for metabolic scaling. More importantly, these data show very clearly that the value of the scaling coefficient relating metabolic power output to body size is not some one magic number, but instead is able to vary over a wide range depending on the relative contributions of aerobic and anaerobic pathways of ATP generation to the supply of power to the locomotory musculature. If studies of the scaling of metabolism are to provide a full picture of scaling phenomena, therefore, the anaerobic contribution to metabolic power must be studied as well as the contributions made by aerobic processes.

Acknowledgments. These studies were supported by National Science Foundation grants PCM78-04321, PCM80-01949, and PCM83-00983 to G.N. Somero, and OCE76-10407, OCE78-08933, and OCE81-10154 to J.J. Childress, Travel funds for G.N. Somero were provided by the John Dove Isaacs Chair in Natural Philosophy.

References

Bainbridge R (1961) Problems of fish locomotion. Symp Zool Soc Lond 5 3–13

Brett JR (1965) The relation of size to rate of oxygen consumption and sustained swimming speed of sockeye salmon *(Oncorhynchus nerka)*. J Fish Res Board Can 22.1491–1501

Brett JR, Groves TDD (1979) Physiological energetics. In: Hoar WA, Randall DJ, Brett JR (eds) Fish physiology, vol VIII. Academic, New York, pp 279–352

Childress JJ, Somero GN (1979) Depth-related enzymic activities in muscle, brain and heart of deep-living pelagic marine teleosts. Mar Biol 52:273–283

Emmett B, Hochachka PW (1981) Scaling of oxidative and glycolytic enzymes in mammals. Respir Physiol 45:261–272

Hand SC, Somero GN (1983) Energy metabolism pathways of hydrothermal vent animals: adaptations to a food-rich and sulfide-rich deep-sea environment. Biol Bull (Woods Hole) 165:167–181

Heusner AA (1982) Energy metabolism and body size. I. Is the 0.75 mass exponent of Kleiber's equation a statistical artifact? Respir Physiol 48:1–12

McMahon TA (1984) Muscles, reflexes, and locomotion. Princeton University Press, Princeton, NJ, pp 235–296

Peters RH (1983) The ecological implications of body size. Cambridge University Press, Cambridge

Schmidt-Nielsen K (1983) Animal physiology: adaptation and environment. Cambridge University Press, Cambridge, pp 201–213

Siebenaller JF, Somero GN, Haedrich RL (1982) Biochemical characteristics of macrourid fishes differing in their depths of distribution. Biol Bull (Woods Hole) 163:240–249

Smith RJ (1980) Rethinking allometry. J Theor Biol 87:97–111

Smith RJ (1984) Allometric scaling in comparative biology: problems of concept and method. Am J Physiol 246:R152–R160

Somero GN, Childress JJ (1980) A violation of the metabolism-size scaling paradigm: activities of glycolytic enzymes in muscle increase in larger-size fish. Physiol Zool 53:322–337

Sullivan KM, Somero GN (1980) Enzyme activities of fish skeletal muscle and brain as influenced by depth of occurrence and habits of feeding and locomotion. Mar Biol 60:91–99

Swezey RR, Somero GN (1982) Skeletal muscle actin content is strongly conserved in fishes having different depths of distribution and capacities of locomotion. Mar Biol Lett 3:307–315

Torres JJ, Belman BW, Childress JJ (1979) Oxygen consumption rates of midwater fishes off California. Deep-Sea Res 26A:185–197

Webb PW (1977) Effects of size on performance and energetics of fish. In: Pedley TJ (ed) Scale effects in animal locomotion. Academic, New York, pp 315–332

Wu TY (1977) Introduction to the scaling of aquatic animal locomotion. In: Pedley TJ (ed) Scale effects in animal locomotion. Academic, New York, pp 203–232

Enzyme Catalysed Fluxes in Metabolic Systems. Why Control of Such Fluxes is Shared Among All Components of the System

J.W. PORTEOUS[1]

1 Introduction: Attempting to Understand Metabolic Control

Attempting to understand the control or regulation of molecular events in complex, intact, biological systems is undoubtedly the most important endeavour in modern biochemistry, genetics, pharmacology, and physiology. This endeavour forms the common ground among these otherwise rather disparate facets of research in biology. Each biological system investigated poses its particular problems for experimentation and none are more severe than those facing the biochemist, geneticist, pharmacologist or physiologist attempting to understand the control of metabolism during exercise. But before we consider particular problems, it seems wise to look at those that are common to all biological systems; and to ask whether current endeavours to understand control of metabolism are directed in the most fruitful direction.

1.1 The First of Two Logical Moves

The first necessary move towards understanding metabolic control involves discovering the routes of catabolism and anabolism of as many substrates as possible in a variety of species and tissues; then drawing metabolic maps in order to summarise our efforts. Such maps will indicate translocation (vectorial) events across membranes as well as so-called non-vectorial (scalar) chemical transformations. In the very limited sense that they show the stoichiometries of translocations and transformations, our maps are quantitative; but in every other respect they are purely qualitative statements of routes that might be followed by certain substrates under environmental conditions that are not stated on the maps (Porteous 1983a). Our maps do not, and cannot, show how and where metabolic regulation occurs.

A relatively recent extension of this mapping of metabolic routes has involved examining isolated parts of the metabolic system in detail. ("Isolated" here means physically extracted from the system for the purposes of examination; or examined in situ without paying attention to the quantitative effects of the part examined on the activities of the whole system.) Consequently, certain parts of our metabolic maps can now be drawn on a larger scale. Thus we now know a great deal about co-operative assemblies of oligomeric proteins, allosteric enzymes, feedback loops, covalent modifi-

1 Department of Biochemistry, University of Aberdeen, Marischal College, Aberdeen AB9 1AS, Scotland, Great Britain

Circulation, Respiration, and Metabolism
(ed. by R. Gilles)

cation of proteins, operon systems, promoter genes, introns, exons, transposons, replicons, splicing mechanisms, oncogenes, and genes associated with differentiation and development. Equally, and perhaps more immediately germane to muscle metabolism and exercise metabolism generally, we now know much about membrane architecture, membrane-located enzymes, symporters and antiporters, membrane-located receptors, and ion-gating mechanisms.

To each and every one of these more detailed parts of our metabolic maps there has been ascribed a control function. We perhaps stand in danger of discovering a greater number of "controls" than there are events to control! More importantly, this loose use of the terms "control" and "regulation" has debased the meaning of the words and has lead to confusion and fruitless dispute.

There are three related objections to this gratuitous use of "control" and "regulation". First, it is illogical to suppose that the properties of an isolated part of any system will necessarily be displayed (even qualitatively) as a property of the system when the part is embedded and functioning in the intact system; an enzyme which is demonstrably controllable when examined in isolation does not necessarily display measurable control of events in the intact metabolic system, either in theory or in practice (Kacser and Burns 1973; Rapoport et al. 1976). Second, the word control has been attached in a non-systematic way to each of the detailed mechanisms and components listed above. In some instances it is implied (but seldom stated explicitly) that the rate of reaction is supposed to be controlled, in others that the concentration of a product is thought to be controlled, in yet others it is implied that a metabolic event is simply "switched on" or "turned off". Third, the majority of claims about metabolic control are stated in a purely qualitatively way; we are seldom if ever told by how much a given flux or metabolite concentration will vary when the activity of an alleged "regulatory", "controlling", "key", "pacemaker" or "rate-limiting" enzyme changes by a given fractional amount.

1.2 The Second Logical Step

To overcome the first of these objections we need to undertake a modelling exercise, basing our model of metabolic systems on the information recorded in the maps already available (the model turns out to be very simple). Next, in order to surmount the second objection, we need a rigorously derived theoretical treatment of the model which will predict how a metabolic system will behave when it is perturbed (for example, by changing the load on or demand from the system; or by changing an enzyme activity within the system). Any such theory should reveal and define system parameters which, when measured by experiment, will provide quantitative indices of metabolic control, thereby eliminating the current habit of using metaphors as substitutes for measurements (as indicated in the third objection above).

Each of these three steps has already been taken by Heinrich and Rapoport (1973, 1974) and by Kacser and Burns (1973, 1981). The approach used by the two independent laboratories was slightly different; the conclusions reached about the nature and origin of metabolic control were identical. The different nomenclatures originally used have now been amalgamated by agreement among several investigators; this common nomenclature will be used here. The second logical step towards an understanding of metabolic control therefore consists of reading the papers published by the Berlin and Edinburgh laboratories and the subsequent papers from the Amsterdam laboratories of Groen, Tager, Westerhoff, and colleagues [see Kacser (1983), Heinrich and Rapoport (1983), Tager et al. (1983), Porteous (1983a,b), Westerhoff et al. (1984), and Westerhoff and Chen (1984) for a comprehensive list of earlier references].

The next section contains brief descriptions of the model metabolic system, the theory derived from the model and the control coefficients that emerge from the theoretical treatment; it emphasises the significant advances in our understanding of metabolic control that have so far been achieved by this systematic and quantitative approach to metabolic control and then asks whether the Kacser and Burns/Heinrich and Rapoport concepts might be applied to investigations of the control of metabolism during exercise.

2 Model, Theory, and Experiment

2.1 A Model of Any Metabolising Cell, Tissue or Organism

Examination of the metabolic maps alluded to above, reveals three features of the functional organization of all cells. The first is that reversible, non-covalent, associations of certain macromolecules is a prerequisite for their involvement in metabolic processes. Examples are seen in the DNA duplex, in the assembly of homologous and heterologous polypeptides to form oligomeric proteins, and in the architecture of membranes. The second (and now universal) feature is that each specific molecule- or ion-dependent step in any scalar or vectorial metabolic process involves an initial, reversible, concentration-dependent, non-covalent association between a protein [an enzyme (E), carrier (C) or receptor (R)] and a specific substrate, product, cofactor, cobsubstrate or effector (S_0, S_1, S_2, etc.). These initial, reversible, associations can be diagrammed as follows:

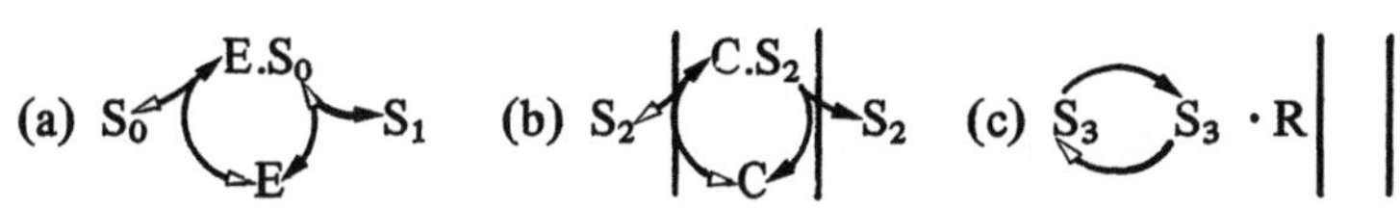

(1)

The carrier (C) and receptor (R) may be membrane-located as drawn, or not. The third, related, generalisation – and the most important for an understanding of metabolic control – is that all metabolic pathways (whether linear, branched or cyclic) are structured according to a simple pattern. Every enzyme (or carrier) is always flanked by related solutes (substrates and products) with which it binds as described in Eq. (1); and every solute (except the first and last in a sequence) is flanked by enzymes (or by an enzyme and a carrier). Figure 1 describes this functional organization and will be recognised as a biochemist's map of part of a metabolising cell. Cosubstrates [(m) and (n) in Fig. 1], cofactors (such as metal cations) and effectors (including hormones and neurotransmitters) are simply additional solutes that interact, as before, with certain enzymes, carriers and receptors. The pathways of DNA replication and transcription, of mRNA translation and enzyme-catalysed "cascades" are a part of this general pattern of organisation equally with pathways concerned (for ex-

For an unsaturated system of enzymes and carriers, the flux equation (Kacser and Burns 1973, 1981) is:

$$F_n = \left(\frac{S_0 - S_n}{K^0_{eq} \cdot K^1_{eq} \cdot K^2_{eq} \cdot \text{----} K^n_{eq}} \right) \Bigg/ \left(\frac{K^0_m}{V^0_{max}} + \frac{K^1_m}{V^1_{max} \cdot K^0_{eq}} + \frac{K^2_m}{V^2_{max} \cdot K^0_{eq} \cdot K^1_{eq}} + \text{----} \right) \tag{3}$$

where the superscript numerals identify each step in the model system (Sect. 2.1) and (------) indicates that all subsequent terms are to be included until each step is accounted for. (An alternative formulation of the product $K^0_{ea} \cdot K^1_{ea} \cdot K^2_{ea} \cdot \text{---} K^n_{ea}$ would be K^{0n}_{eq}.)

In the stationary steady state (non-growing systems) and in the expanding steady state (growing systems), the concentrations of the external solutes (S_0 and S_n in the model system, Sect. 2.1) will remain constant provided we make the external pool large enough; the whole of the numerator term of Eq. (3) can thus be written as a constant (C). There are as many terms in the denominator of Eq. (3) as there are carrier-facilitated or enzyme-catalysed steps in the metabolic system. Each of these denominator terms comprises the mathematical product of (1) the reciprocal of a genetically determined parameter (V_{max}) that characterises the velocity of the facilitated translocation or catalysed transformation, and (2) the ratio of the genetically defined parameter (K_m) and a thermodynamic constant or constants (K_{eq}'s). The general form of Eq. (3) is thus.

$$F_n = \frac{C}{1/\xi_0 + 1/\xi_1 + 1/\xi_2 \text{-----}} \tag{4}$$

where the (ξ) terms represent the activities of the individual carriers and enzymes. It is then easy to show (Kacser and Burns 1981) that the response of the flux (F_n) to a change in activity or concentration of *any one* enzyme or carrier in the system is non-linear (unless the pathway contains only one carrier or enzyme) and becomes increasingly non-linear as the number of enzymes and carriers in the system is increased (Fig. 2a).

Figure 2b then represents the general form of the response of a flux through any system of enzymes and carriers to a change in the activity or concentration of any one carrier or enzyme in the system. If the variable enzyme or carrier had an initial activity (x), a small change in activity would produce a substantial change in the flux through the system in which this enzyme or carrier was embedded (Fig. 2b). If the initial activity happened to be (z), a small change in activity would result in only a very small (perhaps undetectable) change in flux. At intermediate initial values of the variable catalyst [e.g. (y)], the responsiveness of the systemic property (the flux) to a change in one of the components of the system would take an intermediate value (Fig. 2b). To be useful, these qualitative descriptions must be put in quantitative form. This can be achieved if we determine the tangential slope on the response curve (Fig. 2b) at the point corresponding to the existing activity of the variable enzyme or carrier. In experimental practice, this can be done in one of several ways (Sects. 2.2.8, 2.2.9).

that brings about any change in enzyme or carrier activity is thus irrelevant to a consideration of the *effects* of that change on the associated local velocity. We know (from the enzymologists' studies of isolated enzymes) that at a constant substrate concentration, any change in enzyme activity or concentration will bring about a proportional change in local velocity. In order to understand control of metabolism in a *system* of enzymes and carriers (such as that in our model), we have then to ask how a change in the concentration or activity of any one enzyme or carrier will affect those properties of the system that we can measure directly, i.e. (1) the overall flux (F_n) and (2) the concentrations of the intermediate metabolites (S_0, S_1, S_2, etc.).

It is these questions that the theoretical treatment of Kacser and Burns (1973, 1981) seeks to answer. We can anticipate one of the answers here by stating that in contrast to a local velocity (v), the flux (F_n) does *not* respond in a proportional fashion to a change in the activity or concentration of any one enzyme or carrier in the system of enzymes and carriers; a system (Fig. 1 and model system above) possesses properties not exhibited by an isolated component of the system. Failure to recognise this fundamental point is one of several sources of grave error in so much of the existing literature on metabolic control.

2.2 The Theory of Metabolic Control

The treatment here is selective. A full account is given by Kacser and Burns (1973, 1979, 1981) and some aspects are reviewed by Kacser (1983). These papers should be consulted.

2.2.1 A General Flux Equation for a System of Carrier-Facilitated Translocations and Enyzme-Catalysed Chemical Transformations

For a reversible enzyme-catalysed reaction [Eq. (1a)] the non-linear Michaelis-Menten-Henri equation can be written as a linear equation if we assume unsaturation:

$$v = \frac{V^f_{max}}{K^f_m}\left(S_0 - \frac{S_1}{K_{eq}}\right) \qquad (2)$$

where V^f_{max} and K^f_m are enzyme parameters associated with the "forward" reaction $S_0 \rightarrow S_1$ and K_{eq} is the thermodynamic equilibrium constant for the same reaction.

The assumption is justified because: (1) experience indicates that unsaturation is a common situation in vivo (but may not be true in many instances where investigators have employed physiologically unrealistic conditions in vitro!); (2) application of [Eq. (2)] to each of the reactions catalysed by a system of enzymes (e.g. the model system of Sect. 2.1) gives a set of linear simultaneous equations which can be very simply solved to yield a steady state flux equation (Kacser and Burns 1973, 1981); we can then reach otherwise inaccessible conclusions about the kinetic behaviour of unsaturated systems of enzymes and carriers [the kinetics of translocation via carriers will be of the same kind as the kinetics of transformation by an enzyme; cf. Eq. (1a) and (1b)]. It is then possible to extend this treatment to saturated systems (Kacser and Burns 1973, 1979; Rapoport and Heinrich 1975) and to show that the conclusions summarised in Sects. 2.2.2 to 2.2.8 remain valid.

For an unsaturated system of enzymes and carriers, the flux equation (Kacser and Burns 1973, 1981) is:

$$F_n = \left(\frac{S_0 - S_n}{K_{eq}^0 \cdot K_{eq}^1 \cdot K_{eq}^2 \cdot \text{----} K_{eq}^n}\right) \Bigg/ \left(\frac{K_m^0}{V_{max}^0} + \frac{K_m^1}{V_{max}^1 \cdot K_{eq}^0} + \frac{K_m^2}{V_{max}^2 \cdot K_{eq}^0 \cdot K_{eq}^1} + \text{----}\right) \quad (3)$$

where the superscript numerals identify each step in the model system (Sect. 2.1) and (------) indicates that all subsequent terms are to be included until each step is accounted for. (An alternative formulation of the product $K_{ea}^0 \cdot K_{ea}^1 \cdot K_{ea}^2 \cdot \text{---} K_{ea}^n$ would be K_{eq}^{0n}.)

In the stationary steady state (non-growing systems) and in the expanding steady state (growing systems), the concentrations of the external solutes (S_0 and S_n in the model system, Sect. 2.1) will remain constant provided we make the external pool large enough; the whole of the numerator term of Eq. (3) can thus be written as a constant (C). There are as many terms in the denominator of Eq. (3) as there are carrier-facilitated or enzyme-catalysed steps in the metabolic system. Each of these denominator terms comprises the mathematical product of (1) the reciprocal of a genetically determined parameter (V_{max}) that characterises the velocity of the facilitated translocation or catalysed transformation, and (2) the ratio of the genetically defined parameter (K_m) and a thermodynamic constant or constants (K_{eq}'s). The general form of Eq. (3) is thus.

$$F_n = \frac{C}{1/\xi_0 + 1/\xi_1 + 1/\xi_2 \text{-----}} \quad (4)$$

where the (ξ) terms represent the activities of the individual carriers and enzymes. It is then easy to show (Kacser and Burns 1981) that the response of the flux (F_n) to a change in activity or concentration of *any one* enzyme or carrier in the system is non-linear (unless the pathway contains only one carrier or enzyme) and becomes increasingly non-linear as the number of enzymes and carriers in the system is increased (Fig. 2a).

Figure 2b then represents the general form of the response of a flux through any system of enzymes and carriers to a change in the activity or concentration of any one carrier or enzyme in the system. If the variable enzyme or carrier had an initial activity (x), a small change in activity would produce a substantial change in the flux through the system in which this enzyme or carrier was embedded (Fig. 2b). If the initial activity happened to be (z), a small change in activity would result in only a very small (perhaps undetectable) change in flux. At intermediate initial values of the variable catalyst [e.g. (y)], the responsiveness of the systemic property (the flux) to a change in one of the components of the system would take an intermediate value (Fig. 2b). To be useful, these qualitative descriptions must be put in quantitative form. This can be achieved if we determine the tangential slope on the response curve (Fig. 2b) at the point corresponding to the existing activity of the variable enzyme or carrier. In experimental practice, this can be done in one of several ways (Sects. 2.2.8, 2.2.9).

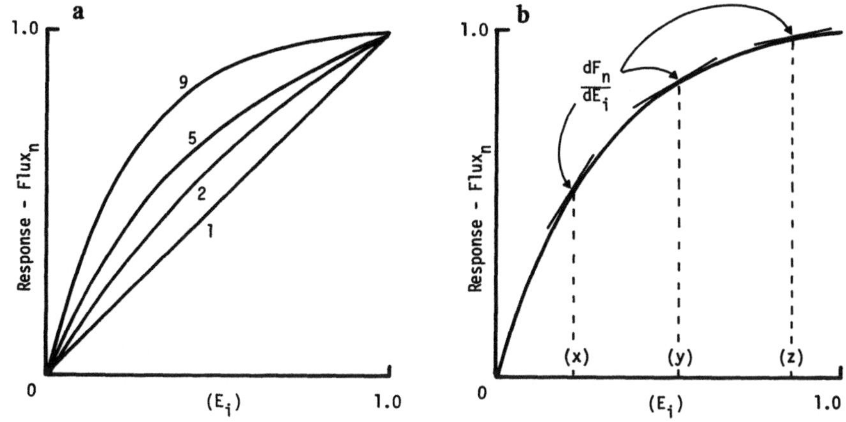

Fig. 2a,b. The response of the flux (*Flux$_n$*) through a system of enzymes when the concentration or activity of any one enzyme (*E$_i$*) is varied. The number of enzymes in the system is indicated against each plot; both scales have been normalised to 1.0 for convenience only. (Redrawn from Kacser and Burns 1981.) **b** The general form of the flux response to a variation in the concentration or activity of any one enzyme in a metabolic system. The tangential slope at any point on the curve corresponding to the existing activity of the variable enzyme (e.g. x, y, z) will give a quantitative measure of control of the specified flux by the particular enzyme in the system under defined experimental conditions. Direct experimental determination of the value of the tangential slope necessitates making successive small change in the activity or concentration of the variable enzyme; larger changes will give poorer estimates; very large changes will yield false information. (Redrawn from Kacser and Burns 1981.) The plot applies equally to translocation of solutes across membranes by substrate-specific carriers

2.2.2 The Flux Control Coefficient (C_E^F) [Formerly Called the Sensitivity Coefficient by Kacser and Burns, and the Control Strength by Heinrich and Rapoport]

The algebraic equivalent of the experimental determination of the tangential slope to the response curve of Fig. 2b is achieved by partial differentiation of the flux equation [Eq. (3)] with respect to the variable enzyme (E_i) to yield $\partial F_n/\partial V_i$. The absolute values of such a slope would depend on the units employed to measure (F_n) and (V_i); Kacser and Burns (1973, 1981) eliminated this inconvenience by defining the flux control coefficient: $C_{E_i}^{F_n} = \frac{\partial F_n}{F_n} \Big/ \frac{\partial E_i}{E_i}$ (i.e. the fractional change in a specified flux divided by the fractional change in a named enzyme) $= \frac{\partial F_n}{\partial E_i} \cdot \frac{E_i}{F_n}$ (i.e. the tangential slope of Fig. 2b multiplied by a scaling factor) $= \partial\,(1\mathrm{n}F_n)/\partial(1\mathrm{n}E_i)$. Thus, $C_{E_i}^{F_n}$ is a true dimensionless coefficient and an exact measure of the control by any one enzyme (E_i) of the flux (F_n) through a metabolic system. In a real cell there will be several fluxes that could be measured (Fig. 1). For any one flux, there will be as many flux control coefficients as there are enzymes and carriers *in the whole system*. Partial differentiation of Eq. (3) then gives, for example:

$$\text{(a)}\ C_{E_1}^{F_n} = \frac{K_m^1/V_{max}^1 \cdot K_{eq}^0}{\text{(denominator of Eq. 3)}} \quad \text{and} \quad \text{(b)}\ C_{E_2}^{F_n} = \frac{K_m^2/V_{max}^2 \cdot K_{eq}^0 \cdot K_{eq}^1}{\text{(denominator of Eq. 3)}} \qquad (5)$$

and similarly for each other enzyme and carrier in the metabolic system.

2.2.3 The Summation Theorem for Flux Control Coefficients

Algebraic statements about two of these coefficients are given in Eq. (5a) and (5b) for a specified flux (F_n); the numerator term for each coefficient is one of the terms of the denominator and this is true of every other flux control coefficient attached to any one flux. It follows that the sum of all flux control coefficients associated with any one flux in a metabolising system of any complexity must be 1.0 (Kacser and Burns 1973, 1979, 1981; Heinrich and Rapoport 1974, 1983). This summation theorem ($\sum_{i=1}^{n} C_{E_i}^{F_n} = 1.0$) does not impose limits on the magnitude of individual flux control coefficients; they could be negative or greater than 1.0, but their sum will always be 1.0. Concerning the application of this summation theorem to growing as well as to non-growing systems, and the origin and importance of negative C_E^F values, see Kacser and Burns (1973), Flint et al. (1981), and Kacser (1983).

Since there are at least 1,000 enzymes and carriers in even the simplest cell, the summation theorem for flux control coefficients shows that the *average* C_E^F value will be of the order of 0.001. That is, on average, a fractional change in the activity of an enzyme embedded in a metabolic system will have an effect on a specified flux that is 0.001 of the effect on velocity we would see if the same change in activity occurred in a reaction catalysed by the isolated enzyme; the same argument applies, *mutatis mutandis*, to flux control coefficients attached to carriers. Some individual enzymes and carriers will have attached to them C_E^F values below the average, and some will be above the average. It is among the latter that we must look (by experimental determination of the C_E^F values) for extraordinary contributions to the "control" of any chosen flux; but labels like "key" or "pacemaker" enzyme (or any other such metaphors) have no useful function in studying metabolic control unless they refer to a measured coefficient (just as "highly alkaline" conveys no useful meaning unless it refers to a measured pH, e.g. pH 9.4).

2.2.4 The Meaning of the Flux Control Coefficient (C_E^F)

The denominator of Eq. (3) contains one term for each carrier and enzyme in the whole metabolising system; the equation thus states explicitly that each and every carrier and enzyme contributes quantitatively to any chosen flux (F_n). This is true no matter how complex the functional organisation of the metabolic system may be. It is inconceivable, for example, that enzymes E_4, E_7, and E_8 (Fig. 1) play no role in determining the magnitude of the flux from S_o to S_p. Since equations analogous to Eq. (3) can be written for every other flux in the same system, simply by rewriting the numerator term in an appropriate way, it is always true that every flux is determined quantitatively by the activities of every carrier and every enzyme in the whole system.

Each flux control coefficient is related to the flux equation [Eq. (3)] as indicated in two examples [Eq. (5a,b)]. By the same arguments just employed, a flux control coefficient is a *systemic parameter*. That is, it is an exact measure of the contribution of a specified enzyme or carrier to a given flux when a particular system is metabolising under known conditions; the flux control coefficient (C_E^F) is not an inherent property of the enzyme (or carrier) itself. The absolute values of a set of flux control coefficients for a given metabolic system may vary and redistribute among the enzymes and carriers (within limits imposed by the summation theorem of Sect. 2.2.3) as the system moves, for whatever reason, from one stationary state to another. The failure to recognise that the "control" exerted by any individual enzyme on a systemic property (such as a flux or the concentrations of metabolites) is itself a systemic parameter and not an inherent property of the enzyme has been a source of misconception, and controversy over interpretation of results, in the existing literature on metabolic control.

2.2.5 The Practical Significance of the Flux Control Coefficient (C_E^F)

Suppose that the C_E^F values for five enzymes (A, H, P, R, W) had been determined for flux (F_p) in a particular tissue, under defined conditions, as: – < 0.001 (A), 0.2 (H), 0.01 (P), 0.015 (R), and 0.1 (W). The sum is 0.325; the remaining enzymes and carriers in the system must then have had attached to them C_E^F values summing to 0.675. Enzyme (H) contributes, within the system, more than any other enzyme yet characterised, to setting the observed value of flux (F_p). There remains a chance (see Sect. 2.2.3) that a C_E^F value greater than 0.2 will yet be found to be attached to some other enzyme or carrier in the same system sustaining the same flux under the same conditions. If, in the meantime, another laboratory measures a different flux (F_q) in the same system and finds a set of C_E^F values for enzymes A, H, P, R, W which differ from those report for flux (F_p), we would be interested, but not surprised, and would not enter into time-wasting controversy; if the other laboratory's findings are for the same flux (F_p), consultation over the conflicting results would be called for.

2.2.6 The Solute Control Coefficient (C_E^S) [Formerly Called the Control Matrix by Heinrich and Rapoport and the Substrate Sensitivity Coefficient by Kacser and Burns]

Just as any flux sustained by a metabolising system is a property of the whole system, so the steady state concentrations of intermediate metabolites [S_0, S_1, etc. and co-substrates, such as (m) and (n), of Fig. 1] are systemic functions. Accordingly, solute control coefficients can be defined: $C_{E_i}^{S_j} = \frac{\partial S_j}{S_j} \Big/ \frac{\partial E_i}{E_i}$. They have a significance and utility analogous to that of the flux control coefficients; the solute control coefficients sum to zero (Heinrich and Rapoport 1974; Rapoport and Heinrich 1975; Kacser and Burns 1979, 1981).

2.2.7 The Elasticity Coefficient (ϵ_S^v) [Formerly Called the Effector Strength by Heinrich and Rapoport]

In contrast to the flux control and solute control coefficients (Sects. 2.2.2 to 2.2.6), the elasticity coefficient is a *local* parameter, not a systemic character. It is a quantifiable index of the response of a local velocity (v) to a small change in the concentration of any solute (S) that interacts with a specified enzyme or carrier so as to modify its activity (Kacser and Burns 1973, 1979; Kacser 1983; Heinrich and Rapoport 1974). The definition extends, *mutatis mutandis*, to effector interactions with a receptor [Eq. (1c) in Sect. 2.1] if that event modifies the activity of an enzyme associated with the receptor. The formal definition is: $\epsilon_{S_j}^{v_i} = \frac{\partial v_i}{v_i} \Big/ \frac{\partial S_j}{S_i}$. The coefficient can be positive or negative, greater or less than 1.0.

There are as many elasticity coefficients attached to each enzyme, carrier or receptor as there are interacting solutes (substrates, cosubstrates, cofactors, effectors) which modify the activities of these entities. Thus, enzyme (E_1) of the model system (Sect. 2.1) will have three elasticity coefficients attached to it: $\epsilon_{S_0}^{v_1}$, $\epsilon_{S_1}^{v_1}$, $\epsilon_{S_3}^{v_1}$. Since a change in substrate and/or product concentration may change a local volocity, the general statement for a change in local volocity at enzyme (E_3) in the model system (Sect. 2.1) would be:

$$\frac{\delta v_3}{v_3} = \epsilon_{S_2}^{v_3} \cdot \frac{\delta S_2}{S_2} + \epsilon_{S_3}^{v_3} \cdot \frac{\delta S_3}{S_3} \ . \tag{6}$$

2.2.8 The Connectivity Theorem

The sum of the mathematical products of (1) the flux control coefficients attached to individual members of any set of enzymes and (2) the appropriate elasticity coefficients attached to any solutes that interact with *each* enzyme of the set (but solely with these enzymes) is always zero (Kacser and Burns 1973). Thus, for the feedback loop of the model system (Sect. 2.1), (S_3) interacts with, and only with: (E_1), (E_3), and (E_4). So:

$$C_{E_1}^{F_n} \cdot \epsilon_{S_3}^{v_1} + C_{E_3}^{F_n} \cdot \epsilon_{S_3}^{v_3} + C_{E_4}^{F_n} \cdot \epsilon_{S_3}^{v_4} = 0 \ . \tag{7}$$

It follows from this particular example that the (allosterically) controllable enzyme (E_1) does not necessarily control flux (F_n) but does play a role in buffering the solute concentrations within the feedback loop; "control" of flux may then pass to an enzyme outside the loop (Kacser and Burns 1973).

Apart from this illuminating example, the connectivity theorem is profoundly important for an understanding of metabolic control. It states quantitatively what is qualitatively obvious on inspection of Fig. 1 (Sect. 2.1): any flux will depend on the activities of all enzymes and carriers in the system; and the local velocity catalysed by each enzyme and carrier depends on the concentration-dependent interactions between it and its substrates, products, cosubstrates, cofactors and effectors. The connectivity theorem also has great practical importance as illustrated in Sect. 2.2.9.

2.2.9 Experimental Determination of Flux Control Coefficients (C_E^F)

Recall (Fig. 2b) that the experimental determination of a flux control coefficient necessarily involves modulating a particular enzyme (or carrier) activity or concentration by a small amount in the intact metabolic system and measuring any consequential change in a chosen flux (see Fig. 1).

Two methods are available for the direct determination of the absolute value of individual flux control coefficients. The first involves isolating a series of mutants (or constructing a series of heterokaryons) for a particular enzyme or carrier, measuring the chosen flux in each of the otherwise isogenic isolates or constructs, and plotting the measured fluxes against the measured enzyme activities (Kacser and Burns 1981).

The method is not particularly well suited to accurate measurement of C_E^F values but provided confirmation (for each of eight enzymes in four widely differing biological systems) of the theoretical prediction (Fig. 2a) that fluxes respond non-linearly to a change in any one enzyme activity. A more accurate method of determining C_E^F values involves gradual titration of a particular enzyme or carrier in the intact system with a specific inhibitor, again measuring the chosen flux at each stage of inhibition of carrier or enzyme. This method has been remarkably successful in confirming several predictions of the Kacser and Burns/Heinrich and Rapoport theory when applied to respiring mitochondria, a notoriously complex metabolic system. The flux control coefficients (C_E^F) for several steps in the system were determined, thus settling long-standing conflicts of opinion as to the identity of alleged rate-limiting steps; and it was shown (in conformity with a prediction of the theory) that the magnitudes of the flux control coefficients were redistributed as the steady state respiration rate changed, no one catalysed step at any time taking on the characteristics of a "rate-limiter", "bottleneck" or "pacemaker" [see Tager et al. (1983) for a summary; and Groen (1984) for other applications].

Two other methods yield *ratios* of the flux control coefficients. Kacser and Burns (1973) established several relationships for the ratios of flux control coefficients (Table 1). It should thus be possible to calculate ratios of C_E^F values (but *not* absolute values) attached to the enzymes of several pathways. These ratios are likely to be approximate because it is not certain that K_m and V_{max} values determined under the

Table 1. Relationships between (a) the ratios of flux control coefficients (C_E^F) attached to various enzymes in a segment of an unsaturated metabolic pathway and (b) the ratios of terms containing (1) the substrate and product concentrations of the catalysed reaction, the K_{eq} for this reaction and the K_{eq} values for any earlier reactions in the pathway; or (2) the thermodynamic disequilibrium for the reaction catalysed and the disequilibria for any earlier reactions; or (3) kinetic parameters of the enzyme concerned and the K_{eq} values for any earlier reactions in the pathway[a]

Segment of metabolic pathway	$\longrightarrow S_0 \xrightarrow{E_1} S_1$		$\xrightarrow{E_2} S_2$		$\xrightarrow{E_3} S_3 \dashrightarrow (F_n)$
a Ratios of flux control coefficients	$C_{E_1}^{F_n}$	:	$C_{E_2}^{F_n}$	:	$C_{E_3}^{F_n}$
b (1) Equivalent ratios	$S_0 - \frac{S_1}{K_{eq_1}}$	:	$\left(S_1 - \frac{S_2}{K_{eq_2}}\right) \cdot \frac{1}{K_{eq_1}}$	:	$\left(S_2 - \frac{S_3}{K_{eq_3}}\right) \cdot \frac{1}{K_{eq_1}} \cdot \frac{1}{K_{eq_2}}$
(2) Equivalent ratios	$1 - \rho_1$	:	$(1 - \rho_2)\,\rho_1$	:	$(1 - \rho_3)\,\rho_1 \cdot \rho_2$
(3) Equivalent ratios	$\frac{K_m^1}{V_{max}^1}$	:	$\frac{K_m^2}{V_{max}^2} \cdot \frac{1}{K_{eq_1}}$	:	$\frac{K_m^3}{V_{max}^3} \cdot \frac{1}{K_{eq_1}} \cdot \frac{1}{K_{eq_2}}$

[a] Note that the absolute value for any one flux control coefficient (C_E^F) cannot be calculated from the terms given in any of the rows (b1, b2, b3); see the text for further details. (Information collected together from Kacser and Burns 1973)

usual conditions in vitro, reflect the values that would obtain in vivo. Nevertheless, such calculations would be a step in the right direction. The relationships shown in Table 1 reveal several invalidities in current qualitative concepts on metabolic control [see Kacser and Burns (1973), for details].

Kacser and Burns (1979) have suggested a method for determining the ratios of flux control coefficients by experiment on intact systems. Measurements are made of the steady state concentrations of metabolites, and of a chosen flux. The system is then perturbed (where the perturbation is applied in the system and how, is irrelevant) and the new steady state metabolite concentrations and flux determined. Another perturbation yields a third set of steady state determinations. Consider just one step ($S_2 \rightarrow S_3$) in the model system (Sect. 2.1), recall Eq. (6) (Sect. 2.2.7), and remember that at steady state, a local velocity is the same as the measured flux. We can then write for the first change from the original steady state: (a), and for the second (b):

$$\text{(a)}\ \frac{\delta F_n'}{F_n} = \epsilon_{S_2}^{v_3} \cdot \frac{\delta S_2'}{S_2} + \epsilon_{S_3}^{v_3} \cdot \frac{\delta S_3'}{S_3} \qquad \text{(b)}\ \frac{\delta F_n''}{F_n} = \epsilon_{S_2}^{v_3} \cdot \frac{\delta S_2''}{S_2} + \epsilon_{S_3}^{v_3} \cdot \frac{\delta S_3''}{S_3}\,. \tag{8}$$

The two simultaneous equations will solve for the two unknowns, $\epsilon_{S_2}^{v_3}$ and $\epsilon_{S_3}^{v_3}$. The same procedure can be repeated at every step at which solute concentrations were measured, so yielding a series of values of individual elasticity coefficients. The connectivity theorem (Sect. 2.2.8) gives, for the steps involving (S_1) and (S_2) of the model system:

$$C_{E_1}^{F_n} \cdot \epsilon_{S_1}^{v_1} + C_{E_2}^{F_n} \cdot \epsilon_{S_1}^{v_2} = 0 \quad \text{or} \quad C_{E_1}^{F_n} \Big/ C_{E_2}^{F_n} = - \epsilon_{S_1}^{v_2} \Big/ \epsilon_{S_1}^{v_1} \tag{9a}$$

$$C_{E_2}^{F_n} \cdot \epsilon_{S_2}^{v_2} + C_{E_3}^{F_n} \cdot \epsilon_{S_2}^{v_3} = 0 \quad \text{or} \quad C_{E_2}^{F_n} \Big/ C_{E_3}^{F_n} = - \epsilon_{S_2}^{v_3} \Big/ \epsilon_{S_2}^{v_2}\,. \tag{9b}$$

By "walking" along the pathway, inserting ratios of elasticity coefficients into appropriate equations [analogous to Eq. (9a,b)], it would be possible to arrive at ratios of (C_E^F) values for the pathway. Determination of the absolute value of just one coefficient would then allow calculation of the absolute values of all coefficients for which ratio values were available. The summation theorem (Sect. 2.2.3) allows us to allocate an "overall flux control coefficient" to any segments of pathways that had not been or could not be assayed (Sect. 2.2.5). Elasticity coefficients determined in this way can be used (by a different calculation) to determine (C_E^S) values (Kacser, personal communication). Note that this method employs the intact metabolising system, does not involve determination of absolute concentrations (since the same sample size can be used for each successive analysis), and involves those kinds of analyses which are routinely used in biochemistry. It is an approach which might commend itself to those interested in metabolic controls associated with exercise – especially to those who use biopsy-needle sampling techniques.

2.2.10 The Effect of Large Changes in a Component of a Metabolic System

It has been assumed so far that the metabolic system was in a steady state and that small changes could be imposed when determining flux control coefficients experimentally (Fig. 2b and Sect. 2.2.9). Is this approach then applicable to systems which suffer large changes of activity of a component, e.g. during a change from a mild to a severely exercised condition? Two points may be made. First, provided the system is in a steady state in the two conditions, the treatment outlined above is valid for each condition. Second, Kacser has shown (personal communication) that if one enzyme suffers a large change in activity so that the flux control coefficient attached to it is altered by a factor (α), then each of the C_E^F values originally attached to all the other enzymes and carriers will change by a factor (β) so that the summation theorem (Sect. 2.2.3) is now satisfied by: $\alpha \cdot C_{E_i}^{Fn} + \beta \Sigma C_{E_j}^{Fn} = 1.0$, where E_i is the variable enzyme and E_j represents the unchanged activities of the remaining enzymes and carriers. Given that the original C_E^F values were known, and that $C_{E_i}^{Fn}$ for the new steady state is determined, all the remaining new C_E^F values can be calculated for unsaturated systems.

2.2.11 Common Sense in the Laboratory

There can be no doubts about the validity of the theoretical treatment of metabolic control outlined above, or about its positive contribution to the experimental elucidation of metabolic control. But, if every enzyme and carrier makes some quantitative contribution to the control of metabolism, common sense dictates that we should attempt to reduce several thousand possible determinations of individual control coefficients to a manageable number. A thorough understanding of the theory will then provide a safer guide to experimentation than will intuition or teleology. Confirmation of this claim may be most readily obtained by experimental determination of the control coefficients attached to those enzymes previously thought on insecure grounds to be "rate-limiting" or similar enzymes in the intact metabolic system. Examples will be found in the literature references quoted of alleged "pacemaker" enyzmes which do in fact make a substantial quantitative contribution to control of metabolism in an intact system, but which scarcely merit the qualitative term previously appended to them; other examples will be found of alleged "bottle-neck" enzymes that in fact make a quantitatively negligible contribution to metabolic control in the system. An independent approach would be to determine the control coefficients attached to those enzymes and carriers lying directly on a flux pathway of particular interest (e.g. $S_0 \rightarrow S_p$ of Fig. 1); this will at least limit the number of initial experiments. As pointed out in Sects. 2.2.2–2.2.6, there can be no reason to suppose that quantitatively important enzymes and carriers do not lie elsewhere in the metabolic system; but once a few quantitatively important enzymes and carriers have been discovered, the number of equally or more important catalysts yet to be discovered may very quickly diminish (Sect. 2.2.5). Further guidance in the search for those enzymes and carriers which it would be most profitable to assess in early experiments may be obtained by considering information given in Table 1 (assuming appropriate parameters and constants are already available or can be determined in preliminary experiments). An equally potent approach also depends on prior information about individual segments of a metabolic pathway and upon a sound understanding of control theory; it consists of "simplifying" the metabolic pathway(s) by application of two principles of (metabolic) model reduction (Heinrich and Rapoport 1974; Rapoport et al. 1976).

3 Conclusions

"... when you can measure what you are speaking about, and express it in numbers, you know something about it; ... when you cannot express it in numbers, your knowledge is of a meagre and unsatisfactory kind ..." [Kelvin 1889]

It is a tradition among many biochemists to regard theory with suspicion, or even derision. I have pleaded (Porteous 1983a,b) that in the absence of sound theory, our experiments may be misconceived and our interpretation of results invalidated. The theoretical treatments of metabolic control by Heinrich and Rapoport and Kacser and Burns summarised and advocated here, are sound, coherent and comprehensive. They have lead directly to novel experimentation with human erythrocytes, human heterokaryon cell lines, *drosophila*, rat liver cells and mitochondria, yeast, *Neurospora crassa* and whole mice. It has been argued that the Kacser and Burns/Heinrich and Rapoport approach to experiments aimed at elucidating metabolic control is inapplicable to tissues such as liver and muscle because of problems associated with compartitioning of metabolites. This argument is misplaced. The problem is not peculiar to the experimental approach advocated here; indeed this approach may help to solve some of these problems in suitable systems (see Groen 1984). The over-riding merit of the Kacser and Burns/Heinrich and Rapoport approach is that it gives a quantitative, universally applicable and easily comprehended meaning to metabolic control; it offers an opportunity for more profitable experimentation. Kelvin would have approved.

Acknowledgements. I am indebted to Bert Groen, Henrik Kacser, Sam Rapoport, Joseph Tager, and Hans Westerhoff for discussions and valuable personal communications. I thank Tom Boyde for an invitation to present a lecture series on this and related topics in the University of Hong Kong and for facilities for drafting this article; and the Croucher Foundation for financial support. Any opinions expressed in this article are those of the author; they do not necessarily coincide with those of other authors listed in the bibliography.

References

Flint HJ, Tateson RW, Bartelmess IB, Porteous DJ, Donachie WD, Kacser H (1981) Control of flux in the arginine pathway of *Neurospora crassa*. Biochem J 200:231–246

Groen AK (1984) Quantification of control in studies on intermediary metabolism. Doctorate Thesis, University of Amsterdam

Heinrich R, Rapoport TA (1973) Linear theory of enzymatic chains; its application for the analysis of the cross-over theorem and of glycolysis of human erythrocytes. Acta Biol Med Ger 31:479–494

Heinrich R, Rapoport TA (1974) A linear steady-state treatment of enzymatic chains. Eur J Biochem 42:89–95

Heinrich R, Rapoport SM (1983) The utility of mathematical models for the understanding of metabolic systems. Biochem Soc Trans 11:31–35

Kacser H (1983) The control of enzymes in vivo: elasticity analysis of the steady state. Biochem Soc Trans 11:35–40

Kacser H, Burns J (1973) The control of flux. Symp Soc Exp Biol 27:65–104

Kacser H, Burns J (1979) Molecular democracy: who shares the controls? Biochem Soc Trans 7: 1149–1160

Kacser H, Burns J (1981) The molecular basis of dominance. Genetics 97:639–666
Kelvin Lord (Thomson W) (1889) Popular lectures and addresses, vol 1, p 73. MacMillan, London
Porteous JW (1983a) The aims and scope of the colloquium on catalysis and modulation in metabolic systems. Biochem Soc Trans 11:29–31
Porteous JW (1983b) Sound practice follows from sound theory – the control analysis of Kacser and Burns evaluated. Trends Biochem Sci 8:200–202
Rapoport TA, Heinrich R (1975) Mathematical analysis of multienzyme systems. I. Modelling of the glycolysis of human erythrocytes. Biosystems 7:120–129
Rapoport TA, Heinrich H, Rapoport SM (1976) A minimal comprehensive model describing steady states, quasi-steady states and time-dependent processes. Biochem J 154:449–469
Tager JM, Groen AK, Wanders RJW, Duszynski J, Westerhoff HV, Vervoorn RC (1983) Control of mitochondrial respiration. Biochem Soc Trans 11:40–43
Westerhoff HV, Chen Y-D (1984) How do enzyme activities control metabolite concentrations? An additional theorem in the theory of metabolic control. Eur J Biochem 142:425–430
Westerhoff HV, Groen AK, Wanders RJW (1984) Modern theories of metabolic control and their application. Biosci Rep 4:1–22

Symposium IV

Comparative Aspects of Erythrocyte Metabolism

Organizer R. ISAACKS

Metabolism of Invertebrate Red Cells: A Vacuum in Our Knowledge

C.P. MANGUM[1] and N.A. MAURO[2]

1 Introduction

When Dr. Isaacks asked us for a contribution on metabolism in invertebrate red cells, we agreed on the condition of a subtitle emphasizing the scarcity of information on the subject. We cannot even address items of such considerable interest as the sites of formation and destruction of these cells, their life history, and the biosynthesis of their most important constituent, the O_2 carrier. To our knowledge not one intensive investigation of any one of these topics in any one phylum outside of the chordates has ever been reported, although casual observations are frequently mentioned in studies of cell ultrastructure (e.g., Fontaine and Lambert 1973; Boilly 1974; Fontaine and Hall 1981). Our task must be to describe what little is known about their structure and physiology in hopes of stimulating investigation in the future, and to summarize some exploratory studies on metabolism of RBCs in two representative species, the annelid bloodworm *Glycera dibranchiata* and the blood clam *Noetia ponderosa*.

2 Structure of Invertebrate RBCs and PBCs

Invertebrates RBCs and PBCs (pink, or hemerythrin-containing, blood cells) are almost as unlike one another as the nine phyla in which they are found (Table 1). They are invariably nucleated, sometimes even bi- or multinucleated. Earlier reports of anucleate PBCs have proven to be incorrect (Terwilliger et al. 1983 and pers. comm.). Because the nucleus is not very compressible, it creates a bulge in the middle of the otherwise biconcave disk. Often these cells are elliptical in outline and in some of the echinoderms they are even elongate (Hetzel 1963), which raises interesting questions about their flow properties. They are usually larger in at least one dimension than mammalian RBCs. In terms of fit their size is not a disadvantage because they do not circulate in small bore tubes analogous to mammalian capillaries, but rather in larger channels or, more typically, in open coelomic spaces. The viscosity of the bloods containing them, however, has not been determined.

1 Department of Biology, College of William and Mary, Williamsburg, VA 23185, USA
2 Department of Biology, Hartwick College, Oneonta, NY 13820, USA

Circulation, Respiration, and Metabolism
(ed. by R. Gilles)

Table 1. Anatomical properties of invertebrate RBCs and PBCs

	Priapulida	Phoronida	Brachiopoda	Annelida	Nemertina	Echiura	Sipunculida	Mollusca	Echinodermata
Diameter (μm)	?	10–12	15–20	2–40	5–7	10–50	6–123	8–25	4–23
O_2 Carrier	Hr	Hb	Hr	Hb, Hr	Hb	Hb	Hr	Hb	Hb
Nucleus	?	+	+	+	+	+	+	+	+
Mitochondria	?	+	+	+	+	+	+	+	+
Endoplasmic reticulum	?	?	?	+	+	?	+	?	+
Glycogen granules	?	?	?	+	+	+	+	+	+
Marginal band system	?	?	?	+	?	?	+	+	?
Golgi apparatus	?	+	?	+	?	?	Rare	?	+
Pinocytotic vesicles	?	+	?	+	+	?	+	+	+
Membrane deformability	?	?	?	Variable	?	?	Variable	?	Variable
Sources[a]		a,b	c	d,e,f,g,h, i,j	a,k	l	d,m,n,o,p,q	d,e,r,s,t,u	d,v,w,x,y

[a]
a E. Ruppert (pers. comm.)
b Emie (1982)
c Storch and Welsch (1976)
d Dawson (1933)
e Cowden (1966)
f Dales (1964)
g Schumacher and Seamonds (1972)
h Sean and Boilly (1980)
i Boilly (1974)
j N.B. and R.C. Terwilliger (pers. comm.)
k Vernet (1979
l Ochi (1975a)
m Stang-Voss (1970)
n Ochi (1975b)
o Nemhauser et al. (1983)
p Terwilliger et al. (1983)
q Ochi (1977)
r Baba (1940)
s Cohen and Nemhauser (1980)
t Freadman and Mangum (1976)
u N.A. Mauro (unpubl. obs.)
v Hetzel (1963)
w Fontaine and Lambert (1973)
x Fontaine and Hall (1981)

Most investigators of ultrastructure have been able to locate well-formed mitochondria, the density of which is quite variable, and less well-formed elements of an endoplasmic reticulum. As pointed out by Fontaine and Lambert (1973), the presence of presumably active organelles such as mitochondria and polyribosomes in mature RBCs suggests that their life history is quite different from that of the nucleated RBC in the lower vertebrates, in which these organelles disappear with maturation of the cell.

With one interesting exception RBCs and PBCs contain O_2 carriers built of loosely linked subunits, each with a single active site. While many aspects of the structure of the Hbs closely resembles those of the vertebrate Hbs, the size of the native multiple varies from monomeric to octameric, sometimes within the RBCs of a single individual (summarized by Mangum 1985). The exception is the lamellibranch mollusk *Barbatia reeveana*, whose RBCs contain not only a conventional tetramer, but also a 430×10^3 d Hb built of subunits containing two covalently linked O_2 binding domains (Grinich and Terwilliger 1980). Otherwise this kind of Hb is found free in the circulation, not inside an RBC, and its intracellular location in the *Barbatia* suggests that its distinctive structure may not be as fundamental as once thought.

The presence of iron containing granules in the cytoplasm of invertebrate RBCs and PBCs (Table 1) is often cited in support of the hypothesis that Hb or hemerythrin (Hr) synthesis is carried out within the cell. Direct investigation (see below) has shown that this idea is true, but it may or may not be related to the iron granules. Thorough examination indicates that at least in some species the iron exists in the form of a protohematin (Baumberger and Michaelis 1931; Dales 1964; R. Kashian, pers. comm.), raising the possibility that it is a product of Hb degradation.

3 Functions of Invertebrate RBCs and PBCs

The O_2 transport function of many invertebrate RBCs and PBCs is very clear (e.g., Hoffmann and Mangum 1970; Mangum and Kondon 1975; Mangum et al. 1975; Wells and Dales 1975; Mangum 1977; Wells et al. 1980; Pritchard and White 1981). At least in the simpler members of various groups it is also clear that a critical respiratory function, perhaps even the primitive one, is O_2 storage for periods when ambient O_2 is not available (Mangum 1977, 1985). O_2 storage is possible because of the large volume of extracellular body fluid and the relatively high hematocrit (up to 18% in the sipunculid *Xenophon mundanus*; Wells 1982).

The ample evidence of the respiratory importance of the annelid and molluskan RBC and the sipunculid PBC may obscure the uncertainty of its function in groups which have not been seriously investigated. The very low hematocrit in animals, such as priapulids (Weber et al. 1979), brachiopods, phoronids, and echinoderms (C.P. Mangum, unpubl. obs.) raises a very real question of respiratory significance. Even in a few species of annelids the respiratory function of the RBC may be secondary and another function, viz. supplying nutrients to the gametes that circulate along with them in the same body compartment, may be primary (Dales 1964; Mangum et al. 1975).

4 Metabolism

4.1 Protein Synthesis

On several occasions the low density of cell organelles in some RBCs and PBCs has led to the conclusion that these cells lead a brief life during which they are not very active metabolically and that, after maturation, they do not continue to synthesize their O_2 carrier (Cowden 1966; Schumacher and Seamonds 1972; Sean and Boilly 1980). This image of the annelid RBC is also supported by low ratios of protein and RNA to DNA in the chromatin (Shaw and Seamonds 1972), presumably relative not to the nucleated RBCs of the vertebrates, but to various mammalian tissues. More direct investigation, however, reveals a very different image of the invertebrate RBC. Not only do both RNA and protein synthesis clearly occur, but Hb is among the proteins synthesized (Hoffmann and Mangum 1970; Shafie et al. 1976).

4.2 Aerobic Metabolism

Is the level of aerobic metabolism in fact low? Not really, unless the comparison is made with exceptionally active tissues, such as insect flight muscle or the cephalopod siphuncle (Table 2). In relation to other tissues in the same kinds of animals the level in the RBC is perhaps best described as average.

Table 2. Rates of O_2 uptake ($\mu l\ g^{-1}$ dry wt. h^{-1}) by invertebrate RBCs and PBCs in comparison with other tissues

Species and tissues	Temp. (°C)	Rate	Source
Annelida			
Glycera dibranchiata RBCs	23	289	Mangum and Carhart (1972)
Amphitrite ornata RBCs	23	677	Mangum et al. (1975)
Enoplobranchus sanguineus RBCs	23	445	Mangum et al. (1975)
Chaetopterus sp. body wall muscle	25	116	Bliss and Skinner (1961)
Mollusca			
Noetia ponderosa RBCs	23	233	Freadman and Mangum (1976)
N. ponderosa adductor muscle			
Red	28	588	Hopkins (1930)
White	28	288	
Nautilus siphuncle	25	4,360	Mangum and Towle (1982)
Sipunculida			
Phascolopsis gouldi PBC	23	108	Mangum and Kondon (1975)
Arthropoda			
Belostoma sp. flight muscle		4,230	Bliss and Skinner (1961)
Chordata			
Elasmobranch RBC	20	127	Altman and Ditmer (1961)
Teleost RBC	20	71	
Amphibian RBC	20	68	
Reptile RBC	25	71	
Bird RBC	20	50	
Mammal RBC	20	23	

Ochi (1975a) reported that O_2 consumption by echiuroid RBCs is not stimulated by exogenous glucose. Several lines of recent evidence, all indirect, suggest that the major pathway of intermediary metabolism in at least annelid and molluscan RBCs is the tricarboxylic acid cycle (TCA) and not the Embden-Meyerhof (EM) or pentose phosphate (PPO_4) pathways.

First, if the PPO_4 pathway were predominant, as in mammalian RBCs, one would expect the temperature dependence of O_2 consumption to be low, which is typical of that pathway (Table 3; Hochachka and Hayes 1962; Mauro and Mangum 1982). But it is not low. It is somewhat greater than the thermal behavior of avian RBCs, which utilize both the EM and PPO_4 pathways, and a lot greater than that of teleost RBCs. Teleost RBCs have the enzymes that participate in both pathways, and the relative contributions of each are unknown (Bachand and Leray 1975). However, the activities of the PPO_4 enzymes are actually higher in teleosts than in most mammalian RBCs and, in one species, O_2 consumption is essentially independent of temperature above 20 °C (Table 3).

Second, the relative rates of CO_2 production from labeled substrate are greater following the addition of exogenous glutamate, which is deaminated and then oxidized via the TCA cycle, than in the presence of glucose, which feeds into either glycolysis or the PPO_4 pathway (Table 4). Nonetheless the rates of uptake of glucose by annelid

Table 3. Temperature dependence of O_2 uptake ($\mu l\ ml^{-1}$ RBCs h^{-1}) by vertebrate and invertebrate RBCs

Species	Temperature range (°C)	Q_{10}
Noetia ponderosa[a]	22–32	2.0
Salmo gairdneri[b]	20–35	0.95
Chicken[a]	22–37	1.8
Rat[a]	22–37	1.4
Man[a]	22–37	1.4

[a] N.A. Mauro and R.E. Isaacks, unpubl. obs.
[b] Eddy (1977)

Table 4. Relative rates of $^{14}CO_2$ uptake and production from ^{14}C-labeled substrate by vertebrate and invertebrate RBCs[a]

	Rat	Chicken	*Noetia ponderosa*	*Glycera dibranchiata*
Glucose				
Uptake	1.0	?	1.5	2.5
CO_2 production	1.0	2.6	6.3	6.0
Glutamate				
Uptake	1.0	1.8	8.4	4.9
CO_2 production	1.0 (Trace)	1.0 (Trace)	200	450

[a] Values were determined as μm substrate ml^{-1} RBC and dpm ml^{-1} RBC during a 1 h incubation period at 22 °C, and are expressed in relation to the rate obtained for rat RBCs. Data collected by N.A. Mauro

Table 5. Levels of ATP (determined as μM ml^{-1} RBC and expressed relative to levels in human RBCs) in RBCs

Species	ATP
Glycera dibranchiata[a]	6.3
G. rouxi[b]	4.5
Noetia ponderosa[a]	5.7
Chimaera[c]	1.1
Elasmobranchs[c,d,e]	0.5 – 5.0
Teleosts[c,d,e]	0.3 – 7.6
Amphibians[c]	2.8 – 3.9
Reptiles[c,d]	0.7 –10.0
Birds[c,d]	0.4 – 4.1
Man[c]	1.0
Other mammals[c,f]	0.03– 2.0

[a] N.A. Mauro and R.E. Isaacks (unpubl. obs.)
[b] Weber and Heidemann (1976)
[c] Bartlett (1980)
[d] Isaacks and Harkness (1980)
[e] Johansen et al. (1978)
[f] Kim et al. (1984)

and mollusk RBCs (as well as glutamate) are high relative to those invertebrates (Table 4). These experiments might be regarded as somewhat misleading since the experimental temperature was 22 °C for the warm-blooded vertebrate RBCs as well as the annelid and mollusk RBCs. (Nonetheless the results suggest considerable thermal adaptation of the annelid and molluscan RBC.) However, when the measurements in Table 4 were repeated at 37 °C the rat RBC was still unable to oxidize glutamate and the value for the chicken was an order of magnitude more or lower than those for *Glycera* and *Noetia* at 22 °C (N.A. Mauro, unpubl. obs.).

Lastly, the levels of ATP in molluscan and annelid RBCs are fairly high, implicating the efficient production expected of the TCA cycle (Table 5).

4.3 Anaerobic Metabolism

Not only do invertebrate RBCs have at least an average capacity for aerobic metabolism, those investigated thus far lack the ability to metabolize in the absence of O_2. As reported earlier (Mangum and Carhart 1972; Mangum et al. 1975), annelid RBCs lyse almost immediately when the PO_2 is lowered to zero. Hemolysis occurs when either whole sera or washed cells are used; therefore, it is unlikely to be due to the depletion of an essential substrate which is obtained in vivo from an exogenous source. Hemolysis (25%) of air equilibrated *Glycera* RBCs also occurs in the presence of KCN (10^{-3} *M*; N.A. Mauro, unpubl. obs.). While less information is available for *Noetia*, its RBCs also lyse when held in the absence of O_2 for more than a few minutes (M.A. Freadman, pers. comm.). This strict O_2 dependence of osmotic fragility is conspicuously untrue of the RBCs of some 17 species of teleost and elasmobranch fishes which were held in the absence of O_2 for periods up to several hours with no detectable lysis (mentioned by Johansen et al. 1978; also C.P. Mangum, unpubl. obs.).

The contrast is somewhat surprising from an adaptive point of view, since aquatic annelids are far more likely to become hypoxic than the fishes studied earlier, many of which were air-breathers. It is also surprising in view of the ability of other annelid tissues to metabolize anaerobically with no loss of integrity. No doubt the avoidance of hemolysis is one of the reasons why coelomic fluid PO_2 in *Glycera* does not fall to zero during prolonged hypoxia even though lactate accumulates in it, indicating that other tissues have switched on anaerobic metabolism (Mangum 1977).

Sipunculid PBCs may differ. At low temperature and pH (10 °C, 6.65) these cells completely switch off aerobic metabolism at a PO_2 as high as 19 mmHg and they remain shut down for several hours with no hemolysis (Mangum and Kondon 1975). Anaerobic metabolism has not been demonstrated directly, however.

In summary, the two invertebrate RBCs studied so far have a well-developed metabolism in which the TCA cycle appears to predominate, and the available information indicates a poorly developed anaerobic metabolism.

4.4 Enzymes Not Associated with Intermediary Metabolism

Our knowledge of the activities of other enzymatic systems that we associate with RBCs is equally rudimentary. Manwell (1977) demonstrated the presence in sipunculid PBCs of superoxide dismutase, which inhibits autoxidation of the O_2 carrier, and NADH diaphorase, which reduces the fully oxidized form back to the functional valence state. To our knowledge no comparable information on invertebrate RBCs has been reported, and once again they may be different. In our experience the formation of metHr *in cellulo* does not occur, while the autoxidation of Hb in intact annelid (but not echuroid) RBCs and even in intact animals is common (C.P. Mangum, unpubl. obs.).

R.P. Henry (pers. comm.) has found a surprising diversity in the activity of carbonic anhydrase (CA), the enzyme that facilitates movements of CO_2 into and out of the RBC. Activity is very high in sipunculid PBCs, moderate in annelid RBCs, and extremely low in molluskan RBCs. The physiological meaning of this diversity is unclear. CO_2 transport and its effect on O_2 binding have not been investigated in these species, although CA activity in the blood of annelids that lack RBCs and instead have extracellular Hbs has been reported on several occasions (Wells 1973; Wells et al. 1980).

4.5 Why Do We Want to Know?

The interest in metabolism of invertebrate RBCs is fundamental. Not only are they believed to represent the primitive ancestral condition of the RBC, but they also differ in enough structural and functional features to indicate that in terms of their metabolic organization they are not merely carbon copies of nucleated vertebrate RBCs, much less mammalian RBCs. From an evolutionary point of view the somewhat heterogeneous class, the invertebrate RBC or PBC, is in fact the archetype and its vertebrate counterpart is the atypical derivative.

4.6 Why Don't We Want to Know?

It is our belief, however, that an understanding of metabolic mechanisms that control respiratory properties of the blood by means of a sensitive feedback regulation may not be a viable goal of the investigation of invertebrate RBC metabolism. A considerable body of information in the literature, much of it relatively early, indicates that purifying invertebrate Hbs or even extracting them from the RBC without stripping alters O_2 binding. In each case in which the evidence has been reexamined, however, negative conclusions have been reached (reviewed by Mangum et al. 1983; Mangum 1985). The available information suggests that organic PO_4 modulation of HbO_2 may have arisen within the vertebrates. The importance of this question would seem to mandate reexamination of the remaining exception as well as investigation of RBCs in which the possibility has not been tested.

To make the point one last time, at least some PBCs may differ. Not only does the O_2 binding of sipunculid and priapulid (but not brachiopod) PBCs differ from that of stripped Hrs (Manwell 1960; Mangum and Kondon 1975; Mangum 1985), but HrO_2 binding is also influenced by a variety of inorganic ions almost certainly found within the cell (de Waal and Wilkins 1976; Petrou et al. 1981). One of these is Cl^- which, in view of the high levels of carbonic anhydrase activity (R.P. Henry, pers. comm.), is highly likely to be a physiological variable. However, O_2 binding in vivo has not been shown to respond adaptively to physiological stimuli and, thus, we cannot call them "modulators". The importance of this question also mandates serious investigation.

References

Altman PL, Ditmer DS (1961) Blood and other body fluids. Fed Am Soc Exp Biol, Bethesda MD

Baba K (1940) The mechanisms of absorption and excretion in a solenogastre, *Epimenia verrucosa* (Nierstrasz), studied by means of injection methods. J Dept Agric Kyusyu Imperial Univ 6: 119–166

Bachand L, Leray C (1975) Erythrocyte metabolism in the yellow perch (*Perca flavescens* Mitchill). I. Glycolytic enzymes. Comp Biochem Physiol 50B:567–570

Bartlett GR (1980) Phosphate compounds in vertebrate red blood cells. Am Zool 20:103–114

Baumberger JP, Michaelis L (1931) The blood pigments of *Urechis caupo*. Biol Bull 61:417–421

Bliss DE, Skinner DM (1961) Tissue respiration in invertebrates. Am Mus Nat Hist, New York

Boilly B (1974) Ultrastructure dès hematies anuclèees de *Magelona papillicornis* F. Muller (annèlide polychète). J Microsc 19:47–58

Cohen WD, Nemhauser I (1980) Association of centrioles with the marginal band of a molluscan erythrocyte. J Cell Biol 86:286–291

Cowden RN (1966) A cytochemical study of the nucleated hemoglobin-containing erythrocytes of *Glycera americana*. Trans Am Microsc Soc 85:45–63

Dales RP (1964) The coelomocytes of the terebellid polychaete *Amphitrite johnstoni*. Q J Microsc Sci 105:263–279

Dawson A (1933) Supravital studies on the colored corpuscles of several marine invertebrates. Biol Bull 64:233–242

de Waal DJA, Wilkins RG (1976) Kinetics of the hemerythrin-oxygen interaction. J Biol Chem 251:2339–2343

Eddy FB (1977) Oxygen uptake by rainbow trout blood, *Salmo gairdneri*. J Fish Biol 10:87–90

Emie CC (1982) The biology of Phoronida. Adv Mar Biol 19:1–89

Fontaine AR, Hall BD (1981) The hemocyte of the holothurian *Eupentacta guinguesemita*: ultrastructure and maturation. Can J Zool 59:1884–1891

Fontaine AR, Lambert P (1973) The fine structure of the haemocyte of the holothurian *Cucumaria miniata* (Brandt). Can J Zool 51:323–332

Freadman MA, Mangum CP (1976) The function of hemoglobin in the arcid clam *Noetia ponderosa*. I. Oxygenation in vitro and in vivo. Comp Biochem Physiol 53A:173–179

Grinich NP, Terwilliger RC (1980) The quaternary structure of an unusual high-molecular weight intracellular hemoglobin from the bivalve mollusc *Barbatia reeveana*. Biochem J 189:1–8

Hetzel HR (1963) Studies on holothurian coelomocytes. I. A survey of coelomocyte types. Biol Bull 125:289–301

Hochachka PW, Hayes FR (1962) The effect of temperature acclimation on pathways of glucose metabolism in the trout. Can J Zool 40:261–270

Hoffmann RJ, Mangum CP (1970) The function of coelomic cell hemoglobin in the polychaete *Glycera dibranchiata*. Comp Biochem Physiol 36:211–228

Hopkins HS (1930) Age differences and the respiration in muscle tissues of mollusks. J Exp Zool 56:209–239

Isaacks RE, Harkness DR (1980) Erythrocyte organic phosphate and hemoglobin function in birds, reptiles and fishes. Am Zool 20:115–129

Johansen K, Mangum CP, Lykkeboe G (1978) Respiratory properties of the blood of Amazon fishes. Can J Zool 56:898–906

Kim HD, Zeidler RB, Sallis J, Nicol S, Isaacks RE (1984) Metabolic properties of low ATP erythrocytes of the monotremes. FEBS 167:83–87

Mangum CP (1977) The annelid hemoglobins: a dichotomy in structure and function. In: Reish DJ, Fauchald K (eds) Essays in memory of Dr. Olga Hartman, Allan Hancock Found., Spec. Publ., Univ So Calif Press, Los Angeles, p 407

Mangum CP (1985) Oxygen transport in the invertebrates. Am J Physiol 248:R505–R514

Mangum CP, Carhart JA (1972) Oxygen equilibrium of coelomic cell hemoglobin from the bloodworm *Glycera dibranchiata*. Comp Biochem Physiol 43A:949–957

Mangum CP, Kondon M (1975) The role of coelomic cell hemerythrin in the sipunculid worm *Phascolopsis gouldi*. Comp Biochem Physiol 50A:777–786

Mangum CP, Towle DW (1982) The *Nautilus* siphuncle as an ion pump. Pac Sci 36:273–282

Mangum CP, Woodin BL, Bonaventura C, Sullivan B, Bonaventura J (1975) The role of coelomic and vascular hemoglobins in the annelid family Terebellidae. Comp Biochem Physiol 51A: 281–294

Mangum CP, Terwilliger RC, Terwilliger NB, Hall R (1983) Oxygen binding of intact coelomic cells and extracted hemoglobin of the echiuran *Urechis caupo*. Comp Biochem Physiol 76A: 253–257

Manwell C (1960) Histological specificity of respiratory pigments. II. Oxygen transfer systems involving hemerythrins in sipunculid worms of different ecologies. Comp Biochem Physiol 1:277–285

Manwell C (1977) Superoxide dismutase and NADH diaphorase in haemerthrocytes of sipunculans. Comp Biochem Physiol 58B:331–338

Mauro NA, Mangum CP (1982) The role of the blood in the temperature dependence of oxidative metabolism in decapod crustaceans. I. Intraspecific responses to seasonal differences in temperature. J Exp Zool 219:179–188

Nemhauer I, Ornberg R, Cohen WD (1980) Marginal bands in blood cells of invertebrates. J Ultrastruct Res 70:308–317

Ochi O (1975a) The erythrocyte and its pigment in echiurans *Urechis unicinctus* and *Ikedosoma gogoshimense*. In: Rice ME, Todorovic M (eds) Proc Int Symp Biol Sipuncula Echiura, vol 2. US Natl Mus, Washington, p 197

Ochi O (1975b) An electron microscope study on the coelomic cells of some Japanese Sipuncula. In: Rice ME, Todorovic M (eds) Proc Int Symp Biol Sipuncula Echiura, vol 1. US Natl Mus, Washington, p 219

Ochi O (1977) X-ray microanalysis on the erythrocytes of sipunculids. Cell Struct Funct 2:51–54

Petrou AL, Armstrong FA, Sykes AG, Harrington PC, Wilkins RG (1981) Kinetics of the equilibrium of oxygen with monomeric and octameric hemerythrin from *Themiste zostericola*. Biochim Biophys Acta 670:370–384

Pritchard A, White FN (1981) Metabolism and oxygen transport in the innkeeper *Urechis caupo*. Physiol Zool 54:44–54

Schumacher HR, Seamonds B (1972) Fine structure of erythrocytes of the common bloodworm *Glycera dibranchiata*. Cytologia 37:359–363

Sean KE, Boilly B (1980) Aspects ultrastructuraux et cytochimiques des hématies nucléees de deux annélides polychètes *Notomastus latericeus* Sars et *Glycera convoluta* Keferstein. Can J Zool 58:589–597

Shaw LM, Seamonds B (1972) Characterization of chromatin prepared from the erythrocytes of the common bloodworm *Glycera dibranchiata*. Life Sci 11:259–266

Shafie SM, Vinogradov SN, Larson L, McCormick JJ (1976) RNA and protein synthesis in the nucleated erythrocytes of *Glycera dibranchiata*. Comp Biochem Physiol 53A:85–88

Stang-Voss C (1970) Zur Ultrastruktur der Blutzellen wirbelloser Tiere. II. Über die Blutzellen von *Golfingia gouldi* (Sipunculidae). Z Zellforsch 106:200–208

Storch V, Welsch U (1976) Elektromikroskopische und enzymhistochemische Untersuchungen über Lophophor und Tentakeln von *Lingula unguis* L. (Brachiopoda). Zool Jahrb Anat Bd 96:225–237

Terwilliger NB, Terwilliger RC, Schabtach E (1983) Two populations of hemerythrin-containing cells in the sipunculan *Themiste dyscritum*. Am Zool 24:1025

Vernet G (1979) Fine structure of the nemertean worm *Lineus lacteus* red blood corpuscles. Cytobios 24:43–46

Weber RE, Heidemann W (1976) The coelomic haemoglobin from the bloodworm *Glycera rouxii*. Molecular and oxygenation properties. Comp Biochem Physiol 57A:151 155

Weber RE, Fange R, Rasmussen K (1979) Respiratory significance of priapulid hemerythrin. Mar Biol Lett 1:87–97

Wells RMG (1973) Carbonic anhydrase activity in *Arenicola marina* (L.). Comp Biochem Physiol 46A:325–331

Wells RMG (1982) Respiratory characteristics of the blood pigments of three worms from an intertidal mudflat. N Z J Zool 9:243–248

Wells RMG, Dales RP (1975) Haemoglobin function in *Terebella lapidaria* L., an intertidal terebellid polychaete. J Mar Biol Assoc UK 55:211–220

Wells RMG, Jarvis PJ, Shumway SE (1980) Oxygen uptake, the circulatory system and hemoglobin function in the intertidial polychaete *Terebella haplochaeta* (Ehlers). J Exp Mar Biol Ecol 46:255–277

Sugar Uptake by Red Blood Cells

R.L. INGERMANN[1,3], J.M. BISSONNETTE[1], and R.E. HALL[2]

1 Introduction

Monosaccharides are an important metabolite in many cells. However, owing to their hydrophilic nature, these sugars tend to enter the cell slowly by simple diffusion through the hydrophobic region of the plasma membrane. Consequently, transport mechanisms have evolved which facilitate cellular monosaccharide uptake. Microorganisms have transport system which (1) couple sugar movement to proton translocation (reviewed by Eddy 1982); (2) involve translocation with concomitant phosphorylation (reviewed by Postma and Roseman 1976); and/or (3) transfer sugar by a carrier-mediated, equilizing transport mechanism (Cirillo 1968, 1981). Among phylogenetically higher organisms, the vast majority of work on sugar transport has focused on mammals and has shown that two principle mechanisms of monosaccharide transport exist. One is a concentrative mechanism found in the epithelial lining of the kidney and gut which involves the cotransport of sodium and sugar (Crane 1965; Kinne 1976). The other is found in most other tissues and is an equilizing, sodium-independent monosaccharide transport system (summarized by Stein 1967).

Among the mammals, the principle model system for the study of sodium-independent D-glucose transport is the red blood cell (RBC). D-glucose enters the human RBC very rapidly, primarily via a stereospecific transporter at maximal rates estimated to exceed utilization by over 12,000-fold (Jacquez 1984). In marked contrast, D-glucose uptake by nonprimate adult RBCs is much slower (with the exception of uptake into blood cells from a porpoise and belukha whale, Andreen-Svedberg 1933; D'Angelo 1982). This may be due to disadvantages associated with high concentrations of monosaccharide within the RBC, involving the nonenzymatic glycosylation of cellular proteins (including hemoglobin and cytoskeletal elements) and possible promotion of hemoglobin oxidation (Miller et al. 1980; Higgins et al. 1982; McMillian and Brooks 1982; Thornallêy et al. 1984). In spite of these potential disadvantages, RBCs from most mammalian fetuses examined have very rapid rates of D-glucose uptake which are generally much greater than that of the corresponding adult cells (Widdas 1955; reviewed by Jacquez 1984). This adult-preadult difference in membrane transport is

1 Department of Obstetrics and Gynecology, Oregon Health Sciences University, Portland, OR 97201, USA
2 Department of Biochemistry, Oregon Health Sciences University, Portland, OR 97201, USA
3 Present address: Department of Biological Sciences, University of Idaho, Moscow, ID 83843, USA

Circulation, Respiration, and Metabolism
(ed. by R. Gilles)

principally due to the presence of greater numbers of transporters in the plasma membranes of the fetal cells (Mooney and Young 1978; Kondo and Beutler 1980). An adult-preadult difference in monosaccharide uptake by RBCs has also been found in some anurans (Flores and Frieden 1972).

Two principal hypotheses, which are not mutually exclusive, may account for the occurrence of rapid monosaccharide uptake by preadult RBCs. Rapid uptake may be a consequence of the maturational state of the cell or the organism. The maturation of the mammalian reticulocyte involves energy-dependent degradation of cell organelles and proteins (Müller et al. 1980; Boches and Goldberg 1982). It is conceivable that such degradation involves proteolysis of the glucose transporter such that fewer transporters remain intact and functional as the cell matures. Consequently, the net rate of glucose uptake would decrease with cell maturation. This appears to be the explanation for the maturation-associated changes in amino acid transport in sheep RBCs (Benderoff et al. 1978). Another hypothesis attributes a physiological importance to rapid glucose uptake by RBCs of the mammalian fetus. In addition to the blood plasma, the RBC represents a portion of the blood volume which can effectively carry nutrients from the placenta: if RBC membrane transport is very rapid, and the rate of transport into these cells exceeds the rate of cellular utilization. Thus, the loading of fetal whole blood with glucose may be facilitated by the capability of fetal RBCs to take up glucose rapidly.

To test predictions made by these hypotheses, and to explore whether a mammalian-like transport system is found in lower organisms, we have examined RBC monosaccharide uptake by carrier-mediated and simple diffusion in four organisms taken to represent two ontogenetic states and a wide phylogenetic range: adult and embryonic chicken *(Gallus domesticus)*, adult and fetal sea perch *(Embiotoca lateralis)*, and mature hagfish *(Eptatretus stouti)*, and sipunculan worm *(Themiste dyscrita)*.

2 Methodology

To assess the actual rate of monosaccharide transport under physiological conditions, it is necessary to know the kinetic constants of the transfer processes as well as the concentration gradient across the plasma membrane. In our examination of monosaccharide transport into RBCs from these organisms, we have made the assumption that uptake at a low external sugar concentration (at an initial intracellular concentration of zero) would allow assessment of relative transfer rates. Consequently, we have examined the uptake of monosaccharides at a concentration gradient of 0.1 mM using L-glucose and 3-O-methyl-D-glucose (3OMG). 3OMG is transported by the sodium-independent D-glucose transporter of higher organisms yet is not metabolized (Hillman et al. 1959; LeFevre 1961; Czech et al. 1974). 3OMG is taken up by carrier-mediated, as well as simple diffusion, and thus represents total uptake. L-glucose is taken up only by simple diffusion. The difference in uptake between 3OMG and L-glucose, therefore, represents stereospecific, carrier-mediated transport.

3 Uptake into Adult and Preadult RBCs

Uptake studies were conducted at 37 °C on adult and embryonic RBCs of the chicken (Ingermann et al. 1985b). Although carrier-mediated monosaccharide uptake could be demonstrated (Fig. 1), there was no significant difference in transport between cells from the adult and day 10, 14, and 18 embryos. (Hatch occurs on day 20–21.) There was, however, a three- to fourfold greater rate of uptake by simple diffusion in embryonic RBCs. Nonetheless, total uptake appeared to be relatively slow, reaching half equilibrium in approx. 2 to 3 h.

To assess transport into RBCs from a phylogenetically lower organism and possible maternal-fetal differences, transport into RBCs of the sea perch, *E. lateralis*, was examined. Studies were conducted with cells from adults and midgestation fetuses collected in early spring. Monosaccharide uptake was measured at 10 °C and is shown in Fig. 2. Both fetal and adult RBCs took up 3OMG by a stereospecific mechanism. Uptake by simple diffusion also occurred, however at a much slower rate. Despite a slightly greater rate of simple diffusion into fetal cells, total uptake did not appear to differ between fetal and adult cells. Further, uptake under these experimental conditions was slow, reaching half equilibrium in 6 to 7 h.

The results of studies with adult and embryonic chicken, and adult and fetal sea perch RBCs, thus, appear to be inconsistent with the hypothesis that rapid monosaccharide transport is a consequence of the maturational state of the cell or the organism. Additionally, the sea perch data appear inconsistent with the interpretation of mammalian data which attributes an importance to increasing the fetal blood volume accessible for glucose loading at the maternal-fetal exchange surfaces.

Despite the lack of a maternal-fetal difference in transport rates, the *E. lateralis* data demonstrate that stereospecific uptake of a monosaccharide takes place into the RBCs of a fish. The trout RBC apparently does not have a stereospecific mechanism for hexose transport (Bolis et al. 1971, Bolis and Luly 1972; Bolis 1973). Since trout are relatively primitive and *E. lateralis* are relatively advanced teleosts (Bond 1979),

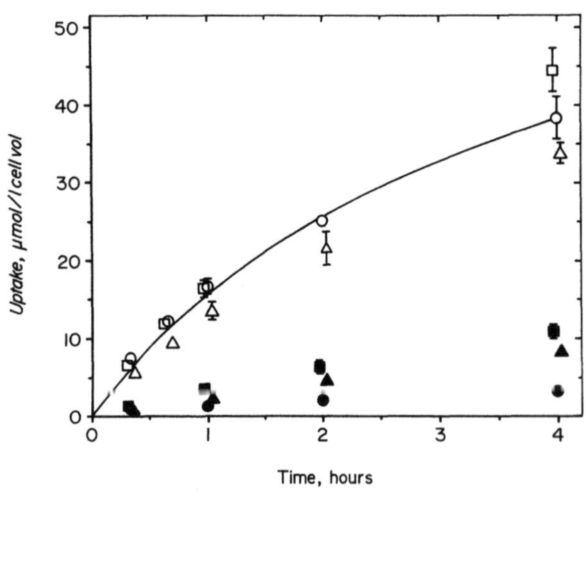

Fig. 1. Carrier-mediated (*open symbols*) and simple diffusion (*filled symbols*) transfer of 3 OMG into RBCs of adult (○), day 10 (□), and day 18 (△) embryonic chickens at 37 °C and at a sugar gradient of 0.1 m*M*. There was no significant difference in carrier-mediated transport between adult and day 10, 14, and 18 embryonic RBCs as assessed by rates of uptake calculated between 20 and 60 min (Ingermann et al. 1985b). Assuming that 75% of the cell volume can equilibrate with sugar, half equilibrium for total uptake occurred in 2 to 3 h. Data represent means ± SD, n = 3–5. *Curve* representing carrier-mediated uptake is fit to data by eye (Figs. 2 and 3, was well)

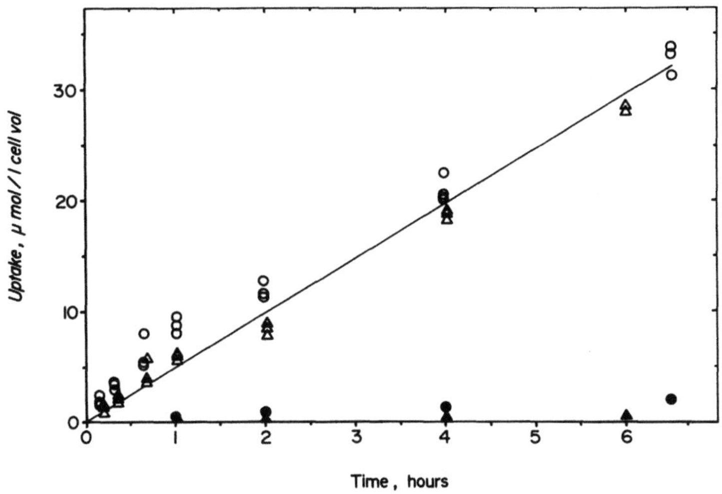

Fig. 2. 3OMG uptake by carrier-mediated (*open symbols*) and simple diffusion transfer (*filled symbols*) for adult (△) and midgestation fetal (○) *Embiotoca lateralis* RBCs at 10 °C. Animals were obtained, identified, and staged as described (Ingermann and Terwilliger 1981). RBCs were collected and washed three times, and experiments conducted in the physiological saline of Cala (1977) (modified solution A: without sugar). Uptake experiments were conducted as described (Ingermann et al. 1984). Individual values are plotted for uptake by carrier-mediated uptake; values for simple diffusion transfer are three individual measurements superimposed upon one another

it appeared that the occurrence of a stereospecific monosaccharide transport system in a circulating blood cell may be a recent evolutionary development.

4 Uptake into Hagfish and Worm RBCs

We examined 3OMG and L-glucose uptake by RBCs from the mature hagfish, *E. stouti*, a representative of the lowest form of true fish and vertebrate, and from a mature invertebrate, the sipunculan worm, *T. dyscrita* to further examine the phylogeny of the RBC monosaccharide transporter (Ingermann et al. 1984, 1985a). Uptake studies into these cells were also conducted with a 0.1 m*M* concentration gradient of 3OMG or L-glucose at 10 °C. Uptake into RBCs from each of these organisms also occurred by a stereospecific mechanism (Fig. 3). However, unlike the transfer in chicken and *E. lateralis* cells which reached half equilibrium in hours, uptake into *E. stouti* and *T. dyscrita* RBCs was much more rapid reaching half equilibrium in about 10 and 20 s, respectively. As observed in chicken and *E. lateralis* RBCs, uptake into *E. stouti* and *T. dyscrita* cells by simple diffusion was very slow: in 4 h it accounted for about 15% and 5% equilibrium, respectively. In spite of our results for *T. dyscrita*, relatively rapid stereospecific uptake of 3OMG does not appear to be universal among invertebrate blood cells. Preliminary and unpublished data from our laboratory has shown that several hours are necessary to reach half equilibrium in RBCs of the polychaete worm, *Pista pacifica*.

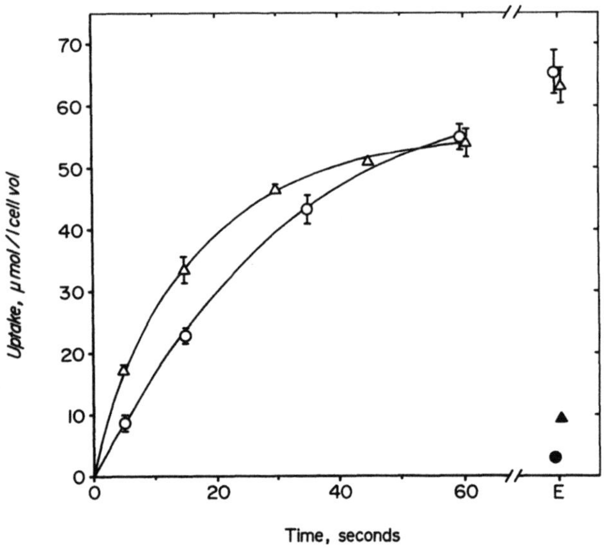

Fig. 3. Uptake of 3OMG by RBCs of *Eptatretus stouti* (△) and *Themiste dyscrita* (○) by stereospecific (*open symbols*) and simple diffusion (*filled symbols*) transfer at 10 °C (from Ingermann et al. 1984, 1985a). *E* is the mean of 1, 2, and 4 h for carrier-mediated transport; it is 4 h for uptake by simple diffusion. Data represent means ± SD, n = 3–8

5 Characterization of the Hagfish and Sipunculan Monosaccharide Transport Systems

In addition to D-glucose and 3OMG, the stereospecific monosaccharide transporter of the human RBC also interacts with and translocates a variety of sugars (LeFevre and Marshall 1958). The human transporter, thus, shows broad specificity for monosaccharide structure. To examine *E. stouti* and *T. dyscrita* blood cell transporters for such structural specificity, a variety of sugars were examined for their ability to compete with 3OMG for uptake (Ingermann et al. 1984, 1985a). Unlabeled 3OMG is effective in reducing labeled 3OMG uptake; while L-glucose, which enters these blood cells very slowly, has very little if any competitive effect on 3OMG transport. The relative degree of interaction between sugar analogs and transporter can, thus, be established and is shown in Table 1. As with the human RBC transporter, the sugar transports system of *E. stouti* and *T. dyscrita* show broad specificities. Further, the specificity patterns are very similar to the sodium-independent monosaccharide transporter of the human RBC.

Sugar uptake into *E. stouti* and *T. dyscrita* blood cells is nonconcentrative, and since α-methyl-D-glucoside has little if any interaction with these transporters, monosaccharide transport appears unlike the sodium-dependent transport systems of mammalian kidney and gut epithelia. We have also found direct evidence that the *T. dyscrita* transport system is sodium-independent (Ingermann et al. 1985a).

Monosaccharide transport is also characterized by the inhibitory effects of various chemical agents, two of which are phlorizin and phloretin. Sodium-dependent transport systems of mammals tend to be very sensitive to phlorizin and less so to phloretin. Sodium-independent transport systems, such as that of the human RBC are much more sensitive to phloretin than phlorizin. Similarly, the monosaccharide transport systems of *E. stouti* and *T. dyscrita* are more sensitive to phloretin than phlorizin (Table 2).

Table 1. Interaction of D-glucose analogs with the sugar transporter of the red blood cell

Human RBC[a]	*Eptatretus stouti*[b]	*Themiste dyscrita*[c]
In decreasing order of interaction with the transporter:		
2-Deoxy-D-glucose	3-O-methyl-D-glucose	3-O-methyl-D-glucose
α-D-glucose	α-D-glucose	2-Deoxy-D-glucose
β-D-glucose	2-Deoxy-D-glucose	α-D-glucose
3-O-methyl-D-glucose	β-D-glucose	β-D-glucose
	D-galactose	
D-mannose	D-mannose	D-galactose
D-galactose	D-xylose	D-mannose
D-xylose		
		D-xylose
		α-Methyl-D-glucoside
		D-fructose
Essentially unreactive with transporter:		
α-Methyl-D-glucoside	D-fructose	D-sorbitol
D-sorbitol	L-glucose	L-glucose
D-fructose	α-Methyl-D-glucoside	Sucrose
L-glucose	D-sorbitol	
Sucrose		

[a] Kozawa (1914), LeFevre and Marshall (1958), Barnett et al. (1973), Carter-Su et al. (1982)
[b] Ingermann et al. (1984)
[c] Ingermann et al. (1985a)

Table 2. Effect of inhibitors on sugar transport into the red blood cell

Human RBC[a] K_I at 37 °C		Conc.	10 °C *Eptatretus stouti* % Control Uptake[b]	10 °C *Themiste dyscrita* % Control Uptake[b]
		5 μM	72 ± 5	98 ± 4
6 μM	*Phloretin*	50	8 ± 1	33 ± 3
		500	1	5
		5 μM	92 ± 10	117 ± 5
500 μM	*Phlorizin*	50	59 ± 4	98 ± 6
		500	8	26 ± 1

[a] LeFevre (1961)
[b] 3-O-methyl-D-glucose uptake, mean ± SD, n = 4 (Ingermann et al. 1984, 1985a)

Based on estimated concentrations necessary to half inhibit transport, the inhibition constant of phloretin for both *E. stouti* and *T. dyscrita* at 10 °C are within a factor of about six relative to that of the human transporter at 37 °C. The influence of phlorizin on *T. dyscrita* transport is only about twofold different than that of the human transporter. In contrast, phlorizin appears to be more effective in inhibiting *E. stouti*, relative to human RBC transport by over one order of magnitude.

A very valuable tool in the study of sugar transport is the fungal metabolite, cytochalasin B. Cytochalasin B has little effect on sodium-dependent sugar transport (Hopfer

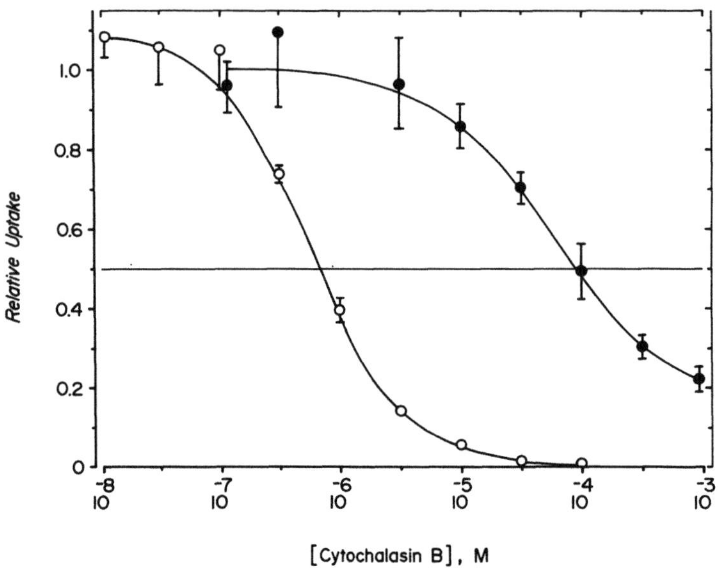

Fig. 4. Influence of cytochalasin B on 3 OMG uptake by *Eptatretus stouti* (○) and *Themiste dyscrita* (●) RBCs (from Ingermann et al. 1984, 1985a). Data represent mean ± SD, n = 4 (except for *E. stouti* at [Cytochalasin B] = 10^{-5} *M*, n = 3)

et al. 1976), however, it is a very potent inhibitor of human sodium-independent monosaccharide transport and half inhibition occurs at a concentration of less than 1 μM (Taverna and Langdon 1973; Jung and Rampal 1977; Golden and Rhoden 1978). Cytochalasin B is a potent inhibitor of carrier-mediated 3OMG uptake in *E. stouti* cells with half inhibition also occurring at less than 1 μM (Fig. 4). In contrast, uptake by *T. dyscrita* cells is much less sensitive to cytochalasin B, with half inhibition occurring at approx. 0.1 m*M* (Fig. 4). The apparent high inhibition constant of the *T. dyscrita* transporter is not without precedent, as transport across the basolateral plasma membrane of rat intestinal epithelial cells (Hopfer et al. 1976), and transport out of the squid axon (Baker and Carruthers 1981) show similar values for inhibition by cytochalasin B.

6 Concluding Remarks

6.1 Physiology

An overview of sugar transport into RBCs representing a broad phylogenetic range of organisms reveals a marked difference in transport rates when examined at an initial concentration gradient of 0.1 m*M*. There was no difference in stereospecific transfer rates between embryonic and adult chicken cells, nor between fetal and adult *E. lateralis* cells. In both organisms total uptake reached half equilibrium in several hours, suggesting that the RBC volume is not an important component of the total blood volume which delivers sugar to the other tissues of the whole organism. The hematocrit of the midgestation *E. lateralis* fetus is 46% (Ingermann and Terwilliger 1982). The blood cell volume, thus, represents a potential increase of 60%–65% in the sugar-carrying volume of the whole blood. Conceivably, sugar metabolism may be relatively unimportant to the fetal organism or the detrimental aspects of high intraerythrocytic glucose may outweigh the advantages associated with increasing the glucose available blood volume. These results appear inconsistent with the mammalian findings. Certainly, more organisms need to be examined to establish whether the lack of transport differences between adult and preadult RBCs seen in the chicken and *E. lateralis* are characteristic of RBC membrane transport among the lower vertebrates.

In contrast to uptake attaining half equilibrium in hours, uptake by RBCs of *E. stouti* and *T. dyscrita* reached half equilibrium in about 20 s under the experimental condition used. The relatively rapid uptake of sugar into *E. stouti* and *T. dyscrita* cells suggests a physiological advantage for rapid transfer independent of ontogenetic or fetal-maternal relationships. Such an advantage may be related to the life strategies of these organisms. The hagfish is exposed to hypoxia while burrowing in muddy sediment and during feeding within the carcass of a dead or moribund fish. The tissues of the hagfish appear to be very resistant to anoxia and apparently rely on anaerobic carbohydrate metabolism (Hansen and Sidell 1983). The sipunculan worm is also tolerant of hypoxia. Ricketts and Calvin (1968) state "[sipunculan worms] can apparently withstand living for a week or so in seawater with no detectable oxygen, and even survive several days in something like mineral oil". The ability of RBCs to take up sugar rapidly may, therefore, insure that this cell can generate cellular energy under anaerobic conditions. Rapid uptake and subsequent release may also allow RBCs to function in delivering sugar to other tissues, thereby supporting an organism's tolerance to hypoxia. This conjecture may explain the apparent ability of RBCs from the diving mammals, a porpoise (a sick individual of unidentified species, Andreen-Svedberg 1933) and the belukha whale (D'Angelo 1982), to take up sugar rapidly. Finally, a sugar transport rate which exceeds cellular metabolic needs may allow the RBC to function as an energy storage tissue, as well as an oxygen delivery tissue. In fact, glycogen rosettes or granules similar in appearance to those of mammalian hepatic cells have been observed in the RBCs of *T. dyscrita* (Terwilliger et al. 1985).

The RBC adenosine triphosphate (ATP) concentrations in adult and day 10 embryonic chickens are different: embryonic cell levels are about ten times the concentration of the adult cells (Bartlett and Borgese 1976). Furthermore, RBCs of fetal *E. lateralis* contain about 25%–30% of the adult intraerythrocytic ATP molar concentration (Ingermann and Terwilliger 1981, 1982). The adult-preadult difference in intraerythrocytic ATP concentrations of chicken and *E. lateralis*, therefore, appear to be independent of any ontogenetic difference in sugar transport. In addition, the ATP concentration in hagfish (Bartlett 1982) and adult chicken RBCs are comparable despite the apparently great difference in their rates of sugar uptake. Intraerythrocytic ATP is, therefore, probably not closely related to rates of membrane carrier-mediated sugar transport in these organisms.

6.2 Biochemistry

At equilibrium, the sugar levels within the *E. stouti* and *T. dyscrita* RBCs do not exceed the sugar concentration of the extracellular medium (Fig. 3). The results, thus, suggest that carrier-mediated sugar transport into these cells occurs by a nonconcentrative, equilizing mechanism. Further, uptake by *T. dyscrita* cells is independent of an experimentally induced sodium gradient. The nonconcentrative sugar transfer in these cells is, thus, similar to transport into mammalian RBCs.

The monosaccharide transport systems of the blood cells of *E. stouti* and *T. dyscrita* show broad specificities with respect to the structures of saccharides that interact with the transport system. The pattern of this specificity is similar among these organisms and the human sodium-independent transporter. These results suggest that there is a similarity of functional groups on the extracellular face of the transport system which recognize and interact with the sugar.

Sugar transport into human, *E. stouti*, and *T. dyscrita* RBCs is inhibited by phloretin, and the half inhibition constants for transport into these cells differ by less than one order of magnitude. Since phloretin is thought to inhibit sugar transport of the human RBC by binding to the transporter at the external sugar binding site (Krupka and Deves 1981), the data suggest that these transporters are structurally similar at their extracellular surface. The sensitivity of transport inhibition by phlorizin in human and *T. dyscrita* RBCs differs by a factor of about two. The transport system of *E. stouti* RBCs is more sensitive to phlorizin than that of human cells by about one order of magnitude. The molecular mechanism by which phlorizin inhibits sodium-independent glucose transport is not known, the nature of structural similarities on this basis is, therefore, unclear. Nonetheless, the general similarities of the effects of phloretin and phlorizin on sugar transport in cells of phylogenetically diverse organisms suggest structural similarities among these transporter systems. Cytochalasin B is a very potent inhibitor of human sodium-independent sugar transport. Transport into *E. stouti* RBCs is similarly sensitive to cytochalasin B, while transport into *T. dyscrita* RBCs is less sensitive by two orders of magnitude. Cytochalasin B appears to inhibit the sugar transport of the human RBC by competitively binding to the D-glucose binding site on the cytoplasmic face of the transporter (Griffin et al. 1982). It seems likely that the sugar binding sites on the cytoplasmic side of the transporters of human and *E. stouti* RBCs are structurally very similar. The *T. dyscrita* data support the findings of Hopfer et al. (1976) and Baker and Carruthers (1981) which suggest that the high affinity cytochalasin B binding site is not an obligatory part of the sodium-independent sugar transporter.

Yeast has an equalizing, carrier-mediated sugar transport system which shows some properties of the mammalian sodium-independent transporter (Cirillo 1968). However, the yeast transporter differs from the mammalian transporter in its ability to interact with fructose (Cirillo 1968, 1981). It also differs in its insensitivity to phloretin, phlorizin, and cytochalasin B (V.P. Cirillo, personal communication). Therefore, whether mammalian-like sodium-independent sugar transporters exist in other simple eukaryotes remains to be determined. However, Carruthers (1983) has shown sodium-independent sugar uptake by invertebrate muscle fibers has properties similar to those of mammalian muscle as well as RBCs. Further, stereospecific sugar uptake into the RBCs of phylogenetically diverse organisms: man, hagfish, and a sipunculid worm, shows marked similarities with respect to (1) equalizing, rather than concentrative uptake; (2) competition among D-glucose analogs; and (3) similarities with respect to the effects of inhibitors. These results suggest an evolutionary conservation of the structural and functional aspects of an important cellular process. Further research is required to establish the extent of this phylogenetic conservation.

Acknowledgement. This work was supported in part by a grant from the Oregon Affiliate of the American Heart Association and by grant 1 R23 HD18108-01 from the United States Public Health Service.

References

Andreen-Svedberg A (1933) On the distribution of sugar between plasma and corpuscles in animal and human blood. Skand Arch f. Physiol 66:113–190

Baker PF, Carruthers A (1981) Sugar transport in giant axons of *Loligo*. J Physiol 316:481–502

Barnett JEG, Holman GD, Munday KA (1973) Structural requirements for binding to the sugar-transport system of the human erythrocyte. Biochem J 131:211–221

Bartlett GR (1982) Phosphates in red cells of a hagfish and a lamprey. Comp Biochem Physiol 73A:141–145

Bartlett GR, Borgese TA (1976) Phosphate compounds in red cells of the chicken and duck embryo and hatchling. Comp Biochem Physiol 55A:207–210

Benderoff S, Blostein R, Johnstone RM (1978) Changes in amino acid transport during red cell maturation. Membr Biochem 1:89–106

Boches FS, Goldberg AL (1982) Role for the adenosine triphosphate-dependent proteolytic pathway in reticulocyte maturation. Science 215:978–980

Bolis L (1973) Comparative transport of sugars across red blood cells. In: Bolis L, Schmidt-Nielsen K, Maddrell SHP (eds) Comparative physiology. North Holland, Holland, p 583

Bolis L, Luly P (1972) Monosaccharide permeability in brown trout *Salmo trutta* L. erythrocytes. In: Bolis L, Keynes RD, Wilbrandt W (eds) Role of membranes in secretory processes. North Holland, Holland, p 215

Bolis L, Luly P, Baroncelli V (1971) D(+)-Glucose permeability in brown trout *Salmo trutta* L. erythrocytes. J Fish Biol 3:273–275

Bond CE (1979) Biology of fishes. WB Saunders, Philadelphia

Cala PM (1977) Volume regulation by flounder red blood cells: the role of the membrane potential. J Exp Zool 199:339–344

Carruthers A (1983) Sugar transport in giant barnacle muscle fibers. J Physiol 336:377–396

Carter-Su C, Pessin JE, Mora R, Gitomer W, Czech MP (1982) Photoaffinity labeling of the human erythrocyte D-glucose transporter. J Biol Chem 257:5419–5425

Cirillo VP (1968) Relationship between sugar structure and competition for the sugar transport system in bakers' yeast. J Bacteriol 95:603–611

Cirillo VP (1981) Unresolved questions on the mechanism of glucose transport in baker's yeast. In: Stewart GG, Russell I (eds) Current developments in yeast research. Pergamon, New York, p 299

Crane RK (1965) Na^+-dependent transport in the intestine and other animal tissues. Fed Proc 24: 1000–1006

Czech MP, Lawrence JC Jr, Lynn WS (1974) Hexose transport in isolated brown fat cells, a model system for investigating insulin action on membrane transport. J Biol Chem 249:5421–5427

D'Angelo G (1982) Evidence for an erythrocyte glucose transport system in the belukha whale, *Delphinapterus leucas*. Cetology 42:1–9

Eddy AA (1982) Mechanisms of solute transport in selected eukaryotic micro-organisms. Adv Microb Physiol 23:1–78

Flores G, Frieden E (1972) Hemolytic effect of phenylhydrazine during amphibian metamorphosis. Dev Biol 27:406–418

Golden SM, Rhoden V (1978) Reconstitution and "transport specificity fraction" of the human erythrocyte glucose transport system; a new approach for identification and isolation of membrane proteins. J Biol Chem 253:2575–2583

Griffin JF, Rampal AL, Jung CY (1982) Inhibition of glucose transport in human erythrocytes by cytochalasins: a model based on diffraction studies. Proc Natl Acad Sci USA 79:3759–3763

Hansen CA, Sidell BD (1983) Atlantic hagfish cardiac muscle: metabolic basis of tolerance to anoxia. Am J Physiol 244:R356–R362

Higgins PJ, Garlick RL, Bunn HF (1982) Glycosylated hemoglobin in human and animal red cells, role of glucose premeability. Diabetes 31:743–748

Hillman RS, Landau BR, Ashmore J (1959) Structural specificity of hexose penetration of rabbit erythrocytes. Am J Physiol 196:1277–1281

Hopfer U, Sigrist-Nelson K, Amman E, Murer H (1976) Differences in neutral amino acid and glucose transport between brush border and basolateral plasma membrane of intestinal epithelial cells. J Cell Physiol 89:805–810

Ingermann RL, Terwilliger RC (1981) Intraerythrocytic organic phosphates of fetal and adult seaperch *(Embiotoca lateralis)*: their role in maternal-fetal oxygen transport. J Comp Physiol 144: 253–259

Ingermann RL, Terwilliger RC (1982) Blood parameters and facilitation of maternal-fetal oxygen transfer in a viviparous fish *(Embiotoca lateralis)*. Comp Biochem Physiol 73A:497–501

Ingermann RL, Hall RE, Bissonnette JM, Terwilliger RC (1984) Monosaccharide transport into erythrocytes of the Pacific hagfish, *Eptatretus stouti*. Mol Physiol 6:311–320

Ingermann RL, Hall RE, Bissonnette JM, Terwilliger RC (1985a) Monosaccharide transport into hemocytes of a sipunculan worm *(Themiste dyscrita)*. Am J Physiol 249:R139–R144

Ingermann RL, Stock MK, Metcalfe J, Bissonnette JM (1985b) Monosaccharide uptake by erythrocytes of the embryonic and adult chicken. Comp Biochem Physiol 80A:369–372

Jacquez JA (1984) Red blood cell as glucose carrier: significance for placental and cerebral glucose transfer. Am J Physiol 246:R289–R298

Jung CY, Rampal AL (1977) Cytochalasin B binding sites and glucose transport carrier in human erythrocyte ghosts. J Biol Chem 252:5456–5463

Kinne R (1976) Properties of the glucose transport system in the renal brush border membrane. Curr Top Membr Transp 8:209–267

Kondo T, Beutler E (1980) Developmental changes in glucose transport of guinea pig erythrocytes. J Clin Invest 65:1–4

Kozawa S (1914) Beiträge zum arteigenen Verhalten der roten Blutkörperchen. III. Artdifferenzen in der Durchlässigkeit der roten Blutkörperchen. Biochem Z 60:231–256

Krupka RM, Deves R (1981) An experimental test for cyclic versus linear transport models. The mechanism of glucose and choline transport in erythrocytes. J Biol Chem 256:5410–5416

LeFevre PG (1961) Sugar transport in the red blood cells: structure-activity relationships in substrates and antagonists. Pharmacol Rev 13:39–70

LeFevre PG, Marshall JK (1958) Conformational specificity in a biological sugar transport system. Am J Physiol 194:333–337

McMillan DE, Brooks SM (1982) Erythrocyte spectrin glucosylation in diabetes. Diabetes 31 (suppl 3):64–69

Miller JA, Gravallese E, Bunn HF (1980) Nonenzymatic glycosylation of erythrocyte membrane proteins, relevance to diabetes. J Clin Invest 65:896–901

Mooney NA, Young JD (1978) Nucleoside and glucose transport in erythrocytes from new-born lambs. J Physiol 284:229–239

Müller M, Dubiel W, Rathmann J, Rapoport S (1980) Determination and characteristics of energy-dependent proteolysis in rabbit reticulocytes. Eur J Biochem 109:405–410

Postma PW, Roseman S (1976) The bacterial phosphoenolpyruvate: sugar phosphotransferase system. Biochim Biophys Acta 457:213–257

Ricketts EF, Calvin J (1968) Between Pacific tides, 4th edn. Stanford University, Stanford, California

Stein WD (1967) The movement of molecules across cell membranes. Academic, New York

Taverna RD, Langdon RG (1973) Reversible association of cytochalasin B with the human erythrocyte membrane; inhibition of glucose transport and the stoichiometry of cytochalasin B binding. Biochim Biophys Acta 323:207–219

Terwilliger NB, Terwilliger RC, Schabtach E (1985) Intracellular respiratory proteins of Sipuncula, Echiura, and Annelida. In: Cohen WD (ed) Blood cells of marine invertebrates, experimental systems in cell biology and comparative physiology. Alan R Liss, New York, p 193

Thornalley PJ, Wolff SP, Crabbe MJC, Stern A (1984) The oxidation of oxyhaemoglobin by glyceraldehyde and other simple monosaccharides. Biochem J 217:615–622

Widdas WF (1955) Hexose permeability of foetal erythrocytes. J Physiol 127:318–327

The Relationship Between Erythrocyte Phosphate Metabolism, Carbon Dioxide, and pH on Blood Oxygen Affinity in Birds

R.E. ISAACKS[1]

1 Introduction

In recent years, we have become increasingly aware of the remarkable diversity in the different organic phosphate compounds in erythrocytes, many of which have been shown to modulate hemoglobin oxygenation. For example: 2,3-bisphosphoglycerate (2,3-P_2-glycerate) previously considered a characteristic only of mammalian erythrocytes, has been observed as a major constituent of the red cells of embryos of birds and reptiles and in the red cells of the armored catfish (Borgese and Lampert 1975; Isaacks and Harkness 1975, 1980); inositol tetrakisphosphate (inositol-P_4) was found to be the major organic phosphate in erythrocytes of the mature ostrich (Isaacks and Harkness 1980); inositol pentakisphosphate (inositol-P_5) previously considered a characteristic of avian erythrocytes has been observed in erythrocytes of several species of fishes and sea turtles (Bartlett 1980; Borgese and Nagel 1978; Isaacks and Harkness 1980; Rapoport and Guest 1941); inositol bisphosphate (inositol-P_2) has been found in erythrocytes of the South American *(Lepidosiren paradoxa)* and African *(Protopterus aethiopicus)* lungfish (Bartlett 1980; Isaacks and Harkness 1980), but its role as a modifier of hemoglobin function has not been verified; and in some of our very recent studies the red blood cells of monotremes, the egg-laying mammals (echidna and duckbill platypus), were found nearly devoid of adenosine triphosphate (ATP) (Kim et al. 1981; Isaacks et al. 1984), only traces were present (0.03 and 0.06 m*M*, respectively).

The red blood cell in mammals possesses an active glycolytic system and a relationship exists between metabolism and function of the red cell in oxygen transport. Specifically, the relationship in red cells of most mammals involves binding of 2,3-P_2-glycerate, a glycolytic intermediate, with deoxyhemoglobin, modulating hemoglobin function and facilitating oxygen delivery in the capillary beds within the tissues (Benesch and Benesch 1967; Chanutin and Curnish 1967). In addition, the allosteric properties of hemoglobins are such that increases in $[H^+]$, CO_2, and temperature, features present in the tissue capillary beds also enhance the release of oxygen.

The in vivo levels of 2,3-P_2-glycerate in human erythrocytes are known to increase above normal during hypoxia from altitude or from pathological conditions, such as anemia and cardiac or pulmonary insufficiency, but presently no satisfactory way is

1 Research Laboratories, Veterans Administration Medical Center and Department of Medicine, University of Miami, Miami, FL 33125, USA

Circulation, Respiration, and Metabolism
(ed. by R. Gilles)

available to manipulate the in vivo concentration of this compound. In human erythrocytes, the levels of 2,3-P_2-glycerate and ATP, which also influences hemoglobin binding, can be depleted after in vitro incubation at 37 °C for 12 to 24 h (Lian et al. 1971); red cells incubated with glycolate also lose 2,3-P_2-glycerate rapidly without changes in ATP concentrations (Rose 1976). The concentration of 2,3-P_2-glycerate can be enriched several-fold in human red blood cells by incubating with inosine, pyruvate, and inorganic phosphate (Lian et al. 1971). Importantly, during depletion and enrichment of erythrocyte 2,3-P_2-glycerate concentrations, there is a corresponding increase and decrease in blood oxygen affinity (Lian et al. 1971).

The presence of 2,3-P_2-glycerate as the major organic phosphate (4–6 m*M*) in the erythrocytes of the chick embryo during the last week of embryonic development and its rapid disappearance from circulating red blood cells shortly after hatching, suggests that it functions as a transitory modulator of hemoglobin function during this period (Isaacks and Harkness 1975, 1980; Borgese and Lampert 1975; Bartlett 1980). As 2,3-P_2-glycerate disappears from the red blood cells of the young chick, inositol-P_5 and ATP concentrations begin to accumulate rapidly corresponding with marked decreases in blood oxygen affinity (Isaacks and Harkness 1980). Inositol-P_5 binds to avian and human hemoglobin, reducing oxygen affinity and is, in fact, the most effective hemoglobin modulator occurring in animal red blood cells (Vandecasserie et al. 1973; Isaacks and Harkness 1980), presumably functioning in birds in a manner similar to that of 2,3-P_2-glycerate in the mammalian system.

But in order to understand and appreciate the similarities, if any, in the mechanisms involved in regulating blood oxygen affinity and hemoglobin oxygenation in these classes of animals, two questions seem apparent from a comparative point of view. First, is the synthesis and degradation of inositol-P_5 in the bird red cell related closely to glycolytic metabolism and secondly, is the concentration of organic phosphates, particularly inositol-P_5, readily adaptable to changing oxygen needs of the bird?

Results from recent, as well as earlier (Wells 1976), studies indicate that the synthesis of inositol-P_5 is not closely related to glycolytic metabolism and further that the chicken red blood cell is unable to catabolize inositol-P_5 once it is formed as suggested from unchanged concentrations of inositol-P_5 after prolonged incubation of the cells (Isaacks et al. 1982b).

It was reported (Jones et al. 1978) that inositol P_5 concentrations rose rapidly in the chicken red cell in response to acute anemia and those workers suggested that the levels of inositol-P_5 were metabolically controlled in response to changing needs of the chicken for oxygen just as 2,3-P_2-glycerate is regulated in erythrocytes of mammals (Jones et al. 1978). We have recently reported (Isaacks et al. 1983) results contrasting sharply with those of Jones et al. (1978). In summary, we produced five age-dependent density populations of chicken erythrocytes from anemic birds (Isaacks et al. 1983). The reticulocyte fraction contained approx. 60% of the level of inositol-P_5 found in the mature red blood cell fraction with synthesis and final accumulation of inositol-P_5 occurring in the retics in circulation. Further, no cell fraction had a concentration of inositol-P_5 greater than that found in mature erythrocytes.

Consequently, it appears that once inositol-P_5 is formed within the erythrocyte, its concentration remains relatively stable. Physiologically, this implies that adaptation to altered requirements for oxygen in birds may not occur by regulation of inositol-P_5

concentration in an analogous manner to changes in 2,3-P_2-glycerate in erythrocytes of mammals. Alternatively, it is possible that additional factor(s), such as CO_2 and [H^+], alone or in concert with inositol-P_5, are required to enhance regulation of oxygen delivery during oxygen deficit in birds, if indeed it is required.

Presented here are results of recent studies, attempting to assess the effects of five concentrations of CO_2 ranging from 0% to 14% at four concentrations of [H^+], ranging from pH near 6.8 to near 8.0, upon the oxyhemoglobin dissociation curves of erythrocyte suspensions from 18-day embryos (4–6 m*M* 2,3-P_2-glycerate and about 0.5 m*M* inositol-P_5), from 2-, 5-, 8-, and 14-day chicks (2,3-P_2-glycerate has rapidly disappeared and inositol-P_5 has accumulated), and from mature chickens (no 2,3-P_2-glycerate, but 3–4 m*M* inositol-P_5).

2 Methods and Materials

Embryos were obtained from fertile eggs of White Leghorn hens after incubating at 99.5 °F in a Humidaire, Model 50, incubator (The Humidaire Co., Wayne Street, New Madison, Ohio). The eggs were opened at the blunt pole on day 18 of incubation and the embryos removed. Blood samples were withdrawn from the embryos by cardiac puncture into a heparinized tuberculin syringe and pooled in heparinized tubes held in an ice bath. Blood samples were also taken by cardiac puncture from 2-, 5-, 8-, and 14-day chicks and from mature chickens in heparinized syringes and transferred to heparinized tubes held in an ice bath.

In order to provide fresh blood for the numerous oxygen dissociation curves on suspensions of intact erythrocytes at five concentrations of CO_2 and four pH's, fertile eggs (two dozen) were placed in the incubator on four consecutive days and the series of five concentrations of CO_2 at one of the selected pH's was performed on the same day. Similarly, oxygen dissociation curves on suspensions of intact erythrocytes from mature chickens were performed at five concentrations of CO_2 and at one of the four pH's on four consecutive days. The oxygen dissociation curves were performed on suspensions of intact erythrocytes at each designated age of chick at five concentrations of CO_2 and at pH near 7.4 only.

Cylinders of certified and analyzed mixtures of CO_2 were purchased from Matheson, Division of Will Ross, Morrow, Georgia. Varying the composition or partial pressure of CO_2 in the reaction vessel at a constant pH required adjustments in the concentration of sodium bicarbonate in the buffer as determined by the Henderson-Hasselbach equation. Thus, the composition of the basic buffer consisted of 0.015 *M* sodium succinate in 0.16 *M* sodium phosphate with appropriate amounts of sodium bicarbonate to give a series of solutions at four pH's, either 6.8, 7.1, 7.4, or 8.0, with five concentrations of CO_2, 0% (0 torr), 6.28% (45 torr), 8.73% (62 torr), 11.75% (84 torr), and 14.64% (104 torr) at each pH (for details, see Isaacks et al., submitted). The actual experimental pH was determined on each sample after completion of the dissociation curves and the mean and standard deviation calculated within each pH series as shown in Table 1 (6.77 ± 0.03, 7.08 ± 0.02, 7.41 ± 0.03, and 7.96 ± 0.04 for 18-day chick embryo, 6.74 ± 0.02, 7.09 ± 0.02, 7.4 and 7.96 ± 0.04 for adult chicken). The con-

Table 1. Effect of CO_2 and pH on oxygen affinity of 18-day chick embryo, adult chicken, and human red cell suspensions[a]

pH	CO_2 (%)	18-Day embryo		Adult chicken		Human	
		P_{50}	n	P_{50}	n	P_{50}	n
6.8	0	38.2	4.4	74.6	6.3	54.1	3.2
	6	38.1	3.3	61.6	4.1	54.9	2.9
	8	38.7	2.8	64.4	4.0	56.7	3.1
	11	40.6	3.0	70.3	3.7	56.6	3.1
	14	42.0	3.2	66.0	3.3	52.7	3.3
7.1	0	23.8	3.3	45.1	5.8	43.1	3.1
	6	26.0	2.9	48.1	3.9	37.2	3.1
	8	27.1	2.5	47.9	3.4	37.3	3.2
	11	33.1	2.9	54.5	3.8	37.1	3.0
	14	34.1	2.7	54.3	3.7	30.6	3.0
7.4	0	16.6	2.9	25.4	4.3	28.0	3.0
	6	23.8	3.1	34.4	3.7	26.2	2.8
	8	22.6	2.7	38.3	3.4	28.1	2.7
	11	30.1	3.6	40.7	3.3	31.4	3.0
	14	35.0	3.6	42.0	3.1	33.5	3.2
8.0	0	9.4	2.4	19.4	3.4	13.1	2.6
	6	17.3	2.7	29.7	3.2	22.8	2.6
	8	20.6	2.5	30.8	3.1	25.4	2.7
	11	22.8	2.5	32.1	3.0	25.7	2.8
	14	22.9	4.9	33.2	3.6	27.6	2.7

[a] Isaacks et al. 1985 (in press)

centration of $[H^+]$ at pH 6.8, 7.1, 7.4, and 8.0 is 15.88, 7.94, 3.97, and 1×10^{-8} $M\,l^{-1}$, respectively, thus, a decrease of 1.2 pH units in these experiments amounts to about 16-fold increase in $[H^+]$ concentration.

The P_{50} (millimeters of mercury or torr at which 50% of the hemoglobin is saturated with oxygen) of intact erythrocyte suspensions was determined in a volume of 0.25 ml of whole blood in a closed jacketed reaction vessel (10 ml) at 37 °C using a modification of the method of Longmuir and Chow (1970) as described previously (Lian et al. 1971). The variability of this method in our laboratory is about ± 1 mmHg.

3 Results

3.1 Effect of CO_2 and pH on Oxygen Affinity of Erythrocyte Suspensions from Embryos and Adult Chickens

The P_{50} of erythrocyte suspensions from 18-day chick embryos and adult chickens increased with increasing concentration of either carbon dioxide or hydrogen ion (Table 1), except for adult bird cell suspensions at pH 6.8 where the effect of CO_2 was inconsistent (Table 1). The ΔP_{50}'s were greater with increases in concentrations of CO_2 from 0% to 14% at a pH near 7.4 in cell suspensions from both the 18-day embryos (18.4 torr, Table 1) and adult birds (16.6 torr, Table 1); the effect on P_{50}'s decreased at pH's above and below 7.4 (Table 1). In fact, there is a greater effect on ΔP_{50}'s from increasing hydrogen ion concentration from pH 8.0 to pH 6.8 in cell suspensions from 18-day embryos (28.8 torr at 0% CO_2) and adult chickens (55.1 torr at 0% CO_2) than from increases in CO_2 concentration at any given pH (Table 1).

The Hill coefficient, n, in the absence of CO_2 at pH 6.8 in the 18-day chick cell suspensions was 4.4 and in the absence of CO_2 at pH 6.8, 7.1, and 7.4 in the adult chicken cell suspensions was 6.3, 5.8, and 4.3, respectively (Table 1).

The P_{50}'s of erythrocyte suspensions from the human performed simultaneously with the series of chicken experiments showed increases with increasing CO_2 concentration at pH 7.4 and 8.0, but showed a reversal in the pattern of response to CO_2 at pH 7.1 and little response to CO_2 at pH 6.8 (Table 1). By far the greater effect on P_{50} was noted in human erythrocyte suspensions from increasing hydrogen ion concentration.

3.2 Effect of CO_2 on Blood Oxygen Affinity of Young Chicks

The P_{50} of erythrocyte suspensions from 2-, 5-, 8-, and 14-day chicks increased with increasing concentration of CO_2 at each age, but the effect of 0% to 14% CO_2 on P_{50} had diminished somewhat with age of the young chick at 14 days (Table 2). There was little difference in the response of oxygen affinity to 14% CO_2 between the four ages of chicks (Table 2). Even though the erythrocytes of the 2-day chick have only about 35% (1.25 m*M*) of the concentration of inositol-P_5 found in erythrocytes of mature birds, 14% CO_2 appears to increase the P_{50} about 13–14 torr (Fig. 1) above normal whole blood values for the 2-day chick (32.8 torr; Isaacks and Harkness 1980) and about 5–6 torr above normal values at 14 days. The effect of 6% CO_2 on P_{50} of the young chick erythrocyte suspensions appears very similar to normal P_{50} values reported previously during this period of chick development (37.1 and 36.5 torr in 5- and 8-day chicks; Isaacks and Harkness 1980). The effects of 8% and 11% CO_2 on P_{50} of cell suspensions are intermediate between that observed from 14% CO_2 and the near physiological normal noted at the 6% CO_2 level (Fig. 1; Table 2). However, the absence of CO_2 in the cell suspensions of the young chicks depresses the oxygen affinity about 13–14 torr (Table 2) from the normal value of 32.8 torr in the 2-day chick (Isaacks

Table 2. Effect of CO_2 on oxygen affinity of young chick red cell suspensions at pH 7.4[a]

CO_2 (%)	2 Days		5 Days		8 Days		14 Days	
	P_{50}	n	P_{50}	n	P_{50}	n	P_{50}	n
0	19.4	3.1	23.4	3.1	26.1	2.5	27.1	2.9
6	31.8	3.1	34.7	3.5	40.4	3.2	37.8	3.4
8	36.6	3.1	40.4	2.9	39.0	3.0	39.9	3.4
11	37.0	2.8	42.0	3.8	46.6	3.9	44.1	3.2
14	46.6	3.3	45.5	3.3	48.6	3.6	45.6	3.5

[a] Isaacks et al. 1985 (in press)

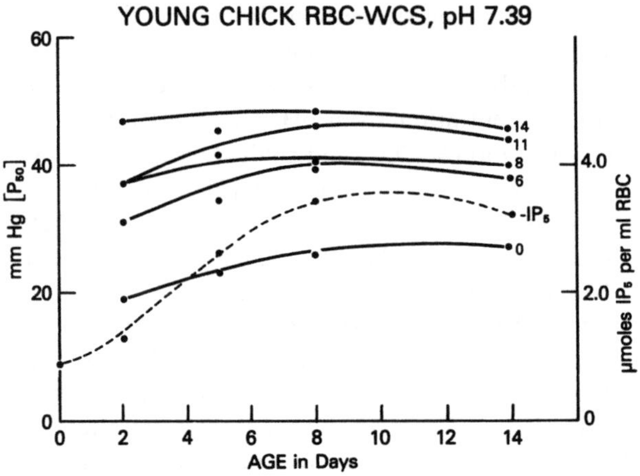

Fig. 1. Correlation of the effect of percentage CO_2 concentration [indicated at each line on the oxygen affinity (P_{50}) of whole cell suspension of chicks with chick age and concentration of inositol-P_5 (IP_5)] (Isaacks et al. 1985, in press)

and Harkness 1980) and the oxygen affinity remains depressed by 9–10 torr from normal values at 14 days even though inositol-P_5 concentration is approaching that of red cells of the mature bird (3.5–4.0 m*M*; Fig. 1).

3.3 Influence of CO_2 on the Bohr Effect Factor of Erythrocyte Suspensions from 18-Day Chick Embryos and Adult Chickens

The Bohr effect was calculated at four pH's over the pH range of about 6.8 to 8.0 and at five concentrations of CO_2 ranging from 0% to 14% (Fig. 2; Table 3). When log P_{50} was plotted against pH for a given concentration of CO_2 at each pH, a linear regression equation was derived for each level of CO_2 with the slope of the line, M, representing the Bohr factor (Table 3).

The Bohr effect factor ($\Delta\log P_{50}/\Delta pH$) for the equations in the absence of CO_2, the usual conditions for determining the Bohr factor, was − 0.508 and − 0.479 for the cell suspensions from 18-day chick embryos and adult birds, respectively (Table 3). These values agree well with Bohr factor values from geese (− 0.50; Danzer and Cohn 1967), herring gull (− 0.48; Clausen et al. 1971), turkey (− 0.53; Isaacks and Harkness

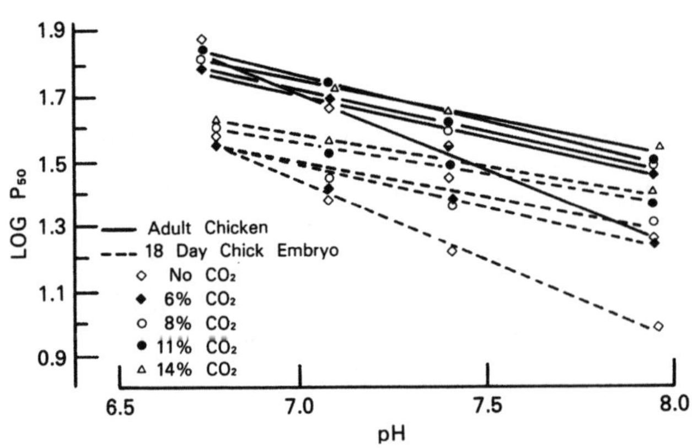

Fig. 2. Specific CO_2 effect on the Bohr factor of embryo and adult chicken whole cell suspension (Isaacks et al. 1985)

Table 3. Specific CO_2 effect on the Bohr factor of embryo and adult chicken red cell suspensions[a]

Regression equation of the plot of log P_{50} vs pH is $Y = A + M \cdot X$

CO_2 (%)	18 Day embryo Computer-derived equation	ΔP_{50}/Unit pH
0	$\text{Log } P_{50} = 5.0007 - 0.50841 \cdot \text{pH}$	19.08
6	$\text{Log } P_{50} = 3.3865 - 0.27154 \cdot \text{pH}$	14.23
8	$\text{Log } P_{50} = 3.0310 - 0.22025 \cdot \text{pH}$	12.27
11	$\text{Log } P_{50} = 3.0412 - 0.20892 \cdot \text{pH}$	14.48
14	$\text{Log } P_{50} = 2.9811 - 0.20393 \cdot \text{pH}$	13.41
	Adult chicken	
0	$\text{Log } P_{50} = 5.0656 - 0.47978 \cdot \text{pH}$	34.07
6	$\text{Log } P_{50} = 3.5645 - 0.2665 \cdot \text{pH}$	22.93
8	$\text{Log } P_{50} = 3.5382 - 0.26017 \cdot \text{pH}$	23.49
11	$\text{Log } P_{50} = 3.7465 - 0.28398 \cdot \text{pH}$	27.53
14	$\text{Log } P_{50} = 3.5002 - 0.25023 \cdot \text{pH}$	24.55

[a] Isaacks et al. 1985 (in press)

1980), and ostrich (− 0.41; Isaacks and Harkness 1980). The Bohr factor in the presence of CO_2 is markedly reduced in both the cell suspensions of 18-day chick embryos and of adult birds, indicating less influence of pH on oxygen affinity in the presence of CO_2 (Fig. 2; Table 3).

4 Discussion

Understanding the mechanisms and interrelationships of the factors contributing to regulation of blood oxygen affinity and oxygen transport and delivery in birds is an unfinished challenge. Despite the presence and concentration of inositol-P_5 in avian red cells and its ability to modulate hemoglobin oxygenation its role in regulating blood oxygen affinity is somewhat of an enigma.

Results from previous work indicated that phosphate modulators had little effect upon the mechanisms regulating hemoglobin function and whole blood oxygen affinity in the loggerhead and green sea turtles (Isaacks and Harkness 1980). Further studies indicated that blood oxygen affinities and hemoglobin function in these two species of marine turtles are altered markedly by CO_2 and to a lesser degree by pH (Isaacks et al. 1982a). Similar results have been reported on blood of the crocodile *(Crocodylus porosus)*, which has low concentrations of organic phosphates in its red cells, its hemoglobin is not modulated by organic phosphates, but appears to be regulated by CO_2 and protons (Bauer and Jelkman 1977; Grigg and Gruca 1979). These effects of CO_2 and $[H^+]$ may have physiological significance in regulating blood oxygen affinity in these species (Isaacks et al. 1982a). Particularly, since the turtles are capable of long migrations, deep dives, prolonged periods of submergence (Felger et al.

1976), and apparently tolerate anoxia very well (Belkin 1963; Millen et al. 1964; Bentley and Lutz 1979).

Without the evidence that inositol-P_5 regulation of hemoglobin function in birds is metabolically controlled, the possibilities exist that CO_2 and $[H^+]$ could have a role in regulation of blood oxygen affinity in the bird. The mean value for arterial pCO_2 and pH in a number of species of birds at rest was 28 mm CO_2 and 7.5, respectively, which is a lower pCO_2 and slightly higher pH than that observed in most mammals (Calder and Schmidt-Nielsen 1968). Heat stress in these birds resulted in hypocapnia and alkalosis (an average decrease of 10–12 mm CO_2 and an increase of about 0.2 pH units), suggesting a more efficient ventilation control in birds than in mammals. In fact, carbon dioxide, particularly in the chicken and probably all birds, is far more important than oxygen in controlling ventilation with important sensory systems, one in the lung and several associated with the circulatory system, that control ventilation to keep carbon dioxide concentrations reasonably constant (Burger 1980). The unusual respiratory system of birds no doubt contributes to their ability to maintain constant carbon dioxide concentrations. Unlike the mammalian lung, which expands to accept the incoming air, the avian lung is relatively rigid and does not need to move much because inspired air flows unidirectionally through the parabronchi into both the lungs and posterior air sacs (caudal group) and upon expiration, air from the anterior air sacs (cranial group) is moved out and the remaining air in the posterior air sac passes into the lungs (Fedde 1980). In effect, the air sacs function as a double-bellows system moving air in a continuous stream through the lungs. Aside from a constant flow of excess air, birds no doubt derive advantages from this system of respiration during migrations, high altitude flying, and in maintaining reasonably constant carbon dioxide concentrations.

Carbon dioxide is acidic and is the acid produced in the greatest quantity in the body and, consequently, it and $[H^+]$ need to be buffered to maintain a reasonable range of physiological pH. The most important buffer of carbon dioxide in the blood is hemoglobin, where it binds at the imino group of histidine and is then transported. The imino group loses some of its affinity for carbon dioxide when hemoglobin is oxygenated in the lungs. The release of protons from oxyhemoglobin converts bicarbonate to water and carbon dioxide, which is then expired from the lungs. The variations in carbon dioxide concentration in blood determines the variation of pH in the gas exchange region because all of the changes in acid content is attributable to changes in carbon dioxide content.

In the data presented here, increasing the CO_2 content from 0% to 14%, equivalent to 0 to 104 torr, reduced oxygen affinity of erythrocyte suspensions from 18-day embryos, young chicks, and mature chickens, particularly at pH near 7.4. Similar effects of 5, 20, 40, and 80 torr pCO_2 on the oxygen dissociation curves of chicken hemoglobin were reported previously, but the pH was not indicated (Burger 1980). Physiologically, less than 5 torr CO_2 is never found except at the beginning of a parabronchus ventilated with inhaled air and 80 torr CO_2 is higher than generally found except in experimental conditions (Burger 1980).

Further, in contrast to our previous findings on sea turtle cell suspensions (Isaacks et al. 1982a), our results indicated a greater decrease in oxygen affinity from increasing hydrogen ion concentration in these cell suspensions than from increasing CO_2

concentration. From these studies we have determined for the first time the CO_2 Bohr effect in both the 18-day embryo and adult chicken (Table 3; Fig. 2), indicating that in the absence of CO_2 the $\Delta P_{50}/\Delta 0.1$ pH is 1.9 and 3.4 torr in the embryo and adult cell suspensions, respectively. In the presence of CO_2, regardless of the concentration, the $\Delta P_{50}/\Delta 0.1$ pH is about 1.35 and 2.45 torr for the embryo and adult chicken blood. Although the curves for the embryo and adult chicken blood are nearly parallel at each CO_2 concentration, and the hemoglobins are comparable except for a little hatching hemoglobin in embryo blood, the difference in P_{50} is undoubtedly due to the presence of 2,3-P_2-glycerate in the embryo cells and to the presence of inositol-P_5 in the adult bird cells (Fig. 2).

Interestingly, the effects of the higher CO_2 concentrations at pH 7.4 clearly decrease blood oxygen affinity in the cell suspensions of the young chicks before inositol-P_5 has reached mature cell concentrations (2-day chick). This effect is maintained in the cell suspensions of the 14-day chick, but to a lesser extent. The absence of CO_2 in the cell suspensions of the young chicks further supports the notion that CO_2 at or above physiological concentrations ($\sim$ 4%) enhances the reduction in blood oxygen affinity. The fact that by 14 days of age the higher concentrations of CO_2 continue to decrease blood oxygen affinity by 6–8 torr when inositol-P_5 is near maximum concentration would seem to rule out a complete, if any, competition of CO_2 with inositol-P_5 for binding sites of deoxyhemoglobin in the chicken. The same effect of CO_2 was seen in cell suspensions of the adult bird at pH 7.4 (Table 2).

Competition between 2,3-P_2-glycerate and CO_2 for the binding sites of human deoxyhemoglobin tetramer have been demonstrated (Bauer 1974) and the binding of inositol hexakisphosphate (inositol-P_6) to human deoxyhemoglobin is similar to that of 2,3-P_2-glycerate (Arnone and Perutz 1974). Further studies indicated a similar competition between inositol-P_6 and CO_2 for binding sites in human deoxyhemoglobin and implied a similar reaction for inositol-P_5 and CO_2 in chicken hemoglobin (Benesch et al. 1968; Wells 1976). However, the binding site in the dyad axis of the chicken hemoglobin molecule includes six amino acids, Val 1β, His 2β, Lys 82β, Arg 135β, His 139β, and Arg 143β (Matsuda et al. 1973), with twelve basic groups available for interacting with phosphates and/or possibly other negatively charged groups as compared with four amino acids, Val 1β, His 2β, Lys 82β, and His 143β with eight basis groups in the binding site of human hemoglobin (Arnone and Perutz 1974). Consequently, the additional basic groups at the binding site of chicken hemoglobin could be sufficient for simultaneous binding of inositol-P_5 and CO_2 without strict competition by either.

The data we have presented indicate that increasing $[H^+]$ and CO_2 concentrations markedly affects the P_{50} of avian blood, particularly at pH 7.4. Exactly how much change in either component would be required to affect blood oxygen affinity under physiological conditions is unknown. Whether whole blood oxygen affinity needs to be regulated in birds is not yet clear, but our data coupled with those of Calder and Schmidt-Nielsen (1968), demonstrating hypocapnia and alkalosis from heat stressed birds would suggest that the oxygen affinity of blood is increased (P_{50} decreased) under these conditions. However, cells already containing a potent hemoglobin modulator, which is resistant to changes in concentration, may require only subtle changes in $[H^+]$ or CO_2 to enhance an already efficient respiratory system.

References

Arnone A, Perutz MF (1974) Structure of inositol hexaphosphate-human deoxyhaemoglobin complex. Nature 249:34–36

Bartlett GR (1980) Phosphate compounds in vertebrate red blood cells. Am Zool 20:103–114

Bauer C (1974) On the respiratory function of haemoglobin. Rev Physiol Biochem Pharmacol 70: 1–31

Bauer C, Jelkman W (1977) Carbon dioxide governs the oxygen affinity of crocodile blood. Nature 269:825–827

Belkin DA (1963) Anoxia: Tolerance in reptiles. Science 139:492–493

Benesch R, Benesch RE (1967) The effect of organic phosphates from the human erythrocyte on the allosteric properties of hemoglobin. Biochem Biophys Res Commun 26:162–167

Benesch R, Benesch RE, Yu CI (1968) Reciprocal binding of O_2 and diphosphoglycerate by human hemoglobin. Proc Natl Acad Sci USA 59:526–532

Bentley TB, Lutz PL (1979) Diving anoxia and nitrogen breathing anoxia in the marine loggerhead turtle. Am Zool 19(3):982

Borgese TA, Lampert LM (1975) Duck red cell 2,3-diphosphoglycerate: Its presence in the embryo and its disappearance in the adult. Biochem Biophys Res Commun 65:822–827

Borgese TA, Nagel RL (1978) Inositol pentaphosphate in fish red blood cells. J Exp Zool 205: 133–140

Burger RE (1980) Respiratory gas exchange and control in the chicken. Poult Sci 59:2654–2665

Calder WA, Schmidt-Nielsen K (1968) Panting and blood carbon dioxide in birds. Am J Physiol 215:477–482

Chanutin A, Curnish RR (1967) Effect of organic and inorganic phosphates on the oxygen equilibrium of human erythrocytes. Arch Biochem Biophys 121:96–102

Clausen G, Sanson R, Storesund A (1971) The HbO_2 dissociation curve of the fulmar and the herring gull. Respir Physiol 12:66–70

Danzer LA, Cohn JE (1967) The dissociation curve for goose blood. Respir Physiol 3:302–306

Fedde MR (1980) Structure and gas-flow pattern in the avian respiratory system. Poult Sci 59: 2642–2653

Felger RS, Cliffton K, Regal PJ (1976) Winter dormancy in Sea Turtles: Independent discovery and exploitation in the Gulf of California by two local cultures. Science 191:283–285

Grigg GC, Gruca M (1979) Possible adaptive significance of low red cell organic phosphates in crocodiles. J Exp Zool 209(1):161–167

Isaacks RE, Harkness DR (1975) 2,3-Diphosphoglycerate in erythrocytes of chick embryos. Science 189:393–394

Isaacks RE, Harkness DR (1980) Erythrocyte organic phosphates and hemoglobin function in birds, reptiles, and fishes. Am Zool 20:115–129

Isaacks RE, Harkness DR, White JR (1982a) Regulation of hemoglobin function and whole blood oxygen affinity by carbon dioxide and pH in the Loggerhead *(Caretta caretta)* and green sea turtle *(Chelonia mydas mydas)*. Hemoglobin 6(6):549–568

Isaacks RE, Kim CY, Johnson AE Jr, Goldman PH, Harkness DR (1982b) Studies on avian erythrocyte metabolism. XII. The synthesis and degradation of inositol pentakis (dihydrogen phosphate). Poult Sci 61:2271–2281

Isaacks RE, Kim CY, Liu HL, Goldman PH, Johnson AE Jr, Harkness DR (1983) Studies on avian erythrocyte metabolism. XIII. Changing organic phosphate composition in age-dependent density populations of chicken erythrocytes. Poult Sci 62:1639–1646

Isaacks RE, Nicols SC, Sallis JD, Zeidler RB, Kim HD (1984) Erythrocyte phosphates and hemoglobin function in monotremes and some marsupials. Am J Physiol 246:R236–R241

Isaacks RE, Goldman PH, Kim CY, Harkness DR (1985) Studies in avian erythrocyte metabolism. XIV. Effect of CO_2 and PH on P_{50} in the chicken. Am J Physiol (in press)

Jones SR, Smith JE, Board PB (1978) Changes in erythrocyte metabolism following acute blood loss in chickens. Poult Sci 57:1667–1674

Kim HD, Zeidler RB, Sallis JD, Nicol SC, Isaacks RE (1981) Adenosine triphosphate-deficient erythrocytes of the egg-laying mammal, echidna (Tachyglossus aculeatus). Science 213:1517–1519

Lian CY, Roth S, Harkness DR (1971) The effect of alteration of intracellular 2,3-DPG concentration upon oxygen binding of intact erythrocytes containing normal and mutant hemoglobins. Biochem Biophys Res Commun 45:151–158

Longmuir IS, Chow J (1970) Rapid method for determining effect of agents on oxyhemoglobin dissociation curves. J Appl Physiol 28:343–345

Matsuda G, Maita T, Mizuno K, Ota H (1973) Amino acid sequence of a beta-chain of A II component of adult chicken haemoglobin. Nature [New Biol] 244:244

Millen JR, Murdaugh jun HV, Baver CB, Robin ED (1964) Circulatory adaptation to diving in the freshwater turtle. Science 145:591–593

Rapoport S, Guest GM (1941) Distribution of acid-soluble phosphorus in blood cells of various vertebrates. J Biol Chem 138:269–282

Rose ZB (1976) A procedure for decreasing the level of 2,3-biphosphoglycerate in red cells in vitro. Biochem Biophys Res Commun 73:1011–1017

Vandecasserie C, Paul C, Schnek AG et al. (1973) Oxygen affinity of avian hemoglobins. Comp Biochem Physiol 44A:711–718

Wells RMC (1976) The oxygen affinity of chicken hemoglobin in whole blood and erythrocyte suspensions. Respir Physiol 27:21–31

ATP Metabolism in Mammalian Red Blood Cells

H.D. KIM[1]

1 Introduction

The pioneering work by Harris (1941) and Danowski (1941) showed that glucose was indispensable for K uptake by erythrocytes, thereby setting the stage for the establishment of the connection between metabolism and active cation transport (Whittam and Ager 1965). Nearly 3 decades later, another metabolic link to a cellular function has been demonstrated. The epoch-making discovery reported independently and simultaneously by Chanutin and Curnish (1967) and Benesch and Benesch (1967) laid the foundation for a close relationship between metabolism and oxygen affinity of hemoglobin. Comparative and developmental studies have continued to unravel the fascinating evolutionary selection of a wide variety of metabolites now known to influence the oxygen affinity of hemoglobin. This interesting topic will be discussed by other speakers in this symposium. In this paper, we will examine ATP metabolism in red cells of three mammalian species from a comparative point of view.

2 Pig Red Blood Cells

To begin with, we will consider the red cells of the adult pig, which exhibit no capacity to use glucose (Kolotivola and Engelhardt 1937; Kim and McManus 1971a,b). Interestingly, the fetal pig red cells have a carrier-mediated glucose transport system capable of glycolysis (Widdas 1955; Zeidler et al. 1976; Kim et al. 1973). The intricate transitory changes in glucose transport kinetics taking place during the postnatal period can be largely accounted for by the emergence of glucose impermeable cells into the circulation (Kim and Luthra 1977). Even in the adult stage, glucose transport is transitorily retained in the reticulocytes. It is in the course of reticulocyte maturation to erythrocytes that glucose transport is lost (Zeidler and Kim 1982), rendering the pig cell nonglycolytic.

Although adult pig cells can not utilize glucose, the cells are amply endowed with metabolic capacity for a variety of substrates including ribose (Kim and McManus 1971a,b), deoxyribose (Kim and McManus 1971a,b), dihydroxyacetone (McManus

1 Department of Pharmacology, University of Missouri-Columbia, School of Medicine, Columbia, MO 65201, USA

Circulation, Respiration, and Metabolism
(ed. by R. Gilles)

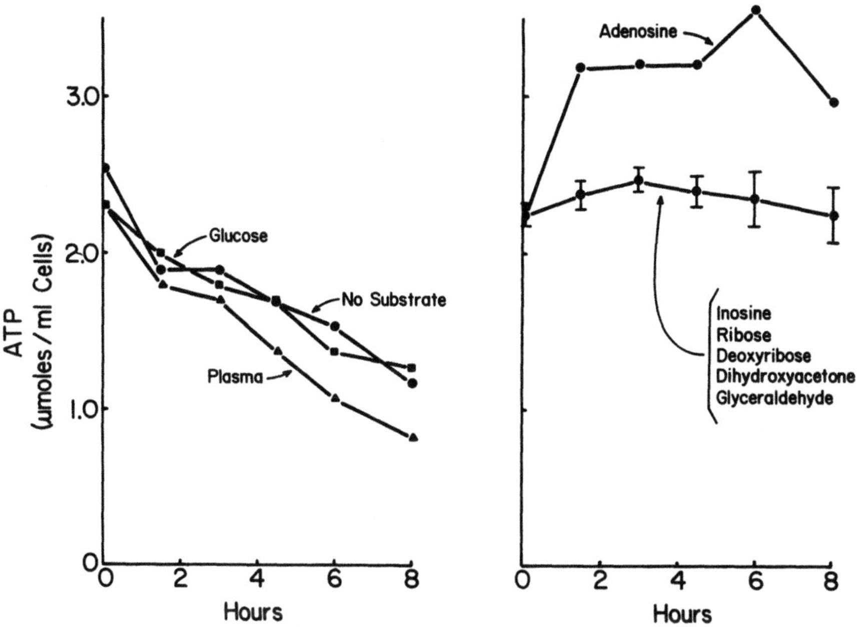

Fig. 1. ATP maintenance of pig red cells by a variety of metabolic substrates. Red cells were suspended to give a hematocryt of 20% in a balanced salt solution of the following compositions (m*M*): 90 NaCl; 5 KCl; 20 phosphate buffer, pH 7.4; 30 glycylglycine/$MgCO_3$ buffer, pH 7.4. All substrate concentration: 5 m*M*, except ribose, 3 m*M*. Temperature 37 °C. (From Kim et al. 1980; reproduced with permission from Elsevier/North-Holland Publishing Company)

1974), glyceraldehyde (McManus 1974), adenosine (Kirschner and Harding 1958), and inosine (Kim et al. 1980). The difficulty in assigning any one of these substrates as in vivo energy source stems from the fact that pig red cells suspended in their own plasma fail to maintain ATP levels (Kim et al. 1980). Figure 1 summarizes ATP maintenance under these conditions. As expected, the cells fare no better in glucose than in plasma or in a balanced salt medium containing no substrate. In every case, ATP fell with a half time of 6–8 h at 37 °C. In sharp contrast, the aforementioned substrates provide adequate measure against ATP deterioration. In the presence of adenosine, even a net synthesis of ATP took place.

These findings led Watts et al. (1979) and Kim et al. (1980) to focus on the liver as a likely energy source. Kim et al. (1980) found that an hepatectomy of the adult animal resulted in an ATP deterioration characteristic of starving cells. Moreover, coincubation of the hepatocytes isolated from 7-day-old piglets with adult pig red cells led to a net synthesis of ATP both in the freshly drawn red blood cells as well in energy depleted cells.

Watts et al. (1979) again using postnatal piglets, collected liver perfusate which was then tested for its efficacy with respect to ATP maintenance. In agreement with the results of the hepatocyte experiments, the cells suspended in the liver perfusate exhibited a net synthesis of ATP. These findings constitute reasonable evidence to postulate for the involvement of adenosine, since it is the only substrate capable of eliciting a net ATP synthesis. Expectation to the contrary, chemical analysis of the liver perfusate showed the presence of inosine, but not adenosine. However, it is possible that the identity of adenosine may have been masked by abundant adenosine deaminase (Ma and Fisher 1969) which could have been leaked out from the liver during perfusion.

Independently and simultaneously, Jarvis et al. (1980) suggested a physiological role for inosine on the basis of the membrane transport kinetic experiments. These authors postulate that the inosine present in the plasma, amounting to 2-11 μmol l^{-1} (Jarvis et al. 1980), is one order of magnitude larger than needed to meet the minimal requirements of cellular sustenance. Incidently, it should be stressed that the notion pertaining to a potential physiological role of a substrate at low concentration was first suggested by McManus (1974). He found an adequate ATP maintenance by pig cells in 10 μmol l^{-1} dihydroxyacetone which was replenished by an infusion calculated to match the cellular consumption rate.

We are currently evaluating the effect of low substrate concentrations on the red cell energetics. It is of considerable interest to ascertain whether the liver is the sole organ site elaborating inosine and whether inosine is the sole physiological energy source for pig red cells.

Needless to say, the metabolic interaction between the liver and red cells has long been established. In addition to inosine described herein, it is thought that the liver produces adenine (Henderson and LePage 1959) and adenosine (Lerner and Lowy 1974), which are in part used by the red cells (Lerner and Lowy 1974) and in part delivered by the red cells from the liver to other tissues, such as bone marrow (Lajtha and Vane 1958). These findings are consistent with observations on the rapid turnover of adenine nucleotides seen in red cells (Mager et al. 1967).

3 Cow Red Blood Cells

Next to the pig cells, cow cells have the lowest glycolytic rate among mammalian cells thus far tested (Laris 1958). It should be pointed out that glycolytic rates are conventionally determined by suspending cells in a suitable salt solution fortified with glucose. However, cow red cells suspended in their own plasma can exhibit much higher glycolytic capacity than in balanced salt media (BSS) (Seider and Kim 1979a, b). These findings suggest that glycolytic rate determined in vitro may not be an adequate reflection of the metabolic capacity of the red cells in vivo.

Since calf cells, which have a higher glycolytic rate than cow cells, use glucose in plasma at the same rate as in BSS, this metabolic property evidently develops as the animal grows older. We found that the cow cell glycolysis is nearly 50% lower in BSS than in plasma, suggesting that the plasma contains putative agents capable of sustaining a higher cellular metabolism. Indeed, dialysis or charcoal treatment of plasma renders it ineffective with respect to glycolytic stimulation (Seider and Kim 1979a, b) as shown in Fig. 2.

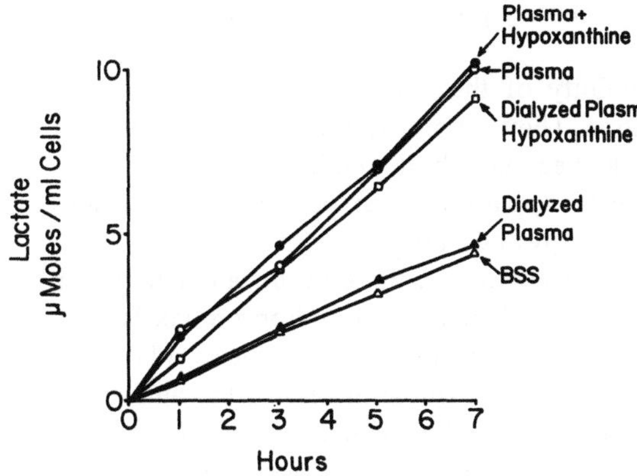

Fig. 2. Effect of plasma on cow cell glycolysis. Plasma was dialyzed for 24 h against buffered balanced salt solution. After dialysis, 5 m*M* glucose or 1 m*M* hypoxanthine was added, depending upon experiment. In 24 h dialysis experiment, fresh blood was drawn from same animal that provided plasma the day earlier. (From Seider and Kim 1979b, reproduced with permission from American Physiology Society)

Moreover, we found that glycolysis can be activated by a variety of purine compounds, including inosine, adenosine, adenine, hypoxanthine, xanthine, and uracil. While inosine at lower concentrations stimulates the metabolism, it inhibits glycolysis at higher concentrations. Although the biphasic actions of inosine are difficult to understand, what is interesting about inosine is the finding that the cow cells are impermeable to inosine (Duhm 1974).

Since the stimulatory agents need not enter the cells, it would seem reasonable to postulate that the metabolic response may be mediated somehow by membrane receptors akin to well characterized purinergic receptors in excitable membranes (Burnstock 1976).

In addition to cow red cells, human cells also apparently respond to externally added agents. Kashket and Denstedt (1958) reported that ADP-supplemented external medium resulted in increased human red cell glycolysis. Ford and Omachi (1972) found that cyclic AMP also elicits a glycolytic activation in human cells. In a preliminary report, Brewer (1971) described an increased glycolytic flux in red cells of rhesus monkey receiving intraperitoneal injection of epinephrine.

External adenine nucleotide is known to cause a dramatic change in membrane permeability to cations. Parker et al. (1977) showed that when ATP is added to external media, a prompt increase in both Na and K permeability of dog cells takes place. Perhaps, what we are dealing with here is a compensatory metabolic activation in response to an increased cation leak. This is an interesting possibility worth pursuing.

In an attempt to ascertain the underlying mechanism, we tested glycolytic enzymes for their response to purines (Seider and Kim 1979a, b). Of the three glycolytic enzymes, hexokinase exhibited increased enzyme activity in the presence of adenine, hypoxanthine, and inosine. The activities of phosphofructokinase, and pyruvate kinase did not change in the presence of purines. Interestingly, epinephrine-mediated glycolytic flux in rhesus monkey red cells is also thought to involve hexokinase as well as phosphofructokinase (Brewer 1971).

When the cow plasma was assayed for presence of purine, we found xanthine and hypoxanthine to be less than 1 μM and no inosine. The reconstitution of balanced salt solution with 1 μM hypoxanthine resulted in only a partial stimulation of glycolysis. More work will be needed to fully identify putative substances in cow plasma responsible for metabolic stimulation.

Another point of interest is the specificity of the plasma effect. The cow cells respond to none other but their own plasma. They can not respond to plasma derived from other species, including calf, cat, rat, guinea pig, rabbit, and human.

In any case, it is clear that the glycolytic apparatus must be operating at a substantially higher rate in vivo than the basal or "ground" rate.

Concomitant with adenosine stimulation of glycolysis, there occurs a net synthesis of ATP by cow cells (Seider and Kim 1979a). Incidently, it was in horse red cells lacking adenosine deaminase that McManus (1974) discovered a net synthesis ATP from adenosine and glucose. He reasoned that although the ribose moiety of the nucleoside could not be metabolized, adenosine itself could yield adenine nucleotides by the adenosine kinase and adenylate kinase routes. The resulting ADP would then require glycolytic intermediates formed from glucose or other alternate carbon source in producing a net ATP synthesis.

4 Echidna Red Blood Cells

In addition to the red cells of cow and horse, sheep cells can also carry out ATP synthesis (McManus 1974; Seider and Kim 1979a). In these cells, ATP content nearly doubles. However, the most spectacular example of this sort comes from red cells of the echidna which can increase their ATP content by approx. 20-fold (Kim et al. 1984a). The result is summarized in Fig. 3a.

It is clear that although neither glucose nor adenosine echidna alone causes ATP synthesis, the combination of the two substrates leads to the formation of an astounding amount of ATP. This dramatic situation stems in part from the uniquely low ATP content seen in freshly drawn red cells of the echidna (Kim et al. 1981). Table 1 compares the red cell ATP content of the monotreme and of the snake to illustrate the existence of a wide range in concentration of as much as 500-fold. These fascinating findings raise several challenging questions.

Is there another high energy compound which might substitute for ATP? An ion-exchange column chromatography of phosphorylated compounds from red cells of the echidna shows a remarkable lack of UV-absorbing compounds as shown in Fig. 4. However, the presence of other glycolytic intermediates is evident. 2,3-DPG is present in both monotreme in large amount. We found that the monotreme hemoglobins stripped of organic phosphates increase P_{50} value substantially upon addition of 2,3-DPG (Isaacks et al. 1984). Interestingly, addition of ATP or inositol pentaphosphate causes a similar increase in P_{50} values.

To what extent are such ATP dependent processes as glycolysis and active cation transport present?

Both monotreme cells also have an impressive capacity to use glucose (Kim et al. 1984a). Moreover, Fig 3b demonstrates the presence of the same type of metabolic response which we described earlier in cow red cells. Lactate production in adenosine

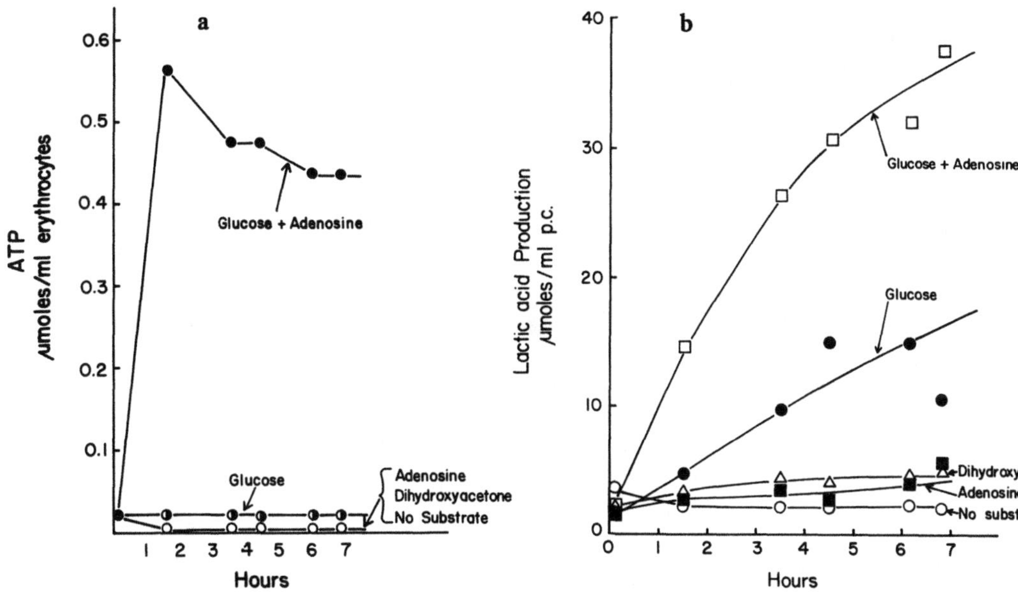

Fig. 3. a. ATP content of the echidna cells suspended in various substrates. **b.** Metabolic response of erythrocytes from the short-beaked echidna to a variety of substrates. (From Kim et al. 1984a, reproduced with permission from the Federation of European Biochemical Societies)

plus glucose is much larger than the sum of individual substrates. These findings make it clear that a trace amount of ATP present in these cells is apparently sufficient to phosphorylate glucose.

The presence of active K transport in the monotreme was determined using Rb^{86} as K analog (Kim et al. 1984b). The ionic composition of the monotreme cells was found to lie in between the low K and high K mammalian cells in that Na and K are present roughly in equal proportions. In each case, there was a steep cation gradient across the membranes. Echidna cells exhibit a small, but measurable active Rb influx, the magnitude of which is comparable to that of the low K sheep cells (Tosteson and Hoffman 1960). Compared with echidna, platypus cells have a higher active Rb influx, the magnitude of which is slightly smaller than that of HK sheep cells.

It is evident that the low ATP content seen in monotreme cells is also sufficient to support active cation transport.

Table 1. ATP content of vertebrate erythrocytes

Species	ATP μmol ml^{-1} cells	Reference
Echidna	0.03	Kim et al. 1984
Platypus	0.06	Isaacks et al. 1984
Human	1.0–1.5	Beutler 1971
Snake	15.00	Rapoport and Guest 1941

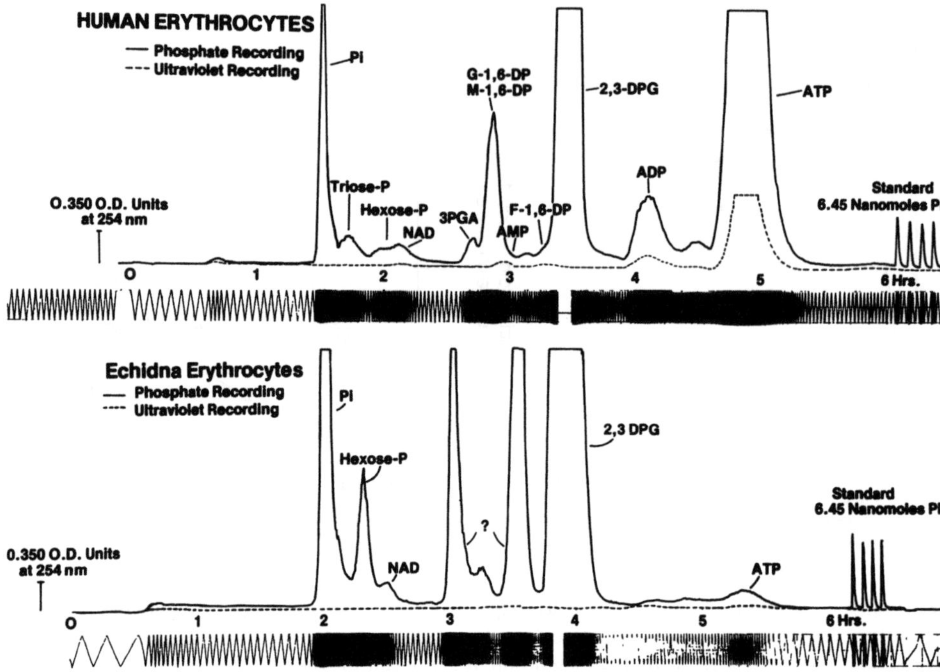

Fig. 4. Phosphorylated compounds in human and echidna erythrocytes. Human cells (0.225 ml) and echidna cells (0.220 ml) were extracted and analyzed by anion-exchange column chromatography. (———) Phosphate recording; (– – – –) ultraviolet recording. For details, see text. *Abbreviations: O.D.*, optical density; *Triose-P*, triose phosphate; *Hexose-P*, hexose phosphate; P_i, inorganic phosphate; *NAD*, nicotinamide adenine dinucleotide; *G-1,6-DP*, glucose 1,6-diphosphate; *M-1,6-DP*, mannose 1,6-diphosphate; *F-1,6-DP*, fructose 1,6-diphosphate; *AMP*, adenosine 5′-monophosphate; *ADP*, adenosine 5′-diphosphate; *ATP*, adenosine 5′-triphosphate; *2,3-DPG*, 2,3-diphosphoglycerate; and *3PGA*, 3-phosphoglycerate. (From Kim et al. 1981, reproduced with permission from American Association for the Advancement of Science)

Table 2 summarizes comparative data on Na + K + Mg ATPase activities in red cells from several mammals. Unexpectedly, the monotreme membranes display spectacularly high ATPase activities. Platypus ATPase activity is some 50-fold higher than the horse ATPase exhibiting the lowest activity. It should be stressed that the enzyme activities were determined in the presence of an excess ATP. When ATP concentrations were varied, we found typical saturation kinetics as shown in Fig. 5. While the V_{max} of the monotreme enzyme was two orders of magnitude larger than that of human enzyme, K_m values for ATP of both monotreme, varying from 0.4 to 0.6 m*M* was similar to that of the human ATPase (Post et al. 1960). We interpret these results to mean that the ATPase with a high V_{max} compensates for low cellular ATP. Thus, monotreme cells may operate with an ion pump activity, perhaps approaching that of other mammalian cells having much higher ATP content, but lower V_{max} of Na + K + Mg ATPase.

In this symposium which is devoted to comparative physiology of erythrocytes, it is fitting to observe that we have seen intriguing concepts emerging from investigations

Table 2. Na + K + Mg ATPase activity of mammalian red blood cell membranes [a]

Species	Number of determinations	Total activity $\mu mol\ mg^{-1}\ h^{-1}$	With ouabain	Ouabain sensitive
Platypus	No. 1	13.18	1.52	11.66
	No. 3	11.97	1.00	10.97
Echidna	n = 5	7.64 ± 1.36	1.09 ± 0.24	6.54 ± 1.02
Guinea pig	n = 3	0.99 ± 0.04	0.40 ± 0.02	0.59 ± 0.04
Human	n = 7	0.64 ± 0.04	0.21 ± 0.02	0.43 ± 0.03
Pig	n = 5	0.70 ± 0.06	0.28 ± 0.01	0.42 ± 0.06
Rabbit	n = 3	1.08 ± 0.16	0.70 ± 0.07	0.38 ± 0.10
Goat	n = 5	0.45 ± 0.04	0.14 ± 0.01	0.31 ± 0.03
Horse	n = 3	0.39 ± 0.11	0.17 ± 0.05	0.22 ± 0.09

[a] The reaction mixture was composed of 5 m*M* KCl, 150 m*M* NaCl, 2 m*M* $MgCl_2$, 2 m*M* vanadate free ATP, 0.2 m*M* EGTA, 20 m*M* Tris buffer, pH 8.0 either with or without 0.1 m*M* ouabain. The reaction was initiated by the addition of saponin-treated membrane protein at the final concentration of 0.04 to 1.10 $\mu g\ ml^{-1}$ at 38 °C and incubated for 30 min to 2 h depending upon the enzyme activity. The reaction was terminated by the addition of chloroform-methanol and inorganic phosphate was determined. Mean ± SEM; n = number of determinations. (From Kim et al. 1984b, reproduced with permission from Academic Press, Inc.)

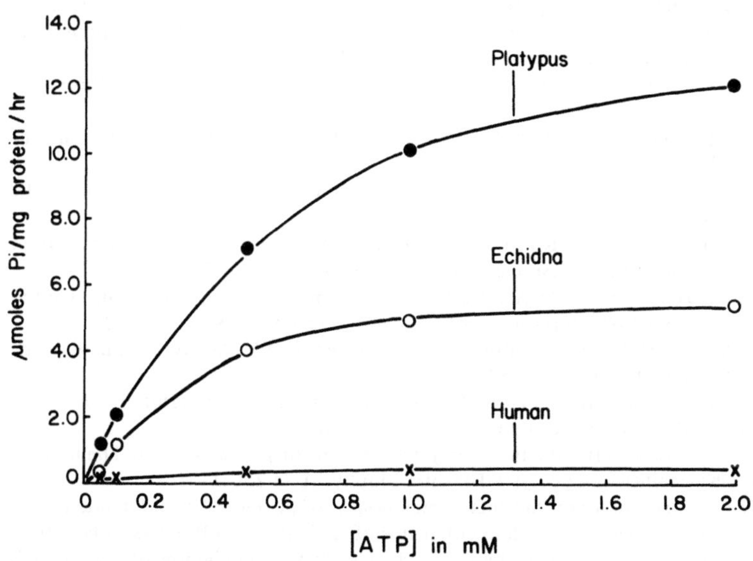

Fig. 5. Effect of ATP on Na + K + Mg ATPase of the monotreme and human erythrocyte membranes. $MgCl_2$ was held at 2 m*M*, while vanadate-free Na_2ATP was varied. The incubation medium was buffered with 20 m*M* Tris-Cl, pH 8.0 at room temperature. ATPase reaction was carried out at 38 °C. *Data points* represent duplicate determination of the membranes from two animals of each monotreme. (From Kim et al. 1984b, reproduced with permission from Academic Press, Inc.)

with comparative approaches. We explored the metabolic interaction of red cells and the liver. We gathered evidence to show that the operation of metabolic apparatus in vivo may be much different than previously thought. Finally, we described an incredibly efficient adaptation to a vanishing low ATP concentration in the maintenance of cellular functions.

Acknowledgments. I would like to thank Drs. R.B. Zeidler and R.E. Isaacks for their stimulating collaborations. Thanks are also due to Genie Eckenfels for her expert typing. Supported in part by NIH grant AM33456 and NSF grant PCM815821.

References

Benesch R, Benesch RE (1967) The effect of organic phosphates from the human erythrocyte on the allosteric properties of hemoglobin. Biochem Biophys Res Commun 26:162–167

Beutler E (1971) Red cell metabolism. A manual of biochemical methods. Grune & Stratton, New York, pp 1–146

Brewer GJ (1971) In: Rorth M, Astrup P (eds) Oxygen affinity of hemoglobin and red cell acid base status. Proceedings of the Alfred Benson Symposium IV held at the premesis of the Royal Danish Academy of Sciences and Letters, Copenhagen, May 17–22, p 609

Burnstock G (1976) Purinergic receptors. J Theor Biol 62:491–503

Chanutin A, Curnish RR (1967) Effect of organic and inorganic phosphates on the oxygen equilibrium of human erythrocytes. Arch Biochem Biophys 121:96–102

Danowski TS (1941) The transfer of potassium across the human blood cell membrane. J Biol Chem 139:693–705

Duhm J (1974) Inosine permeability and purine nucleoside phosphorylase activity as limiting factors for the synthesis of 2,3-diphosphoglycerate from inosine, pyruvate, and inorganic phosphate in erythrocytes of various mammalian species. Biochim Biophys Acta 343:89–100

Ford DL, Omachi A (1972) Influence of cyclic 3′,5′-AMP on glycolysis in human erythrocytes. Biochim Biophys Acta 279:587–592

Harris EJ (1941) The influence of the metabolism of human erythrocytes on their potassium content. J Biol Chem 141:579–595

Henderson JF, LePage GA (1959) Transport of adenine-8-^{14}C among mouse tissues by blood cells. J Biol Chem 234:3219–3223

Isaacks RE, Nicol S, Sallis J, Zeidler R, Kim HD (1984) Erythrocyte phosphates and hemoglobin function in monotremes and some marsupials. Am J Physiol 246:R236–R241

Jarvis SM, Young JD, Ansay M, Archibold AL, Harkness RA, Simmonds RJ (1980) Is inosine the physiological energy source of pig erythrocytes? Biochim Biophys Acta 597:183–188

Kashket S, Denstedt OF (1958) The metabolism of the erythrocyte. XV. Adenylate kinase of the erythrocyte. Can J Biochem Physiol 36:1057–1064

Kim HD, Luthra MG (1977) Pig reticulocytes. III. Glucose permeability in naturally occurring reticulocytes and red cells from newborn piglets. J Gen Physiol 70:171

Kim HD, McManus TJ (1971a) Studies on the energy metabolism of pig red cells. I. The limiting role of membrane permeability in glycolysis. Biochim Biophys Acta 230:1–II

Kim HD, McManus TJ (1971b) Studies on the energy metabolism of pig red cells. II. Lactate formation from free ribose and deoxyribose with maintenance of ATP. Biochim Biophys Acta 230:12

Kim HD, McManus TJ, Bartlett GR (1973) Transitory changes in the metabolism of pig red cells during neonatal development. In: Gerlack E, Moser K, Deutch E, Williams W (eds) Erythrocytes, thrombocytes, leukocytes: Recent advances in membrane and metabolic research. Thieme, Stuttgart, p 146

Kim HD, Watts RP, Luthra MG, Schwalbe CR, Conner RT, Brendel K (1980) A symbiotic relationship of energy metabolism between a 'non-glycolytic' mammalian red cell and the liver. Biochim Biophys Acta 589:256–263

Kim HD, Zeidler RB, Sallis JD, Nicol SC, Isaacks RE (1981) Adenosine triphosphate-deficient erythrocytes of the egg-laying mammal, echidna (Tachyglossus aculeatus). Science 213:1517–1519

Kim HD, Zeidler RB, Sallis JD, Nicol SC, Isaacks RE (1984a) Metabolic properties of low ATP erythrocytes of the monotremes. FEBS Lett 167:83–87

Kim HD, Baird M, Sallis JD, Nicol SC, Isaacks RE (1984b) Active cation transport and Na + K + Mg ATPase of the monotreme erythrocytes. Biochem Biophys Res Commun 119:1161–1167

Kirschner LB, Harding N (1958) The effect of adenosine on phosphate esters and sodium extrusion in swine erythrocytes. Arch Biochem Biophys 77:54–61

Kolotilova A, Engelhardt W (1937) Permeability of sugar distribution and glycolysis in red blood cells. Biokhimiya 2:387 (CA 31:5419)

Lajtha LG, Vane JR (1958) Dependence of bone marrow cells on the liver for purine supply. Nature 182:191–192

Laris PC (1958) Permeability and utilization of glucose in mammalian erythrocytes. J Cell Comp Physiol 51:273–307

Lerner MH, Lowy BA (1974) The formation of adenosine in rabbit liver and its possible role as a direct precursor of erythrocyte adenine nucleotides. J Biol Chem 249:959–966

Ma P, Fisher J (1969) Comparative studies of mammalian adenosine deaminases – some distinctive properties in higher mammals. Comp Biochem Physiol 31:771–781

Mager J, Hershko A, Zeithlin-Beck R, Shoshani T, Razin A (1967) Turnover of purine nucleotides in rabbit erythrocytes. I. Studies in vivo. Biochim Biophys Acta 149:50–58

McManus TJ (1974) In: Greenwalt TJ, Jamieson (eds) Alternate pathways for metabolism: A comparative view. The human red cell in vitro. Grune & Stratton, New York, pp 49–63

Parker JC, Castronova V, Goldinger JM (1977) Dog red blood cells: Na and K diffusion potentials with extracellular ATP. J Gen Physiol 69:417–430

Post RL, Merritt CR, Kinsolving CR, Albright CD (1960) Membrane adenosine triphosphatase as a participant in the active transport of sodium and potassium in the human erythrocyte. J Biol Chem 235:1796–1802

Rapoport S, Guest GM (1941) Distribution of acid-soluble phosphorus in the blood cells of various vertebrates. J Biol Chem 138:269–283

Seider MJ, Kim HD (1979a) Cow red blood cells. I. Effect of purines, pyrimidines, and nucleosides in bovine red cell gylcolysis. Am J Physiol 236(5):C255–C261

Seider MJ, Kim HD (1979b) Cow red blood cells. II. Stimulation of bovine red cell glycolysis by plasma. Am J Physiol 236(5):C262–C267

Tosteson DC, Hoffman JF (1960) Regulation of cell volume by active cation transport in high and low potassium sheep cells. J Gen Physiol 44:169–194

Watts RP, Brendel K, Luthra MG, Kim HD (1979) Inosine from liver as a possible energy source for pig red blood cells. Life Sci 25:1577–1582

Whittam R, Ager ME (1965) The connection between active cation transport and metabolism in erythrocytes. Biochem J 97:214–227

Widdas W (1955) Hexose permeability of fetal erythrocytes. J Physiol 127:318–327

Zeidler RB, Kim HD (1982) Pig reticulocytes. IV. In vitro maturation of naturally occurring reticulocytes with permeability loss to glucose. J Cell Physiol 112:360–366

Zeidler RB, Lee P, Kim HD (1976) Kinetics of 3-O-methyl glucose transport in red blood cells of newborn pigs. J Gen Physiol 67:67–80

Erythropoiesis: Cellular and Molecular Mechanisms

V.M. INGRAM[1]

1 Introduction

Our discussion will be concerned with a comparison of the cellular and molecular changes during the differentiation of the erythropoietic system in vertebrates, with relatively little mention of invertebrate systems because much less is known about them.

When looking at vertebrate erythropoiesis one is struck by the similarities between the schemes observed amongst different groups of animals. Differences are observed, of course, but they seem to be of secondary importance.

For example, in all cases there are two major erythrocyte lines – a primary (larval, embryonic) and a definitive (fetal, adult) cell line. These are usually morphologically distinct and usually contain different globin polypeptide chains. The change from primitive to definitive erythrocytes, therefore, involves not only a change of cell type with changes in the proteins that make up the structure of a cell, but also a change in globin gene activities. We do not know when in development the split between primitive and definitive erythropoiesis occurs, much less what the mechanism is.

Many, perhaps most vertebrates, show a further switch of erythropoiesis *within* the definitive cell population. This secondary switch is superimposed on the original switch to definitive cells. It involves a quantitative change in the expression of certain globin genes, but sometimes – as in the human (Hb F to Hb A) – the change is so great as to be almost qualitative. This secondary switch may also involve genes other than the globin genes; for example, in the human case a red cell surface antigen changes from i in fetal cells to I in adult cells. A similar situation exists in the chicken (Weise and Ingram 1976). The molecular events involved in this switch of globin gene activation is not understood. Neither is it clear that the secondary switch is truly within a single cell population or whether there is a series of (similar) cell populations with intermediate degrees of switching. In the newly hatched chick there seem to be at least two morphologically distinguishable red cell populations; however, the differences reported are really very small.

During the development of these three kinds of hemoglobin – we talk of "hemoglobin switching". However, it is believed that the *structure of the various hemoglobin genes* remains the same during development, but that the *expression of the hemoglobin genes* is regulated. Similarly the different red cell populations that are seen during

1 Department of Biology, Massachusetts Institute of Technology, Cambridge, MA 02139, USA

Circulation, Respiration, and Metabolism
(ed. by R. Gilles)

development are also controlled as to when in development they are turned on and when they are turned off. Superimposed in the mammal on this developmental control of red cell production is the day by day control of the red cell *number* to be produced and this is brought about by the hormone erythropoietin. These radically different control situations will be discussed.

2 Invertebrates

The smaller invertebrates with aerobic metabolism have no need for oxygen-carrying pigments under most circumstances; but if they live in stagnant pond water, as does the water flea, then a need for such pigments is present and their synthesis is induced. *Daphnia* contains a dissolved hemoglobin of mol. wt. 34,000 in its hemolymph (Fox 1953). Oxygen deprivation stimulates hemoglobin production six- to ninefold during a 10 day period, and remains high if oxygen continues to be in short supply. Hemoglobin production decreases again to the low starting levels when oxygen is once more abundant. It is important to realize that this hemoglobin, in contrast with the vertebrate situation, is not contained inside red cells, but is dissolved in the "plasma", indicating a very different mechanism for control. Therefore, it should be possible in this animal to study the hemoglobin control mechanism without having to worry about the control of red cell production. Since *Daphnia* is parthenogenetic, clones of the animal are easily obtained, making the analysis of gene structure and gene expression much easier. It is strange that this appealing system has to my knowledge not been exploited.

Certain large invertebrates have respiratory pigments. In some mollusks the protein pigment, a hemocyanin, forms an aggregate of several million molecular weight, dissolved in the circulating hemolymph. Other mollusks do have suspended cells in their celomic fluids which contain respiratory protein pigments. Some insect larvae, on the other hand, that live in the stomach of grazing animals (an oxygen poor environment), have large amounts of dissolved low molecular weight hemoglobins. The mechanisms of control, molecular or cellular, involved in any of these animals are obscure.

3 Vertebrates

3.1 Hemoglobin Switching in a Fish (Rainbow Trout)

Primary or "larval" erythrocytes in this fish are easily distinguishable from the definitive adult cells, because the former are round and remain so, while the definitive cells are clearly oval. Both remain nucleated when mature, as do all but the mammalian erythrocytes. The site of erythropoiesis changes from blood islands in the larval "intermediate cell mass" to the kidney and spleen (Iuchi and Yamamoto 1983). This is a typical switch of site. At the same time the hemoglobins switch from the larval types to the adult types, again a typical switch seen also in the frog, the chicken, and in man. In this fish there is a documented switch in red cell membrane glycolipids (globoside to ganglioside) which is associated with the change in cell type.

3.2 Hemoglobin Switching in an Amphibian

During metamorphosis of a frog there is typically not only a changeover from larval or tadpole hemoglobins to adult frog hemoglobins, but there is also an abrupt change in erythrocyte population. As in the fish and the chicken, the larval or embryonic erythrocytes can be distinguished morphologically from the later adult cells. This is so whether metamorphosis is natural or induced artificially through the administration of thyroxine. To quote Dorn and Broyles (1982). "The expression of pre- and post-switch hemoglobins in separate erythrocytes during a developmental transition would imply that the regulation of hemoglobin switching operates at an early stage in hemopoietic cell differentiation (Chapman and Tobin 1979). Thus, immunocytochemical studies using antisera specific to different hemoglobin types have been done during hemoglobin switching of humans (Farquhar et al. 1980), mice (Brotherton et al. 1979), chickens (Chapman and Tobin 1979), and several amphibians (Maniatis and Ingram 1971; Benbassat 1974; Jurd and Maclean 1970; Flavin et al. 1981). The degree to which pre- and postswitch hemoglobins coexist in the same cell during hemoglobin switching seems to vary from species to species, even within the amphibia. There have been conflicting reports of the metamorphic hemoglobin switch of *Rana catesbeiana* – the first studied and prototypic example of hemoglobin switching in animals (McCutcheon 1936)."

In the case of the bullfrog, *R. catesbeiana*, the system has been elegantly defined by the recent work of Dorn and Broyles (1982). They have "purified newly differentiating cells from the blood of metamorphosing tadpoles ... These new cells have an immature morphology, are very active in the synthesis of *adult* hemoglobin and contain no detectable tadpole hemoglobin. The tadpole cells have no detectable adult hemoglobin, are synthetically inactive, increase in density during the switch and are then cleared

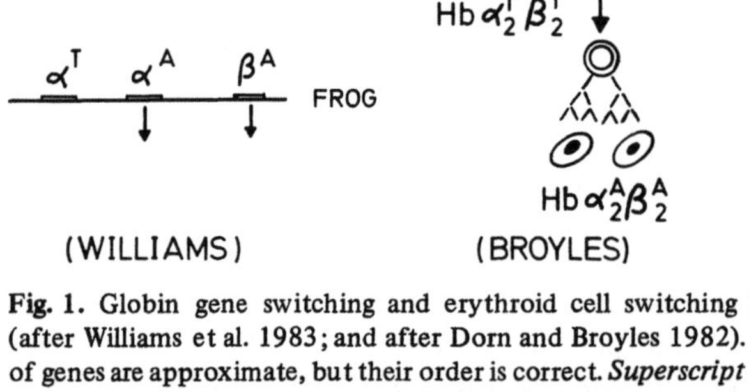

Fig. 1. Globin gene switching and erythroid cell switching in the amphibian, *Rana catesbeiana* (after Williams et al. 1983; and after Dorn and Broyles 1982). In the gene switching panel, spacings of genes are approximate, but their order is correct. *Superscript T* identifies a gene which is expressed only in tadpole erythrocytes; *superscript A* identifies a gene expressed only in adult frog erythrocytes. In the cell switching panel, immature erythroblasts of either lineage are indicated by having large, active nuclei synthesizing either tadpole-specific globins or frog-specific globins, but never both. Mature erythrocytes are shown with small condensed nuclei

from the circulation." In other words, switching in this instance is a matter of (1) stimulating the production of a brand new cell line, and (2) activating specifically the adult globin genes; both events are presumably caused by the recognized hormonal changes within the metamorphosing tadpole. The molecular mechanism of these events is unknown.

It is not necessary to postulate separate mechanisms for shutting off tadpole erythrocyte production and tadpole globin gene expression, since at least in *R. catesbeiana* metamorphosis occurs at a time when there is neither (tadpole) erythrocyte production nor (tadpole) globin synthesis.

The Fig. 1 shows the scheme just discussed. The change in control of globin gene activation in the frog is particularly interesting. Unlike the higher vertebrates - the chicken and the mammal - amphibian α and β globin genes seem to be located on the *same* chromosome next to each other. At least that is the case for some of the globin genes of the frog *Xenopus levis* (Patient et al. 1980). In contrast, the mammalian α and β globin gene clusters occur on different chromosomes, as is also the case in chickens. It will be interesting to see whether the molecular mechanism which switches from tadpole genes to frog genes is different from the mammalian mechanism, because in the frog the α and β genes involved are on the same chromosome, while in the chicken and the mammal they are on separate chromosomes.

3.3 Hemoglobin Switching in a Bird

Two morphologically distinct erythrocyte populations are elaborated during the *chick* embryo's development which lasts 21 days (Dantschakoff W 1908; Bruns and Ingram 1973). First, primitive erythrocytes appear as a cohort, but without a self-perpetuating stem cell population. Although these cells persist for approx. 2 weeks, by day 5 of embryonic development definitive erythrocytes appear; since these do have a self-perpetuating stem cell population, they become the dominant erythrocyte type of the late embryo and of the adult chicken. Other birds seem to have a similar pattern of red cell development. Characteristic and distinct hemoglobins are found in the primitive and definitive cells (Bruns and Ingram 1973) (Fig. 2).

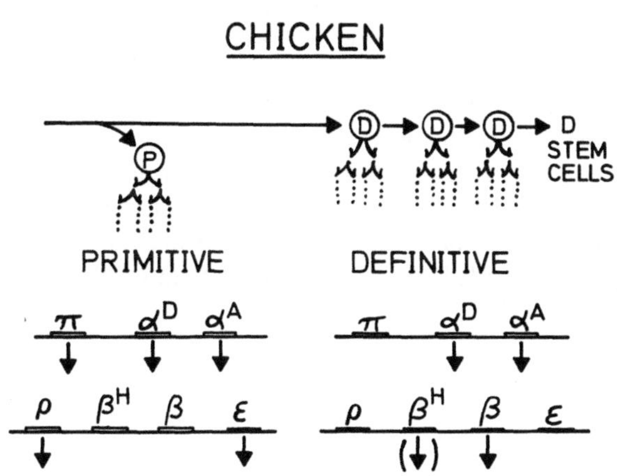

Fig. 2. Erythroid cell switching and globin gene switching in an avian system, the chicken *Gallus gallus*. The gene β^H is a globin gene expressed as a β-like globin, but only in definitive cells and only just before and just after hatching

The primary mesenchyme is the site where precursors to the erythrocytes first appear amongst the group of migrating cells found between epiblast (ectoderm) and hypoblast (endoderm). This migration is part of the process of gastrulation. Descendants of these mesenchymal cells become part of the yolk sac which is the dominant erythropoietic organ in the chick embryo. Late in embryonic life and after hatching, the bone marrow is the site for erythropoiesis, as it is in most adult vertebrates. Correlated with this change of erythropoietic site is a changeover (Bruns and Ingram 1973) in the morphological type of red cell, from primitive to definitive cells.

When total numbers of primitive and of definitive cells per embryo are plotted as a function of days of embryonic development (Bruns and Ingram 1973), it is clear that the primitive cells are a transient population and that the definitive cells are a rapidly expanding, permanent population. In the chicken the assignment of embryonic hemoglobins to the primitive cells and of adult hemoglobins to definitive cells is correct, but during the period of changeover there seems to be simultaneous production within at least some of the earliest definitive cells of β-globin chains (adult) and of the π and ρ peptide chains of hemoglobin P (embryonic) (Tobin et al. 1979). It is not known how many cells at the intermediate stage of switching respond in this way and what the proportions of adult and embryonic chains within a particular chain might be. Apparently there is a similar period of "overlap" during the changeover from primitive to definitive red cells in the embryonic mouse (Chui et al. 1979).

In the chicken there is another developmentally regulated switch in the ratio of major to minor adult hemoglobins in the definitive erythrocyte population very late in embryonic life and after hatching (Bruns and Ingram 1973). This phenomenon is independent of the appearance and disappearance of hemoglobin H (or β^H globin chain) production around the time of hatching. When the adult hemoglobins of the chicken, hemoglobin A and hemoglobin D first appear, more hemoglobin D is produced, yet hemoglobin D is the minor (25%) component in the adult. This is a change in the ratios of α^A and α^D chains, since both hemoglobins have the same β-globin chain. It is not known whether this change in ratio occurs within cells of a single definitive population or whether it is due to a change in the numbers of two different cell populations, each with its own characteristic proportion of the two globin chains. This change in ratio probably does not parallel the change from γ to β globin chain productions in the human fetus, because the chicken phenomenon is a modulation of genes in the α globin locus and the human changes modulation of genes in the β globin gene cluster on a different chromosome.

In addition to the changes in globin chain type there are other alterations as primitive cells are replaced by definitive cells. Surface glycoproteins differ in the two erythrocyte populations (Sanders 1968; Weise and Ingram 1976); the differences are both qualitative and quantitative. Then there is a developmentally regulated change in the synthesis of the chick erythrocyte-specific histone H5. Although this occurs in both primitive and definitive red cells, synthesis of H5 occurs much later than synthesis of the universal histones H1-H4 (Moss et al. 1973).

Primitive and definitive cells make α-like and β-like globin chains:

	α-like	β-like
Primitive	π, α^D, α^A	ρ, ϵ
Definitive	α^D, α^A	$(\beta^H), \beta$

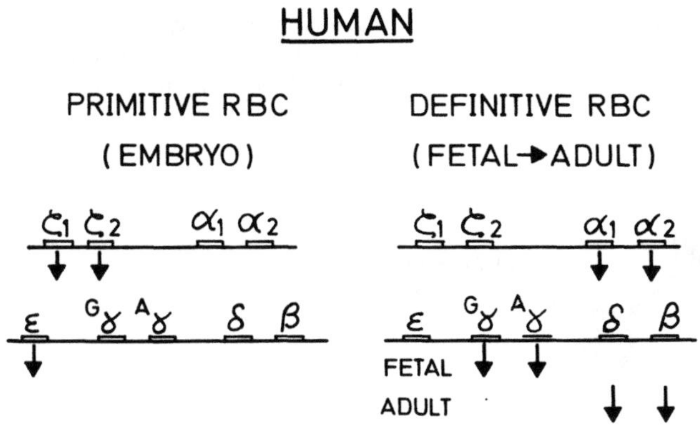

Fig. 3. Erythroid cell switching and globin gene switching in a mammal, *Homo sapiens*. Globin genes α_1 and α_2 are separate genes with identical globin products. The late fetal switch from hemoglobin F (using α genes and γ genes) to hemoglobins A and A_2 (using β and δ genes as well as α) is shown on the *right*

Culture of primary mesenchyme indicates (Keane et al. 1979) that the α-like chains are turned on first, particularly the π-chain. Among the β-like chains made in this culture system the ϵ chain appears first. The temporal sequence for the β-like globin chains in the embryo itself is: ϵ and ρ in the primitive cells, followed in the definitive cells by β^H and β, followed by β only. The order of the corresponding genes, however, reading from the 5′ end is: ρ, β^H, β, ϵ. The first and the last gene are activated together in the early embryo. This is quite different from the situation in the human, where the spatial arrangement 5′ to 3′ parallels the temporal sequence of expression during development (see Fig. 3). The way chicken globin genes are activated during development is clearly different from the mammalian mechanism, at least this is true for the human case. The chicken α globin genes follow in their developmental sequence the 5′–3′ "rule".

So far a search for chicken erythropoietin has been unsuccessful (Coll and Ingram 1978). Stimulating factors have been purified from anemic chicken serum and they do stimulate both colony and burst formation, but they turned out to be similar or identical with transferrin.

3.4 Hemoglobin Switching in a Mammal

We will use the human case to illustrate hemoglobin switching in a mammal. As shown in Fig. 4, the erythrocyte populations can again be divided into primitive and definitive cells. The primitive erythrocytes of the human embryo remain nucleated when mature, as do those of other mammals and all the red cells of lower vertebrates. On the other hand, the human (mammalian) definitive erythrocytes lose their nuclei during maturation, unlike those of lower vertebrates. Here is a physiologically important difference between mammals and lower vertebrates, since the nonnucleated red cell is much more deformable and, therefore, more efficient in penetrating into narrow capillaries.

Next we see that the primitive erythrocyte activates a distinct set of α-like and β-like globin genes and that separate sets of genes are active in the definitive cell. Furthermore, there is a very important developmental sequence superimposed on the definitive erythrocyte population: the change from fetal (HbF) to adult (HbA) hemoglobin production. This involves shutting down γ-globin chain synthesis and activating β-globin genes. There is evidence (Papayannopoulou et al. 1984) that extracellular factors can

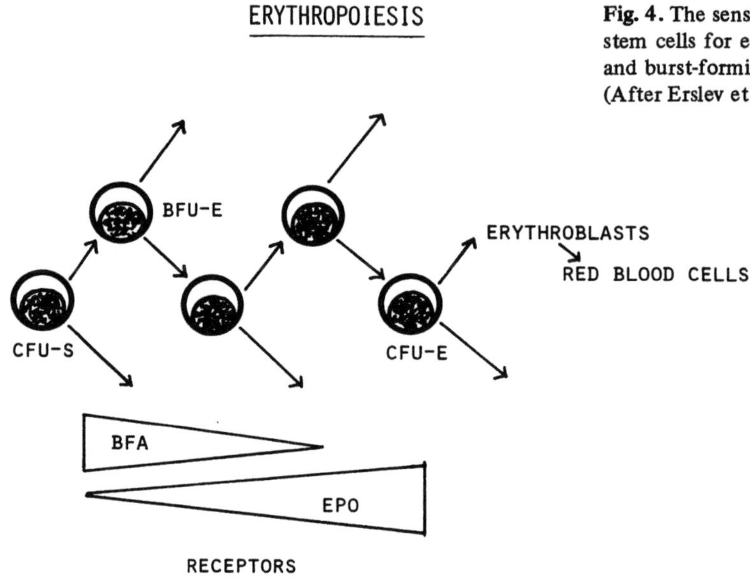

Fig. 4. The sensitivity of mammalian stem cells for erythropoietin (Epo) and burst-forming activity (BFA). (After Erslev et al.)

produce such a shutting down of γ-chain synthesis and possibly activating β-chain synthesis and that such a factor can be found in sheep serum. In addition, there are noticeable differences between the definitive cells carrying HbF in the fetus and those carrying HbA in the adult. For example, the mean corpuscular volume decreases by 25% during the first 3 months after birth, there is a change in one of the surface antigens of the red cell from predominantly **i** on fetal cells to **I** on adult cells, all within the definitive cell population.

What causes the switch from primitive to definitive erythrocytes in the human embryo is unknown, as it is in other mammals and lower vertebrates. More is known about the change within definitive cells from fetal to adult red cell and from fetal to adult hemoglobins. For example, it is postulated (Stamatoyannopoulos et al. 1983) that the progenitor erythroid cells differ from each other in the degree to which they are "programmed" to make HbF only, a mixture of HbF and HbA, or HbA only. In this sense they form a developmental sequence, so that erythroid precursors in the embryo have not developed very far and are still in the HbF "program"; such cells would then mature into HbF-containing fetal red cells. Later the precursors have developed further and are now programmed to make HbF + HbA, the state of things at birth. Later still the precursors have developed still more and will mature into adult cells making HbA. The adult individual in this view contains in the bone marrow a mixture of precursor cells which are more or less differentiated, but under normal demands in the bone marrow for red cell maturation only the most mature precursors undergo final maturation and, therefore, only adult mature cells with HbA are formed. But the potential for causing less differentiated precursors to mature into Hb-containing red cells still exists; it is displayed during periods of acute crisis when the demand for red cells is exceptionally heavy or in certain hereditary hemoglobinopathies – thalassemias, hereditary persistance of fetal hemoglobin – when HbF-containing red cells are produced. The demand for red cells is modulated by the hormone erythropoietin, which will be discussed later. This regulates the number of red cells to be produced. The normal intracellular switching mechanism from fetal to adult hemoglobin ($^G\gamma$, $^A\gamma$ gene activation to δ, β gene activation) is not understood, but there are some straws in the wind.

The indications for an extracellular, therefore *trans*-acting factor are interesting (Papayannopoulou et al. 1984).

Explanations for a γ to δ β switch have been sought amongst certain partial gene deletions in the β-globin gine cluster (Forget et al. 1983; Fritsch et al. 1979). Some deletions lead to a phenotype in which HbF is made in red cells of the adult, leading to a totally benign condition, hereditary persistence of fetal hemoglobin (HPFH). Other deletions produce the phenotype of β^0-thalassemia, a serious hemoyltic anemia. In the first case there has been no switch from fetal to adult globin synthesis, so no adult globin is made, but fetal globin is made and in adequate amounts. In the other, the switch from fetal to adult globin production occurs, but since the adult globin genes are deleted no adult globin can be made and a severe anemia results. However, an examination of the extent of the deletions does not allow one to deduce that any particular region of the β-like globin gene cluster is the site of information for the fetal to adult switch.

The transfection of DNA-containing chimeric globin genes into cells, followed by expression of those genes has given rise to interesting speculations (Axel 1984), Chimeric genes are constructed as follows: the promoter of one globin gene (α or β) is joined to the structural portion of another globin gene. The chimeric genes and controls are transfected into cells and the expression of the genes which they carry is measured. The expression of these genes in MEL (murine erythroleukemia) cells is particularly interesting; it looks as if there is a *dominant* regulatory element within the *structural* part of the gene; the promoter is of lesser importance. Possibly the regulatory element might be in the intervening sequence of the structural gene. The overall model suggests a two-step regulation of gene activation:

1. Depression into an active configuration.
2. Influence of a positive regulatory element in the DNA of the structural gene responding to a tissue-specific activator or factor. Presumably in the above experiments both α and β genes are introduced in the "active" configuration; α does not require a second step, β does. This is a provocative hypothesis and further results are eagerly awaited.

4 Control of Red Cell Production

In mammals the number of erythrocytes produced by the adult marrow is regulated by the hormone *erythropoietin*. The hormone also regulates red cell production in the human fetus during the third trimester, the bone marrow phase of the fetal erythropoiesis. It is not known whether fetal red cells are regulated during the earlier liver phase of fetal definitive erythrocyte production. The even earlier embryonic primitive red cells are probably not affected by erythropoietin, since they are not a self-perpetuating stem cell population, but a cohort of cells. However, evidence on this point is not available. In goats, sheep, rats, and rabbits erythropoietin regulates fetal erythropoiesis.

Under normal conditions the total number of red cells is fairly tightly controlled both by red cell loss through aging and removal and by control of red cell production through a feedback mechanism; red cell production is stimulated by low tissue oxygen tension. The mechanism seems to be that the serum of anemic and hypoxic individuals contains higher than usual levels of the hormone erythropoietin, produced in response to the low oxygen tension by the kidney. The site within the kidney and the precise mechanism which produces the hormone is still not understood. The hormone, a glycoprotein with a mol. wt. of 39,000 has been purified (Miyake et al. 1977), but another species of mol. wt. 23–27,000 has also been recognized. Erythropoietin acts, it is thought, on certain sensitive stem cells, presumably both at the burst-forming level of erythroid differentiation (BFU–E) and at the later erythroid-colony-forming stem cell (CFU–E), the latter being much more sensitive to low erythropoietin (Erslev et al. 1980). It is clear that erythropoietin stimulates both cell division (the numbers increase) and the rate of cell maturation to form hemoglobinized cells; the mechanism whereby this occurs is unknown. Yet the erythropoietin effect can be demonstrated both in vivo and in culture. Erythropoietin may not be the only factor regulating numbers of erythrocytes formed, but it is probably the most important. Interestingly, it is known that other and different factors regulate the production of granulocytes and megakaryocytes; while the factors involved are different, the overall mechanism seems similar.

5 Conclusions

Amongst the invertebrates there seems to be great diversity of oxygen-carrying pigments, including some which resemble hemoglobins; there also seems to be no general rule as to whether cells functionally resembling vertebrate erythrocytes are used or not.

The vertebrates form a much more homogeneous group. Generally, the red cell populations can be divided into early primitive erythrocytes and later, self-perpetuating definitive erythrocytes. Different sets of globin genes are active in primitive and definitive cells, but sometimes some genes are active in both. Amongst the mammals and to some extent in birds there is a later, second changeover of active globin genes within the definitive cell population, apparently accompanied by other changes in definitive cell structure. Finally, in mammals the production of erythrocytes in the late fetus and in the adult is regulated by the hormone erythropoietin.

Hardly anything is known about the mechanisms which bring about these complex and obviously well-regulated changes, except that in frogs the changes from primitive to definitive cells (accompanied by the globin changes) is triggered during metamorphosis by thyroxine; this fact is interesting, but does not yet give us a molecular mechanism. In the late human switch from fetal γ-globin gene activation to adult β-globin gene activation might involve an extracellular humoral factor and might require a particular DNA sequence within the structural gene which is to be "turned on". Much further work is needed before we can really describe the important molecular events of erythropoiesis.

References

Axel R (1984) Cooley's Anemia Symposium, New York City, NY

Benbasset J (1974) J Cell Sci 16:143–156

Brotherton TW, Chui DHK, Gauldie J, Patterson M (1979) Proc Natl Acad Sci USA 76:2853–2857

Bruns GAP, Ingram VM (1973) The erythroid cells and hemoglobins of the chick embryo. Philos Trans R Soc Lond B Biol Sci 266:225–305

Chapman BS, Tobin AJ (1979) Dev Biol 69:375–387

Chui DHK, Brotherton TW, Gauldie J (1979) Hemoglobin ontogeny in fetal mice; adult hemoglobin in yolk sac derived erythrocytes. In: Stamatoyannopoulos G, Nienhuis AW (eds) Cellular and molecular regulation of hemoglobin switching. Grune & Stratton, New York, pp 213–226

Coll J, Ingram VM (1978) The stimulation of heme accumulation and erythroid colony formation in cultures of chick bone marrow cells by chick plasma. J Cell Biol 76:184–190

Dantschakoff W (1908) Untersuchungen über die Entwicklung des Blutes und Bindegewebes bei den Vögeln. I. Die ersten Blutzellen beim Hühnerembryo und der Dottersack als blutbildendes Organ. Anat Hefte 37:471–589

Dorn AR, Broyles RH (1982) Erythrocyte differentiation during the metamorphic hemoglobin switch of Rana catesbeiana. Proc Natl Acad Sci USA 79:5592–5596

Erslev AJ, Caro J, Birgegard G, Silver R, Miller O (1980) The biogenesis of erythropoietin. Exp Hematol, vol 8, suppl 8:1–13

Farquhar MM, Papayannopoulou T, Brice M, Kan YW, Stamatoyannopoulos G (1980) Dev Biol 80:64–78

Flavin M, Blouquit Y, Duprat AM, Deparis P, Tonthat H, Rosa J (1981) In: Stamatoyannopoulos G, Nienhuis AW (eds) Hemoglobins in development and differentiation. Alan R Liss, New York, pp 215–221

Forget BG, Tuan D, Newman MV, Feingold EA, Collins F, Fukumaki Y, Jagadeeswaran P, Weissman SM (1983) Molecular studies of mutations that increase Hb F production in man. In: Stamatoyannopoulos G, Nienhuis AW (eds) Globin gene expression and hematopoietic differentiation. Alan R Liss, New York, pp 65–76

Fox HM (1953) Factors influencing hemoglobin synthesis by Daphnia. Proc R Soc Lond B Biol Sci 141:179–189

Fritsch EF, Lawn RM, Maniatis T (1979) Characterisation of deletions which affect the expression of fetal globin genes in man. Nature 279:598–603

Ingram V (1981) Hemoglobin switching in amphibians and birds. In: Stamatoyannopoulos G, Nienhuis AW (eds) Hemoglobins in development and differentiation. Alan R Liss, New York, pp 147–160

Iuchi I, Yamamoto M (1983) Erythropoiesis in the developing rainbow trout, *Salmo gairdneri irideus*: histochemical and immunochemical detection of erythropoietic organs. J Exp Zool 226:409–417

Jurd RD, Maclean N (1970) J Embryol Exp Morphol 23:299–309

Keane RW, Lindblad PC, Pierik LT, Ingram VM (1979) Isolation and transformation of primary mesenchymal cells of the chick embryo. Cell 17:801–811

Maniatis GM, Ingram VM (1971) J Cell Biol 49:372–379, 380–389, 390–404

McCutcheon FH (1936) J Cell Comp Physiol 8:63–81

Miyake T, Kung CKH, Goldwasser E (1977) Purification of human erythropoietin. J Biol Chem 252:5558–5564

Moss BA, Joyce WG, Ingram VM (1973) Histones in chick embryonic erythropoiesis. J Biol Chem 248:1025–1031

Papayannopoulou T, Tatsis B, Kurachi S, Nakamoto B, Stamatoyannopoulos G (1984) A haemoglobin switching activity modulates hereditary persistence of fetal haemoglobin. Nature 309: 71–73

Patient RK, Elkington JA, Kay RM, Williams JG (1980) Internal organization of the major adult α- and β-globin genes of *X. laevis*. Cell 21:565–574

Sanders BG (1968) Developmental disappearance of a fowl red blood cell antigen. J Exp Zool 167: 165–178

Stamatoyannopoulos G, Papayannopoulou T, Nakamoto B, Kurachi S (1983) Hemoglobin switching activity. In: Stamatoyannopoulos G, Nienhuis AW (eds) Globin gene expression and hematopoietic differentiation. Alan R Liss, New York, pp 347–364

Tobin AJ, Chapman BS, Hansen DA, Lasky L, Selvig SE (1979) Regulation of embryonic and adult hemoglobin synthesis in chickens. In: Stamatoyannopoulos G, Nienhuis AW (eds) Cellular and molecular regulation of hemoglobin switching. Grune & Stratton, New York, pp 205–212

Weise MJ, Ingram VM (1976) Proteins and glycoproteins of membranes from developing chick red cells. J Biol Chem 251:6667–6673

Williams JG, Bendig MM, Patient RK, Banville D, Greaves DR, Mahbubani H (1983) Replication, methylation and expression of X. laevis globin genes injected into fertilized Xenopus eggs. In: Stamatoyannopoulos G, Nienhuis AW (eds) Globin gene expression and hematopoietic differentiation. Alan R Liss, New York, pp 27–38

Mechanisms of the Maturation of the Reticulocyte

S.M. RAPOPORT [1]

1 Characteristics of the Reticulocyte

Differentiation is generally conceived to be a program of gene expression, primarily determined by regulation of transcriptional processes. In some cell types, however, including those constituting the erythron, one can distinguish a subprogram which involves practically exclusively posttranscriptional events. This phase of differentiation may be designated as maturation. During the differentiation of the erythron the activity of the nucleus ceases after the stage of the basophilic erythroblast. There ensues pyknosis of the nucleus with cessation of new formation and export of ribonucleic acids. The pyknotic nucleus of the late erythroblast (normoblast) is expelled in the mammalia, but is retained in an inactive form in all other vertebrates. The transition from the basophilic erythroblast to the normoblast defines clearly the end of the transscriptional phase of differentiation and the beginning of its maturational phase. The penultimate stage of the differentiation process, represented by the reticulocyte, is well-defined by an inactive or absent nucleus, on the one hand, and on the other, by the presence of functional mitochondria, ribosomes as well as of several active membrane transporters and receptors. Under normal circumstances the period of differentiation from stem cell to mature erythrocyte is about 7 days, half of which is accounted for by the reticulocyte stage. Therefore, one can distinguish reticulocytes of differing maturity. The maturity may be gauged by their hemoglobin concentration, which increases with maturity, by their decreasing basophilia which is equivalent to their amount of RNA, or by the decline of the number and the functional state of the mitochondria (for review see Rapoport et al. 1974). Among other characteristics there may be mentioned the absence of lipoxygenase in the most immature reticulocytes. The last stage of maturation, the transition to the mature erythrocyte involves the complete degradation of mitochondria and ribosomes, as well as changes of the cell membrane with selective decay of enzymes of the cytosol. Consequently, respiration and protein synthesis are abolished, glycolysis and various transport processes are strongly reduced.

A comparison of constituents and metabolic pathways between reticulocytes and erythrocytes of rabbits is given in Table 1.

The reticulocytes of the blood under normal steady state conditions represent an advanced stage of maturation, with low oxygen consumption and a high degree of uncoupling. In man even during anemia with reticulocytosis the reticulocyte usually ap-

1 Institut für Physiologie und Biologische Chemie, Humboldt Universität, 104 Berlin, GDR

Circulation, Respiration, and Metabolism
(ed. by R. Gilles)

Table 1. A comparison of the main constituents and metabolic pathways between reticulocytes and erythrocytes of rabbits

Constituent		Reticulocyte	Erythrocyte	Remarks
RNA	mg ml^{-1}	25– 4	0.3	
Hemoglobin	mg ml^{-1}	300–200	330	
Non-Hb Prot.	mg ml^{-1}	45	15	
Lipid	mg ml^{-1}	9	5	
Pathway				
RNA synthesis		–	–	
Protein synthesis		+	–	
Heme synthesis		+	–	
Purine synthesis		+	–	Some enzymes remain
Pyrimidine synthesis		+	–	
Glycogen synthesis		–	–	
Glycogenolysis		–	–	
Lipid synthesis		(+)	–	Weak in reticulocytes: only phosphatidic acid synthesized in erythrocytes
Respiration		+	–	
Citrate cycle		+	–	Some enzyme remain
Glycolysis		++	+	decreased
Gluconeogenesis		–	–	
OPP pathway		+	+	
Fatty acid breakdown		+	–	
Amino acid catabolism		+	–	Some enzymes remain
Active transport amino acids		++	(+)	Some carriers remain
Active transport cations		++	+	
Receptors		++	(+)	Decreased, some remain
Adenyl cyclase		+	–	

Table 2. Comparison of reticulocytes of man and rabbit values in 10^{-3} μM min^{-1} ml^{-1} cells

	Man (10)	Rabbit (21)
Cytc.-Oxyd.	24 ± 5	270 ± 40
$NADH_2$-Cytc.-Red.	240 ± 30	330 ± 40
O_2 Consumption	30 ± 4	110 ± 9
Reti (%)	28.5 ± 3.4	30.6 ± 2.4

pears to be highly mature (Richter-Rapoport et al. 1977). In addition, there may be species differences as shown in Table 2.

The activities of cytochrome oxidase and the succinate-cytosolic system are in human reticulocytes about one order of magnitude lower than in rabbit reticulocytes, whereas the activities of the $NADH_2^-$ Cytc reductase system are comparable. The decline of the activities of the components of the respiratory chain corresponds to their

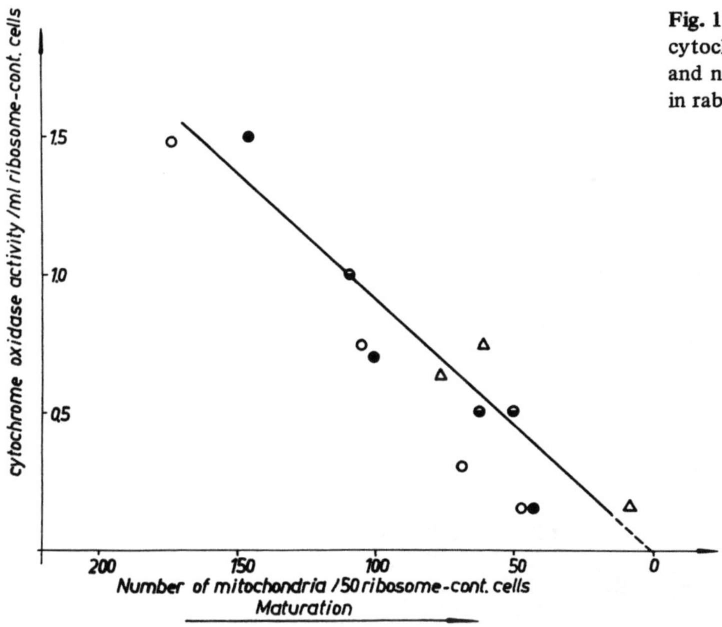

Fig. 1. Correlation between cytochrome oxidase activity and number of mitochondria in rabbit reticulocytes

rate of desintegration of mitochondria. As shown in Fig. 1 there is a strict correlation between the decline of cytochrome oxidase and the number of mitochondria.

The decrease of enzymic activities is not restricted to mitochondria, but also involves enzymes occurring in the cytosol and in the plasma membrane.

2 The Metabolism of the Reticulocyte

Let me briefly discuss the metabolism of the reticulocyte. In a glucose-containing salt medium glucose is the main substrate and the main source of ATP generation, whereas under endogenous conditions amino and fatty acids, arising mostly from the breakdown of the mitochondria furnish a major share. In addition, there is ribose made available by the breakdown of ATP and ribonucleic acids. The fate of glucose is depicted in Fig. 2.

It may be seen that more than half may be accounted for by the formation of lactate, which furnishs about one-fifth of the 180 m*M* ATP formed per liter and hour (Siems et al. 1982). A sizeable percentage is represented by the formation of serine, by two parallel pathways via hydroxypyruvate and phosphoserine, respectively, for which glucose contributes the C-skeleton (Rapoport et al. 1980). The oxidative pentose phosphate pathway, on the other hand, is of minor importance.

Respiration accounts for 80% of the total ATP production of 120–200 mmol l^{-1} reticulocyte and h^{-1}, the rest being contributed by glycolysis. The respiratory substrate is 85% glucose with fatty acids and amino acids contributing the remainder. The total

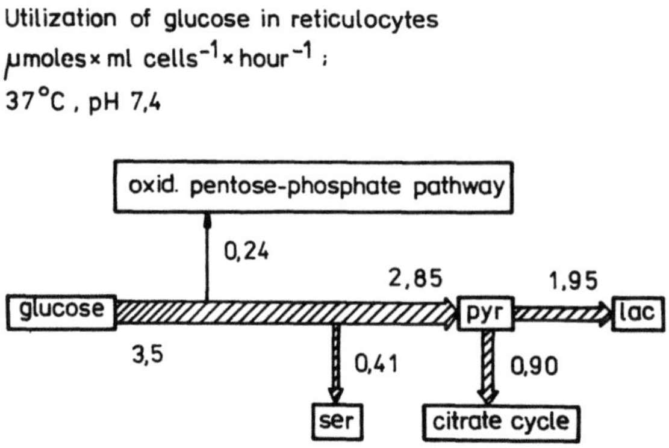

Fig. 2. Utilization of glucose in rabbit reticulocytes (mmol l^{-1} cells^{-1} h^{-1})

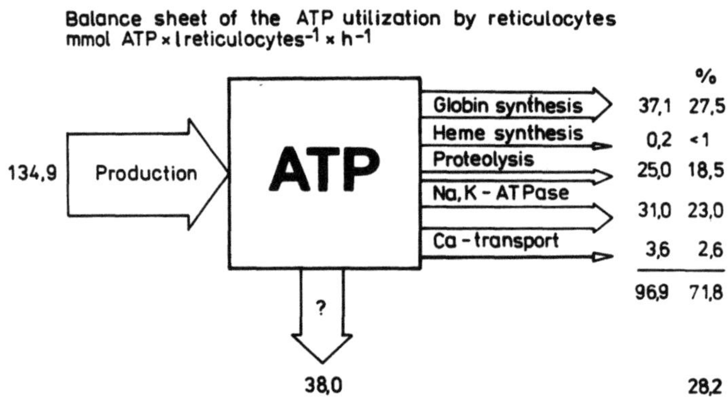

Fig. 3. Balance sheet of the ATP utilization by rabbit reticulocytes (mmol ATP l^{-1} reticulocytes^{-1} h^{-1})

ATP production is about two orders higher than in erythrocytes. What about the ATP consuming processes in reticulocytes?

A synopsis of our results is presented in Fig. 3. There are three major consumers of ATP: first protein synthesis with about 30%, secondly Na, K, ATPase with about 25%, and thirdly ATP-dependent proteolysis with 15%, whereas 30% are sofar unaccounted for. These processes do not compete with each other for ATP and appear to be regulated independently (Siems et al. 1984).

3 LOX and the Program of Maturation

What are the mechanisms underlying the phenomena of maturation? We focussed our attention on the factors instrumental in the destruction of mitochondria, on the assumption that the reduction of ATP formation with the consequent loss of protein synthesis and other ATP-requiring processes plays a key role in the transition to the erythrocyte. We found that an erythroid cell-specific lipoxygenase (LOX) is the primary destructive agent (Schewe et al. 1975). Other types of LOX are distributed widely in different tissues of man and animals and are considered to be of importance as cellular signals, and in pathological processes, such as inflammation and asthma. The specific function of the reticulocyte enzyme appears to be the oxidative damage to mitochondria, which may be considered a teleonomic key event in the maturation of red cells.

4 Properties and Actions of Reticulocyte Lipoxygenase (LOX)

The enzyme from rabbit reticulocytes is so far the only lipoxygenase of animal origin to be purified and characterized (Rapoport et al. 1979). Some properties are summarized in Table 3.

It consists of a single polypeptide chain and contains 1 Fe mol^{-1}, properties which appear to be common to other lipoxygenases. The C-terminal heterogeneity is probably due to the endogenous posttranslational action of exopeptidases. The enzyme has a high affinity for both O_2 and its organic substrates.

The reticulocyte enzyme has the exclusive property to attack intact phospholipids and even biological membranes. It may be classified as a n-6 - LOX which exhibits also an affinity for the n-9 position (Kühn et al. 1983). The enzyme catalyzes two consecutive reactions, namely, (1) oxygenation, which leads to the formation of hydroperoxy fatty acids and (2) hydroperoxidase activity by which the products of the first reaction are converted into various products, including hydroxy and epoxyhydroxy compounds as well as ketodienoic acids (Rapoport et al. 1984). A further feature is the self-inactivation of the enzyme.

Several actions of the reticulocyte LOX may be involved in a variety of maturational changes. They are listed in Table 4.

LOX causes rupture and lysis of mitochondrial membranes depending on the functional state of mitochondria, whereas cytoplasmic membranes are much more resistant to its attack (Fritsch et al. 1979).

LOX exerts two types of inhibitory actions on the respiratory chain. One is exerted on cytochrome oxidase and is limited to the phospholipids (Wiesner et al. 1981). The other occurs at the sites between the proximate FeS centers and ubiquinone (Schewe

Table 3. Characteristics of reticulocyte LOX

Mol. wt.	78,500
N-terminus	Gly
C-terminus	His, Ile, Asp
Structure	1 Chain
Substrates	Free and esterified polyenic acids
Positional specificity	n-6 $\gg$ n-9
Inactivators	Polyacetylenic, hydroperoxy-fatty acids
pH Optimum	≈ 8
K_m (Linoleate)	12 μM
K_m (O_2)	7 μM

Table 4. Actions of reticulocyte LOX

1.	Lysis of mitochondrial membrane
2.	Inactivation of the respiratory chain at the level of complexes I and II
3.	Inhibition of the cytochrome oxidase
4.	Uncoupling of oxidative phosphorylation
5.	Cooxidative destruction of the Fe-S centers of the mitochondrial outer membrane
6.	Cooxidative inactivation of SH-enzymes
7.	Self-inactivation by lipohydroperoxides
8.	Triggering of ATP-dependent proteolysis

et al. 1981), and apparently also involves proteins. The attack on the outer mitochondrial membrane and the inactivation of cytosolic SH enzymes are caused by cooxidation, thus, accounting for their decline during maturation.

The self-inactivation of LOX accounts partly for its disappearance during maturation. Its reaction products, the hydroperoxy fatty acids, cause the inactivation of the enzyme, which is due to the oxidation of one, out of 13 methionines, which is probably located in the active center of the enzyme (Rapoport et al. 1984).

There are several types of evidence for the functional role of LOX in vivo. For one, it can be shown with selective inhibitors that it accounts for about 5% of the total and for about 25% of the nonrespiratory oxygen uptake of reticulocytes (Salzmann et al. 1985). Secondly, we demonstrated the production of pentane, a breakdown product of the hydroperoxidase reaction of LOX in intact reticulocytes (Salzmann et al. 1984). Thirdly, salicyl hydroxamate (SHAM) a selective inhibitor of LOX, inhibited the decline of cytochrome oxidase during maturation in vitro.

There are three questions to be asked: for one, is the LOX found in rabbit reticulocytes specific for erythroid cells, secondly, does the maturation in the normal steady state proceed according to the same mechanism, and thirdly, can the results be generalized and do they apply to other species?

The answer to all three questions appears to be affirmative. Collaborative work of P. Harrison from the Beatson Institute in Glasgow and B. Thiele from our institute led to the isolation of a clone of cDNA derived from the LOX mRNA. By hybridization experiments with the cDNA it was found that LOX mRNA occurs only in bone marrow and the peripheral blood, thus, confirming earlier immunological data. The mRNA is found in the bone marrow and to a small extent in the peripheral blood of nonanemic animals. The occurrence of LOX or its mRNA has been demonstrated both in rats and mice and in earlier work was indicated even in the reticulocytes of chicken (Augustin and Rapoport 1959) as shown in Fig. 4.

5 Biological Dynamics of LOX

Diverse stimuli can induce the outpouring of reticulocytes from the bone marrow, thereby abbreviating their stay there, while prolonging their maturation time in the peripheral blood. Intense stimulation, such as drastic bleeding or hemolysis, induce the

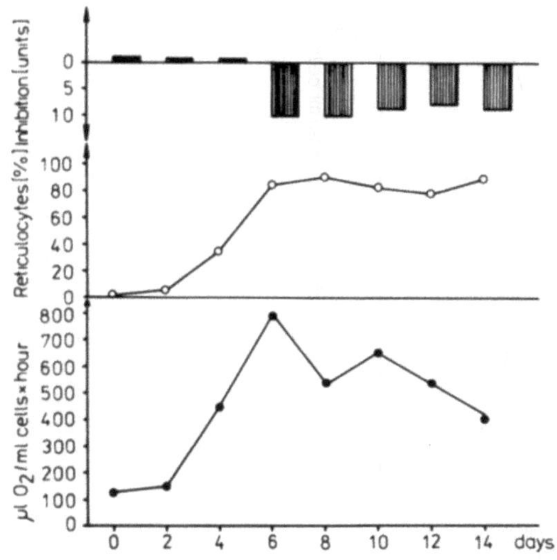

Fig. 4. Oxygen consumption, reticulocytosis, and respiratory-chain inhibitory activity of chicken red cells during an anemia

appearance of immature reticulocytes, including those which arise from skipped divisions. During the course of an experimental anemia various populations of reticulocytes succeed each other in close sequence, which differ among themselves in various characteristics. At about the sixth day the megaloreticulocytes which arise from skipped divisions predominate. At that time respiration which reaches its peak on the fourth day reaches a somewhat lower plateau. LOX is absent in normal blood. It appears generally on the fourth day and increases greatly thereafter during continuous bleeding. After bleeding is discontinued LOX may persist for a long time, even after the reticulocyte count has returned to normal. This time course is presented in Fig. 5.

In a complex study on various characteristics of density-fractionated cells during an experimental bleeding anemia in rabbits it was found that young ribosome-rich cells as well as bone marrow lacked LOX, which reached its maximal activity at the very time of the steepest decline of mitochondrial and other maturation-dependent parameters (Wiesner et al. 1973), such as cytochrome oxidase, ϑ-ALA synthetase, glutamin-

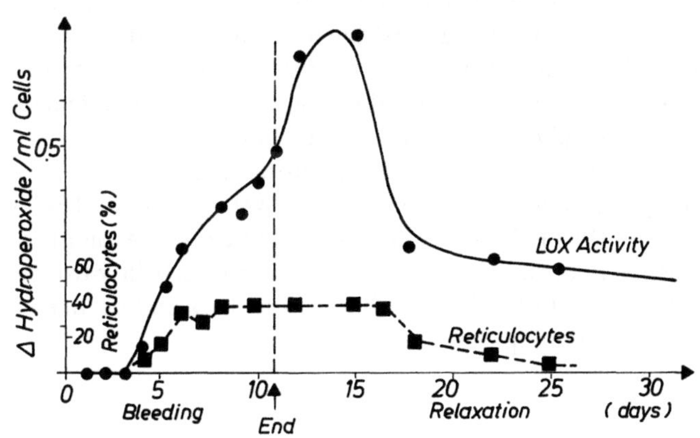

Fig. 5. Reticulocytosis and LOX activity during an anemia of rabbits

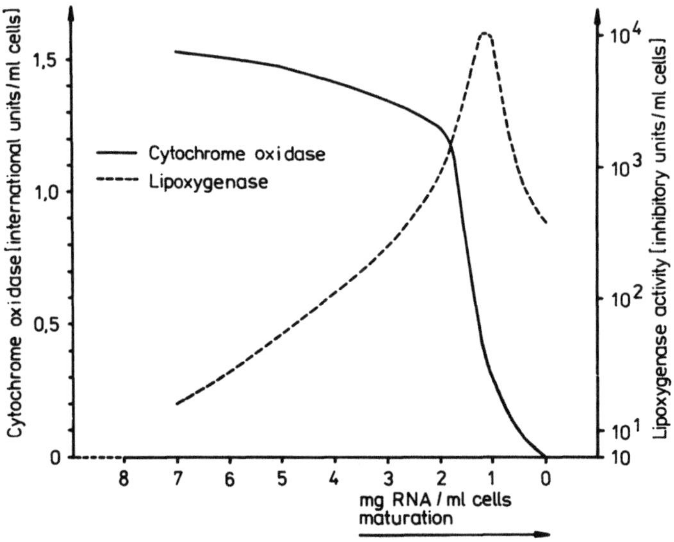

Fig. 6. Biological dynamics of cytochrome oxidase and LOX during maturation of reticulocytes

ase malate dehydrogenase, desoxyribonuclease, and pyrophosphatase. Subsequently the LOX disappears again owing to self-inactivation and degradation. This time course is depicted in Fig. 6.

Further work demonstrated that despite the fact that LOX mRNA is present in both bone marrow and immature reticulocytes. LOX synthesis occurs only in more mature reticulocytes. The reason for this circumstance is the masking of the mRNA in form of translationally inactive mRNP particles. The unmasking is assumed to be due to limited proteolysis (Thiele et al. 1979, 1982).

6 The Role of ATP-Dependent Proteolysis

Reticulocytes contain an ubiquitin-ATP-dependent system which accounts for more than 90% of the total proteolysis and is directed against the mitochondria (Müller et al. 1980). We have also found an ATP-dependent proteolytic system in liver mitochondria (Rapoport et al. 1982, 1983). It is conceivable that the widely distributed system of ATP-dependent proteolysis may generally be selective for mitochondria.

What about the relation between the actions of LOX and proteolysis?

Suppression of LOX activity, be it by anaerobiosis, or SHAM, prevents ATP-dependent proteolysis (Dubiel et al. 1981). Thus, the attack by LOX precedes and triggers the subsequent ATP-dependent proteolysis. Maturation experiments in vitro demonstrated clearly the inhibitory effect of SHAM on both proteolysis and decline of cytochrome oxidase activity as shown in Fig. 7.

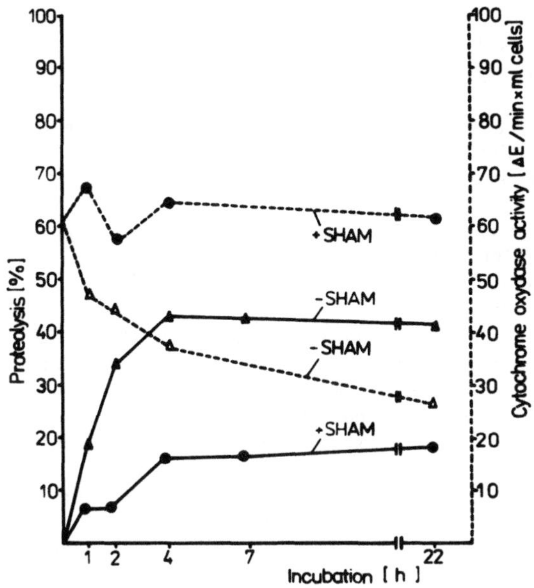

Fig. 7. The effect of SHAM on ATP-dependent proteolysis and cytochrome oxidase activity during in vitro maturation of rabbit reticulocytes at 37 °C

7 Conclusion

The main events of the degradation of mitochondria in reticulocytes may be summarized as follows.

The first event is the unmasking of LOX mRNA and the synthesis of LOX protein. Secondly, the LOX attacks the mitochondrial membranes and inactivates the respiratory chain, at the same time suffering self-inactivation. Now the way is open for the third step, the ubiquitin-ATP-dependent proteolysis. Finally, the breakdown is completed by other proteinases and phospholipases in mitochondria and in the cytosol.

References

Augustin HW, Rapoport S (1959) Über Atmung und Succinatoxydasesystem bei reifen und jugendlichen Hühnererythrozyten. Acta Biol Med Ger 3:433–449

Dubiel W, Müller M, Rapoport S (1981) ATP-dependent proteolysis of reticulocyte mitochondria is preceded by the attack of lipoxygenase. Biochem Int 3:165–171

Fritsch B, Maretzki D, Hiebsch CH, Schewe T, Rapoport S (1979) Zur Selektivität der Wirkung der Lipoxygenase aus Kaninchenretikulozyten auf Mitochondrien- und Erythrozytenmembranen. Acta Biol Med Ger 38:1315–1321

Kühn H, Wiesner R, Schewe T, Rapoport SM (1983) Reticulocyte lipoxygenase exhibits both n-6 and n-9 activities. FEBS Lett 153:353–356

Müller M, Dubiel W, Rathmann J, Rapoport S (1980) Determination and characteristics of energy dependent protolysis in rabbit reticulocytes. Eur J Biochem 109:405–410

Rapoport SM, Rosenthal S, Schewe T, Schultze M, Müller M (1974) The metabolism of the reticulocyte. In: Yoshikawa H, Rapoport SM (eds) Cellular and molecular biology of erythrocytes. University of Tokyo Press, Tokyo, p 93

Rapoport SM, Schewe T, Wiesner R, Halangk W, Ludwig P, Janicke-Höhne M, Tannert CH, Hiebsch CH, Klatt D (1979) The lipoxygenase of reticulocytes. Purification, characterization and biological dynamics of the lipoxygenase; its identity with the respiratory inhibitors of the reticulocyte. Eur J Biochem 96:545–561

Rapoport S, Müller M, Dumdey R, Rathmann J (1980) Nitrogen economy and the metabolism of serine and glycine in reticulocytes of rabbits. Eur J Biochem 108:449–455

Rapoport S, Dubiel W, Müller M (1982) Characteristics of an ATP-dependent proteolytic system of rat liver mitochondria. FEBS Lett 147:93–96

Rapoport S, Dubiel W, Müller M (1983) Calcium exerts an indirect effect on ATP-dependent proteolysis of rat liver mitochondria. FEBS Lett 160:134–136

Rapoport S, Härtel B, Hausdorf G (1984) Methionine sulfoxide formation – the cause of self-inactivation of reticulocyte lipoxygenase. Eur J Biochem 139:573–576

Richter-Rapoport SKN, Dumdey R, Hiebsch CH, Thamm R, Uerlings I, Rapoport S (1977) Charakterisierung von Retikulozyten des Menschen: Atmung, Pasteur-Effekt und elektronenmikroskopische Befunde an Mitochondrien. Acta Biol Med Ger 36:53–64

Salzmann U, Kühn H, Schewe T, Rapoport SM (1984) Pentane formation during the anaerobic reactions of reticulocyte lipoxygenase. Comparison with lipoxygenases from soybeans and green pea seeds. Biochem Bioph Acta 795:535–542

Salzmann U, Ludwig P, Schewe T, Rapoport SM (1985) The share of lipoxygenase in the antimycin-resistant oxygen uptake of intact rabbit reticulocytes. Biomed Biochim Acta 44:213–221

Schewe P, Halangk W, Hiebsch CH, Rapoport SM (1975) A lipoxygenase in rabbit reticulocytes which attacks phospholipids and intact mitochondria. FEBS Lett 60:1

Schewe T, Albracht SPJ, Ludwig P (1981b) On the site of action of the inhibition of the mitochondrial respiratory chain by lipoxygenase. Biochim Biophys Acta 636:210

Siems W, Müller M, Dumdey R, Holzhütter HG, Rathmann J, Rapoport SM (1982) Quantification of pathways of glucose utilization and balance of energy metabolism of rabbit reticulocytes. Eur J Biochem 124:567–576

Siems W, Dubiel W, Dumdey R, Müller M, Rapoport SM (1984) Accounting for the ATP-consuming processes in rabbit reticulocytes. Eur J Biochem 139:101–107

Thiele BJ, Belkner J, Andree H, Rapoport TA, Rapoport SM (1979) Synthesis of a lipoxygenase in reticulocytes. Synthesis of non-globin-proteins in rabbit erythroid cells. Eur J Biochem 96: 563–569

Thiele BJ, Andree H, Höhne M, Rapoport SM (1982) Translational repression of lipoxygenase mRNA in reticulocytes. Eur J Biochem 129:133–141

Wiesner R, Rosenthal S, Hiebsch CH (1973) Leitkriterien der Retikulozytenreifung. II. Das Verhalten von Zytochromoxydase und Hemmstoff F der Atmungskette bei der Retikulozytenreifung. Acta Biol Med Ger 30:631–646

Wiesner R, Schewe T, Ludwig P, Rapoport SM (1981) Reversibility of the inhibition of the cytochrome c oxidase by reticulocyte lipoxygenase. FEBS Lett 123:123

Symposium V

Comparative Cardiac Metabolism in Ectotherms

Organizers B. SIDELL and W. DRIEDZIC

Molluscan Circulation: Haemodynamics and the Heart

P.J.S. SMITH[1]

1 The Anatomy

The anatomy of the cardiovascular system in molluscs is highly variable, although with the exception of the Scaphopoda, all the molluscan classes have atria draining the venous blood to a central ventricular heart or hearts. This major pump propels the blood round a system of arteries, some with clearly elastic properties (for example, see Shadwick and Gosline 1981). Blood sinuses are invariably present although the cephalopods also have a peripheral bed of exchange vessels (Browning 1982). Venous vessels, frequently contractile, return deoxygenated blood to the gills and systemic heart. In decapod cephalopods there is a well developed accessory branchial heart (Johansen and Martin 1962; Smith 1982).

The Scaphopoda or tusk shells, have neither vessels nor a clearly differentiated ventricle. The circulatory system is made up entirely of sinuses, an area near the anus being contractile. Oxygen uptake takes place across the wall of the mantle cavity (Pelseneer 1906). In the Polyplacophora two atria empty into a medial ventricle via a variable number of ostia, whereas in the Monoplacophora there are two ventricles fed by two atria each (Fig. 1) and the heart empties into a single major vessel which in turn discharges into an open vascular system of sinuses (Fig. 1). In the bivalves and gastropods the ventricle is fed by one or two atria. Higher blood pressures recorded in the gastropod *Helix*, has led Jones (1983) to suggest that here the system is "more closed" and the term "openness" with regard to bivalves and particularly gastropods is an oversimplification. Although the tetrabranchiate *Nautilus* retains a low pressure open system with four atria feeding a central ventricle, no other group of molluscs attains the complexity of the dibranchiate cephalophod, either peripherally or at the level of the main cardiac organs (Fig. 2).

The molluscan heart is usually contained within a pericardium and indeed in the Aplacophora, the rudimentary ventricle and atrium are clearly derived from an invagination of the pericardial membrane (Hyman 1967). In cephalopods, however, the pericardium is much reduced (Marthy 1968) with the systemic heart and some major veins being contained in the now massive renal sacs.

Normally in molluscs, there is no cardiac arterial supply, the myocardium being either very thin or composed of numerous trabeculae (illustrated for *Busycon canali-*

1 A.F.R.C. Unit of Insect Neurophysiology and Pharmacology, Department of Zoology, University of Cambridge, Downing Street, Cambridge CB2 3EJ, Great Britain

Circulation, Respiration, and Metabolism
(ed. by R. Gilles)

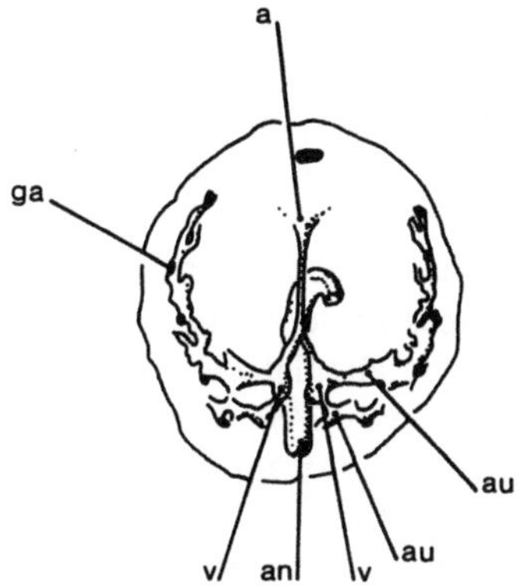

Fig. 1. Monoplacophoran circulatory system (*Neopilina galathea*: after Martin and Johansen 1965). *a*, aorta; *an*, anus; *au*, auricle; *ga*, entrance of the artery from the gill; *v*, ventricle

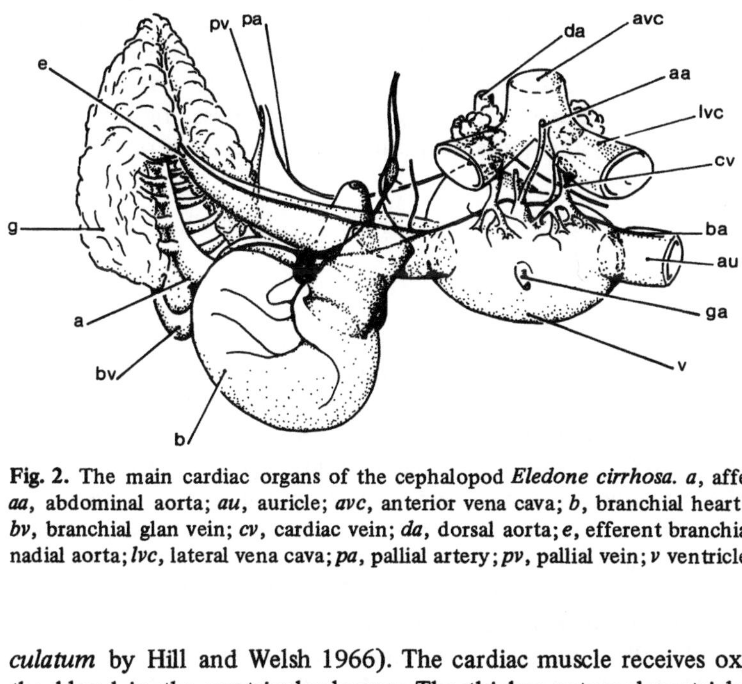

Fig. 2. The main cardiac organs of the cephalopod *Eledone cirrhosa*. *a*, afferent branchial vessel; *aa*, abdominal aorta; *au*, auricle; *avc*, anterior vena cava; *b*, branchial heart; *ba*, branchial artery; *bv*, branchial glan vein; *cv*, cardiac vein; *da*, dorsal aorta; *e*, efferent branchial vessel; *g*, gill; *ga*, gonadial aorta; *lvc*, lateral vena cava; *pa*, pallial artery; *pv*, pallial vein; *v* ventricle

culatum by Hill and Welsh 1966). The cardiac muscle receives oxygen directly from the blood in the ventricular lumen. The thicker octopod ventricle shares the general molluscan feature of lacking an endothelium, but has well-developed ventricular arteries draining into the lateral venae cavae through cardiac venous vessels (Fig. 2 and Smith and Boyle 1983).

2 The Role of the Circulatory System

The molluscan cardiovascular system functions in the transport of gas and metabolites, in renal physiology and in locomotion. The ventricle is a major site of ultrafiltration (Martin 1983) with the size of the renopericardial canal affecting the efficiency of ventricle refilling (Dale 1974). As the molluscs are soft bodied, the circulatory system is both affected by body movements and can play a role in executing these movements (Trueman et al. 1966). Although blood pressures can be affected by movement, the heart is protected from the more extreme pressure fluctuation by valves (Alexandrowicz 1962; Brand 1972; Bourne and Redmond 1977).

The circulatory system clearly has a role in respiratory gas transport but the work by Booth and Magnum (1978) and Bekius (1972) introduces some confusion. The former showed that in highly oxygenated water, blocking the circulatory system of *Modiolus demissus*, a bivalve, reduced the O_2 consumption by only 14%. The primary route of O_2 uptake is direct and the blood "does not have a major respiratory role". Experiments by Taylor and Brand (1975) lead to a similar conclusion for *Arctica islandica*. Both these species lack a respiratory pigment and Booth and Mangum conclude that in such species "the circulating body fluids must be involved primarily in other physiological processes such as the delivery of nutrients, salt and water balance and the excretion of metabolic waste". This conclusion applies to some degree throughout the bivalves (see also Famme 1981). The gastropod *Haliotis* can remain alive for 5 days after the removal of the heart (Bekius 1972) leading Bekius to propose that the circulatory system is probably aided by other pulsatile movements such as those of the buccal mass.

3 Venous Pressures

The magnitude of returning venous pressure is an important factor in cardiac physiology and the aneural control of the heart. The end diastolic stretch of the cardiac muscle controls the cardiac output by reflex. Where measured, returning venous blood seldom has a pressure above 10 cm H_2O (Table 1). One exception is *Octopus dofleini* (Johansen and Martin 1962) although lower pressures have been recorded from *O. vulgaris*. In both cases the pressure waves are complicated by being superimposed on those generated by the contraction of the mantle. In *O. vulgaris* the mantle contraction only generates a pressure of 1–2 cm H_2O (Wells and Smith 1985). However, the difference in the diastolic pressure level between the two species requires confirmation.

The data from Duval (1983), using a closed loop measuring system on terrestrial slugs, are confusing (Table 1b). Blood pressure records from the auricle of *Limax* (Duval 1983) show a pulse amplitude similar to previous studies (4 cm H_2O). The contradiction is in both the auricular and ventricular diastolic pressure levels (29 cm H_2O). Although other studies have used open loop systems, a standing pressure of such magnitude should not be lost. Further study is needed.

4 Intraventricular and Aortic Pressures

The pressures generated by ventricular contractions are about as varied among the species as is the anatomy. Several points remain to be clarified. Florey and Cahill

Table 1. Circulatory pressures, blood volumes and ventricular contraction rates for bivalves, gastropods and cephalopods (pressures in cm H_2O)

Species (Ref.)[a]		Atrial pressure Systolic / diastolic	Ventricular pressure Systolic / diastolic	Heart rate (beats min^{-1})	Blood volume % wet. wt.
a) BIVALVES					(Minus shell)
Anodonta anatina	(4)	1.5 / 0.7	2 –4 / 0.6	20 (R.T.)	–
Anodonta cygnea	(13)	1.5 / 0.6	4.8 / 1.1	–	55 (14)
Mya arenaria	(7)	–	3.0 / 0 –1.5	17 (11–12 °C)	–
Tresus nutalli	(7)	0.2 / 0	1.5–3.0 / 0	8 (18–20 °C)	–
Saxidomus giganteus	(7)	–	6.0 / 0	8 (11–12 °C)	–
b) GASTROPODS	: Prosobranche				(Minus shell)
Haliotis corrugata	(21)	2.2 / 1.1	8.8 / 5.9	21 (15 °C?)	–
			(Aortic pressure)		
Patella vulgata	(9)	2.5–3.5 / 1 –2.5	5 / 2.5	21	66 (11)
	: Pulmonates				
Lymnaea stagnalis (aquatic)	(5)	4 –5 / 3 –4	10 / 4	15– 40 (15)	34 (18)
Helix pomatia	(10)	6 –8 / 4 –5	24 / 6	60	–
Limax pseudoflavus	(6)	54 / 29.3	74.5 / 51.9	60 (20 °C)	–
Deroceras reticulatum	(6)	22 / 12.2	35.2 / 21	88–108 (20 °C)	–
		Efferent branchial vessel pressure Systolic / diastolic	Ventricular and aortic pressure Systolic / diastolic	Heart rate (beats min^{-1})	Blood volume % wet. wt.
c) CEPHALOPODS					
Nautilus pompilius	(3)	2.5 / 0.5	28 /1 35 /16	5–18 (17 °C)	–
Octopus dofleini	(8)	10 –25 / 5 –15	– 45–70/25–50	8–18 (7–9 °C)	–
Octopus vulgaris	(19)	3.0 / 0.5 (17)	– 25–50/10–15	45	5.8 (12)
Loligo pealei	(1)	5.6 / 0.1	76.7/0.7 73.2 /27.7	95 (19–22 °C)	–

[a] Sources: (1) Bourne (1982); (2) Bourne and Redmond (1977); (3) Bourne et al. (1978); (4) Brand (1972); (5) Dale (1974); (6) Duval (1983); (7) Florey and Cahill (1977); (8) Johansen and Martin (1962); (9) Jones (1970); (10) Jones (1971); (11) Jones and Trueman (1970); (12) Martin et al. (1958); (13) Peggs (in Jones 1984); (14) Potts (1954); (15) Schwartzkopff (1954); (16) Smith and Davis (1965); (17) Smith and Wells (unpublished); (18) Soffe et al. (1978); (19) Wells (1979)

(1977) found normal pressure in the ventricle of *Tresus*, *Saxidomus*, and *Mya*, higher than reported in earlier studies (see also Jones 1983). As these authors note the differences are probably attributable to the technique used for measurement. The pressure is important if ultrafiltration is to occur across the bivalve ventricle. [The colloid osmotic pressure of Lamellibranch blood is 1.3 cm H_2O (Mangum and Johansen 1975)]. Coleoid cephalopods clearly have the highest pressures (with the exception of Duval's data) and the lowest blood volumes (Table 1c) a result to be expected if the vascular system is closed. *Nautilus* retains a more open system (Griffin 1900) and has on average lower aortic pressures (Bourne et al. 1978).

When the range of blood pressures encountered in the phylum is examined, that in the terrestrial pulmonate *Helix* is found to be markedly higher than the aquatic gastropods (Table 1b). This may be due to *Helix* possessing a more closed system with higher peripheral resistances (Jones 1983) but could also reflect the use of the haemolymph as a hydrostatic skeleton since a land animal lacks the supportive water medium (Jones 1983).

The discussion of Duval (1983) in the previous section also applies to the ventricular pressures.

5 Cardiac Cycle

5.1 In Vitro Studies

The different classes of molluscs form an ascending series where the role of the cardiovascular system assumes a greater importance in the delivery of oxygen to the tissues. In the cephalopods, where circulation is no longer greatly influenced by body movements, many developments are superficially convergent with the vertebrates (Packard 1972). Clearly one of the most interesting questions to ask is whether the increased importance of the cardiovascular system is reflected in the development and specialization of the cardiac muscle and its performance. Unfortunately this is one area of research which has been largely ignored. The isolated molluscan ventricle has been used extensively in pharmacological studies (see Lever and Boer 1983). Most, however, have observed on contraction rate and 'force', which are both dependent on the degree of stretch (Nomura 1963; Almqvist 1973, Hill and Irisawa 1967). Stretch alters the rise time of the cardiac muscle prepotential resulting in a beat rate increase. There is also an increase in the duration of the action potential plateau giving a force increase (Nomura 1963). In spike-plateau action potentials the force of contraction is related to the Ca^{2+} controlled Na^{+} dependent plateau amplitude (for review see Jones 1983). In a perfused *Busycon* heart the form of the action potential changes with perfusion conditions and the resultant output (Figs. 3 and 5; Smith 1985).

The isolated ventricle of the octopod *Eledone cirrhosa* shows a clear relationship between heart rate, stroke volume and the level of an artificial venous return pressure (Fig. 4). Heart rate increases linearly throughout the pressure range despite the high levels at the upper limit (Table 1). As in the fish, large changes in perfusion pressure are required to produce significant increases in heart rate (Jones and Randall 1978). Although stroke volume is affected by changes in input and back pressure on the heart,

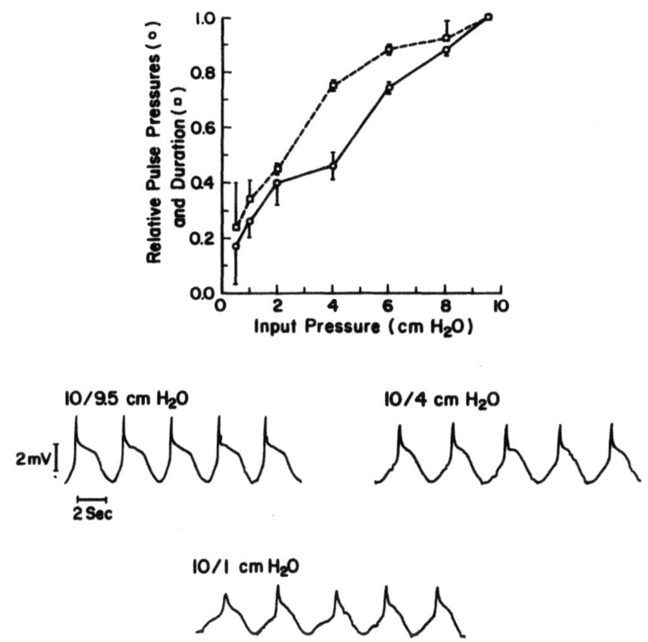

Fig. 3. Records made with a suction electrode on the surface of a beating, isolated and perfused heart from the whelk *Busycon canaliculatum* (20 °C). The amplitude and duration of the simultaneously measured aortic pressure pulse is also shown (mean values from five measurements). The output back pressure on the heart (P_2 in Fig. 5) was held at 10 cm H_2O and the input pressure varied. Note that the fast spike duration and plateau duration change with input pressure, as does the rise time of the prepotential (Smith 1985)

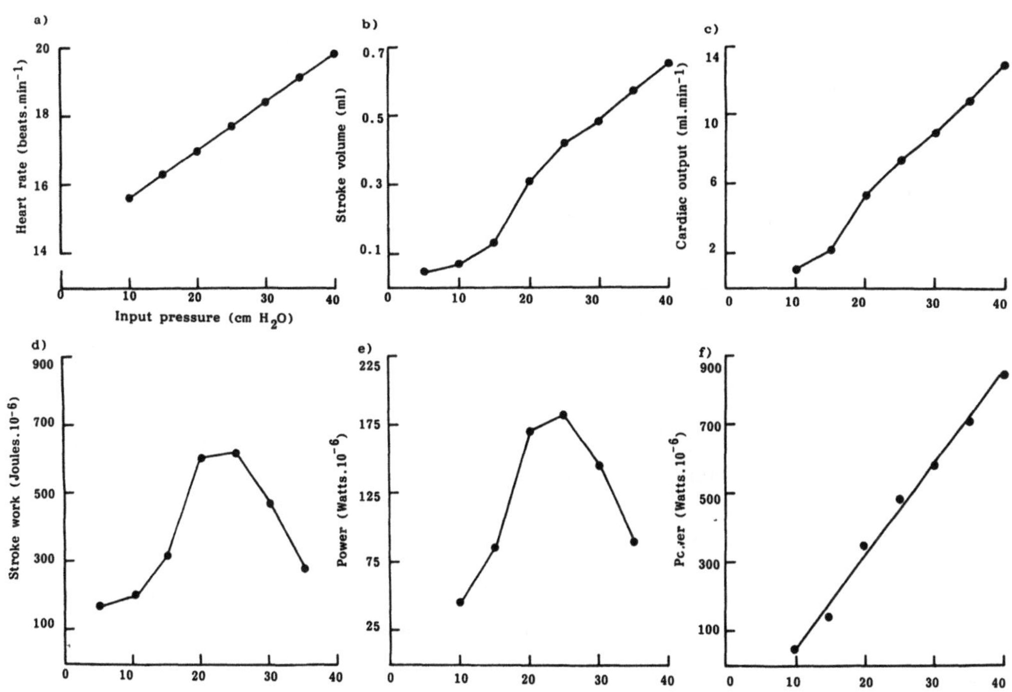

Fig. 4a–f. Response parameters from the isolated ventricle of the octopus *Eledone cirrhosa* to variation of the input pressure (10 °C). The output back pressure was constant at 40 cm H_2O. The results are mean values from three preparations perfused as described by Smith (1981)

heart rate is not affected by the latter (Smith 1981). As stroke volume is affected by pressures on both sides of the heart, the effects of end diastolic pressure and the pressure difference across the heart are confused. Smith (1981) showed that the former is important between 0 and 20 cm H_2O, whereas at higher pressures it is the pressure difference across the heart which regulates output.

Stroke work can be calculated from the stroke volume and the pressure difference across the heart. [If stroke volume is expressed in cubic metres and the pressure difference in newton/metre^2 (Pascals: 98.1 Nm^{-2} = 1 cm H_2O) their product is in Joules).] Above the pressure of 20 cm H_2O the ventricle works less as the stroke volume increase does not compensate for the drop in the pressure difference.

Ventricular power can be calculated by relating the stroke work to the frequency of contraction [Power (Watts) = Js^{-1}).] The ventricular powers (Fig. 4) are calculated from the results of animals between 700 g and 1 kg and are therefore approximately equal to Watts 10^{-6} kg^{-1}. A 900 g *E. cirrhosa* has a ventricle weight of approx. 1.25 g [Boyle and Knobloch 1982: the power output in relation to heart weight is 32 to 136 mWatts kg^{-1} ventricular muscle over an input pressure range of 5–20 cm H_2O. This compares with 45 Watts kg^{-1} for human cardiac muscle (Weis-Fogh and Alexander 1977).] For the octopus measurements output back pressure was set at 40 cm H_2O, a level based on the data from *Octopus dofleini* (Table 1). This level would now appear too high so the power outputs are probably artificially low.

The stroke volume of the isolated heart of *Busycon canaliculatum* (Gastropoda) behaves in a similar way to octopus on alteration of input pressure (Fig. 5; Smith 1985). Over the pressure ranges used, output back pressure (P_2) had no effect on output. Heart rate is almost independent of pressure between 10 and 3 cm H_2O but begins to fall off below these levels. This corresponds to the appearance of an irregular myogram recording (Fig. 3). As the octopus heart frequently failed to contract at input pressures less than 5 cm H_2O, the *Busycon* ventricle is clearly capable of working better at these lower pressures. Both work and power curves, however, are greatly reduced at input pressures of less than 1 cm H_2O. In *Mercenaria mercenaria*, a bivalve, the heart still works well at an input pressure of 0.75 cm H_2O (Fig. 5; Smith 1985). The power output of the *Busycon* heart ranges from 18 to 26 mWatts kg^{-1} ventricular muscle over an input pressure range of 2–3.5 cm H_2O (P_2 = 6 cm H_2O). *Mercenaria* has a similar power range, 13 to 29 mWatts kg^{-1} ventricular muscle with input pressures of 0.25–0.75 cm H_2O (P_2 = 2 cm H_2O). *Mercenaria* cardiac muscle is clearly adapted to work at still lower pressure levels.

Smith and Hill (in preparation) have shown that the putative cardiac excitatory transmitter 5-hydroxytryptamine and the peptide FMRFamide elevate the cardiac output from the isolated *Busycon* heart primarily by increasing stroke volume. The most striking effect of both is on the flow rate, as the output pressure pulse increases is amplitude but is decreased in duration.

The only other power values for a molluscan ventricle are from Herold's (1975) study on *Helix*. He calculated mechanical power by the formula:

Power (Watts) = K.h. V.N.

K is the product of the perfusion fluids specific weight and its acceleration (9.81 m s^{-2}), h is the height of the perfusion fluid in the cannula (m), V is the stroke volume (μl), and N is the contraction frequency in hertz.

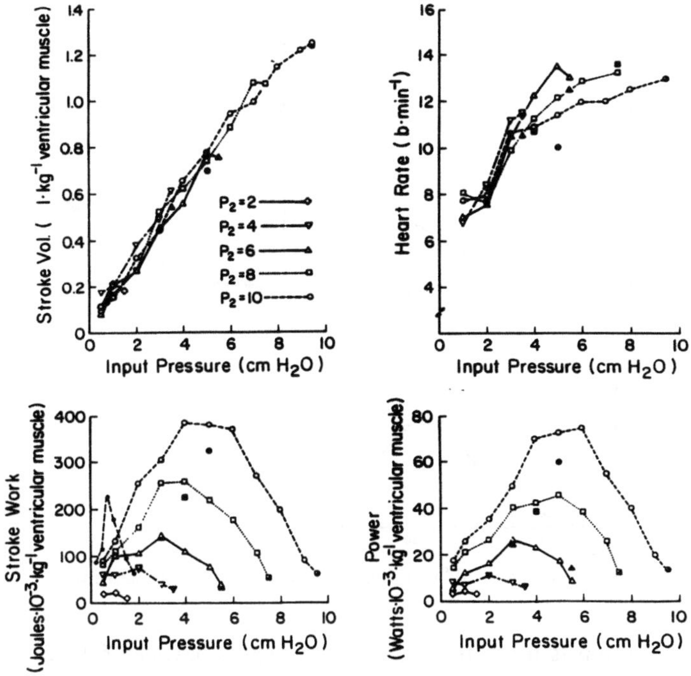

Fig. 5. Response parameters from the isolated heart of *B. canaliculatum* (20 °C; means from four preparations with an average heart weight of 253 mg). The measurement were made at five different levels of output back pressure (P_2). One stroke work curve is given for the heart of the bivalve, *Mercenaria mercenaria* (0———0; 20 °C, mean values from two preparations at one output pressure of 2 cm H_2O. (The mean heart weight was 68 mg.) (Smith 1985)

As Herold (1975) perfused his preparation by a Straub cannula the difference between the input and output pressure levels is ill-defined and the meaning of the resulting power values remains unclear. If the octopus data are used to calculate power by this formula it is clearly different from the results discussed above (Fig. 4e,f). The power of the heart is confused with the potential energy of the system. Bearing in mind the reservation about this approach to calculating power, Herold (1975) found values two orders of magnitude lower than octopus. When ventricular weight is taken into account and hearts perfused at low input pressures (6 cm H_2O), a value of 103 mWatts kg^{-1} ventricular muscle results for *Helix*.

The efficiency of the *Helix* ventricle is affected by the ionic composition and the PO_2 of the perfusion fluid (Herold 1975). Temperature can also affect cardiac output as Duval (1983) has shown for *Limax pseudoflavus*, in vivo. In *Eledone* a temperature change of 10 °C has a considerable effect on in vitro ventricular performance (Fig. 6). Heart rate increases but there is a reduction in stroke volume. In this context the distribution of this species is interesting as it extends from the North Atlantic to the Mediterranean over which a considerable temperature range could be ecnountered. Some compensation in the cardiac metabolism might be expected. However, although in the North Sea *E. cirrhosa* is common in shallow sub-littoral waters (Boyle and Knob-

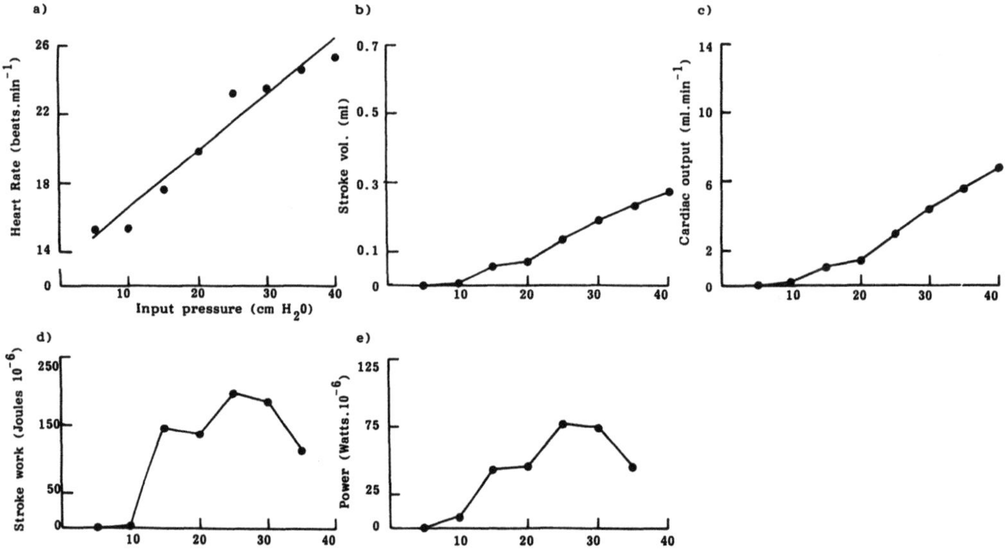

Fig. 6a–e. As in Fig. 4 but at a temperature of 18° –20 °C. (Mean values from three preparations)

loch 1982) its southern distribution is restricted primarily to depths between 60 and 120 m (Mangold-Wirz 1963) where presumably temperatures are lower.

5.2 In Vivo Studies

Although the exact function of cardiac output is species dependent, in many cases a major role is to deliver oxygen to the tissues during exercise. For the cephalopods oxygen delivery is clearly a primary function of the cardiovascular system, and this is one reason why more data are available for exercise in this group. It is also relatively easy to control and quantify (O'Dor 1982; Wells et al. 1983).

At rest the stroke volume of a 500 g *O. vulgaris* is approx. 0.5 ml with heart rates around 45 beats min^{-1} (Wells 1979). Aortic diastolic and systolic pressures are about 20 and 36 cm H_2O respectively. There are no intraventricular pressure measurements for *Octopus* but assuming it relates to aortic pressure as in *Haliotus* (Bourne and Redmond 1977) and *Helix* (Jones 1971) an approximate 'work-loop' can be drawn (Fig. 7). The area of this loop equals the potential energy component of stroke work (Mountcastle 1974).

During exercise the oxygen consumption of an octopus goes up about three times (Wells et al. 1983). Heart rate increases by 16% (Wells 1979). Unlike the mammal there seems to be no extra capacity in *Octopus* blood for loading or unloading oxygen during exercise (Houlihan, Duthie, Smith, and Wells, in preparation). Stroke volume must therefore increase by around 2.4 times. This is within the capability of an isolated heart (Smith 1981 and unpublished data). When the work loop is redrawn, for exercise, its area increases dramatically (five times the resting area).

The shape of the work-loop in octopus is quite different from the mammal (Figs. 7 and 8) and more similar to the fish (Randall 1970). In the human loop the majority of

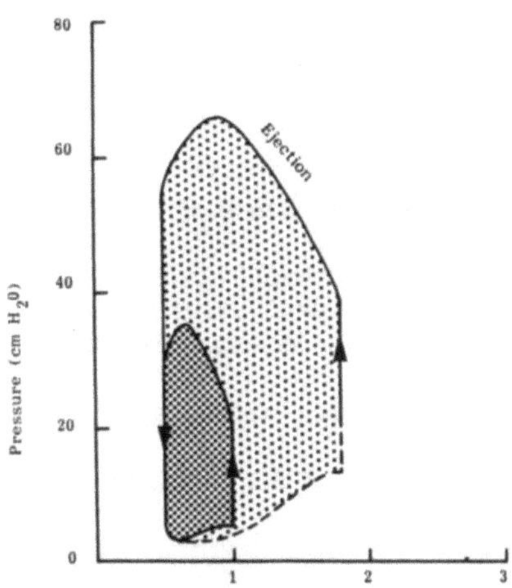

Fig. 7

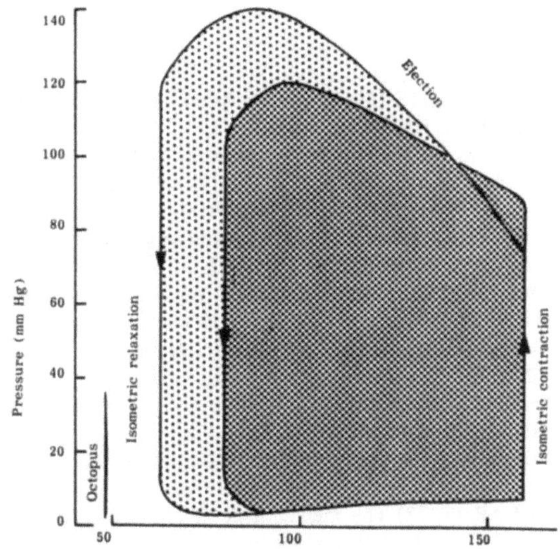

Fig. 8

Fig. 7. Work-loop for a 500 g *Octopus vulgaris* at rest (*dense stipple*) and immediately after exercise (*light stipple*). The position of both loops relative to end diastolic volume is not known

Fig. 8. Work-loop for a man at rest (*dense stipple*) and during exercise (*light stipple*). The data is from Mountcastle (1974). The octopus resting work-loop is also shown at the same scale

the pressure increase is taken up by the isometric component, with the aortic pressure wave is generally smaller (Mountcastle 1974). The systolic pressure in the dorsal aorta of octopus can be almost twice the diastolic (36.3 cm H_2O/20 cm H_2O; calculated from Wells 1979). The work loop is less dominated by the isometric phase. Also, when a mammal exercises the peripheral resistance goes down, reflected by a drop in the diastolic pressure. This is clearly not so in octopus and may, by analogy, imply an increase in peripheral resistance. In mammals the area increase of the work loop during exercise is only about 30% although oxygen consumption increases by five or six times (data from Mountcastle 1974). This is partly because an increase in heart rate accommodates a higher level of cardiac output. When this is taken into account the power output in octopus is seven times the resting level whereas in man it is only 2.5. Increasing cardiac output by stepping up stroke volume, at a relatively constant heart rate, is the same strategy used by the teleost during exercise (Jones and Randall 1978). It would now be timely and of great interest to examine the consequence of these differences on the physiology of the cardiac muscle, muscles similar in function but quite different in performance.

Acknowledgements. The previously unpublished material reported in this paper was financed by a Royal Society Study Visit Grant and by a NSF Grant No. PCM 830809 to Dr. R.B. Hill, Rhode Island, whose support I gratefully acknowledge. I would also like to thank Dr. C. Ellington, Dr. I.G. Priede, and Dr. G. Smith for advice and criticism on the work and manuscript.

References

Alexandrowicz JS (1962) An accessory organ of the circulatory system in *Sepia* and *Loligo*. J Mar Biol Assoc UK 42:405–418

Almqvist M (1973) Dynamic properties of stretch-induced chronotropy in the isolated heart of the snail *Helix pomatia*. Acta Physiol Scand 87:39A–40A

Bekius R (1972) The circulatory system of *Lymnaea stagnalis* (L). Neth J Zool 22:1–58

Booth CE, Mangum CP (1978) Oxygen uptake and transport in the lamellibranch mollusc *Modiolus demissus*. Physiol Zool 51:17–32

Bourne GB (1982) Blood pressure in the squid *Loligo pealei*. Comp Biochem Physiol 72A:23–27

Bourne GB, Redmond JR (1977) Hemodynamics of the Pink Abalone, *Haliotis corrugata* (Mollusca, Gastropoda) 1. Pressure relations and pressure gradients in intact animals. J Exp Biol 200:9–16

Bourne GB, Redmond JR, Johansen K (1978) Some aspects of hemodynamics in *Nautilus pompilius*. J Exp Zool 205:63–70

Boyle PR, Knobloch D (1982) On growth of the octopus *Eledone cirrhosa*. J Mar Biol Assoc UK 62:277–296

Brand AR (1972) The mechanisms of blood circulation in *Anodonta anatina* L. (Bivalvia, Unionidae). J Exp Biol 56:361–379

Browning J (1982) The density and dimensions of exchange vessels in *Octopus pallidus*. J Zool (Lond) 196:569–579

Dale B (1974) The eco-physiological significance of the circulatory mechanics of *Lymnaea stagnalis* L. Comp Biochem Physiol A47:1105–1113

Duval A (1983) Heartbeat and blood pressure in terrestrial slugs. Can J Zool 61:987–992

Famme P (1981) Haemolymph circulation as a respiratory parameter in the mussel, *Mytilus edulis* L. Comp Biochem Physiol A69:243–247

Florey E, Cahill MA (1977) Haemodynamics in lamellibranch molluscs: confirmation of constant-volume mechanism of auricular refilling. Remarks on the heart as site of ultrafiltration. Comp Biochem Physiol A57:47–52

Griffin LE (1900) The anatomy of *Nautilus pompilius*. Mem Natl Acad Sci 8:103–197

Herold JP (1975) Myocardial efficiency in the isolated ventricle of the snail, *Helix pomatia* L. Comp Biochem Physiol 52A:435–440

Hill RB, Irisawa H (1967) The immediate effect of changed perfusion pressure and the subsequent adaptation in the isolated ventricle of the marine gastropod *Rapana thomasiana*. Life Sci 6: 1691–1697

Hill RB, Welsh JH (1966) Heart, circulation and blood cells. In: Wilbur KM, Yonge CM (eds) Physiology of mollusca, vol II. Academic, New York, p 125

Hyman LH (1967) The invertebrates, vol 6. McGraw Hill, New York

Johansen K, Martin AW (1962) Circulation in the cephalopod *Octopus dofleini*. Comp Biochem Physiol 5:161–176

Jones DR, Randall DJ (1978) The respiratory and circulatory systems during exercise. In: Hoar WS, Randall DJ (eds) Fish physiology, vol VII. Academic, New York, p 425

Jones HD (1970) Hydrostatic pressures within the heart and pericardium of *Patella vulgata* L. Comp Biochem Physiol 34:263–272

Jones HD (1971) Circulatory pressures in *Helix pomatia* L. Comp Biochem Physiol 39A:289–295

Jones HD (1983) The circulatory systems of gastropods and bivalves. In: Saleuddin ASM, Wilbur KM (eds) The mollusca, vol 5, part 2. Academic, New York, p 189

Jones HD, Truman ER (1970) Locomotion of the limpet, *Patella vulgata* L. J Exp Biol 52:201–216

Lever J, Boer HH (eds) (1983) Molluscan neuro-endocrinology. North-Holland, Amsterdam

Mangold-Wirz K (1963) Biologie des Cephalopodes benthiques et nectoniques de la Mer Catalane. Suppl 13, Vie Milieu

Mangum CP, Johansen K (1975) The colloid osmotic pressures of invertebrate body fluids. J Exp Biol 63:661–671

Marthy HJ (1968) Die Organogenese des Coelomsystems von *Octopus vulgaris* Lam. Rev Suisse Zool 75:723–763

Martin AW (1983) Excretion. In: Saleuddin ASM, Wilbur KM (eds) The mollusca, vol 5, part 5. Academic, New York, p 353

Martin AW, Johansen K (1965) Adaptations of the circulation in invertebrates. In: Hamilton WF, Dow P (eds) Handbook of physiology, vol III. American Physiological Society, Washington, p 2545–2581

Martin AW, Harrison FM, Huston MJ, Stewart DM (1958) The blood volumes of some representative molluscs. J Exp Biol 35:260–279

Mountcastle VB (ed) (1974) Medical physiology, vol 2. C.V. Mosby, Saint Louis

Nomura H (1963) The effect of stretching on the intracellular action potential from the cardiac muscle fibre of the marine mollusc, *Dolabella auricularia*. Sci Rep Tokyo Kyoiku Daigaku B11: 153–165

O'Dor RK (1982) Respiratory metabolism and swimming performance of the squid *Loligo opalescens*. Can J Fish Aquat Sci 39:580–587

Packard A (1972) Cephalopods and fish: The limits of convergence. Biol Rev 47:241–307

Pelseneer P (1906) Mollusca. In: Lankester ER (ed) A treatise on zoology, part V. Adams & Charles Black, London

Potts WTW (1954) The rate of urine production of *Anodonta cygnea*. J Exp Biol 31:614–618

Randall DJ (1970) The circulatory system. In: Hoar WS, Randall DJ (eds) Fish physiology, vol IV. Academic, New York, p 133

Schwartzkopff J (1954) Über die Leistung der isolierten Herzen der Weinbergschnecke (*Helix pomatia* L.) im künstlichen Kreislauf. Z Vgl Physiol 36:543–594

Shadwick RE, Gosline J (1981) Elastic arteries in invertebrates: Mechanics of the *Octopus* aorta. Science 213:759–761

Smith LS, Davis JC (1965) Haemodynamics in *Tresus nuttali* and certain other bivalves. J Exp Biol 43:171–181

Smith PJS (1981) The role of venous pressure in the regulation of output from the heart of the octopus, *Eledone cirrhosa* (Lam.). J Exp Biol 93:243–255

Smith PJS (1982) The contribution of the branchial heart to the accessory branchial pump in the Octopoda. J Exp Biol 98:229–238

Smith PJS (1985) Cardiac performance in response to loading pressures in *Busycon canaliculatum* (Gastropoda) and *Mercenaria mercenaria* (Bivalvia). J Exp Biol 119 (in press)

Smith PJS, Boyle PR (1983) The cardiac innervation of *Eledone cirrhosa* (Lamarck) (Mollusca: Cephalopoda). Phil Trans R Soc Lond B300:493–511

Soffe SR, Benjamin PR, Slade CT (1978) Effects of environmental osmolarity on blood composition and light microscope appearance of neurosecretory neurones in the snail *Lymnaea stagnalis* (L.). Comp Biochem Physiol 61A:577–584

Taylor AC, Brand AR (1975) Effects of hypoxia and body size on the oxygen consumption of the bivalve *Artica islandica* (L). J Exp Mar Biol Ecol 19:187–196

Trueman ER, Brand AR, David P (1966) Burrowing in bivalves. J Exp Biol 44:460–492

Weis-Fogh T, Alexander RMcN (1977) The sustained power output obtained from striated muscle. In: Pedley TJ (ed) Scale effects in animal locomotion. Academic, New York, p 520

Wells MJ (1979) The heartbeat of *Octopus vulgaris*. J Exp Biol 78:87–104

Wells MJ (1983) Circulation in cephalopods. In: Saleuddin ASM, Wilbur KM (eds) The mollusca, vol 5, part 2. Academic, New York, p 239

Wells MJ, Smith PJS (1985) The ventilation cycle in octopus. J Exp Biol 116:375–383

Wells MJ, O'Dor RK, Mangold K, Wells J (1983) Oxygen consumption in movement by *Octopus*. Mar Behav Physiol 9:289–303

Cardiac Energy Metabolism in Relation to Work Demand and Habitat in Bivalve and Gastropod Mollusks

W. R. ELLINGTON[1]

1 Introduction

Bivalve and gastropod mollusks show considerable diversity in terms of habitat, general behavior, and degree of activity. Some species are sluggish or sessile, while others, such as scallops and sea hares display a high degree of locomotory activity. The cardiovascular systems of bivalve and gastropod mollusks show the same general functional arrangement, although there is considerable variation in the structure of the heart (Hill and Welsh 1966). Animals in these two groups display a broad range of hemodynamic properties (Jones 1983). The circulatory system may participate in a wide variety of processes including water and ion balance, gas transport, and locomotion (Smith 1985) as well as nutrient translocation. In general, heart rates and pressure development are low especially when viewed in terms of cardiovascular function in cephalopods (Smith 1985).

Given the great diversity of life-style and habitat of bivalve and gastropod mollusks, it is of interest as to whether these differences are reflected in patterns of cardiac energy metabolism in these animals. This paper probes aspects of energy metabolism in relation to life-style with emphasis being placed on marine species.

2 Ultrastructural Correlates of Metabolic Function

The ultrastructure of bivalve (Watts et al. 1981) and gastropod (Sanger 1979) hearts has been investigated extensively. Ventricles tend to show relatively high mitochondrial densities, especially when compared to phasic muscles, such as the translucent adductor muscle of bivalves (Nicaise and Amsellem 1983). Mitochondria are found in rows along the contractile elements as is the case of the ventricle of the gastropod *Helix aspersa* (North 1963) or scattered throughout the muscle fiber (Hawkins et al. 1980). The overall structure of mitochondria and number of cristae is highly variable.

In many muscle systems mitochondrial density is strongly correlated with the ability to perform sustained contractile activity (Josephson 1975) and in the case of hearts, the resting cardiac work rate (Hoppeler et al. 1984). Some differences have been noted in mitochondrial densities in bivalve ventricles. For instance, mitochondrial densities in

1 Department of Biological Science, Florida State University, Tallahassee, FL 32306, USA

Circulation, Respiration, and Metabolism
(ed. by R. Gilles)

the ventricle of the oyster *Crassostrea virginica* (Hawkins et al. 1980) are somewhat higher than corresponding densities in the clam *Mercenaria mercenaria* (Kelly and Hayes 1968). However, these differences do not appear to be well correlated with work demand on the heart since both are sluggish or sessile species. Dykens and Mangum (1979) have compared mitochondrial density and structure in the ventricles of a bivalve (*M. mercenaria*) and a cephalopod (squid *Lolliguncula brevis*). The squid maintains a 25-fold higher heart rate and this is correlated with a much higher mitochondrial density. Individual squid heart mitochondria show a larger number of cristae per organelle than mitochondria from the bivalve. Thus, in the case of the bivalve vs cephalopod there is a good correlation between mitochondrial density and work demand (Dykens and Mangum 1979).

A striking feature of the ultrastructure of bivalve and gastropod ventricle cells is the abundance of glycogen particles (Kelly and Hayes 1968; Irisawa et al. 1973; Hawkins et al. 1980; Watts et al. 1981). These particles typically are present as α-rosettes with diameters ranging from 500–600 Å (Kelly and Hayes 1968; Watts et al. 1981). Glycogen particles can be found scattered throughout the fiber, but are often observed concentrated in the vicinity of mitochondria as is the case in *M. mercenaria* (Kelly and Hayes 1968) and in the whelk *Busycon contrarium* (Chih and Ellington, unpublished).

The total glycogen content in the ventricles of a number of marine bivalve and gastropod mollusks is listed in Table 1. Glycogen content is extremely high, especially in bivalve ventricles, ranging up to approx. 22% of the total dry weight of the tissue. These levels of glycogen are, in general, considerably higher than corresponding levels in fish hearts and three to eight times higher than what has been observed in the mammalian heart (Table 1). Hochachka (1980) has indicated that α-rosettes represent an efficient form of glycogen packaging, and that these granules are rarely found in vertebrate heart or red muscle. The high levels of glycogen in the ventricles of bivalves and gastropods indicate that there is a potential for a strong reliance on carbohydrate metabolism under certain physiological circumstances.

Table 1. Levels of glycogen in the ventricles of certain bivalve and gastropod mollusks[a]

Species		μmol Hexose g^{-1} wet wt.
Bivalves		
Crassostrea virginica	(oyster)	95.66 ± 24.39 (n = 3)
Mercenaria mercenaria	(clam)	240.87 ± 63.52 (n = 4)
Macrocallista nimbosa	(clam)	211.58 ± 32.98 (n = 4)
Gastropods		
Melongena corona	(conch)	142.46 ± 16.02 (n = 4)
Busycon contrarium	(whelk)	103.11 ± 11.31 (n = 3)
Fasciolaria tulipa	(tulip shell)	116.80 ± 6.95 (n = 3)
Fish hearts[b]		34.0 – 86.1
Rat heart[c]		27.5

[a] Glycogen levels were determined by the method of Keppler and Decker (1974). Each value represents a mean ± 1 SD

[b] Driedzic and Stewart 1982

[c] Goldspink 1983

3 Myoglobin

Myoglobin is widely distributed in the muscles of gastropods and bivalves, and typically has a high affinity for oxygen with P_{50} values ranging from 0.5–5 mmHg (Read 1966). This pigment is present in the ventricles of a number of bivalves and gastropods but is totally absent in others (Table 2). When present in gastropods, myoglobin content appears to be much higher than the content of bivalve ventricles. In fish there appears to be a positive correlation between heart myoglobin content and habitat and degree of activity (Giovane et al. 1980). That is, active, pelagic species appear to have a much higher myoglobin content. No such correlation between ventricle myoglobin content and life-style is evident in bivalve and gastropod mollusks (Table 2). For instance, two gastropods (*Busycon spiratum* and *Fasciolaria tulipa*) occupy the same microhabitat, and qualitatively appear to display the same degree of activity, but differ radically in ventricle myoglobin content.

Myoglobin content of muscle is often observed to be positively correlated with capacity for aerobic energy metabolism (Lawrie 1953). In the case of bivalve and gastropod ventricles, myoglobin levels are not strongly correlated with activities of citrate synthase, an indicator enzyme for overall capacity for Krebs cycle activity (Table 2). This lack of correlation with capacity for aerobic energy metabolism has been observed in certain fish hearts where there is a fivefold range of myoglobin content (Driedzic and Stewart 1982). The role of myoglobin in the facilitation of oxygen diffusion into mammalian muscle is well established (Cole 1983). Furthermore, Driedzic et al. (1982) have shown that myoglobin has a protective effect in certain fish hearts during hypoxic perfusions. In view of the high myoglobin content of some bivalve and gastropod ventricles and the lack of vascularization of the ventricular myocardium (Smith 1985), it is likely that myoglobin assumes a role of facilitation of oxygen diffusion in these hearts.

4 Aspects of Mitochondrial Function

In comparison to other molluscan muscles, the hearts of bivalve and gastropod mollusks display substantially greater rates of oxygen consumption (Nomura 1950; Mangum and Polites 1980), and this correlates well with the higher mitochondrial densities. Mitochondria from the ventricles of the clam *M. mercenaria* (Ballantyne and Storey 1983) and the whelk *B. contrarium* (Chih and Ellington, unpublished) have been isolated and characterized. Both studies provide evidence for the existence of the malate/aspartate and α-glycerophosphate shuttles. Substrate preferences of clam and whelk heart mitochondria were very similar. Intermediates of fatty acid metabolism were poor substrates or were not oxidized at all (Ballantyne and Storey 1983; Chih and Ellington, unpublished). Succinate, malate, and pyruvate were good substrates, although the presence of a small amount of malate was necessary for pyruvate oxidation (Table 3). In addition, proline, aspartate, glutamate, and ornithine were also readily oxidized by clam and whelk ventricle mitochondria.

Table 2. Levels of myoglobin and citrate synthase activities in the ventricles of certain bivalve and gastropod mollusks (each value represents a mean ± SD)

Species	Habitat	Activity level	Ventricle weight (mg)	Myoglobin content[a] (nmol subunit g^{-1} wet wt.)	Citrate synthase[b] (μmol min^{-1} g^{-1} wet wt.) at 25 °C
Bivalves					
Crassostrea virginica (oyster)	Subtidal/intertidal	Sessile	26– 51	<0.5 (n = 4)	9.69 ± 1.08 (n = 3)
Mercenaria mercenaria (clam)	Subtidal/intertidal	Sluggish burrower	111–179	16.61 ± 6.23 (n = 4)	19.58 ± 4.89 (n = 3)
Macrocallista nimbosa (clam)	Subtidal in well-drained sediments	Rapid burrower	61– 78	9.04 ± 2.92 (n = 6)	23.62 ± 1.42 (n = 3)
Argopecten irradians (bay scallop)	Subtidal	Active swimmer	10– 21	<0.5 (n = 3)	12.38 ± 1.03 (n = 3)
Gastropods					
Melongena corona (conch)	Subtidal/intertidal	Low activity	19– 36	111.69 ± 21.33 (n = 8)	7.34 ± 0.65 (n = 3)
Busycon contrarium (whelk)	Subtidal/intertidal	Low activity	73–139	68.66 ± 16.10 (n = 8)	3.38 ± 0.20 (n = 3)
Busycon spiratum (whelk)	Subtidal/intertidal	Active	94–140	44.58 ± 22.34 (n = 4)	–
Fasciolaria tulipa (tulip shell)	Subtidal/intertidal	Active	95–127	<0.5 (n = 4)	5.90 ± 0.47 (n = 3)

[a] Assayed according to the procedure of Sidell (1980) using the molar absorptivity coefficient for mammalian oxymyoglobin
[b] Assayed according to the procedure of Sugden and Newsholme (1975) using crude, uncentrifuged extracts of ventricles

Table 3. State 3 respiration in mitochondria from the clam *Mercenaria mercenaria* (Ballantyne and Storey 1983) and the whelk *Busycon contrarium* (Chih and Ellington, unpublished) (rates are expressed as a percentage of succinate oxidation)

Substrate	*M. mercenaria*	*B. contrarium*
Succinate	100	100
Malate[a]	97	80
Pyruvate[b]	99	45
Proline[a]	125	78
Aspartate[a]	73	33
Glutamate[a]	72	50
Ornithine[a]	85	54
αGlycerophosphate	47	49

[a] Assayed in the presence of 0.1–0.3 m*M* pyruvate
[b] Assayed in the presence of 0.1–0.3 m*M* malate

Glycogen and certain free amino acids represent important endogenous substrates of aerobic energy metabolism in the ventricles of bivalves and gastropods. The role of exogenous substrates from the hemolymph is unclear. Significant levels of glucose (Livingstone and de Zwaan 1983) and free amino acids (Pierce 1971) are present in the hemolymph. Furthermore, isolated ventricles readily take up and metabolize radiolabeled glucose and aspartate (Collicutt and Hochachka 1977; Gäde and Ellington 1983).

5 Role of Arginine Phosphate

Conventional enzyme-linked metabolite assays (Ellington 1981) as well as phosphorus nuclear magnetic resonance studies (Ellington 1983) have shown that the levels of arginine phosphate are comparatively high in the ventricle of the whelk *B. contrarium*. Arginine phosphokinase activities are high on the order of 100 μmol min^{-1} g^{-1} wet wt. at 25 °C, and the mass action ratio approaches unity (Ellington, unpublished). It is likely that arginine phosphokinase catalyzes a reaction near equilibrium with its substrates in the whelk ventricle. Arginine phosphate appears to function as a reservoir of high energy to minimize changes in the ATP/ADP ratio under certain physiological conditions. This role is readily evident when whelk ventricles were subjected to anoxia plus ischemia (Ellington 1983). There was a linear decay of arginine phosphate levels, while ATP levels remained relatively constant. Diminished ATP production under these conditions was offset by a net utilization of arginine phosphate (Ellington 1983). Arginine phosphokinase in *B. contrarium* ventricle is located in the cytoplasmic compartment (Ellington, unpublished), and as a consequence there is no potential for a phosphagen shuttle system in this tissue.

Preliminary phosphorus nuclear magnetic resonance studies of the ventricles of the clams *M. mercenaria* and *M. nimbosa* have shown that arginine phosphate levels are considerably lower than levels in the whelk ventricle (Ellington, unpublished). The significance of this difference is unclear.

6 Glycolysis

The deposition of large amounts of glycogen in the ventricles of gastropod and bivalve mollusks (Table 1) indicates that carbohydrates may be important substrates of energy metabolism under certain circumstances. A survey of the activities of key enzymes of glycolysis reveals a high capacity for carbohydrate utilization (Table 4). Phosphofructokinase (PFK) is thought to be the major rate limiting enzyme in the glycolytic pathway. PFK activities in the ventricles of a number of bivalve and gastropod mollusks are similar. Furthermore, these enzyme activities are in the same range as PFK activities observed for the phasic adductor muscle of the scallop *Placopecten magellanicus* (Table 4). The phasic muscle of *P. magellanicus* is capable of relatively high rates of glycolysis especially during recovery following contractile activity (Livingstone et al. 1981).

The ability of a tissue to utilize exogenous glucose is limited by the rate of glucose transport and its phosphorylation by hexokinase. Rapidly contracting, phasic muscles utilize endogenous energy sources (arginine phosphate, glycogen), and hexokinase activities tend to be low (Table 4; see also Zammit and Newsholme 1976). In contrast, hexokinase activities in ventricles of bivalves and gastropods are higher (Table 4) implying enhanced capacity for glucose utilization. Thus, hemolymph glucose may be an important substrate for cardiac energy metabolism.

Phosphofructokinase and hexokinase activities in the ventricles of bivalve and gastropod mollusks listed in Table 4 are quite similar. Differences in life-style and habitat are not reflected in patterns of enzyme activity in these tissues.

Table 4. Activities of phosphofructokinase (PFK), hexokinase (HEX), lactate dehydrogenase (LDH), octopine dehydrogenase (ODH), alanopine dehydrogenase (ALNDH), and strombine dehydrogenase (STRDH) in ventricles of certain bivalve and gastropod mollusks[a]

	PFK	HEX	LDH	ODH	ALNDH	STRDH
Bivalves						
Crassostrea virginica	4.36	–	1.50	0.14	1.06	1.91
Mercenaria mercenaria	1.62	1.13	2.67	0	60.96	68.33
Macrocallista nimbosa	4.60	–	4.22	15.37	50.37	44.68
Argopecten irradians	3.61	1.42	4.22	20.50	0	0
Gastropods						
Melongena corona	3.74	–	7.83	75.56	25.15	9.24
Busycon contrarium	4.85	1.32	24.12	269.93	58.68	74.36
Busycon spiratum	5.54	0.63	77.04	365.15	112.25	77.71
Fasciolaria tulipa	5.69	0.99	0	188.10	86.41	76.68
Scallop fast adductor muscle *(Placopecten)*[b]	2.69	0.10	0.20	25.70	0.10	–

[a] PFK and HEX were assayed according to the procedures of de Zwaan et al. (1981). Extracts for LDH, ODH, ALNDH, and STRDH assays were first passed through a Sephadex G-25 column and then assayed according to the procedures of Ellington (1981)

[b] de Zwaan et al. (1980)

Pyruvate oxidoreductases (lactate dehydrogenase, "opine" dehydrogenases) catalyze the terminal step of glycolysis under conditions of relatively high energy demands in muscles (Livingstone et al. 1983). Lactate dehydrogenase activity is widely distributed in the ventricles of bivalve and gastropod mollusks (Table 4). However, activities are low when compared to those of "opine" dehydrogenases. Octopine dehydrogenase (pyruvate + arginine + NADH ⇌ octopine + NAD) is present in the ventricles of gastropod mollusks (Table 4) at activities which are among the highest yet reported (Zammit and Newsholme 1976; Livingstone et al. 1983). ODH activities in bivalve ventricles are considerably lower. Alanopine dehydrogenase (pyruvate + alanine + NADH ⇌ Alanopine + NAD) is present in the ventricles of the two clams and in the four gastropod species (Table 4). The activities are also among the highest yet reported for mollusks (Livingstone et al. 1983). Strombine dehydrogenase (pyruvate + glycine + NADH ⇌ strombine + NAD) activities in these ventricles are in the same range as the activities of alanopine dehydrogenase.

It is evident that opine dehydrogenase activities are generally high in the ventricles of marine gastropods. Livingstone et al. (1983) suggested that the distribution of these enzymes is related to the rate of energy demand of the tissue and the availability of amino acid substrates. The use of these enzymes is often found restricted to periods of contractile activity or periods of recovery following contractile activity or anoxia (de Zwaan and Dando 1984). In mollusks, opine dehydrogenases are found primarily in marine species where free amino acid levels are high (Livingstone et al. 1983). It is difficult to rationalize the high activities of opine dehydrogenases in gastropod ventricles on the basis of the above. It is assumed that energy demands on these ventricles are relatively constant and low. Furthermore, free amino acid substrate levels may not be particularly high. The levels of free arginine, alanine, and glycine under normoxic conditions in the ventricle of the whelk *B. contrarium* were 1.2, 3.8, and 0.6 μmol g^{-1} wet wt, respectively (Ellington, unpublished). A possible role of these enzymes during recovery from anoxia cannot be totally discounted.

7 Energy Metabolism During Hypoxia

The typical cardiac response to exposure of the whole animal to declining oxygen tensions is a marked bradycardia (Brand and Roberts 1973; Deaton and Mangum 1978). However, in vitro heart preparations of bivalve and gastropod mollusks often display rather different responses to hypoxia. The isolated heart of the bivalve *Tapes watlingi* maintained near normal rates and force of contraction for extended time periods under anoxic conditions (Jamieson and de Rome 1979). Similar results were observed with the isolated perfused ventricle of the whelk *Busycon contrarium*, although contractile performance under hypoxic conditions was considerably more variable (Ellington 1981).

The sources of energy during anoxia have been investigated extensively (Gäde and Ellington 1983). Arginine phosphate utilization is important in the ventricle of *B. contrarium* (Ellington 1981), although the rates of utilization are low with the half time for decay of the entire pool being 5-6 h (Ellington 1983). Glycolytic energy production involves the simultaneous utilization of glycogen and aspartate leading to the accumulation of alanine and succinate (Collicutt and Hochachka 1977; Baginski and Pierce 1978; Ellington 1981; Foreman and Ellington 1983). The overall stoichiometric equation for aspartate fermentation as suggested by de Zwaan (1983) is as follows:

$$\text{glycogen} + 2\ \text{aspartate} + 4.71\ \text{ADP} \rightleftharpoons 1.71\ \text{succinate} + 1.14\ CO_2 + 2\ \text{alanine} + 4.71\ \text{ATP}\,.$$

This leads to an alanine/succinate accumulation ratio of slightly greater than one. Actual alanine/succinate accumulation ratios in bivalve and gastropod hearts are greater than unity and can be as high as three (Gäde and Ellington 1983). In the case of the ventricle of *B. contrarium*, the lower than expected succinate accumulations cannot be accounted for on the basis of release of succinate into the medium or by the conversion of succinate into propionate (Graham and Ellington, unpublished).

De Zwaan et al. (1983) proposed the existence of an aspartate-alanine hydrogen translocase system in the adductor muscle of *Mytilus edulis* in which part of the conversion of aspartate to alanine is coupled to the translocation of reducing equivalents into the mitochondrion during anoxia. Central to this system is the decarboxylation of malate (derived from aspartate) by a mitochondrial malic enzyme (de Zwaan et al. 1983). There is convincing isotopic evidence for malic enzyme activity in bivalve (Collicutt and Hochachka 1977) and gastropod (Gäde and Ellington 1983) ventricles as well as direct measurement of enzyme activity in ruptured mitochondria from the ventricle of *B. contrarium* (Ellington, unpublished). Thus, it seems likely that a similar hydrogen transport system is operating in these hearts.

Opines are minor end products of energy metabolism during anoxia in bivalve and gastropod ventricles. In the case of the isolated ventricle of the oyster *C. virginica* alanopine/strombine accumulated to only a minor extent during 90 min of anoxia (Foreman and Ellington 1983). Perfusion of isolated ventricles of *B. contrarium* for 6 h under anoxia conditions resulted in small accumulations of octopine and alanopine/strombine. Subsequent analysis of alanopine/strombine levels by HPLC revealed that alanopine was the major compound accumulated in both *C. virginica* and *B. contrarium* ventricles (Fiore et al. 1984). Opine production can be greatly augmented by blocking aspartate utilization using the aminotransferase inhibitor aminooxyacetate (Foreman and Ellington 1983; Graham and Ellington 1984). Aminooxyacetate greatly reduces the rate of succinate production and virtually blocks alanine accumulation. Under these circumstances, there was a threefold increase in alanopine accumulation in *C. virginica* ventricle (Foreman and Ellington 1983) and a fivefold increase in octopine accumulation in *B. contrarium* ventricle (Graham and Ellington 1984).

An important feature of energy metabolism during anoxia is the degree of intracellular pH change in ventricles. In the ventricle of the whelk *B. contrarium* the rates of accumulation of alanine and succinate during anoxia were low, on the order of 10 to 20 nmol $\text{min}^{-1}\ \text{g}^{-1}$ wet wt (Ellington 1983). Intracellular pH changes, as measured using phosphorus nuclear magnetic resonance techniques, were quite low, as the intracellular pH fell in the range of 0.25 to 0.35 units over a period of 6 h. The modest anoxic acidifications in the whelk ventricle can be attributed to the relatively low rates of energy metabolism and to active/passive pH regulating processes in the tissue. Nonbicarbonate buffering capacity in the ventricle of *B. contrarium* is relatively high (Graham, unpublished).

8 Summary and Conclusions

An analysis of aspects of energy metabolism in the ventricles of marine bivalve and gastropod mollusks reveals great similarities in spite of large differences in degree of activity and characteristics of the microhabitat. Where differences are evident, these are generally not well correlated with the life-style of the animal. Thus, work demand and habitat do not appear to have been important selective factors in the development of the overall metabolic capabilities of the ventricles of these marine bivalve and gastropod mollusks. The presence of a large metabolic scope would not be that essential given the low work demands placed on the hearts.

Acknowledgments. Some of my work described in this paper was supported by the U.S. National Science Foundation (PCM-8202370). I thank two of my students L.R. Robinson and C.P. Chih for performing some of the enzyme and metabolite assays.

References

Baginski RM, Pierce SK (1978) A comparison of amino acid accumulation during high salinity adaptation with anaerobic metabolism in the ribbed muscle *Modiolus demissus*. J Exp Zool 203:419–428

Ballantyne JS, Storey KB (1983) Mitochondria from the ventricle of the marine clam *Mercenaria mercenaria*: substrate preferences and pH and salt concentration on proline oxidation. Comp Biochem Physiol 76B:133–138

Brand AR, Roberts D (1973) The cardiac responses of the scallop *Pecten maximus* L. to respiratory stress. J Exp Mar Biol Ecol 13:29–43

Cole RP (1983) Skeletal muscle function in hypoxia: effect of alteration of intracellular myoglobin. Respir Physiol 58:1–14

Collicutt JM, Hochachka PW (1977) The anaerobic oyster heart: coupling of glucose and aspartate fermentation. J Comp Physiol 115:147–157

Deaton LE, Mangum CP (1978) The cardiac response of the ponderous ark clam, *Noetia ponderosa*, to reduced oxygen levels. Comp Biochem Physiol 59A:229–230

Driedzic WR, Stewart JM (1982) Myoglobin content and activities of enzymes of energy metabolism in red and white fish hearts. J Comp Physiol 149:67–73

Driedzic WR, Stewart JM, Scott DL (1982) The protective effect of myoglobin during hypoxic perfusion of isolated fish hearts. J Mol Cell Cardiol 14:673–677

Dykens JA, Mangum CP (1979) The design of cardiac muscle and the mode of metabolism in molluscs. Comp Biochem Physiol 62A:549–554

Ellington WR (1981) Energy metabolism during hypoxia in the isolated, perfused ventricle of the whelk, *Busycon contrarium*. J Comp Physiol 142:457–464

Ellington WR (1983) Phosphorus nuclear magnetic resonance studies of energy metabolism in molluscan tissues: Effect of anoxia and ischemia on the intracellular pH and high energy phosphates in the ventricle of the whelk *Busycon contrarium*. J Comp Physiol 153:159–166

Fiore GB, Nicchitta CV, Ellington WR (1984) High-performance liquid chromatographic separation and quantification of alanopine and strombine in crude tissue extracts. Anal Biochem 139:413–417

Foreman RA, Ellington WR (1983) Effects of inhibitors and substrate supplementation on anaerobic energy metabolism in the ventricle of the oyster, *Crassostrea virginica*. Comp Biochem Physiol 74B:543–547

Gäde G, Ellington WR (1983) The anaerobic molluscan heart: Adaptation to environmental anoxia. Comparison with energy metabolism in vertebrate hearts. Comp Biochem Physiol 76A:615–620

Giovane A, Greco G, Maresia A, Tota B (1980) Myoglobin in the heart of tuna and other fishes. Experientia 36:219–220

Goldspink G (1983) Alterations in myofibril size and structure during growth, exercise and changes in environmental temperature. In: Peachy LD (ed) Handbook of physiology. American Physiological Society, Bethesda, p 539

Graham RA, Ellington WR (1984) Intracellular pH change and the qualitative nature of anaerobic end products in molluscan cardiac muscle. Am Zool 24:134A

Greenwalt DE, Bishop SH (1980) Effect of aminotransferase inhibitors on the pattern of free amino acid accumulation in isolated mussel hearts subjected to hyperosmotic stress. Physiol Zool 53:262–269

Hawkins WE, Howse HD, Sarphie TG (1980) Ultrastructure of the heart of the oyster *Crassostrea virginica*. J Submicrosc Cytol 12:359–374

Hill RB, Welsh JH (1966) Heart, circulation and blood cells. In: Wilbur KM, Yonge CM (eds) The physiology of mollusca, vol II. Academic Press, New York, p 125

Hochachka PW (1980) Living without oxygen. Harvard University Press, Cambridge

Hoppeler H, Lindstedt SL, Claassen H, Taylor CR, Mathieu O, Weibel ER (1984) Scaling of mitochondria volume in heart to body mass. Respir Physiol 55:131–137

Irisawa H, Irisawa A, Shigeto N (1973) Physiological and morphological correlation of the functional syncytium in the bivalve myocardium. Comp Biochem Physiol 44A:207–219

Jamieson DD, de Rome P (1979) Energy metabolism in the heart of the molluscs *Tapes watlingi*. Comp Biochem Physiol 63B:399–405

Jones HD (1983) The circulatory systems of gastropods and bivalves. In: Saleuddin ASM, Wilbur KM (eds) The mollusca, vol 5, part 2. Academic, New York, p 189

Josephson RK (1975) Extensive and intensive factors determining the performance of striated muscle. J Exp Zool 194:135–154

Kelly RE, Hayes RL (1968) The ultrastructure of smooth cardiac muscle in the clam, *Venus mercenaria*. J Morph 127:163–176

Keppler D, Decker K (1974) Glycogen-determination with amyloglucosidase. In: Bergmeyer HU (ed) Methods of enzymatic analysis. Academic Press, New York, p 1127

Lawrie RA (1953) The activity of the cytochrome system in muscle and its relation to myoglobin. Biochem J 55:298–305

Livingstone DR, de Zwaan A (1983) Carbohydrate metabolism of gastropods. In: Hochachka PW (ed) The mollusca, vol I. Academic, New York, p 177

Livingstone DR, de Zwaan A, Thompson RJ (1981) Aerobic metabolism, octopine production and phosphoarginine as sources of energy in the phasic and catch adductor muscles of the giant scallop *Placopecten magellanicus* during swimming and the subsequent recovery period. Comp Biochem Physiol 70B:35–44

Livingstone DR, de Zwaan A, Leopold M, Marteijn E (1983) Studies on the phylogenetic distribution of pyruvate oxidoreductases. Biochem Syst Ecol II:415–425

Mangum CP, Polites G (1980) Oxygen uptake and transport in the prosobranch molluscs *Busycon canaliculatum* I. Gas exchange and the response to hypoxia. Biol Bull 158:77–90

Nicaise G, Amsellem J (1983) Cytology of muscle and neuromuscular junction. In: Saleuddin ASM, Wilbur KM (eds) The mollusca, vol 4, part 1. Academic, New York, p 1

Nomura S (1950) Energetics of the heart muscle of the oyster, work performed and oxygen consumption. Sci Rep Tohoku Univ 18:279–285

North RJ (1963) The fine structure of the myofibers in the heart of the snail *Helix aspersa*. J Ultrastruc Res 8:206–218

Pierce SK (1971) A source of solute for volume regulation in marine mussels. Comp Biochem Physiol 38A:619–635

Read KRH (1966) Molluscan hemoglobin and myoglobin. In: Wilbur KM, Yonge CM (eds) The physiology of molluscan, vol II. Academic, New York, p 209

Sanger JW (1979) Cardiac fine structure in selected arthropods and molluscs. Am Zool 19:9–27

Sidell BD (1980) Response of goldfish *(Carassius auratus)* muscle to acclimation temperature: alterations in biochemistry and proportions of different fiber types. Physiol Zool 53:98–107

Smith PJS (1985) Molluscan circulation: haemodynamics and the heart. In: Gilles R (ed) Proc 1st Int Congr Comp Physiol Biochem, vol 1. Springer, Berlin Heidelberg New York

Sugden PH, Newsholme EA (1975) Activities of citrate synthase, NAD linked and NADP linked isocitrate dehydrogenase, glutamate dehydrogenase, aspartate aminotransferase and alanine aminotransferase in nervous tissues from vertebrates and invertebrates. Biochem J 150:105–111

Watts JA, Koch RA, Greenberg MJ, Pierce SK (1981) Ultrastructure of the heart of the marine mussel *Geukensia demissia*. J Morph 170:301–319

Zammit VA, Newsholme EA (1976) The maximum activities of hexokinase, phosphorylase, phosphofructokinase, glycerolphosphate dehydrogenase, lactate dehydrogenase, phosphoenolpyruvate carboxykinase, nucleoside diphosphate kinase, glutamate-oxaloacetate transaminase in relation to carbohydrate utilization in muscles from invertebrates. Biochem J 160:447–462

de Zwaan A (1983) Carbohydrate catabolism in bivalves. In: Hochachka PW (ed) The mollusca, vol I. Academic, New York, p 137

de Zwaan A, Dando PR (1984) Phosphoenolpyruvate metabolism in bivalve molluscs. Mol Physiol 5:285–312

de Zwaan A, Thompson RJ, Livingstone DR (1980) Physiological and biochemical aspects of the valve snap and valve closure responses in the giant scallop, *Placopecten magellanicus* II. Biochemistry. J Comp Physiol 137:105–114

de Zwaan A, de Bont AMT, Hemelraad J (1983) The role of phosphoenolpyruvate carboxykinase in the anaerobic metabolism of the sea mussel *Mytilus edulis* L. J Comp Physiol 153:267–274

Role of Free Amino Acids in the Oxidative Metabolism of Cephalopod Hearts

U. HOEGER and T.P. MOMMSEN [1]

1 Introduction

The development of an efficient pump system was of critical importance in the evolution of circulatory systems. They designate a limiting factor to the rate of oxygen transport as well as the exchange of metabolites between tissues. Among invertebrates, cephalopods are unique in matching the circulatory system to their outstanding physical performance; in contrast to all other molluscs, they display closed circulatory systems and hemolymph circulation is further improved by a pair of branchial hearts supporting the function of the systemic heart. Cephalopods such as the pelagic squids are outstanding high speed swimmers able to perform long migrations. In this review, we shall discuss some metabolic characteristics of the heart of these highly specialized invertebrates. Most of our discussion will focus on studies of the squid heart; however, data from other cephalopods and molluscs will be included for comparison.

A characteristic feature of the cephalopod heart is its predominantly aerobic working mode. This is evident by its positioning in the hemolymph circuit, its ample supply of oxygen and mitochondria (Dykens and Mangum 1979) and a concomitant high activity of citrate synthase (80 U g^{-1} tissue; 15 °C; Mommsen and Hochachka 1981). The activity of this mitochondrial enzyme falls into the same range as the activity found in mouse and rat hearts (100–150 U g^{-1} tissue, 25 °C; Alp and Newsholme 1976). Except for the overall aerobic working capability, however, cephalopod hearts show an entirely different metabolic organization in that they rely largely on amino acids as substrates for their oxidative metabolism.

2 Cephalopod Heart Metabolism: Possible Oxidative Substrates

2.1 Free Fatty Acids

Free fatty acids constitute the preferred oxidative substrates in mammalian hearts with glucose and lactate supplying additional portions of the metabolic energy (Neely and Morgan 1977; Spitzer 1974). Studies carried out on squid hearts, in contrast,

1 Department of Zoology, University of British Columbia, Vancouver, British Columbia, Canada V6T 2A9

Circulation, Respiration, and Metabolism
(ed. by R. Gilles)

failed to demonstrate the same importance of lipids in this tissue. First, structural phospholipids are the main lipids present in squid tissue (Shchepkin et al. 1976) with the exception of the liver (Boucaud-Camou 1971), and second, there is no evidence for lipid storage per se from electron micrographs (Storey and Storey 1983). Third, all enzymes of lipid metabolism measured in heart are present only in low activities (succinyl-CoA transferase: 0.6 U g^{-1}, carnitine acetyl-CoA transferase: < 0.02 U g^{-1}; Ballantyne et al. 1981) or are altogether absent (thiolase; Storey, unpubl., β-OH-butyrate dehydrogenase, Ballantyne et al. 1981). The same observations hold true for isolated squid heart mitochondria (Mommsen and Hochachka 1981). In comparison, much higher levels of free fatty acid catabolizing enzymes are present in mammalian hearts. For instance, 3-OH-CoA dehydrogenase in rat heart (Emmett 1981) is about 50 times more active than in the systemic heart of squid (Ballantyne et al. 1981; Mommsen et al. 1983). Therefore, on the basis of both substrate availability and enzyme equipment, lipids appear to be a very minor source for the energy metabolism of the cephalopod heart.

2.2 Carbohydrates

Glycogen concentrations in squid heart (*Illex illecebrosus*: 3–28 m*M* glycosyl U g^{-1} tissue; Hoeger, unpubl.) are in the same range as in the tissues of several other cephalopods (up to 24 m*M* glycosyl U g^{-1} tissue; Storey and Storey 1978, 1983) and thus generally low. These data have to be considered with some caution, since a rapid depletion of glycogen may have occurred as the animals struggled during capture.

The use of glycogen as endogenous fuel for heart steady state metabolism requires a balance between both synthesizing and catabolizing pathways. In heart (and other tissues) of *Loligo opalescens* only negligible activities of the glycogen mobilizing enzyme, glycogen phosphorylase, are found. The same applies to one enzyme controlling gluconeogenesis, phosphoenolpyruvate carboxykinase (Ballantyne et al. 1981). In tissues of two other squid, *Todarodes sagittatus* and *Loligo forbesi*, measurable activities (about 10 U g^{-1}; 25 °C) of glycogen phosphorylase were apparent (Zammit and Newsholme 1976). This raises the possibility that the squid heart might have some capability for utilization of endogenous glycogen. The incorporation of label from ^{14}C-precursors (proline, glutamate, glucose) into glycogen has been demonstrated in *Octopus macropus* and *Nautilus pompilius* (Hochachka and Fields 1982 Fields and Hochachka 1982). In both animals, however, incorporation of label into glycogen was less in the heart than in other tissues.

In the ventricles of bivalves, glycogen granules are easily visible in electron micrographs (Watts et al. 1981). For many molluscs, such large glycogen reserves are important for energy production under extended hypoxic or even anoxic conditions (Gäde and Ellington 1983), a situation unlikely to occur in cephalopods taking their pelagic life-style into account.

With respect to the utilization of exogenous glucose, it is interesting to note that glycolytic enzyme activities in the heart of *Loligo opalescens* (Ballantyne et al. 1981) are higher than in other tissues and are in fact only surpassed by the brain. Hexokinase is abundant in both tissues (Ballantyne et al. 1981; Mommsen et al. 1983). Thus it is possible that the heart is supplied with exogenous carbohydrate from the hemolymph, where glucose is present in a concentration of 0.5–1 m*M* (Hochachka et al. 1978; Storey et al. 1979). The degree to which glucose *versus* amino acids is utilized as fuel for the oxidative metabolism of heart seems to be species dependent; in *Nautilus*, glucose is preferred over amino acids, whereas in *Octopus* and squid, amino acids are preferentially utilized (Fields and Hochachka 1982; Hochachka and Fields 1982; Mommsen et al. 1982). This situation is apparent in heart as well as in other tissues and may reflect differences in the general metabolic organization within the cephalopods. It could have some correlation with the peak or steady state energy demand on the system. Judged from the relevant enzyme activities, however, it seems that gluconeogenic potential of squid tissues is low. This fact already puts a limit on glucose turnover and maximally sustainable rates of its utilization.

2.3 Amino Acids

The high concentrations of free amino acids in marine molluscs are thought to play an important role in osmoregulation (Bishop et al. 1983). This cannot be the case for cephalopods considering their stenohaline environment. While arginine glutamate and the ubiquitous alanine and glycine are the major tissue amino acids for marine molluscs, high levels of proline are unique to cephalopods (Storey et al. 1979; Ballantyne et al. 1981; Mommsen et al. 1982, 1983). Amino acid catabolism proceeds through three major pathways – direct oxidation, transamination, deamination (for review see Bishop et al. 1983). In the following we will therefore focus on those amino acids found to be of characteristic importance to the cephalopod heart (see Fig. 1).

Proline. The squid heart relies on proline as an oxidative substrate more than any other tissue. This is exemplified best if the ratios for $^{14}CO_2$ production rates for this amino acid are calculated over those from glucose for different tissues (Fig. 2). In the much less active *Nautilus*, these ratios are generally lower, but heart still reveals the highest value for all tissues investigated (Fields and Hochachka 1982).

As depicted in Fig. 1 (enzyme 1), the first step in the catabolism of proline involves its oxidation with molecular oxygen to the intermediate pyrroline-5-carboxylate (P5C). Just as in other organisms, proline oxidase is associated with the outer mitochondrial membrane. It probably represents the rate limiting step in the utilization of proline (Brunner and Neupert 1969). Subsequently, P5C is further oxidized to glutamate, via the unstable intermediate glutamate-γ-semialdehyde, and then funnelled into the Krebs cycle by either transamination or deamination involving glutamate dehydrogenase.

Besides its central role in amino acid metabolism (Fig. 1), P5C may also act as an important regulatory intermediate, since it exerts a pronounced stimulatory effect on the activity of the pentose-phosphate shunt (Phang et al. 1982). If this stimulation proves true for the cephalopod heart as well, it would give helpful insights into the regulatory interrelationship between proline and glucose metabolism.

In many aspects of proline utilization, the squid heart resembles beetle flight muscle (Weeda et al. 1980): (1) in both tissues, transamination is favoured over deamination; (2) isolated mitochondria require pyruvate for the oxidation of proline; and (3) this process is inhibited by the transaminase inhibitor aminooxyacetate (Mommsen and Hochachka 1981).

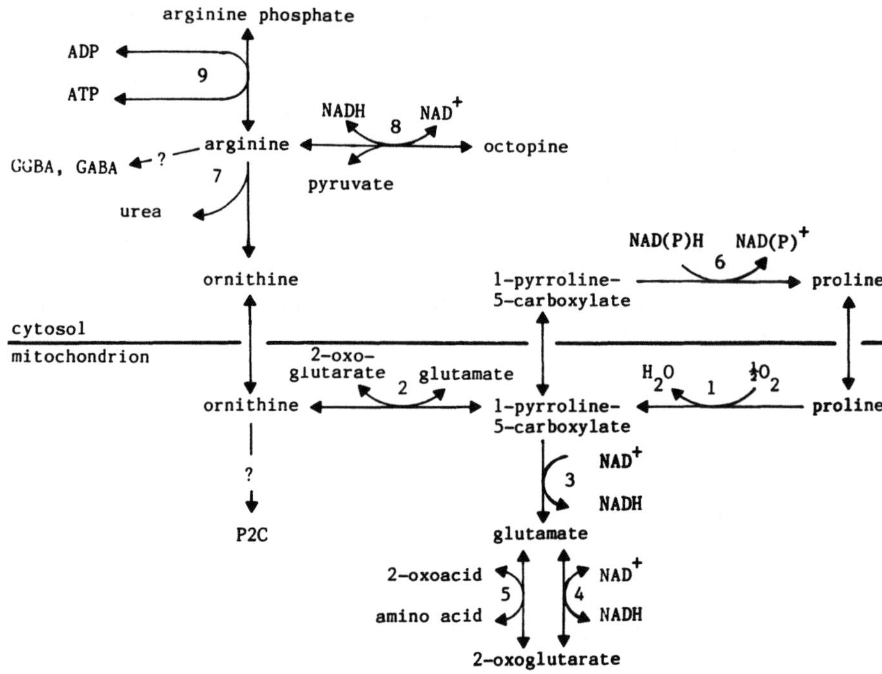

Fig. 1. Metabolic pathways of amino acid interconversions in cephalopod heart (after Mommsen and Hochachka 1981 and Adams and Frank 1980). The *numbers* indicate the following enzymes: *1* proline oxidase; *2* ornithine aminotransferase; *3* Δ'-pyrroline-5-carboxylate dehydrogenase; *4* glutamate dehydrogenase; *5* aspartate or alanine aminotransferase; *6* Δ'-pyrroline-5-carboxylate reductase; *7* arginase; *8* octopine dehydrogenase; *9* phosphoarginine kinase. *GGBA*: γ-guanidino butyramide; *GABA*: γ-amino butyric acid; *P2C*: pyrroline-2-carboxylate (2-keto-aminovalerate)

At the level of 2-oxoglutarate, the integration of proline-derived carbon in the Krebs cycle may proceed in two ways: first, it may function to augment the pool of Krebs cycle intermediates – a function well described for insect muscles (Sacktor 1975) as well as for the squid heart (Mommsen and Hochachka 1981); second, proline may be completely oxidized by a pathway which requires reentry of proline-derived carbon into the Krebs cycle via pyruvate. For the squid heart, the presence of the latter route is supported by the reported transfer of proline-derived label into the pyruvate/alanine pool (Mommsen et al. 1983).

The cytosolic fraction of the squid heart contains the enzyme P5C reductase (Fig. 1, enzyme 6) which reconverts the intermediate P5C into proline. This situation could potentially lead to futile cycling between proline oxidation on one side and proline synthesis on the other (see Fig. 1). Regulatory control, however, is facilitated by the different compartmentation of the enzymes involved. We postulate that proline oxidation will be favoured during periods of high aerobic energy output, while proline formation will prevail under conditions of low energy demand on the heart and while the cytosol is partially reduced.

In mammals, proline oxidase is strongly inhibited by L-lactate, an enzyme property which has been ascribed an important regulatory function (Kowaloff et al. 1977).

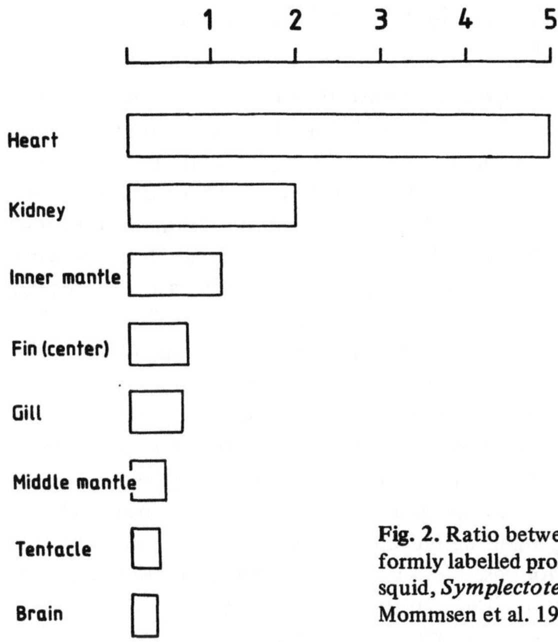

Fig. 2. Ratio between $^{14}CO_2$-production rates from uniformly labelled proline and glucose in different tissues of the squid, *Symplectoteuthis oualaniensis* (modified after Mommsen et al. 1982)

While L-lactate does not normally accumulate in cephalopods, squid proline oxidase is also strongly inhibited by L-lactate, but not by D-lactate (the more physiological isomer for molluscs), arginine or octopine (Mommsen and Hoeger, unpublished). Octopine, which assumes a role similar to L-lactate in vertebrates (i.e. it functions as electron acceptor during hypoxia), does not regulate proline oxidase in squid heart at all. This observation is not surprising, considering the aerobic working mode of the squid heart. Other mechanisms must therefore exist which direct the flux of carbons towards synthesis or catabolism of proline.

It has already been pointed out that proline concentrations in cephalopod heart are substantially higher than either glutamate or ornithine, its potential precursors. One advantage of accumulating proline is its weak charge, which eases its transport across membranes (Meyer 1977). Furthermore, low concentrations of glutamate are to be expected, since this acidic amino acid is known to serve as a neurotransmitter in cephalopods (Bone et al. 1982). In the highly aerobic squid heart, proline seems also to be best suited to supply carbons in moments of high energy needs, since it already supplies a reduced nucleotide during its oxidation to glutamate.

The presence of all the above mentioned enzymes has been confirmed for the cephalopod heart. In the scheme shown in Fig. 1, P5C assumes a central position, linking arginine and proline metabolism. While in rat and at least one insect, P5C can also be synthesized via the glutamate reductase pathway (Ross et al. 1977; Wadano 1981), the squid heart does not possess the enzymic machinery for this pathway (Mommsen et al. 1983). Therefore, the squid heart has to rely on P5C supplied by the ornithine aminotransferase for in situ proline synthesis. Potential sources are the endogenous or exogenous pools of ornithine and arginine. Besides proline synthesis within heart tissue itself, it could also be transported to the heart by the circulatory system after synthesis in other tissues especially kidney or gill.

Glutamate. The free glutamate pool designates a substrate for both the glutamate dehydrogenase and the diverse aminotransferases. In cephalopod heart the importance of the glutamate dehydrogenase reaction (Fig. 1, enzyme 4) in glutamate metabolism is evident from recent studies on squid *(Loligo pealii)* which revealed some characteristics entirely different from mammalian enzymes. When the forward (glutamate forming) and reverse (ammonia forming) directions of the enzyme were compared, the enzyme from squid mantle showed a ratio of 1:1 (Storey et al. 1978). In contrast, mammalian glutamate dehydrogenases revealed ratios of about 1:15 (cf. Storey and Storey 1978). Other observations were the large differences between individuals apparent in the above ratio of glutamate dehydrogenase reaction in different tissues of the squid *Illex illecebrosus* and a clam (*Panopea generosa*; Hoeger and Hochachka, in preparation). These findings suggest the presence of different isozymes regulating the direction of flux through the glutamate dehydrogenase reaction according to the physiological state of the tissue. The observed variations also imply alterations in the mitochondrial properties because in the molluscs, glutamate dehydrogenase is exclusively located in this compartment (Sollock et al. 1979; Mommsen and Hochachka 1981). Mammals, in contrast, possess both cytoplasmic and mitochondrial forms of the enzyme.

In the cephalopod heart, the kinetic properties of glutamate dehydrogenase would poise it correctly for the rapid transfer of glutamate (arising from proline oxidase and P5C dehydrogenase reactions) into the pool of Krebs cycle intermediates. On the other hand, it should be recalled that the oxidation of glutamate by isolated heart mitochondria is inhibited by the transaminase inhibitor aminooxyacetate (Mommsen and Hochachka 1981). Such inhibition indicates that at least in vitro glutamate oxidation proceeds mainly via transamination (Fig. 1, enzymes 1, 3, 5).

Arginine and Ornithine. In fish protein, arginine constitutes the most abundant amino acid (Cowey et al. 1962) and therefore comprises a major part of the dietary amino acid in piscivorous cephalopods such as pelagic squid.

Arginine is the substrate of several important metabolic reactions in the cephalopod heart. After phosphorylation by phosphoarginine kinase (Fig. 1, enzyme 9) it serves as an endogenous energy store in the form of arginine-phosphate. In heart, phosphoarginine kinase is less than half as active as in other muscular tissues of squid. Further observations on squid heart also reveal that phosphagen and arginine stores are limited (below 10 μmol g^{-1} tissue; Ballantyne et al. 1981; Mommsen et al. 1983). From the scanty data available it seems premature at this point to evaluate the overall importance of phosphagen to heart metabolism.

The catabolism of arginine in squid heart can proceed along two different routes. In vitro the most important pathway involves an L-amino acid oxidase (LAO). This enzyme is responsible for the transfer of large portions of ^{14}C-arginine derived label into γ-guanidino butyrate (Mommsen et al. 1983). The same enzyme can possibly act also on ornithine, yielding the intermediate 2-ketoaminovalerate (Mommsen et al. 1983), which would lead to synthesis of proline without involving ornithine aminotransferase. In similar studies on another species of squid, labelled γ-aminobutyrate was detected (Mommsen et al. 1982) which might indicate the presence of enzymes necessary to further degrade γ-guanidinobutyrate. In squid, LAO is localized predominantly in the heart, although it also occurs in gill and kidney (Mommsen et al. 1982, 1983; cf. Bishop et al. 1983 for other molluscs).

Smaller portions of squid heart arginine are converted into ornithine and urea in a non-oxidative step catalyzed by the enzyme arginase. As in *Octopus* (Gaston and Campbell 1966) this enzyme is most active in the kidney rather than the heart. The resulting urea is excreted by the kidneys and possibly also across the gills. Ornithine, the other product, can be utilized as substrate by ornithine aminotransferase (Fig. 1, enzyme 2). This enzyme, which is quite active in squid heart, links arginine metabolism to the proline/glutamate pathway through its product P5C. The overall importance of this transaminase to squid heart metabolism was demonstrated by the use of the transaminase inhibitor aminooxyacetate, which inhibited the oxidation of both glutamate and ornithine by isolated squid heart mitochondria (Mommsen and Hochachka 1981).

Ornithine serves as a good oxidative substrate for isolated mitochondria, but is present in the squid heart only in low concentrations (below 1 m*M*). Of course ornithine could be supplied to the heart from exogenous sources, especially the kidney (cf. Hochachka et al. 1983). The assessment of the ultimate importance of ornithine to squid heart metabolism obviously requires further detailed study on turnover, flux direction and tissue uptake of this basic amino acid.

Generally, production rates of $^{14}CO_2$ from uniformly labelled arginine in squid heart are an order of magnitude lower than those from proline. This is not surprising, since in the absence of urease activity in squid tissues generally (Mommsen et al. 1983), arginine is positioned further away from CO_2 liberating reactions than proline (cf. Fig. 1). Furthermore, arginine did not serve as an oxidative substrate for isolated mitochondria, while ornithine did (Mommsen and Hochachka 1981). This observation suggests that ornithine derived from arginine breakdown in situ contributes little if anything to the oxidative metabolism of the heart. As in the case of ornithine, further experimentation is needed to analyze the real importance of arginine, its turnover, metabolism and diverse functions in the squid heart.

If injected into the vena cava of *Sepia*, ^{14}C-arginine gives rise to labelled ornithine and citrulline in heart, which in turn leads to subsequent de novo synthesis of arginine, also in the heart (Hochachka et al. 1983). In the cuttlefish, a functional urea cycle may therefore be operative in heart as well as in other tissues, while in a squid *(Illex)* the whole enzymatic machinery for the cycle seems to be present in the systemic heart only (Mommsen et al. 1983). At this point, we prefer to refrain from speculating about the in vivo function of the urea cycle in cephalopod hearts, apart from its well-known excretory role.

2.4 Octopine

Arginine also constitutes the precursor for octopine, an imino acid which numerous cephalopods accumulate in non-cardiac tissue to maintain glycolytic flux under periods of anaerobic work (Gäde 1980). The reaction, catalyzed by octopine dehydrogenase (Fig. 1, enzyme 8), is easily reversible in vivo, and in the cephalopod heart, the importance of octopine lies in its function as possible oxidative substrate.

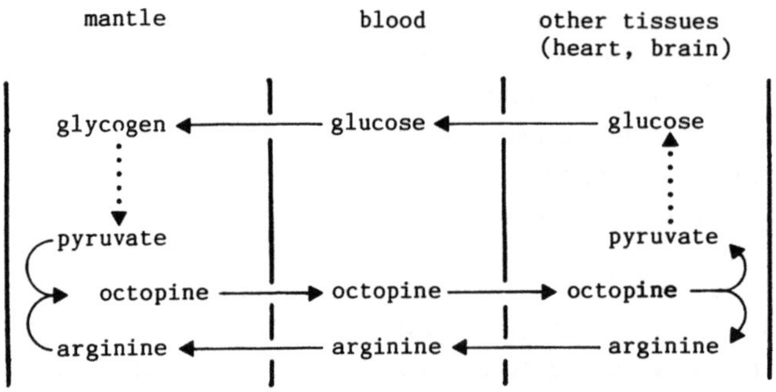

Fig. 3. Proposed exchange of metabolites between cephalopod tissues (after Storey and Storey 1983). Octopine produced anaerobically in the mantle is released in the circulation and taken up by the heart in which it is oxidized. The resulting pyruvate is used either in the Krebs cycle or for glyconeogenic reactions and the newly formed glucose can be channelled back into the mantle

In *Sepia* ^{14}C-octopine is rapidly removed from the hemolymph and actively catabolized in the ventricles as opposed to mantle tissue where octopine is oxidized slowly (Storey and Storey 1979a). These findings, together with the presence of tissue-specific isozymes of octopine dehydrogenase (Storey and Storey 1979b), suggest a situation in which octopine, accumulated in the muscle under anaerobic exercise, is released into the circulation during recovery. Subsequently, octopine will be taken up by the heart where it serves as an oxidative substrate (Storey and Storey 1983). In an additional study, it was confirmed in *Sepia* heart that most of the label from octopine was spread into other metabolites, especially proline (Hochachka et al. 1983). Storey and Storey (1983) further suggest that the breakdown products of octopine oxidation in heart could be recycled back into the mantle via the circulatory system; arginine would leave the heart directly whereas pyruvate would be utilized for gluconeogenesis first. After export, glucose would then enter the mantle to replenish its glycogen stores (Fig. 3). This situation appears analogous to the Cori cycle in vertebrates. It depends, however, on the washout kinetics of octopine from the muscular tissues and its uptake rate by the heart which remains to be investigated. Also, the participation of the cephalopod heart in this cycle depends on its glyconeogenic capabilities. The synthesis of glycogen from various precursors in the heart of *Octopus* and the presence of glycogen in squid heart were demonstrated (see Sect. 2.2). In another squid *(Loligo opalescens)*, however, only minute activities of two key gluconeogenic enyzmes, phosphoenolpyruvate carboxykinase and fructose-bisphosphatase, were found (Ballantyne et al. 1981).

References

Adams E, Frank L (1980) Metabolism of proline and hydroxyproline. Annu Rev Biochem 49: 1005–1061

Alp PR, Newsholme EA (1976) Activities of citrate synthase and NAD$^+$ and NADP$^+$ linked isocitrate dehydrogenase in muscle from vertebrates and invertebrates. Biochem J 154:689–700

Ballantyne JS, Hochachka PW, Mommsen TP (1981) Studies on the metabolism of the migratory squid, *Loligo opalescens*: enzymes of tissues and heart mitochondria. Mar Biol Lett 2:75–85

Bishop SW, Ellis LL, Burcham JL (1983) Amino acid metabolism in molluscs. In: Wilbur KM (ed) The mollusca, vol I. Academic, New York, p 244

Bone Q, Packard A, Pulsford AL (1982) Cholinergic innervation of muscle fibers in squid. J Mar Biol Assoc UK 62:193–199

Boucaud-Camou E (1971) Constituants lipidiques du foie de *Sepia officinalis*. Mar Biol 8:66–69

Brunner G, Neupert W (1969) Localization of proline oxidase and '-pyrroline-5-carboxylic acid dehydrogenase in rat liver. FEBS Lett 3:283–286

Cowey CB, Daisley KW, Parry G (1962) Study of amino acids, free or as components of protein, and some B vitamins in the tissue of the Atlantic salmon, *Salmo salar*, during spawning migration. Comp Biochem Physiol 7:29–38

Dykens JA, Mangum CP (1979) The design of cardiac muscle and the mode of metabolism in molluscs. Comp Biochem Physiol 62A:549–554

Emmett B (1981) Metabolic biochemistry of the wandering shrew, *Sorex vagrans*. Master's thesis, University of British Columbia, p 72

Fields JHA, Hochachka PW (1982) Glucose and proline metabolism in *Nautilus*. Pac Sci 36:337–341

Gäde G (1980) Biological role of octopine formation in marine molluscs. Mar Biol Lett 1:121–135

Gäde G, Ellington WR (1983) The anaerobic molluscan heart: adaptation to environmental anoxia. Comparison with energy metabolism in vertebrate hearts. Comp Biochem Physiol 76A:615–620

Gaston S, Campbell JW (1966) Distribution of arginase activity in molluscs. Comp Biochem Physiol 17:259–270

Hochachka PW, Fields JHA (1982) Arginine, glutamate, and proline as substrates for oxidation and for glyconeogenesis in cephalopod tissues. Pac Sci 36:325–335

Hochachka PW, French CJ, Meredith J (1978) Metabolic and ultrastructural organization in *Nautilus* muscles. J Exp Zool 205:51–62

Hochachka PW, Mommsen TP, Storey J, Storey KB, Johansen K, French CJ (1983) The relationship between arginine and proline metabolism in cephalopods. Mar Biol Lett 4:1–21

Kowaloff EM, Phang JM, Granger AS, Downing SJ (1977) Regulation of proline oxidase activity by lactate. Proc Natl Acad Sci USA 74:5368–5371

Meyer J (1977) Proline transport in rat liver mitochondria. Arch Biochem Biophys 178:387–395

Mommsen TP, Hochachka PW (1981) Respiratory and enzymatic properties of squid heart mitochondria. Eur J Biochem 120:345–350

Mommsen TP, French CJ, Emmett B, Hochachka PW (1982) The fate of arginine and proline carbon in squid tissues. Pac Sci 36:343–348

Mommsen TP, Hochachka PW, French CJ (1983) Metabolism of arginine, proline, and ornithine in tissues of the squid, *Illex illecebrosus*. Can J Zool 61:1835–1846

Neely JR, Morgan HE (1977) Relationship between carbohydrate and lipid metabolism and the energy balance of heart muscle. Annu Rev Physiol 36:413–459

Phang JM, Downing SJ, Yeh GC, Smith RJ, Williams JA, Hagedorn CH (1982) Stimulation of the hexosemonophosphate – pentose pathway by pyrroline-5-carboxylate in cultured cells. J Cell Physiol 110:255–261

Ross G, Dunn G, Jones ME (1977) Ornithine synthesis from glutamate in rat intestinal mucosa homogenates: evidence for the reduction of glutamate to gamma-glutamyl semialdehyde. Biochem Biophys Res Comm 85:140–147

Sacktor B (1975) Biochemistry of insect flight. In: Candy DJ, Kilby BA (eds) Insect biochemistry and function. Chapman & Hall, London, p 1

Shchepkin VY, Shul'man GY, Sigayeva TG (1976) Tissue lipids in Mediterranean squid of different ecology. Gidrobiol Zh 12:76–79

Sollock RL, Vorhaben JE, Campbell JW (1979) Transaminase reactions and glutamate dehydrogenase in gastropod tissues. J Comp Physiol A 129:129–135

Spitzer JJ (1974) Effect of lactate infusion on canine myocardial free fatty acid metabolism in vivo. Am J Physiol 226:213–217

Storey KB, Storey JM (1978) Energy metabolism in the mantle muscle of the squid *Loligo pealii*. J Comp Physiol 123:169–175

Storey KB, Storey JM (1979a) Octopine metabolism in the cuttlefish, *Sepia officinalis*: octopine production by muscle and its role as an aerobic substrate for nonmuscular tissues. J Comp Physiol B 131:311–319

Storey KB, Storey JM (1979b) Kinetic characterization of tissue-specific isozymes of octopine dehydrogenase from mantle muscle and brain of *Sepia officinalis*. Eur J Biochem 93:545–552

Storey KB, Storey JM (1983) Carbohydrate metabolism in caphalopod molluscs. In: Wilbur KM (ed) The mollusca, vol I. Academic, New York, p 92

Storey KB, Fields JHA, Hochachka PW (1978) Purification and properties of glutamate dehydrogenase from the mantle muscle of the squid, *Loligo pealii*. Role of the enzyme in energy production from amino acids. J Exp Zool 205:111–118

Storey KB, Storey JM, Johansen K, Hochachka PW (1979) Octopine metabolism in *Sepia officinalis*: effect of hypoxia and metabolite load on the blood levels of octopine and related compounds. Can J Zool 57:2331–2336

Wadano A (1981) Proline synthesis from glutamate in the mitochondria isolated from a blowfly, *Aldrichina grahami*. Experientia 36:1028–1029

Watts JA, Koch RA, Greenberg MJ, Pierce SK (1981) Ultrastructure of the heart of the marine mussel, *Geukensia demissa*. J Morphol 170:301–319

Weeda E, deKort CAD, Beenakkers AMT (1980) Oxidation of proline and pyruvate by flight muscle mitochondria of the Colorado beetle, *Leptinotarsa decemlineata* Say. Insect Biochem 10:305–311

Zammit VA, Newsholme EA (1976) The maximum activities of hexokinase, phosphofructokinase, glycerol phosphate dehydrogenase, lactate dehydrogenase, octopine dehydrogenase, phosphoenolpyruvate carboxykinase, nucleoside diphosphokinase, glutamate oxaloacetate transaminase and arginine kinase in relation to carbohydrate utilization in muscles from marine invertebrates. Biochem J 160:447–462

Cardiovascular and Hemodynamic Energetics of Fishes

T. FARRELL[1]

1 Introduction

The aim of this paper is to given an overview of the cardiovascular energetics of fish. I will limit myself to teleost fish and will focus on more recent findings. The first section of the talk considers the heart as a pump and deals with the flow and pressure generated by the ventricle. The second section presents data on how myocardial O_2 consumption is related to power output of the heart. The last section addresses the importance of temperature and catecholamines in regulating cardiac performance.

I would like to acknowledge the assistance and collaboration of Ken MacLeod, Bill Driedzic, Tom Hart, Sheila Wood, Mark Graham, and Louise Milligan for some of the data and ideas presented here.

2 The Heart as a Pump

The teleost heart has a single ventricle, which pumps blood into the ventral aorta, a single atrium and a sinus venosus, which acts as a reservoir for venous blood. The cardiac pacemaker is located at the sino-atrial junction.

Power output of the heart is determined primarily by the product of cardiac output ($\dot{V}_b$) and ventral aortic pressure (P_{va}). Power output is a measure of the heart's ability to work as a pump, that is, it's ability to provide sufficient blood flow (cardiac output) to meet the metabolic needs of the tissues, and to generate sufficient pressure to overcome vascular resistance. Power output is about 1 mW g^{-1} ventricle wet weight in benthic fish such as the sea raven and lingcod and about 1.7 mW g^{-1} in active fish such as the trout. Power output increases to about 7.4 mW g^{-1} at the critical swimming speed in trout (Kiceniuk and Jones 1977) and in experimental restrained tuna it is about 12 mW g^{-1} (Jones and Brill, personal communication). In comparison, power output of the human heart at 37 °C is about 5.4 mW g^{-1} and can increase to 50 mW g^{-1} in trained atheletes (Gibbs and Chapman 1979).

The ability of the teleost heart to vary its power output is dependent upon a number of intrinsic and extrinsic mechanisms which ultimately affect P_{va} and $\dot{V}_b$. The factors which affect P_{va} and $\dot{V}_b$ are now discussed.

1 Department of Biological Sciences, Simon Fraser University, Burnaby, British Columbia, Canada V5A 1S6

Circulation, Respiration, and Metabolism
(ed. by R. Gilles)

2.1 Blood Pressure

Blood pressure is determined by the product of blood flow and vascular resistance. In teleosts, the heart pumps venous blood through the gills before oxygenated blood is distributed to the various segments of the systemic circuit. Vascular resistance is, therefore, readily subdivided into gill and systemic resistances (R_g and R_s). Thus, for a given $\dot{V}_b$, P_{va} is set by the sum of R_g and R_s, which are in-series, and P_{va} is indicative of the pressure work done by the heart. The dorsal aortic pressure, P_{da}, is set by R_s and is indicative of the blood pressure needed to distribute $\dot{V}_b$ through the various, in-parallel systemic circulations. The mean P_{va} is normally of the order of 35 to 65 cmH_2O, with the corresponding P_{da} being about 8 to 20 cmH_2O lower. This indicates that R_s is about three times greater than R_g.

Both R_s and R_g are regulated by neural and humoral influences, which produce important changes in P_{va} (Wood 1974; Wood and Shelton 1975; Pettersson and Nilsson 1979). The teleost heart can cope with physiological changes in P_{va} through homeometric regulation. This intrinsic mechanism enables the heart to vary pressure development to accomodate changes in R_g and R_s without changing $\dot{V}_b$ (Farrell et al. 1982).

Few studies have measured the central venous blood pressure (P_{cv}) in teleosts (Chan and Chow 1976; Kiceniuk and Jones 1977; Farrell, unpublished data). All indications are that P_{cv} is normally less than 2 cmH_2O above ambient pressure. The P_{cv} represents the filling pressure of the heart.

2.2 Cardiac Output

The majority of direct measurements of $\dot{V}_b$ range from 6 to 30 ml min^{-1} kg^{-1} for teleosts in temperate water (see Farrell 1984 for references). $\dot{V}_b$ increases several-fold during sustained swimming in trout (Randall et al. 1967; Kiceniuk and Jones 1977) and in experimentally restrained tuna $\dot{V}_b$ is about 90 ml min^{-1} kg^{-1} at 20 °C (Jones and Brill, personal communication). Cardiac output is the product of heart rate (f_H) and stroke volume (SV_H), both of which are regulated by independent mechanisms (Fig. 1).

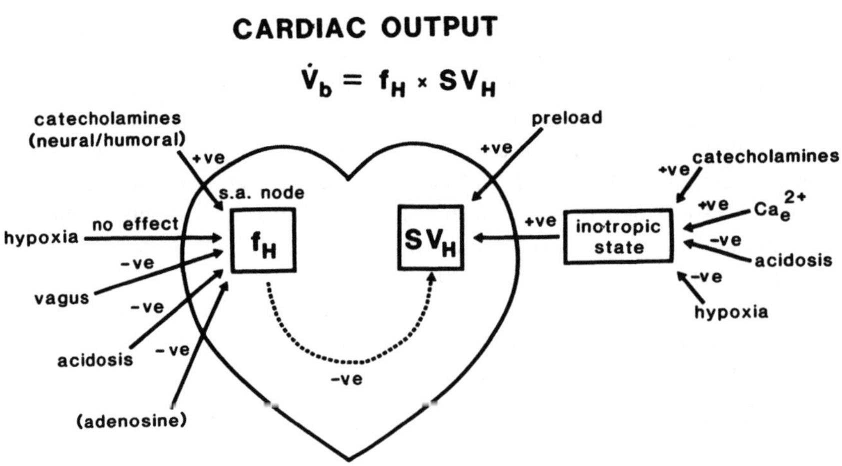

Fig. 1. A summary of the factors which affect cardiac output ($\dot{V}_b$) in teleost fish

2.2.1 Heart Rate

The intrinsic f_H is set by the discharge frequency of the sino-atrial pacemaker (Huang 1973; Saito 1973). The intrinsic f_H is in turn modulated by several neural and humoral mechanisms. A primary control mechanism is the inhibitory vagal tone which causes bradycardia (see Laurent et al. 1983, for references). In resting fish, f_H is somewhat lower than the pacemaker frequency because of a resting vagal tone to the heart. Thus, the resting fish can increase f_H by removing this resting vagal tone. Tachycardia is also produced by circulating catecholamines and/or adrenergic innervation when present (Laurent et al. 1983). Hypercapnic acidosis and extracellular adenosine both decrease f_H (Farrell et al. 1983; Farrell, unpublished data). A moderate level of hypoxia (P_{O_2} > 40 mmHg) does not affect the intrinsic f_H (Farrell et al. 1985), even though environmental hypoxia causes a reflex bradycardia through an increase in vagal tone (see for example Randall and Smith 1967; Holeton and Randall 1967; Smith and Jones 1978; Daxboeck and Holeton 1978; Wood and Shelton 1980b; Farrell 1982).

2.2.2 Stroke Volume

Studies with perfused hearts have clearly established the importance of preload in determining SV_H through the Starling mechanism (Farrell et al. 1982, 1984). Work with trout, salmon, sea raven and ocean pout hearts has shown that small changes in preload (< 2 cmH_2O) produce formidable changes in $\dot{V}_b$ (Fig. 2). This response occurs without

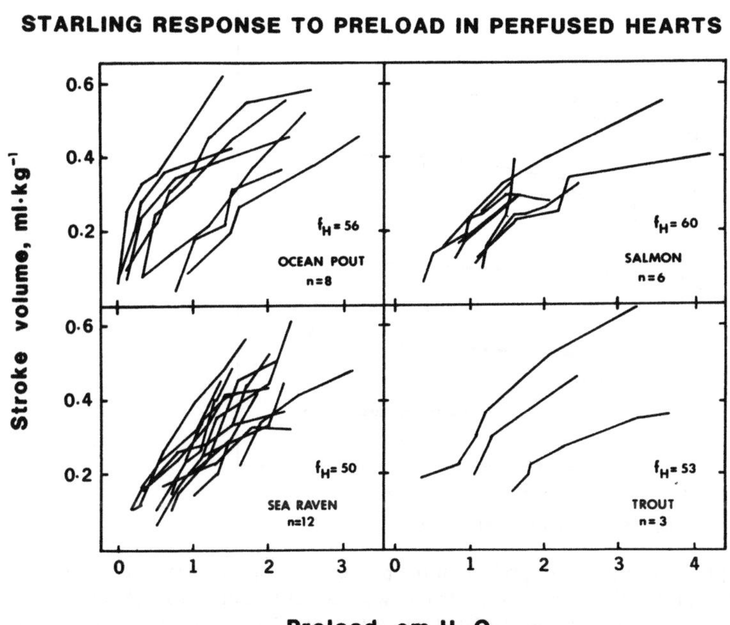

Fig. 2. The effect of preload on stroke volume (Starling response) in teleosts. The average heart rate (f_H, bpm) is indicated for n preparations. Data for sea raven and ocean pout were adapted from Farrell et al. 1982, 1983, respectively. The salmon and trout data are preliminary observations by the author

a major change in f_H and at a constant diastolic afterload. The important differences between these data and previous demonstrations of preload effects (Bennion 1968; Stuart et al. 1983) are (1) the preloads were physiological, (2) the heart generated a physiological power output; and (3) f_H did not increase with preload.

In addition to preload, the inotropic state of the myocardium affects SV_H. Catecholamines and extracellular Ca^{2+} improve the heart's inotropic state (Gesser et al. 1982; Gesser and Poupa 1983; Farrell et al. 1983, 1984). Acidosis and hypoxia have negative inotropic effects (Gesser and Poupa 1983; Turner and Driedzic 1980; Farrell et al. 1983, 1985). However, a marked change in the inotropic state of the myocardium does not necessarily alter the resting performance of the heart considerably. Instead, the change in inotropic state is manifest as a modification of the Starling response. For example, severe extracellular acidosis (pH 7.4) reduces the resting $\dot{V}_b$ by only 15% to 20%, but the Starling response to preload is reduced over threefold (Fig. 3). A similar situation exists when the heart is exposed to hypoxia (Farrell et al. 1985). Such observations indicate that the teleost heart has an intrinsic reserve for maintaining its *resting* work level.

SV_H is also dependent on f_H because SV_H is in part determined the time available for ventricular filling ($1/f_H$). The importance of this is that whenever f_H changes the

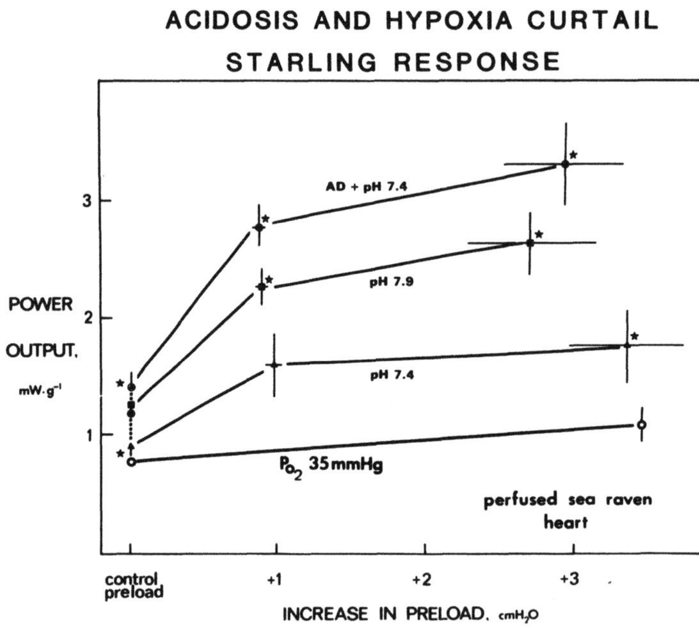

Fig. 3. The control (■, pH 7.9) response to preload in the perfused sea raven heart increases power output twofold. Acidosis (▲, pH 7.4) reduces the control power output by about 15% (*broken line*), and markedly reduces the Starling response (*solid line*). Addition of 0.1 μM adrenaline during acidosis (▲, AD + pH 7.4) restores the control power output (*broken line*), and the Starling response (*solid line*). Progressive hypoxia (o, P_O, 35 mmHg) also curtails the Starling response. Data adapted from Farrell et al. 1984a and Farrell 1984b. * Denotes significant difference from the respective control

counteracting change in SV_H will tend to regulate $\dot{V}_b$ at a given level. This type of regulation has been observed in perfused hearts (Farrell et al. 1983) and probably occurs in vivo during environmental hypoxia when there is a marked bradycardia and little in change $\dot{V}_b$.

3 Myocardial Oxygen Uptake

The only studies of myocardial O_2 consumption ($\dot{M}_{O_2}$) are with teleosts without a separate coronary circulation, i.e. they rely on the venous blood being pumped through the heart for their O_2 supply. For such fish the following general points have been established: (1) the mechanical efficiency of the heart is about 15%; an efficiency comparable to the mammalian myocardium. (2) The anaerobic contribution to metabolism is < 5%. (3) Calculations indicate that at rest, less than 4% of the available O_2 is extracted from the venous blood during its passage through the heart and so venous blood provides an adequate O_2 supply. (4) The O_2 requirement of the heart in the resting fish is less than 1% of the standard O_2 uptake of the fish (Driedzic et al. 1983; Farrell et al. 1985).

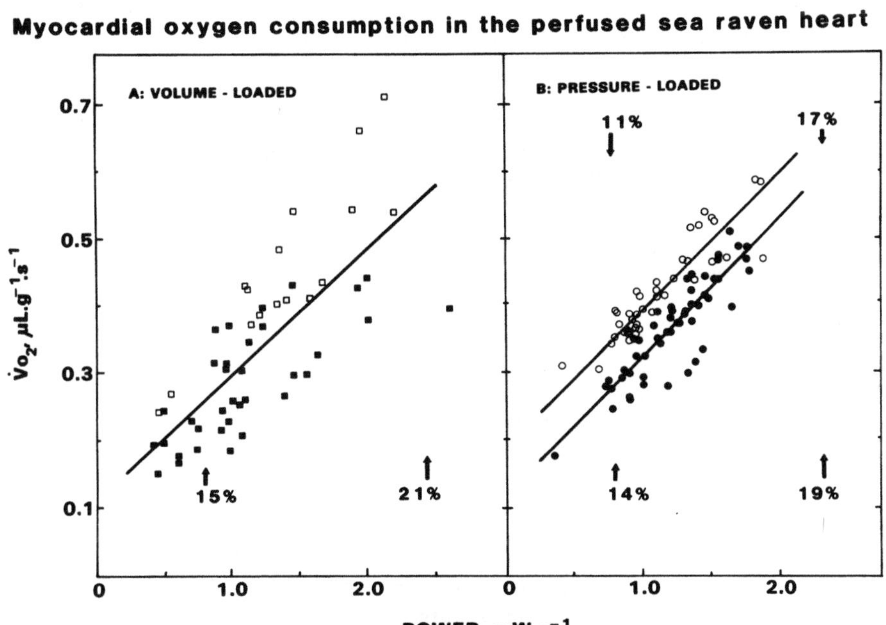

Fig. 4. The linear relationship between myocardial oxygen consumption ($\dot{V}_{O_2}$) and power output of the perfused sea raven heart when $\dot{V}_b$ and pressure development are increased independently, volume-loaded and pressure-loaded, respectively. Mechanical efficiency, at comparable points on each line, is indicated by the percentages. *Open* and *closed symbols* represent summer and winter fish, respectively. Data adapted from Farrell et al. 1985

It is clear from the above that the heart can do extra work by increasing $\dot{V}_b$ and P_{va}. What then are the effects of additional pressure work or additional flow work on the myocardial O_2 requirements? When power output of the perfused sea raven heart is increased there is a linear increase in $\dot{M}_{O_2}$ and an improvement in mechanical efficiency (Fig. 4). This is true whether the heart does extra pressure work when P_{va} is raised, or extra flow work when $\dot{V}_b$ is raised by increasing preload (P_{va} also increases slightly when $\dot{V}_b$ is increased). The difference between the two work challenges lies in the O_2 extraction from the perfusate; O_2 extraction increases with additional pressure work, but actually decreases when $\dot{V}_b$ is increased. These findings are qualitatively similar to those for the mammalian heart which is dependent on a coronary circulation. Whether these data from the sea raven are applicable to other fish species and whether f_H affects myocardial $\dot{M}_{O_2}$ are unknown.

4 Temperature Effects

Temperature effects on the neural control of f_H are established (see Farrell 1984, for review). However, I wish to focus briefly on the important effects of temperature on the intrinsic properties of the heart. Studies with perfused hearts have demonstrated positive chronotropic and inotropic responses to temperature (Bennion 1968; Graham and Farrell 1984). This means that as the metabolic needs of the tissues increase with temperature, $\dot{V}_b$ can increase intrinsically to meet these needs.

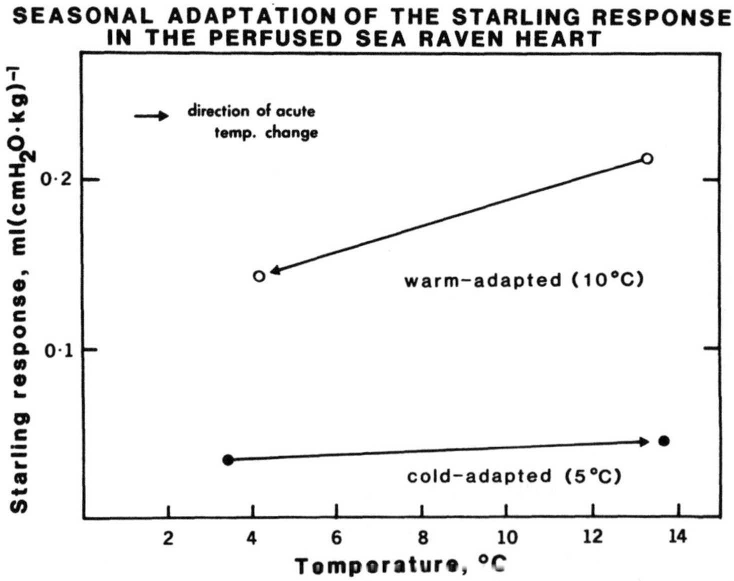

Fig. 5. The attenuation of the Starling response (change in SV_H/change in preload) with an acute and a seasonal decrease in water temperature. Adapted from Graham and Farrell 1984

Perhaps more unusual is the observation of a seasonal adaptation to temperature. Both the intrinsic f_H and the inotropic capabilities of the sea raven adapted to cold conditions (5 °C) were significantly lower than those of warm-adapted fish (10 °C) that were acutely cooled to a lower temperature (3 °C). Conversely, the inotropic capabilities of fish adapted to warm conditions were greater than those of cold-adapted fish that were acutely warmed to a higher temperature (13 °C) (Fig. 5). The significance of this inverse compensation of the heart's intrinsic capabilities with temperature may be related to the quiescent state of these fish during winter when the water temperature is near 0 °C.

5 The Role of Catecholamines

It is clear from the above that catecholamines provide inotropic and chronotropic stimulation of the heart. Until recently, it was unclear under what conditions adrenergic stimulation of the heart was important. This was especially true in view of the equivocal results obtained with adrenergic drug infusions into intact fish (Helgason and Nilsson 1973; Wood and Shelton 1980a; Pettersson and Nilsson 1980; Farrell 1981), and the finding that adrenergic stimulation did not increase the *resting* power output of perfused hearts (Farrell et al. 1982; Stuart et al. 1983). I wish to present data which indicate that catecholamines play an important role in protecting the integrity of the myocardium under adverse and stressful circumstances.

The first point is that circulating catecholamines increase during stress. Stressful exercise also results in as much as a 0.5 pH unit decrease in blood pH and a decrease in venous P_{O_2} (Wood et al. 1977; Graham et al. 1982; Milligan and Farrell 1986). Both the acidosis and hypoxia should impair the inotropic capabilities of the heart based on perfused heart and isolated ventricular strip studies. However, following strenuous exercise in vivo when there is a respiratory acidosis, cardiac performance is enhanced rather than diminished (Neumann et al. 1983). The explanation for this is probably adrenergic stimulation of the myocardium during the acidosis since a physiological level of adrenaline (0.1 μM) restores and even improves on the Starling response of the perfused heart exposed to a severe (pH 7.4) hypercapnic acidosis (Fig. 3). The mechanism(s) underlying the protective effect of adrenaline on the teleost myocardium is (are) unclear at this time.

6 Summary

Instead of a lengthy summary of the many points I have brought to your attention. I will make two general conclusions. (1) It is totally inappropriate to refer to the teleost heart as a "primitive". Many similarities with the mammalian myocardium have been uncovered, such as the Starling response, homeometric regulation, mechanical efficiency, $\dot{M}_{O_2}$ requirements as power output changes, etc. I expect that through the use of preparations which are more physiological we will continue to reveal the intricasy and beauty of the fish heart. (2) Any assessment of cardiac function in fish must monitor the important variables. From the above, I hope it is apparent that $\dot{V}_b$ and P_{va} are the two important variables, followed by P_{cv} and P_{da}.

References

Bennion GR (1968) The control of the function of the heart in teleost fish. M.S. thesis, University of British Columbia, Vancouver, Canada

Chan DKO, Chow PH (1976) The effects of acetylcholine, biogenic amines and other vasoactive agents on the cardiovascular functions of the eel, *Anguilla anguilla*. J Exp Zool 196:13–26

Daxboeck C, Holeton GF (1978) Oxygen receptors in the rainbow trout, *Salmo gairdneri*. Can J Zool 56:1254–1259

Driedzic WR, Scott DL, Farrell AP (1983) Aerobic and anaerobic contributions to energy metabolism in perfused isolated sea raven *(Hemitripterus americanus)* hearts. Can J Zool 61:1880–1883

Farrell AP (1981) Cardiovascular changes in the lingcod *(Ophiodon elongatus)* following adrenergic and cholinergic drug infusions. J Exp Biol 91:293–305

Farrell AP (1982) Cardiovascular changes in the unanesthetised lingcod *(Ophiodon elongatus)* during short-term progressive hypoxia and spontaneous activity. Can J Zool 60:933–941

Farrell AP (1984) A review of cardiac performance in the teleost heart: intrinsic and humoral regulation. Can J Zool 62:523–536

Farrell AP (1985) A protective effect of adrenaline on the acidotic teleost heart. J Exp Biol 116 (in press)

Farrell AP, MacLeod KR, Driedzic WR (1982) The effects of preload, afterload and epinephrine on cardiac performance in the sea raven, *Hemitripterus americanus*. Can J Zool 60:3165–3171

Farrell AP, MacLeod KR, Driedzic WR, Wood S (1983) Cardiac performance during hypercapnic acidosis in the in situ, perfused fish heart. J Exp Biol 107:415–429

Farrell AP, Hart T, Wood S, Driedzic WR (1984) The effect of extracellular calcium and preload on a teleost heart during extracellular hypercapnic acidosis. Can J Zool 62:1429–1435

Farrell AP, Wood S, Hart T, Driedzic WR (1985) Myocardial oxygen consumption in the sea raven, *Hemitripterus americanus*: the effects of volume loading, pressure loading and progressive hypoxia. J Exp Biol 117 (in press)

Gesser H, Poupa O (1983) Acidosis and cardiac muscle contrabiligy: comparative aspects. Comp Biochem Physiol 76(A):559–566

Gesser H, Andresen P, Brams P, Sund-Laursen J (1982) Inotropic effects of adrenaline on the anoxic or hypercapnic myocardium of rainbow trout and eel. J Comp Physiol 147:123–128

Gibbs CL, Chapman JB (1979) Cardiac energetics. In: Berne RM (ed) Handbook of physiology, vol 1, sect 2. Publ Am Physiol Soc, Bethesda Maryland, pp 775–804

Graham MS, Farrell AP (1985) The effects of temperature on cardiac performance in the sea raven. J Exp Biol 118 (in press)

Graham MS, Wood CM, Turner JD (1982) The physiological responses of the rainbow trout to strenous exercise: interactions of water hardness and environmental acidity. Can J Zool 60: 3153–3164

Helgason SS, Nilsson S (1973) Drug effects on pre- and post-bronchial blood pressure and heart rate in a free swimming marine teleost, *Gadus morhua*. Acta Physiol Scand 88:533–540

Holeton GF, Randall DJ (1967) Changes in blood pressure in the rainbow trout during hypoxia. J Exp Biol 46:297–306

Huang TF (1973) The action potential of the myocardial cells of the golden corp. Jpn J Physiol 23:529–540

Kiceniuk JW, Jones DR (1977) The oxygen transport system in trout *(Salmo gairdnerii)* during sustained exercise. J Exp Biol 69:247–260

Laurent P, Holmgren S, Nilsson S (1983) Nervous and humoral control of the fish heart: structure and function. Comp Biochem Physiol 76(A):525–542

Milligan CL, Farrell AP (1986) Extra- and intracellular acid-base status following enforced activity in the sea raven *(Hemitripterus americanus)*. J Comp Physiol (in press)

Neumann P, Holeton GF, Heisler N (1983) Cardiac output and regional blood flow in gills and muscles after strenuous exercise in rainbow trout *(Salmo gairdneri)*. J Exp Biol 105:1–14

Pettersson K, Nilsson S (1979) Nervous control of the branchial vascular resistance of the Atlantic cod, *Gadus morhua*. J Comp Physiol 129:179–183

Pettersson K, Nilsson S (1980) Drug induced changes in cardio-vascular parameters in the Atlantic Cod, *Gadus morhua*. J Comp Physiol 137:131–138

Randall DJ, Smith LS (1967) The effect of environmental factors on circulation and respiration in teleost fish. Hydrobiologica 29:113–124

Randall DJ, Holeton GF, Stevens ED (1967) The exchange of oxygen and carbon dioxide across the gills of rainbow trout. J Exp Biol 46:339–348

Saito T (1973) Effects of vagal stimulation on the pacemaker action potentials of carp heart. Comp Biochem Physiol 44A:191–199

Smith FM, Jones DR (1978) Localization of receptors causing hypoxic bradycardia in trout *(Salmo gairdneri)*. Can J Zool 56:1260–1265

Stuart RE, Hedtke JL, Weber LJ (1983) Physiological and pharmacological investigation of the nonvascularised marine teleost heart with adrenergic and cholinergic agents. Can J Zool 61: 1944–1948

Turner JD, Driedzic WR (1980) Mechanical and metabolic response of the perfused isolated fish hearth to anoxia and acidosis. Can J Zool 58:886–889

Wood CM (1974) A critical examination of the physical and adrenergic factors affecting blood flow through the gills of the rainbow trout. J Exp Biol 60:241–265

Wood CM, Shelton G (1975) Physical and adrenergic factor affecting systemic vascular resistance in the rainbow trout: a comparison with branchial vascular resistance. J Exp Biol 63:505–524

Wood CM, Shelton G (1980a) Cardiovascular dynamics and adrenergic responses of the rainbow trout in vivo. J Exp Biol 87:247–270

Wood CM, Shelton G (1980b) The reflex control of heart rate and cardiac output in the rainbow trout: Interactive influences of hypoxia, hemorrhage and systemic vasomotor tone. J Exp Biol 87:271–284

Wood CM, McMahon BR, McDonald DG (1977) An analysis of changes in blood pH following exhausting activity in the stary flounder *(Platichthys stellatus)*. J Exp Biol 69:173–185

Relationship Between Cardiac Energy Metabolism and Cardiac Work Demand in Fishes

B.D. SIDELL[1] and W.R. DRIEDZIC[2]

1 Introduction

The diversity in both anatomy and physiology of piscine hearts raises a wealth of experimental questions for investigators interested in evolution of cardiac function within the vertebrate subphylum. In particular, a series of questions which focus upon cardiac energy metabolism rank prominently.

The energy metabolism of fish hearts although generally aerobic, has been implicated as being highly dependent on carbohydrate utilization (MacIntyre and Driedzic 1981; Sidell 1983). The homologous tissue of mammals, on the other hand, exhibits a marked preference for fatty acid fuels to support its energy metabolism (Neely and Morgan 1974). Are these and other metabolic differences between the myocardia of lower and higher vertebrates correlated with differences in the energetic demand placed upon the tissues? If so, do these differences begin to emerge within the range of heart work encountered among fish species? We will attempt to address these metabolic questions in this paper.

1.1 Objectives

The specific objectives we hope to achieve are twofold:

1. to describe the general characteristics of cardiac energy metabolism in major groups of fishes; and,
2. to determine whether quantitative or qualitative differences in myocardial metabolism between fish species can be correlated with presumptive cardiac workloads.

The first objective is now realistically attainable. Sufficient physiological and biochemical data have been accumulated to permit a reasonably accurate description of piscine cardiac metabolism (see Sect. 2). The second objective is somewhat more elusive. In order to approach it, we must first assess whether the range of cardiac work demands among fish species is sufficiently broad to predict salient metabolic differences.

1 Department of Zoology, University of Maine at Orono, Orono, ME 04469, USA
2 Department of Biology, Mount Allison University, Sackville, New Brunswick EOA 3CO, Canada

Circulation, Respiration, and Metabolism
(ed. by R. Gilles)

1.2 Resting Cardiac Workloads Among Fishes

Mechanical work of the fish heart has been determined from pressure and flow determinations in intact animals. The majority of studies have reported values for cardiac output and ventral aortic pressure in resting fishes and these data form the best basis for interspecific comparisons of cardiac work (Table 1). Several features of the data deserve mention. First, in terms of fold differences, there is a substantial spectrum of cardiac workloads among fish species. Hearts of hagfish develop only ~5% of the resting power output of hearts from more active teleost species, such as rainbow trout (Table 1). From the broader perspective of the vertebrate subphylum, however, fish hearts still can be considered capable of only relatively low workloads. For example, the work output of even the more competent of fish hearts listed in Table 1 is still < 10% of comparable resting values in mammals (e.g., 43–46 mWatt kg^{-1} body wt. in the rat, Lang et al. 1984). Attempts to correlate metabolic profile of the heart with workload in fishes must be tempered by this perspective.

Second, it is evident that cardiac work in fishes also correlates highly with taxon. The lowest cardiac work demands are found in cyclostomes (at least the hagfishes). Elasmobranch hearts appear capable of only modest work levels, while teleost heart work ranks above these groups, but shows a broad range among species. Significant phylogenetic differences in the physiology of these major fish classes may influence the metabolic profiles of their cardiac tissue and possibly obscure relationships between cardiac work and metabolism. This factor is particularly evident in viewing the metabolism of elasmobranch hearts (see Sect. 2.2), and is a source of some confusion in approaching our second objective.

Table 1. Estimates of resting mechanical power developed by hearts of fishes[a]

	Power (mWatt kg^{-1} body wt.)	Reference
AGNATHA		
Myxine spp.	<0.1[b]	
ELASMOBRANCHII		
Raja binoculata	0.53	Satchell (1971)
Squalus acanthias	0.54	Kent et al. (1980)
TELEOSTEI		
Anguilla anguilla	0.95	Davie and Forster (1980)
Ophiodon elongatus	0.97	Farrell (1982)
Hemitripterus americanus	1.00	Farrell et al. (1982)
Platichthys stellatus	1.63	Wood et al. (1979)
Gadus morhua	1.90	Jones et al. (1974)
Salmo gairdneri	2.60	Wood and Shelton (1980)

[a] Power development is calculated from reported values of weight-specific cardiac output and mean ventral aortic pressure

[b] Power development of hagfish heart assumes cardiac output of the systemic heart to be < 5 ml (kg • min^{-1}) (Bloom et al. 1963) and ventral aortic pressures of 1–8 mmHg (Chapman et al. 1963)

1.3 The Question of Heart Size

There is a greater variation in weight-specific heart size among fishes than any other vertebrate group. That is, different species of fish with equivalent body weights may have hearts which vary by over fourfold in mass (Poupa and Ostadal 1969). Because of the relatively simple two-chambered design of fish hearts, an equivalent range is found in comparing body weight-specific ventricular size (see Santer et al. 1983). Power output of this single ventricle is ultimately responsible for the cardiac work estimated in Table 1 (with the possible exception of elasmobranchs, whose contractile conus arteriosus may contribute significantly to mean ventral aortic pressure). Biochemical measurements of heart muscle expressed as tissue specific activities (e.g., enzyme units g^{-1} heart) thus may be useful to the physiologist interested in comparing characteristics of cardiac muscle between species. However, if these same measurements are to be correlated meaningfully with body weight-specific cardiac workloads, it is necessary to normalize for differences in relative ventricular size (e.g., units of cardiac enzyme activity kg^{-1} body wt.). We will present the latter of these expressions for each species in the present paper, but will also provide measurements of relative ventricular size to permit calculation of the former activity.

2 Cardiac Metabolism in the Major Fish Taxa

2.1 Cyclostomes

The cardiac metabolism of jawless fishes, in particular the myxinoids or hagfishes, is of considerable interest for two reasons. As living members of the ancient class Agnatha, the hagfishes are generally considered the most primitive extant vertebrate group. Features of their cardiac metabolism, therefore, may reflect those of ancestral vertebrates. In addition, the very low vascular pressures (Chapman et al. 1963) and partially open circulatory system of these animals results in the lowest demand of cardiac work among the fishes (cf. Table 1). These factors also presumably account for the presence of several accessory hearts in the myxinoids (portal, cardinal, and caudal; see Hardisty 1979). The hagfish systemic heart is composed entirely of spongy tissue and, lacking a compact epicardium, is also devoid of coronary circulation.

The extremely low work demand imposed upon the myxinoid myocardium is undoubtedly the key factor in the remarkable anoxic tolerance of the tissue. Resting mechanical performance of the hagfish systemic ventricle is refractory to even the most severe anoxic challenges, including poisoning of the respiratory chain by azide or cyanide or hypoxia under nitrogen gas (Hansen and Sidell 1983). Anaerobic energy metabolism of the tissue is apparently by the conventional glycolytic pathway leading to lactate production since hypoxic challenge results in depletion of tissue glycogen and substantial increases of both tissue and blood lactate concentrations (Hansen and Sidell 1983).

The enzymatic profile of the hagfish myocardium is consistent with the tissue's anoxic tolerance (Table 2). Relatively robust activity is found for enzymes of the glycolytic sequence, while indicators of mitochondrial aerobic capacity (e.g., cyto-

Table 2. Biochemical indices from fish ventricle for anaerobic and aerobic energy metabolism and capacities to oxidize alternative metabolic fuels

	Activity[a]								
	Ventricle size (% body weight)	Anaerobic		Aerobic			Fuels		Ref.
		PFK	PK	CS	CO	HK	CPT	3-OHBDH	
AGNATHA									
Myxine glutinosa	0.09 ± 0.01	2.30 ± 0.16	32.4 ± 2.6	6.23 ± 0.58	3.74 ± 0.49	1.53 ± 0.11	0.05 ± 0.01	ND	b
(Atlantic hagfish)	(16)	(8)	(13)	(12)	(19)	(13)	(10)	(9)	
ELASMOBRANCHII									
Raja erinacea	0.055 ± 0.004	2.92 ± 0.21	8.4 ± 0.2	4.82 ± 0.46	10.52 ± 0.61	2.76 ± 0.20	ND	2.30 ± 0.24	c
(little skate)	(6)	(6)	(6)	(6)	(6)	(6)	(6)	(6)	
Squalus acanthias	0.085 ± 0.005	8.85 ± 0.19	69.9 ± 7.5	21.12 ± 0.87	8.54 ± 0.45	3.81 ± 0.18	ND	2.45 ± 0.15	c
(spiny dogfish)	(6)	(6)	(6)	(6)	(6)	(6)	(6)	(6)	
TELEOSTEI									
Hemitripterus americanus	0.069 ± 0.002	1.26 ± 0.11	35.78 ± 2.78	11.64 ± 1.09	34.6 ± 7.8	2.43 ± 0.68	0.11 ± 0.01	<0.05	d
(sea raven)	(7)	(3)	(3)	(3)	(5)	(6)	(7)	(6)	
Gadus morhua	0.079 ± 0.003	7.61 ± 1.17	46.50 ± 0.95	9.58 ± 0.37	30.80 ± 1.83	4.92 ± 0.15	0.35 ± 0.06	<0.05	b
(Atlantic cod)	(10)	(4)	(6)	(6)	(7)	(6)	(6)	(4)	
Dicentrarcus labrax	0.077 ± 0.004	7.35 ± 0.89	36.06 ± 1.94	10.16 ± 0.52	21.07 ± 1.84	6.33 ± 1.23	0.46 ± 0.04	<0.05	c,e
(sea bass)	(7)	(7)	(5)	(5)	(5)	(5)	(6)		
Morone saxatilis	0.124 ± 0.004	11.52 ± 1.03	37.0 ± 2.60	5.94 ± 0.36	48.7 ± 7.4	14.84 ± 1.20	0.65 ± 0.04	<0.05	c
(striped bass)	(7)	(7)	(6)	(6)	(3)	(7)	(7)		

[a] Activity is expressed as units of cardiac enzyme activity per kilogram body weight at 15 °C assay temperature. One unit = 1 μmol of substrate utilized per minute
[b] Hansen and Sidell (1983); also incorporates some revised values from more recent experiments
[c] Driedzic, Sidell, Stowe and Johnston, unpublished data
[d] Driedzic and Stewart (1982); activities were corrected from 10 °C to 15 °C assuming Q_{10} = 1.9 for all enzymes; also incorporate revised values from more recent experiments
[e] Zammit and Newsholme (1979)

chrome oxidase) are low in activity. Both biochemical data and mechanical performance studies strongly implicate carbohydrate oxidation as predominating over β-oxidation of fatty acids in the energy metabolism of the myxinoid heart (see Sect. 3.2).

Virtually nothing is known about cardiac energy metabolism in the other major cyclostome group, the lampreys. Lampreys have a considerably more activity life-style than hagfishes, possess a closed circulation and presumably place greater work demand upon their hearts since both cardiac output and vascular pressures are closer to those found in teleosts (Hardisty 1979). A careful comparison of lamprey and hagfish cardiac metabolism could help resolve whether the unusual metabolic features of the hagfishes are characteristics of their phylogenetic class or more closely correlated with their extremely low-work cardiovascular system.

2.2 Elasmobranchs

Cardiac work capacities of common elasmobranch fishes appear to be intermediate between the cyclostome and teleost groups (Table 1). The elasmobranch heart is obligately aerobic (Driedzic 1978) and characteristically possesses a distinct compact epicardium perfused by coronary circulation, although the degree to which these features are developed varies considerably among species (Santer and Greer-Walker 1980). Ventricular work in cartilagenous fishes is probably overestimated by vascular pressure and flow determinations because of the contribution of the muscular conus arteriosus to aortic pressure development.

The biochemical profile of elasmobranch cardiac muscle (Table 2) is consonant with the tissue's highly aerobic nature. Activity of cytochrome oxidase, from the mitochondrial respiratory chain, is considerably higher than that of the cyclostome tissue and capacity for substantial rates of carbohydrate catabolism is also evident. The limited anaerobic potential of elasmobranch hearts appears to be dependent upon glycolytic metabolism since tissue lactate concentration rises markedly in response to hypoxia (Driedzic 1978). Information derived from both enzymatic and mechanical performance studies indicate that carbohydrates and ketone bodies are effective metabolic fuels for the elasmobranch myocardium, but that the tissue has little capacity for catabolism of fatty acids (see Sect. 3.2; Zammit and Newsholme 1979; Driedzic and Hart 1984).

As is the case for cyclostomes, the available biochemical data on elasmobranch hearts is confined to relatively inactive species of common sharks, skates, and rays. Yet, there are several much more active pelagic species of elasmobranchs whose life history suggests substantially higher cardiac power requirements. Cardiac tissue from these more "athletic" cartilagenous fishes offers many interesting experimental possibilities to the most intrepid comparative physiologists.

2.3 Teleosts

There are more species of teleosts than any other major fish group (20–30,000 known species compared to 700–800 species of Chondrichthyes and 50 Agnathans). The spectrum of habitats and life-styles occupied by teleosts is equally impressive, from tropical to polar regions, both fresh and salt waters, from sedentary benthic or even semiparasitic species to extremely large, powerful, and fast-swimming pelagic predators.

Considerable anatomical variation is also evident among hearts of bony fishes. Utilizing both scanning electron microscopy and light microscopy, Santer and Greer-Walker (1980) were able to identify the presence of a variably-developed compact epicardium with coronary circulation in only 20 of 93 teleost species examined. Because these anatomical features are present in widely dispersed genera of teleosts, a compact myocardial layer appears to be more closely correlated with an active life history of the species than with phylogenetic position (Santer and Greer-Walker 1980).

Although some range in anoxic tolerance of the teleost myocardium has already been noted, anaerobic metabolism cannot long sustain even resting workloads, and normally accounts for only about 2% of ATP regeneration by the tissue (Driedzic 1983). Enzyme activities from the mitochondrial respiratory chain and TCA cycle are robust in these aerobic tissues and range considerably higher than in homologous tissues of elasmobranchs and cyclostomes (Table 2). As in elasmobranchs, enzymes of carbohydrate catabolism are present in high activity and capacity for oxidation of these fuels to CO_2 has been amply demonstrated (e.g., Lanctin et al. 1980). In teleost hearts, we also first encounter development of significant enzymatic capacity for oxidation of fatty acids (Zammit and Newsholme 1979; Table 2), fuels which are capable of sustaining at least resting performance of the tissue (Driedzic and Hart 1984; Sect. 3.2).

3 Relationships Between Metabolic Capacities and Workloads

Two major types of information may be exploited to examine possible quantitative and qualitative relationships between cardiac energy metabolism and power development of fish hearts. The first data base is biochemical. A series of enzymes have been identified which appear to reflect metabolic capacity of central pathways for anaerobic and aerobic metabolism, and pathways for catabolism of the major energetic fuels (carbohydrates, fatty acids, and ketone bodies). Measurements of mechanical performance of fish hearts in response to various exogenously-supplied metabolic substrates or to anoxic challenge provide the second, and most direct source of physiological information.

3.1 Aerobic and Anaerobic Capacities

Capacities for energy metabolism via aerobic pathways generally appear to correlate with power output of fish hearts. The catalytic power of both cytochrome oxidase and citrate synthase, markers of mitochondrial aerobic metabolism, tends to increase in progression from cyclostomes to elasmobranchs and through teleosts as cardiac power demands rise (Table 2). The conclusion to be drawn from these data is simply that cardiac metabolism in fishes is generally aerobic, and capacity for aerobic ATP regeneration is expanded as demand for ATP increases.

A regular pattern relating anaerobic capacity and resting work demand among the fishes is much more elusive. From a biochemical standpoint, the reaction catalyzed by phosphofructokinase (PFK) is considered to be the primary flux-limiting step of anaerobic glycolysis in vertebrate muscles (Crabtree and Newsholme 1972a; Beis and Newsholme 1975). The ratio of cardiac PFK capacity (Table 2) to resting cardiac work

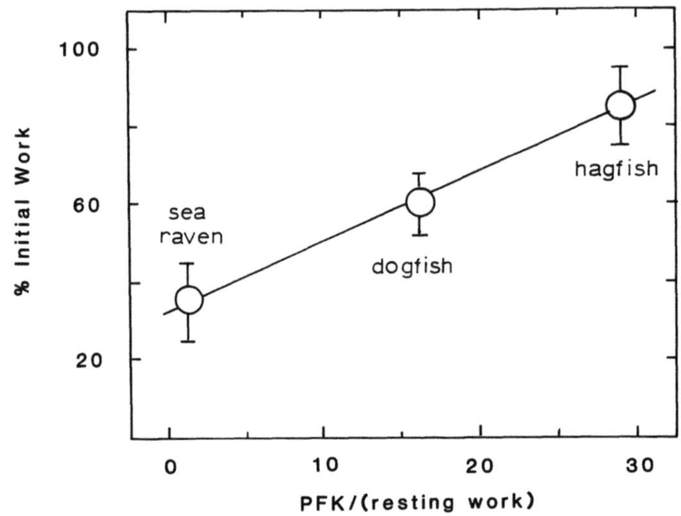

Fig. 1. Relationship between hypoxic tolerance of fish hearts and relative ability of anaerobic metabolism to meet resting workload. Hypoxic tolerances are expressed as percent initial performance of the heart after 50–60 min under N_2 gas (data from: sea raven: Turner and Driedzic 1980; dogfish: Driedzic 1978; hagfish: Hansen and Sidell 1983). Resting work and PFK values are from Tables 1 and 2, respectively

(Table 1), in species where both are known, should give some indication of relative tolerance of their hearts to hypoxic challenge. For the few species of fish where cardiac performance in response to hypoxia has been tested under comparable conditions this relationship appears valid (Fig. 1). Tolerances to hypoxia among the hearts of fishes, however, appear more closely related to the ecology and life history of the species than to absolute levels of cardiac work (e.g., see Gesser and Poupa 1974).

3.2 Preferred Metabolic Fuels

One major characteristic of myocardial metabolism which might vary as a function of cardiac workload is relative reliance on alternative metabolic fuels. Does the preferred metabolic fuel of fish hearts shift, for example, to greater utilization of fatty acids, as cardiac workloads increase? The answer, unfortunately, seems equivocal since taxonomic differences in physiology appear to outweight other factors.

3.2.1 Availability of Blood-Borne Fuels

The presence or absence of coronary circulation notwithstanding, cardiac muscle is arguably the best perfused tissue in the body of fishes. It is upon blood-borne, rather than endogenous metabolic fuels that the myocardium relies to support its continuous pumping work. Plasma concentrations of these major types of metabolic fuel differ very significantly between the fish taxa and their availability markedly influences substrate preferences of hearts from each group.

Carbohydrate, predominantly in the form of glucose, is found in the millimolar range of concentrations in plasma of cyclostomes (Robertson 1974), elasmobranchs, and teleosts (Larsson and Fange 1977; Zammit and Newsholme 1979). However, the plasma of elasmobranchs contains only extremely low levels of fatty acids and triacylglycerols (Zammit and Newsholme 1979) and appears to completely lack albumin capable of binding and transporting fatty acid (Fellows and Hird 1981). Ketone bodies, on the other hand, circulate at significant concentrations in the blood of cartilagenous species while being present in only trace quantities in the blood of teleosts (Zammit and Newsholme 1979).

3.2.2 Comparison of Substrate Preferenda

Each of the above fuel availabilities seem to agree well with the maximum activities of marker enzymes for aerobic metabolism of carbohydrate, fatty acid, and ketone fuels in hearts of fishes. Crabtree and Newsholme (1972a,b) have presented convincing evidence that the maximum rates of oxidation of glucose, fatty acids, and ketone bodies by muscle tissue of vertebrates are best correlated with maximum tissue activities of hexokinase (HK), carnitine palmitoyltransferase (CPT), and 3-hydroxybutyrate dehydrogenase (3-OHBDH), respectively.

Cyclostomes. Energy metabolism of the low-work hearts in Atlantic hagfish appears to be almost totally carbohydrate-based. Hexokinase activity of this tissue is modest but far greater than that of CPT or 3-OHBDH, the latter being undetectable (Table 2). When theoretical ATP yields are calculated from these enzyme activities, an approx. 20-fold greater energy-generating capacity is found for the glucose pathway than for fatty acid oxidation (Table 3). Mechanical studies performed on isolated hagfish ven-

Table 3. Capacities for regeneration of ATP by oxidation of carbohydrates, fats, or ketone bodies calculated from enzyme activities of fish hearts[a]

	μmol ATP (kg • min)$^{-1}$ from Carbohydrate	Fatty acid	Ketone
AGNATHA			
Myxine glutinosa	58 ± 5	3 ± 1	0
ELASMOBRANCHII			
Raja erinacea	105 ± 8	0	61 ± 7
Squalus acanthias	145 ± 7	0	71 ± 4
TELEOSTEI			
Hemitripterus americanus	92 ± 26	7 ± 1	< 2
Gadus morhua	187 ± 6	23 ± 4	< 2
Dicentrarcus labrax	241 ± 2	30 ± 3	< 2
Morone saxatilis	564 ± 46	44 ± 2	< 2

[a] Based upon ATP yields of 38 ATP/glucose, 130 ATP/palmitate, 29 ATP/3-hydroxybutyrate; calculated from activities of HK (carbohydrate), CPT (palmitate), and 3-OHBDH (butyrate), respectively, from Table 1. CPT activity was divided by 2 to account for the position of this enzyme at two places in the fatty acid pathway on either side of the mitochondrial membrane (see Bremer 1983; Zammit and Newsholme 1979)

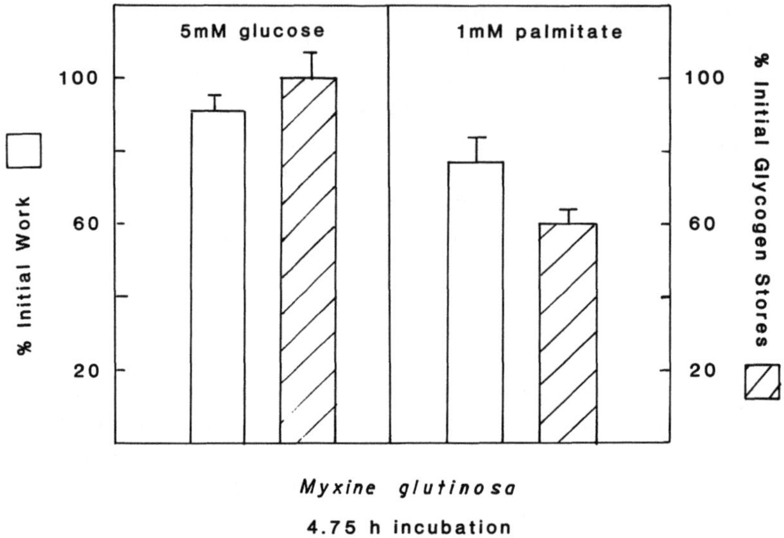

Fig. 2. Ability of exogenous metabolic fuels to support mechanical work and protect glycogen stores of the isolated hagfish ventricle. Mechanical performance when supplied with 1.0 m*M* palmitic acid only is significantly lower than with 5.0 m*M* glucose. Glycogen reserves remain intact in the presence of 5.0 m*M* glucose, but nearly one-half of available glycogen is mobilized when palmitic acid is the sole exogenously-supplied substrate. Incubation times = 4–5 h. (Data from Sidell et al. 1984)

tricles also confirm the predominance of carbohydrate metabolism in the tissue. When 5 m*M* glucose is exogenously provided to the isolated hagfish ventricle, mechanical performance is completely supported without detectable breakdown of endogenous glycogen stores (Fig. 2). Provision of 1 m*M* palmitate alone, however, causes significant decrement in mechanical performance and extensive mobilization of endogenous glycogen stores to meet the energetic shortfall (Fig. 2). Although not tested directly, there is little reason to suspect that ketone bodies could adequately support mechanical work of the cyclostome tissue, since enzymes associated with this pathway are not detectable (Table 2).

Elasmobranchs. The next step up in cardiac power development of fishes is elasmobranch species, but this also corresponds with dramatic shifts in availability of metabolic substrates alluded to previously. Activity of hexokinase in the moderate-work hearts of sharks and skates is significantly increased in comparison to that in cardiac tissue of hagfishes (Table 2). Catalytic capacity for oxidation of fatty acids (as indicated by CPT activity), however, is completely absent in hearts of this fish group. Elasmobranch hearts, however, display the most robust ability for oxidizing ketone bodies among the fishes (as indicated by 3-OHBDH; Table 2, Zammit and Newsholme 1979). Indeed, ATP yield calculated from enzyme activities indicates that the energy-generating capacity of ketone body oxidation in the elasmobranch tissue approaches the potential of carbohydrate oxidation (Table 3). Evidence from mechanical studies corroborates these conclusions.

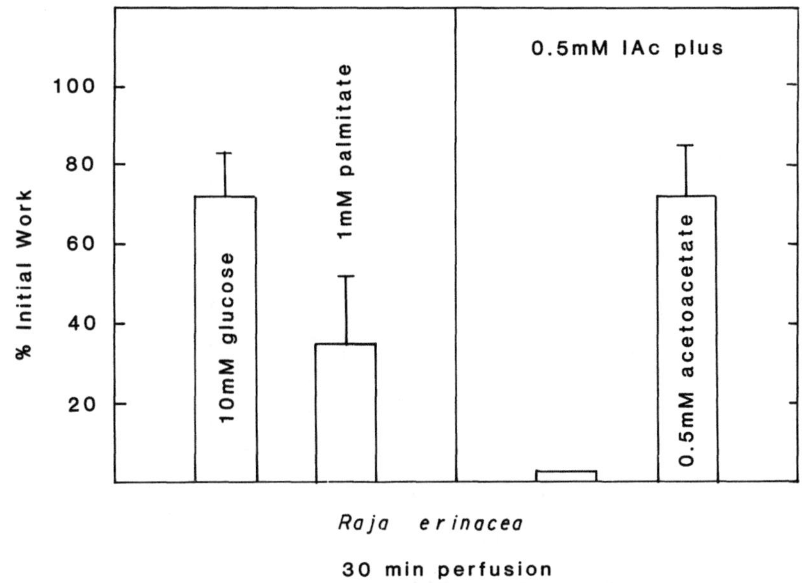

Fig. 3. Ability of exogenous metabolic fuels to support mechanical work in the isolated perfused heart of the little skate. The ketone body, acetoacetate, supports mechanical work effectively when carbohydrate metabolism is blocked by the glycolytic inhibitor, iodoacetic acid (IAc). Mechanical work in the presence of fatty acid, 1 m*M* palmitate, is significantly reduced even when carbohydrate metabolism remains unimpeded. Perfusion time = 30 min. (Data from Driedzic and Hart 1984)

The abilities of glucose, palmitate, and acetoacetate to support physiological levels of work by the isolated perfused skate heart have been assessed recently by Driedzic and Hart (1984). Exogenously supplied glucose or acetoacetate were found to be equally effective in supporting mechanical work of the skate heart, even when hearts receiving the ketone body were glycolytically-poisoned with iodoacetate to prevent mobilization of endogenous glycogen stores (Fig. 3). Moderate levels of fatty acid were not only unable to support mechanical performance of the preparation, but actually caused a decrement in work, even in the absence of glycolytic blockage (Fig. 3).

Teleosts. As cardiac power requirements progressively increase above those of elasmobranchs in the bony fishes, capacity for oxidation of carbohydrate is expanded considerably (Table 2). Calculated energy-generating capacity from glucose oxidation (Table 3) rises roughly in parallel with presumptive cardiac workload. Catalytic potential for oxidizing fatty acids (as judged by CPT) displays an equivalent trend of expansion with increasing cardiac work (Table 2), but consistently shows only 8%–10% of the ATP regenerating capacity of carbohydrate oxidation (Table 3). The magnitude of difference in these biochemical indicators is surprising in light of results from physiological preparations.

Both 10 m*M* glucose and 1 m*M* palmitic acid are equally effective in supporting physiological levels of work by the isolated perfused teleost heart, even when glycogen mobilization in the presence of palmitate is blocked by iodoacetic acid (Driedzic and

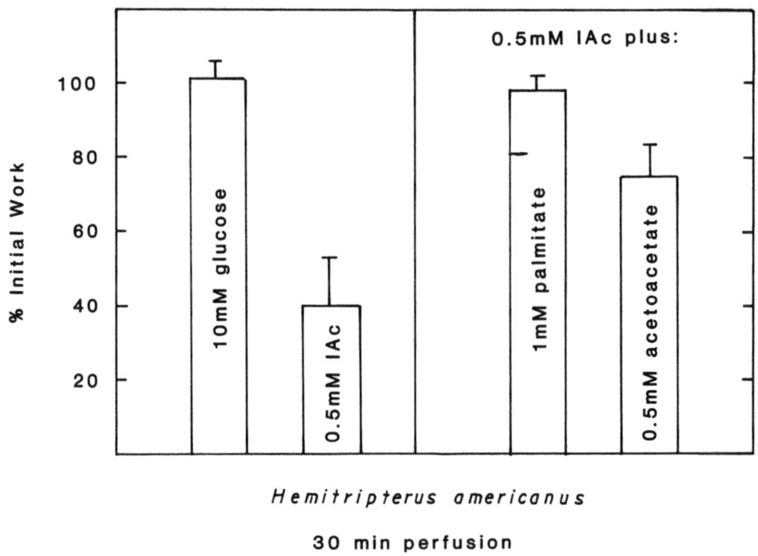

Fig. 4. Ability of exogenous metabolic fuels to support mechanical work in the isolated perfused sea raven heart. Both 10 m*M* glucose and 1 m*M* palmitic acid are effective fuels for supporting mechanical work, even when recruitment of carbohydrate metabolism in the presence of palmitate is blocked by iodoacetic acid (IAc). Mechanical work is significantly reduced when ketone, acetoacetate, is the only exogenous fuel supplied to glycolytically-blocked hearts. Perfusion time = 30 min. (Data from Driedzic and Hart 1984)

Hart 1984; Fig. 4). As expected from biochemical data, the ketone body, acetoacetate, is not an adequate fuel to support heart work in the teleost preparation (Fig. 4). Yet for the first time in our survey, physiological and biochemical data, at least as pertain to quantitative estimates of glucose and fatty acid utilization, appear at variance. The mechanical studies are unequivocal. Is it possible that our biochemical indices have either quantitatively underestimated capacity for oxidizing fatty acids or overestimated that for carbohydrate? The answer possibly is "yes" to both questions.

3.3 Potential Biochemical Pitfalls

Measurement of CPT activity in tissue homogenates assesses the catalytic potential of the long chain fatty acyltransferase of mitochondria to form palmitoylcarnitine from palmitoylCoA and free carnitine. Because this acyltransferase system is capable of operating on a wide class of long chain fatty acylCoA thioesters (see Bremer 1983), there exists the possibility of marked difference in substrate specificity depending on either chain length or degree of unsaturation of the fatty acyl moeity. In other words, although palmitoyl CoA may be an optimal substrate for the transferase in higher vertebrates, other acylCoA substrates may be more suitable for the enzyme from ectothermic tissues. We have recently performed a limited series of experiments that suggest that this is the case.

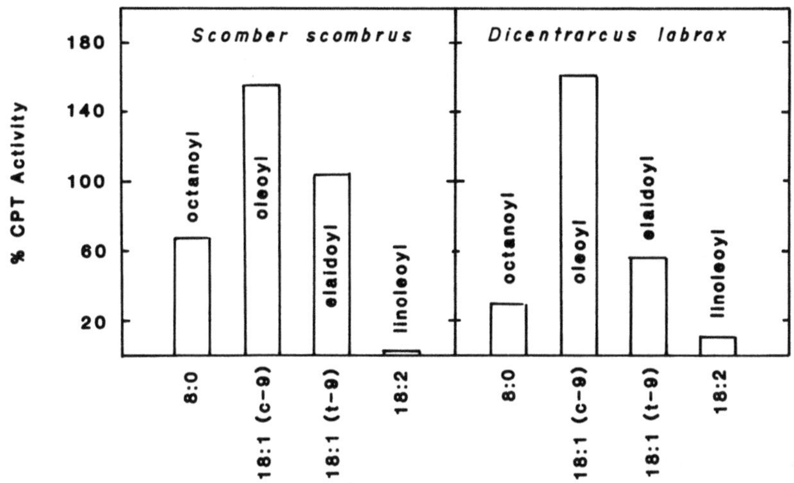

Fig. 5. Substrate selectivity of the long chain acyltransferase system of teleost heart mitochondria. Species examined: mackerel, *Scomber scombrus*, and sea bass, *Dicentrarcus labrax*. Data are presented as percentage of activity when palmitoyl CoA (16:0) is used as substrate. All assays were conducted at 15 °C, pH 8.0. (Sidell and Driedzic, unpublished results)

The long chain acyltransferase system from hearts of at least two teleost species, bass and mackerel, displays 60%–70% greater activity with oleoyl CoA (18:1) as substrate than with palmitoyl CoA (16:0) (Fig. 5). What is particularly surprising is the marked difference in activity of the enzyme between substrates which differ only modestly in their chemical structure. It remains possible that other acylCoA thioesters not yet examined show even greater activity than oleoylCoA. But it is encouraging to note that oleic and not palmitic acid is the predominant component of plasma nonesterified fatty acids in a variety of fishes (see, for example, Fellows et al. 1980). In any case, it appears likely that CPT activity of teleost tissues may significantly underestimate the energy-generating capacity from oxidation of fatty acids.

There is also reason to suspect that hexokinase activity may overestimate the maximum rate at which glucose can be oxidized aerobically by heart muscle. First, the isolated, perfused working rat heart is capable of only approximately one-fifth the rate of glucose uptake predicted by hexokinase activity of the tissue. This result has led England and Randle (1967) to suggest that cardiac hexokinase is inhibited sufficiently in situ by intracellular glucose-6-phosphate levels so that its full catalytic potential is never attained. A second, although admittedly more speculative suggestion can be made.

It is at least theoretically possible that blood-borne glucose could be taken up by the hearts of fishes and anaerobically fermented to lactic acid during hypoxia or extremes of activity. The uniquely high level of perfusion of the myocardium can ensure not only a continuous supply of glucose from the blood, but also extremely effective washout of accumulated lactate from the tissue. Perhaps the energetically less efficient use of blood-borne glucose (2 ATP/glucose) than glycogen (3 ATP/glucosyl unit) as a substrate for anaerobic glycolysis is preferable to depletion of the tissue's "last resort" glycogen reserves. An anaerobic role of hexokinase activity in heart might also explain

Table 4. Differential in hexokinase activity between cardiac and red skeletal muscle of teleosts[a]

	Hexokinase activity[b] [Units • (g wet wt.)$^{-1}$]	
	Heart	Red skeletal muscle
Carassius carassius (carp)	7.49 ± 0.53 (5)	0.58 ± 0.07 (5)
Esox niger (chain pickerel)	2.33 ± 0.25 (9)	0.72 ± 0.13 (9)

[a] All fish were acclimated to 25 °C; assays were performed at 15 °C. Data are from Kleckner and Sidell, 1985 (chain pickerel) and Sidell, Driedzic, and Johnston, unpublished (carp)
[b] One unit of activity = 1 μmol product formed per minute

why the enzyme is so much more active in cardiac tissue than red skeletal muscle of fishes (Table 4). Although otherwise metabolically very similar to heart, red skeletal muscle is less well-perfused than the myocardium, especially at extremely high levels of muscular activity.

If any of the above arguments regarding cardiac CPT and HK activities are correct, the relative importance of carbohydrate and fatty acids as metabolic fuels of the teleost heart potentially are much more similar than our biochemical data would suggest. Such an interpretation is more consistent with the direct physiological results (e.g., Driedzic and Hart 1984). However, these possibilities still do not provide arguments to suggest that major differences in fuel preference exist among the hearts of teleost species. Our working hypothesis at this stage must be that the range of cardiac power demands in bony fishes has not expanded sufficiently to elicit major shifts in preferenda for metabolic fuels.

4 Possible Avenues for Future Work

Each of the earlier discussions in this paper have tried to point out specific gaps in information on cardiac metabolism of the fishes. From a broader perspective, however, it should be equally clear that the comparative survey approach to assessing relationships between heart work and metabolism has many inherent weaknesses. In particular, factors of phylogeny and life history strategy are recurrent sources of confusion which separate us from any broad generalizations. In future efforts alternative experimental approaches may address more clearly the question of how differing cardiac workloads affect heart metabolism in the fishes.

Experimental treatments which significantly alter chronic heart work within individuals of a single species appear particularly promising. Two possible treatments worthy of future study are exercise and temperature. Available data on the subject suggest that power development by fish hearts may increase by up to fivefold during strenuous activity (Kiceniuk and Jones 1977, Jones and Randall 1978). Yet little is known regarding endurance exercise effects on chronic cardiac work or metabolism in fishes.

Environmental temperature is a physical variable which has been manipulated to great advantage in studying the metabolism of fishes at both biochemical (Hazel and Prosser 1974) and whole animal (Fry 1971) levels of organization. Yet, little attention has been paid to the potential effects of temperature on cardiac work or even specifically on cardiac metabolism in eurythermal fishes. The a priori supposition is that cardiac work demand for equivalent perfusion of tissues may increase markedly at cold temperatures because of elevated blood viscosity and lowered vascular compliance. Certainly we have indications from biochemical data that enzymatic capacities of heart tissue in cold-active fishes is greatly affected by thermal acclimation (Kleckner and Sidell 1985).

Finally, we should also mention a basic inequity in the estimates of cardiac power development and metabolic capacity that have been compared throughout this paper. The former physiological measurements have consistently reflected basal or resting rates while the biochemical data estimate maximum available capacities from in vitro measurements of enzymes under optimal conditions. Some effort should be expended to develop comparable data sets for relating these two features of cardiac physiology. Realistic assessment of "resting" metabolic capacities and substrate oxidation rates are probably most accurately approached by studies utilizing isolated perfused hearts and isotopically-labeled metabolites. Such studies are technically difficult and costly to execute. A more intriguing challenge is to identify a biochemical indicator which accurately reflects relative maximum energy demands of fish hearts. One possibility that we are currently exploring is the measurement of ATPase activity of crude cardiac homogenates under conditions which favor activation of the contractile machinery, an approach used successfully in comparative studies of skeletal muscle (Zammit and Newsholme 1976).

5 Summary

1. Resting power development of fish hearts varies over a considerable range, from 0.1 mWatt kg^{-1} body wt. to > 3 mWatt kg^{-1} body wt. However, compared to endothermic vertebrates (e.g., rat: 45 mWatt kg^{-1} body wt.) the fish myocardium is a low-work system.
2. Cardiac power development increases from cyclostomes to elasmobranchs and increases further from sluggish to active teleosts.
3. Capacity for aerobic energy metabolism generally expands through the fishes as cardiac power development increases.
4. Anaerobic tolerance of fish hearts is related more to species' ecology and life history than phyletic position or cardiac workload.
5. Physiological results indicate that resting heart work may be supported by metabolism of carbohydrate in cyclostomes, by either carbohydrate or ketone bodies in elasmobranchs and by either carbohydrate or fatty acids in teleosts. Fuel preferenda appear more related to phylogenetic position than cardiac workload.
6. Biochemical indices suggest a significantly greater capacity for oxidation of carbohydrate than fatty acids in teleost hearts, but may overestimate the true differential.

7. The range of cardiac workloads among fishes is apparently not sufficient to have caused the marked predominance of fatty acid utilization characteristic of myocardial energy metabolism in endothermic vertebrates.

Acknowledgments. The authors' work is supported by grants from the U.S. National Science Foundation, American Heart Association, Maine Affiliate (B.D.S.); and N.S.E.R.C. of Canada, New Brunswick Heart Association (W.R.D.).

References

Beis I, Newsholme EA (1975) The contents of adenine nucleotides, phosphagens and some glycolytic intermediates in resting muscles from vertebrates and invertebrates. Biochem J 152:23–32
Bloom G, Ostland E, Fange R (1963) Functional aspects of cyclostome hearts in relation to recent structural findings. In: Brodal A, Fange R (eds) The biology of *myxine*. Universitatets Forlaget, Oslo, pp 317–339
Bremer JE (1983) Carnitine – Metabolism and functions. Physiol Rev 63:1420–1480
Chapman CB, Jensen D, Wildenthal K (1963) On circulatory control mechanisms in the Pacific hagfish. Circ Res 12:427–440
Crabtree B, Newsholme EA (1972a) The activities of phosphorylase, hexokinase, phosphofructokinase, lactate dehydrogenase and the glycerol-3-phosphate dehydrogenases in muscles from vertebrates and invertebrates. Biochem J 126:49–58
Crabtree B, Newsholme EA (1972b) The activities of lipases and carnitine palmitoyltransferase in muscles from vertebrates and invertebrates. Biochem J 130:697–705
Davie PS, Forster ME (1971) Cardiovascular responses to swimming in eels. Comp Biochem Physiol 67A:367–373
Driedzic WR (1978) Carbohydrate metabolism in the perfused dogfish heart. Physiol Zool 51:42–50
Driedzic WR (1983) The fish heart as a model system for the study of myoglobin. Comp Biochem Physiol 76A:487–493
Driedzic WR, Hart T (1984) Relationship between exogenous fuel availability and performance by teleost and elasmobranch hearts. J Comp Physiol 154:593–599
Driedzic WR, Stewart JM (1982) Myoglobin content and the activities of enzymes and energy metabolism in red and white fish hearts. J Comp Physiol 149:67–73
England PJ, Randle PJ (1967) Effectors of rat heart hexokinases and the control of rates of glucose phosphorylation in the perfused rat heart. Biochem J 105:907–920
Farrell AP (1982) Cardiovascular changes in the unanesthetized lingcod *(Ophiodon elongatus)* during short-term progressive hypoxia and spontaneous activity. Can J Zool 60:933–941
Farrell AP, MacLeod K, Driedzic WR (1982) The effects of preload, after load and epinephrine on cardiac performance in the sea raven *(Hemitripterus americanus)*. Can J Zool 60:3165–3171
Fellows FCI, Hird FJR (1981) Fatty acid binding proteins in the serum of various animals. Comp Biochem Physiol 68B:83–87
Fellows FCI, Hird FJR, McLean RM, Walker TI (1980) A survey of the non-esterified fatty acids and binding proteins in the plasmas of selected animals. Comp Biochem Physiol 67B:593–597
Fry FEJ (1971) The effect of environmental factors on the physiology of fish. In: Hoar WS, Randall DJ (eds) Fish physiology, vol 6. Academic, New York, pp 1–98
Gesser H, Poupa O (1974) Relations between heart muscle enzyme pattern and directly measured tolerance to acute anoxia. Comp Biochem Physiol 48A:97–103
Hansen CA, Sidell BD (1983) Atlantic hagfish cardiac muscle: Metabolic basis of tolerance to anoxia. Am J Physiol 244:R356–R362
Hardisty MW (1979) Biology of the cyclostomes. Chapman and Hall, London, p 396
Hazel JR, Prosser CL (1974) Molecular mechanisms of temperature compensation in poikilotherms. Physiol Rev 54:620–677

Jones DR, Randall DJ (1978) The respiratory and circulatory systems during exercise. In: Hoar WS, Randall DJ (eds) Fish physiology, vol 7. Academic, New York, pp 425–501

Jones DR, Langille BL, Randall DJ, Shelton G (1974) Blood flow in dorsal and ventral aortas of the cod, *Gadus morhua*. Am J Physiol 226:90–95

Kent B, Levy M, Opdyke MB (1980) Effect of acetylcholine on oxygen uptake in the gill of *S. acanthias*. Bull Mt Desert Isl Biol Lab 20:109–112

Kiceniuk J, Jones DR (1977) The oxygen transport system in trout *(Salmo gairdneri)* during sustained exercise. J Exp Biol 69:247–260

Kleckner NW, Sidell BD (1985) Comparison of maximal activities of enzymes from tissues of thermally-acclimated and naturally-acclimatized chain pickerel *(Esox niger)*. Physiol Zool 58: 18–28

Lanctin HP, McMorran L, Driedzic WR (1980) Rates of glucose and lactate oxidation by the perfused isolated trout *(Salmo gairdneri)* heart. Can J Zool 58:1708–1711

Lang CH, Bagby GJ, Ferguson JL, Spitzer JJ (1984) Cardiac output and redistribution of organ blood flow in hypermetabolic sepsis. Am J Physiol 246:R331–R337

Larsson A, Fange R (1977) Cholesterol and free fatty acids (FFA) in the blood of marine fish. Comp Biochem Physiol 57B:191–196

MacIntyre A, Driedzic WR (1981) Activities of enzymes in cardiac energy metabolism. Can J Zool 59:325–328

Neely JR, Morgan HE (1974) Relationship between carbohydrate and lipid metabolism and the energy balance of heart muscle. Annu Rev Physiol 36:413–459

Poupa O, Ostadal B (1969) Experimental cardiomegalies and "cardiomegalies" in free-living animals. Ann NY Acad Sci 156:445–468

Robertson JD (1974) Osmotic and ionic regulation in cyclostomes. In: Florkin M, Sheer BT (eds) Chemical zoology, vol 8. Academic, New York, pp 149–193

Santer RM, Greer-Walker M (1980) Morphological studies on the ventricle of teleost and elasmobranch hearts. J Zool (Lond) 190:259–272

Santer RM, Greer-Walker M, Emerson L, Witthames PR (1983) On the morphology of the heart ventricle in marine teleost fish (Teleostei). Comp Biochem Physiol 76A:453–457

Satchell GH (1971) Circulation in fishes. Cambridge University Press, London

Sidell BD (1983) Cardiac metabolism in the myxinidae: Physiological and phylogenetic considerations. Comp Biochem Physiol 76A:495–505

Sidell BD, Stowe DB, Hansen CA (1984) Carbohydrate is the preferred metabolic fuel of the hagfish *(Myxine glutinosa)* heart. Physiol Zool 57:266–273

Turner JD, Driedzic WR (1980) Mechanical and metabolic response of the perfused isolated fish heart to anoxia and acidosis. Can J Zool 58:886–889

Wood CM, Shelton G (1980) Cardiovascular dynamics and adrenergic responses of the rainbow trout in vivo. J Exp Biol 87:247–270

Wood CM, McMahon BR, MacDonald BG (1979) Respiratory gas exchange in the resting starry flounder, *Platichthys stellatus*: a comparison with other teleosts. J Exp Biol 78:167–179

Zammit VA, Newsholme EA (1976) The maximum activities of hexokinase, phosphorylase, phosphofructokinase, glycerolphosphate dehydrogenases, lactate dehydrogenase, octopine dehydrogenase, PEP carboxykinase, nucleoside diphosphate kinase, glutamate-oxaloacetate transaminase, and arginine kinase in relation to carbohydrate utilization in muscles from marine invertebrates. Biochem J 160:447–462

Zammit VA, Newsholme EA (1979) Activities of enzymes of fat and ketone-body metabolism and effects of starvation on blood concentrations of glucose and fat fuels in teleost and elasmobranch fish. Biochem J 184:313–322

Effects of Hypoxia and Acidosis on Fish Heart Performance

H. GESSER[1]

1 Introduction

The vertebrate cells function with few exceptions optimally under aerobic conditions and at pH values within a close range. The heart muscle is frequently taken as an example of a tissue where this is particularly true.

For some mammals and also for many poikilothermic vertebrates, normal life includes frequent episodes of a diminished oxygen supply as well as of acidosis also as far as the heart is concerned. Heart muscle is called upon to function uninterrupted and it is of great interest to examine to what extent and by what means this function is maintained during such perturbations.

The present paper deals with fishes since a wide variation in oxygenation and acid-base balance of the tissues can be expected in this group. Thus, an increased contractile demand seems to be associated with oxygen deprivation as well as acidosis in the heart muscle of many fishes. One reason for this is the varying supply of coronaries to the heart tissue among fishes (Ostadal and Schiebler 1971; Poupa et al. 1974; Santer and Greer-Walker 1980). Due to the organization of circulation in fishes, the tissue lacking coronaries is nourished by systemic venous blood. However, due to circulatory shunts there may be an admixture of arterial well-oxygenated blood with a low pCO_2 (Randall 1984). In this mixed blood the acidity and the pO_2 can be expected to vary with the state of activity of the fish.

In addition to variations in O_2 supply and acid-base balance in the surroundings of the heart cells due to anatomical conditions, there are also environmental conditions aggravating such stresses. In general, pO_2 and pCO_2 are more variable in water than in air (Dejours 1975).

2 Anoxic Tolerance and Metabolic Capacities

In view of the variation in environmental conditions and in heart tissue vascularization among fish species, it was not surprising to find large differences in the ability of isolated heart tissue to maintain contractility under anaerobic conditions. As an example Fig. 1 shows that the twitch tension of heart tissue of plaice is reduced by 50% after about

1 Department of Zoophysiology, University of Aarhus, 8000 Aarhus C, Denmark

Circulation, Respiration, and Metabolism
(ed. by R. Gilles)

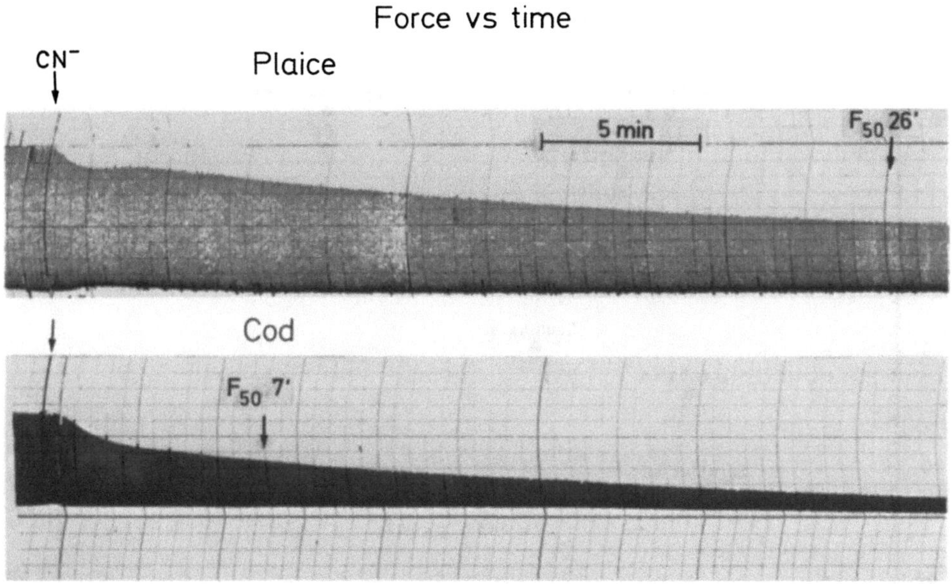

Fig. 1. Effects of cyanide on twitch force development in ventricle strips of plaice and cod. F_{50} denotes the time of anaerobiosis necessary for a 50% reduction of force

30 min exposure, whereas the same change occurs in less than 10 min for the cod. As to the immediate cause for the drop in contractility during oxygen deprivation, several hypotheses exist. The anaerobic energy liberation is maintained solely by glycolysis. Hence, interest was directed to the possibility of a variation in the capacity of this pathway. Pette (1965) has pointed out that the activity of the different enzymes constituting a metabolic pathway, such as glycolysis, appears to be related by constant ratios. In accordance with this the tissue activity of pyruvate kinase and cytochrome oxidase measured in vitro has been shown to give a good estimate of the capacity of the tissue for anaerobic lactate production and oxygen consumption, respectively. Conceivably the loss of force during anoxia depends inversely on the extent to which the lost aerobic energy liberation can be replaced with glycolysis. To estimate this the activity ratio of pyruvate kinase to cytochrome oxidase was calculated. This ratio was compared with the loss of contractile force during anoxia estimated as the time necessary for losing 50% of the preanaerobic force in hearts from seven species of fish (Gesser and Poupa 1974). A strong correlation between these quantities appeared (Fig. 2) indicating that the tolerance to oxygen lack involves an enhanced relative capacity for glycolysis. However, this does not necessarily imply that the loss of contractility under anoxia primarily is a result of a lowered ATP, creatine phosphate content.

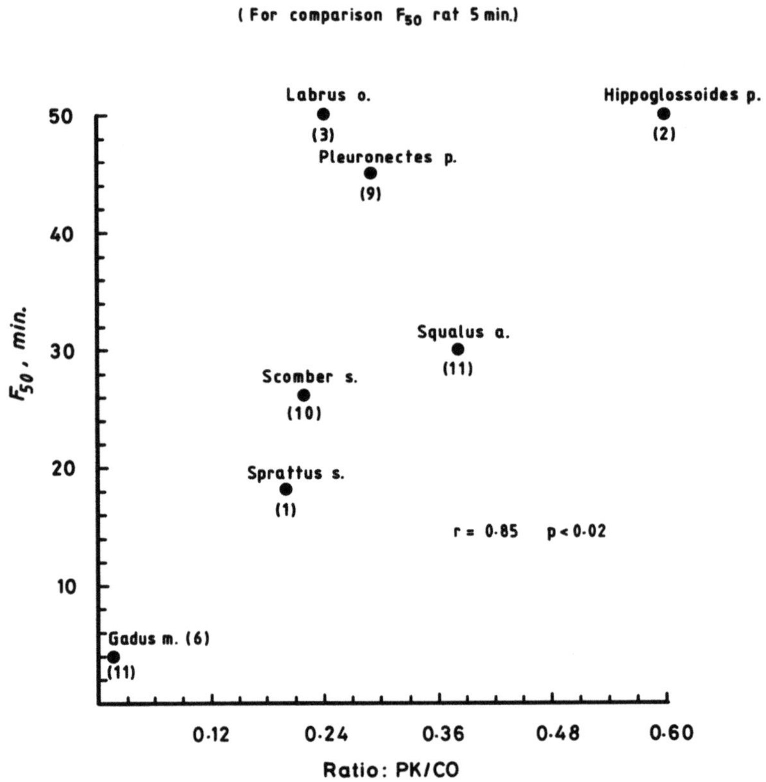

Fig. 2. Relation between anoxic twitch force development, F_{50}, and the activity ratio of pyruvate kinase to cytochrome oxidase (Gesser and Poupa 1974)

3 Hypoxia, Adrenaline, and $[Ca^{2+}]_o$

Conditions like physical stress inflicting oxygen lack upon the heart cells often involve other effects as well. Obviously an adrenergic stimulation may occur. For the mammalian heart muscle deprived of oxygen adrenaline causes at most a marginal enhancement of contractility. Frequently, this is only transient being followed by an enlarged loss of contractility (Bing et al. 1972; Davidson et al. 1974).

Quite another behavior was seen for the fish heart (Gesser et al. 1982). The heart muscle of rainbow trout loss about 60% of contractile force during 30 min of anoxia. Despite this, adrenaline administration exerted the same stimulatory effect in relative terms as during oxygenation. Maximally the twitch force increased by about 150% for the trout heart. In particular, it should be noticed that the stimulatory effect of adrenaline was maintained throughout the anoxia period of 30 min and the subsequent reoxygenation period. Similar results were obtained for the eel heart. Hence, unlike the situation for the mammalian heart, adrenaline does not seem to accelerate any process of contractile deterioration in the oxygen-deprived fish heart.

The mechanism for the stimulation of the heart contractility by adrenaline probably includes an enhanced rate of energy liberation (Williamson 1964) and an enlargement of the Ca^{2+} pool in the electromechanical (EC) coupling (e.g., Niedergerke and Page 1977). During severe physical exercise a further enlargement of this pool may be obtained by an increase in plasma calcium, which means an elevation of the extracellular calcium activity. Thus, Ruben and Bennett (1981) found that the plasma calcium increased in several vertebrates exposed to strenuous exercise. In trout, for instance, this increase amounted to about 70%. Somewhat in contrast, however, a recent study of rainbow trout subjected to a similar treatment showed plasma calcium activity only marginally elevated (Andreasen, to be published).

Like adrenaline administration an elevation in $[Ca^{2+}]_0$ had a different effect on the heart of mammal and fish (Nielsen and Gesser 1983). It had a strong positive inotropic effect on the oxygenated heart muscle of both rat and rainbow trout. During 90 min of oxygen deprivation, however, this effect disappeared for the rat heart, while it became accentuated for the trout heart muscle as measured in relative terms. Similar results were obtained with eel, crucian carp, and roach. Hence, the adrenergic stimulation and a possible increase in $[Ca^{2+}]_0$ may be of importance for the maintenance of an appropriate heart function during stress, for instance, in vertebrates having a heart tissue with a limited coronary supply.

4 Mechanisms for Loss of Contractility During Oxygen Lack

The stimulatory effect of adrenaline and of an increased $[Ca^{2+}]_0$ for the heart of, e.g., rainbow trout and eel during anoxia (Nielsen and Gesser 1984a,b) accentuates the question of what is causing the loss of force during this condition. Furthermore, to what extent does the fish heart differ from the mammalian heart in this respect? Hence, in spite of a decrease in force by about 70% at a $[Ca^{2+}]_0$ within a physiological range, the anaerobic capacity of the trout heart does not appear to be exhausted. This was directly shown since the lactate production already tripled by the onset of anoxia was about doubled when $[Ca^{2+}]_0$ was changed from 1.25 to 5 m*M* (Nielsen and Gesser 1984b). The mechanism for this stimulation by Ca^{2+} is unclear. It cannot be ascribed to a release of phosphofructokinase activity due to lowering of ATP since no change in the tissue ATP concentration could be observed during anoxia with a higher or a low $[Ca^{2+}]_0$. However, the creatine phosphate concentration was reduced from about 1.4 to about 0.2 μmol g^{-1} wt. after 30 min of anoxia (Nielsen and Gesser 1984b). This stresses the possibility of a cellular compartmentalization of ATP. Evidence exists suggesting that there are a mitochondrial and a cytoplasmatic pool of ATP linked together by a creatine-creatine phosphate shuttle (McClellan et al. 1983; Saks et al. 1978). If so, the decrease in creatine phosphate indicates a lacking capacity to rephosphorylate ADP in the compartment containing the contractile system.

The view that diffusion barriers exist for ATP, which are penetrable for creatine phosphate has recently been challenged. This view seems to rest on data which could as well be explained with a kind of facilitated transport of ATP. Here, ATP is close to equilibrium with creatine phosphate throughout the diffusion path. Thus, ATP has the additional possibility to diffuse as creatine phosphate (Meyer et al. 1984). Immediately, however, this model seems hard to apply on the anoxic trout heart, since the concentration of creatine phosphate fell by about 85%, while the ATP concentration was unchanged (Nielsen and Gesser 1984b).

The creatine phosphate concentration does not seem to be the upper limit for the anoxic performance of the trout heart. Despite the fact that an elevation of $[Ca^{2+}]_0$ from 1.25 to 5 m*M* gave about a doubling of anoxic twitch force, no change in the tissue CP level could be documented (Nielsen and Gesser 1984b). Alternatively, however, it is conceivable that an elevation in $[Ca^{2+}]_0$ overcomes a negative inotropic effect of a lowered CP level by increasing the activated portion of the contractile system.

Other possible mechanisms exist for the loss of contractility during anoxia, which do not immediately relate to the energy state. Thus, the action potential duration has been found to decrease in oxygen-deprived heart muscle of guinea pig. This suggests that less Ca^{2+} is made available for the E,-C-coupling (McDonald and MacLeod 1971). However, this possibility does not appear to apply for trout heart muscle, for which the action potential duration was not affected by either removal of oxygen or exposure to 3 m*M* cyanide, in spite of a reduction in twitch force by about 70% (Höglund and Gesser, unpublished).

The anaerobic catabolism results in formation of lactic acid, which due to its low pKa means an acid load at a physiological pH. Hydrogen ions among other negative effects appear to inhibit the stimulatory action of Ca^{2+} on the contractile system in a competitive way (Fabiato and Fabiato 1978). Acidosis as a tentative cause for the anoxic force loss has received much interest, and evidence has been obtained that it is of significance in the mammalian heart muscle (Matthews et al. 1981). A competitive inhibition of the inotropic effect of Ca^{2+} would of course be counteracted by an increased Ca^{2+} availability. However, results obtained do not suggest that acidosis is of importance in the anoxic trout heart. Hence, measurements of pH_i with the DMO method did not reveal any significant acidosis (Nielsen and Gesser 1984b).

Another argument against explaining the anoxia force loss of the trout heart by an intracellular acidosis is given in Fig. 3. Here it is seen that the force decreases more under an anoxic exposure than under exposure to an acidosis obtained by an elevated pCO_2, when $[Ca^{2+}]_0$ is 1.0 m*M*. However, by increasing $[Ca^{2+}]_0$, this difference is reversed. There is no obvious reason to believe that an anoxic acidosis should be less severe at a high than at a low $[Ca^{2+}]_0$, or that the reverse should be true for the acidosis imposed by a high CO_2. Rather, due to the stimulation of the lactic acid production by Ca^{2+}, the negative inotropy of anoxia as compared to that of a high CO_2 should be more expressed at an elevated $[Ca^{2+}]_0$. La Manna et al. (1980) working with the heart of toad likewise obtained results suggesting that the force decrease during anoxia does not correlate with a drop in intracellular pH.

Perhaps there is a difference between the mammalian and the fish heart as to the prime cause of the force loss during anoxia. Hence, it seems to be fairly well documented that anoxia causes a significant acidification of the mammal heart muscle (e.g., Matthews et al. 1981). It should be noted, however, that the lactate content and rate of production is about ten times higher in the rat heart (Hearse and Chain 1972) than in the trout heart (Nielsen and Gesser 1984b). Moreover, the teleost heart cell, in general, appears to be considerably smaller than the mammalian one (Kilarsky 1967). Thus, it

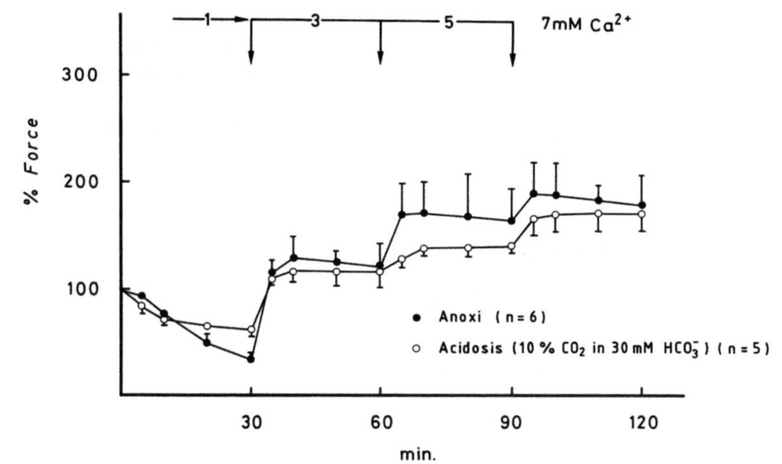

Fig. 3. Twitch force of trout ventricular strips exposed to either acidosis or anoxia with stepwise increases in $[Ca^{2+}]_o$ (Nielsen and Gesser 1984a)

should rid itself more efficiently of lactic acid. Noteworthy in this context is the absence of correlation between the anoxic tolerance and the nonbicarbonate buffer value in hearts of vertebrates (Damm Hansen and Gesser 1980).

Although acidosis does not appear to explain the negative inotropic effect of anoxia in the heart of fish, it may be a potential treat during this condition. This was suggested by the fact that a decrease in the extracellular buffering capacity by a parallel lowering of the CO_2 and bicarbonate levels caused a pronounced impairment of the anoxic tolerance of the isolated heart muscle of carp. This was also seen for the trout heart, although to a lesser extent (Gesser 1977).

Tentatively, these effects are explained with a cellular excretion of H^+-ions by a sarcolemmal chloride-bicarbonate shift (Strome et al. 1976). This would likely be less efficient at a lowered extracellular bicarbonate CO_2. Alternatively or in addition, the intracellular buffering power may be impoverished. Hence, the concentration of undissociated lactate in the cell built up at the cost of a given drop in pH_i would be lowered and in parallel the rate of lactate efflux and of lactate production, i.e., of anaerobic energy liberation (Gesser 1976). Noteworthy, lactate appears to cross the cell membrane in the undissociated form, although only a fraction of 10^{-4}–10^{-3} of the total lactate is undissociated at physiological pHs (Roos 1975).

Commonly, acidosis enhances the negative inotropic effect of anoxia on vertebrate heart muscle. This is not always the case, however. Thus, Fig. 4 shows that the oxygen-deprived eel heart maintained at a $[Ca^{2+}]_o$ of 1.0 m*M* lost less twitch force at an elevated pCO_2 than at a low one. When $[Ca^{2+}]_o$ was increased this relation was reversed (Nielsen and Gesser 1984a). The stimulation of the contractility of the anoxic muscle by acidosis probably involves an intracellular release of calcium (Gesser and Poupa 1978; Nielsen and Gesser 1984a). The relative importance of such a release will probably be greater at a low than at a high $[Ca^{2+}]_o$.

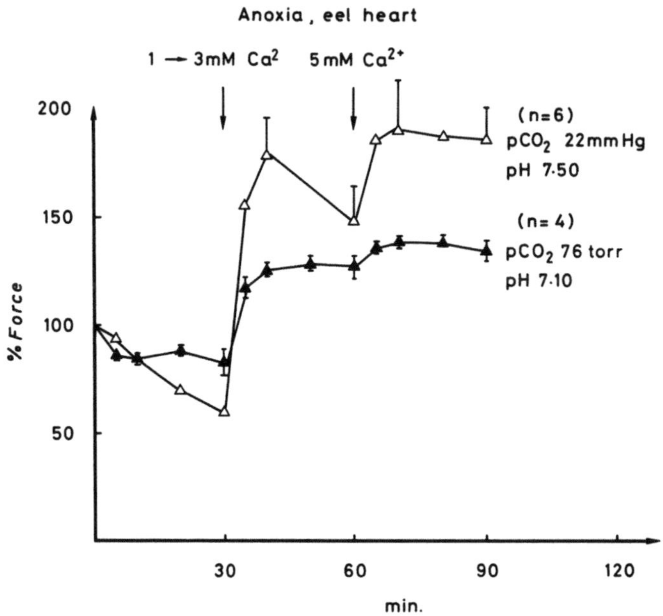

Fig. 4. Twitch force of anoxic eel ventricular strips exposed to either a high or a low pCO_2 with stepwise increases in $[Ca^{2+}]_o$ (Nielsen and Gesser 1984a)

Before the specific effects of acidosis are discussed it should be concluded that the effects of hypoxia on the heart function still are unclear. One feature deserving further attention is the large decrease in creatine phosphate. Here, it should be noted that Kammermaier et al. (1982) ascribe main importance to changes in the free energy release upon ATP hydrolysis in the cytoplasmatic compartment.

5 Effect of Acidosis

Generally, fishes have a lower total CO_2 concentration than air-breathers. The pelagic marine fish easily gets rid of CO_2 through the gills, due to the high solubility of this substance relative to oxygen. Hence, in these fishes the plasma pCO_2 is about 5 mmHg, whereas for instance in man it is about 40 (Dejours 1975). The environment of fishes may present high pCO_2 levels, in some eutroph waters values as high as 50 mmHg have been recorded (Heisler 1982). Furthermore, it should again be recalled that the heart in many fishes to a large extent is nourished by systemic venous blood. During exercise, then, as already pointed out, it will likely be exposed to a pCO_2 elevated by the skeletal muscle activity. CO_2 is formed either directly or by lactic acid reacting with bicarbonate.

The effects of acidosis on the heart muscle have recently been reviewed from a comparative point of view (Gesser and Poupa 1983). Therefore, only the major points will be summarized here.

Figure 5 shows the force development under hypercapnic acidosis for several vertebrates. It should be noted that the experimental conditions vary somewhat. Never-

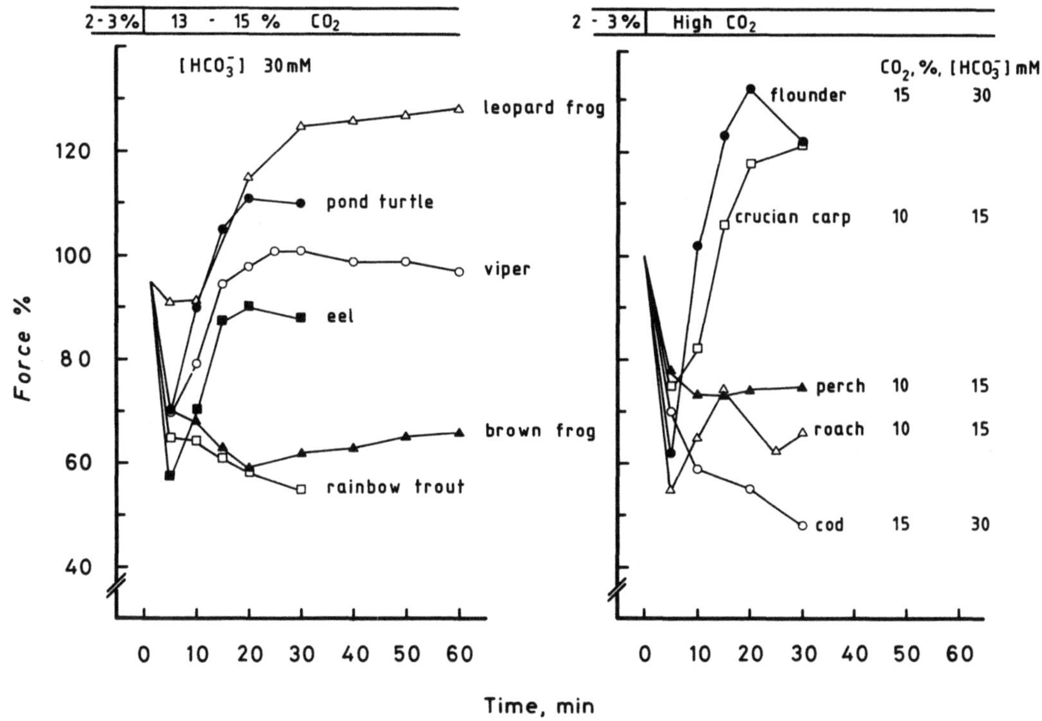

Fig. 5. Effect of hypercapnic acidosis on twitch force of ventricular strips from 11 vertebrate species

theless, it is evident that large differences exist between species in ability of the isolated heart muscle to develop force under acidosis. Two types of response can be observed. Generally, the switch to a high pCO_2 immediately causes a drop in contractile force. For some species, however, there is a subsequent increase in force so that the pre-hypercapnic level is often reached and like for the flounder heart even surpassed.

It has been suggested that the capacity of the heart to function under acidosis is more developed in air- than in water-breathing vertebrates and that this is an adaptation to the general difference in body total CO_2 levels (Gesser and Poupa 1983). In accordance with this the heart muscle of all the pure air-breathers investigated shows the ability to recover force, whereas this is true only for some of the fish hearts (Gesser and Poupa 1983). The high tolerance, for example, of the flounder heart to acidosis suggests, however, that factors like coronary supply, scope of activity, and environment should be taken into account, in addition to the general differences between water- and air-breathers.

The recovery of force during ongoing acidosis appears to involve an enlargement of the Ca^{2+} pool in the E,C-coupling. Evidence exists that intracellular, probably mitochondrial, calcium stores are released (Gesser and Poupa 1978, 1981, 1983). The effects of acidosis on pH_i appear to be independent on the ability to recover force (Gesser and Jørgensen 1982).

6 General Conclusions

A large variation exists among fish species as to the performance of the heart muscle during oxygen lack or acidosis.

A covariation was found between the anoxic performance and the fraction of the total energy liberating capacity being anaerobic.

The fish heart muscle subjected to anoxia shows a reserve capacity for contractility and energy liberation, which can be elicited by an adrenergic impulse or an elevation in $[Ca^{2+}]_0$. In this respect it seems to differ from the mammalian heart muscle.

The cause of the force loss in the anoxic fish heart is unclear, but apparently not an intracellular acidification.

As to the effects of acidosis a capacity to liberate intracellular, probably mitochondrial, calcium stores is likely of main importance for the maintenance of contractility under this condition.

References

Andreasen P (submitted) Free and total calcium concentrations in blood of rainbow trout, *Salmo gairdneri*, during "stress" conditions

Bing OHL, Brooks WW, Messer JV (1972) Effects of isoprotorenol on heart muscle performance during myocardial hypoxia. J Mol Cell Cardiol 4:319–328

Damm Hansen H, Gesser H (1980) Relation between non-bicarbonate buffer value and tolerance to cellular acidosis: A comparative study of myocardial tissue. J Exp Biol 84:161–167

Davidson S, Maroko PR, Braunwald E (1974) Effects of isoprotorenol on contractile function of the ischemic and anoxic heart. Am J Physiol 227:439–443

Dejours P (1975) Principles of comparative physiology. North-Holland, Amsterdam

Ellis D, Thomas RC (1976) Direct measurement of the intracellular pH of mammalian cardiac muscle. J Physiol 262:755–771

Fabiato A, Fabiato F (1978) Effects of pH on the myofilaments and the sarcoplasmatic reticulum of skinned cells from cardiac and skeletal muscles. J Physiol 276:233–255

Gesser H (1976) Significance of the extracellular bicarbonate buffer system to anaerobic glycolysis in hypoxic muscle. Acta Physiol Scand 98:110–115

Gesser H (1977) The effects of hypoxia and reoxygenation on force development in myocardia of carp and rainbow trout: Protective effects of CO_2/HCO_3^-. J Exp Biol 69:199–206

Gesser H, Jørgensen E (1982) pH_i, contractility and Ca-balance under hypercapnic acidosis in the myocardium of different vertebrate species. J Exp Biol 96:405–412

Gesser H, Poupa O (1974) Relations between heart muscle enzyme pattern and directly measured tolerance to anoxia. Comp Biochem Physiol 48:97–104

Gesser H, Poupa O (1975) Lactate as substrate for force development in hearts with different isoenzyme patterns of lactate dehydrogenase. Comp Biochem Physiol 52B:311–313

Gesser H, Poupa O (1978) The role of intracellular Ca^{2+} under hypercapnic acidosis of cardiac muscle: Comparative aspects. J Comp Physiol 127:307–313

Gesser H, Poupa O (1981) Acidosis and Ca^{2+} distribution in myocardial tissue of flounder and rat. J Comp Physiol 143:245–251

Gesser H, Poupa O (1983) Acidosis and cardiac muscle contractility: comparative aspects. Comp Biochem Physiol 76A:559–566

Gesser H, Andresen P, Brams P, Sund-Laursen J (1982) Inotropic effects of adrenaline on the anoxic or hypercapnic myocardium of rainbow trout and eel. J Comp Physiol 147:123–128

Hearse DJ, Chain EB (1972) The role of glucose in the survival and recovery of the anoxic isolated perfused rat heart. Biochem J 128:1125–1133

Heisler N (1982) Transepithelial ion transfer processes as mechanisms for fish acid-base regulation in hypercapnia and lactacidosis. Can J Zool 60:1108–1122

Kammermaier H, Schmidt P, Jüngling E (1982) Free energy change of ATP hydrolysis: a causal factor of early hypoxic failure of the myocardium? J Mol Cell Cardiol 14:267–277

Kilarsky W (1967) The fine structure of striated muscles in teleosts. Z Zellforsch 79:562–580

La Manna JC, Saive JJ, Snow TR (1980) The relative time course of early changes in mitochondrial function and intracellular pH during hypoxia in the isolated toad ventricle strip. Circ Res 46: 755–763

Matthews PM, Radda GK, Taylor DJ (1981) A ^{31}Pnmr study of metabolism in the hypoxic perfused rat heart. Trans Biochem Soc 9:236–237

McClellan G, Weisberg A, Winegard S (1983) Energy transport from mitochondria to myofibril by a creatine phsophate shuttle in cardiac cells. Am J Physiol 245:C423–C427

McDonald TF, MacLeod DP (1971) Anoxia-recovery cycle in ventricular muscle: Action potential duration contractility and ATP content. Pflügers Arch 325:305–322

Meyer RA, Sweeny LH, Kushmerick MJ (1984) A simple analysis of the "phosphocreatine shuttle". Am J Physiol 246:C365–C377

Niedergerke R, Page S (1977) Analysis of catecholamine effect in single atrial trabeculae of the frog heart. Proc R Soc Lond B Biol Sci 197:333–362

Nielsen KE, Gesser H (1983) Effects of $[Ca^{2+}]_o$ on contractility in the anoxic cardiac muscle of mammal and fish. Life Sci 2:1437–1442

Nielsen KE, Gesser H (1984a) Eel and rainbow trout myocardium under anoxia and/or hypercapnic acidosis, with changes in $[Ca^{2+}]_o$ and $[Na^+]_o$. Mol Physiol 5:189–198

Nielsen KE, Gesser H (1984b) Energy metabolism and intracellular pH in trout heart muscle under anoxia and different $[Ca^{2+}]_o$. J Comp Physiol 154(5):523–527

Ostadal B, Schiebler TH (1971) Über die terminale Strombahn in Fischherzen. Z Anat Entwicklungsgesch 134:101–110

Pette D (1965) Plan und Muster im zellulären Stoffwechsel. Naturwissenschaften 52:557–616

Poupa O, Gesser H, Jonsson S, Sullivan L (1974) Coronary-supplied compact shell of ventricular myocardium in Salmonids. Growth and enzyme pattern. Comp Biochem Physiol 48A:85–95

Randall D (1984) In: Johansen K, Burggren W (eds) Shunts in fish gills. Alfred Benzon Symposium 21, Munksgaard (in press)

Roos A (1975) Intracellular pH and distribution of weak acids across the cell membranes. A study of D- and L-lactate and of DMO in rat diaphragm. J Physiol (Lond) 249:1–26

Ruben JA, Bennett AF (1981) Intense exercise, bone structure and blood calcium levels in vertebrates. Nature 291:411–413

Saks VA, Rosenshtrauk LV, Smirnow VN, Chazov EI (1978) Role of creatine phosphokinase in cellular function and metabolism. Can J Physiol Pharmacol 56:691–706

Santer RM, Walker MG (1980) Morphological studies of the ventricle of teleost and elasmobranch hearts. J Zool (Lond) 190:259–272

Strome RD, Clancy RL, Gonzales NC (1976) Myocardial CO_2 buffering: Role of transmembrane transport of H^+ or HCO_3^- ions. Am J Physiol 230:1037–1041

Williamson JR (1964) Metabolic effects of epinephrine in the isolated perfused rat heart. J Biol Chem 239:2721–2729

Symposium VI

Intracellular pH: Role and Regulation

Organizer A. MALAN

Alphastat Regulation of Intracellular Acid-Base State?

R.B. REEVES[1]

1 Introduction

Research of the last 10 years has greatly clarified the regulation of a nonequilibrium intracellular pH by the identification of the active acid extrusion processes principally involved (Roos and Boron 1981). This decade has also broadened our horizons as to what variable is being regulated in the extracellular blood compartment. From studies of how changing body temperature affects the pH of blood, we have learned that the regulation of the extracellular space for most organisms is not one of maintenance of constant pH; instead as shown in Fig. 1, the regulated pH characteristic of each body temperature is found to decrease as temperature increases (Howell et al. 1970; Reeves 1977). Since blood pH now is demonstrated to be only a dependent, secondary variable, the question arises as to what is the primary regulated acid-base variable. An analogous

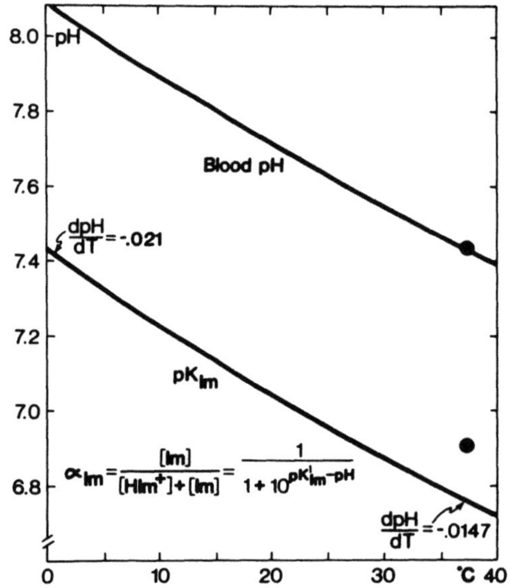

Fig. 1. The change in blood pH compared to the change in the pK of histidyl imidazole with temperature

1 Department of Physiology, School of Medicine, State University of New York, Buffalo, NY 14214, USA

Circulation, Respiration, and Metabolism
(ed. by R. Gilles)

inquiry is apposite for acid-base regulation of the intracellular space; this paper reviews the current status of the identity of the regulated intracellular variable.

2 The Imidazole Alphastat Hypothesis

One hypothesis nominates protein charge state for the regulated acid-base variable of blood. Changes in protein charge state can be assessed as a varying fractional dissociation of histidyl imidazole groups, termed alpha imidazole and defined in Fig. 1 (Reeves 1972). For a wide variety of air-breathing and water-breathing species, extracellular acid-base regulation can be demonstrated to maintain alpha imidazole constant at a level of about 0.8–0.85 as body temperature changes (Reeves 1977; Reeves and Rahn 1979). The term *imidazole alphastat*, used to describe this pattern of acid-base regulation, denotes a regulation that preserves a single value of alpha imidazole. Figure 1 also illustrates the significant change in histidyl imidazole pK with temperature; note that the relationship between pK_{Im} and temperature is not a straight line function, the slope being less steep at high temperatures than at low temperatures. The curve describing the change in blood pH with temperature is a constant ΔpH above the pK_{Im} curve, hence, alpha imidazole for blood is constant. The question addressed in this communication is whether active acid extrusion processes controlling the regulated intracellular state also show a similar regulatory pattern. One test of alphastat control of the intracellular compartment is to ask whether intracellular pH values for a given cell type when measured in vivo over a range of temperatures fall along a curve of constant alpha.

Why might regulation to maintain a constant alpha imidazole have profound importance to the activities of cells, tissues, and organisms? There are 20 amino acids coded for genetically; all proteins are composed of these amino acids. Only one of these, *histidine*, has an R group or side group as shown in Fig. 2 that takes up and gives off a proton in the physiological pH range 6–8 (Edsall Wymann 1958). Hence, when any change in acid-base state of the cytoplasm occurs, it is chiefly histidyl imidazole groups that are titrated. The pK of any particular histidyl imidazole depends on its local electrostatic microenvironment within the protein; charge-charge interaction can greatly affect the pK of the imidazole ring. The range of pK_{Im} values has been found to be 6.2–8.1 for 11 titratable histidine imidazoles of a typical globular protein, deoxy hemoglobin (Matthew et al. 1979). It is this broad distribution of pK_{Im} values in proteins that confers the near-linear titration curve on intracellular protein over the pH

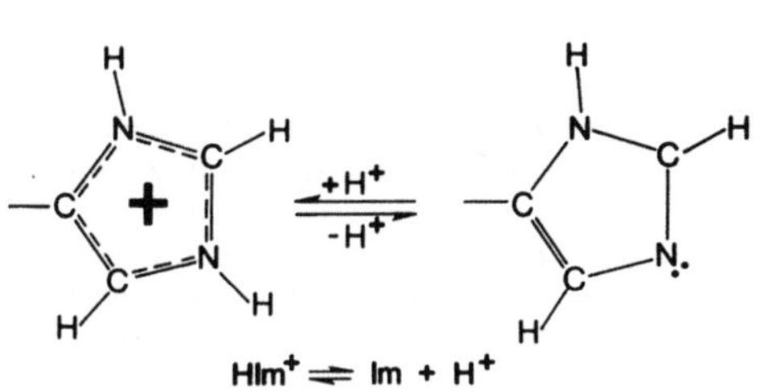

Fig. 2. The proton equilibrium at the imidazole ring of histidine. Note that only the protonated imidazole ring is charged +1. Notation used in the text that corresponds to this equilibrium is shown below

range 6–8 (Malan et al. 1976). Since it is useful to designate an *average global value*, a single pK_{Im} value of 6.96 at 25 °C will be used (Reeves 1977).

Of the greatest importance to biological systems is the large sensitivity of pK_{Im} to temperature change, characterized by an enthalpy of 7 kcal mol^{-1}, which requires that pK_{Im} vary over the biologically relevant temperature range at about −0.019 U deg^{-1} (Reeves 1976). The enthapy of dissociation is fully as variable microscopically as is the pK_{Im}. For a $\Delta pK_{Im}/\Delta T$ generally applicable to whole cell systems an experimentally determined average value for isolated isoionic proteins, hemoglobin, and separately, a mixture of plasma proteins, is used (Reeves 1976).

The conspicuous physiological importance of the ionization state of histidyl imidazole derives from the sensitivity of all protein structure and function to proton titration. Maintenance of constant charge state confers stability of function in the form of conserved affinity constants for substrates, inhibitors, and allosteric modulators, and, hence, is central to the integrity and performance of integrated multienzyme metabolic systems (Somero 1981).

3 Intracellular pH Regulation as Temperature Changes in Air-Breathing Vertebrates

Data from the first published attempt to assess intracellular acid-base state as body temperature varies are shown in Fig. 3 (Malan et al. 1976). Using the weak acid dimethyloxazolidine-dione (DMO) to estimate intracellular pH, we found both in turtles and in frogs that several compartments appeared to behave as alphastat controlled systems. Figure 3 shows measured blood and intracellular points; the dotted lines are least squares straight lines fitted to the data and the heavy curves are those of constant alpha for the value indicated. These data were the first to indicate that (1) actively

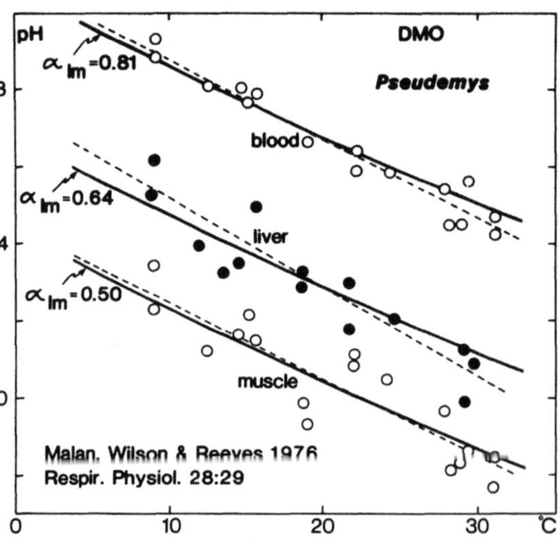

Fig. 3. Blood and intracellular pH data as a function of body temperature in the turtle, *Pseudemys*. *Dotted lines* are least squares straight lines. *Heavy curves* depict constant alpha imidazole value indicated

regulated intracellular pH in vivo had a different value at every body temperature, i.e., pH_i decreased as body temperature rises, and (2) that a constant alpha imidazole characteristic of each cell type was preserved.

As body temperature changes, the animal's central chemoreceptor control of ventilation maintains only that CO_2 partial pressure at each body temperature sufficient to keep alpha imidazole constant (Hitzig 1982). The dependent variable, blood pH, then is fixed at each temperature. No renal compensation is required because plasma strong ion difference is preserved constant (i.e., no change in buffer base occurs). Even though the CO_2 equilibrium curves of blood proteins are very sensitive to temperature change (because of the large effect of temperature on the pK_{Im} of histidyl imidazole), no titration of protein occurs. Plasma bicarbonate concentration and total CO_2 content are constant as body temperature varies.

The specific temperature-controlled CO_2 partial pressure exists for all cellular compartments in the organism, not just for blood. To regulate muscle and liver alpha imidazole it is necessary that active membrane ion pumps in each cell defend the appropriate intracellular strong ion difference (SID or buffer-base concentration). In general, this is done by an active flux of sodium and bicarbonate ions into the cells just sufficient to balance loss of SID from metabolism and leak fluxes. However, of striking adaptive significance is the fact that if the acid-base transition from one body temperature to another in an air-breathing vertebrate occurs by ventilatory control of CO_2 tension, then a constant intracellular alpha imidazole is achieved with no adjustment of intracellular ion concentrations (Reeves and Malan 1976).

DMO studies have a number of well-known limitations, hence, observations on intracellular pH as a function of body temperature made using ^{31}P-NMR spectroscopy are of great interest. Hitzig (unpublished observations) chose two common salamander species, *Plethodon* and *Notopthalmus*, as his experimental material because their body form made them ideal for studies using normal test-tube cuvettes of the NMR spectrometer. He placed his specimen with respect to the NMR field coils so that he measured intracellular pH in the large muscle mass of the pelvic region and base of the tail, hence, these spectra had a prominent absorption for phosphocreatine. The resulting pH_i data show regulation of striated muscle intracellular pH to be one of constant alpha imidazole over a 20° range; in *Notopthalmus* α_{Im} = 0.5 and in *Plethodon*, α_{Im} = 0.6. By utilizing a lungless skin-breathing species, Hitzig made a further observation of some general importance. In *Plethodon* uptake of oxygen through the skin is adequate at lower body temperatures and metabolic rates despite the large diffusion resistance of the skin. At 30°, however, a resting animal in air cannot obtain sufficient oxygen to maintain tissue oxygen tensions and a metabolic acidosis develops culminating in death after about 6 h unless relived. As a consequence of tissue hypoxia, α_{Im} falls precipitously to ca. 0.4. If, however, the air surrounding the animal is replaced with pure oxygen, the intracellular pH recovers to the alpha value seen at lower body temperatures. These observations using a noninvasive technique on unanesthetized, uncatheterized animals, signify that muscle pH_i is controlled to preserve constant protein charge state. However, some conditions, like hypoxia, can seriously compromise that control and disturb normal tissue acid-base values.

This theme is even more elaborately developed in the elegant series of measurements carried out on the desert iguana, *Dipsosaurus* (Bickler 1981, 1982, 1984). Using

DMO methods, Bickler studied not only steady state values of pH_i for myocardium and striated muscle, but he studied the response to sudden changes as well. He observed that in the steady state the iguana preserved intracellular protein charge state, i.e., maintained α_{Im} = 0.72–0.74, provided the temperature range did not exceed the animal's preferred temperature range (32–43 °C). As soon as the animal was cooled to a lower temperature, the hyperventilation required to prevent carbon dioxide stores from building was not achieved and a significant departure from constant alpha was found. If rapid transitions of body temperature were made from 35° to 17°, even greater departures from a curve of constant alpha imidazole were observed.

4 Acid-Base Regulation in the Hierarchy of Regulatory Function

Observations of the nature exemplified by Hitzig's *Plethodon* and Bickler's *Dipsosaurus* measurements suggest that acid-base regulation, both extra- and intracellular, is confined to restricted ranges of body temperature characteristic for each organism; some species may have relatively large ranges over which alphastat control operates, such as *Pseudemys*, while other species like *Dipsosaurus* are quite circumscribed in the zone of regulation. Thus, as Malan (1978) has emphasized, patterns of acid-base regulation as temperature changes may not be uniform over the whole temperature range an organism can tolerate.

5 Intracellular pH Measurements in Water-Breathing Vertebrates

Major weak acid studies on intracellular acid-base regulation as temperature changes have also been carried out on three fish species and are summarized in Table 1. In the case of *Ictalurus* and *Scyliorhinus*, the dpH/dT for both red and white muscle are not significantly different from blood dpH/dT and, hence, parallel the change in blood

Table 1. Blood and tissue (DMO) ^pH/^T (U/°C)

	Ictalurus[a]	Scyliorhinus[b]	Anguilla[c]
Blood	−0.013	−0.015 ad −0.014 juv	−0.008
White muscle	−0.015	−0.018	−0.009
Red muscle	−0.019	−0.033	−0.003
Heart	−0.012*	−0.010*	−0.020*
Liver	–	–	−0.018*
Mean pH_1	−0.016		

* Significantly different from blood
[a] Cameron and Kormanik (1982)
[b] Heisler er al. (1976)
[c] Walsh and Moon (1982)

alpha. The data for myocardium are, however, significantly different from blood, as Malan et al. (1976) also noted in their studies on *Pseudemys* and *Rana*; one important unanswered question about the use of DMO for pH_i determinations is whether a tissue with a large mitochondrial fraction, like myocardial fibers, yields reliable estimates of $\Delta pH_i/\Delta T$. The study on the eel, *Anguilla*, shows a remarkable instance in which two cell compartments, heart and liver, follow alphastat regulation at a time when blood and muscle do not. Comments already made about control over restricted temperature ranges may ultimately be found important for *Anguilla*; at the moment there is no simple interpretation of this species' regulation.

Fish and other water-breathers are forced to utilize a different strategy for primary control of their intracellular pH in response to body temperature changes. These species cannot alter ventilation to adjust carbon dioxide tension because they primarily utilize their ventilation to extract the very meagre amounts of oxygen dissolved in their respiratory medium, water. To control blood acid-base state when temperature changes water-breathers must actively move significant equivalents of ions through the gill epithelium; i.e., to achieve the pH required at lower temperature to keep alpha constant, these species must take up sodium and bicarbonate equivalents from the environment. Moreover, each individual cell as well must carry out the same process if an intracellular constant alpha is to be maintained; hence, rapid changes in temperature may require long periods before individual regulated compartments in the animal come to a new steady acid-base state. Ions must be moved, CO_2 must be loaded in cooling and lost in warming; these processes take time and energy. The air-breather has a transient-free, flexible system in this regard (Reeves and Rahn 1979).

6 Single Cell in Vitro pH_i Control as Temperature Varies

The very first study of control of intracellular acid-base status when temperature changes utilizing intracellular glass microelectrodes has been carried out on the European crayfish *Astacus leptodactylus*, a gill-breather, by Rodeau (1984). Unlike weak acid distribution methods for measuring intracellular pH, the glass electrode provides a measure of the true cytosolic pH. Rodeau's elegant study provided data from extensor muscle fibers of the carpopodite of a walking leg and from an abdominal ganglion neuron. The changes observed in pH_i for both cell types, under conditions of constant extracellular pH and bicarbonate concentration, follow a constant alpha imidazole path (muscle α_{Im} = 0.62; neuron α_{Im} = 0.67). These results agree well with DMO studies on intact crayfish carried out by Malan (1984); interestingly, Malan observed that below 13 °C (6°–13 °C) alphastat control characterized both intra- and extracellular compartments. Above 13 °C (13°–20 °C), the extracellular compartment did not follow alphastat, but the intracellular compartments did. Over the lower temperature range, P_{CO_2} changes from regulated ventilation, were the primary mechanisms for attaining alphastat control intracellularly. At the higher temperature range, active readjustment by cell membrane acid extrusion processes must have been principally responsible. In Rodeau's study a major portion of the regulatory effort is being carried by ion transport mechanisms in the plasmalemma over the whole temperature range.

Note that these observations indicate the existence of some mechanism for adjusting the regulated pH_i set point as temperature changes.

A similar conclusion derives from DMO studies of on isolated hepatocytes from *Anguilla*, the eel (Walsh and Moon 1983). If isolated hepatocytes are suspended at different temperatures in HEPES buffer that has had its pH adjusted to the corresponding in vivo blood pH at that temperature, hepatocytes regulate their internal pH just as they do in vivo, along a path of constant alpha imidazole (α_{Im} = 0.77). Just as in Rodeau's *Astacus*, ion transport by the plasmalemma is required because CO_2 tension was kept low and constant in these hepatocyte measurements. Hence, we are again presented with the question of how individual cells are able to control their intracellular acid-base state to preserve alpha imidazole constant.

7 Possible Mechanisms of Intracellular Imidazole Alphastat Control

It is possible to suggest mechanisms compatible with current knowledge about active acid extrusion processes in the cell membrane that can achieve alphastat regulation. Let me attempt to sketch such a mechanism. Figure 4 summarizes the current paradigm for regulation of intracellular pH (Roos and Boron 1981); for a typical actively regulated cell, the intracellular pH is a nonequilibrium steady state resulting from the equality of acid-loading (left-hand) processes and active acid extrusion process (on the right). Continual acid loading occurs from acidic metabolic products, extracellular weak acids (when present), and leak fluxes of bicarbonate (and protons) down an electrochemical gradient. Active acid extrusion occurs by a sodium for proton antiport and/or a bicarbonate for chloride antiport. These may be separate or combined; in this form the stoichiometry is two acid equivalents extruded per sodium ion and bicarbonate ion equivalent taken up. Acid extrusion is dependent upon substrate concentrations, i.e., extracellular sodium and bicarbonate concentrations. Active acid extrusion is also enhanced by increasing extracellular pH and decreasing intracellular pH (Roos and Boron 1981).

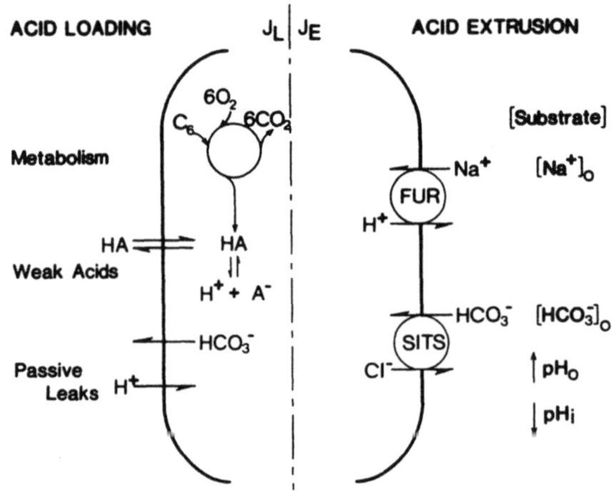

Fig. 4. Determinants of intracellular pH (see text for explanation)

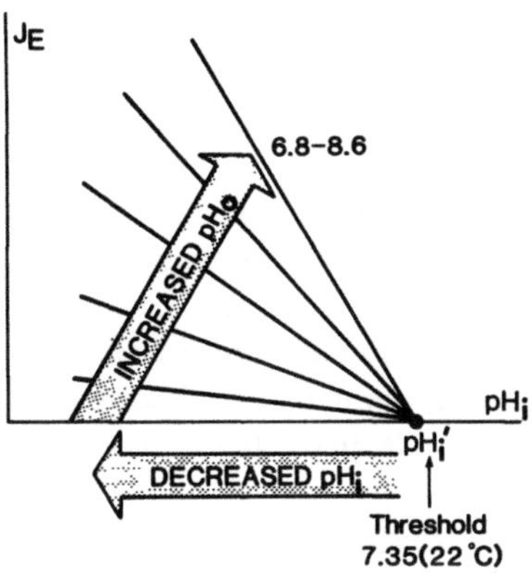

▲

Fig. 5. Kinetic determinants for acid extrusion (J_E) emphasizing differing role of intracellular and extracellular pH

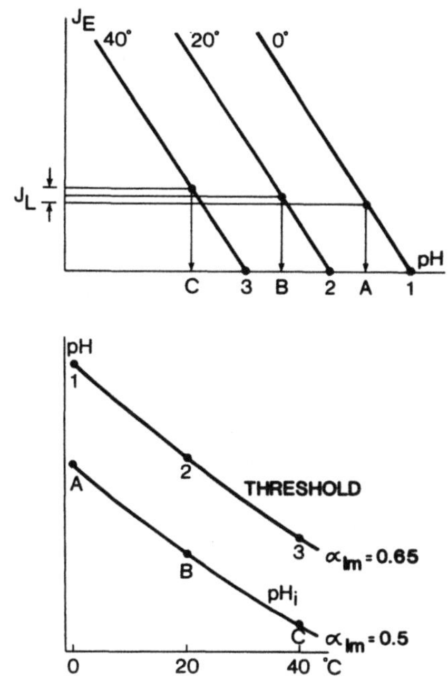

Fig. 6. Proposed effect of change of temperature on slope and threshold of acid extrusion curve. *Lower figure* shows threshold and regulated pH_i values as following separate curves of constant alpha imidazole ▶

Kinetic determinants of this model are shown schematically in Fig. 5 and are based primarily on work carried out on barnacle muscle (Boron et al. 1979). The acid extrusion curves are linear and intercept the abscissa at a common threshold value, pH_i'. Above threshold intracellular pH, no acid extrusion occurs. Acid extrusion increases linearly below threshold. The slope of the extrusion curve is a function of extracellular substrate concentrations, i.e., increased sodium and bicarbonate ion concentrations stimulate acid extrusion. The slope of the extrusion curve is also increased by extracellular alkalosis. From the point of view of putative alphastat control it is important to note the pH ranges over which extracellular pH changes acid extrusion; a range of 6.8 to 8.6 in *Balanus* muscle at 22° (Boron et al. 1979). Similarly, note the value of threshold pH_i', 7.35. These are pH values squarely in the middle of the histidyl imidazole titration range. As a working hypothesis then, the effects of extra- and intracellular pH values on acid extrusion arise from proton binding to histidyl imidazole moieties at independent sites on the corresponding extra- and intracellular portions of the acid-extruding transport protein.

Figure 6 depicts how these mechanisms might participate in preserving alphastat control of the intracellular acid-base state in an air-breathing species. If the proton-binding site responsible for threshold pH_i' is a histidyl imidazole group, we expect that threshold will be found to be importantly shifted by temperature change, as shown below. Analogously, if the extracellular proton sensitive site on the transport protein is an histidyl imidazole, we may reasonably suppose that it is alpha imidazole at that site that is important, not pH per se. If extracellular acid-base regulation maintains extracellular alphastat acid-base conditions, then the slope of the acid extrusion curve will be independent of temperature. Supporting this notion is the demonstrated fact that extracellular bicarbonate ion concentration is constant independent of temperature. The resulting regulated pH_i depends not only on extrusion rate, but also on acid loading; if the result of temperature change on the combined processes of metabolism and passive ion fluxes, about which nothing is now known, is roughly constant, then the steady state regulated pH_i could indeed approach the alphastat condition depicted.

Water-breathers are likely to have more complex variations of this basic mechanism because in these species extracellular substrate, i.e., bicarbonate ion, concentration increases as temperature falls. Nonetheless, this simple extension of the Roos-Boron paradigm offers an attractive hypothesis that makes testable predictions about the temperature behavior of the system.

It is important to recall that a clear-cut parallel of utilizing a specific histidyl imidazole for the general control of tissue alpha imidazole has been shown to occur in the brainstem chemoreceptor that regulates ventilation in air-breathers (Hitzig 1982). Elegant studies of the chemoreceptor response to changing CSF pH in ventriculo-cisternally perfused unanesthetized turtles indicate that the chemoreceptor defends an alpha imidazole set-point by adjustments in ventilation (Hitzig 1982).

8 Summary

The imidazole alphastat hypothesis for control of intracellular acid-base regulation remains a useful one. Quite evidently the physiological regulation observed when temperature changes is not on the whole consistent with pH-stat regulation; as yet no other integrating hypothesis exists.

Many cell types on which intracellular pH measurements have been made with a variety of techniques show regulation consistent with alphastat control. Exceptions remain. Whether these discrepancies are real or due to limitations of technique must yet be established. There is a palpable need for exploration of intracellular pH regulation as a function of temperature on the same cell type with more than one technique.

It is becoming increasingly evident that patterns of acid-base control may not be the same over all temperatures tolerated; acid-base regulation may conflict with other regulations and a hierarchy of regulatory commitments may exist. It is also becoming evident that the intracellular state may follow alphastat control when the extracellular regulation is compromised by environmental limitations.

Finally, the Roos-Boron acid extrusion paradigm suggests obvious tests for possible alphastat control. Experimental data testing whether the proton-sensitive intra- and extracellular sites on the acid-extrusion carrier protein(s) are histidyl imidazole groups are especially wanted.

References

Bickler PA (1981) Effects of temperature on acid-base balance and ventilation in desert iguanas. J Appl Physiol 51:452–460

Bickler PA (1982) Intracellular pH in lizard *Dipsosaurus dorsalis* in relation to changing body temperature. J Appl Physiol 53:1466–1472

Bickler PA (1984) Effects of temperature on acid and base excretion in a lizard, *Dipsosaurus dorsalis*. J Comp Physiol B 154:97–104

Boron WF, McCormick WC, Roos A (1979) pH regulation in barnacle muscle fibers: Dependence on intracellular and extracellular pH. Am J Physiol 237:C185–C193

Boron WF, McCormick WF, Roos A (1981) pH regulation in barnacle muscle fibers: Dependence on extracellular sodium and bicarbonate. Am J Physiol 240:C80–C89

Cameron JN, Kormanik GA (1982) Intracellular and extracellular acid-base status as a function of temperature in the freshwater channel catfish, *Ictalurus punctatus*. J Exp Biol 99:127–142

Edsall JT, Wyman J (1958) Biophysical chemistry. Academic, New York

Heisler N, Weitz H, Weitz AM (1976) Extracellular and intracellular pH with changes of temperature in the dogfish, *Scyliorhinus stellaris*. Respir Physiol 26:249–263

Hitzig BM (1982) Temperature-induced changes in turtle CSF pH and central control of ventilation. Respir Physiol 49:205–222

Howell BJ, Baumgardner F, Bondi K, Rahn H (1970) Acid-base balance in cold-blooded vertebrates as a function of body temperature. Am J Physiol 218:600–616

Malan A (1978) Intracellular acid-base status as a variable temperature in an breathing vertebrates and its representation. Respir Physiol 33:115–119

Malan A (1984) Intracellular pH temperature relationships in a water breather, the crayfish. Mol Physiol (in press)

Malan A, Wilson T, Reeves RB (1976) Intracellular pH in cold-blooded vertebrates as a function of body temperature. Respir Physiol 28:29–47

Matthew JB, Hanania GI, Gurd FRN (1979) Electrostatic effects in hemoglobin: Hydrogen ion equilibria in human deoxy- and oxyhemoglobin A. Biochemistry 18:1919–1928

Reeves RB (1972) An imidazole alphastat hypothesis for vertebrate acidbase regulation: tissue carbon dioxide content and body temperature in bullfrogs. Respir Physiol 14:219–236

Reeves RB (1976) Temperature-induced changes in blood acid-base status; pH and P_{CO_2} in a binary buffer. J Appl Physiol 40:752–761

Reeves RB (1977) The interaction of body temperature and acid-base balance in ectothermic vertebrates. Annu Rev Physiol 39:559–586

Reeves RB, Malan A (1976) Model studies of intracellular acid-base temperature responses in ectotherms. Respir Physiol 28:49–63

Reeves RB, Rahn H (1979) Patterns in acid-base regulation. In: Wood S, Lenfant C (eds) Evolution of respiratory processes: A comparative approach. Marcel Dekker, New York

Rodeau J-L (1984) Effect of temperature on intracellular pH in crayfish neurons and muscle fibers. Am J Physiol 246:C45–C49

Roos A, Boron W (1981) Intracellular pH. Physiol Rev 61:296–434

Somero GN (1981) pH-temperature interactions on proteins: Principles of optimal pH and buffer system design. Mar Biol Lett 2:163–178

Walsh PJ, Moon TW (1982) The influence of temperature on extracellular and intracellular pH in American eel, *Anguilla rostrata*. Respir Physiol 50:129–140

Walsh PJ, Moon TW (1983) Intracellular pH-temperature interactions of hepatocytes from American eels. Am J Physiol 245:R32–R37

Intracellular pH Regulation of Renal-Epithelial Cells

W.F. BORON[1]

1 Introduction

Studying intracellular pH (pH_i) in renal-tubule cells is of interest not only because of the information it may provide on the general problem of pH_i regulation, but because the mechanisms responsible for renal-tubule pH_i regulation may play a crucial role in the transepithelial secretion of acid and alkali. The pH sensitivity of virtually every biological process requires that pH be appropriately regulated in cells and in the extracellular fluid. I emphasize that "appropriate regulation" does not mean that the pH in any compartment must be permanently fixed, but that the pH be adjusted to a value appropriate for the conditions. For example, unfertilized eggs (for review, see Busa and Nuccitelli 1984) and quiescent (i.e., nondividing) cultured fibroblasts (Schuldiner and Rozengurt 1982) have relatively acidic pH_i values. However, when the eggs are fertilized or when the fibroblasts are stimulated to divide by the addition of a growth factor pH_i rises by a few tenths of a pH unit. This intracellular alkalinization, which is thought by some to be causally related to subsequent cell division, represents a controlled shift in the value to which pH_i is regulated.

The fundamental problem of pH_i regulation is to extrude from the cell acid that accumulates as a result of passive fluxes of ions and, possibly, as a result of cellular metabolism. A typical mammalian cell might have a cell voltage of -60 mV (inside negative), and be exposed to an extracellular solution having a pH of 7.4. From these values, we can use the Nernst equation to predict that H^+ should be in electrochemical equilibrium across the cell membrane only when pH_i is about one full unit more acid than the external pH (pH_o). In actual fact, pH_i is generally only a few tenths of a pH unit lower than pH_o. Thus, there is a strong tendency for H^+ to passively enter the cell, thereby lowering pH_i. One can also show (see Roos and Boron 1981) that there is a similar tendency for HCO_3^- to leave the cell. Because $[HCO_3^-] >>> [H^+]$ at physiological pH values, passive fluxes of HCO_3^- are probably more important than those of H^+ for producing chronic intracellular acid loads. Depending upon the metabolic state of the cell, cellular metabolism can also produce a chronic intracellular acid load.

If pH_i is to be stabilized in the face of these tendencies toward intracellular acidification, then acid must be extruded from the cell as rapidly as it accumulates. Elucidating the mechanisms of pH_i regulation is extremely difficult under such steady state condi-

[1] Department of Physiology, Yale University School of Medicine, 333 Cedar Street, New Haven, CT 06510, USA

Circulation, Respiration, and Metabolism
(ed. by R. Gilles)

tions. Much more rewarding has been the approach of studying the response of pH_i to acute intracellular acid loads. Several investigators (for example, R.C. Thomas 1974; Boron and DeWeer 1976) have found that when pH_i is rapidly reduced by any number of means, the cell responds by slowly returning pH_i toward its initial value. This pH_i recovery (i.e., alkalinization) must be due to an active transport process, inasmuch as the direction of H^+ and/or HCO_3^- transport required to explain the pH_i recovery is opposite to that predicted for the passive movements of these ions. To date, two general classes of pH_i-regulating ion transport mechanisms have been identified in animal cells. The first to be studied was the Na-dependent Cl–HCO_3 exchange system of invertebrate cells (Russell and Boron 1976; R.C. Thomas 1977). This transporter exchanges external Na^+ and HCO_3^- (or an equivalent ion) for internal Cl^- and possibly H^+. It is blocked by the stilbene derivatives SITS and DIDS, or by removal of any one of the transported ions. The transporter is nearly shut off at pH_i 7.4, and is gradually stimulated as pH_i falls below this "threshold". Isotopic data suggest that this pH_i dependence is due to an allosteric site on the cytoplasmic surface of the transporter (Boron et al. 1978). The second pH_i-regulating system is the Na–H exchanger that has been identified in several vertebrate cells (see, for example, Aickin and Thomas 1977a). It is inhibited by amiloride, but unaffected by the stilbene derivatives. Its transport rate gradually increases as pH_i falls below an activation threshold (Boron and Boulpaep 1983a), presumably due to an intracellular modifier site analogous to that proposed for the Na-dependent Cl–HCO_3 exchanger (Aronson et al. 1982).

2 pH_i Regulation by Proximal-Tubule Cells of the Salamander

2.1 Methodology

All of the experiments reported in this communication were performed on isolated, perfused renal tubules. The rationale for this is fourfold. In the first place, a tubule can in principal be isolated from any segment of the nephron. The older technique of microperfusing tubules can only be done on nephron segments that border the surface of the kidney. A second major advantage of using isolated tubules is that one has excellent control over the composition of the solutions contacting both the luminal and basolateral (i.e., blood-side) surfaces of the cells. A third reason for working with isolated tubules is that their excellent spatial stability is ideal for impaling with microelectrodes. Finally, isolated tubules are ideal for making optical absorbance measurements.

Our techniques for measuring intracellular ion activities in isolated tubules has been detailed elsewhere (Boron and Boulpaep 1983a); the following is a summary of our methods. Figure 1 is a diagram of the isolated, perfused tubule apparatus, similar to the one originally described by Burg and his co-workers (Burg et al. 1966). Each end of the tubule is held by an assembly of three concentric glass pipettes. In each assembly, the outermost pipette surrounds the tubule, and the middle pipette cannulates it. In the right-hand assembly, the innermost pipette carries the perfusate to the lumen, whereas on the left, the innermost pipette collects fluid that has already traversed the

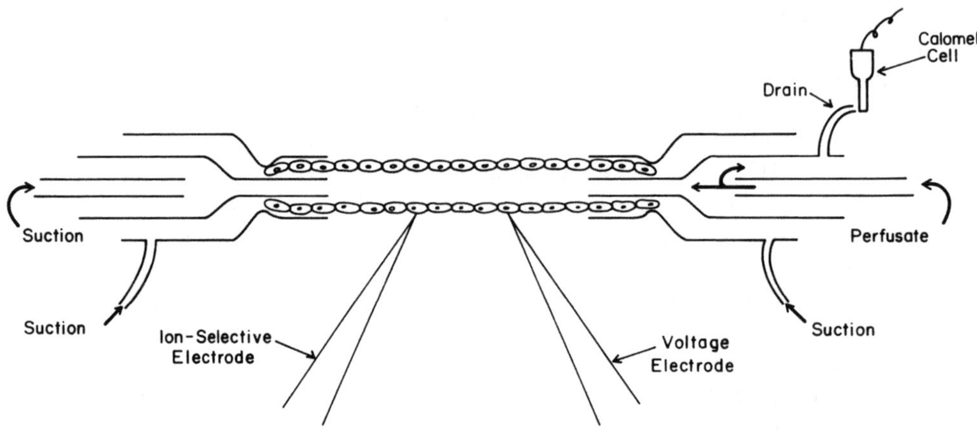

Fig. 1. Isolated, perfused tubule apparatus, *top view*. The tubule (~100 μm diameter for a salamander proximal tubule) is held between two pipette assemblies. It is continuously perfused and superfused. A slight vacuum between the outermost and middle pipette holds the tubule in place. Nearby cells can be impaled with an ion-sensitive (e.g., pH, Na^+ or Cl^-) and with a voltage-sensitive electrode. From Boron and Boulpaep (1983a)

lumen. The salamander proximal tubule is ~100 μm in diameter, and the distance between pipette assemblies 500 to 1,000 μm. The cells are about 20 μm on an edge.

Measurements of intracellular ion activities are made by impaling two separate cells with microelectrodes. One microelectrode is sensitive to pH, Na^+, or Cl^-, whereas the second electrode is sensitive only to voltage. The signals from the electrodes are amplified by high-impedance electrometers and then substracted. The difference signal is linearly related to pH (in the case of a pH-sensitive electrode) or to the logarithm of the ion activity (in the case of Na- and Cl-sensitive electrodes).

These experiments on salamander proximal tubules were performed at room temperature (~22 °C). Nominally HCO_3-free solutions were buffered with 13 m*M* HEPES, usually to pH 7.5. HCO_3^--containing solutions were equilibrated with 1.5% CO_2, usually at pH 7.5 (10 m*M* HCO_3^-).

2.2 Na–H Exchange in Salamander Proximal Tubules

A detailed account of the experiments in this section can be found in Boron and Boulpaep (1983a). Because we suspected that the salamander proximal tubule might have separate transport systems for H^+ and HCO_3^-, we attempted to simplify the interpretation of the data by first studying a possible H^+ transport system in the total absence of HCO_3^-. Accordingly, all of the solutions for the experiments described in this section were nominally HCO_3^- free. As noted in the Sect. 1, the best approach for examining an acid-extruding transport system is to monitor pH after applying an acute intracellular acid load. The simplest way to apply such an acid load to very small cells is to prepulse with Ringer containing m*M* quantities of NH_4^+, a technique first used in squid giant axons (Boron and DeWeer 1976a,b). Typical results are illustrated in

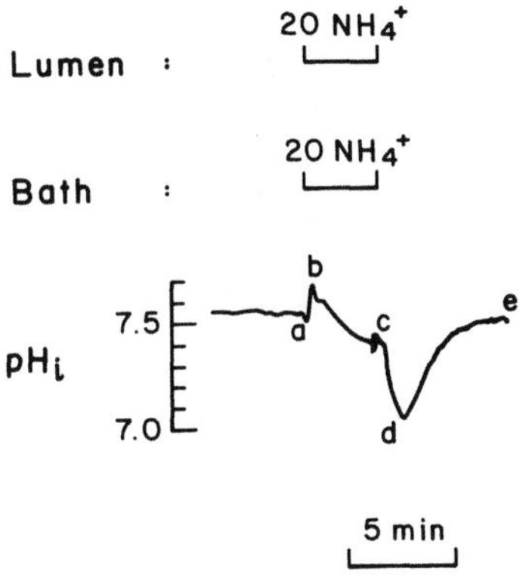

Fig. 2. pH_i recovery from an NH_4^+-induced acid load. The experiment was performed on a salamander proximal tubule at room temperature, measuring pH_i with a recessed-tip pH-sensitive microelectrode. *Bath* is the basolateral (i.e., blood-side) solution. Between *a* and *c*, 20 m*M* NH_4^+ was added to both lumen and bath Ringer (replacing Na^+). Upon removal of the external NH_4^+, pH_i undershot the initial value (*cd*) and then slowly recovered (*de*). The external solutions were nominally HCO_3^--free throughout. From Boron and Boulpaep (1983a)

Fig. 2. When NH_4^+ is simultaneously added to the lumen and bath (i.e., basolateral solution), pH_i rapidly increases (segment *ab*). This alkalinization is due to the passive influx and protonation of NH_3 (which is present in small quantities in NH_4^+-containing solutions). Eventually, however, $[NH_3]_i$ approaches $[NH_3]_o$, and the further net influx of NH_3 is negligible. The influx of the weak acid NH_4^+ continues, however, and leads to a slow fall in pH_i *(bc)* as a small fraction of entering NH_4^+ dissociates to form H^+ and NH_3. When the NH_4^+ and NH_3 are removed from the external solutions, pH_i rapidly falls *(cd)* as the intracellular NH_4^+ which had previously accumulated dissociates into NH_3 (which passively leaves the cell) and H^+ (which is trapped within the cell). The period from *a* to *d* serves only to acid load the cell, and is formally equivalent to injecting the cell with a strong acid, such as HCl. The biologically important portion of the experiment is the cell's response to this acute acid load. As can be seen, pH_i gradually recovers toward its initial value *(de)*, following a time course that is approximately exponential. This pH_i recovery is indicative of an acid-extruding transport system. As noted above, because the passive movements of H^+ and/or HCO_3^- would cause pH_i to fall rather than to recover, this segment-*cd* pH_i recovery must reflect active transport.

To elucidate the mechanism of this pH_i recovery, we examined its sensitivity to the diuretic amiloride, a known inhibitor of Na–H exchange (Aickin and Thomas 1977). As shown in Fig. 3, 2 m*M* amiloride in either the lumen or the bath substantially inhibits the pH_i recovery from an NH_4^+-induced acid load, with the inhibition being the greatest when the amiloride is present in both the lumen and the bath. Other experiments (not shown) have demonstrated that the segment-*de* pH_i recovery is blocked by the total removal of Na^+ from the lumen and the bath. If the Na^+ is removed from only one side, however, the pH_i recovery still occurs, though at a reduced rate. These results, coupled with the observations on amiloride inhibition, suggest that the pH_i recovery is brought about by a Na–H exchanger that is present at both the luminal and basolateral cell membranes. The apparent bilateral distribution of the Na–H exchangers was unexpected, inasmuch as the basolateral Na–H exchanger would work against net acid secretion into the tubule lumen (see below).

To confirm the tentative identification of the transporter as a Na–H exchanger, we performed several additional experiments (not shown). We demonstrated that the period of pH_i recovery from an acid load *(de)* is accompanied by a transient rise of intracellular

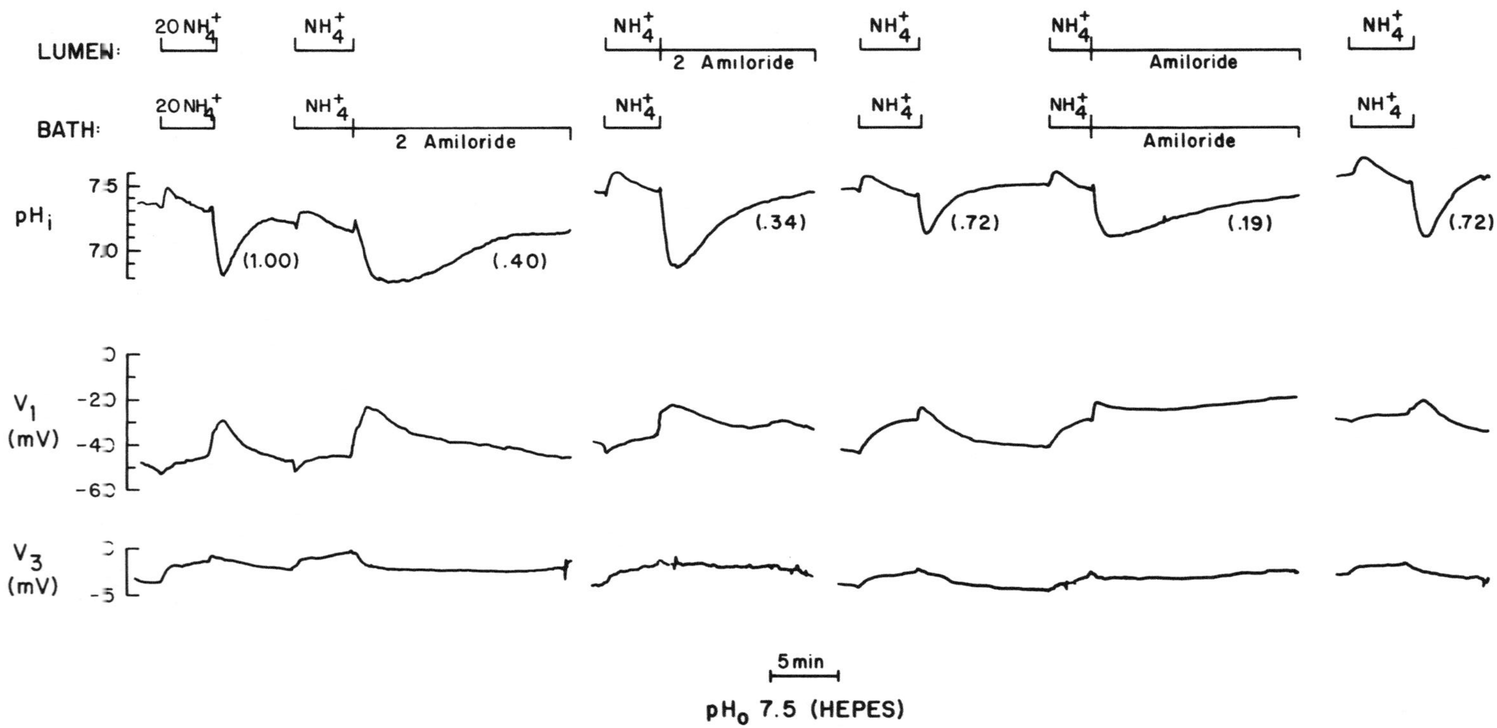

Fig. 3. Effect of amiloride on pH_i recovery from NH_4^+-induced acid load. The design of the experiment was similar to that of Fig. 2, except that the cells were acid loaded six times, with or without 2 m*M* amiloride in the lumen or the bath. V_1 is the cell potential across the basolateral membrane and V_3 is the transepithelial potential, in each case referenced to the basolateral solution. The pH_i recoveries were fitted to a single exponential, using a nonlinear least-squares approach. The rate constant (min^{-1}) is given in *parentheses* for each recovery. From Boron and Boulpaep (1983a)

Na^+ activity, as expected of a Na–H exchanger. Both the pH_i recovery and the rise in Na^+ activity are blocked by reducing pH_o to 6.5. Furthermore, the pH_i recovery has none of the properties expected of a Na-dependent Cl–HCO_3 exchanger. As noted above, the pH_i recovery occurred in the nominal absence of HCO_3^-. In addition, we could detect only very small changes of intracellular Cl^- activity during segment-*de* pH_i recoveries. Finally, the pH_i recoveries were unaffected by exposing the tubule to SITS or by removing the Cl^-. The data, thus indicate that these proximal-tubule cells regulate their pH_i by means of a Na–H exchanger.

2.3 Basolateral HCO_3^- Transport in Salamander Proximal Tubules

To investigate possible HCO_3^- transport systems, we performed a series of experiments in HCO_3^--containing Ringer in which we monitored pH_i while briefly lowering the pH either in the bath (b) or the lumen. The details of these experiments have been published elsewhere (Boron and Boulpaep 1983b). To avoid pH_i changes due to shifts in pCO_2, pH_o was lowered from 7.5 to 6.8 by reducing $[HCO_3^-]_o$ from 10 to 2 m*M* at a constant pCO_2 of 1.5%. As shown in the left panel of Fig. 4, such a reduction in the extracellular pH, when imposed at the basolateral membrane, has two effects: (1) there is large fall of pH_i, and (2) a large and nearly instantaneous depolarization of the basolateral membrane. Furthermore, as shown in the right panel of Fig. 4, the pH_i and voltage changes

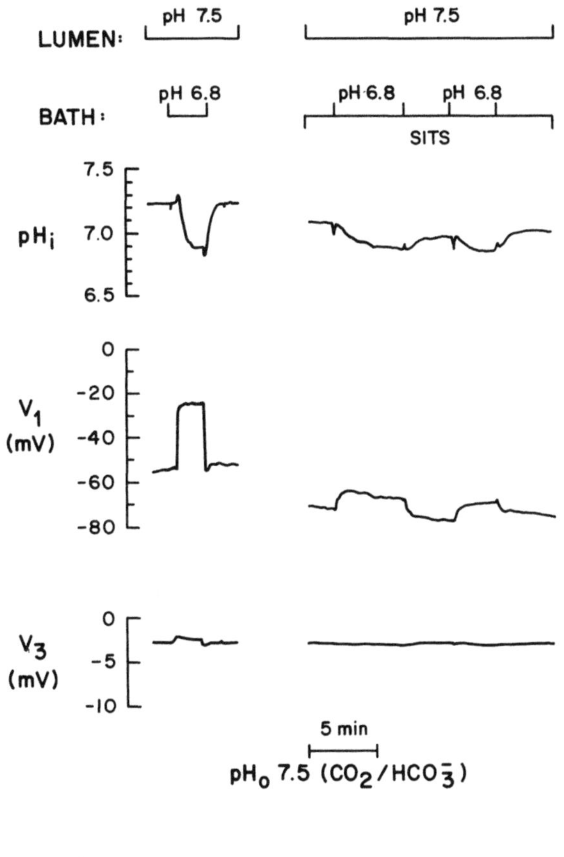

Fig. 4. Effect of reducing basolateral pH and $[HCO_3^-]$. Except at the times indicated by pH 6.8, the luminal and basolateral solutions were buffered to pH 7.5 with 10 m*M* HCO_3^-/1.5% CO_2. Three times the basolateral pH was decreased to 6.8 by lowering $[HCO_3^-]$ to 2 m*M* at constant pCO_2, the last two times, in the presence of 0.5 m*M* SITS. From Boron and Boulpaep (1983b)

are largely blocked by SITS. We found that when a similar reduction of extracellular pH is imposed at the luminal membrane, only small changes occur in pH_i and the basolateral membrane potential (not shown). These results, thus indicate that there is a major HCO_3^- transport system at the basolateral, but not at the luminal membrane. The most straightforward mechanistic explanation for the data presented thus far is that there is a simple HCO_3^- conductance at the basolateral membrane. According to this hypothesis, a reduction in $[HCO_3^-]_b$ would lead to the exit of HCO_3^- (thereby lowering pH_i) and net negative charge (thereby making the cell more positive).

Although the above data are superficially consistent with a simple basolateral conductance to HCO_3^-, it is important to consider other possible explanations. An alternate possibility is that the pH_i changes of Fig. 4 were mediated by an electroneutral Cl–HCO_3 exchanger. However, this model cannot account for the large shifts in cell voltage accompanying the changes in $[HCO_3^-]_b$. Furthermore, we found (not shown) that the pH_i and voltage changes elicited by altering $[HCO_3^-]_b$ do not require Cl^-, nor are they accompanied by substantial changes of intracellular Cl^- activity. These Cl^- data thus rule out a simple Cl–HCO_3 exchanger, as well as the Na-dependent Cl–HCO_3 exchanger described above for invertebrates. A third possibility, though one never previously suggested, is that the movement of HCO_3^- is coupled not to Cl^-, but to Na^+. Figure 5 illustrates the results of an experiment in which basolateral Na^+ was removed at a constant $[HCO_3^-]_b$. The Na^+ removal produces pH_i and voltage changes that are very similar to those induced by altering $[HCO_3^-]_b$ (Fig. 5, left panel). Furthermore, these Na-induced changes in pH_i and voltage are blocked by SITS (Fig. 5, right panel),

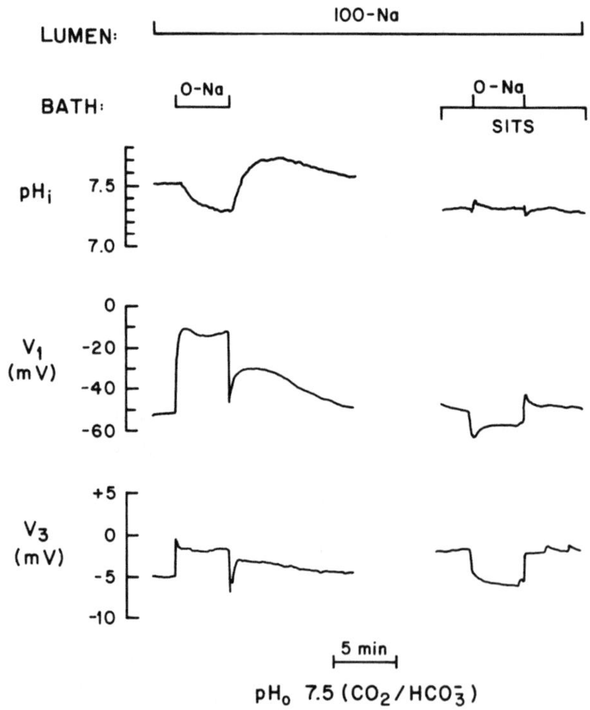

Fig. 5. Effect of removing basolateral Na^+. Except at the indicated times, the tubule was incubated in pH-7.5 HCO_3^-/CO_2 Ringers containing 100 m*M* Na^+. The basolateral Na^+ was twice replaced with N-methyl-D-glucammonium, the second time, in the presence of 0.5 m*M* SITS. From Boron and Boulpaep (1983b)

and require the presence of HCO_3^- (not shown). The SITS-sensitive pH_i changes induced by altering $[Na^+]_b$ are consistent with the hypothesis that HCO_3^- movement across the basolateral membrane is tightly coupled to the movement of Na^+ in the same direction. Thus, the exit of Na/HCO_3 from the cell (and the resultant fall of pH_i) can be elicited by lowering either $[HCO_3^-]_b$ or $[Na^+]_b$. This model leads to the prediction that lowering $[HCO_3^-]_b$ should also lead to a SITS-sensitive decrease of intracellular Na^+ activity. This has indeed been confirmed (Boron and Boulpaep 1983b). In order for this putative Na/HCO_3 cotransport system to account for the SITS-sensitive voltage changes observed upon altering either $[HCO_3^-]_b$ or $[Na^+]_b$, the transporter's stoichiometry would have to be more than one HCO_3^- (or equivalent species) for each Na^+.

More recent experiments (Boron and Fong, unpublished) have shown that transport by the Na/HCO_3 system is reversibly inhibited by the carbonic-anhydrase inhibitor acetazolamide (ACZ). The drug inhibits both the pH_i and voltage changes elicited by altering either $[HCO_3^-]_b$ or $[Na^+]_b$. The half-maximal dose is in the low-μM range. It is not clear whether the entire inhibitory effect of ACZ can be accounted for by its effects on intracellular carbonic anhydrase, or whether there is also a component of direct inhibition of the transporter. If the ACZ is applied when the intracellular Na^+ activity is normal, the effect is to raise pH_i. This indicates that Na/HCO_3 normally exits from the cell. If the ACZ is applied after the intracellular Na^+ activity has been lowered to about half its normal value (by removing luminal Na^+), then the effect is to lower pH_i. This result, which implies that the uninhibited transporter operates backwards under conditions of low internal Na^+, is consistent with a stoichiometry of one Na^+ for every two or more HCO_3^-.

pH_i regulation in the salamander proximal tubule can thus be summarized as follows. The normal efflux of Na/HCO_3 leads to chronic intracellular acid load. The cell's pH_i-regulating system responds to this fall in pH_i by exchanging Na^+ for H^+ across both the luminal and basolateral membranes. The H^+ that happens to be secreted into the lumen constitutes net renal acid secretion. Thus, the universal process of pH_i regulation has, by the introduction of an asymmetric acid-loading pathway (i.e., the basolateral Na/HCO_3 transporter), been transformed into a mechanism capable of transepithelial acid secretion. By definition, the acid secretion should occur to the extents that the cell has need of pH_i regulation. If the acid-loading pathway is blocked either by the removal of CO_2/HCO_3^- (Boron and Boulpaep 1983b) or by the addition of an inhibitor (Boron and Fong, unpublished), the steady state pH_i rises by ~0.2. This rise in pH_i is almost certainly due to Na–H exchange. At this new, alkaline pH_i, however, the Na–H exchanger is nearly inactive (i.e., it has approached its threshold for activation). We would predict that this cell with no need of pH_i regulation is incapable of net acid secretion.

3 An Optical Method for Measuring pH_i in Isolated Renal Tubules

3.1 Methodology

Although it is reasonably easy to use microelectrodes to obtain pH_i measurements on amphibian proximal tubules, the same cannot be said for mammalian tubules. J.R. Chaillet and I, therefore, set out to develop a method for using the absorbance spectrum of a pH-sensitive dye to monitor pH_i in renal tubules. Such an approach would have the advantage of being applicable to extremely small cells. To verify that the dye was functioning properly, however, we decided to first apply it to salamander proximal tubules, a preparation in which we could directly compare the dye to pH-sensitive microelectrodes.

The isolated-tubule apparatus in these experiments was the same as that shown in Fig. 1, except for the introduction of the optical apparatus. The latter is described in detail in a forthcoming paper (Chaillet and Boron 1985). We employed a standard inverted microscope. The source of white light was a 100 W quartz-halogen bulb housed in an air-tight and water-cooled container. A beam splitter diverted a portion of the light to a photodiode, which fed back to the light's power supply. The remainder of the light reached the field aperture, which had two shutter-controlled pinholes. One pinhole allowed a 26 μm diameter spot of light to shine directly on the tubule ("spot on"), and the other, a similar spot of light next to the tubule ("spot off"). The spot on is used to determine the transmitted light intensity, and the spot off, the incident light intensity. After passing through the plane of the tubule, the spot-on or spot-off light was projected onto a diffraction grating. A portion of the resulting spectrum between 400 and 800 nm was then focused on a linear array of 1,024 photodiodes in such a way that red light fell at one end of the array and blue light at the other. The array, interfaced with an 11/23-based computer, was scanned approximately once every second, yielding intensity spectra digitized to 14 bit precision. Incident-light-intensity spectra were obtained with the spot off, and transmitted-light-intensity spectra, with the spot on. The absorbance spectra are calculated as the difference in the logarithms (base 10) of the incident- and transmitted-light-intensities spectra.

3.2 Characteristics of the Intracellular Dye

The intracellular dye chosen for these experiment was 4′,5′-dimethyl-5 (and -6)-carboxyfluorescein (Me_2CF). It is available as the permeant diacetate derivative which enters the cells, and is converted by native intracellular esterases to Me_2CF. Absorbance spectra of intracellular Me_2CF were obtained by subtracting an estimate of the absorbance of the undyed cells from the total absorbance.

The dye was calibrated inside the tubule cells by recording dye absorbance spectra at several fixed values of pH_i. Our approach for establishing a fixed pH_i was essentially the same as that of J.A. Thomas et al. (1979). The tubule cells were exposed to Na-free Ringer containing 10 *μM* of the K–H exchanger nigericin, and 75 m*M* K^+. $[K^+]_o$ was chosen so that the K^+ activity would be the same outside the cell as inside. Because $[K^+]_i = [K^+]_o$, the nigericin should in principal make $pH_i = pH_o$. We found that the pH dependence of the intracellular absorbance spectra similar, but not identical, to that of Me_2CF calibrated in a cuvette. The peak absorbance inside the cell was at ~510 nm, ~5 nm higher than in a cuvette. There was a similar increase in the isosbestic wavelength (at which the absorbance is insensitive to pH) from 470 to 471–477 nm. The ratio of absorbance at the peak to that at the isosbestic wavelength is related to pH by a simple pH-titration curve. In a cuvette the apparent pK is ~6.90, whereas inside the cell, the apparent pK is ~7.35. Because of these differences between the dye's behavior in the cuvette and inside a cell, one would make a substantial error by calculating pH_i from a cuvette calibration curve. We found, however, that if pH_i was calculated from the intracellular calibration curve for Me_2CF, the resultant pH_i (i.e., ~7.44) was very close to that previously measured (Boron and Boulpaep 1983a) with pH-sensitive microelectrodes (i.e., ~7.43).

As a final test of the dye's ability to accurately measure pH_i, we simultaneously monitored pH_i with the dye and with a microelectrode during and after a pulse of 20 m*M* NH_4^+ (as in Fig. 2). We found that the pH_i changes recorded by the two methods were virtually identical. Thus, the absorbance spectrum of Me_2CF can be used not only to calculate accurate steady state values of pH_i, but also to follow rapid pH_i transients.

4 Na–H Exchange in the Rabbit Cortical Collecting Tubule

4.1 Methodology

A detailed account of this study can be found in a forthcoming paper (Chaillet et al. 1985). The isolated-tubule apparatus for these experiments was the same for those on salamander proximal tubules, except that the pipette diameters were appropriately reduced to accommodate the smaller mammalian tubules, and the temperature was raised to 37 °C. The optics were identical to those described in the previous section, with the exception that the spot size was 10 μm instead of 26 μm. Because the purpose of these experiments was to investigate mechanisms of H^+ transport, we only used Ringers solutions that were nominally HCO_3^- free (buffered with 32 m*M* HEPES, gassed with 100% O_2).

4.2 Introduction to the Cortical Collecting Tubule

The cortical collecting tubule (CCT) is made up of two cell types (see Kaissling and Kriz 1979), the principal cells (PC's) and the intercalated cells (IC's). The former make up about two-thirds of total cells (the percentage varying with distance along the CCT). The IC's make up the remaining approximately one-third, and are believed responsible for acid secretion in this segment of the nephron. In particular, the IC's are believed to possess an electrogenic, luminal H^+ pump, similar to the one of the turtle bladder (for review, see Al-Awqati 1978; Steinmetz and Andersen 1982). No information was previously available on mechanisms of pH_i regulation in either of these cell types. As far as possible mechanisms of pH_i regulation by the IC's are concerned, one might expect that the hypothesized luminal H^+ pump could serve the same acid-extruding function that the Na–H exchanger or Na-dependent Cl–HCO_3 exchanger does in other cells. This H^+ pump could regulate pH_i by itself, or in conjunction with another transporter, such as a Na–H exchanger. There is no relevant data on pH_i regulation in the PC's, although I had speculated that pH_i in the PC's might be regulated by a basolateral Na–H exchanger (Boron 1983). The rationale for this was that the Na–H exchanger, the most likely transporter, could not be expected to consistently operate in the forward direction in the CCT, where luminal pH can fall to very low values.

It should be pointed out that our present technology does not permit us to distinguish between PC's and IC's in our pH_i measurements. Inasmuch as our spot of incident light is ~10 μm in diameter, and passes through two layers of cells, we expect that any pH_i measurement is in fact a mean pH_i value of from two to six cells. Although most of our measurements probably include both PC's and IC's, some were probably made on fields that included only PC's.

4.3 Response of the Cells to an Acute Acid Load

The cells of the CCT were acid loaded by a brief pulse with NH_4^+, as described above for the salamander proximal tubule (see Fig. 2). The changes in pH_i were qualitatively the same as in the salamander, though the rate and magnitude of the pH_i changes were much greater. We found that the cells spontaneously recovered from an NH_4^+-induced acid load (i.e., pH_i increased towards its initial value), and that the time course of the pH_i recovery usually had two distinct components. The first component was always present, was rapid, and was blocked by the bilateral removal of Na^+. The second component was not always present, was much slower, and was independent of Na^+. We made no further attempt to study the slow-phase mechanism.

As noted above, the rapid-phase mechanism of pH_i recovery in the CCT is Na^+ dependent. To determine the sideness of this Na^+ dependence, we removed Na^+ from both lumen and bath, acid loaded the cells, and then added the Na^+ back to either the bath or to the lumen. The readdition of the Na^+ to the lumen had only a slight effect on pH_i, leading to a very slow recovery. Returning the Na^+ to the basolateral side of the tubule, however, caused a rapid and complete pH_i recovery. The maximum speed of the recovery was approx. 13 times greater than when the Na^+ was added to the lumen. This leads us to conclude that the transporter responsible for the pH_i recovery is localized primarily, if not exclusively, at the basolateral membrane. In experiments in which we returned varying amounts of Na^+ to the bath, we determined that the apparent K_m for basolateral Na^+ is about 27 mM. This value is within the range reported by others for Na–H exchangers. If this rapid-phase pH_i recovery is indeed caused by a Na–H exchanger, then it should be inhibited by amiloride. We found this to be the case. When $[Na^+]_b$ was 15 mM, amiloride caused a ~60% decrease in the pH_i recovery rate at a concentration of 50 μM, and nearly a 100% inhibition when the concentration of the drug was 1 mM. Such a Na^+ dependence of the inhibition by amiloride is consistent with competition between the drug and Na^+, as has previously been reported for the Na–H exchanger of kidney-derived membrane vesicles (Kinsella and Aronson 1981).

These results thus indicate that the CCT has a basolateral Na–H exchanger. The relative numbers of the PC's and IC's lead us to suggest that the Na–H exchanger is at least in the PC's, and may also be in the IC's as well. A basolateral Na–H exchanger in the PC's would certainly serve the necessary function of pH_i regulation. In addition, this exchanger could also participate in net alkali secretion by the CCT, which is known to occur in CCT's taken from animals fed an alkaline diet (McKinney and Burg 1977; McKinney and Burg 1978, Lombard et al. 1983). If these alkali-secreting tubules were to have a transporter that mediates the efflux of HCO_3^- from the PC into the lumen (e.g., a Cl–HCO_3 exchanger), then this HCO_3^- exit would acid load the cell and thereby stimulate the Na–H to extrude acid across the basolateral membrane. The net effect would be net alkali secretion. Thus, it is possible that a Na–H exchanger could not only regulate pH_i, but also participate in net alkali secretion when the conditions demand it.

5 Conclusions

In this brief communication, I have tried to show that the techniques exist for accurately measuring rapid pH_i transients in renal tubules, and that rapid pH_i transients can be used to identify the various mechanisms of H^+ and HCO_3^- transport in renal tubules. It should be clear that the nephron is a fertile area for research in the area of pH_i regulation. Not only do all the cells of the nephron have to regulate their pH_i for purposes of good housekeeping (as do muscle and nerve cells, for example), but certain cells of the nephron must transport H^+ and/or HCO_3^- for a living. Research in the immediate future will probably be directed in two major directions. First, the nephron will have to be examined segment by segment to describe the H^+ and/or HCO_3^- transport systems present. Second, the regulation of these transport systems will be studied to determine if and how the regulation of H^+ and/or HCO_3^- transport systems is related to the adaptive response of the kidney to various physiological and pathological stresses. It is likely that the regulation of renal H^+ and HCO_3^- transport processes are important for the normal functioning of the kidney, and that elucidating these mechanisms of regulation will have important implications for cells in other organ systems.

Acknowledgment. This work was supported by grants from the National Institutes of Health (AM30344 and AM17433).

References

Aickin CC, Thomas RC (1977a) An investigation of the ionic mechanism of intracellular pH regulation in mouse soleus muscle fibres. J Physiol 273:295–316

Al-Awqati A (1978) H^+ transport in urinary epithelia. Am J Physiol 235:F77–F88

Aronson PS, Nee J, Suhm MA (1982) Modifier role of internal H^+ in activating the Na^+–H^+ exchanger in renal microvillus membrane vesicles. Nature 299:161–163

Boron WF (1983) Transport of H^+ and of ionic weak acids and bases. J Membr Biol 72:1–16

Boron WF, Boulpaep EL (1983a) Intracellular pH regulation in the renal proximal tubule of the salamander: Na–H exchange. J Gen Physiol 81:29–52

Boron WF, Boulpaep EL (1983b) Intracellular pH regulation in the renal proximal tubule of the salamander: basolateral HCO_3^- transport. J Gen Physiol 81:53–94

Boron WF, DeWeer P (1976a) Intracellular pH transients in squid giant axons caused by CO_2, NH_3, and metabolic inhibitors. J Gen Physiol 67:91–112

Boron WF, DeWeer P (1976b) Active proton transport stimulated by CO_2/HCO_3^-, blocked by cyanide. Nature 259:240–241

Boron WF, Russell JM, Brodwick MS, Keifer DW, Roos A (1978) Influence of cyclic AMP on intracellular pH regulation and chloride fluxes in barnacle muscle fibres. Nature 276:511–513

Burg MJ, Grantham J, Abramow M, Orloff J (1966) Preparation and study of fragments of single rabbit nephrons. Am J Physiol 210:1293–1298

Busa WB, Nuccitelli R (1984) Metabolic regulation via intracellular pH. Am J Physiol 246:R409–R438

Chaillet JR, Boron WF (1985) Intracellular calibration of a pH-sensitive dye in isolated perfused salamander proximal tubules. J Gen Physiol, in press

Chaillet JR, Lopes AG, Boron WF (1985) Basolateral Na–H exchange in the rabbit cortical collecting tubule. J Gen Physiol, in press

Kaissling B, Kriz W (1979) Structural analysis of the rabbit kidney. Adv Anat Embryol Cell Biol 56:1–123

Kinsella JL, Aronson PS (1981) Amiloride inhibition of the Na^+–H^+ exchanger in renal microvillus membrane vesicles. Am J Physiol 241:F374–F379

Lombard WE, Kokko JP, Jacobson HR (1983) Bicarbonate transport in cortical and outer medullary collecting tubules. Am J Physiol 244:F289–F296

McKinney TD, Burg MB (1977) Bicarbonate transport by rabbit cortical collecting tubules: Effect of acid and alkali loads in vivo on transport in vitro. J Clin Invest 60:766–768

McKinney TD, Burg MB (1978) Bicarbonate secretion by rabbit cortical collecting tubules in vitro. J Clin Invest 61:1421–1427

Roos A, Boron WF (1981) Intracellular pH. Physiol Rev 61:296–434

Russell JM, Boron WF (1976) Role of chloride transport in regulation of intracellular pH. Nature 264:73–74

Schuldiner S, Rozengurt E (1982) Na^+/H^+ antiport in Swiss 3T3 cells: mitogenic stimulation leads to cytoplasmic alkalinization. Proc Natl Acad Sci USA 79:7778–7782

Steinmetz PR, Andersen OS (1982) Electrogenic proton transport in epithelial membranes. J Membr Biol 65:155–174

Thomas JA, Buchsbaum RN, Zimniak A, Racker E (1979) Intracellular pH measurements in Ehrlich ascites tumor cells utilizing spectroscopic probes generated in situ. Biochemistry 81:220–2218

Thomas RC (1974) Intracellular pH of snail neurones measured with a new pH-sensitive glass microelectrode. J Physiol 238:159–180

Thomas RC (1977) The role of bicarbonate, chloride and sodium ions in the regulation of intracellular in snail neurones. J Physiol 273:317–338

^{31}P NMR Studies of Intracellular pH in Skeletal and Cardiac Muscle

D.G. GADIAN[1]

1 Introduction

Phosphorus nuclear magnetic resonance (^{31}P NMR) was first used to study cellular and tissue metabolism over 10 years ago (Moon and Richards 1973; Hoult et al. 1974), and since then there has been increasing interest in this non-invasive method of studying metabolism in vivo. An important feature of the method is that it provides a measure of intracellular pH. In this article, we shall firstly describe in general terms the type of information that is available from ^{31}P NMR, concentrating in particular on the measurement of pH, and we shall then discuss some specific applications to skeletal and cardiac muscle. Finally, we shall mention some additional possibilities afforded by the use of ^{1}H NMR for metabolic studies.

2 What Can ^{31}P NMR Tell Us About Metabolism

The type of information that ^{31}P NMR can provide is illustrated by Fig. 1, which shows a ^{31}P NMR spectrum obtained from the leg of an anaesthetised rat. The spectrum contains signals that can be assigned to the γ, α, and β phosphates of ATP, phosphocreatine, and inorganic phosphate within the leg muscle. The simplicity of the spectrum reflects the fact that narrow signals are observed only from mobile phosphorus-containing compounds that are present at concentrations of above 0.2–0.5 mM; highly immobilised compounds such as membrane phospholipids produce very broad signals, which often show up in the spectra as a sloping baseline, while compounds that are present at concentrations below 0.2 mM produce weak signals that may be lost in the noise. If ADP were present in mobile form at sufficiently high concentration, it would generate two signals overlapping with the signals from the α and γ phosphates of ATP. It is of considerable interest that ADP generally makes no detectable contribution to the spectra of well-oxygenated tissues, suggesting that the concentration of free ADP is very much lower than the total amounts that are estimated by the technique of freeze-clamping. It is similarly of interest that the concentration of inorganic phosphate as measured by NMR is generally considerably lower than values obtained using other

1 Department of Physics in Relation to Surgery, Royal College of Surgeons of England, Lincoln's Inn Fields, London, Great Britain

Circulation, Respiration, and Metabolism
(ed. by R. Gilles)

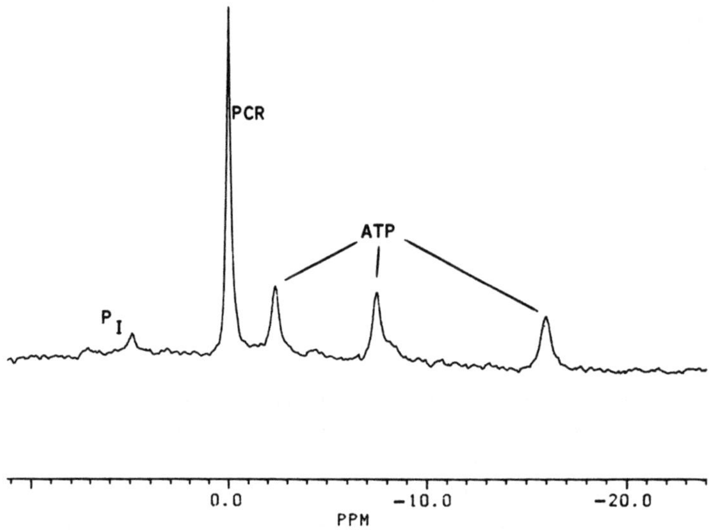

Fig. 1. ^{31}P NMR spectrum obtained at 145.7 MHz from the leg of an anaesthetised rat. The spectrum represents the accumulation of 32 scans repeated at 2 s intervals. Signal assignments are (from *right to left*) β, α, and γ phosphates of ATP, phosphocreatine (PCr), and inorganic phosphate (P_i)

methods. A possible explanation for these discrepancies is that the more traditional invasive methods involve an unavoidable breakdown of high-energy phosphates. In addition, it could be that significant quantities of these metabolites are bound in such a way that the bound fraction generates no detectable signal. For example, it was first suggested by Barany et al. (1975) that the ADP that is bound to muscle myofilaments is too immobilised to generate detectable NMR signals. NAD produces signals that can often be detected as a shoulder just to the right (i.e. to low frequency) of the signal from the α-phosphate of ATP.

A particularly important feature of the spectra is that the frequency of the inorganic phosphate signal is sensitive to pH variations in the normal physiological range. This signal therefore provides a monitor of intracellular pH. The sensitivity to pH arises because the state of ionisation of inorganic phosphate changes in the physiological range (the compound has a pK_a of about 6.75). In the next section, a discussion is given of the principles of pH measurements using NMR.

The frequencies of the ATP signals are sensitive to the binding of divalent metal ions, as was first shown by Cohn and Hughes (1962). On the basis of titrations performed in vitro (see, for example, Gadian et al. 1979), it can be concluded from the spectrum of Fig. 1 that the ATP in the muscle is predominantly complexed to divalent metal ions, presumably Mg^{2+} ions, and similar conclusions regarding the state of ATP have been reached for other tissues. This conclusion is of interest because the state of ATP in vivo has a considerable influence on its biological activity, and also because it enables information to be obtained about the concentration of free Mg^{2+} in vivo. However, the precise value of the free Mg^{2+} concentration, and the accuracy with which it can be estimated, is subject to some discussion (Gupta and Moore 1980; Wu et al. 1981; Gadian and Radda 1981; Garfinkel and Garfinkel 1984).

The concentrations of the various metabolites are, under certain conditions, proportional to the areas of their respective signals, and therefore metabolic processes can be followed simply by monitoring how the signal areas vary with time. However, it should be noted that careful controls are necessary in order to quantify the concentrations of metabolites, particularly in studies where there may be some uncertainty as to precisely where within the sample the signal is coming from (for example, when using surface coils). The spectrum of Fig. 1 was obtained in 1 min, which is a fairly typical accumulation time. However, for some experiments, the time resolution can be greatly enhanced by synchronising the collection of data with physiological activity (Dawson et al. 1977; Fossel et al. 1980).

NMR can readily be used in this way to study the changes in metabolite levels and pH that are associated with muscular contraction, ischaemia, etc., and many examples have been discussed in reviews (Gadian 1982, 1983; Gadian and Radda 1981; Gadian et al. 1982; Griffiths et al. 1982; Ingwall 1982; Meyer et al. 1982; Radda et al. 1984). In Sect. 4 we discuss some of the studies in which the measurement of intracellular pH has been of particular importance, but firstly we shall consider in more detail the principles and potential problems associated with pH measurements using NMR.

3 The Measurement of Intracellular pH

NMR was first used to measure intracellular pH by Moon and Richards (1973) in their studies of red blood cells. Their measurements were based on the observation that 2,3-diphosphoglycerate and inorganic phosphate generate ^{31}P NMR signals whose frequencies are sensitive to pH variations in the normal physiological range. [The frequencies of NMR signals are normally expressed in terms of the parameter known as the chemical shift, with dimensionless units of parts per million (ppm).] In principal, any signal whose frequency is sensitive to pH can provide an indication of pH, but in practice the inorganic phosphate signal is most commonly used, because it is readily observable in the majority of ^{31}P spectra, and because its frequency is particularly sensitive to pH changes in the region of neutrality.

Inorganic phosphate exists mainly as HPO_4^{2-} and $H_2PO_4^-$ around neutral pH. If these two species did not exchange with each other, they would give rise to two signals separated from each other by about 2.4 ppm. In solution, however, the two species exchange with each other very rapidly, and therefore the observed spectrum consists of a single signal, the frequency of which is determined by the relative amounts of the two species; in fact, the frequency measured as a function of pH gives the standard type of pH curve. It should thus be possible to determine intracellular pH simply by measuring the chemical shift of the inorganic phosphate signal in vivo, and determining from a standard titration curve performed on a solution the pH to which this chemical shift corresponds. However, there are several potential difficulties that have to be considered.

Firstly, we must consider whether any factors other than pH can affect the chemical shift of inorganic phosphate in vivo. Certainly, large variations in ionic strength, metal ion binding, and temperature can influence the chemical shift of inorganic phosphate, largely through their effects on the pK_a of this compound. Therefore, it is always important to perform the standard titration curve using a solution whose ionic composition resembles that found in vivo. Of course, there is bound to be some uncertainty regarding the precise ionic environment within the cell, but fortunately control experiments have shown that the errors generated by these uncertainties are likely to be very small. The effects of ionic strength and metal ion binding have been discussed in several papers (see, for example, Gadian et al. 1979; Gillies et al. 1982; Jacobson and Cohen 1981; Jacobus et al. 1982; Roberts et al. 1981).

The binding of inorganic phosphate to macromolecules is another factor that could influence its chemical shift in vivo. However, it has been found that inorganic phosphate titration curves are similar in simple aqueous solution and in homogenised dog heart preparations (see Jacobus et al. 1982). This suggests that phosphate-protein binding interactions have little effect on the observed inorganic phosphate chemical shifts. Additional evidence that such binding has little effect is largely circumstantial, and is based on the similarity between pH values obtained by NMR and by other means. For example, the intracellular pH of resting human skeletal muscle has been measured using ^{31}P NMR to be 7.03, which is in excellent agreement with values that had previously been obtained using other techniques (see Taylor et al. 1983).

An additional problem is that the chemical shift of any signal is measured by comparing the frequency of that signal with that of a reference compound. It is therefore necessary for a suitable reference compound to be present. In the case of ^{31}P NMR studies of muscle, phosphocreatine provides a very convenient reference, for the pK_a of this compound is about 4.6 and its signal is therefore insensitive to pH changes in the region of neutrality. Thus, for many whole tissue studies, the intracellular pH is determined from the chemical shift difference between the inorganic phosphate and phosphocreatine signals. When phosphocreatine is not present, the ^{1}H NMR signal from the water within the tissue can provide an acceptable frequency standard (Ackerman et al. 1981).

A further problem is that there may occasionally be some doubt as to the assignment of the signal that is to be used for pH measurements. For example, one of the signals from 2,3-diphosphoglycerate (2,3-DPG) has a chemical shift that is similar to that of the inorganic phosphate signal, and it was suggested that 2,3-DPG may make a significant contribution to the ^{31}P spectrum of the rat brain (Ackerman et al. 1980). If this were the case, it would be difficult to measure the intracellular pH. However, it has been estimated that the effective concentration of 2,3-DPG is in fact too low to make a significant contribution to the observed brain spectra (Shoubridge et al. 1982).

Finally, there may be some uncertainty as to the distribution of inorganic phosphate within the various intracellular compartments. For example, mitochondria occupy about 40% of the intracellular volume of the heart, and if the inorganic phosphate were located within the mitochondria, we might inadvertently be measuring mitochondrial rather than cytoplasmic pH. Experiments utilising 2-deoxyglucose have indicated that in the rat heart ^{31}P NMR does indeed monitor cytoplasmic pH (Bailey et al. 1981). However, in some recent studies of the rat heart, an additional contribution to the inorganic phosphate signal has been observed, corresponding to a more alkaline environment. It was suggested that this is from mitochondrial inorganic phosphate, and this interesting observation (Garlick et al. 1983) is certainly worthy of further investigation.

Provided that these various problems are adequately addressed, it is clear that ^{31}P NMR provides a convenient and reliable means of estimating intracellular pH. Nevertheless, some uncertainties remain. In particular, the standard titration curve is generally obtained by plotting the chemical shift of inorganic phosphate in solution against the solution pH measured with a combined glass electrode. Illingworth (1981) has pointed out that such electrodes can produce pH measurements that are in error by as much as 0.2 pH units or even more. This could account for the fact that different pH titration curves that have been reported in the literature differ slightly from each other, and in view of this and the other points that have been raised above, it is unwise to hope for absolute accuracy of better than 0.1 pH unit when measuring intracellular pH by NMR. However, changes in pH can often be measured to better than 0.05 pH unit.

As mentioned at the beginning of this section, any signal whose frequency is sensitive to pH can in principle provide an indication of pH, and several compounds other than inorganic phosphate have been used for pH measurements. For example, studies have been performed in which rat hearts were perfused with 2-deoxyglucose (Bailey et al. 1981). In these studies, both the inorganic phosphate and the 2-deoxyglucose 6-phosphate signals provided measurements of intracellular pH, and the values obtained using the two signals were in good agreement with each other under normoxic and ischaemic conditions. This agreement over a large range of pH values provides additional confirmation for the reliability of the NMR technique. The ^{31}P signals of ATP have also been used for measurements of pH in very acidic environments (Casey et al. 1977; Fuldner and Stadler 1982).

^{19}F NMR has also been used to measure intracellular pH in studies where the compounds α-difluoromethylalanine and α-trifluoromethylalanine provide pH markers. The methyl esters of these compounds can be added to cellular suspensions prior to the NMR measurements, and having been taken up by the cells are cleaved to the free acid by endogenous esterases (Deutsch et al. 1982; Kashiwagura et al. 1984). In addition, the intracellular pH can somtimes be deduced from pH-sensitive ^{1}H NMR signals, in particular from carnosine and anserine (Yoshizaki et al. 1981; Arus et al. 1984). In fact, as mentioned at the end of this article, there is increasing interest in the use of ^{1}H NMR for metabolic studies.

It is interesting to note that NMR can provide information about heterogeneity of pH. For example, the inorganic phosphate signal is sometimes split into two components, implying that the sample under observation contains two environments of differing pH. Alternatively, if there is a range of environments of differing pH values, the sample will produce a range of inorganic phosphate signals that are very slightly shifted from each other, the net effect being that the observed signal is broadened. Such effects have been observed in isolated skeletal muscle (Busby et al. 1978; Seeley et al. 1976), and in human forearm muscle during exercise (Taylor et al. 1983).

4 Applications of pH Measurements

In this section, we consider some selected applications of ^{31}P NMR in studies of cardiac and skeletal muscle. An important aspect of these studies is that pH is measured in conjunction with several of the key phosphorus-containing metabolites. This has several implications, as illustrated by the examples described below.

4.1 The Creatine Kinase Reaction – Concentration of Free ADP

Creatine kinase catalyses the reaction

$$\text{Phosphocreatine} + \text{ADP} + \text{H}^+ \rightleftharpoons \text{ATP} + \text{creatine} .$$

Using a technique known as saturation transfer NMR, it has been shown that under many conditions this reaction is close to equilibrium in skeletal and cardiac muscle (Gadian et al. 1981; Nunnally and Hollis 1979; Matthews et al. 1982; Meyer et al. 1982; Shoubridge et al. 1984). Under such conditions, this reaction can be used to deduce the concentration of free ADP, provided that the other reactants can be measured. Of the other reactants, ATP, phosphocreatine, and H^+ can be estimated directly from the ^{31}P NMR spectra, and creatine can be deduced by making use of the data that are available from freeze-clamping studies. An estimate of the ADP concentration can therefore be made, and it is of considerable interest that the values obtained are low – about 5–20 μM in different types of skeletal muscle at rest (Gadian et al. 1981; Meyer et al. 1982; Taylor et al. 1983; Arnold et al. 1984), and rather higher in the perfused heart (Matthews et al. 1982), the precise value in the heart depending on the substrate that is used. These values represent the concentration of ADP that is available to interact with creatine kinase, and therefore the concentration of ADP that is free in solution. The low level of free ADP (it is only a few percent of the total that is measured by other techniques) has important implications in relation to the control of glycolysis (Newsholme and Start 1973) and oxidative phosphorylation (Jacobus et al. 1982).

An additional aspect of these measurements is that the free energy of hydrolysis of ATP can be determined in vivo, because the NMR studies permit measurement of ATP and its products of hydrolysis. A specific application is described below in Sect. 4.2, but in general it should be noted that as a result of the low levels of free ADP and inorganic phosphate in tissues, the free energy of hydrolysis of ATP is higher than many of the values previously reported.

4.2 Muscular Exercise

^{31}P NMR is ideally suited to investigating the metabolic events associated with exercise, for mechanical function and metabolic state can be monitored continuously and non-invasively during rest, exercise, and recovery. The biochemical basis of fatigue has been studied by ^{31}P NMR in frog gastrocnemius muscles maintained at 4 °C under anaerobic conditions (Dawson et al. 1978, 1980a,b). The muscles were stimulated repetitively over a period of about 75 min, and the force measurements revealed the typical signs of fatigue; the force gradually declined in magnitude, and the rate of mechanical relaxation became progressively slower. The spectra indicated the expected decline in phosphocreatine, increase in inorganic phosphate, and decline in intracellular pH. The specific role of acidification in the development of fatigue was difficult to assess, because the pH decline was accompanied by many other metabolic changes. But one of the possible mechanisms that would be consistent with the observations is that the build up of ADP, inorganic phosphate, and H^+ (i.e. the products of ATP hydrolysis) could, by product inhibition, slow down the rate of ATP hydrolysis at the cross bridges and hence cause a reduction in the force development. It was also suggested that the decline in free energy of ATP hydrolysis that is observed to accompany fatigue causes a reduction in the rate of Ca^{2+} uptake into the sarcoplasmic reticulum, which in turn causes a slowing down in the rate of mechanical relaxation (Dawson et al. 1980b).

In these studies, the pH measurements were used to estimate the quantity of lactic acid that was produced, and hence to estimate the rate of glycogen breakdown. Interesting conclusions could therefore be reached about the control of glycogen breakdown in vivo. With the development of wide bore magnets, it is now possible to perform similar studies on human muscle, and studies of forearm muscle metabolism have been described both of control subjects and of patients with muscle disease (for a review see Radda et al. 1984). We shall firstly describe briefly some of the results from control subjects reported by Taylor et al. (1983), concentrating in particular on the pH measurements.

The intracellular pH of the forearm muscle at rest is measured to be 7.03 ± 0.03 (SD, n = 33), which is in excellent agreement with measurements on human muscle made using other techniques. During exercise, the intracellular pH in the finger flexor muscle can reach a value as low as 6.0, and one subject was able to exercise for several minutes at this low pH. These results suggest that contrary to previous belief, glycolysis, and therefore the enzyme phosphofructokinase, must be at least partially active even at pH 6.0. Another interesting observation was that there was a lack of metabolic recovery during a 6 min ischaemic period following exercise, whereas normal recovery is half completed in about 1 min. This indicates that glycolysis is switched off at the end of exercise, despite increases in the concentrations of AMP and ADP. A similar observation had been reported by Harris et al. (1976) following exercise to fatigue, and had been explained in terms of the inhibition of glycolysis by lactic acidosis (Hultman et al. 1981). However, in the NMR studies, the pH decline in some subjects was small, and could not account for the lack of glycolytic activity following exercise. A more likely explanation is that Ca^{2+} ions are required for the activation of phosphorylase kinase and that only the *a* form of phosphorylase is active. A similar conclusion was reached from the NMR studies of frog gastrocnemius muscles described above. However, these conclusions contrast with results obtained from the ischaemic rat heart (Bailey et al. 1981) and from the hamstring muscles of mice with a phosphorylase kinase deficiency (Rahim et al. 1980), which show that in these systems phosphorylase *b* is also active.

A striking deviation from normal pH behaviour is observed in patients with McArdle's syndrome (a defect of glycogen metabolism). The intracellular pH in the forearm muscle of these patients becomes alkaline during exercise, rather than acid (Ross et al. 1981), which is totally consistent with the inability of these patients to generate lactic acid from glycogen.

4.3 Ischaemia in Cardiac Muscle

^{31}P NMR has been extensively used to study the metabolic events associated with ischaemia and recovery in cardiac muscle. As an example of these studies, it has been shown that in perfused rat hearts, the decrease in pH during ischaemia (from about

7.05 to 6.2 in a 15 min period) is considerably reduced by the presence of the buffer bis-Tris-propane in the perfusion medium (Garlick et al. 1979). This appears to have a significant protective effect on the ischaemic tissue. In further studies, Jacobus et al. (1982) have studied the role of intracellular pH in the regulation of myocardial contractility. Their results demonstrate a fairly tight coupling between pH and contractility in the normal heart during respiratory acidosis. However, during total global ischaemia, the extent of acidification cannot fully account for the observed contractile depression; other metabolic and/or physiological mechanisms must also contribute to the effect.

The decline in pH during ischaemia can be interpreted in terms of the anaerobic breakdown of glycogen (Garlick et al. 1979), and therefore, as mentioned in Sect. 4.2, provides a method of studying the rate of glycogenolysis. Infusion of insulin into rat hearts prior to ischaemia causes an increase in the rate and extent of acidosis during the period of no flow, while the rate of ATP depletion is decreased. This has been interpreted in terms of an increased content and accessibility of glycogen as a result of the insulin treatment (Bailey et al. 1982). Moreover, these studies show that glycolysis and hence phosphofructokinase activity are maintained in the ischaemic heart, despite a decrease in intracellular pH to below 6.0. This maintenance of phosphofructokinase activity at low pH is in agreement with the human forearm studies discussed above.

4.4 The Use of Nuclei Other Than ^{31}P

In Sect. 3, it was mentioned that ^{1}H and ^{19}F NMR can sometimes be used for pH measurements. There is increasing interest in the use of ^{1}H NMR for metabolic studies, and following earlier work on cellular suspensions (Brown et al. 1977), studies of isolated muscle (Yoshizaki et al. 1981; Arus et al. 1984), and of the brain of anaesthetised animals (Behar et al. 1983) have now been described. These studies illustrate the scope for using ^{1}H NMR for measurements of intracellular pH (by means of signals from carnosine or anserine), and lactate, and the joint use of ^{1}H and ^{31}P NMR could prove to be particularly useful for metabolic studies.

Acknowledgments. The author was previously at the University of Oxford, where many of the studies described here were performed. He thanks the Rank Foundation and Picker International for support at the Royal College of Surgeons of England.

References

Ackerman JJH, Grove TH, Wong GG, Gadian DG, Radda GK (1980) Mapping of metabolites in whole animals by ^{31}P NMR using surface coils. Nature 283:167–170

Ackerman JJH, Gadian DG, Radda GK, Wong GG (1981) Observation of ^{1}H NMR signals with receiver coils tuned for other nuclides. The optimisation of B_0 homogeneity and a multinuclear chemical shift reference. J Magn Reson 42:498–500

Arnold DL, Matthews PM, Radda GK (1984) Metabolic recovery after exercise and the assessment of mitochondrial function in vivo in human skeletal muscle by means of ^{31}P NMR. Magn Reson Med 1:307–315

Arus C, Barany M, Westler WM, Markley JL (1984) ^{1}H NMR of intact muscle at 11 T. FEBS Lett 165:231–237

Bailey IA, Williams SR, Radda GK, Gadian DG (1981) The activity of phosphorylase in total global ischaemia in the rat heart: a ^{31}P NMR study. Biochem J 196:171–178

Bailey IA, Radda GK, Seymour A-ML, Williams SR (1982) The effects of insulin on myocardial metabolism and acidosis in normoxia and ischaemia. Biochim Biophys Acta 720:17–27

Barany M, Barany K, Burt CT, Glonek T, Myers TC (1975) Structural changes in myosin during contraction and the state of ATP in the intact frog muscle. J Supramol Struct 3:125–140

Behar KL, den Hollander JA, Stromski ME, Ogino T, Schulman RG, Petroff OAC, Prichard JW (1983) High-resolution ^{1}H nuclear magnetic resonance study of cerebral hypoxia in vivo. Proc Natl Acad Sci USA 80:4945–4948

Brown FF, Campbell ID, Kuchel PW, Rabenstein DC (1977) Human erythrocyte metabolism studies by ^{1}H spin echo NMR. FEBS Lett 82:12–16

Busby SJW, Gadian DG, Radda GK, Richards RE, Seeley PJ (1978) Phosphorus nuclear magnetic resonance studies of compartmentation in muscle. Biochem J 170:103–114

Casey RP, Njus D, Radda GK, Sehr PA (1977) Active proton uptake by chromaffin granules: observation by amine distribution and phosphorus-31 nuclear magnetic resonance techniques. Biochemistry 16:972–977

Cohn M, Hughes TR (1962) Nuclear magnetic resonance spectra of adenosine di- and tri-phosphate. II. Effect of complexing with divalent metal ions. J Biol Chem 237:176–181

Dawson MJ, Gadian DG, Wilkie DR (1977) Contraction and recovery of living muscles studied by ^{31}P nuclear magnetic resonance. J Physiol 267:703–735

Dawson MJ, Gadian DG, Wilkie DR (1978) Muscular fatigue investigated by phosphorus nuclear magnetic resonance. Nature 274:861–866

Dawson MJ, Gadian DG, Wilkie DR (1980a) Mechanical relaxation rate and metabolism studied in fatiguing muscle by phosphorus nuclear magnetic resonance (^{31}P NMR). J Physiol 299:465–484

Dawson MJ, Gadian DG, Wilkie DR (1980b) Studies of the biochemistry of contracting and relaxing muscle by the use of ^{31}P NMR in conjunction with other techniques. Philos Trans R Soc Lond B Biol Sci 289:445–455

Deutsch C, Taylor JS, Wilson DF (1982) Regulation of intracellular pH by human peripheral blood lymphocytes as measured by ^{19}F NMR. Proc Natl Acad Sci USA 79:7944–7948

Fossel ET, Morgan HE, Ingwall JS (1980) Measurement of changes in high-energy phosphates in the cardiac cycle by using gated ^{31}P nuclear magnetic resonance. Proc Natl Acad Sci USA 77: 3654–3658

Fuldner HH, Stadler H (1982) ^{31}P-NMR analysis of synaptic vesicles: status of ATP and internal pH. Eur. J Biochem 121:519–524

Gadian DG (1982) Nuclear magnetic resonance and its applications to living systems. Oxford University Press, Oxford

Gadian DG (1983) Whole organ metabolism studied by NMR. Ann Rev Biophys Bioeng 12:69–89

Gadian DG, Radda GK (1981) NMR studies of tissue metabolism. Ann Rev Biochem 50:69–83

Gadian DG, Radda GK, Richards RE, Seeley PJ (1979) ^{31}P NMR in living tissue: the road from a promising to an important tool in biology. In: Shulman RG (ed) Biological applications of magnetic resonance. Academic, New York, pp 463–535

Gadian DG, Radda GK, Brown TR, Chance EM, Dawson MJ, Wilkie DR (1981) The activity of creatine kinase in frog skeletal muscle studied by saturation transfer nuclear magnetic resonance. Biochem J 196:215–228

Gadian DG, Radda GK, Dawson MJ, Wilkie DR (1982) pH measurements of cardiac and skeletal muscle using ^{31}P NMR. In: Nuccitelli R, Deamer DW (eds) Intracellular pH: its measurement, regulation and utilization in cellular functions. Alan R. Liss Inc. USA, pp 61–77

Garfinkel L, Garfinkel D (1984) Calculation of free-Mg^{2+} concentration in adenosine 5′-triphosphate containing solutions in vitro and in vivo. Biochemistry 23:3547–3552

Garlick PB, Radda GK, Seeley PJ (1979) Studies of acidosis in the ischaemic heart by phosphorus nuclear magnetic resonance. Biochem J 184:547–554

Garlick PB, Brown TR, Sullivan RH, Ugurbil K (1983) Observation of a second phosphate pool in the perfused heart by ^{31}P NMR; is this the mitochondrial phosphate? J Mol Cell Cardiol 15: 855–858

Gillies RJ, Alger JR, den Hollander JA, Shulman RG (1982) Intracellular pH measured by NMR: methods and results. In: Nuccitelli R, Deamer DW (eds) Intracellular pH: its measurement, regulation and utilization in cellular functions. Alan R. Liss, USA, pp 79–104

Griffiths JR, Iles RA, Stevens AN (1982) NMR studies of metabolism in living tissue. Prog NMR Spectros 15:49–200

Gupta RK, Moore RD (1980) ^{31}P NMR studies of intracellular free Mg^{2+} in intact frog skeletal muscle. J Biol Chem 255:3987–3993

Harris RC, Edwards RHT, Hultman E, Nordesjo L-O, Nylind B, Sahlin K (1976) The time course of phosphocreatine resynthesis during recovery of the quadriceps muscle in man. Pflügers Arch Eur J Physiol 367:137–142

Hoult DI, Busby SJW, Gadian DG, Radda GK, Richards RE, Seeley PJ (1974) Observations of tissue metabolites using ^{31}P nuclear magnetic resonance. Nature 252:285–287

Hultman E, Sjoholm H, Sahlin K, Edstrom L (1981) Glycolytic and oxidative energy metabolism and contraction characteristics of intact human muscle. In: Ciba Foundation Symposium 82, Pitman Medical, London, pp 19–40

Illingworth JA (1981) A common source of error in pH measurements. Biochem J 195:259–262

Ingwall JS (1982) Phosphorus nuclear magnetic resonance spectroscopy of cardiac and skeletal muscles. Am J Physiol 242:H729–H744

Jacobson L, Cohen JS (1981) Improved technique for investigation of cell metabolism by ^{31}P NMR spectroscopy. Biosci Rep 1:141–150

Jacobus WE, Moreadith RW, Vandegaer KM (1982) Mitochondrial respiratory control: evidence against the regulation of respiration by extramitochondrial phosphorylation potentials or by ATP/ADP ratios. J Biol Chem 257:2397–2402

Jacobus WE, Pores IH, Lucas SK, Kallman CH, Weisfeldt ML, Flaherty JT (1982) The role of intracellular pH in the control of normal and ischemic myocardial contractility: a ^{31}P nuclear magnetic resonance and mass spectrometry study. In: Nuccitelli R, Deamer DW (eds) Intracellular pH: its measurement, regulation and utilization in cellular functions. Alan R. Liss, USA, pp 537–565

Kashiwagura T, Deutsch CJ, Taylor J, Erecinska M, Wilson DF (1984) Dependence of gluconeogenesis, urea synthesis, and energy metabolism of hepatocytes on intracellular pH. J Biol Chem 259:237–243

Matthews PM, Bland JL, Gadian DG, Radda GK (1982) A ^{31}P-NMR saturation transfer study of the regulation of creatine kinase in the rat heart. Biochim Biophys Acta 721:312–320

Meyer RA, Kushmerick MJ, Brown TR (1982) Application of ^{31}P NMR spectroscopy to the study of striated muscle metabolism. Am J Physiol 242:C1–C11

Moon RB, Richards JH (1973) Determination of intracellular pH by ^{31}P nuclear magnetic resonance. J Biol Chem 284:7276–7278

Newsholme EA, Start C (1973) Regulation in metabolism. John Wiley, London

Nunnally RL, Hollis DP (1979) Adenosine triphosphate compartmentation in living hearts; a phosphorus nuclear magnetic resonance saturation transfer study. Biochemistry 18:3642–3646

Radda GK, Bore PJ, Rajagopalan B (1984) Clinical aspects of ^{31}P NMR spectroscopy. Br Med Bull 40:155–159

Rahim ZHA, Perrett D, Lutaya G, Griffiths JR (1980) Metabolic adaption in phosphorylase kinase deficiency. Biochem J 186:331–341

Roberts JKM, Wade-Jardetzky N, Jardetzky O (1981) Intracellular pH measurements by ^{31}P nuclear magnetic resonance. Influence of factors other than pH on ^{31}P chemical shifts. Biochemistry 20:5389–5394

Ross BD, Radda GK, Gadian DG, Rocker G, Esiri M, Falconer-Smith J (1981) Examination of a case of suspected McArdle's syndrome by ^{31}P nuclear magnetic resonance. N Eng J Med 304: 1338–1342

Seeley PJ, Busby SJW, Gadian DG, Radda GK, Richards RE (1976) A new approach to compartmentation in muscle. Biochem Soc Trans 4:62–64

Shoubridge EA, Briggs RW, Radda GK (1982) ^{31}P NMR saturation transfer measurements of the steady state rates of creatine kinase and ATP synthetase in the rat brain. FEBS Lett 140:288–292

Shoubridge EA, Bland JL, Radda GK (1984) Regulation of creatine kinase during steady-state isometric twitch contraction in rat skeletal muscle. Biochim Biophys Acta 805:72–78

Taylor DJ, Bore PJ, Styles P, Gadian DG, Radda GK (1983) Bioenergetics of intact human muscle: a ^{31}P nuclear magnetic resonance study. Mol Biol Med 1:77–94

Wu ST, Pieper GM, Salhany JM, Eliot RS (1981) Measurement of free magnesium in perfused and ischemic arrested heart muscle. A quantitative phosphorus-31 nuclear magnetic resonance and multiequilibria analysis. Biochemistry 20:7399–7403

Yoshizaki K, Seo Y, Nishikawa H (1981) High resolution proton magnetic resonance spectra of muscle. Biochim Biophys Acta 678:283–291

The Role of Intracellular pH in Hormone Action

R.D. Moore[1]

1 Introduction

pH is the measure of the "effective" concentration of protons. In fact, pH is defined in terms of this effective concentration, or thermodynamic activity, of protons: pH $\equiv$ $-\log \alpha_H$ where α_H is the thermodynamic activity of the proton. This definition is especially useful since it means that the pH is directly proportional to the energy, or Gibbs free energy, available from protons to do work. A familiar example provided by this is the fact that the voltage generated by a pH electrode is *directly proportional* to the pH sensed by that electrode. In a similar manner, in the living cell, pH is directly proportional to the ability of the proton to do work or to affect the cellular machinery.

As discussed below, one of the several advantages of the proton is that it's properties make it uniquely suited to relate the global activities of the whole cell to individual molecular events. Thus, it is not surprising that the proton turns out to be involved in the action of those hormones, such as growth factors and insulin, which have multiple actions upon the cell.

1.1 Intracellular pH and Insulin

Perhaps the first clear suggestion that the proton, as expressed by intracellular pH, pH_i, is an intracellular regulator of metabolism was made by Trivedi and Danforth in 1966. By 1984, pH_i was implicated in the action of several hormones. The role of pH_i in hormone action is best established in growth factors (L'Allemain et al. 1984) such as: serum (Frelin et al. 1983), epithelial growth factor (EGF) (Rothenberg et al. 1983), and insulin (see below). In addition, pH_i has been implicated in the action of glucocorticoids (Freiberg et al. 1982; Kinsella et al. 1984), glucagon (Fenton et al. 1978), and parathyroid hormone (Cohn et al. 1983). Insulin was probably the first hormone which was implicated with changes in pH_i and was the first hormone shown to change pH_i by activation of Na:H exchange.

The suggestion that pH_i may be the intracellular signal for insulin was first advanced by Manchester in 1970 and was based upon the sharp pH-profile of some intracellular enzymes, especially phosphofructokinase. A kinetic analysis, based upon a physical

1 Biophysics Laboratory, State University of New York, Plattsburgh, NY 12901, USA and Dept. of Physiology and Biophysics, College of Medicine, University of Vermont, Burlington, VT 05405, USA

Circulation, Respiration, and Metabolism
(ed. by R. Gilles)

model of the (Na^++K^+)-pump, led Moore to suggest in 1973 that insulin might be increasing the rate of proton extrusion from muscle and that changes in pH_i are part of the signaling system in insulin action. Manchester (1970) had reported three experiments on rat diaphragm muscle which demonstrated an insulin-induced increase in pH_i of 0.04, 0.07, and 0.15 pH units.

1.1.1 Effect of Insulin upon pH_i in Amphibian Tissue

Moore and co-workers have confirmed, in frog skeletal muscle, the prediction that in vitro addition of insulin increases pH_i, using both the weak acid [^{14}C] 5,5-dimethyl-oxazolidine-2,4-dione (DMO) (Moore 1977; Moore 1979; Moore et al. 1979; Moore 1981; Fidelman et al. 1982) and the noninvasive technique of ^{31}P-NMR (Moore and Gupta 1980). Neither growth hormone nor albumin, at the same concentration (2 μM) as insulin, affect pH_i indicating that the change in pH_i produced by insulin is not a nonspecific protein effect (Moore 1979). In the absence of CO_2/HCO_3^-, the magnitude of the effect of insulin is +0.16 ± 0.03 U when determined with DMO (Moore 1979) and +0.16 ± 0.05 when determined with ^{31}P-NMR using the difference between the resonance peaks of intracellular inorganic phosphate and phosphocreatine (Moore and Gupta 1980). In the presence of CO_2/HCO_3^-, the effect of insulin is about +0.13 ± 0.02 (Fidelman et al. 1982).

More recently, Putnam and Roos (1983) have used pH-sensitive microelectrodes to confirm the elevation of pH_i when frog semitendinosus muscle is exposed to 1 mU insulin ml^{-1} in the presence of 0.1% bovine serum albumin. By 50 min after addition of this concentration of insulin, pH_i had risen by about 0.08 ± 0.01, while the plasma membrane potential increased in magnitude by 5.4 ± 1.9 mV. Although in the absence of albumin, insulin (400 mU ml^{-1}) did not increase pH_i recovery after 5% CO_2 acidification, when Ringer K^+ was increased from 2.5 to 15 mM, the hormone nearly tripled recovery of pH_i after the acid load. This recovery was inhibited by 1 mM amiloride leading these investigators to also suggest that insulin can activate Na:H exchange in frog skeletal muscle and, thus, stimulate extrusion of acid.

The effect of insulin upon pH_i in amphibians is not limited to skeletal muscle. Morrill (Morrill et al. 1983) and co-workers have used ^{31}P-NMR to follow the effect of insulin upon pH_i in frog prophase arrested oocytes. In frog Ringer without albumin, control pH_i is 7.38 and 0.1 to 10 μM insulin elevates pH_i to between 7.75 and 7.8 over a 1 to 2 h period. Presoaking the oocytes in Na^+-free Ringer for 30 to 60 min lowers control pH_i to 7.25 and blocks the effect of insulin upon pH_i. Addition of Na^+ restores the effect of insulin to elevate pH_i. One mM amiloride blocks not only the elevation of pH_i by insulin, but also blocks the insulin stimulation of cell division.

1.1.2 Effect of Insulin upon pH_i in Mammalian Tissue

Since this work, there have been two reports, both based upon ^{31}P-NMR determination of pH_i, that insulin does not affect pH_i in mammalian muscle. In rat heart, insulin (approx. 3 mU ml^{-1}, or 20 nM) does not affect pH_i as measured by ^{31}P-NMR using the resonance peak of 2-deoxyglucose-6-phosphate (Bailey et al. 1982). The possibility has not been ruled out that the presence of the nonphysiological agent 2-deoxyglucose alters the insulin response. Meyer et al. (1983) have measured pH_i in perfused cat soleus and biceps brachii using the ^{31}P-NMR resonance peak of intracellular inorganic phosphate and 2-deoxyglucose-phosphate and reported that in the presence of high (250 mU min^{-1}) levels of insulin, pH_i does not change.

In both of these studies which report no effect of insulin upon pH_i, the concentrations of insulin was unusually high. It is an old observation that large elevations of the concentration of some hormones reverse the effect. Another possibility is that when the experiments began, sufficient insulin may still have been present upon the hormone's receptor to activate the acid extrusion mechanism. Alternatively, even if the insulin receptor becomes unoccupied by the hormone, it may be that it takes additional time before the stimulated acid extrusion mechanism returns to a base-line activity.

On the other hand, Podo (Podo et al. 1982) has also used ^{31}P-NMR and reported that in the isolated rat diaphragm insulin increases pH_i as determined by the resonance peak of intracellular inorganic phosphate. The muscles were preincubated in the presence or absence of insulin in a Warburg respirometer containing Ringer with glucose-6-phosphate. pH_i was then determined by next placing the muscles in the NMR spectrometer at 4 °C. Those muscles preincubated with insulin had pH_i values which averaged 0.15 higher than that of the controls.

Perhaps the most decisive test of the *physiological* significance of a hormone effect is to demonstrate the predicted consequences of *reduced* levels of the hormone. If a physiological function of insulin is to elevate pH_i, hypoinsulinemia sufficiently moderate as not to produce metabolic acidosis, as reflected by blood pH, should nevertheless produce an observable decrease in pH_i. Recently, we (Brunder et al. 1983) have used ^{14}C-DMO to determine in vivo pH_i in soleus muscles of rats made diabetic by injection of streptozotocin (SZ) 65 or 75 mg kg^{-1}. SZ-injected rats show classical signs of diabetes, i.e., elevated (greater than twofold) plasma glucose and reduced (by about 50%) immunoreactive insulin levels. However, as reflected by blood pH, none of the diabetic rats had metabolic acidosis. Intracellular Na^+ was elevated as observed previously. pH_i was significantly *decreased* by 0.07 ± 0.024 (SE) and by 0.127 ± 0.031 7 days after 65 and 75 mg SZ kg^{-1}, respectively.

Clancy et al. (1983) have reported similar results in rats made diabetic with alloxan or SZ. Although blood pH was significantly depressed (to 7.07) in rats diabetic for 2 days, blood pH had returned to normal by 7 and 28 days. In rats diabetic for 7 days, pH_i was decreased by 0.28 in cardiac muscle, by 0.23 in skeletal muscle, and by 0.16 in liver. After 28 days, pH_i was decreased by about the same amount in cardiac and skeletal muscle, but the decrease in liver was now 0.24. Administration of insulin for 4 to 5 h restored pH_i of cardiac muscle and skeletal muscle to normal, while blood pH remained depressed. In hemidiaphragm preparations from normal and from 2 day diabetic rats, in vitro administration of 100 mU insulin ml^{-1} increased pH_i by 0.1 to 0.25 U and this effect was blocked by amiloride.

Moreover, there is evidence that insulin can elevate pH_i in other mammalian cells. Moolenaar et al. (1983) used an internalized fluorescent pH_i indicator to continuously follow pH_i in diploid human fibroblasts. Insulin is known to act synergistically with growth factors like EGF in stimulating DNA synthesis. Dialyzed fetal calf serum (FCS) elevates pH_i by about 0.2 pH units within 15–20 min. Depleted FCS, which lacks mitogenic activity, has a negligible effect upon pH_i. The polypeptide growth factor EGF also increases pH_i by about 0.1 pH units after 20 min. Although insulin (8×10^{-7} *M*) by itself does not produce a significant elevation of pH_i, in the presence of EGF the hormone does produce a statistically significant elevation of pH_i. When the pH_i elevation due to EGF has reached a steady state, addition of insulin produces a further rise of pH_i by almost 0.1 pH units. Moolenaar et al. (1983) concluded that both the effect of EGF and insulin upon pH_i are due to stimulation of Na:H exchange because in these cells pH_i recovery following an acid load (1) is blocked by amiloride, (2) is accompanied by an increase in amiloride-sensitive ^{22}Na influx and H^+ efflux, and (3) depends upon the extracellular Na^+ concentration.

1.2 Summary

The evidence seems to indicate beyond reasonable doubt that insulin increase pH_i in amphibian tissue. In frog skeletal muscle, this has been confirmed by three totally different techniques: ^{14}C-DMO, ^{31}P-NMR determination of the difference between inorganic phosphate and phosphocreatine peaks, and pH-sensitive microelectrodes. That this effect is not limited to amphibian muscle is indicated by the finding that insulin elevates pH_i in frog oocytes.

The evidence that insulin exerts this effect upon pH_i in mammalian cells is not yet as clear. However, in spite of two negative reports of the effect of insulin upon muscle, two other laboratories report that they have observed in vitro addition of insulin to rat diaphragm elevates pH_i as measured by ^{31}P-NMR in one study and by ^{14}C-DMO in the other. Moreover, two groups also report that diabetic rats which are not keto-acidotic (as reflected by normal blood pH), nevertheless have abnormally low values of pH_i.

The finding by Moolenaar et al. (1983) that insulin does elevate pH_i in human fibroblasts, but only in the presence of EGF is especially provocative. It may well be that, at least in mammals, the presence of some other serum factor(s) is required for insulin to elevate pH_i. If so, addition of insulin to tissues incubated or perfused in Ringer containing only ions and purified albumin would not be expected to affect pH_i.

2 Mechanism of Effect of Hormones on pH_i

2.1 Regulation of pH_i

It is now generally recognized that the proton is not at equilibrium across the plasma membrane. For example, Kostyuk and Sorokina (1961) pointed out that in frog sartorius muscle, the equilibrium pH_i (calculated from the measured pH_0 of 7.4 and V_m of -90 mV) is about 5.9, far lower than the value they observed, 7.1, using pH-sensitive microelectrodes. This equilibrium persists in the face of the general tendency of metabolism to produce protons (Roos and Boron 1981). Physicochemical and organellar buffering and biochemical proton sources and sinks can produce short-term effects upon the amount of free intracellular protons produced by the acidifying effects of metabolism. However, the central role in maintaining pH_i above its equilibrium level is due to transport systems which remove acid from the cell, i.e., extrude H^+ and/or accumulate HCO_3^- or OH^-.

In squid axons, snail neurons, and barnacle muscle, acid extrusion is achieved by coupling the entry of Na^+ and HCO_3^- to the exit of Cl^- and possibly H^+. In crayfish neurons, amphibian proximal tubular cells, and amphibian skeletal muscle, Na:H exchange is used to extrude acid. In mammalian muscle, both Na:H exchange and Cl^-:HCO_3^- exchange operate in parallel. For complete discussion, see Roos and Boron (1981).

2.2 Na:H Exchange

So far, where it exists, the evidence concerning the mechanism whereby hormones elevate pH_i, indicates that the effect is due to activation of Na:H exchange. With some significant exceptions to be discussed below, much of the data pertaining to the mechanism whereby hormones affects pH_i has been obtained in studies using frog skeletal muscle.

The Na:H exchange mechanism operates by using the energy made available by Na^+ moving down its free energy gradient, ΔG_{Na}, to drive the proton up its own free energy gradient, ΔG_H. Therefore, the average free energy change, $\langle \Delta G \rangle_{Na:H}$, for an Na:H exchange mechanism that couples Na^+ influx to proton efflux is the sum of the free energy required to transport n Na^+ ions inward and that required to transport m H^+ ions outward:

$$\langle \Delta G \rangle_{Na:H} = n\Delta G_{Na} + (-m\Delta G_H)$$

or

$$\langle \Delta G \rangle_{Na:H} = n\,[eV_m + kT \ln \frac{\gamma_{Na,i}[Na^+]_i}{\gamma_{Na,o}[Na+]_o}] - m\,[eV_m + kT \ln \frac{\alpha_{H,i}}{\alpha_{H,o}}] \qquad (1)$$

where e = the protonic charge, k = Boltzmann constant, T = absolute temperature, V_m = membrane potential, $\gamma_{Na,i}$ ($\gamma_{Na,o}$) = intracellular (extracellular) Na^+ activity coefficient, $[Na^+]_i$ ($[Na^+]_o$) = intracellular (extracellular) Na^+ concentration, and $\alpha_{H,i}$ ($\alpha_{H,o}$) = intracellular (extracellular) H^+ activity. This mechanism will have sufficient energy available from ΔG_{Na} to transport protons outward when $\langle \Delta G \rangle_{Na:H} < 0$. If $[Na^+]_o$ is decreased sufficiently to null the average free energy, i.e., $\langle \Delta G \rangle_{Na:H} = 0$, this system should not transport protons. If $[Na^+]_o$ is decreased below this null point so that $\langle \Delta G \rangle_{Na:H} > 0$, activation of the system should transport protons inward. Under physiological conditions $\Delta G_{Na} < 0$, $\Delta G_H < 0$, and $\langle \Delta G \rangle_{Na:H} < 0$.

2.3 Criteria for Activation of Na:H Exchange

Therefore, the hypothesis that the increase in pH_i caused by a hormone is due to stimulation of Na:H exchange yields the following four predictions:

1) The elevation of pH_i should be associated with an increased influx of Na^+. In the presence of sufficient ouabain to inhibit possible stimulation of the $(Na^+ + K^+)$-pump by the hormone or by an increase in intracellular Na^+, this increased Na^+ influx would be manifest by an increase in intracellular Na^+ which should be correlated with the increase in pH_i produced by the hormone. Interpretation of ^{22}Na influx is less certain due to the presence of Na:Na exchange in some cells, especially frog skeletal muscle.

2) In frog sartorius, $[Na^+]_i$ is about 7–8 m*M*. For $pH_o = 7.4$, Eq. (5) (assuming $n/m = 1$) indicates that decreasing $[Na^+]_o$ to about 6.8 m*M* should null the free energy for Na:H exchange ($\langle\Delta G\rangle_{Na:H} = 0$) and at that value of $[Na^+]_o$, the hormone should have no effect upon pH_i.

3) Lowering $[Na^+]_o$ still further reverses the sign of $\langle\Delta G\rangle_{Na:H}$ and, thus, reverses the direction of Na:H exchange. Therefore, removing extracellular Na^+ should convert the action of the hormone from an increase to a decrease in pH_i.

4) Because the diuretic drug amiloride (3,5-diamino-6-chloropyrazinoyl-guanidine) blocks Na:H exchange (Benos 1982) this drug should block all the above effects of the hormone in question, i.e., it should block the effect of the hormone upon both H^+ efflux and the associated Na^+ influx and upon the decrease in pH_i produced by insulin in Na-free Ringer.

2.4 Application to Insulin

The effect of insulin on pH_i is not blocked by 10^{-3} *M* ouabain (Moore 1981). Considerable evidence now indicates that the change in pH_i by insulin is due to activation of Na:H exchange. As just indicated for any hormone, the test of whether the hormone is activating Na:H exchange is obtained by *confirmation of criteria* listed just above.

1) This criterion was satisfied by the observation that in the presence of 10^{-3} *M* ouabain, the change in pH_i produced by insulin in each frog sartorius muscle is positively correlated ($r = 0.689$, $P < 0.01$) with net Na^+ influx as reflected by the elevation in Na_i^+ produced by the hormone in the same muscle (Moore 1981).

Consistent with these results is the observation by Clausen and Kohn (1977) that in the presence of 10^{-3} *M* ouabain, insulin produced a significant 30% to 40% increase in $^{22}Na^+$ influx in rat soleus muscle. In the presence of ouabain, insulin also increases $^{22}Na^+$ uptake by hepatocytes (Fehlmann and Freychet 1981). However, in rat adipocytes preincubated 15 min with 10^{-3} *M* ouabain, insulin (3 n*M*) does not increase $^{22}Na^+$ uptake (Resh et al. 1980).

2) In Ringer containing CO_2 and bicarbonate, decreasing $[Na^+]_o$ to this calculated value (a 15-fold reduction) completely blocked ($P > 0.05$) the effect of insulin upon pH_i in frog skeletal muscle (Moore 1979) confirming this prediction. This prediction is also supported by the finding that depleting frog oocytes of Na^+ by soaking in Na^+-free Ringer (which by decreasing the Na^+ driving force should bring $\langle\Delta G\rangle_{Na:H}$ towards zero) blocks the effect of insulin upon pH_i (Morrill et al. 1983).

3) When either Mg^{2+} or choline is used to replace the Na^+ in the Ringer [actually about 0.12 m*M* Na^+, see (Moore 1981)], the effect of insulin upon pH_i is converted to a statistically significant *decrease* ($P < 0.005$) (Moore 1981). This decrease in pH_i produced by insulin in Mg^{2+}-Ringer has been confirmed using the noninvasive method of ^{31}P-NMR to measure pH_i (Moore and Gupta 1980).

Of equal importance, the magnitude of ΔG_{Na} is sufficient to move H^+ against its energy gradient, as is indicated by the fact that $\langle\Delta G\rangle_{Na:H}$ always has the proper sign for the observed flux of H^+ (Moore 1981).

It is possible that all of the above results could be due to the operation of a Na^+–CO_3^{2-} co-transport system (Funder et al. 1978). However, the elevation of pH_i by insulin in HCO_3^--free Ringer rules out this possibility (Moore 1979, 1981) and also argues against the effect being due to HCO_3^- exchange.

4) In the presence of 5% CO_2/30 m*M* HCO_3^-, 0.5 m*M* amiloride does block the elevation by insulin of pH_i in frog sartorius muscle (Moore et al. 1979). In an identical Ringer but lacking amiloride, insulin significantly ($P < 0.001$) increases pH_i by 0.096 ± 0.016 (Moore 1977). In Ringer lacking CO_2/HCO_3^-, amiloride still blocks the elevation of pH_i by insulin (Moore 1981).

The increase in Na_i^+ due to exposure of muscles to insulin for 90 min in the presence of both 1 m*M* ouabain and 0.5 m*M* amiloride is essentially zero ($P > 0.5$) (Moore 1981).

Finally, in muscles placed in Ringer in which Na^+ is replaced by osmotically equivalent amounts of Mg^{2+}, 0.5 m*M* amiloride inhibits the decrease in pH_i produced by insulin (Moore 1981).

These effects of amiloride are not limited to amphibian skeletal muscle as the elevation of pH_i in frog oocytes by insulin is also blocked by this drug (Morrill et al. 1983). Moreover, the elevation of pH_i in rat hemidiaphragms by in vitro addition of insulin is also blocked by (5×10^{-4} *M*) amiloride (Clancy et al. 1983).

Of considerable importance is the finding that 1 m*M* amiloride inhibits the stimulation by insulin of pH_i recovery which occurs in acid-loaded frog semitendinous muscles in the presence of 15 m*M* K_o^+ (Putnam and Roos 1983). This is especially important because this clearly demonstrates that the effect of amiloride is to block the stimulation by insulin of *acid transport* (as opposed to metabolic or buffer-induced pH_i changes).

There is no evidence that the change in pH_i is secondary to metabolic changes. To the contrary, insulin stimulates both glycolysis and ATP hydrolysis by increased activity of the (Na^++K^+)-pump. In the cell, over 90% of ATP^{4-} is bound to Mg^{2+} (Gupta and Moore 1980): it is the $MgATP^{2-}$ form which is hydrolyzed by cell processes, such as the (Na^++K^+)-ATPase. Although glycolysis does not produce intracellular H^+ under anaerobic conditions (Busa and Nuccitelli 1984), the accompanying hydrolysis of ATP does, according to the reaction (at $pH \geqslant 8$):

$$MgATP^{2-} + H_2O \rightarrow MgADP^{1-} + P_i^{2-} + H^+ . \tag{2}$$

Even at lower pH values, the combined effect of glycolysis and hydrolysis of the ATP produced, is still to produce protons according to the reaction [see Busa and Nuccitelli (Busa and Nuccitelli 1984)]:

$$\text{glucose} \rightarrow 2\ \text{lactate}^{1-} + 2\ H^+ . \tag{3}$$

Accordingly, under anaerobic conditions, if insulin does not stimulate acid extrusion, the hormone would be expected to produce a *decrease* in pH_i due to stimulation of ATP hydrolysis by the (Na^++K^+)-pump. Thus, the insulin-induced increase in pH_i which occurs in the face of increased production of acid under anoxic conditions (Fidelman et al. 1982) together with the studies of Putnam and Roos (1983) on recovery from acid-loaded cells, adds considerable weight to the argument that at least in frog muscle, this hormone can stimulate acid extrusion by Na:H exchange.

2.5 Conclusions

It seems reasonably well established that in frog skeletal muscle and oocytes, insulin can stimulate the Na:H exchange system in the plasma membrane and that amiloride blocks the stimulation by insulin of this transport system.

The inconsistency of reports about the effect of insulin upon pH_i in mammalian tissues is not yet resolved. The fact remains that three different laboratories report that insulin can elevate pH_i in mammalian tissue. Moreover, one of these groups plus our own have observed the expected decrease in pH_i in nonketoacidotic diabetic rats. Possible explanations for the failure of two other groups to observe an elevation of pH_i by insulin include: the use of very high concentrations of insulin, possible lack of a necessary "cofactor", and in one case the acidifying action of phosphorylation of 2-deoxyglucose (Bailey et al. 1982). Another factor to consider is that the intracellular buffering power reported for rat skeletal muscle (about 68 mmol pH^{-1} l^{-1}) and rat heart (51 and 77 mmol pH^{-1} l^{-1}) is twice that reported for frog sartorius muscle (35 mmol pH^{-1} l^{-1}) (Roos and Boron 1981) thus an effect of insulin upon pH_i in mammals might be significantly slower than in amphibians.

3 Effects of Changes in pH_i upon Cell Function

3.1 The Example of Insulin

Several lines of evidence indicate that at least in frog skeletal muscle, the effect of insulin upon glycolysis is mediated by a change in pH_i which is caused by activation of Na:H exchange by the hormone.

If the action of insulin upon glycolysis is due to Na:H exchange, the Na^+ concentration, or activity, component of $\langle\Delta G\rangle_{Na:H}$ should determine both the magnitude as well as the *direction* of the acute action of insulin upon glycolysis. This has been shown to be the case. That merely lowering $[Na^+]_0$ *reverses* the action of insulin on glycolytic flux indicates that ionic phenomena play the predominant role in mediating the acute action of insulin upon glycolysis. The confirmation of this particular prediction provides especially powerful support for the model and strongly implies a direct, functional relationship between the Na^+ concentration gradient and the intracellular signal that mediates the insulin effect upon glycolysis.

The thesis that the immediate signal which mediates this insulin effect is the change in pH_i is supported by the finding that there is a correlation between changes in pH_i, whether induced by insulin or by changes in CO_2 levels, and changes in glycolytic flux. Moreover, the plot of substrate vs rate indicate that the effect of changes in pH_i is to effect the activity of phosphofructokinase, (PFK), the pacemaker or rate-limiting enzyme (Karpatkin et al. 1964) of glycolysis. This is consistent with the extreme sensitivity (Trivedi et al. 1966) to small changes in pH of PFK; phosphofructokinase isolated from frog skeletal muscle can be maximally activated by pH elevations as small as 0.1 to 0.2 U (Trivedi et al. 1966), i.e., a 20% to 37% decrease in H^+ activity.

The elevation of pH_i may also play a role in insulin action upon several other cell functions, including protein synthesis, nucleic acid synthesis, and cell division. Moreover, insulin probably stimulates type A amino acid transport by increased synthesis of new transport sites and there is reason to suspect that this may be mediated at least in part by an increase in pH_i (Moore 1983). A provocative finding is that in rat adipocytes alkaline pH causes a minor, but definite enhancement of hexose transport in the absence of insulin (Sonne et al. 1981).

4 Role of Intracellular pH in Hormone Action

Elevation of pH_i not only appears to be involved in insulin action, but a growing body of evidence suggests that elevation of pH_i is part of the signal system for several stimuli which trigger anabolic cell functions, such as DNA synthesis (Winkler and Steinhardt 1981) and/or protein synthesis (Winkler and Steinhardt 1981; Winkler 1982) (see Busa and Nuccitelli 1984). In the action of mitogens, elevation of pH_i appears to play a "permissive" role as other signals are also required.

4.1 How Can a Change in pH_i Effect Cell Function?

Allosteric effects upon enzyme activity (1) immediately comes to mind as a way in which pH_i can regulate cell function. Such an effect would be in essence a *structural* (allosteric means change in shape) changes. However, there are four other ways in which pH_i can affect cell function and each of these is a *process*, e.g., (2) intracellular signal systems, (3) membrane transport systems, (4) metabolic reactions per se, (5) cell energetics (thermodynamics).

4.1.1 Enzyme Activity and Response to Allosteric Modifiers

So far the best example of an effect upon enzyme activity is provided by phosphofructokinase. The effect of pH on this enzyme is highly cooperative, resulting in a very steep pH profile (see above).

In addition to effects of pH upon enzyme activity per se (i.e., the proton acting as an allosteric modifier), changes in pH can effect the binding of effector molecules, thus, modifying the response of an enzyme to other signals. A few such effects are known (Busa and Nuccitelli 1984).

There is also reason to suspect that pH_i effects specific protein phosphorylation and dephosphorylation reactions. These reactions probably are involved in amplification of many effects of insulin and other hormones.

Initiation of DNA synthesis by insulin and other growth factors is preceeded by phosphorylation of ribosomal S6 proteins and by amiloride-sensitive, Na^+-dependent H^+ efflux. Either amiloride or dissipation of H^+ gradients, with DNP (dinitrophenol) or CCCP (carbonyl cyanide m-chlorophenyl hydrazone), blocks both the phosphorylation of ribosomal S6 proteins and the amiloride-sensitive Na^+-dependent H^+ efflux (Winkler 1982). Moreover, in isolated hepatocytes, within 30 to 40 min, insulin increases phosphorylation of two proteins fractions and decreases phosphorylation of four others and all of these phosphorylation effects are inhibited by amiloride (Pouyssegur et al. 1982). These studies are not yet conclusive since amiloride enters the cell (Le Cam et al. 1982), pH_i was not measured, nor were changes in pH_i by physical-chemical methods used to mimic the biochemical effects. Nevertheless, the most likely interpretation is that an elevation of pH_i is part of the signal, whereby insulin triggers changes in the phosphorylation state of key cell proteins.

4.1.2 Possible Interactions Between pH_i and Other Intracellular Signals

There are several interactions between the level of pH_i and other intracellular ionic signals and also molecular second messengers. These interactions suggest the existence of novel ionic feedback mechanisms, both positive and negative, of probable biological significance.

Interactions Between pH_i and Intracellular Ca^{2+}. Intracellular H^+ and Ca^{2+} "buffer" each other by competition for the same binding sites and by Ca:H exchange across the inner mitochondrial membrane. Calmodulin mediates the Ca^{2+} responses of several important enzymes and is one of the most primitive of all proteins. This Ca^{2+}-binding protein is also very pH sensitive. Busa and Nuccitelli (1984) have pointed out that because of the pH sensitivity of the Ca^{2+} dissociation constant of calmodulin, an increase of pH_i by 0.5 U would have the same effect as a fivefold increase in free Ca_i^{2+} at constant pH_i. They suggest that calmodulin might more accurately be considered a Ca^{2+}/H^+-binding protein, and may serve as a sensor of *both* $[Ca^{2+}]_i$ *and* pH_i changes.

Possible Interactions with Cyclic AMP. There is some evidence that cAMP elevates pH_i by stimulation of the Cl^-:HCO_3^- acid extrusion mechanism (Busa and Nuccitelli 1984). Conversely, circumstantial evidence suggesting that moderate changes in pH_i may regulate $[cAMP]_i$. The cAMP-mediated stimulation of lipolysis and calorigenesis is markedly depressed by acidosis and increased by alkalosis (Busa and Nuccitelli 1984).

Based upon these and other considerations, Busa and Nuccitelli (1984) have suggested that "H^+, Ca^{2+}, and cAMP might all function interdependently as "synarchic messengers" (Rasmussen 1981) with pH_i providing a metabolic context within which, e.g., a hormonal stimulus might have rather different consequences at two different pH_is – –".

4.1.3 Membrane Transport Systems

Changes in pH can affect membrane transport by two different mechanisms. Of course, the change in pH_i could be coupled thermodynamically to the transport of another substance by a membrane exchange system which couples the movement of the proton to that of another ion or molecule. In addition, the effect can be direct to modify the transport system or, in some cases, modify the number of transport systems in the plasma membrane.

Skou (1982) has shown that (Na^++K^+)-ATPase is quite sensitive to changes in pH. An elevation of pH increases the affinity of the enzyme system for Na^+. This suggests that in the intact cell, activity of the (Na^++K^+)-pump may be increased by an elevation of pH_i. In the case of insulin, although insulin probably also activates the (Na^++K^+)-pump through intramembrane processes (Moore 1983), the pH sensitivity of the (Na^++K^+)-pump would provide a positive feed-back loop.

Another positive feed-back loop is suggested by the fact that an increase in pH_i increases Na:Ca exchange (Moore 1983).

4.1.4 Metabolic Processes in Which the Proton is Part of the Reaction

The hydrolysis of ATP [Eq. (4)] is an example of a chemical reaction in which protons are either a substrate or a product. Therefore, simply on the basis of the law of mass action, a change in pH will effect such reactions. Conversely, such reactions can effect pH_i.

4.1.5 Regulation of Energy State of the Cell

Intracellular pH can influence the energy state of the cell by effecting the free energy of hydrolysis of ATP, ΔG_{AT}. Whether or not ΔG_{AT} is effected depends on how the hormone effects the other reactants and products, such as ATP, ADP, P_i. In the case of insulin, the hormone stimulates production of the "energy currency of the cell", ATP, by stimulating both glycolysis and oxidative phosphorylation. This hormone also decreases the level of intracellular P_i (Bailey et al. 1982).

The proton is part of the chemical reaction for hydrolysis of MgATP as expressed in [Eq. (4)]:

$$MgATP^{2-} + H_2O \rightarrow MgADP^{1-} + P_i^{2-} + H^+ . \qquad (4)$$

Therefore, it is clear that the free energy, ΔG_{MgATP}, available from this chemical reaction is a function of the proton activity:

$$\Delta G_{MgATP} = \Delta G^0_{ATP} - RT \ln \left(\frac{\alpha_{MgATP}}{\alpha_{MgADP} \cdot \alpha_{Pi} \cdot \alpha_H} \right) . \qquad (5)$$

Thus, the increase in pH_i produced by insulin results in an increase in the free energy, ΔG_{MgATP}, available from the hydrolysis of MgATP.

Since pH is defined as the negative logarithm of the hydrogen ion activity, α_H:

$$pH \equiv - \log_{10} \alpha_H$$

the free energy available from MgATP hydrolysis in the cell may be written explicitly as a function of the intracellular pH:

$$\Delta G_{MgATP} = \Delta G^0_{MgATP} - RT \ln \left(\frac{\alpha_{MgATP}}{\alpha_{MgADP} \cdot \alpha_{Pi}} \right) - 2.3026\ RT \cdot pH_i . \qquad (6)$$

In textbooks, the term 2.3026 RT $\cdot$ pH is incorporated into the standard free energy, ΔG^0_{ATP}, thus leading to the statement that this term is pH dependent. However, when the equation is expressed in its more complete form, as above, ΔG^0_{ATP} becomes more nearly a true constant.

From [Eq. (6)] is it apparent that at 37 °C, an elevation of pH_i typical of insulin, say from 7.15 to 7.3, would of itself result in an additional 0.212 Kcal mol^{-1} available from ATP hydrolysis. This is only about 2% of the total available, but because of the ability of this particular hormone to decrease intracellular P_i (Bailey et al. 1982), the increase in energy available from ATP hydrolysis could be still greater. This increased energy is not a violation of the first law of thermodynamics because it takes work to change the environment of the cell: for example, energy is required to elevate pH_i.

Thus, insulin can not only vary the rate of energy delivery via ATP synthesis, but also regulate the energy *intensity*, i.e., ΔG/molecule of ATP hydrolyzed. This provides one more opportunity for insulin to exert a cascade of regulatory influences.

This suggests the concept that pH_i plays a role in not only the biochemical, but the biophysical *energy state* of the whole cell. In an analogy, the cell becomes somewhat like an atom in an excited state. In contrast to energizing an atom or molecule, it takes several minutes to change the energy state of the whole cell. For example, because of intracellular buffering, as well as the energy required to extrude protons, the effect of insulin, for example, upon pH_i is relatively slow, taking over 1 h to reach maximum. However, once in the higher energy state, the "energized cell" is poised for energy-demanding chores, such as synthesis of macromolecules, cell division, or performing other forms of work. Thus, in this view, by taking the cell into a higher biophysical energy state, the effect of hormones, such as insulin, can prepare anabolic processes in the cell to respond to more specialized intracellular signals, other than pH_i, such as Ca^{2+} or messenger peptides.

5 Unique Properties of the Proton for Intracellular Regulation

It seems reasonable to assume that the very first cells to appear on this planet needed to regulate very basic or fundamental functions, for example, the ability to vary the rate of energy flow within the cell, and the related problem of taking the cell from a state of relative dormancy to one of activity. To do this, something in those first cells would be required to act as an intracellular signal and those first cells would have to have used whatever was at hand. It would be much later that the development of the more complex functions required for differentiation would appear and, thus, require the need for intracellular signals of increased specificity, and, therefore, more complex structure, i.e., molecular "second messengers". As Busa and Nuccitelli (1984) have pointed out: "it seems safe to assume that, just as it does today, from the very beginnings of life on earth biochemistry has involved weakly ionized compounds and has relied heavily on acid-catalysis: regulation of pH_i therefore provided a powerful means of regulating the metabolism of early cells *without requiring the evolution of special receptor molecules*." From these considerations one may begin to infer that the first primitive metabolic functions were controlled in part by protons.

For a substance to be an intracellular signal:

1. it must bind tightly to organic macromolecules in order to stay bound and to have sufficient energy of binding to induce conformational changes;
2. it must have specificity for certain sites on these macromolecules; and
3. its levels within the cell must be regulatable.

5.1 The Proton as an Intracellular Signal in the First Cell

Of those substances dissolved in the primitive ocean, as today most were ions and in all likelihood the same ions as in today's seas. Of these, Ca^{2+} and H^+ (1) have the highest field strength and, therefore, the highest energies of binding to macromolecules; (2) have specificities of binding which vary with the structure of the macromolecule; and (3) could have been regulated provided appropriate transport systems were present in the primitive membrane.

It is not without significance that the intracellular thermodynamic activities of the proton and of Ca^{2+} ($\sim 10^{-7}$ *M* for both) are not only of the same order of magnitude, but are both much less than that of other cations, including not only Na^+ and K^+, but also Mg^{2+} (Gupta and Moore 1980). The fact that both H^+ and Ca^{2+} bind tightly to macromolecules would require that their thermodynamic activity be kept low if changes in such activity were to play a role in regulation of the conformational state, and therefore, function, of macromolecules and macromolecular complexes. The fact that the specificity of these two cations overlaps paves the way for their interaction, and therefore, for the immediate development of some complexity of regulation in the early cells.

5.2 Unique Physical Properties of the Proton

The proton has certain features unique among all ions. The fact that in an aqueous environment its mobility exceeds that of any other ion by an order of magnitude makes it ideal not only for rapid signaling, but also for producing a more pervasive, environmental, change in the cell interior than possible with molecular signals (Moore et al. 1982). Moreover, by nature of its small (when unhydrated) size, the proton may participate in quantum mechanical tunneling, whereas the probability for this with other ions or with molecules is vanishingly small.

5.3 The Proton and Energy Transducing Systems

The proton plays a key role in cellular energy transduction. Not only is it part of the reaction, and therefore, a determinant of the energy released, of ATP hydrolysis (see above), but it also plays a central role in ATP synthesis by chemiosmotic mechanisms. By using the same currency for both energy transduction and information transduction, the earliest cells would have immediately provided opportunities for close coordination of these two critical processes. As Busa and Nuccitelli (1984) have pointed out, the proton is especially suited to both sense and coordinate energy processing systems. In their words, "the apparent potential of pH_i to communicate information regarding cellular *energy balance* to enzymes and structures which may share no other common effector further emphasizes the *integrative* character of pH_i - precisely the charactieristic required of a central effector of metabolism." It is not surprising that the proton, or pH_i, can both sense and regulate the energy state of the cell when one remembers that the proton is *part of* the energy state.

5.4 The Proton as a Metabolic Regulator

In their 1966 paper, Trivedi and Danforth had remarked that " it would be surprising if physiological regulation of enzyme activity by pH were not a widespread and perhaps primitive method of control."

Insulin is a primitive hormone, existing widely throughout the animal kingdom. Therefore, it would be expected to effect very primitive, functions, such as energy flow, essential to survival of the cell. Such primitive, or general, functions would likewise be expected to use a primitive intracellular signal system.

A corollary of the thesis that pH_i is a normal regulatory factor in the cell is that there must be a pattern to the pH profiles of enzymes. Otherwise, changes in pH_i *would* produce metabolic chaos. In the course of evolution, enzymes with inappropriate responses to pH changes would of necessity have been selected against. But pH profiles per se are not sufficient to identify possible control points for ΔpH_i. Because of the complex nature of allosteric enzymes, pH may not necessarily effect the activity of the enzyme directly (i.e., its pH profile may be flat), but as mentioned above may effect the interaction of another allosteric effector with that enzyme. Such examples are known (Moore et al. 1982).

In view of the fundamental differences in physical behavior of protons and discrete molecules, it would perhaps be wise to confine the term second messenger to molecules. This would emphasize the profound differences between these two types of signal systems. Molecular signals represent discrete signals analogous to letters, whereas the proton is a less discrete factor producing a more pervasive, or environmental, and therefore, general effect. As an analogy, a molecular second messenger signal is similar to a letter being sent into a city, it finds a specific address. A change in pH_i, or proton activity, on the other hand, is more like fog rolling into the city, there is no specific address, but almost *everything* is effected. Letters, or molecules, are specific discrete messages. Fog, like pH, is like an *atmospheric* change - the effect is less specific, but more pervasive.

If we assume that H^+ is perhaps the first intracellular signal, one might expect that during evolution molecular second messengers would have been added to enable insulin and other hormones to assume the ability to influence more specific regulatory processes as they evolved.

References

Bailey IA, Radda GK, Seymour AL, Williams SR (1982) The effects of insulin on myocardial metabolism and acidosis in normoxia and ischaemia. Biochim Biophys Acta 720:17–27

Benos DJ (1982) Amiloride: a molecular probe of sodium transport in tissues and cells. Am J Physiol 242:C131–C145

Brunder DG, Oleynek JJ, Moore RD (1983) In vivo measurement of intracellular pH in soleus muscles of streptozotocin-diabetic rats. J Gen Physiol 82:154

Busa WB, Nuccitelli R (1984) Metabolic regulation via intracellular pH. Am J Physiol 246:R409–R438

Clancy RL, Gonzalez NC, Shaban M, Cassmeyer V (1983) Acid-base balance in diabetic diaphragms. Meet Fed Proc 42(3):477

Clausen T, Kohn PG (1977) The effect of insulin on the transport of sodium and potassium in rat soleus muscle. J Physiol (Lond) 265:19–42

Cohn DE, Klahr S, Hammerman MR (1983) Metabolic acidosis and parathyroidectomy increase Na^+–H^+ exchange in brush border vesicles. Am J Physiol 245:F217–F222

Fehlmann M, Freychet P (1981) Insulin and glucagon stimulation of (Na^+-H^+)-ATPase transport activity in isolated rat hepatocytes. J Biol Chem 256:7449–7453

Fenton RA, Gonzalez NC, Clancy RL (1978) The effect of dibutyryl cyclic AMP and glucagon on the myocardial cell pH. Respir Physiol 32:213–223

Fidelman ML, Seeholzer SH, Walsh KB, Moore RD (1982) Intracellular pH mediates action of insulin on glycolysis in frog skeletal muscle. Am J Physiol 242:c87–c93

Freiberg JM, Kinsella J, Sacktor B (1982) Glucocorticoids increase the Na^+-H^+ exchange and decrease the Na^+ gradient dependent phosphate-uptake systems in renal brush border membrane vesicles. Proc Natl Acad Sci USA 79:4932–4936

Frelin C, Vigne P, Lazdunski M (1983) The amiloride-sensitive Na^+/H^+ antiport in 3T3 fibroblasts. J Biol Chem 258:6272–6276

Funder J, Tosteson DC, Wieth JO (1978) Effects of bicarbonate on lithium transport in human red cells. J Gen Physiol 71:721–746

Gupta RK, Moore RD (1980) ^{31}P NMR studies of intracellular free Mg^{2+} in intact frog skeletal muscle. J Biol Chem 255:3987–3993

Karpatkin S, Helmreich E, Cori CF (1964) Regulation of glycolysis in muscle. J Biol Chem 239: 3139–3145

Kinsella J, Cujdik T, Sacktor B (1984) Na^+-H^+ exchange activity in renal brush border membrane vesicles in response to metabolic acidosis: the role of glucocorticoids. Proc Natl Acad Sci USA 81:630–634

Kostyuk PG, Sorokina ZA (1961) In: Kleinzeller A, Kotyk A (eds) Membrane transport metabolism. Academic, New York, pp 193–203

L'Allemain G, Paris S, Pouyssegur J (1984) Growth factor action and intracellular pH regulation in fibroblasts. J Biol Chem 259:5809–5815

Le Cam A, Auberger P, Sampson M (1982) Insulin enhances protein phosphorylation in isolated hepatocytes by inhibiting an amiloride sensitive phosphotase. Biochem Biophys Res Commun 106:1062–1070

Manchester KL (1970) Speculations on the mechanism of action of insulin. Hormones 1:342–351

Meyer RA, Kushmerick MJ, Dillon PF, Brown (1983) Lack of insulin effect in intracellular pH in mammalian skeletal muscle. Fed Proc 42:a1248

Moolenaar WH, Tsien RY, van de Saag PT, de Laat SW (1983) Na^+/H^+ exchange and cytoplasmic pH in the action of growth factors in human fibroblasts. Nature 304:645–648

Moore RD (1973) Effect of insulin upon the sodium pump in frog skeletal muscle. J Physiol (Lond) 232:23–45

Moore RD (1977) Effect of insulin upon intracellular pH. Biophys J 17:259a

Moore RD (1979) Elevation of intracellular pH by insulin in frog skeletal muscle. Biochem Biophys Res Commun 91:900–904

Moore RD, Fidelman ML, Seeholzer SH (1979) Correlation between insulin action upon glycolysis and change in intracellular pH. Biochem Biophys Res Commun 91:905–910

Moore RD (1981) Stimulation of Na:H exchange by insulin. Biophys J 33:203–210

Moore RD (1983) Effects of insulin upon ion transport. Biochem Biophys Acta 737:1–49

Moore RD, Gupta RK (1980) Effect of insulin on intracellular pH as observed by ^{31}P NMR spectroscopy. Int J Quantum Chem, Quantum Biol Symp 7:83–92

Moore RD, Fidelman ML, Hansen JC, Otis JN (1982) The role of intracellular pH in insulin action. In: Nuccitelli R, Deamer DW (eds) Intracellular pH: Its measurement, regulation, and utilization in cellular functions. Alan R. Liss, New York, pp 385–416

Morrill G, Kostellow A, Weinstein SP, Gupta RJ (1983) Hormone induced changes in intracellular Na and pH during the first meiotic division in amphibian oocytes. Fed Proc 42:1791

Podo R, Carpinelli G, D'Agnolo G (1982) International Conference on Magnetic Resonance in Biological Systems, Stanford, Aug 29–Sept 3, p 14

Pouyssegur J, Chambard JC, Paris S (1982) In: Boynton AL, McKeehan WL, Whitfield JF (eds) Symposium on ions, cell proliferation and cancer. Academic, New York, pp 205–218

Putnam RW, Roos A (1983) Insulin effects on pH_i and V_m of frog muscle. Physiologist 26(4):A70

Rasmussen H (1981) Calcium and cAMP as synarchic messengers. Wiley, New York

Resh MD, Nemenoff RA, Guidotti G (1980) Insulin stimulation of (Na^+, K^+)-adenosine triphosphotase-dependent $^{86}Rb^+$ uptake in rat adipocytes. J Biol Chem 255:10938–10945

Roos A, Boron WF (1981) Intracellular pH. Physiol Rev 61:296–433

Rothenberg P, Glasen L, Schlesinger P, Cassel D (1983) Activation of Na^+/H^+ exchange by epidermal growth factor elevates intracellular pH in A431 cells. J Biol Chem 258:12644–12653

Skou JC (1982) The $[Na^+ + K^+]$-ATPase: Coupling of the reaction with ATP to the reaction with Na^+ and K^+. Ann NY Acad Sci 402(II):169–184

Sonne O, Gliemann J, Linde S (1981) Effect of pH on binding kinetics and biological effect of insulin in rat adipocytes. J Biol Chem 256:6250–6254

Straus DS (1981) Effects of insulin on cellular growth and proliferation. Life Sci 29:2131–2139

Trivedi B, Danforth WH (1966) Effect of pH on the kinetics of frog muscle fructokinase. J Biol Chem 241:4110–4114

Winkler MM (1982) Regulation of protein synthesis in sea urchin eggs by intracellular pH. In: Nuccitelli R, Deamer DW (eds) Intracellular pH: Its measurement, regulation, and utilization in cellular functions. Alan R. Liss, New York, pp 325–340

Winkler MM, Steinhardt RA (1981) Activation of protein synthesis in a sea urchin cell-free system. Dev Biol 84:432–443

Intracellular pH in Response to Ambient Changes: Homeostatic or Adaptive Responses

A. MALAN[1]

1 Introduction

For comparative physiologists interested in the responses of organisms to environmental conditions, basic studies on intracellular pH raise two major questions: (1) Does ionic regulation of intracellular pH (or α-imidazole) occur in vivo, and if so, how does its contribution compare with that of ventilatory regulation? (2) Do controlled changes of intracellular pH contribute to adaptive responses through the modulatory role of pH on cell functions?

2 Homeostatic Responses

2.1 Natural Fluctuations of Ambient Factors

Many ambient factors may cause acid-base changes in body fluids, either directly (inspired P_{CO_2}, temperature) or indirectly (inspired P_{O_2}, especially in water breathers, water ionic composition, diet, etc.) (review in Truchot 1981). In air, for instance, fractional concentration of CO_2 varies from the normal level of 0.0003 (0.03%) up to 0.0135 (13.5%) in the burrows of mammals (Williams and Rausch 1973; Withers 1975), and 0.065 (6.5%) in the nests of burrowing birds (White et al. 1978). In general, the aquatic medium is even more variable with respect to acid-base influencing factors, expect perhaps for temperature. For instance, diurnal variations in photosynthesis may cause water P_{O_2} to fluctuate between near anoxia at night to near saturation in the afternoon in shallow waters in summer (Jones 1961; Garey and Rahn 1970; Kramer et al. 1978). In rock pools on the seashore, water P_{CO_2}, pH, P_{O_2}, temperature, etc. all present wide diurnal fluctuations; for P_{CO_2}, the amplitude excedes three orders of magnitude (Truchot and Duhamel-Jouve 1980).

2.2 Acidosis Due to Hypercapnia or to Water Hyperoxia

In a hypercapnic environment, ionic regulation is nearly the only possible mechanism for pH homeostasis. It can take place either at the level of the cell membrane, to regulate

1 CNRS, Laboratoire de Physiologie Respiratoire, 23, Rue Becquerel, 67087 Strasbourg, France

Circulation, Respiration, and Metabolism
(ed. by R. Gilles)

intracellular (cytosolic) pH, pH_i, or at the blood-medium interface (kidney, gill, etc.) to adjust blood pH. Ionic homeostasis of pH_i in response to hypercapnia has long been documented in laboratory mammals (Messeter and Sjesjö 1971; Wood and Schaefer 1978, review in Roos and Boron 1981). In recent years, experimenters have become more aware of the need to study animals in naturally occurring conditions, and studies have largely been extended to water breathers. In such animals, extracellular respiratory acidosis currently occurs as a result either of ambient hypercapnia, or of a hypoventilatory response to ambient hyperoxia (Truchot 1975; Dejours and Beekenkamp 1977; review in Truchot 1981). The latter observation has led to put in doubt the importance of pH homeostasis for aquatic animals: in the freshwater crayfish, for instance, the extracellular respiratory acidosis brought about by exposure to hyperoxic water remains almost uncompensated, even after several months (Dejours and Beekenkamp 1977; Dejours and Armand 1980; Sinha and Dejours 1980). Lack of compensation of extracellular hypercapnia has also been observed in the air-breathing fish *Synbranchus* on emersion (Heisler 1980, 1982), and in the amphibian *Siren* on exposure to poorly-ionized hypercapnic water (Heisler et al. 1982).

Even in the absence or incompleteness of ionic ("metabolic") compensation of extracellular hypercapnic acidosis, pH homeostasis takes place at the intracellular level in *Synbranchus, Siren* (loc cit) as well as in the toad *Bufo marinus* (Toews and Heisler 1982) and in the freshwater crayfish *Astacus leptodactylus* (Gaillard and Malan 1983). In the latter, after 24 h hypercapnia or 48 h hyperoxia both resulting in a threefold increase in hemolymph P_{CO_2}, intracellular pH is fully restored in the nerve cord, heart, and claw muscle; by that time, restoration is still partial (or not significant) in the abdominal muscles (Fig. 1), in which it will reach completeness only after 5 to 7 days (Gaillard and Malan, unpublished). This illustrates the variability between organs as concerns pH_i regulation.

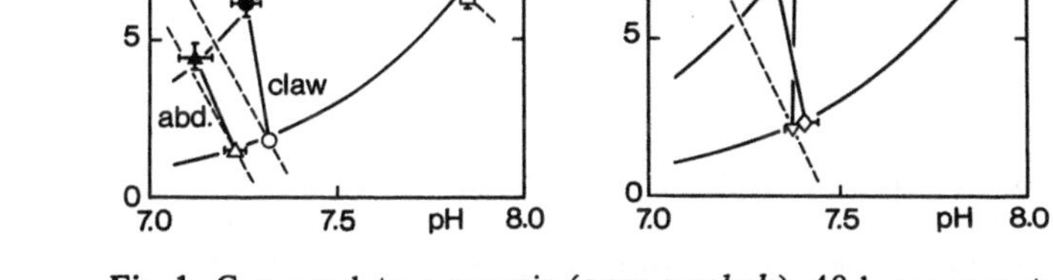

Fig. 1. Compared to normoxia (*open symbols*), 48 h exposure to hyperoxia (*closed symbols*) induced a respiratory acidosis in crayfish, *Astacus leptodactylus*, at 13 °C. Ionic (metabolic) compensation was slight in the hemolymph, and still indetectable in abdominal muscle. In nerve cord and heart, and to a lesser extent in claw muscle, ionic exchanges restored intracellular pH to its control value (DMO data). Hypercapnic exposure (24 h) gave similar results. *Dotted line:* nonbicarbonate buffer line. Mean ± SE. (Redrawn from Gaillard and Malan 1983)

When it is present, as in the dogfish (Heisler et al. 1976; Heisler and Neumann 1977) or in the rainbow trout (Höbe et al. 1984), metabolic compensation of extracellular pH probably facilitates intracellular pH regulation: the extrusion of acidic equivalents from the cell in response to intracellular acidification is enhanced by increasing extracellular pH or extracellular bicarbonate concentration (Boron et al. 1979, 1981; Rodeau 1982).

2.3 Temperature-Related Acid-Base Loads

In order to see the interplay of ventilatory and ionic regulation of pH, another ambient factor than hypercapnia has to be studied. In response to temperature changes, most air-breathing ectotherms achieve alphastat regulation of extra- and intracellular pH by ventilatory adjustment of P_{CO_2}, the ionic gradients across membranes being kept constant (Reeves 1972; Malan et al. 1976; Reeves and Malan 1976, Reeves and Rahn 1979; Reeves, this symposium). Aquatic animals, however, can adjust their blood P_{CO_2} by ventilation only over a restricted range, because of the constraints imposed by the low solubility of oxygen (Rahn and Baumgardner 1972; see below). This probably explains most of the deviations from extracellular alphastat regulation observed in water breathers (Heisler 1980; Truchot 1981; Dejours and Armand 1983).

When a crayfish is transferred from 13° to 6 °C, it increases its relative ventilation like a turtle, and hemolymph P_{CO_2} decreases, while pH increases in parallel with imidazole pK' (Fig. 2, left; Gaillard and Malan 1985). This offsets the increase in CO_2 solubility and, therefore, CO_2 concentration does not vary significantly. On the contrary, when going from 13° to 20 °C, the animal reaches a point where in order to in-

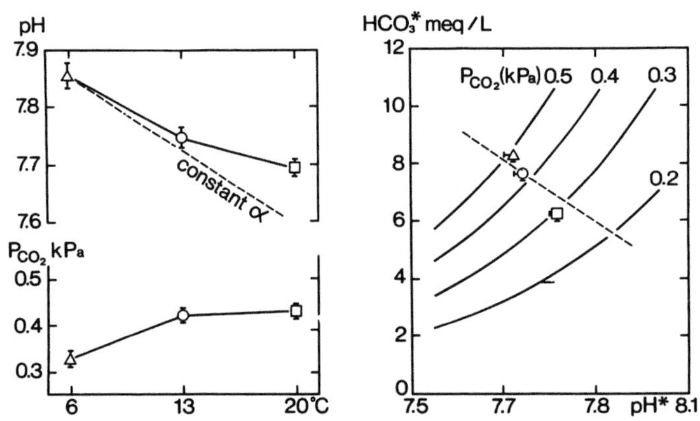

Fig. 2. In vivo acid-base vs temperature relationships of crayfish hemolymph. *Left panel:* data measured at animal temperature (mean ± SE, n = 12). From 13° to 6 °C, P_{CO_2} was reduced, and pH varied in parallel with imidazole pK' (*broken line*), thus, keeping α-imidazole constant. From 13° to 20 °C, P_{CO_2} levelled off and pH decreased less than imidazole pK'. *Right panel:* same data corrected to 13 °C on a bicarbonate (plus carbonate) – pH diagram, by simulation of closed-system temperature changes. Δ: 6 °C data; ○: 13 °C data; □: 20 °C data. *Dotted line:* nonbicarbonate buffer line. From 13° to 6 °C, no deviation from closed-system conditions was observed. From 13° to 20 °C, a respiratory alkalosis developed, with perhaps a partial metabolic compensation. (Redrawn from Gaillard and Malan 1985)

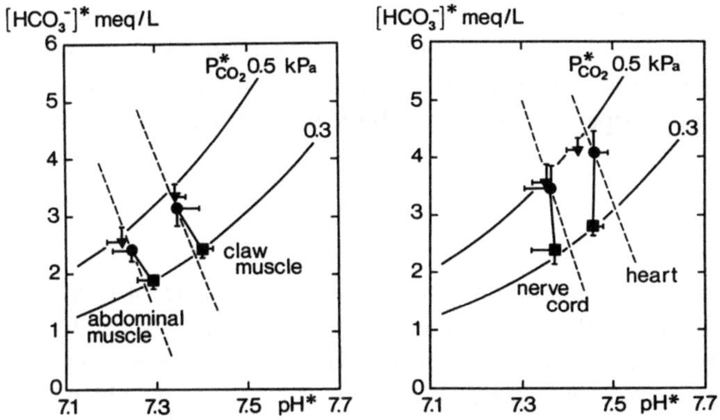

Fig. 3. Same experiment as Fig. 2, temperature-corrected intracellular data (DMO pH). ▼ : 6 °C; ●: 13 °C; ■: 20 °C. From 13° to 6 °C, no deviation from closed-system conditions was observed in any tissue. From 13° to 20 °C, an uncompensated respiratory alkalosis developed in the muscles (*left panel*); in the heart and nerve cord, constancy of temperature-corrected pH, pH* (and α-imidazole) was achieved by ionic exchanges (*right panel*). (Redrawn from Gaillard and Malan 1985)

crease its P_{CO_2} further, he should extract more than 100% of the available oxygen. As a consequence, hemolymph CO_2 concentration decreases, and pH decreases less than imidazole pK'. Is this a respiratory alkalosis? On Fig. 2, right panel, the same data has been temperature-corrected (Malan 1977, 1978a) to eliminate temperature effects on acid-base variables and to show only the changes in chemical factors, like C_{CO_2} and strong ion difference (Stewart 1978). It is now clear that from 13° to 6 °C, by increasing its relative ventilation, the animal keeps constant temperature-corrected acid-base variables, respectively, pH* and $[HCO_3^-]^*$; while from 13° to 20 °C a respiratory alkalosis develops (owing to the buffer composition of body fluids, a constant pH* also closely corresponds to a constant α-imidazole; Malan 1977).

Due to the P_{CO_2} adjustment, temperature-corrected intracellular pH, pH_i^*, stays almost constant from 13° to 6 °C in all tissues studied. From 13° to 20 °C, a respiratory alkalosis develops. In the heart and nerve cord, it is fully compensated by ionic intracellular pH regulation and pH_i^* (and α-imidazole) remains unchanged. In muscles, in which pH_i regulation in response to hypercapnia at 13 °C was slower, no compensation is yet to be seen after 48 h (Fig. 3). This illustrates the alternative use of ventilatory and ionic regulation of intracellular pH, here rather α-imidazole. At temperatures below 13 °C, ventilatory control of P_{CO_2} provides a fast and efficient means of adjusting pH_i in all tissues as temperature varies. When this fails, above 13 °C, ionic regulation of pH_i takes over, at least in a certain proportion of tissues.

3 Adaptive Responses

Other cases do not correspond to the maintenance of a fixed value of pH or α-imidazole, but to a systematic acidification correlated with, and presumably part of, an adaptive response to adverse environmental conditions. Two such examples have been studied in this respect: mammalian hibernation and the cryptobiosis of brine shrimp larvae. Examples of intracellular alkalinizations involved in hormone action, cell division, and egg fertilization will be found in the papers by Moore and by Steinhardt (this symposium).

3.1 Hibernation of Mammals

A major characteristic of mammalian hibernation is the controlled cycling of body temperature which takes place during the winter season, and allows the animal to survive several months on the body fat or food stores accumulated during the previous summer. Periodically, body temperature is lowered down to a value close to ambient and metabolic rate is reduced to about 1/30 of basal rate. After a few days or weeks, however, depending on the species, the animal will rewarm spontaneously back to 37 °C, stay normothermic for a short while, and then cool again. The reason for these spontaneous arousals is still unknown, but the rewarming process does require a considerable amount of heat, which is derived both from shivering and from nonshivering thermogenesis. At the same body temperature of, e.g., 18 °C, the metabolic rate thus varies by a factor of 40:1, twice as much as in a high level human athlete. Do acid-base changes contribute to this exceptional metabolic scope?

Deep hibernation corresponds to a sizeable respiratory acidosis (Malan et al. 1973), with a fourfold increase of arterial P_{CO_2} (temperature-corrected). In blood, brain, diaphragm, and skeletal muscle, tissues which together represent most of body intracellular water, this acidosis is uncompensated (Fig. 4; DMO data; Malan et al. 1981;

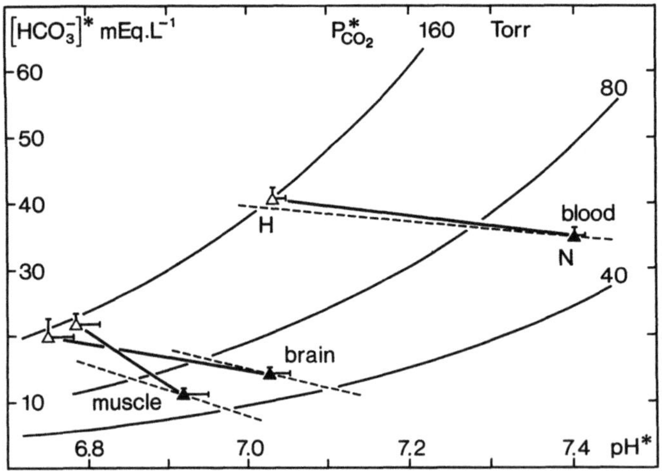

Fig. 4. Temperature-corrected acid-base data of intracellular fluid of skeletal muscle and brain (DMO pH), and of (venous) blood in European hamsters, hibernating (*open symbols*), or not (*closed symbols*). Standard temperature is 37 °C. Mean hibernation body temperature was 10 °C. *Broken lines* are the nonbicarbonate buffer lines. In hibernation, a respiratory acidosis took place in blood and in these tissues. (Redrawn from Malan 1982)

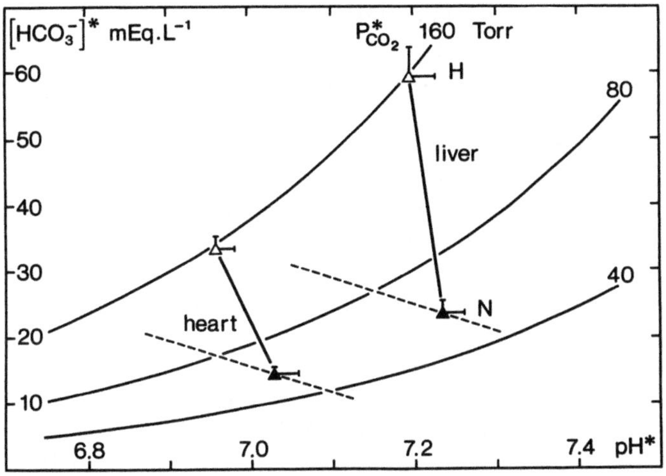

Fig. 5. Same as Fig. 4 for heart and liver. In hibernation, ionic exchanges tended to bring intracellular pH* back to the normothermic value

Malan 1982). The lack of ionic compensation of the acidosis is particularly noticeable for brain, which is known to have an efficient homeostasis of pH_i in normothermy (Messeter and Sjesjö 1971). In liver, however, and to a lesser extent in heart, ionic regulation tends to restore intracellular pH*, i.e., intracellular α-imidazole (Fig. 5). This may be related to the importance of these two organs for survival. Again, this illustrates the interplay of ventilatory and ionic control of intracellular pH. Relative hypoventilation during entrance into hibernation (Snapp and Heller 1981) increases temperature-corrected P_{CO_2} throughout the body. On this acidotic background, ionic regulation permits a local modulation of pHi, probably related to tissue function.

It has been proposed earlier (Malan 1978b, 1980) that acidosis might contribute to at least two features of hibernation, inhibition of thermogenesis (hence the increased metabolic scope) and inhibition of brain structures involved in temperature regulation. The organs in which ionic regulation of pH_i tends to maintain a constant α-imidazole, i.e., liver and heart, would escape such an inhibition. In muscle and brain, an obvious candidate for inhibition by acidosis is the major rate-limiting enzyme of glycolysis, phosphofructokinase (cf. Moore, this symposium). As in other vertebrates, the enzyme from a hibernator's muscle is highly sensitive to acid inhibition (Hand and Somero 1983). When temperature is lowered, the pH-inactivation curve is shifted towards higher pH, as expected from a phenomenon controlled by the titration of imidazole groups in the protein. Combining this information with our own pH_i data, one finds that on going from normothermy to hibernation, the enzyme must be nearly 60% inactivated in the muscle; in the liver, owing to ionic regulation of pH, inactivation is probably negligible and glycolysis can still proceed at a normal rate.

Brown adipose tissue is the major site of nonshivering thermogenesis in hibernating mammals. Heat generation is elicited by norepinephrine released by sympathetic terminals. In isolated cells, at temperatures ranging from 15° to 37 °C, thermogenic response to norepinephrine is reduced up to 75% by increasing P_{CO_2} and reducing medium pH to values equivalent to those found in hibernation (Malan 1984). The inhibition is at least partly intracellular (ibid); it probably involves the facilitation by acidosis of the inhibition of uncoupling by nucleotides previously observed on isolated mitochondria (Nicholls 1976).

Hypercapnia also depresses the firing rate of temperature-sensitive (also temperature-insensitive) neurons of the preoptic area of the hypothalamus, an area involved in temperature detection and regulation (Wünnenberg and Baltruschat 1982). This effect may be responsible for the lowering by hypercapnia of the body temperature threshold for shivering (Schaefer and Wünnenberg 1976). Respiratory acidosis thus probably contributes to the downward shift of the regulated body temperature observed in hibernation (Heller et al. 1978).

Another argument in favor of an inhibitory role of intracellular respiratory acidosis in hibernation is provided by the three- to fourfold increase in relative ventilation which takes place at the very beginning of arousal (Malan et al. 1973). Within 15 to 45 min, arterial P_{CO_2} is divided by half. Most of the thermogenic effort takes place only later on, i.e., at an arterial $P^*_{CO_2}$ ranging between 60 and 80 torr. This reduction of $P^*_{CO_2}$ is enough to revert almost completely the acidotic inhibition of brown adipocytes (Malan 1984). It probably also allows for full glycolytic rate and for the upward resetting of temperature responses; all these are obvious permissive conditions for the strenuous thermogenic effort of arousal.

3.2 Cryptobiosis

A somewhat similar pattern of inhibition by low pH has been found by Busa et al. (1982) in the cryptobiosis of the encysted larvae of the brine shrimp *Artemia salina*. Cryptobiosis normally occurs in response to dehydration, but it can be induced in the laboratory in rehydrated cysts by anaerobiosis. Intracellular pH has been determined in the cysts by nuclear magnetic resonance of ^{31}P. The induction of cryptobiosis (and the corresponding near suppression of metabolic rate) is associated with a large drop in

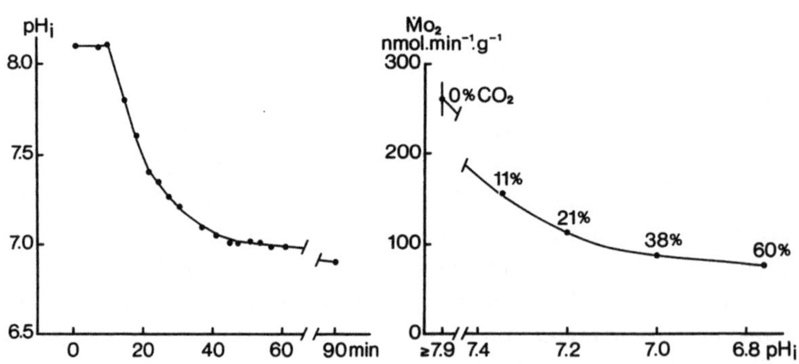

Fig. 6. *Left panel:* Acidification of *Artemia* cysts during the induction of cryptobiosis by aerobic to anaerobic transition. Intracellular pH measured by NMR. *Right panel:* Dependence of oxygen consumption of *Artemia* cysts on intracellular pH. The cysts were exposed to graded increases of P_{CO_2} by equilibrating the superfusion fluid with gas mixtures with the indicated CO_2 concentrations (P_{O_2} was constant). (Redrawn from Busa et al. 1982 and Busa and Crowe 1983)

pH_i, from 7.9 to 6.3 (Fig. 6). Here also acidosis probably plays a causative role: high CO_2 lowers pHi (from 7.9 to 6.8) and reduces metabolic rate in a pHi-dependent manner (Busa and Crowe 1983).

4 Conclusion

In vivo like in vitro, ionic exchange mechanisms at the cell membrane level play an important role in the homeostasis of intracellular pH in response to ambient changes, in combination with the ventilatory control of P_{CO_2}. The latter provides a fast and economical means of adjusting pH over most body compartments simultaneously, while ionic control permits local modulation of pH_i according to specific needs (e.g., in hibernation). Ionic regulation of pH_i also serves as a second line of defense, taking over when regulation of P_{CO_2} becomes ineffective, like in crayfish tissues when temperature rises above 13 °C.

The same mechanisms can also modify intracellular pH as part of an integrated adaptive response to environmental changes, such as hibernation or cryptobiosis.

References

Boron WF, McCormick WC, Roos A (1979) pH regulation in barnacle muscle fibers: dependence on intracellular and extracellular pH. Am J Physiol 237:C185–C193

Boron WF, McCormick WC, Roos A (1981) pH regulation in barnacle muscle fibers: dependence on extracellular sodium and bicarbonate. Am J Physiol 240:C80–C89

Busa WB, Crowe JH (1983) Intracellular pH regulates transition between dormancy and development of brine shrimp (*Artemia salina*) embryos. Science 221:366–368

Busa WB, Crowe JH, Matson GB (1982) Intracellular pH and the metabolic status of dormant and developing *Artemia* embryos. Arch Biochem Biophys 216:711–718

Dejours P, Armand J (1980) Hemolymph acid-base balance of the crayfish *Astacus leptodactylus* as a function of the oxygenation and the acid-base balance of the ambient water. Respir Physiol 41:1–11

Dejours P, Armand J (1983) Acid-base balance of crayfish hemolymph: effects of simultaneous changes of ambient temperature and water oxygenation. J Comp Physiol 149:463–468

Dejours P, Beekenkamp H (1977) Crayfish respiration as a function of water oxygenation. Respir Physiol 30:241–251

Gaillard S, Malan A (1983) Intracellular pH regulation in response to ambient hyperoxia or hypercapnia in the crayfish. Mol Physiol 4:231–243

Gaillard S, Malan A (1985) Intracellular pH-temperature relationships in a water breather, the crayfish. Mol Physiol 7:1–16

Garey WF, Rahn H (1970) Gas tensions in tissues of trout and carp exposed to diurnal changes in oxygen tension of the water. J Exp Biol 52:575–582

Hand SC, Somero GN (1983) Phosphofructokinase of the hibernator *Citellus beecheyi*: Temperature and pH regulation of activity via influences on the tetramer-dimer equilibrium. Physiol Zool 56:380–388

Heisler N (1980) Regulation of the acid-base status in fishes. In: Ali M (ed) Environmental physiology of fishes. Plenum, New York, p 123

Heisler N (1982) Intracellular and extracellular acid-base regulation in the tropical fresh-water teleost fish *Synbranchus marmoratus* in response to the transition from water breathing to air breathing. J Exp Biol 99:9–28

Heisler N, Neumann P (1977) Influence of sea-water pH upon bicarbonate uptake induced by hypercapnia in an elasmobranch fish (*Scyliorhinus stellaris*). Pflügers Arch 368:Suppl R19

Heisler N, Weitz H, Weitz AM (1976) Hypercapnia and resultant bicarbonate transfer processes in an elasmobranch fish (*Scyliorhinus stellaris*). Bull Eur Physiopathol Respir 12:77–85

Heisler N, Forcht G, Ultsch GR, Anderson JF (1982) Acid-base regulation in response to environmental hypercapnia in two aquatic salamanders, *Siren lacertina* and *Amphiuma means*. Respir Physiol 49:141–158

Heller HC, Walker GM, Florant GL, Glotzbach SF, Berger RJ (1978) Sleep and hibernation: Electrophysiological and thermoregulatory homologies. In: Wang LC, Hudson JW (eds) Strategies in cold: Natural torpidity and thermogenesis. Academic, New York, p 225

Höbe H, Wood CM, Wheatly M (1984) The mechanisms of acid-base and ionoregulation in the freshwater rainbow trout during environmental hyperoxia and subsequent normoxia. I. Extra- and intracellular acid-base status. Respir Physiol 55:139–154

Jones JD (1961) Aspects of respiration in *Planorbis corneus* L. and *Limnaea stagnalis* L. (Gastropoda: Pulmonata). Comp Biochem Physiol 4:1–29

Kramer DL, Lindsey CC, Moodie GEE, Stevens ED (1978) The fishes and the aquatic environment of the central Amazon basin, with particular reference to respiratory patterns. Can J Zool 56: 717–729

Malan A (1977) Blood acid-base state at a variable temperature. A graphical representation. Respir Physiol 31:259–275

Malan A (1978a) Intracellular acid-base state at a variable temperature in air-breathing vertebrates and its representation. Respir Physiol 33:115–119

Malan A (1978b) Hibernation as a model for studies on thermogenesis and its control. In: Girardier L, Seydoux J (eds) Effectors of thermogenesis. Birkhäuser, Basel, p 303

Malan A (1980) Enzyme regulation, metabolic rate and acid-base state in hibernation. In: Gilles R (ed) Animals and environmental fitness. Pergamon, Oxford, p 487

Malan A, Lyman CP, Willis JS, Wang LCH (1982) Respiration and acid-base state in hibernation. In: Hibernation and torpor in mammals and birds. Academic, New York, p 237

Malan A (1985) Effet du pH sur la réponse de l'adipocyte brun à la noradrénaline. Son rôle dans l'hibernation. J Physiol (Paris) 79:80a

Malan A, Arens H, Waechter A (1973) Pulmonary respiration and acid-base state in hibernating marmots and hamsters. Respir Physiol 17:45–61

Malan A, Wilson TL, Reeves RB (1976) Intracellular pH in cold-blooded vertebrates as a function of body temperature. Respir Physiol 28:29–47

Malan A, Rodeau JL, Daull F (1981) Intracellular pH in hibernating hamsters. Cryobiology 18: 100–101

Messeter K, Siesjö BK (1971) The intracellular pH' in the brain in acute and sustained hypercapnia. Acta Physiol Scand 83:210–219

Nicholls DG (1976) Hamster brown-adipose-tissue mitochondria. Purine nucleotide control of the ion conductance of the inner membrane, the nature of the ion binding site. Eur J Biochem 62: 223–228

Rahn H, Baumgardner FW (1972) Temperature and acid-base regulation in fish. Respir Physiol 14: 171–182

Reeves RB (1972) An imidazole alphastat hypothesis for vertebrate acid-base regulation: tissue carbon dioxide content and body temperature in bullfrogs. Respir Physiol 14:219–236

Reeves RB, Malan A (1976) Model studies of intracellular acid-base temperature responses in ectotherms. Respir Physiol 28:49–63

Reeves RB, Rahn H (1979) Patterns in vertebrate acid-base regulation. In: Wood SC, Lenfant C (eds) Evolution of respiratory processes. Dekker, New York, p 225

Rodeau JL (1982) L'état acide-base intracellulaire: Analyse théorique appliquée à l'érythrocyte des Mammifères et étude expérimentale des cellules nerveusess et musculaires des Crustacés. Thèse Etat Sciences, Strasbourg

Roos A, Boron WF (1981) Intracellular pH. Physiol Rev 61:296–434

Schaefer KE, Wünnenberg W (1976) Threshold temperatures for shivering in acute and chronic hypercapnia. J Appl Physiol 41:67–70

Sinha NP, Dejours P (1980) Ventilation and blood acid-base balance of the crayfish as functions of water oxygenation (40–1500 Torr). Comp Biochem Physiol 65A:427–432

Snapp BD, Heller HC (1981) Suppression of metabolism during hibernation in ground squirrels (*Citellus lateralis*). Physiol Zool 54:297–307

Stewart PA (1978) Independent and dependent variables of acid-base control. Respir Physiol 33: 9–26

Toews DP, Heisler N (1982) The effects of hypercapnia on intracellular and extracellular acid-base status in the toad *Bufo marinus*. J Exp Biol 97:79–86

Truchot JP (1975) Changements de l'état acide-base du sang en fonction de l'oxygénation de l'eau chez le crabe *Carcinus maenas* (L.). J Physiol (Paris) 70:583–592

Truchot JP (1981) L'équilibre acido-basique extracellulaire et sa régulation dans les divers groupes animaux. (J Physiol Paris) 77:529–580

Truchot JP, Duhamel-Jouve A (1980) Oxygen and carbon dioxide in the marine intertidal environment: diurnal and tidal changes in rockpools. Respir Physiol 39:241–254

White FN, Bartholomew GA, Kinney JL (1978) Physiological and ecological correlates of tunnel nesting in the European bee-eater, *Merops apiaster*. Physiol Zool 51:140–154

Williams DD, Rausch RL (1973) Seasonal carbon dioxide and oxygen concentrations in the dens of hibernating mammals (Sciuridae). Comp Biochem Physiol 44A:1227–1235

Withers PC (1975) A comparison of respiratory adaptations of a semi-fossorial and a surface dwelling Australian rodent. J Comp Physiol 98:193–203

Wood SC, Schaefer KE (1978) Regulation of intracellular pH in lungs and other tissues during hypercapnia. J Appl Physiol 45:115–118

Wünnenberg W, Baltruschat D (1982) Temperature regulation of golden hamsters during acute hypercapnia. J Therm Biol 7:83–86

The Activation of Protein Synthesis by Intracellular pH

R.A. STEINHARDT[1] and M.M. WINKLER[2]

1 Ionic Activation at Fertilization

Studies on the activation of development at fertilization of the sea urchin egg have suggested that increases in free calcium and increases in intracellular pH are essential steps in triggering cell division (Whitaker and Steinhardt 1982; Steinhardt 1982). To link an ionic change to a subsequent event, such as cell division, it is necessary to demonstrate three things. First, one must be able to show that the ionic change occurs naturally and always accompanies the event in question. Second, artificially inducing the ionic change should lead to the induction of the event. Third, blocking the ionic change to the natural stimulus should block the subsequent event. All three conditions have been demonstrated for linking increases in free intracellular calcium and increases in intracellular pH to the activation of cell division at fertilization in the sea urchin egg.

The first ionic change at fertilization is an increase in free intracellular calcium which by itself is both necessary and sufficient to activate all following events with the single exception of the actual mechanics of cell division which requires the centriole contributed by a fertilizing sperm. Merely increasing free calcium by the application of the calcium ionophore A23187 results in the cortical secretion reaction, the increase in oxygen uptake, the rapid acceleration of protein synthesis, and the initiation of DNA synthesis (Steinhardt and Epel 1974). Cell division will follow if a centriole is provided by hypertonic shock treatment (Brandriff et al. 1975). Naturally occurring increases in intracellular free calcium upon fertilization were also shown (Steinhardt et al. 1977). Finally, blocking the increase in free calcium by means of EGTA injections blocked the cortical reaction, although whether increases in macromolecular synthesis occurred was not tested in the small number of eggs that could be treated in this fashion (Zucker and Steinhardt 1978).

One of the consequences of the rise in free calcium is the activation of a sodium-hydrogen exchange (Johnson et al. 1976) which elevates the intracellular pH (Shen and Steinhardt 1978). Elevation of the intracellular pH by artificial means, such as the application of penetrating weak bases, had already been shown to induce protein synthesis and DNA synthesis (Steinhardt and Mazia 1973; Epel et al. 1974; Mazia and Ruby 1974). Blocking the rise in intracellular pH by removing sodium or by artificially lowering intracellular pH with weak penetrating acids will halt cell division and macro-

1 Department of Zoology, University of California, Berkeley, CA 94720, USA
2 Department of Zoology, University of Texas, Austin, TX 78712, USA

Circulation, Respiration, and Metabolism
(ed. by R. Gilles)

molecular synthesis (Chambers 1976; Johnson et al. 1976; Grainger et al. 1979; Shen and Steinhardt 1978; Winkler et al. 1980). There is an absolute requirement for the intracellular pH to increase for protein synthesis and DNA synthesis (Whitaker and Steinhardt 1982).

2 Mechanism of the Intracellular pH Rise

How is the sodium-hydrogen exchange activated? By applying high pressure to eggs during fertilization it has been possible to separate sodium-dependent acid efflux from the insertion of new membrane during the cortical secretion reaction (Schmidt and Epel 1983). Therefore, the increase in free calcium does not activate the sodium-hydrogen exchange by means of the cortical reaction. Instead there is good indirect evidence that the sodium-hydrogen exchange is activated by protein kinase C. The tumor promotor TPA activates sodium-hydrogen exchange and raises the pH in unfertilized sea urchin eggs (Swann and Whitaker 1984) just as had been earlier shown for vertebrate cells in culture (Burns and Rozengurt 1983). TPA mimics diacylglycerol and activates protein kinase C (Nishizuka 1983). Diacylglycerol is one of the breakdown products of phosphoinositol hydrolysis which accompanies fertilization (Turner et al. 1984; Whitaker and Aitchison 1984). Another breakdown product inositol-triphosphate may be responsible for the calcium release from intracellular stores which results in the rise in free calcium (Whitaker and Irvine 1984). Both calcium and the diacylglycerol will activate protein kinase C and the results with applied TPA suggest this is sufficient to stimulate the sodium-hydrogen antiport. TPA by itself will not change the calcium levels in the early period when protein synthesis is already accelerating (Poenie et al. 1985).

3 The Relationship Between Intracellular pH and Protein Synthesis in Intact Eggs

Figure 1 shows the relation between the acceleration in protein synthesis and the increase in intracellular pH in eggs that have been fertilized or stimulated by treatment with the weak base ammonia (NH_4Cl in seawater at pH 8). It can be clearly seen that in ammonia-treated eggs protein synthesis lags behind the acceleration seen at fertilization, although both start up within a few minutes. The entire difference between the rate of acceleration of protein synthesis from these two treatments can be accounted for by the increase in free calcium which follows fertilization (see Fig. 2), but which is missing from the early period of ammonia treatment (Poenie et al. 1985). Figure 2 also clearly demonstrates that the increase in intracellular pH is an absolute requirement for the acceleration of protein synthesis, the increase in free intracellular calcium by itself not being able to affect protein synthesis rate unless the intracellular pH has been elevated to the permissive range above pH 7.0.

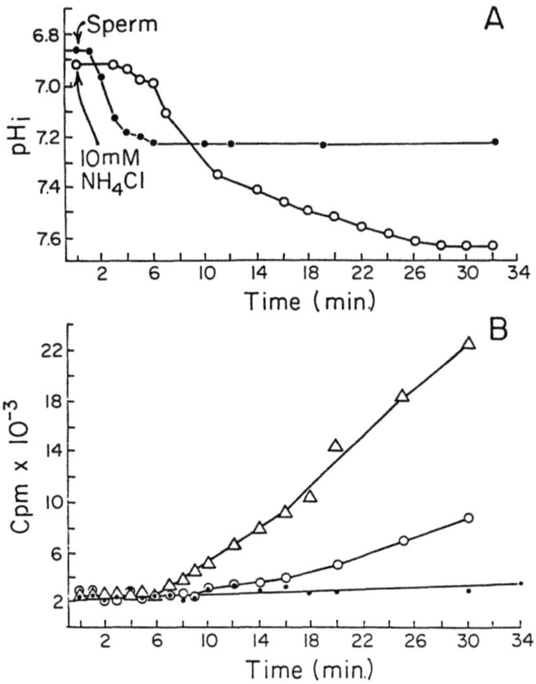

Fig. 1. A The change in intracellular pH following fertilization or ammonia activation of ***Lytechinus pictus*** eggs. The increase in intracellular pH was monitored continuously with Thomas type recessed-tip pH microelectrodes. **B** The increase in the rate of protein synthesis following fertilization or ammonia activation. (△) fertilized. (○) 10 m*M* NH_4 Cl-treated. Eggs were preloaded with ^{3}H valine and sampled for TCA insoluble counts

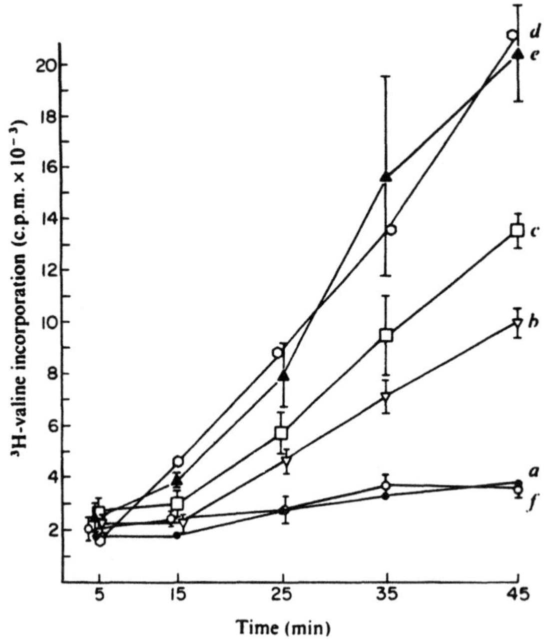

Fig. 2. ^{3}H Valine incorporation after activation in various ionic media. *a* Eggs activated by 2.5 *μM* A23187 in zero-sodium seawater – calcium release without a pH rise. *b* Eggs activated by 10 m*M* NH_4 Cl in zero-calcium sea-water – a pH rise without an early calcium release or entry. *c* As in (*b*) with some calcium entry from normal seawater. *d* Eggs activated by 2.5 *μM* A23187 in zero-sodium seawater plus 10 m*M* NH_4 Cl – a totally artificial calcium release and pH rise. *e* Normal fertilization. *f* Unfertilized

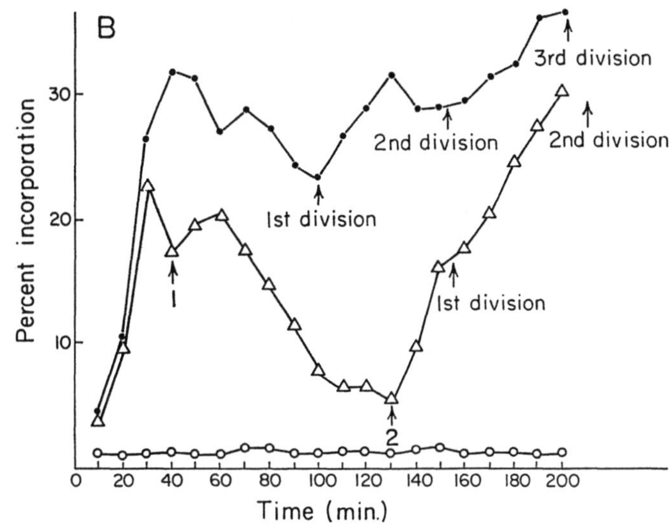

Fig. 3. Lowering the intracellular pH with acetate. At 30 min postfertilization, one batch of eggs is placed in seawater containing 10 m*M* Na-acetate at pH 6.5 (*arrow 1*). At 130 min the acetate solution is replaced with normal seawater (*arrow 2*). Pulse labeling of percent incorporation of ^3H valine into protein. Events of the cell cycle are also scored

Lowering the intracellular pH will attenuate the protein synthesis rate even after the rapid acceleration that takes place at fertilization. This is usually accomplished by treating the eggs with weak penetrating acids at low external pH (Grainger et al. 1979). The lower intracellular pH will delay and halt the cell cycle as well. However, the block produced by lowering the intracellular pH after the initial rise at fertilization has different effects than the absolute block produced by not allowing the intracellular pH to rise in the first place. Dropping the intracellular pH after the initial acceleration will not immediately stop protein synthesis, acting only gradually and reaching the unfertilized rate only after a considerable delay (Fig. 3). This suggests that mRNA already recruited into polysomes may continue to be translated at some definite rate even at the lower intracellular pH.

4 The Contribution of Free Intracellular Calcium to Protein Synthesis Rate

Provided the permanent increase in intracellular pH follows the transient increase in free calcium at fertilization, that increase in calcium ion has a large influence on protein synthesis rate. A completely artificial calcium transient (from ionophore) followed by an artificial pH rise (ammonia treatment in zero-sodium seawater) will accelerate protein synthesis just as well as normal fertilization (Fig. 2). However, the calcium rise by itself has no effect (Winkler et al. 1980). These results point out two important features of ionic regulation in these cells which cannot be emphasized enough. First, although a pH rise is essential it does not by itself complete the activation process, the calcium transient is doing a large part of the job in ways for which the pH rise cannot substitute. Second, the calcium ion cannot do its part in accelerating protein synthesis unless the intracellular pH has reached the permissive range. A similar dual ionic control of the initiation of DNA synthesis has also been found in this system in which even tighter control is exerted in that neither the pH rise nor the calcium transient has an effect unless both are present (Whitaker and Steinhardt 1981). The idea of a *permissive range* for intracellular pH is the single most important concept to emerge from these studies and one which will be generalized to include other cell systems and other pairs of regulatory molecules.

5 The Relationship Between pH and Protein Synthesis in Vitro: Steps in Protein Synthesis Directly Altered by Intracellular pH

We have previously described a cell-free system for protein synthesis which preserves many of the features of the intact egg or early embryo (Winkler and Steinhardt 1981). The system is capable of de novo initiation and synthesizes high molecular weight products. The system is sensitive to pH being up to 20 times more active at pH 7.4 than at pH 6.9. The elongation rate is twofold greater at the higher pH and is comparable to those measured in vivo (Brandis and Raff 1979; Hille and Albers 1979; Goustin and Wilt 1981). The polysome profiles are similar at the two pHs (Winkler and Steinhardt 1981). These results indicate that for mRNAs in polysomes, initiation and elongation occur at near physiological rates, i.e., the in vitro system mimics the in vivo system quite closely. It also argues that the low synthetic rate observed at pH 6.9 is not simply impaired initiation caused by a deviation from the optimal pH, but rather reflects a physiological relevant control.

Adding exogenous mRNA will not stimulate the cell-free translation system to produce more total protein which implies that some element or elements of the translational machinery are limiting (Winkler et al. 1985). To assess the potential to initiate protein synthesis, the level of 43S initiation complexes was measured at pH 6.9 and 7.4 using labeled initiator tRNA (^{35S}Met-tRNA). Initiation complexes consist of a 40S ribosomal subunit, the initiator tRNA, GTP, and a number of initiation factors. The tRNA in the 43S complex freely exchange with the pool of unbound tRNA (Darnbrough et al. 1973). Thus, the level of initiation complexes can be measured by adding labeled tRNA and analyzing the binding to 40S ribosomal units by sucrose gradient centrifugation as shown in Fig. 4. In these experiments elongation has been inhibited by the addition of sparsomycin. This allows time for the labeled Met-tRNA pool to equilibrate with initiation complexes without depleting the pool of labeled tRNA by ongoing protein synthesis. Figure 4 shows that the level of initiation complexes is about three times higher at pH 7.4 than 6.9, hence, the potential to initiate protein synthesis is higher at the more alkaline pH. This higher capacity to initiate complements the faster elongation rate also found at the alkaline pH.

However, these steps in protein synthesis directly affected by pH cannot by themselves account for most of the stimulation even if we allowed ourselves the simple, but

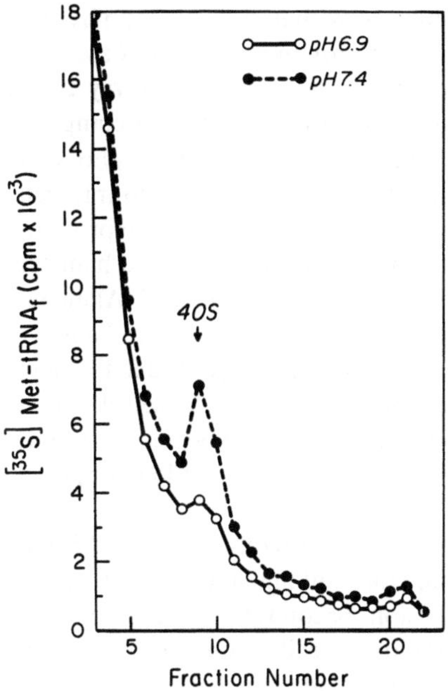

Fig. 4. Comparison of 43S Met-tRNA initiation complexes at pH 6.9 and 7.4. ^{35}S Met-tRNA is added to the cell-free translation at the same time as 50 μM sparsomycin to inhibit elongation. Samples were incubated for 5 min at 20 °C and 25 μl aliquots were diluted with 125 μl ice cold buffer (40 mM NaCl, 100 mM K-acetate, 0.57 mM $MgCl_2$, 1 mM EGTA, 1 mM DTT, 10 mM Hepes, pH 7.4 and 0.2% glutaldehyde. Samples were layered onto 15%–40% (wt/vol) sucrose gradients, made with the buffer described above except that glutaldehyde was omitted, and centrifuged at 50,000 rpm in a Beckman SW-56 rotor for 2.5 h. Fractions (0.2 ml) were collected directly into scintillation vials and counted with an aqueous fluor

misleading arithmetic 2 × 3 = 6 and then merely sought some other factor to account for a three- or fourfold further stimulation. Until we actually know the rate-limiting step in each situation it is pointless to compare numbers of this short. What is known is that the bulk of the mRNA is not available for recruitment into polysomes either in vitro or in vivo and merely raising the pH does not allow for a significant portion of the mRNA to shift into polysomes in our in vitro system (Winkler et al. 1985).

6 Do Increases in Intracellular pH Account for the Stimulation of Cell Proliferation in General in Response to Mitogen or Growth Factors?

A general theory to explain cell growth can be pieced together from the work on polyphosphoinositide hydrolysis and the sodium-hydrogen antiport (Sect. 2). Mitogen or growth factor reacts with receptors at the plasma membrane giving rise to polyphosphoinositide hydrolysis. The breakdown products, diacylglycerol and inositol 1,4 5-triphosphate both act directly or indirectly to activate protein kinase C. Diacylglycerol alters the affinity for Ca^{2+} and activates the kinase at resting free-calcium levels. Inositol triphosphate acts to release calcium from intracellular stores. The protein kinase C, in turn, phosphorylates a number of targets, among them the sodium-hydrogen antiport. The sodium-hydrogen antiport exhibits an increased affinity for intracellular protons and the intracellular pH is forced up. The higher intracellular pH activates transcription of a cellular analog of an oncogene which causes a cascade of gene activations resulting in cell proliferation. The two main posts on which this line of speculation has rested are the ability of amiloride analogs to inhibit growth *and* the sodium-hydrogen antiport and the ability of phorbol esters to stimulate protein kinase C *and* sodium-hydrogen exchange in some cells. The calcium transient in this simple pathway is redundant, but there are other more serious anomalies to deal with. The most complete and careful study of intracellular pH and mitogenesis in human peripheral lymphocytes has shown that these cells do not undergo an alkalinization in response to mitogen (Deutsch et al. 1984) even through they exhibit a sodium requirement for the proliferative response. A second serious problem arises from the fact that Chinese hamster lung fibroblasts are only amiloride-sensitive in the bicarbonate-free media (L'Allemain et al. 1984). Both anomalies suggest the stimulation of sodium-hydrogen exchange is a mechanism to help keep the intracellular pH in a permissive range under the acid load of growth metabolism rather than a key step on the pathway to cell division. Amiloride blockage of the sodium-hydrogen antiport is then only expected to inhibit growth if other pH regulatory mechanisms cannot keep up with the acid load imposed on the cell.

How has the sea urchin shed light on understanding cell growth in general? First, the sea urchin was the material used for the original work linking increases in free calcium and intracellular pH to stimulation of cell division. Second, the concept of sodium-hydrogen exchange itself and its sensitivity to amiloride really dates from the 1976 paper by Johnson, Epel, and Paul on sea urchin eggs. If it had been discovered before, it had gone unnoticed. However, we should look carefully at the sea urchin story as we have done here before generalizing to other systems. The sea urchin egg is at a low intracellular pH (6.8–7.0) and must alkalinize before the calcium signal can result in cell proliferation. The generalization for all cell types, which should stand up to future tests, is that intracellular pH must be kept in a permissive range if the program for cell division is to go forward. Notwithstanding that essential requirement, the key stimulus for proliferation still seems to be the increase in free intracellular calcium levels.

7 New Directions; The Links Between Protein Synthesis and Calcium Homeostasis and Cell Growth

In more recent work to be reported elsewhere, it has become clear that there are direct links between protein synthesis and the levels at which the intracellular free calcium is set. Agents, such as ammonia and TPA, which alkalinize the egg, at first do not affect free-calicum levels, but after a considerable delay (40 min or more) the calcium levels go up a few-fold to the higher levels characteristic of the activated cell state (Poenie et al. 1985). Other agents which inhibit protein synthesis, such as emetine, severely inhibit the calcium pump-down after fertilization. These observations promise to open up a whole new approach to understanding calcium homeostasis in cells; once again we have reason to be grateful for the sea urchin egg for helping us find new ideas and new approaches to understanding cell regulation.

Acknowledgments. The work described here has been supported in part by grants from the National Science Foundation (USA) to R.A.S. and a grant from the National Institutes of Health to M.M.W.

References

Brandis JW, Raff RA (1979) Elevation of protein synthesis in a complex response to fertilization. Nature 278:467–469

Brandriff B, Hinegardner RT, Steinhardt RA (1975) Development and life cycle of the parthenogenetically activated sea-urchin embryo. J Exp Zool 192:13–24

Burns CP, Rozengurt E (1983) Serum, platelet-derived growth factor, vasopressin and phorbol esters increase intracellular pH in Swiss 3T3 cells. Biochem Biophys Res Commun 116:931–938

Chambers EL (1976) Na is essential for activation of the inseminated sea urchin egg. J Exp Zool 197:149–154

Darnbrough C, Legon S, Hunt T, Jackson RJ (1973) Initiation of protein synthesis: Evidence for messenger RNA-independent binding of methionyl-transfer RNA to the 40 S ribosomal subunit. J Mol Biol 76:379–403

Deutsch C, Taylor JS, Price M (1984) pH homeostasis in human lymphocytes: Modulation by ions and mitogen. J Cell Biol 98:885–893

Epel D, Steinhardt RA, Humphreys T, Mazia D (1974) An analysis of the partial metabolic depression of sea urchin eggs by ammonia: The existence of independent pathways. Dev Biol 40:245–255

Goustin AS, Wilt FH (1981) Protein synthesis, polyribosomes, and peptide elongation in early development of *Strongylocentrotus purpuratus*. Dev Biol 82:32–40

Grainger JL, Winkler MM, Shen SS, Steinhardt RA (1979) Intracellular pH controls protein synthesis rate in the sea urchin egg and early embryo. Dev Biol 68:396–406

Hille MB, Albers AA (1979) Polypeptide chain elongation increases after fertilization of sea urchin eggs. Nature 278:469–471

Johnson JD, Epel D, Paul M (1976) Intracellular pH and activation of sea urchin eggs after fertilization. Nature 262:661–664

L'Allemain G, Paris S, Pouysségur J (1984) Growth factor action and intracellular pH regulation in fibroblasts. Evidence for a major role of the Na^+/H^+ antiport. J Biol Chem 259:5809–5815

Mazia D, Ruby A (1974) DNA synthesis turned on in unfertilized sea urchin eggs by treatment with NH_4OH. Exp Cell Res 85:167–172

Nishizuka Y (1983) Calcium, phospholipid turnover and transmembrane signalling. Philos Trans R Soc London B Biol Sci 302:101–112

Poenie M, Alderton J, Tsien RA, Steinhardt RA (1985) Nature 315:147–149

Schmidt T, Epel D (1983) High hydrostatic pressure and the dissection of fertilization responses. I. The relationship between cortical granule exocytosis and proton efflux during fertilization of the sea urchin egg. Exp Cell Res 146:235–248

Shen SS, Steinhardt RA (1978) Direct measurement of intracellular pH during metabolic depression at fertilization and ammonia activation of the sea urchin egg. Nature 272:253–254

Steinhardt RA (1982) Ionic logic in the activation of the cell cycle. In: Boynton A, McKeehan W, Whitfield W (eds) Ions, cell proliferation and cancer. Academic, New York, p 311

Steinhardt RA, Epel D (1974) Activation of sea-urchin eggs by a calcium ionophore. Proc Natl Acad Sci USA 71:1915–1919

Steinhardt RA, Mazia D (1973) Development of K^+-conductance and membrane potentials in unfertilized sea urchin eggs after exposure to NH_4OH. Nature 241:400–401

Steinhardt RA, Zucker R, Schatten G (1977) Intracellular calcium release at fertilization in the sea urchin egg. Dev Biol 58:185–196

Swann K, Whitaker M (1984) The tumour promoter TPA stimulates protein synthesis in sea-urchin eggs by activating the plasma membrane sodium-hydrogen ion exchange. J Physiol 353:86P

Turner P, Sheetz M, Jaffe LA (1984) Nature 310:414–415

Whitaker M, Aitchison M (1984) Nature (in press)

Whitaker M, Irvine R (1984) Nature 312:636–639

Whitaker MJ, Steinhardt RA (1981) The relation between the increase in reduced nicotinamide nucleotides and the initiation of DNA synthesis in sea urchin eggs. Cell 25:95–103

Whitaker MJ, Steinhardt RA (1982) Ionic regulation of egg activation. Q Rev Biophys 15:593–666

Winkler MM, Steinhardt RA (1981) Activation of protein synthesis in a sea urchin cell free system. Dev Biol 84:432–439

Winkler MM, Steinhardt RA, Grainger JL, Minning L (1980) Dual ionic controls for the activation of protein synthesis at fertilization. Nature 287:558–560

Winkler MM, Nelson E, Lashbrook C, Hershey J (1985) Dev Biol 107:290–300

Zucker RS, Steinhardt RA (1978) Prevention of the cortical reaction in fertilized sea urchin eggs by injection of calcium-chelating ligands. Biochem Biophys Acta 541:459–466

Regulatory Mechanisms of Intracellular pH in Excitable Cells

A. DE HEMPTINNE [1]

In the last decade, a number of technical developments have permitted to obtain reproducible, precise, and coherent measurements of the value of the intracellular pH (pH_i) in different cell types. Several reviews have been recently published in which techniques used to measure pH_i are compared and critically evaluated (Poole-Wilson 1978; Malan 1981; Roos and Boron 1981; Nuccitelli and Deamer 1982; Busa and Nuccitelli 1984; Thomas 1984).

The main conclusion one can draw from the growing number of papers which report measurements of pH_i in physiological and steady state condition, is that pH_i is close to neutrality, ranging usually between 7 and 7.5.

Considering the value of the extracellular pH [2] (pH_0) which is normally close to 7.4 and the negative value of the intracellular membrane potential (V) as measured in excitable cells, it appears that the equilibrium potential for H^+ (E_{H^+}) does not correspond to V in steady state conditions. pH_i is considerably more alkaline than expected from thermodynamic equlibrium distribution of the H^+ ions which can be calculated from the Nernst equation: $E_{H^+} = (pH_i - pH_0)$ 2.3 RT/F, where R, T, and F have their usual thermodynamic meaning.

In order (1) to maintain the relative alkaline steady state pH_i and (2) to recover the relative alkaline pH_i value after having perturbed pH_i in acid direction (Thomas 1984), the existence of an active acid extrusion mechanism has been postulated (Thomas 1976; Boron and De Weer 1976).

The ionic requirements for the operation of the postulated H^+ extrusion system have been extensively investigated in both invertebrate and vertebrate preparations.

In both snail neurones and barnacle muscle fibers, regulation of pH_i requires the presence of both Na^+ and HCO_3^- in the extracellular medium and also Cl^- in the intracellular compartment. Removal of one of these ions inhibits the pH_i regulation system (Thomas 1977; Boron et al. 1981). This is illustrated in the left panel of Fig. 1. Inhibition is also observed in the presence of the stilbene derivative SITS (Russell and Boron 1976; Thomas 1976; Boron 1977). These observations are consistent with a carrier model for acid extrusion as reproduced in the left panel of Fig. 1, of which different, but amost equivalent versions, have been presented. All models are compatible with electroneutrality in pH_i regulation and dependence on external Na^+, HCO_3^-, and internal Cl^- (Boron 1983).

---

1 Laboratory of Normal and Pathological Physiology, University of Gent, 9000 Gent, Belgium

2 pH_0 should be measured at the surface of the cells because in metabolically active preparations, surface pH can be significantly more acid than extracellular bulk pH (de Hemptinne 1980)

Circulation, Respiration, and Metabolism
(ed. by R. Gilles)

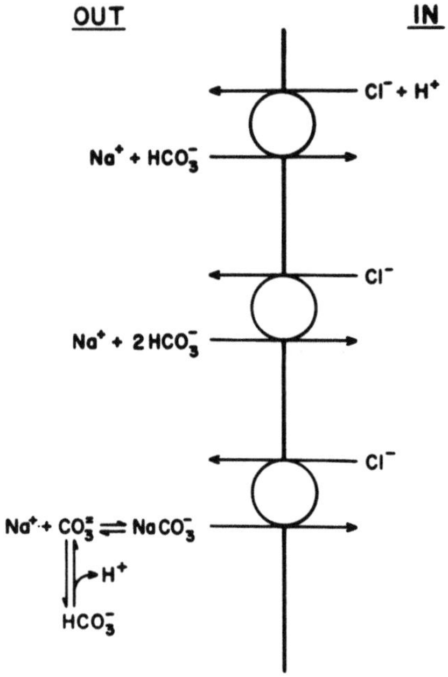

Fig. 1. *Right panel:* Acid extrusion rate expressed as a function of the extracellular Na^+ concentration in the presence of 10 m*M* (*top trace*) and 2.5 m*M* HCO_3^- (*bottom trace*), respectively. *Left panel:* Three models for electroneutral exchange of ions involved in pH_i regulation. (Reproduced from Boron et al. 1981; Roos and Boron 1981)

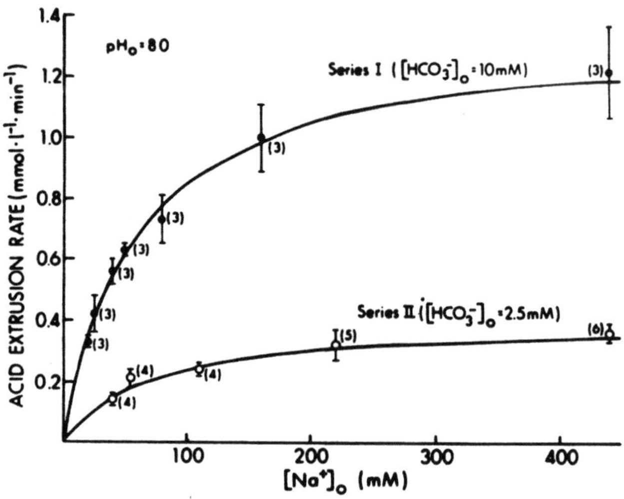

In mammalian preparations, such as sheep cardiac Purkinje fiber, acid extrusion is apparently affected neither by SITS, nor by the nominal absence of external HCO_3^- and internal Cl^- (Fig. 2) (Vaughan-Jones 1979; Vanheel et al. 1984). Removal of external Na^+ or addition of the diuretic amiloride (3,5-diamino-6-chloro-N-(diamino-methylene) pyrazine-carboxamide) inhibits the H^+ extrusion system as illustrated in Fig. 3 (Deitmer and Ellis 1980). This is compatible with a model for acid extrusion which exchanges extracellular Na^+ for intracellular H^+ in an electroneutral way. In the mammalian soleus muscle, besides the postulated Na^+-H^+ exchange, a SITS-sensitive Cl^-/HCO_3^- exchange seems to participate for about 20% in the acid extrusion mechanism (Aickin and Thomas 1977).

In sheep cardiac Purkinje fiber, experimental evidence has also been presented which indicates that the presence of external Cl^- is not necessary as a mediator for H^+ extrusion, but causes instead a continuous intracellular acid load. This is compatible with extrusion of intracellular HCO_3^-, or perhaps OH^- when HCO_3^- is not available, in exchange for external Cl^- (Vaughan-Jones 1979; Vanheel et al. 1984).

Recent experiments have provided indications of a close relationship between pH_i and pCa_i. An increase of the intracellular Ca^{2+} activity causes a fall of pH_i and vice versa (Bers and Ellis 1982). Current investigations are under way to test whether a possible "apparent" regulation of pH_i could rely on a precise and efficient regulation of pCa_i.

To what extent is the steady state value of pH_i influenced by metabolic production of acids, such as CO_2 and lactic acid? It is well documented that pH_i becomes acid

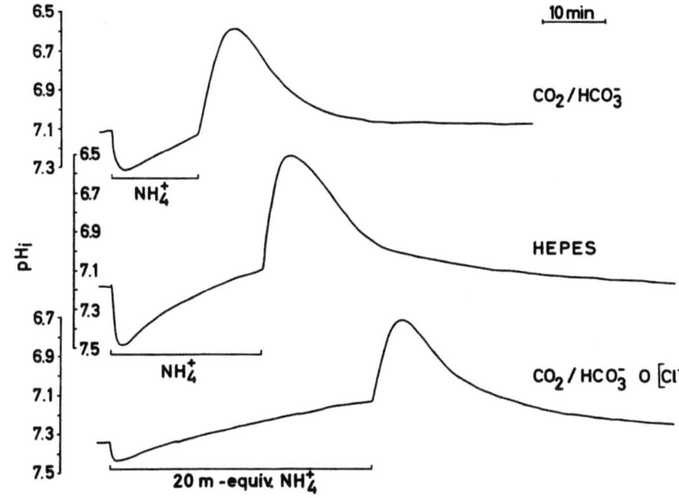

Fig. 2. Effect of addition and removal of NH_4^+ (20 m*M*) on pH_i in the presence of both HCO_3^- and Cl^- (*top trace*), in the nominal absence of HCO_3^- (superfusion solution buffered with HEPES) (*middle trace*) and in the absence of Cl^- (*bottom trace*). Notice the intracellular acidification caused by removal of NH_4^+ followed by the recovery process of pH_i which implies net acid extrusion

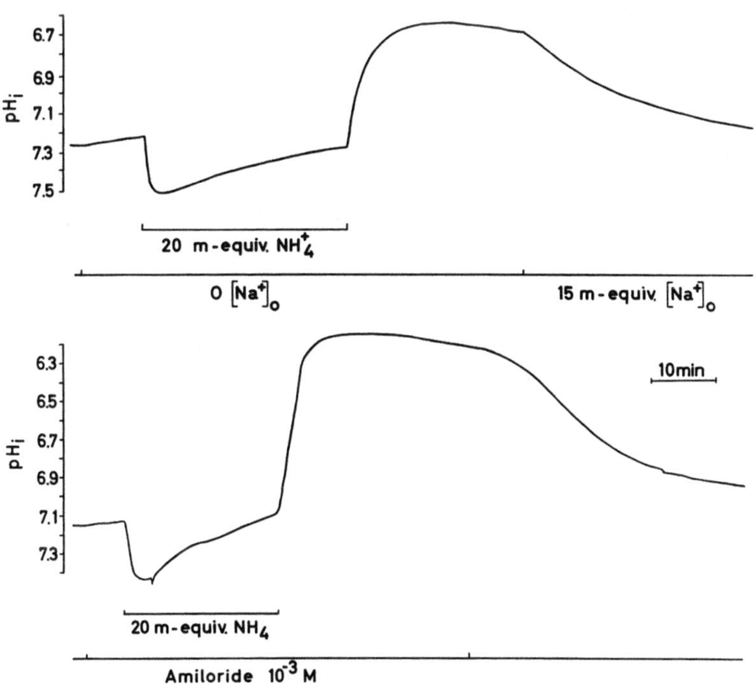

Fig. 3. Effect of addition and removal of NH_4^+ on pH_i in the absence of extracellular Na^+ (*top trace*), in the presence of amiloride 10^{-3} *M* (*bottom trace*). Notice the sustained intracellular acidification caused by removal of NH_4^+. Recovery from intracellular acidification is seen in the presence of 15 mEq. Na^+ and after removal of amiloride

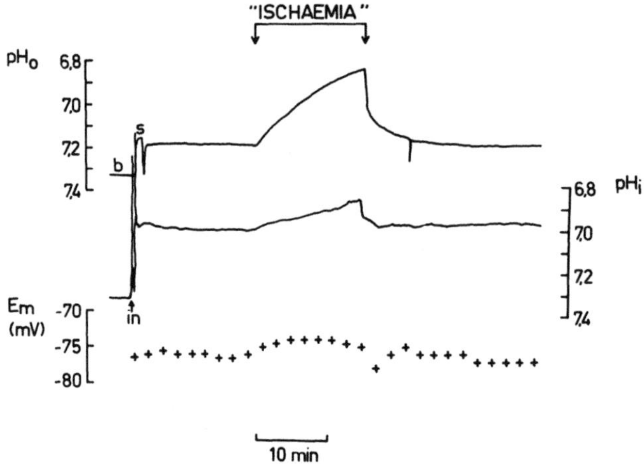

Fig. 4. Effect of simulated ischaemia in rat soleus muscle on extracellular pH, intracellular pH and transmembrane potential. Label *b* and *S* indicate, respectively, bulk solution and fiber surface. At *in* a muscle cell was impaled. (Reproduced from de Hemptinne and Huguenin 1984, Fig. 4)

when circulation of blood is impaired. Also in vitro conditions, simulating "ischemia" causes intracellular acidosis. This is also associated with extracellular surface acidosis when CO_2 and lactic acid accumulate in the interstitial compartment. In this condition, as shown in Fig. 4, extracellular acidosis can be much greater than the intracellular acidosis as expected from the greater buffer capacity of the intracellular compartment (de Hemptinne and Huguenin 1984).

When a muscle preparation is normally superfused, the cells are surrounded by an unstirred layer of fluid through which metabolically produced acids diffuse after leaving the cells. This accounts for the observation that the surface pH is slightly more acid than bulk pH (Dubuisson 1937; Caldwell 1958; Distèche 1960; de Hemptinne 1980). When muscle respiration is inhibited by cyanide, both the intracellular and surface compartments show a progressively increasing acidosis resulting from increased lactic acid production. Acidosis is, however, preceded by a transient surface and intracellular alkalosis (Fig. 5) which is presumably related to the immediate depression of CO_2 production. Blocking both respiration and glycolysis with cyanide and iodoacetic acid greatly reduces the pH gradient between the cell surface and the bulk solution (de Hemptinne and Huguenin 1984). Similar effects are also seen as a result of cooling.

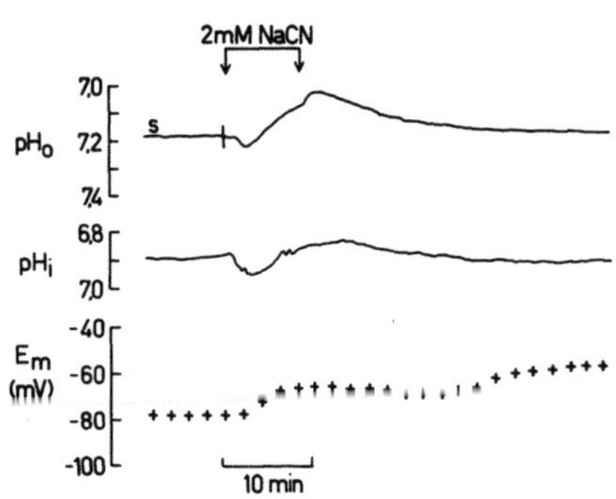

Fig. 5. Effect of NaCN on extracellular (surface) pH, intracellular pH and transmembrane potential measured in rat soleus muscle. (Reproduced from de Hemptinne and Huguenin 1984, Fig. 5)

References

Aickin CC, Thomas RC (1977) An investigation of the ionic mechanism of intracellular pH regulation in mouse soleus muscle fibres. J Physiol (Lond) 273:295–316

Bers DM, Ellis D (1982) Intracellular calcium and sodium activity in sheep heart Purkinje fibres. Effects of changes of external sodium and intracellular pH. Pflügers Arch 393:171–178

Boron W (1977) Intracellular pH transients in giant barnacle muscle fibers. Am J Physiol 233: C61–C73

Boron W (1983) Transport of H^+ and of ionic weak acids and bases. J Membr Biol 72:1–16

Boron WF, De Weer P (1976) Intracellular pH transients in squid giant axon caused by CO_2, NH_3 and metabolic inhibitors. J Gen Physiol 67:91–112

Boron W, McCormick W, Roos A (1981) pH regulation in barnacle muscle fibers: Dependence on extracellular sodium and bicarbonate. Am J Physiol 240:C80–C89

Busa WB, Nuccitelli R (1984) Metabolic regulation via intracellular pH. Am J Physiol 246:R409–R438

Caldwell PC (1958) Studies on the internal pH of large muscle and nerve fibres. J Physiol (Lond) 142:22–62

de Hemptinne A (1980) Intracellular and surface pH in skeletal and cardiac muscle measured with a double-barrelled pH microelectrode. Pflügers Arch 386:121–126

de Hemptinne A, Huguenin F (1984) The influence of muscle respiration and glycolysis on surface and intracellular pH in fibres of the rat soleus muscle. J Physiol (Lond) 347:581–592

Deitmer JW, Ellis D (1980) Interactions between the regulation of intracellular pH and sodium activity of sheep cardiac Purkinje fibres. J Physiol (Lond) 304:471–488

Distèche A (1960) Contribution à l'étude des échanges d'ions hydrogène au cours du cycle de la contraction musculaire. Mem Acad R Med XXXII, 1–69

Dubuisson M (1937) Untersuchungen über die Reaktionsänderung des Muskels im Verlauf der Tätigkeit. Pflügers Arch 239:314–326

Malan A (1981) L'Etat acide-base intracellulaire. J Physiol (Paris) 77:581–596

Nuccitelli R, Deamer D (1982) Intracellular pH: its measurement, regulation and utilization in cellular functions. Kroc Found Ser, vol 15. Alan Liss, New York

Poole-Wilson PA (1978) Measurement of myocardial intracellular pH in pathological states. J Mol Cell Cardiol 10:511–526

Roos A, Boron W (1981) Intracellular pH. Physiol Rev 61:296–434

Russell JM, Boron WF (1976) Role of chloride transport in regulation of intracellular pH. Nature 264:73–74

Thomas RC (1976) Ionic mechanism of the H pump in a snail neurone. Nature 262:54–55

Thomas RC (1977) The role of bicarbonate, chloride and sodium ions in the regulation of intracellular pH in snail neurones. J Physiol (Lond) 273:317–338

Thomas RC (1984) Review lecture: experimental displacement of intracellular pH and the mechanism of its subsequent recovery. J Physiol (Lond) 354:3P–22P

Vanheel B, de Hemptinne A, Leusen I (1984) Analysis of Cl^-–HCO_3^- exchange during recovery from intracellular acidosis in cardiac Purkinje strands. Am J Physiol 246:C391–C400

Vaughan-Jones RD (1979) Regulation of chloride in quiescent sheep heart Purkinje fibres studied using intracellular chloride and pH sensitive microelectrodes. J Physiol (Lond) 295:111–137

Symposium VII

Comparative Aspects of Adaptation to Cold

Organizer L.C.H. WANG

Seasonal Acclimation and Thermogenesis

G. HELDMAIER[1], H. BÖCKLER[1], A. BUCHBERGER[1], G.R. LYNCH[1,2], W. PUCHALSKI[1], S. STEINLECHNER[1], and H. WIESINGER[1]

1 Introduction

Low ambient temperatures and the shortage of food during winter are confronting all endotherms with severe challenges of thermoregulation and threaten their energy balance. In order to cope with cold mammals, birds, and endothermic insects must possess a capacity for thermoregulatory heat production large enough to compensate all heat lost to the environment. This is especially true for small mammals where thermal insulation of fur is rather limited, and they have to rely primarily on heat production (Scholander et al. 1950; Hart 1965).

From a great number of laboratory studies we know that the poor cold tolerance of small mammals does improve during several weeks of cold treatment (Hart and Heroux 1953; Hart 1971; Heroux 1961; Kreider 1972). Most of these studies assumed that similar improvements of cold tolerance may also occur in small mammals living in their natural environment during winter. However, very little information is available on seasonal acclimation of mammals in their natural environment or when they were kept under seasonally changing environmental conditions in the laboratory. Seasonal changes in cold tolerance and thermogenesis have so far only been studied in wild Norway rats (*Rattus norvegicus*, Hart and Heroux 1963), white-footed mice (*Peromyscus sp.*, Hart and Heroux 1953; Wickler 1980), in hares (*Lepus americanus*, Feist and Rosenmann 1975), in Arctic voles (*Clethrionomys rutilus*, Rosenmann et al. 1975; Feist and Morrison 1981), and in the Djungarian hamster (*Phodopus sungorus*, Heldmaier et al. 1982a).

Although only few species have been studied, the general concepts and physiological mechanisms of seasonal acclimation are discernable, and same basic questions can already be answered:

1. Which physiological mechanism is responsible for seasonal changes in cold tolerance?
2. Which environmental signals are used for control of seasonal acclimation?
3. Which endocrine or neural mechanisms are involved in seasonal acclimation, i.e., are conveying environmental signals to the thermoregulatory system?

1 Fachbereich Biologie, Philipps-Universität, 3550 Marburg, FRG
2 Present address: Department of Biology, Wesleyan University, Middletown, CT 06457, USA

Circulation, Respiration, and Metabolism
(ed. by R. Gilles)

2 Cold Tolerance and Thermogenesis

The most detailed information is available for the Djungarian hamster, where physiological and biochemical mechanisms of seasonal acclimation have been studied extensively. This species proved to be an ideal object for studies on seasonal acclimation. It inhabits the Siberian steppe, an area with most extreme seasonal changes in ambient temperature, often being higher than +35° in the summer and falling below - 35 °C during winter. Furthermore, this species shows rather obvious seasonal changes in fur coloration, body weight, and reproductive activity, thus, allowing rapid assessment of its seasonal state.

To determine the cold tolerance of Djungarian hamsters we placed them in a climate chamber at thermoneutrality and lowered ambient temperature in 10 °C steps until the cold limit was reached (smaller steps close to the cold limit). At each temperature level the hamsters were kept for about 1 h to obtain resting metabolic rate. The lowest ambient temperature which could be reached was - 83 °C. During this exposure heat production, body temperature (by radio transmitter), and the electromyogram of hamsters was recorded continuously and the cold limit was defined as the final ambient temperature at which the hamsters no longer increased heat production (HPmax = maximum thermoregulatory heat production) and became hypothermic. As can be seen from two individual records in Fig. 1 the cold limit of the Djungarian hamster living at thermoneutrality during summer was found at - 24 °C, whereas the winter acclimated hamster could tolerate ambient temperature below - 60 °C [mean values - 26.1 + - 2.2 (n = 7) and - 68.5 + - 1.9 (n = 8), respectively, see Table 1].

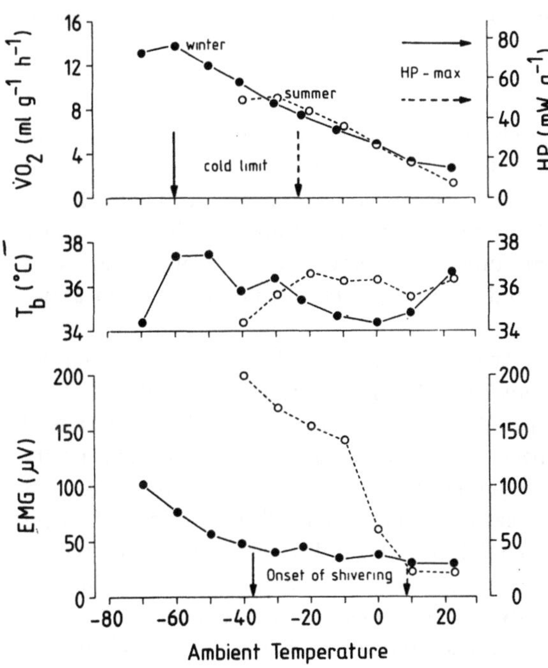

Fig. 1. Heat production, body temperature, and shivering of a summer (○) and a winter (●) acclimated Djungarian hamster, *Phodopus sungorus*, during cold tolerance test. HPmax = maximum thermoregulatory heat production of resting hamsters. The cold limit was calculated as the intercept of HPmax with the slope of heat production. The onset of shivering was calculated as the intercept of the slope in EMG-integral with the basal level of EMG-integral. Body temperature was recorded with intraperitoneally implanted transmitters

Table 1. Seasonal changes of maximum thermoregulatory heat production (HPmax) and the cold limit in small mammals

	Body mass (g)	HPmax ($mW\ g^{-1}$)	Cold limit (°C) T_a	Ref.
Peromyscus maniculatus				
Summer	21.3	62.0[a]	- 12	Hart and Heroux 1953
Winter	19.7	87.0[a]	- 32	
Peromyscus leucopus				
Summer	19	63.8	- 5	Wickler 1980
Winter	18	107.8[b]	- 40[b]	
Clethrionomys rutilus				
Summer	28	74.3[b]	- 40[b]	Rosenmann et al. 1975
Winter	15	145.2[b]	- 74[b]	
Phodopus sungorus[d]				
Summer	41.1	45.5	- 26.1	This study
Winter	26.3	94.6	- 68.5	
Rattus norvegicus				
Summer	250	20.4	- 20	Hart and Heroux 1963
Winter	281	30.3	- 45	
Lepus americanus				
Summer	1,544	18.7[c]	- 74[c]	Feist and Rosenmann 1975
Winter	1,506	23.1[c]	-157[c]	

[a] Estimated from conductance and metabolic rate at 1–2 °C T_a
[b] HPmax was measured during short exposures at –7° to –11 °C in a helium/oxygen mixture and the cold limit was extrapolated for conditions in air
[c] In *Lepus* HPmax was further increased by shaving, oiling, or wetting the fur
[d] Data reported for *Phodopus* are similar to previous reports (Heldmaier et al. 1982a), but were repeated for this study with an improved climate chamber allowing cold exposures as low as –83 °C

Similar changes in cold tolerance have been observed in other species of small mammals (Table 1). Further direct measurements, i.e., cold exposure in air, were performed only in *Peromyscus* and *Rattus*, and both species showed a less pronounced cold tolerance as compared to *Phodopus*. All other data were extrapolated from moderate cold exposures in a helium/oxygen mixture which may produce more exaggerated values for HPmax and cold limit as compared to resting metabolic rate in direct measurements.

The primary reason for the improved tolerance is an extended capacity for thermoregulatory heat production, and in *Phodopus* HPmax increased from about 50 to 80 $mW\ g^{-1}$ in the two individuals compared in Fig. 1 [mean values 45.5 + - 2.9 (n = 7) and 94.6 + - 3.5 (n = 8) $mW\ g^{-1}$, respectively]. This seasonal increase in the capacity for thermoregulatory heat production is primarily due to nonshivering thermogenesis (NST), the predominant pathway for heat production in small mammals (Heldmaier 1971). The NST maximum was found to be 30.0 + - 1.9 $mW\ g^{-1}$ in summer and 67.3 + - 3.5 $mW\ g^{-1}$ in winter-acclimated hamsters as determined by the noradrenaline test (Böckler et al. 1982). The increase in NST by 37.3 $mW\ g^{-1}$ may explain 80% of the total increase in HPmax during winter. A similar coincidence between seasonal changes in NST and HPmax has also been described for *Clethrionomys rutilus* (Feist and Morrison 1981).

Small mammals preferably use NST during cold exposure and shivering thermogenesis is only activated when the cold load exceeds the heating capacity of NST (Brück and Wünnenberg 1966; Jansky 1973). This is also valid for seasonal acclimation. EMG records in Fig. 1 show that at a moderate cold load heat was first produced preferably by NST and shivering started at +8 °C in the summer-acclimated and at - 38 °C in the winter-acclimated hamster [mean values for onset of shivering 6.2 + - 1.5 (n = 7) and - 34.2 + - 1.6 (n = 16) °C T_a] (Böckler and Heldmaier 1983).

The effect of the greater heating capacity on cold tolerance in winter-acclimated hamsters is assisted by reduced body weight and by a voluntary hypothermia at moderate cold load. The latter is demonstrated by T_b records in Fig. 1. The summer-acclimated hamster maintained a body temperature of 36 °C throughout the entire range of cold exposure except a drop in T_b at the cold limit. The winter acclimated hamster lowered his body temperature to 34 °C during moderate cold exposures, but when T_a fell below - 20 °C it increased its body temperature to about 38 °C until the cold limit was reached. The same pattern of thermoregulatory changes in T_b was observed in all winter-acclimated hamsters and mean values for T_b in moderate cold were 34.6 + - 0.3 (0 °C T_a) and in severe cold 36.9 + - 0.4 °C (- 50 °C T_a). This slight hypothermia in moderate cold reduces heat loss to the environment and may save about 15% of heat requirements for thermoregulation.

Seasonal reduction of body mass from 40.7 to 27.9 g for summer to winter, respectively, plays an important role for cold tolerance in the hamsters reported above. Reductions of body weight in winter were found in other species of small mammals, too (Table 1), although changes are less pronounced than in *Phodopus*. At first glance this sounds paradoxical, since smaller-sized mammals are more sensitive to cold and may store less energy reserves than larger mammals. However, this is only partly true in Djungarian hamsters, since thermal insulation of winter fur is greatly improved, thus, compensating the effect of the enlarged relative surface area due to their smaller mass (Heldmaier and Steinlechner 1981a). Therefore, a smaller-sized body must be heated and even with the same capacity for heat production small winter-acclimated hamsters would tolerate much lower ambient temperatures. All data on heat production presented before were given per unit body weight to compensate for changes in body weight. If total heat production per animal is calculated the seasonal differences are slightly less, but still significantly elevated during winter, i.e., NST maximum is 1,216 + - 103 mW during summer and increases to 1,864 + - 78 mW in winter. Thus, the smaller-sized body of winter-acclimated hamster is equipped with a greater total potential for NST and the combination of these properties is responsible for the admirable cold tolerance of Djungarian hamsters.

The most obvious organ site for NST is the brown adipose tissue (BAT). However, since BAT amounts to only a few percent of body mass its quantitative contribution to total NST has been a matter of controversy and estimates ranged between 10% and 80% of total NST (Smith and Horwitz 1969; Jansky 1973; Foster and Frydman 1978). In vivo measurements of blood flow in rats with radioactive labeled microspheres indeed supported the view that up to 80% of total NST may be produced by BAT (Foster and Frydman 1978). This is supported by measurements of blood flow in the Djungarian hamster, and our data suggest that the quantitative contribution largely depends upon the state of acclimation.

Hamsters living at thermoneutrality increase blood flow in BAT by 3.02 ml min^{-1} (X 18) and reduced blood flow through intestinal organs and muscle tissue during NST. Hamsters kept outdoors are redistributing blood flow in a similar manner, but further increase cardiac output from 24.3 to 38.8 ml min^{-1} and this excess cardiac output is mainly used for blood supply to BAT [blood flow increased by 9.06 ml min^{-1} (X 28)]. These data suggest that in vivo metabolic rate of BAT is increasing during NST 28 and 48 times, respectively; indicating that about two-thirds of total NST is concentrated in BAT of cold-adapted hamsters, whereas in non-cold adapted hamsters BAT contributes only one-third to total NST (Puchalski et al., unpublished). Similar percentages were also obtained in experiments with partial removal of BAT, where NST was measured before and after the operation, thus, allowing an estimate of the contribution of BAT during NST (Heldmaier et al. 1983).

These findings suggest that changes in thermogenic capacity are based on an improved heating capacity of BAT, whereas other sources of heat remain unchanged. BAT can, thus, be regarded as the organ site for seasonal thermogenic acclimation and we should expect seasonal changes of its biochemical and thermogenic properties. The content of mitochondria in BAT increased from about 12 mg mitochondrial protein per total BAT in hamsters living at thermoneutrality during summer to about 91 mg mitochondrial protein in winter-acclimated hamsters (Rafael et al. 1981). This sevenfold increase is larger than the increase in total NST, illustrating that seasonal thermogenic improvements are indeed concentrated in BAT.

3 Environmental Cues for Thermogenic Acclimation

Several weeks of chronic cold exposure or treatment with intermittent cold exposures in the laboratory caused an improvement of NST in all newborn and adult mammals investigated so far (for a review see Smith and Horwitz 1969; Chaffee and Roberts 1971; Heldmaier 1971; Jansky 1973). This suggests that decreasing ambient temperatures during autumn might be adequate to initiate seasonal improvements of thermogenic capacity. However, in the temperate and subantarctic climate zones the ambient temperature is rather variable and, thus, not a highly reliable environmental signal for seasonal acclimation. During summer and early fall constant high ambient temperatures may be followed by drop in T_a within a few days. Such sudden outbreaks of cold are too fast for thermogenic adaptation, which always requires about 3 weeks for completion (Jansky et al. 1967), and it needs to occur only once to wipe out the population of small mammals. Therefore, it seems reasonable to use more reliable environmental signals for seasonal acclimation, as may be provided by seasonal changes in photoperiod.

It has been demonstrated in several species of rodents that an exposure to short photoperiod at thermoneutrality stimulates BAT or enhances the capacity for NST, e.g., in *Mesocricetus auratus* (Hoffman et al. 1965), *Peromyscus leucopus* (Lynch 1973; Lynch and Epstein 1976; Lynch et al. 1978), *Phodopus sungorus* (Heldmaier and Hoffmann 1974; Heldmaier et al. 1981b), *Praomys natalensis* and *Rhabdomys pumilio* (Haim and Fourie 1980), and *Microtus ochrogaster* (Wunder, in press). In white rats and in the red-backed vole *Clethrionomys rutilus* apparently no thermotrophic response to short photoperiod was found (Hagelstein and Folk 1978; Feist and Morrison 1981; Kott and Horwitz 1982). However, rats and voles were kept in L:D 12:12 prior to any experiment, which is close to or already below the critical photoperiod and may have interfered with a possible thermotrophic response to short photoperiod. In contrast to this lacking response at room temperature, a thermotrophic response of rats to short photoperiod was observed when they were kept in a cold environment (Hagelstein and Folk 1978).

In *Phodopus sungorus* 8 weeks of exposure to a short photoperiod (8:12 L:D) increased total NST capacity to about 1,000 mW as compared to 731 mW in hamsters living in long photoperiod (Heldmaier et al. 1981b). This was accompanied by an increase in respiratory properties of BAT, whereby the activity of cytochrome oxidase in BAT and the total amount of mitochondria in BAT increased by nearly 200% (Table 2). The amplitude of these changes was about the same as observed in hamsters exposed to seasonal changes in natural photoperiod (Heldmaier et al. 1982a; Steinlechner and Heldmaier 1982; Rafael et al. 1981).

This clearly shows that photoperiod is effective on the molecular basis of heat generation in BAT. A dominant portion of heat in BAT is generated by uncoupled oxidation in mitochondria, which are exclusively equipped with a heat dissipating protein (GDP-binding protein, Thermogenin, Cannon et al. 1978; Desautels and Himms-Hagen 1979; Cannon et al. 1982; Nicholls and Locke 1984). Table 2 shows that the occurrence of total thermogenin is also enhanced in Djungarian hamsters exposed to short photoperiod, concluding that heat generating properties are specifically controlled by the photoperiod.

Table 2. Effects of photoperiod and brain implants with melatonin on nonshivering thermogenesis, biochemical and adrenergic properties of BAT in Djungarian hamsters kept at thermoneutrality

	Long photoperiod (16:8 L:D)	+ Melatonin implants	Short photoperiod (8:16)
NST capacity (mW)	731 +- 55	1,026 +- 45	1,084 +- 94
($mW\ g^{-1}$)	19.3 +- 1.6	27.6 +- 1.8	39.3 +- 3.3
Mitochondrial protein in total BAT (mg)	7.8 +- 1.1	20.6 +- 2.8	21.6 +- 2.3
Activity of cytochrome oxidase in total BAT (U)	47.5 +- 9.4	106.3 +- 13.5	113.3 +- 108
GDP binding of BAT mitochondria (pmol GDP mg^{-1} mito.prot.)	273 +- 25	830 +- 46	307 +- 24
(nmol GDP/total BAT)	2.3 +- 0.4	18.2 +- 2.8	7.1 +- 1.1
Tyrosinehydroxylase in total BAT (nmol l-Dopa h^{-1})	20.5 +- 1.7	40.2 +- 3.8	30.3 +- 4.3
Noradrenaline content of total BAT (ng)	1,209 +- 129	1,980 +- 237	1,176 +- 86

The thermotrophic action of short photoperiod is qualitatively identical to seasonal acclimation or to cold adaptation in every detail investigated so far. Quantitatively thermogenic improvements by short photoperiod were about 40% to 60% of the changes observed in seasonal acclimation or during laboratory cold adaptation (Heldmaier et al. 1982b). The seasonal shortening of the photoperiod may, thus, provide a signal for seasonal thermogenic adaptation, prior to any experience of winter cold. Subsequent cold exposures may further improve thermogenic capacity to obtain maximum cold tolerance during midwinter. This suggests a substitutive role of photoperiod and cold, whereby both can act as environmental cues for seasonal acclimation, but the thermotrophic action of short photoperiod in fall simply anticipates the adaptive action of winter cold.

However, at the present point of knowledge we cannot completely exclude an interaction between both cues. We found a greater NST capacity in Djungarian hamsters adapted to cold in short photoperiod as compared to cold adaptation in long photoperiod (Heldmaier et al. 1981b). This is most obvious when the effect of a chronic exposure to moderate cold (15 °C) is compared in both photoperiods. It is further supported by the fact that the cold limit of cold-adapted hamsters during summer (natural photoperiod) was to be - 49 °C, whereas in natural short photoperiod in winter the cold limit was - 68 °C. This suggests that besides the substitutive role of both environmental cues, the short photoperiod may further facilitate thermogenic adaptation to cold.

4 Endocrine Mechanisms Involved in Seasonal Control of Thermogenesis

The influence of the photoperiod on reproduction is transmitted via the pineal gland. This gland synthesizes and releases melatonin mainly during the hours of darkness and this synthesis is inhibited when light falls on the eyes. Thus, the amount or diurnal pattern of melatonin may provide a suitable endocrine signal for the photoperiod, whereby more melatonin is to be expected in short photoperiod (for review see Hoffmann 1979; Reiter 1980; Goldman 1983). Injections or implantations of exogenous melatonin may, thus, mimic the effect of short photoperiod, and cause an improvement of thermogenic capacity. This has been shown in Djungarian hamsters, where a chronic release of melatonin from subcutaneous implants or daily injections with melatonin caused the same increase in NST and respiratory properties of BAT as observed during exposure to short photoperiod (Heldmaier et al. 1981b; Holtorf et al., unpublished). These experiments further showed that not merely a large amount of melatonin will cause a thermotrophic response, but the diurnal distribution patterns is of critical importance since daily injections of melatonin were only effective when given in the late afternoon, whereas injections of melatonin in the morning or at noon were ineffective (Holtorf et al., unpublished). Similar diurnal changes in sensitivity towards melatonin were also reported for its effect on gonadal activity (Tamarkin et al. 1976; Richardson et al. 1981).

The diurnal pattern of sensitivity towards melatonin as well as the unsuccessful attempts to demonstrate receptors for melatonin in BAT, suggest that this hormone

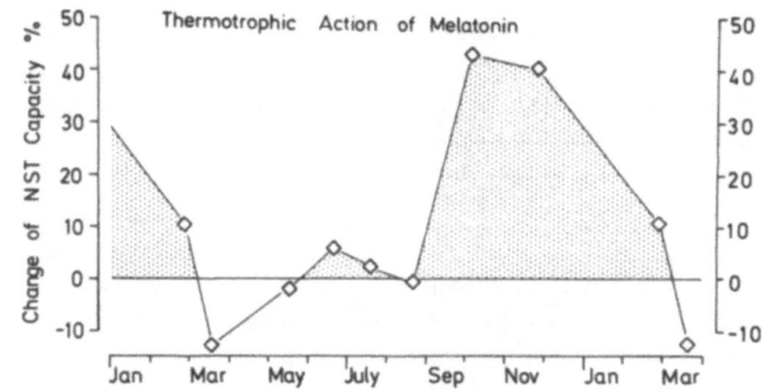

Fig. 2. Djungarian hamsters, *Phodopus sungorus*, were implanted with melatonin at different times of the year (silastic capsules, subcutaneously, Steinlechner and Heldmaier 1982). Immediately before the implantation they were transferred from natural photoperiod to an artificial long period (L:D 16:8). Nonshivering thermogenesis was measured after 8 weeks of treatment with melatonin and presented as percent of the initial value

does not act on peripheral thermogenic organs directly. This view is also supported by the fact that the thermotrophic effect of melatonin strongly depends upon the photoperiodic history of the experimental animals (Steinlechner and Heldmaier 1982). We kept Djungarian hamsters in seasonally changing photoperiod at thermoneutrality and implanted hamsters with melatonin at different times of the year (Fig. 2). Melatonin was ineffective at those times of the year when the hamsters were refractory to short photoperiod, and only during fall when hamsters were sensitive to short photoperiod (by previous experience of long photoperiod) we could elicit a thermotrophic response. Therefore, it appears more likely to assume that melatonin is acting on central structures and, thus, indirectly influences peripheral organs through a more complex pathway of signal transmission. Lynch and co-workers have shown that small amounts of melatonin implanted into the hypothalamic region of *Peromyscus leucopus* cause gonadal regression like short photoperiod. The same amount of melatonin given peripherally was ineffective, suggesting that the action site of melatonin for photoperiodic time measurement is to be located in the brain (Glass and Lynch 1981).

We used such brain implants of melatonin to test for the action site of melatonin for thermogenic adaptation. Djungarian hamsters living in long photoperiod received stainless steel cannula containing 20 μg of melatonin suspended in bees wax, stereotaxically implanted into the hypothalamic area. This treatment in fact elevated NST capacity and mitochondrial protein in BAT to the same extent as short photoperiod (Table 2). However, it had a specific and exaggerating effect on GDP binding of BAT mitochondria which increased to about 830 pmol mg^{-1} mitochondrial protein. These values were higher than in 5 °C cold exposed hamsters and were so far only found when hamsters were chronically exposed to severe cold at − 5 °C (unpublished observation). A similar increase was observed in the activity of tyrosinehydroxylase, the rate-limiting enzyme for the synthesis of catecholamines. The content of catecholamines in BAT was largely enhanced in hamsters treated with melatonin brain implants, although we previously observed no change in the level of catecholamines during cold adaptation or photoperiod acclimation (Table 2). Obviously, central stimulation by

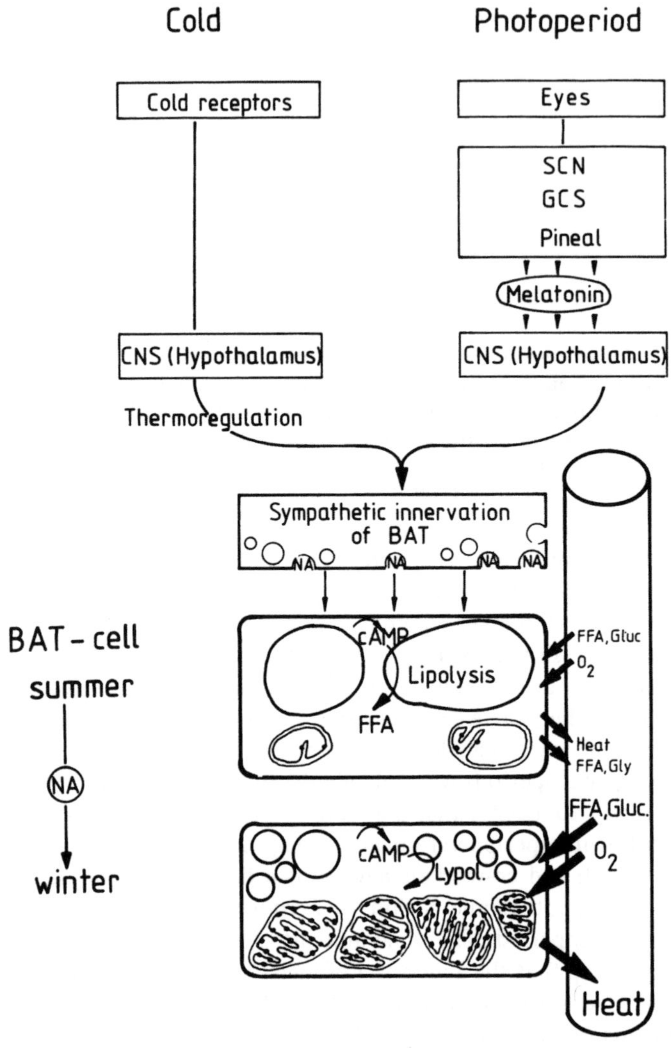

Fig. 3. Pathway of environmental signal transmission for seasonal thermogenic acclimation. Cold is sensed by cold receptors and via the usual pathway for acute control of thermoregulatory heat production in brown adipose tissue is stimulated through its sympathetic innervation for nonshivering thermogenesis (NST) (β-adrenergic receptors, lipolysis, and heat dissipation from uncoupled oxidation in mitochondria). Permanent or repeated use of this pathway will improve thermogenic properties of BAT cells ("capacity training"). For photoperiodic time measurement light is sensed by the retina and conveyed through the retinohypothalamic tract, in conjunction with the suprachiasmatic nucleus (SCN), to the superior cervical ganglion (GCS), innervating the pineal gland. Melatonin produced according to the lighting condition stimulates, via some unknown central structures, the sympathetic innervation of BAT cells, thus, causing improvement of their thermogenic properties

melatonin effected synthesis and turnover of catecholamines in BAT and had an exaggerating effect on GDP binding of mitochondria. The latter effect is of a rather peculiar specificity since only the GDP binding of mitochondria was maximal, whereas the amount of mitochondria in BAT showed only a moderate increase following melatonin, as seen in short photoperiod.

This indicates that the thermotrophic response to short photoperiod and melatonin is transmitted via the sympathetic innervation of BAT. The same pathway is also suggestive for the thermogenic improvements during cold adaptation. Chronic release of noradrenaline from miniosmotic pumps implanted into rats or pharmacological long-term stimulation of β-adrenergic receptors caused biochemical changes in BAT comparable to cold adaptation (Mory et al. 1984). Short daily cold exposures and daily injections of noradrenaline with the same acute effects on thermogenesis caused identical thermogenic adaptation (Heldmaier and Jablonka, unpublished). This is in contrast to earlier experiments where treatment with noradrenaline was not found as effective as compared to cold adaption (LeBlanc and Villemaire 1970), but this difference may be due to the circumstance that chronic cold exposure and daily injections of noradrenaline, i.e., qualitatively and quantitatively differing treatments, were compared. We could further demonstrate that catecholamines are also involved in the thermotrophic response to short photoperiod, since short days as well as melatonin enhanced the synthesis of catecholamines in the adrenals and BAT (Buchberger et al. 1983, Table 2).

From the present evidence we suggest that seasonal thermogenic acclimation in small mammals is controlled by a cooperative cueing of photoperiod and ambient temperature, as described above. The molecular basis of this acclimation is an increased amount of mitochondrial protein and a better equipment of these mitochondria with the GDP-binding protein which facilitates heat dissipation by uncoupled oxidation (Fig. 3). The endocrine and/or neural mechanisms by which the environmental information is conveyed to BAT are not yet fully understood. The coincidence between photoperiodic control of reproduction and thermogenesis suggests a common mechanism, i.e., light is perceived through the retina, and retinal information is conveyed through fibers of the retinohypothalamic tract in conjunction with the suprachiasmatic nucleus to the superior cervical ganglion, which in turn sends fibers back into the brain innervating the pineal gland. The light-dependent synthesis and release of melatonin (or peptides) from this gland stimulates brain structures in the hypothalamic area which may activate synthesis and turnover of catecholamines in the sympathetic innervation of BAT.

Environmental cold is measured by peripheral (or central) cold receptors, and this information is integrated in thermoregulatory areas in the hypothalamus, causing an acute stimulation of BAT. This acute stimulation is also transmitted by the sympathetic innervation to BAT, there increasing turnover and synthesis of catecholamines. The final link of thermotrophic stimulation appears to be common for photoperiod and cold adaptation. Whether it is already combined at the central level and the same brain structures are involved in control of photoperiod and temperature acclimation has to be clarified by further research.

Acknowledgment. This research was supported by the Deutsche Forschungsgemeinschaft, Schwerpunktprogramm „Temperaturregulation und Adaptation" (He 990).

References

Böckler H, Heldmaier G (1983) Interaction of shivering and nonshivering thermogenesis during cold exposure in seasonally acclimatized Djungarian hamsters (Phodopus sungorus). J Therm Biol 8:97–98

Böckler H, Steinlechner St, Heldmaier G (1982) Complete cold substitution of noradrenaline-induced thermogenesis in the Djungarian hamster. Phodopus sungorus. Experienta 38:261–262

Brück K, Wünnenberg W (1966) Beziehung zwischen Thermogenese im „braunen" Fettgewebe, Temperatur im cervicalen Anteil des Vertebralkanals und Kältezittern. Pflügers Arch 290:167–183

Buchberger A, Heldmaier G, Steinlechner St, Latteier B (1983) Photoperiod and temperature effects on adrenal tyrosinehydroxylase and its relation to nonshivering thermogenesis. Pflügers Arch 399:79–82

Cannon B, Nedergaard J, Romert L, Sundin U, Svartengren J (1978) The biochemical mechanism of thermogenesis in brown adipose tissue. In: Wang LCH, Hudson JW (eds) Strategies in cold. Academic, New York, pp 567–617

Cannon B, Hedin A, Nedergaard J (1982) Exclusive occurence of thermogenin antigen in brown adipose tissue. Febs Lett 150:129–132

Chaffee RRJ, Roberts JC (1971) Temperature acclimation in birds and mammals. Annu Rev Physiol 33:155–202

Desautels M, Himms-Hagen J (1979) Roles of noradrenaline and protein synthesis in the cold-induced increase in purine nucleotide nucleotide binding by rat brown adipose tissue mitochondria. Can J Biochem 57:968–976

Feist DD, Morrison PR (1981) Seasonal changes in metabolic capacity and norepinephrine thermogenesis in the Alaskan red-backed vole: environmental cues and annual differences (1981). Comp Biochem Physiol 69A:697–700

Feist DD, Rosenmann M (1975) Seasonal sympatho-adrenal and metabolic responses to cold in the Alaskan snowshoe hare (Lepus americanus macfarlani). Comp Biochem Physiol 51A:449–455

Foster DD, Frydman ML (1978) Nonshivering thermogenesis in the rat. II. Measurements of blood flow with microspheres point to brown adipose tissue as the dominant site of calorigenesis induced by noradrenaline. Can J Physiol Pharmacol 56:110–122

Glass JD, Lynch GR (1981) Melatonin: identification of sites of antigonadal action in mouse brain. Science 124:821–823

Goldman BD (1983) The physiology of melatonin in mammals. In: Reiter R (ed) Pineal research reviews. Allen R. Liss, New York, 1:145–182

Hagelstein KA, Folk GF Jr (1978) Effects of photoperiod, cold acclimation and melatonin on the white rat. Comp Biochem Physiol 62C:225–229

Haim A, Fourie FR (1980) Heat production in cold and long scotophase acclimated and winter acclimated rodents. Int J Biometeorol 24:231–236

Hart JS (1965) Seasonal changes in insulation of the fur. Can J Zool 34:53–57

Hart JS (1971) Rodents. In: Whittow GC (ed) Comparative physiology of thermoregulation, vol II. Academic, New York, pp 1–149

Hart JS, Heroux O (1953) A comparison of some seasonal and temperature-induced changes in Peromyscus: cold resistance, metabolism, and pelage insulation. Can J Zool 31:528–534

Hart JS, Heroux O (1963) Seasonal acclimatization in wild rats (Ratus norvegicus). Can J Zool 41: 711–716

Heldmaier G (1971) Zitterfreie Wärmebildung und Körpergröße bei Säugetieren. Z Vgl Physiol 73: 222–248

Heldmaier G, Hoffmann K (1974) Melatonin stimulates growth of brown adipose tissue. Nature 247:224–225

Heldmaier G, Steinlechner St (1981a) Seasonal control of energy requirements for thermoregulation in the Djungarian hamster (Phodopus sungorus), living in natural photoperiod. J Comp Physiol 142:429–437

Heldmaier G, Steinlechner St, Rafael J, Vsiansky P (1981b) Photoperiodic control and effects of melatonin on nonshivering thermogenesis and brown adipose tissue. Science 212:917–919

Heldmaier G, Steinlechner St, Rafael J (1982a) Nonshivering thermogenesis and cold resistance during seasonal acclimatization in the Djungarian hamster. J Com Physiol 149:1–9

Heldmaier G, Steinlechner St, Rafael J, Latteier B (1982b) Photoperiod and ambient temperature as environmental cues for seasonal thermogenic adaptation in the Djungarian hamster, Phodopus sungorus. Int J Biometeorol 26:339–345

Heldmaier G, Buchberger A, Seidl K (1983) Contribution of brown adipose tissue to thermoregulatory heat production in the Djungarian hamster. J Therm Biol 8:413–415

Heroux O (1961) Comparison between seasonal and thermal acclimation in white rats. V. Metabolic and cardiovascular response to noradrenaline. Can J Biochem Physiol 39:1829–1836

Hoffman RA, Hester RJ, Towns C (1965) Effect of light and temperature on the endocrine system of the golden hamster, Mesocricetus auratus waterhouse. Comp Biochem Physiol 15:525–533

Hoffmann K (1979) Photoperiod, pineal, melatonin and reproduction in hamsters. In: Kappers JA, Pevet P (eds) The pineal gland of vertebrates including man, vol 52. Elsevier, North-Holland Biomedical, pp 397–415

Jansky L (1973) Nonshivering thermogenesis and its thermoregulatory significance. Biol Rev 48: 85–132

Jansky L, Bartunkova R, Zeisberger E (1967) Acclimation of the white rat to cold: noradrenaline thermogenesis. Physiol Bohemoslov 16:366–371

Kott KS, Horwitz BA (1982) Photoperiod and pinealectomized do not affect cold-induced deposition of brown tissue in the long-evans rat. Cryobiology 20:100–105

Kreider MB (1972) Stimulus for metabolic acclimation to cold: intensity versus duration. Proc Int Symp Environ Physiol (Bioenergetics), FASEB

LeBlanc J, Villemaire A (1970) Thyroxine and noradrenaline on noradrenaline sensitivity, cold resistance, and brown fat. Am J Physiol 218:1742–1745

Lynch GR (1973) Seasonal changes in thermogenesis, organ weights, and body composition in the white-footed mouse, Peromyscus leucopus. Oecologia 13:363–376

Lynch GR, Epstein AL (1976) Melatonin induced changes in gonads, pelage and thermogenic characters in the white-footed mouse, Peromyscus leucopus. Comp Biochem Physiol 53C:67–68

Lynch GR, White SE, Grundel R, Berger MS (1978) Effects of photoperiod, melatonin administration and thyroid block on spontaneous daily torpor and temperature regulation in the white-footed mouse, Peromyscus leucopus. J Comp Physiol 125:157–163

Mory G, Bouillaud F, Combes-George M, Ricquier D (1984) Noradrenaline controls the concentration of the uncoupling protein in brown adipose tissue. FEBS Lett 166:393–397

Nicholls DG, Locke RM (1984) Thermogenic mechanisms in brown fat. Physiol Rev 64:1–64

Rafael J, Vsiansky P, Heldmaier G (1981) Adaptive changes in brown adipose tissue of Djungarian hamsters. Acta Univ Carol Biol 1979:327–330

Reiter RJ (1980) The pineal gland: a regulator of regulators. Prog Psychobiol Physiol Psychol 9: 323–356

Richardson BA, Vaughan MK, Brainard GC, Huerter JJ, De los Santos, Reiter R (1981) Influence of morning injections on the antigoandotrophic effects of afternoon melatonin administration in male and female hamsters. Neuroendocrinology 33:112–117

Rosenmann M, Morrison P, Feist D (1975) Seasonal changes in the metabolic capacity of red-backed voles. Physiol Zool 48:303–310

Scholander PF, Hock R, Walters V, Johnson F, Irving L (1950) Heat regulation in some arctic and tropical mammals and birds. Biol Bull 99:237–257

Smith RE, Horwitz BA (1969) Brown fat and thermogenesis. Physiol Rev 49:330–425

Steinlechner St, Heldmaier G (1982) Role of photoperiod and melatonin in seasonal acclimatization of the Djungarian hamster, Phodopus sungorus. Int J Biometeorol 26:329–337

Tamarkin L, Westrom WK, Hamill AI, Goldman BD (1976) Effect of melatonin on the reproductive systems of male and female Syrian hamsters: a diurnal rhythm in sensitivity to melatonin. Endocrinology 99:1534–1541

Wickler StJ (1980) Maximal thermogenic capacity and body temperatures of white-footed mice, Peromyscus, in summer and winter. Physiol Zool 53:338–346

Wunder BA (in press) Strategies for and environmental cueing mechanisms of seasonal changes in thermoregulatory parameters of small mammals. In: Merritt J (ed) Winter ecology of small mammals. Carnegie Mus Nat Hist Spec Publ

Biochemical Mechanisms of Thermogenesis

B. CANNON and J. NEDERGAARD[1]

1 Nonshivering Thermogenesis is Brown-Fat Thermogenesis

Today there is no doubt that facultative nonshivering thermogenesis is synonymous with the heat generated from brown adipose tissue. The questions presently addressed concerning the mechanism of thermogenesis are related to the *acute* cellular regulation of the process, as mediated both via α_1- and β-adrenergic receptors, and to the control of the *adaptive* changes within the tissue occurring during acclimation to cold (or to certain diets).

In the following we shall briefly review some of the present discussions in these areas; some more detailed and specialized reviews can be found elsewhere, e.g., on the brown fat cell as such (Nedergaard and Lindberg 1982), on brown-adipose-tissue mitochondria (Nicholls 1976, 1979; Lindberg et al. 1981; Nedergaard and Cannon 1984a), on the question of lipid metabolism (McCormack 1982), on the adaptive processes (Barnard et al. 1980), or on the tissue as a whole (Cannon and Nedergaard 1985a,b; Nicholls and Locke 1984; Rothwell and Stock 1984). Recently, two books have also been published, edited by Girardier and Stock (1983) and Trayhurn and Nicholls (1986).

We shall here refer to brown fat as being in an *"active"* or *"inactive"* state, and by this we refer to the short-term, acute, facultative shifts occurring in the degree of thermogenic activity of the tissue, actually taking place within minutes. This is, of course, what happens within the animal as a response to different intensities of nervous stimulation (with accompanying norepinephrine release), changing from less than two impulses per second in the inactive state, to about eight impulses in the active state (Nijima et al. 1984). This activation is also what can be followed in isolated brown fat cells after the addition of norepinephrine (Sects. 2 and 3), or what should be followed in the isolated mitochondria by the addition of the still disputed intracellular "mediator" of thermogenesis (Sect. 2.2).

Similarly, we shall refer to brown fat as being in a *recruited* or *atrophied* state, indicating by this the adaptive changes in the tissue. Thus, the recruited state is that seen in (small) mammals acclimated to cold, in newborn mammals, and in animals which are fed certain types of food, notably the so-called cafeteria diets. We assume that these recruited states have a high degree of commonality; they may even be identical on the biochemical plane, but this is presently under discussion (see, e.g., Sect. 2.2.2). (We

1 The Wenner-Gren Institute, University of Stockholm, S-10691 Stockholm, Sweden

Circulation, Respiration, and Metabolism
(ed. by R. Gilles)

avoid the word "hypertrophied" when describing the recruited state, because this term is too easily associated with simply an increased wet weight; an increased wet weight generally occurs in the recruited state, but unfortunately also often in the atrophied state.) The questions related to the acquirement of the recruited state are discussed in Sect. 4.

Thus, whereas it is the degree of "recruitment" which sets the limits for what the tissue can actually produce in terms of heat, it is the "activity state" which determines how much of the capacity is used in each instance.

2 The β-Adrenergic Pathways for Regulation of Tissue Activity

By the term "the β-adrenergic pathways" we refer to the traditionally studied pathways leading from activation of β-adrenergic receptors by *norepinephrine* to both the procuration of substrate for thermogenesis *and* to the stimulation of the mitochondrial thermogenic mechanisms, i.e., the two necessary conditions for thermogenesis: the fuel and the ignition.

There are reports that hormones other than catecholamines may influence thermogenesis. Although most of these hormone effects are poorly characterized, it does, however, seem possible that *glucagon* can induce thermogenesis in rabbit (Heim and Hull 1966), mouse (Doi and Kuroshima 1982), and rat brown adipose tissue (Kuroshima and Yahata 1979), but not in hamsters (Nedergaard and Lindberg 1982). The concentrations necessary in vitro are, however, much higher than those found in the circulation. *Adenosine* can partially inhibit catecholamine-stimulated thermogenesis in brown fat cells (Schimmel and McCarthy 1984; Szillart and Bukowiecki 1983), but the main function of adenosine may be to be the mediator of the increased blood flow through the tissue, which is apparently not a β-adrenergic effect in itself (Foster and Depocas 1980).

There are a large number of typical β_1-adrenergic receptors on brown fat cells (Svoboda et al. 1979), and their coupling to the stimulation of adenylate cyclase for the production of cyclic-AMP is dependent upon the physiological state of the animal, with, perhaps surprisingly, a lower degree of coupling being observed in the cold-acclimated state (Svartengren et al. 1982; Svoboda et al. 1984a,b). This means that more norepinephrine (in the animal: a higher rate of nervous stimulation) would be needed in the cold. An alternative (but so far unconfirmed) explanation for this discrepancy between norepinephrine binding to the receptor and its ability to stimulate adenylate cyclase would be that the cyclic-AMP production was *not* coupled to the plentiful β_1-type receptors in the tissue, but rather resulted from the stimulation of a totally different class of β-receptors. There is pharmacological evidence for the existence both in white and brown fat (Arch et al. 1984) of a "new" species of β-receptors (sometimes called β_3-receptors) which could be those linked to adenylate cyclase stimulation. If this is the case, we are left with a large number of β_1-receptors in the tissue, the function of which would then be unknown or nonexistent.

Irrespective of which receptors are involved, an increase in cyclic-AMP level ensues (Fain et al. 1973; Pettersson and Vallin 1976); this is the first step in the process leading to the procuration of substrate for thermogenesis.

2.1 The Procuration of Substrate

The major part of the substrate for thermogenesis is generated via β-adrenergic pathways which lead both to a degradation of intracellular triglycerides to free fatty acids and to an increased uptake of lipids from the circulation (via activation of lipoprotein lipase) (see Fig. 3).

2.1.1 Mobilization of Intracellularly Stored Substrate

About 30% of brown adipose tissue in the recruited state is triglyceride (Nedergaard and Cannon 1984b), and this triglyceride is the fuel for the acute phase of thermogenesis. This triglyceride is very quickly mobilized in the activated state (Fig. 1), and there is no reason to assume that this mobilization of intracellularly stored substrate does not occur in the traditional way for cyclic-AMP-dependent processes. Thus, there

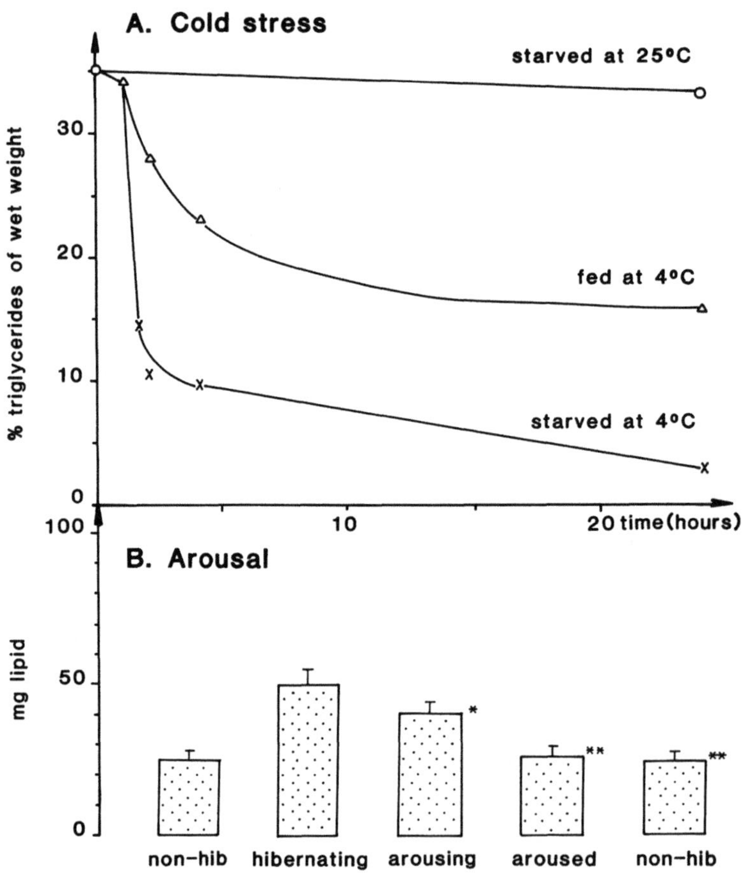

Fig. 1A,B. Mobilization of intracellular substrate in brown adipose tissue of **(A)** cold-stressed hamsters and **(B)** hamsters arousing from hibernation. Adapted from Lindberg et al. (1976) and Nedergaard and Cannon (1984b)

is an increased activity of protein kinase(s) in the active state (Skala and Knight 1977), and although this has not been experimentally demonstrated, a hormone-sensitive lipase is probably stimulated by the protein kinases. The breakdown of triglycerides leads to a release of fatty acids within the cells; it was initially assumed that all fatty acids were combusted within the cells (Hull and Segall 1966), but experiments with isolated cells from animals with brown adipose tissue in both the recruited state and in the atrophied state (Bieber et al. 1975; Nedergaard 1982), as well as direct measurements of blood leaving the tissue (Hardman and Hull 1970; Portet et al. 1974) indicate that fatty acids are released from the cells, even when nonshivering thermogenesis occurs. These released fatty acids may, e.g., be fuel for shivering thermogenesis in other tissues (Joel 1965; Nedergaard and Cannon 1984b); a significant fraction of the mobilized fatty acids do, however, become the fuel for nonshivering thermogenesis within the tissue (Fig. 3).

However, it can also been seen in Fig. 1 that these intracellular stores of substrate do not last long; after only a few hours in the active state, there must be another route for the continuous supply of fuel for thermogenesis. It is seen from Fig. 1A that food can replenish the lipid stores of the tissue – but how is this accomplished?

2.1.2 Activation of Lipoprotein Lipase

Within the first few hours in the cold, i.e., within exactly the same time interval as the decrease in triglyceride content, there is a very rapid increase in the activity of lipoprotein lipase activity (Carneheim et al. 1984) (Fig. 2). This increase in the enzyme that hydrolyzes the triglycerides found in the circulating very-low-density lipoproteins and chylomicrons and, thus, delivers fatty acids to the tissue, is probably the explanation for the ability of the tissue to continue to produce heat after the intracellular triglyceride depots have been emptied. As seen in Fig. 2, after only 4 h in the cold, there is a fourfold increase in the activity of lipoprotein lipase, and this increase is apparent-

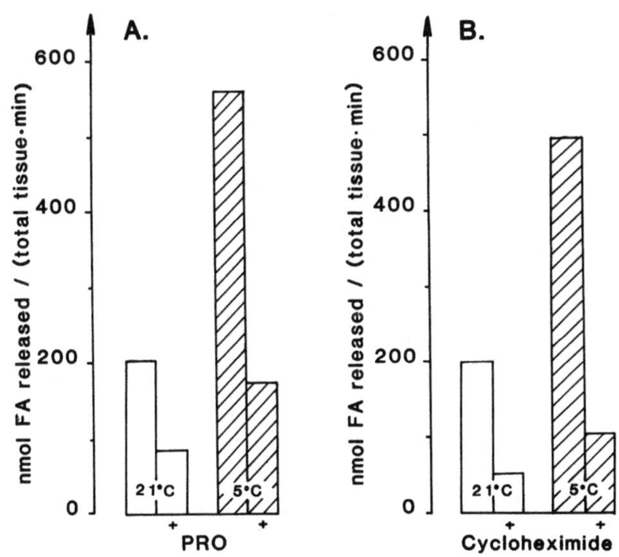

Fig. 2A,B. The activation of lipoprotein lipase by cold and its inhibition by the β-adrenergic blocker propranolol (**A**) and the protein synthesis inhibitor cycloheximide (**B**). Adapted from Carneheim et al. (1984, 1985b)

ly fully dependent upon the synthesis of new lipoprotein lipase enzymes, as it is completely abolished by the protein synthesis inhibitor cycloheximide (Fig. 2). In in vivo experiments it has been found that this increase in lipoprotein lipase activity can be mimicked by the injection of β-adrenergic agonists, and that the increase caused by cold can be totally abolished by the injection of the β-adrenergic blocker propranolol, indicating that the increase in lipoprotein lipase is a direct effect of β-adrenergic stimulation of the tissue (Carneheim et al. 1984) (Figs. 2 and 3). This adrenergic stimulation of lipoprotein lipase is apparently specific for brown adipose tissue; in white adipose tissue, lipoprotein lipase is stimulated by insulin.

The activation of lipoprotein lipase has the effect that fatty acids in the tissue tend to reflect the composition of the diet, especially in the activated state (Carneheim et al. 1985a). Thus, e.g., diets high in erucic acid ($C_{20:1}$) will lead to high amounts of this long-chain fatty acid being found in brown fat (Utne et al. 1977). Erucoyl esters are inhibitory to the oxidation of other substrates (Alexson et al. 1985a,b), and it is possible that the peroxisomal proliferation seen during acclimation to cold (Nedergaard et al. 1980) functions to enable the degradation of, e.g., erucoyl esters, as long-chain acyl esters are preferentially degraded in the peroxisomes (as compared to the mitochondria) (Alexson and Cannon 1984).

2.1.3 Activation of Lipid Synthesis

We find it reasonable to assume that the quantitatively significant amount of fatty acids needed for prolonged thermogenesis is delivered to the tissue by the action of lipoprotein lipase. It has, however, long been known that the tissue is capable of fatty acid synthesis (e.g., Angel 1969), and it was pointed, out both by Angel and by McCormack and Denton (1977) that the capacity for fatty acid synthesis in the tissue was unusually high. This high rate of lipid synthesis has generally been observed under conditions where thermogenesis is also evoked [e.g., in the cold-stressed rat (Trayhurn 1979)], and although there are exceptions to this rule [e.g., cold-acclimated hamster (Trayhurn 1981)], the general coincidence of these two opposing pathways of fatty acid metabolism has been difficult to understand. At least three different theories have been put forward:

1. that this could be a *futile cycle* leading to the purposeless synthesis and degradation of fatty acid, but using ATP (Trayhurn 1979), and thereby stimulating thermogenesis. However, as often pointed out, the capacity of the ATP-synthase in brown adipose tissue is, in general, so low that the oxygen consumption to which it is coupled cannot explain the high rates of thermogenesis observed (Lindberg et al. 1976; Cannon and Vogel 1977).
2. that – as suggested by Cooney and Newsholme (1984) – brown adipose tissue may have developed from a lipidogenic tissue, which would initially have functioned to synthesize fatty acids from carbohydrates. In this synthesizing process, ATP molecules are produced (due to the production of NADPH), and such a high capacity lipidogenic tissue would, therefore, be faced with the problem how to degrade ATP. The unique ability for the mitochondria to become "uncoupled" would then initially have been developed to fulfill this demand; later in evolution (concurrent with the development of homeothermia in mammals) the main function of the tissue would change to be thermogenic (utilizing the "uncoupling" property of the mitochondria);

the high lipidogenic capacity would remain associated with the tissue, but should be considered to be mainly or only a vestigial trait. This is an interesting hypothesis, but it is, as all evolutionary hypotheses, naturally somewhat difficult to prove or disprove.

3. that the regulation of the degree of uncoupling of the mitochondria (i.e., the activity of thermogenin) is regulated by cytosolic fatty acids or their derivatives. Thus, if, e.g., carbohydrates are to be used as fuel for thermogenesis, the "signal" for thermogenin activation cannot be created, unless some of the carbohydrate is transformed to lipid (Cannon and Nedergaard 1985a,b). Thus, *positive modulator production* would then be the function of the fatty acid synthesis.

However, to date, no positive effect of sympathetic activation has been directly demonstrated.

2.2 The Regulation of Thermogenin Activity

The ability to produce heat is associated with a brown-fat specific protein, thermogenin, found in the mitochondrial membrane. This allows substrate to be combusted without ATP synthesis. The development of this concept has been described in detail (Lindberg et al. 1981), and we shall here only summarize that thermogenin is a purine-nucleotide (ATP, GTP, ADP, GDP) binding protein, which is found in the inner membrane of brown-adipose-tissue mitochondria. It is associated with a high Cl^- and proton permeability, which is diminished by the binding of purine nucleotides to the protein. It has a mol wt. of 32,000, but probably exists in the membrane as a dimer, with one nucleotide binding site per dimer (Fig. 3).

A more detailed description of thermogenin [which is also known as the uncoupling protein (UCP), the 32,000 protein, the nucleotide-binding protein (NbP), the GDP-binding protein, and the "proton conductance pathway"] can be found in more specialized reviews, such as Nicholls (1976, 1979) or Nedergaard and Cannon (1985a).

Further, thermogenin has now been isolated (Lin and Klingenberg 1980, 1982; Klingenberg 1984), and within a short time, it should be possible to know its amino acid sequence. Experiments to incorporate thermogenin into liposomes to investigate whether in its isolated form it has the properties observed in brown-fat mitochondria have not been too successful; a recent report incidates that it is occasionally possible to obtain a GDP-sensitive incorporation (Bouillaud et al. 1983).

2.2.1 The Question of the Positive Modulator

As nonshivering thermogenesis is facultative, the activity of thermogenin is necessarily regulated, but the mechanism for this is still controversial (cf. Fig. 3).

As pointed out by Cannon et al. (1973), it is not very likely that it is the ATP or ADP (or GDP or GTP) levels around the mitochondria which regulate the degree of activation of thermogenin. This is mainly because cytosolic levels of adenine nucleotides are rather high (millimolar) as compared to the affinity of thermogenin for these nucleotides (micromolar); thus, the nucleotide binding site on thermogenin is probably saturated in the in vivo conditions, and changes in the ATP/ADP levels do not per se have much effect on thermogenin activation (Cannon et al. 1973; LaNoue et al. 1982). Rather, free fatty acids (or their acyl-CoA derivatives) have been implicated, because their level is increased during thermogenesis (Cannon et al. 1973). There have also been other candidates, but free fatty acids and their derivatives have the principal advantage that they (as discussed in Sect. 2.1 above) are created as a direct effect of catecholamine stimulation of intracellular or extracellular lipases. Further, the addition of free fatty acids to brown fat cells can *mimic* the effect of norepinephrine addition (Reed and Fain 1968; Prusiner et al. 1968a,b; Bukowiecki et al. 1981). We, therefore, consider it reasonable to assume that no signal other than a stimulated lipolysis is necessary to stimulate thermogenesis [i.e., the so-called minimal hypothesis (Bieber et al. 1975)].

It is still, however, not fully understood how this happens at the molecular level. The two candidates mainly discussed presently are free-fatty acids and acyl-CoAs.

Acyl-CoA esters have the theoretical advantage that they compete with ATP/GDP for the nucleotide binding site on thermogenin (Cannon et al. 1977; Strieleman et al. 1983), and that they apparently can activate thermogenin by so doing [i.e., they can increase Cl^- permeability in mitochondria in which this has been inhibited by GDP (Cannon et al. 1977)]. Acyl-CoAs are found associated with the mitochondria when they are isolated, and a higher level is found when the tissue is in an activated state (Normann and Flatmark 1984), but what this means physiologically is uncertain. There is no convincing evidence that they can actually lead to an increased thermogenesis in isolated mitochondria, nor that they can diminish the membrane potential of the mitochondria.

Free fatty acids were as early as 1969 by Rafael et al. observed to be able to uncouple brown-fat mitochondria. There is an increased sensitivity to free-fatty acids in mitochondria from cold-acclimated animals (Locke et al. 1982a,b; Barre et al. 1985), but in reality, the sensitivity is only 3–9 times higher than for, e.g., liver mitochondria (Barre et al. 1985). Free fatty acids do *not* compete with GDP for the binding site of thermogenin (Rial and Nicholls 1983), and they do not seem to be able to induce an increased chloride permeability (Nicholls and Lindberg 1973), but they do increase the proton permeability in isolated mitochondria (Rial et al. 1983). That this is via interaction with thermogenin is only indirectly shown; the best demonstration is that GDP can interfere with the increased proton permeability caused by fatty acid addition (Rial et al. 1983).

Thus, we cannot today say with certainty which is the intracellular mediator.

2.2.2 The Question of the Existence of a "Masked" Form of Thermogenin

It has been discussed that an "unmasking" mechanism for thermogenin occurs under certain circumstances.

The experimental background for this suggestion is summarized in Table 1. In Table 1, a series of physiological situations associated with a changed (i.e., in most cases increased) level of (^{3}H)GDP binding to the mitochondria is listed. As indicated above, thermogenin has a mol. wt. of 32,000, and it would, therefore, be imagined that an increased (^{3}H)GDP binding would always be associated with an increased peak height at the 32 kDa region on SDS polyacrylamide gel electrophoresis of brown-adipose-tissue mitochondria. However, as seen in Table 1, this is not always the case: in a number of situations, increased GDP binding was observed without concomitant apparent increase in peak height at 32,000.

Table 1. Relationship between three methods for thermogenin measurement[a]

Physiological stimulus	(^{3}H)GDP binding Increase	32 kD Area Increase	Thermogenin antigen Increase
Cold acclimation of rats	Yes[1–3]	Yes[2]	Yes[4,5]
Cold acclimation of mice	Yes[6]	Yes[7]	Yes[8]
Cold acclimation of hamsters	Yes[9,10,11,12]	No[9,22]	Yes[11,12]
Cold stress of rats (1 day)	Yes[2,5]	Yes[2] or no[13]	Yes[5]
Cold stress of rats (1 h)	Yes[2,13] or no[5]	No[2,13]	No[5]
Cafeteria feeding	Yes[14,15,16]	No[15]	Yes[16]
Lactation	Decrease[17]	?	Decrease[17]
Ob/ob mouse	Decrease[6,18]	No decrease[6,7]	Decrease[18]
Surgical denervation + cold	Yes[19]	No[19]	?
Surgical denervation + cafeteria	Yes[19]	No[19]	?
Norepinephrine injection	Yes[2] or no[3]	No[2]	?
Norepinephrine pumps	Yes[20]	Yes[20]	?
Pheochromocytoma in mice	Yes[21]	Yes[21]	Yes[21]
Cold stress + prot. synt. inh.	Yes[13]	No[13]	?

[a] The papers referred to above are (1): Cannon et al. 1978; (2): Desautels et al. 1978; (3): Sundin and Cannon 1980; (4): Lean et al. 1983; (5): Nedergaard and Cannon 1985; (6): Himms-Hagen and Desautels 1978; (7): Hogan and Himms-Hagen 1980; (8): Ashwell et al. 1983a; (9): Himms-Hagen and Gwilliam 1980; (10): Sundin 1981; (11): Trayhurn et al. 1983; (12): Nedergaard and Cannon 1984c; (13): Desautels and Himms-Hagen 1979; (14): Brooks et al. 1980; (15): Himms-Hagen et al. 1981; (16): Nedergaard et al. 1984; (17): Ashwell et al. 1983b; (18): Ashwell et al. 1983c; (19): Himms-Hagen and Park 1984; (20): Mory et al. 1984; (21) Ricquier et al. 1983; (22): Ricquier et al. 1979

From this type of observation, a masked form of thermogenin and an unmasking process was suggested, initially by Desautels et al. (1978). Thus, the masked form of thermogenin [characterized by an inability to bind (^{3}H)GDP and probably also by being nonfunctional in thermogenesis], would in the initial stages of thermogenesis undergo some kind of structural change which would allow it to become functional and enable it to bind (^{3}H)GDP. The existence of such a masked form must indeed be postulated to account for the observations summarized in the first two columns of Table 1.

In recent years it has become possible to obtain antibodies to thermogenin (Cannon et al. 1982; Lean et al. 1983; Ricquier et al. 1983), and by the use of these antibodies, quantitative specific measurement methods for thermogenin have been developed: enzyme-linked immunosorbent assay (ELISA) (Cannon et al. 1982; Nedergaard and Cannon 1985) and radioimmunoassay (RIA) (Lean et al. 1983) methods. With the use of these quantitative, specific methods, the question of an unmasked form of thermogenin has been reinvestigated.

As seen in Table 1, in all cases where an increased peak height had been observed on gel electrophoresis, an increased content of thermogenin antigen – as determined by the immunological methods – has also been observed. Further, in all cases so far examined, the immunological methods have been able to detect an increase in thermogenin *even* when the 32 kD peak height has not been clearly elevated.

It must be noted that in some of the observed states where a discrepancy between 32 kD peak height and (^{3}H)GDP binding has been observed, there have not as yet been investigations with the immunological methods, but in general, the evidence would presently tend to favor the contention that a masked form of thermogenin does not exist.

This then means that we have only two regulatory mechanisms for thermogenin: the acute (Sect. 2.2.1) and the adaptive (see Sect. 4).

3 The α-Adrenergic Pathways

A minor but not insignificant part (about 20%) of total thermogenesis in isolated brown fat cells results from the activation of α_1-adrenergic pathways (Mohell et al. 1981, 1983a; Schimmel et al. 1983). Similarly, there is evidence from in vivo experiments that α_1-adrenergic pathways have a synergistic effect on thermogenesis (Foster 1984). The cellular processes involved are not known in detail, but a plasma membrane depolarization, mobilization of intracellular Ca^{2+}, increased phosphatidyl-inositol turnover, and an activation of Ca^{2+}-dependent K^+ channels seem to be involved. We shall here briefly summarize what is known about these processes (see Fig. 4).

3.1 The Presence of α_1-Adrenergic Receptors

Brown adipose tissue cells possess α_1-adrenergic receptors (Mohell et al. 1983b) (Fig. 4). It is noteworthy that the density of these receptors (as well as the ratio α_1/β_1-receptors) is increased under conditions when the tissue is activated. This is at least true for

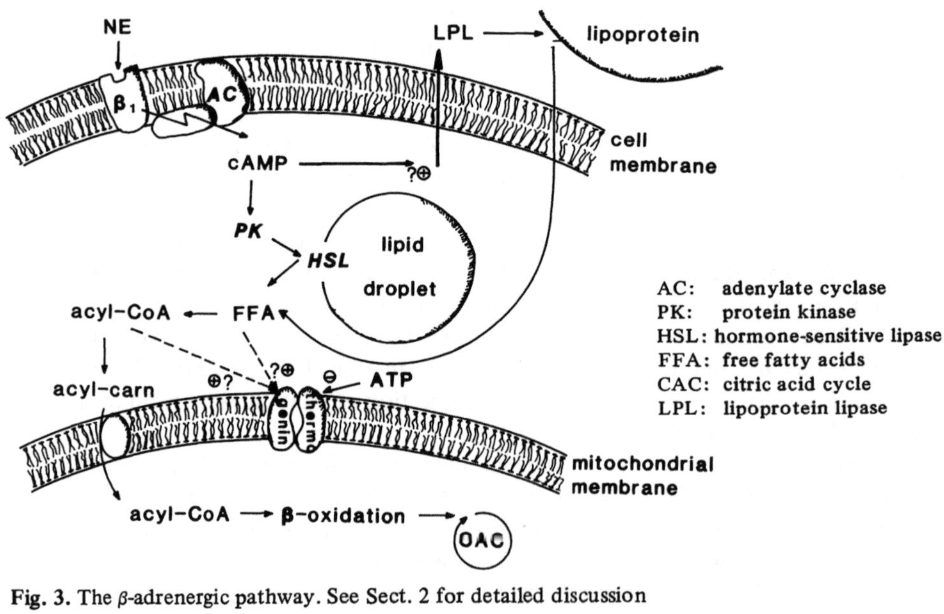

Fig. 3. The β-adrenergic pathway. See Sect. 2 for detailed discussion

both cold-acclimated (Raasmaja et al. 1984) and cafeteria-fed (Raasmaja et al. 1985) rats. This increase is especially intriguing because situations with an intensified nervous stimulation and an accompanying increase in norepinephrine turnover – as is the case here (Kennedy et al. 1977; Landsberg and Young 1983) – are normally associated with a decreased receptor number (so-called desensitization).

In general, α_1-adrenergic stimulation is associated with an increased turnover of phosphatidyl-inositol, and this has also been observed in brown adipose tissue (Garcia-Sainz et al. 1980; Strunecka et al. 1981; Mohell et al. 1984). The role of this increased turnover in the mediation of α_1-adrenergic stimulation is debated, and at least a fraction of the turnover is dependent upon extracellular Ca^{2+} (Mohell et al. 1984) and, thus, cannot be a primary effect of hormone/receptor interaction. However, some of the increased phosphatidylinositol turnover may be associated with the formation of inositol-*tris*-phosphate, which in several tissues has been discussed as being a primary mediator of α_1-adrenergic effects.

3.2 A Membrane Depolarization

The first observable response to a nervous stimulation in brown fat cells is a membrane depolarization (Seydoux and Girardier 1978). This membrane depolarization was initially observed by Girardier et al. (1968) and was for some time discussed as being directly involved in the thermogenic process due to the indirect activation of Na^+K^--ATPase and the ensuing utilization of ATP (Horwitz 1978).

The membrane depolarization (which brings the membrane potential down from about −60 mV to about −25 mV) is caused by an increased membrane permeability (Horowitz et al. 1971), probably especially for Na^+. The initial component of the membrane depolarization is α-adrenergic in nature (Girardier and Schneider-Picard 1983).

The functional significance of the membrane depolarization is not well understood today. We shall below elaborate one possible hypothesis; the fact that the brown fat cells are electrically coupled (Sheridan 1971), and that this coupling apparently increases when the tissue is in the recruited state (Schneider-Picard et al. 1980, 1984), would tend to indicate that these membrane phenomena play an important role in the tissue.

3.3 An Increased Cytosolic Ca^{2+} Level

In general, α_1-adrenergic stimulation is associated with an increase in cytosolic Ca^{2+} levels. This has not been directly observed in brown fat cells, but there is a mobilization of Ca^{2+} from intracellular stores (Connolly et al. 1984). As this mobilization can be mimicked by the addition of the mitochondrial uncoupler FCCP, it is likely that the Ca^{2+} is mobilized from the mitochondria. Further, as the mobilization is dependent upon the presence of Na^+ in the extracellular space (Connolly et al. 1984), we have suggested (Nedergaard et al. 1984) that the release of Ca^{2+} is due to an activation of the Na^+/Ca^{2+} exchange mechanism which is active in these mitochondria (Al-Shaikhaly et al. 1979). Thus, it may be imagined that the increased membrane permeability observed after norepinephrine addition (Horowitz et al. 1971) leads to an increase in cytosolic Na^+ which is sufficient high to induce the mitochondria to release sufficient

Ca^{2+} to increase the cytosolic level (Nedergaard 1981, 1983) (Fig. 4). It must, however, be stressed that there are presently no direct measurements of an increased Na^+ concentration in α_1-stimulated cells.

3.4 An Activation of a Ca^{2+}-Dependent K^+ Channel

One effect of the increased cytosolic Ca^{2+} level is an activation of Ca^{2+}-dependent K^+ channels (Fig. 4). These channels can be opened by any manipulation which leads to an increase in cytosolic Ca^{2+}, and the observed K^+ efflux (Girardier et al. 1968; Nånberg et al. 1984) can be fully inhibited by the specific inhibitor apamin (Nånberg et al. 1985). The opening of these K^+ channels seems to have a function in the maintenance of the α_1-adrenergic effect (Nånberg et al. 1985).

3.5 An α_1-Adrenergic Activation of Proliferative Processes?

It is unlikely that the α_1-adrenergic pathway (Fig. 4) has as its sole function to promote itself. Other effects, probably related to the increased cytosolic Ca^{2+} levels, have been described, such as increased cyclic-GMP levels (Skala and Knight 1979) or increased T_4 to T_3 conversion (Silva and Larsen 1983) (Fig. 4).

We think that it is possible that the α_1-adrenergic pathway is in some way related to the recruitment of the tissue. Indirect evidence for this can, e.g., be found in the increase in α_1-receptors in the recruited state or in the increased cyclic-GMP dependent protein kinase activity (Skala and Knight 1979). A deeper molecular understanding of these problems is, however, not presently at hand.

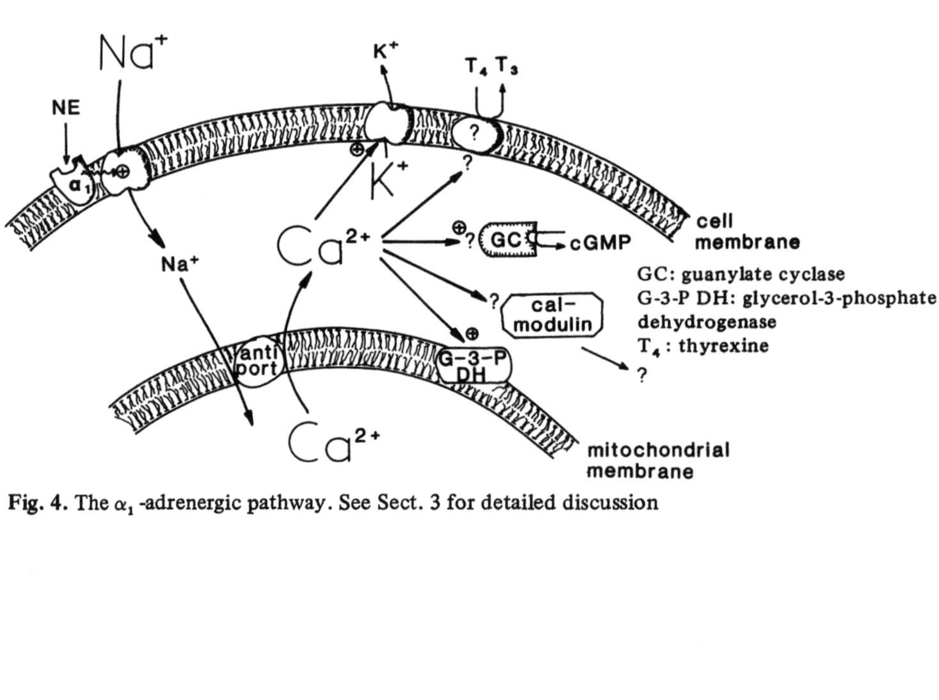

Fig. 4. The α_1-adrenergic pathway. See Sect. 3 for detailed discussion

4 The Adaptation Process

Our knowledge about the adaptive changes [the recruitment and atrophy of the tissue as a response to (or even in anticipation of) different physiological situations] is much more limited than our knowledge of the acute changes involved in thermogenesis activation (which in itself, as seen above, is by far complete).

It is probable that certain regions in the anterior part of the hypothalamus are involved; these regions would be partly or fully identical to those regulating the acute activity of the tissue.

One major question in this respect is whether it is simply the same signal which leads to both a stimulation of the recruitment of the tissue and to the acute activation of the tissue, i.e., norepinephrine released from the sympathetic nervous system.

It is clear that an activated sympathetic nervous system is necessary for the adaptation process (Mory et al. 1982). Although it has been considered for a long time that catecholamine treatment did not bring about changes identical to those seen in the recruited state, recent developments tend to indicate that it is possible (although not certain) that a chronic adrenergic stimulation is a sufficient signal to bring about the recruitment of thermogenic capacity seen, e.g., in acclimation to cold (Mory et al. 1984).

The molecular events taking place in this system, those that lead to cell multiplication (Cameron and Smith 1964) and to cell differentiation are still totally unknown. It would seem that there are cells in the tissue which although undifferentiated are already committed (Nechad et al. 1983), but how the impressive transformation from near dormancy to proliferation and differentiation is brought about is a question for the future.

References

Alexson S, Cannon B (1984) A direct comparison between peroxisomal and mitochondrial preferences for fatty acyl beta-oxidation predicts channelling of medium-chain and very-long-chain unsaturated fatty acids to peroxisomes. Biochim Biophys Acta 796:1–10

Alexson S, Nedergaard J, Cannon B (1985a) Inhibition of acetyl-carnitine oxidation by erucoyl-carnitine in rat brown-adipose-tissue mitochondria is due to CoA sequestration. Biochim Biophys Acta 834:149–158

Alexson S, Nedergaard J, Cannon B (1985b) Partial protection against erucoyl-carnitine inhibition due to high CoA levels in hamster brown-adipose-tissue mitochondria. Comp Biochem Physiol (in press)

Al-Shaikhaly MHM, Nedergaard J, Cannon B (1979) Sodium-induced calcium release from mitochondria in brown adipose tissue. Proc Natl Acad Sci USA 76:2350–2353

Angel A (1969) Brown adipose cells: spontaneous mobilization of endogenously synthesized lipid. Science 163:288–290

Arch JRS, Ainsworth AT, Thody VE, Wilson C, Wilson S (1984) Atypical beta-adrenoceptor on brown adipocytes as target for anti-obesity drugs. Nature 309:163–165

Ashwell M, Jennings G, Richard D, Stirling DM, Trayhurn P (1983a) Effect of acclimation temperature on the concentration of the mitochondrial "uncoupling" protein measured by radioimmunoassay in mouse brown adipose tissue. FEBS Lett 161:108–112

Ashwell M, Jennings G, Trayhurn P (1983b) Measurement by radioimmunoassay of the mitochondrial uncoupling (GDP binding) protein from brown adipose tissue of lactating mice. Proc Nutr Soc 160A

Ashwell M, Jennings G, Trayhurn P (1983c) Evidence from radioimmunoassay for a decreased concentration of mitochondrial "uncoupling" protein from brown adipose tissue of genetically obese (ob/ob) mice. Biochim Soc Trans 11:727–728

Barnard T, Mory G, Nechad M (1980) Biogenic amines and the trophic response of brown adipose tissue. In: Parvez S, Parvez H (eds) Biogenic amines in development. Elsevier-North Holland, Amsterdam, pp 391–439

Barre H, Nedergaard J, Cannon B (1985) Uncoupling effects of free fatty acids in brown adipose tissue. Submitted for publication

Bieber LL, Pettersson B, Lindberg O (1975) Studies on norepinephrine-induced efflux of free fatty acid from hamster brown adipose tissue cells. Eur J Biochem 58:375–381

Bouillaud F, Ricquier D, Gulik-Krzywicki T, Gary-Bobo CM (1983) The possible proton translocation activity of the mitochondrial uncoupling protein of brown adipose tissue. Reconstitution studies in liposomes. FEBS Lett 164:272–276

Brooks SL, Rothwell NJ, Stock MJ, Goodbody AE, Trayhurn P (1980) Increased proton conductance pathway in brown adipose tissue mitochondria of rats exhibiting diet-induced thermogenesis. Nature 286:274–276

Bukowiecki LJ, Follea N, Lupien J, Paradis A (1981) Metabolic relationships between lipolysis and respiration in rat brown adipocytes. J Biol Chem 256:12840–12848

Cameron IL, Smith RE (1964) Cytological responses of brown fat tissue in cold-exposed rats. J Cell Biol 23:89–100

Cannon B, Nedergaard J (1985a) The biochemistry of an "inefficient" tissue: brown adipose tissue. Essays Biochem 20:110–164

Cannon B, Nedergaard J (1985b) Brown adipose tissue. The molecular mechanisms controlling activity and thermogenesis. In: Van R, Cryer A (eds) New perspectives in adipose tissue. Butterworth, London, pp 233–270

Cannon B, Vogel G (1977) The mitochondrial ATPase of brown adipose tissue. Purification and comparison with the mitochondrial ATPase from beef heart. FEBS Lett 76:284–289

Cannon B, Nedergaard J, Romert L, Sundin U, Svartengren J (1978) The biochemical mechanism of thermogenesis in brown adipose tissue. In: Wang LL, Hudson J (ed) Strategies in cold: Natural torpidity and thermogenesis. Academic Press, New York, pp 567–594

Cannon B, Nicholls DG, Lindberg O (1973) Purine nucleotides and fatty acids in energy coupling of mitochondria from brown adipose tissue. In: Azzone GF et al. (eds) Mechanisms in bioenergetics. Academic, New York, pp 357–364

Cannon B, Hedin A, Nedergaard J (1982) Exclusive occurrence of thermogenin antigen in brown adipose tissue. FEBS Lett 150:129–132

Carneheim C, Nedergaard J, Cannon B (1984) Beta-adrenergic stimulation of lipoprotein lipase activity in rat brown adipose tissue during acclimation to cold. Am J Physiol 246:E327–E333

Carneheim C, Nedergaard J, Cannon B (1985a) Fatty acids in the lipids of brown adipose tissue of the hibernating and arousing hamster. Submitted for publication

Carneheim C, Nedergaard J, Cannon B (1985b) Transcription-dependent increase in lipoprotein lipase activity in rats. Submitted for publication

Connolly E, Nånberg E, Nedergaard J (1984) Na^+ dependent, alpha-adrenergic mobilization of intracellular (mitochondrial) Ca^{2+} in brown adipocytes. Eur J Biochem 141:187–193

Cooney GJ, Newsholme EA (1984) Does brown adipose tissue have a metabolic role in the rat? Trends Biochem Sci 9:303–305

Desautels M, Himms-Hagen J (1979) Roles of noradrenaline and protein synthesis in the cold-induced increase in purine nucleotide binding by rat brown adipose tissue mitochondria. Can J Biochem 57:968–976

Desautels M, Zaror-Behrens G, Himms-Hagen J (1978) Increased purine nucleotide binding, altered polypeptide composition, and thermogenesis in brown adipose tissue mitochondria of cold acclimated rats. Can J Biochem 56:378–383

Doi K, Kuroshima A (1982) Thermogenic response to glucagon in cold-acclimated mice. Jpn J Physiol 32:377–385

Fain JN, Jacops MD, Clement-Cormier YC (1973) Interrelationship of cyclic AMP, lipolysis, and respiration in brown fat cells. Am J Physiol 224:346–351

Foster DO (1984) Auxilary role of alpha-adrenoceptors in brown adipose tissue thermogenesis. In: Hales RHS (ed) Thermal physiology. Raven, New York, pp 201–204

Foster DO, Depocas F (1980) Evidence against noradrenergic regulation of vasodilation in rat brown adipose tissue. Can J Physiol Pharmacol 58:1418–1425

Garcia-Sainz JA, Hasler AK, Fain JN (1980) $Alpha_1$-adrenergic activation of phosphatidylinositol labeling in isolated brown fat cells. Biochem Pharmacol 29:3330–3333

Girardier L, Schneider-Picard G (1983) Alpha- and beta-adrenergic mediation of membrane potential changes and metabolism in rat brown adipose tissue. J Physiol 335:629–641

Girardier L, Stock MJ (eds) (1983) Mammalian thermogenesis. Chapman and Hall, London

Girardier L, Seydoux J, Clausen T (1968) Membrane potential of brown adipose tissue. A suggested mechanism for the regulation of thermogenesis. J Gen Physiol 52:925–940

Hardman MJ, Hull D (1970) Fat metabolism in brown adipose tissue in vivo. J Physiol (Lond) 206: 263–273

Heim T, Hull D (1966) The effect of propanolol on the calorigenic response in brown adipose tissue of new-born rabbits to catecholamines, glucagen, corticotrophin and cold exposure. J Physiol 187:271–283

Himms-Hagen J, Desautels M (1978) A mitochondrial defect in brown adipose tissue of the obese (ob/ob) mouse: reduced binding of purine nucleotides and a failure to respond to cold by an increase in binding. Biochem Biophys Res Commun 83:628–634

Himms-Hagen J, Gwilliam C (1980) Abnormal brown adipose tissue in hamsters with muscular dystrophy. Am J Physiol 239:C18–C22

Himms-Hagen J, Park IRA (1984) Role of the sympathetic innevation in the growth and maintenance of brown adipose tissue in cold-adapted and cafeteria-fed rats. In: Hales JRS (ed) Thermal physiology. Raven Press, New York, pp 193–196

Himms-Hagen J, Triandafillou J, Gwilliam C (1981) Brown adipose tissue of cafeteria-fed rats. Am J Physiol 241:E116–E120

Hogan S, Himms-Hagen J (1980) Abnormal brown adipose tissue in obese (ob/ob) mice: response to acclimation to cold. Am J Physiol 239:E301–E309

Horwitz BA (1978) Plasma membrane involvement in brown fat thermogenesis. In: Girardier L, Seydoux J (eds) Effectors of thermogenesis. Exper Suppl 32. Birkhäuser, Basel, pp 19–23

Horowitz JM, Horwitz BA, Smith RE (1971) Effect in vivo of norepinephrine on the membrane resistance of brown fat cells. Experientia 27:1419–1421

Hull D, Segall MM (1966) Distinction of brown from white adipose tissue. Nature 212:469–472

Joel CD (1965) The physiological role of brown adipose tissue. In: Renold AE, Cahill GF Jr (eds) Adipose tissue. Am Physiol Soc, Washington, pp 59–85 (Handbook of physiology, sect 5)

Kennedy DR, Hammond RP, Hamolsky MW (1977) Thyroid cold acclimation influences on norepinephrine metabolism in brown fat. Am J Physiol 232:E565–E569

Klingenberg M (1984) Characteristics of the uncoupling protein from brown fat mitochondria. Biochem Soc Transac 12:390–393

Kuroshima A, Yahata T (1979) Thermogenic response of brown adipocytes to noradrenaline and glucagon in heat-acclimated and cold-acclimated rats. Jpn J Physiol 29:683–690

Landsberg L, Young JB (1983) Autonomic regulation of thermogenesis. In: Girardier L, Stock MJ (eds) Mammalian thermogenesis. Chapman and Hall, London, pp 99–140

LaNoue KF, Koch CD, Meditz RB (1982) Mechanism of action of norepinephrine in hamster brown adipocytes. J Biol Chem 257:13740–13748

Lean MEJ, Branch WJ, James WPT, Jennings G, Ashwell M (1983) Measurement of rat brown adipose tissue mitochondrial uncoupling protein by radioimmunoassay: increased concentration after cold acclimation. Biosci Rep 3:61–71

Lin CS, Klingenberg M (1980) Isolation of the uncoupling protein from brown adipose tissue mitochondria. FEBS Lett 113:299–303

Lin CS, Klingenberg M (1982) Characteristics of the isolated purine nucleotide binding protein from brown fat mitochondria. Biochemistry 21:2950–2956

Lindberg O, Bieber LL, Houstek J (1976) Brown adipose tissue metabolism: an attempt to apply results from in vitro experiments on tissue in vivo. In: Jansky L, Musacchia XJ (eds) Regulation of depressed metabolism and thermogenesis. Thomas, Springfield, Ohio, pp 117–136

Lindberg O, Cannon B, Nedergaard J (1981) Thermogenic mitochondria. In: Lee CP et al. (eds) Mitochondria and microsomes. Addison-Wesley, Reading, pp 93–119

Locke R, Rial E, Scott ID, Nicholls DG (1982a) Fatty acids as acute regulators of the proton conductance of hamster brown-fat mitochondria. Eur J Biochem 129:373–380

Locke RM, Rial E, Nicholls DG (1982b) The acute regulation of mitochondrial proton conductance in cells and mitochondria from the brown fat of cold-adapted and warm-adapted guinea pigs. Eur J Biochem 129:381–387

McCormack JG (1982) The regulation of fatty acid synthesis in brown adipose tissue by insulin. Prog Lipid Res 21:195–223

McCormack JG, Denton RM (1977) Evidence that fatty acid synthesis in the intercapsular brown adipose tissue of cold-adapted rats is increased in vivo by insulin by mechanisms involving parallel activation of pyruvate dehydrogenase and acetyl-coenzyme A carboxylase. Biochem J 166:627–630

Mohell N, Nedergaard J, Cannon B (1981) An attempt to differentiate between alpha- and beta-adrenergic respiratory responses in hamster brown fat cells. Adv Physiol Sci 32:495–497

Mohell N, Nedergaard J, Cannon B (1983a) Quantitative differentiation of alpha- and beta-adrenergic respiratory responses in isolated hamster brown fat cells: evidence for the presence of an alpha$_1$-adrenergic component. Eur J Pharmacol 93:183–193

Mohell N, Svartengren J, Cannon B (1983b) Identification of (^3H)prazosin binding sites in brown adipose tissue as alpha$_1$-adrenergic receptors. Eur J Pharmacol 92:15–25

Mohell N, Wallace M, Fain JN (1984) Alpha$_1$-adrenergic stimulation of phosphatidylinositol turnover and respiration of brown fat cells. Mol Pharmacol 25:64–69

Mory G, Ricquier D, Nechad M, Hemon P (1982) Impairment of trophic response of brown fat to cold in guanethidine-treated rats. Am J Physiol 242:C159–C165

Mory G, Bouillaud F, Combes-George M, Ricquier D (1984) Noradrenaline controls the concentration of the uncoupling protein in brown-adipose tissue. FEBS Lett 166:393–396

Nånberg E, Nedergaard J, Cannon B (1984) Alpha-adrenergic effects on $^{86}Rb^+$ (K^+) potentials and fluxes in brown fat cells. Biochim Biophys Acta 804:291–300

Nånberg E, Connolly E, Nedergaard J (1985) Presence of a Ca^{2+}-dependent K^+ channel in brown adipocytes. Possible role in maintenance of alpha$_1$-adrenergic stimulation. Biochim Biophys Acta 844:42–49

Nechad M, Kuusela P, Carneheim C, Björntorp P, Nedergaard J, Cannon B (1983) Development of brown fat cells in monolayer culture. I. Morphological and biochemical distinction from white fat cells in culture. Exp Cell Res 149:105–118

Nedergaard J (1981) Effects of cations on brown adipose tissue in relation to possible metabolic consequences of membrane depolarisation. Eur J Biochem 114:159–167

Nedergaard J (1982) Catecholamine sensitivity in brown fat cells from cold-adapted hamsters and rats. Am J Physiol 242:C250–C257

Nedergaard J (1983) The relationship between extra-mitochondrial Ca^{2+} concentration, respiratory rate, and membrane potential in rat brown-adipose-tissue mitochondria. Eur J Biochem 133:185–191

Nedergaard J, Cannon B (1984a) Thermogenic mitochondria. In: Ernster L (ed) New comprehensive biochemistry (bioenergetics). Elsevier, Amsterdam, pp 291–314

Nedergaard J, Cannon B (1984b) Preferential utilization of brown adipose tissue lipids during arousal from hibernation in the golden hamster. Am J Physiol 247:R506–R512

Nedergaard J, Cannon B (1984c) Regulation of thermogenin expression and activity in brown adipose tissue. In: Hales JRS (ed) Thermal physiology. Raven, New York, pp 169–173

Nedergaard J, Cannon B (1985) (^3H)GDP binding and thermogenin amount in brown adipose tissue mitochondria from cold-exposed rats. Am J Physiol 248:C365–C371

Nedergaard J, Alexson S, Cannon B (1980) Cold adaptation in the rat: increased brown fat peroxisomal beta-oxidation relative to maximal mitochondrial oxidative capacity. Am J Physiol (Cell Physiol) 239:C208–C216

Nedergaard J, Lindberg O (1982) The brown fat cell. Int Rev Cytol 74:187–286

Nedergaard J, Connolly E, Nånberg E, Mohell N (1984) A possible physiological role of the Na^+/Ca^{2+} exchange mechanism of brown fat mitochondria in the mediation of alpha$_1$-adrenergic signals. Biochem Soc Trans 12:393–396

Nicholls DG (1976) The bioenergetics of brown adipose tissue mitochondria. FEBS Lett 61:103–110

Nicholls DG (1979) Brown adipose tissue mitochondria. Biochim Biophys Acta 549:1–29

Nicholls DG, Lindberg O (1973) Brown-adipose-tissue mitochondria. The influence of albumin and nucleotides on passive ion permeabilities. Eur J Biochem 37:523–530

Nijima A, Rohner-Jeanrenaud F, Jeanrenaud B (1984) Effect of cold stimulation on the efferent discharges of nerves innervating interscapular brown adipose tissue in the rat. In: Hales JRS (ed) Thermal physiology. Raven, New York, pp 189–192

Normann PT, Flatmark T (1984) Increase in mitochondrial content of long-chain acyl-CoA in brown adipose tissue during cold-acclimation. Biochim Biophys Acta 794:225–233

Pettersson B, Vallin I (1976) Norepinephrine-induced shift in levels of adenosine 3′:5′-monophosphate and ATP parallel to increased respiratory rate and lipolysis in isolated hamster brown fat cells. Eur J Biochem 62:383–390

Portet R, Laury MC, Mertin R, Senault C, Hluszko MT, Chevillard L (1974) Hormonal stimulation of substrate utilization in brown adipose tissue of cold acclimated rats. Proc Soc Exp Biol Med 147:807–812

Prusiner SB, Cannon B, Ching TM, Lindberg O (1968a) Oxidative metabolism in cells isolated from brown adipose tissue. 2. Catecholamine-regulated respiratory control. Eur J Biochem 7:51–57

Prusiner SB, Cannon B, Lindberg O (1968b) Oxidative metabolism in cells isolated from brown adipose tissue. I. Catecholamine and fatty acid stimulation of respiration. Eur J Biochem 6: 15:22

Raasmaja A, Mohell N, Nedergaard J (1984) Increased alpha$_1$-receptor density in brown adipose tissue of cold-acclimated rats and hamsters. Eur J Pharmacol 106:489–498

Raasmaja A, Mohell N, Nedergaard J (1985) Increased alpha$_1$-adrenergic receptor density in brown adipose tissue of cafeteria-fed rats. Biosci Rep 4:851–859

Rafael J, Ludolph HJ, Hohorst HJ (1969) Mitochondrien aus braunem Fettgewebe: Entkopplung der Atmungskettenphosphorylierung durch langkettige Fettsäuren und Rekopplung durch Guanosintriphosphat. Hoppe-Seyler's Z Physiol Chem 350:1121–1131

Reed N, Fain JN (1968) Potassium-dependent stimulation of respiration in brown fat cells by fatty acids and lipolytic agents. J Biol Chem 243:6077–6083

Rial E, Nicholls DG (1983) The regulation of the proton conductance of brown fat mitochondria. Identification of functional and non-functional nucleotide-binding sites. FEBS Lett 161:284–288

Rial E, Poustie A, Nicholls DG (1983) Brown adipose tissue mitochondria: the regulation of the 32,000 M_V uncoupling protein by fatty acids and purine nucleotides. Eur J Biochem 173: 197–203

Ricquier D, Mory G, Hemon P (1979) Changes induced by cold adaption in the brown adipose tissue from several species of rodents, with special reference to the mitochondrial components. Can J Biochem 57:1262–1266

Ricquier D, Barlet JP, Garel JM, Combes-Georges M, Dubois MP (1983) An immunological study of the uncoupling protein of brown adipose tissue mitochondria. Biochem J 210:859–866

Ricquier D, Mory G, Nechad M, Combes-George M, Tribault D (1983) Development and activation of brown fat in rats with pheochromocytoma PC12 tumors. Am J Physiol 245:C172–C177

Rothwell NJ, Stock MJ (1984) Brown adipose tissue. Rec Adv Physiol 10:349–384

Schimmel RJ, McCarthy L (1984) Role of adenosine as an endogenous regulator of respiration in hamster brown adipocytes. Am J Physiol 246:C301–C307

Schimmel RJ, McCarthy L, McMahon KK (1983) Alpha$_1$-adrenergic stimulation of hamster brown adipose respiration. Am J Physiol 244:C362–C368

Schneider-Picard G, Carpentier JL, Orci L (1980) Quantitative evaluation of gap junctions during development of the brown adipose tissue. J Lipid Res 21:600–607

Schneider-Picard G, Carpentier JL, Girardier L (1984) Quantitative evaluation of gap junctions in rat brown adipose tissue after cold acclimation. J Membr Biol 78:85–89

Seydoux J, Girardier L (1978) Control of brown fat thermogenesis by the sympathetic nervous system. In: Seydoux J, Girardier L (eds) Effectors of thermogenesis, exper suppl 32. Birkhäuser, Basel, pp 153–167

Sheridan JD (1971) Electrical coupling between fat cells in newt fat body and mouse brown fat. J Cell Biol 50:795–803

Silva JE, Larsen PR (1983) Adrenergic activation of triiodothyronine in brown adipose tissue. Nature 305:712–713

Skala JP, Knight BL (1977) Protein kinases in brown adipose tissue of developing rats. J Biol Chem 252:1064–1070

Skala JP, Knight BL (1979) Cyclic GMP and cyclic GMP-dependent protein kinase in brown adipose tissue of developing rats. Biochim Biophys Acta 582:122–131

Strieleman PJ, Elson CE, Shrago E (1983) Modulation of GDP binding to brown adipose tissue mitochondria by coenzyme A thioesters. Fed Proc 42:1324

Strunecka A, Olivierusova L, Kubista V, Drahota Z (1981) Effect of cold stress and norepinephrine on the turnover of phospholipids in brown adipose tissue of the golden hamster (Mesocricetus auratus). Physiol Bohemoslov 30:307–313

Sundin U (1981) Brown adipose tissue. Control of heat production. Development during ontogeny and cold adaption. University of Stockholm, Stockholm

Sundin U, Cannon B (1980) GDP-binding to the brown fat mitochondria of developing and cold-adapted rats. Comp Biochem Physiol 65B:463–471

Svartengren J, Svoboda P, Cannon B (1982) Desensitisation of beta-adrenergic responsiveness in vivo. Decreased coupling between receptors and adenylate cyclase in isolated brown fat cells. Eur J Biochem 128:481–488

Svartengren J, Svoboda P, Drahota Z, Cannon B (1984) The molecular basis for adrenergic desensitization in hamster brown adipose tissue: uncoupling of adenylate cyclase activation. Comp Biochem Physiol 78C:159–170

Svoboda P, Svartengren J, Snochowski M, Houstek J, Cannon B (1979) High number of high-affinity binding sites for (–)-(^{3}H)dihydroalprenolol on isolated hamster brown-fat cells. Eur J Biochem 102:203–210

Svoboda P, Svartengren J, Naprstek J, Jirmanova Z (1984a) The functional and structural reorganisation of the plasma membranes of brown adipose tissue induced by cold acclimation of the hamster. I. Changes in catecholamine-sensitive adenylate cyclase activity. Molec Physiol 5:197–210

Svoboda P, Svartengren J, Drahota Z (1984b) The functional and structural reorganisation of the plasma membranes of brown adipose tissue induced by cold acclimation of the hamster. II. The beta-adrenergic receptor. Molec Physiol 5:211–220

Szillart D, Bukowiecki LJ (1983) Control of brown adipose tissue lipolysis and respiration by adenosine. Am J Physiol 245:E555–E559

Trayhurn P (1979) Fatty acid synthesis in vivo in brown adipose tissue, liver and white adipose tissue of the cold-acclimated rat. FEBS Lett 104:13–16

Trayhurn P (1981) Fatty acid synthesis in mouse brown adipose tissue. The influence of environmental temperature on the proportion of whole-body fatty acid synthesis in brown adipose tissue and the liver. Biochim Biophys Acta 664:549–560

Trayhurn P, Nicholls DG (eds) (1986) Brown adipose tissue. Edward Arnold, London (in press)

Trayhurn P, Richards D, Jennings G, Ashwell M (1983) Adaptive change in the concentration of the mitochondrial 'uncoupling' protein in brown adipose tissue of hamsters acclimated at different temperatures. Biosci Rep 3:1077–1084

Utne F, Njaa LR, Braekkan OR, Lambertsen G, Julshamn K (1977) Hydrogenated marine fat, its influence on rat tissue lipids, compared to fish oil, rapeseed oil and lard. Fisk Dir Ser Ernaer 2: 23–41

Neural Control of Mammalian Hibernation

H.C. HELLER and T.S. KILDUFF [1]

1 Introduction

Hibernation is the most profound, nonpathological change of behavioral state occurring in mammals. It requires dramatic adjustments in all physiological systems (Lyman et al. 1982). We might study any one of a number of physiological control systems to understand how homeostasis is maintained and, therefore, how animals survive throughout the hibernation cycle. To understand the mechanisms of induction and control of the hibernation cycle, however, we should focus on those physiological control systems which appear to be prime movers in the generation of hibernation. These appear to be the neural systems controlling arousal state changes and regulating body temperature. We now know that the entrance is accompanied by a progressively lowered hypothalamic set point for T_b regulation and that a set point is continuously present during a hibernation bout (Heller et al. 1978). The set point is also lower during euthermic slow wave sleep (SWS) than during wakefulness (Glotzbach and Heller 1976), suggesting possible homology between torpor and SWS. Electrophysiological studies support this contention as hibernation is entered through sleep and a bout of torpor consists mostly of SWS with REM sleep being virtually absent when T_b falls below 27 °C (Walker et al. 1977). Therefore, it seems entirely likely that shallow torpor and, subsequently, deep hibernation could have evolved as a result of selective pressures favoring the energy savings from a lowering of the regulated T_b during SWS (Heller et al. 1978). Hibernation is, thus, a valuable model system for the study of mechanisms controlling mammalian arousal states and the physiological alterations which accompany them.

This profound change of arousal state which we call hibernation and the less extreme euthermic sleep states are examples of widespread modulation of neuronal activity in the nervous system. This is a broader use of the concept of neuronal modulation than is currently common in neurobiology. Neuromodulation is frequently used in a more restrictive sense to refer to the potentiating action of a neuropeptide on the responsiveness of postsynaptic cells to a neurotransmitter. In contrast to classical neurotransmitter action, such peptides are effective at extremely low concentrations and have actions which last from minutes to hours. These properties of long-lasting effectiveness at low concentration and widespread distribution are precisely those which one might expect for a substance involved in arousal state changes.

1 Department of Biological Sciences, Stanford University, Stanford, CA 94305, USA

Circulation, Respiration, and Metabolism
(ed. by R. Gilles)

To investigate the neural circuits which modulate thermoregulatory and arousal state controlling mechanisms during hibernation, we undertook a broad survey of regional neural activity throughout the central nervous system as a function of events in the hibernation cycle. The technique we chose for this study is the autoradiographic 2-deoxyglucose method (Sokoloff et al. 1977) because it: (1) enables observations to be made without disrupting the arousal state of the animal, and (2) enables simultaneous comparison of activity of all neural structures on the same scale whether it be absolute or relative. The 2-deoxyglucose (2DG) method is based on the fact that 2DG is taken up by actively metabolizing cells and is phosphorylated like glucose, but its further metabolism proceeds very slowly if at all. The 2DG thus appears to be enzymatically trapped and its rate of intracellular accumulation is proportional to the metabolic rate of those cells. A radioactive label on the 2DG enables quantification of relative metabolic rates of specific regions of the CNS through autoradiography and densitometry or through scintillation counting.

2 Methods

Golden-mantled ground squirrels *(Citellus lateralis)* were used in our experiments. They were chronically catheterized in the jugular vein and received a subcutaneous thermocouple reentrant tube at least 1 week prior to an experiment. The animals were held in rooms at 5 °C and on a photoperiod of LD 12:12. They were caged individually, each with a nest box, nesting material, and food and water ad libitum. Experiments were performed in a metabolism chamber in darkness at 5 °C. A thermocouple was placed into the reentrant tube and the jugular catheter was connected to a syringe on the outside of the metabolism chamber and surrounding temperature controlled chamber. Body temperature and metabolic rate were continuously measured. Injection (150 μCi kg^{-1}) and incubation of the [^{14}C] 2DG occurred during euthermia, entrance (T_b = 30°-25 °C, 25°-20 °C, and 20°-15 °C), deep (T_b = 6°-7 °C) or shallow (T_b = 20 °C and 15 °C) hibernation, arousal from hibernation, and during forced hypothermia (T_b = 6°-7 °C) induced by the halothane-heliox method. Incubation times were adjusted to compensate for the influence of T_b on metabolism. The animal was sacrificed at the end of the incubation period by an overdose of sodium pentobarbital delivered through the catheter. Its brain and spinal cord were rapidly removed, frozen, sectioned, and autoradiographed. At least eight-five neural structures were identified on the autoradiographs and the relative 2DG uptake (R2DGU) of each structure for each condition was computed as the ratio of the optical density (O.D.) of the structure on the autoradiographs to the O.D. of the optic tract on the same set of autoradiographs:

$$R2DGU_{structure} = \frac{O.D._{structure}}{O.D._{optic\ tract}} .$$

Another technique which we have employed to reveal possible sights of synthesis of neuropeptides was to inject into the animal through a chronic i.v. cannula as described above, 1-[^{14}C]leucine (100 μCi kg^{-1}) at specific stages in the hibernation cycle (Kennedy et al. 1981). Incubation, tissue treatment, and autoradiographic procedures were the same as for the 2DG experiments. Metabolism of [^{14}C] leucine when labeled in the #1 carbon position is relatively simple; it is either respired as [^{14}C] CO_2 or incorporated into protein (Kennedy et al. 1981). Areas of high optical density on the resulting autoradiographs are sites of accumulation of the labeled leucine and, therefore, probable incorporation of the labeled amino acid into translational products.

3 Results and Discussion

3.1 Deep Hibernation

Autoradiographs of brains of hibernating squirrels were remarkably homogeneous in comparison to those of euthermic animals. This was reflected in the distribution of R2DGU values in the two experimental conditions; the values from hibernating animals were clustered around a lower mean value than were the values from the euthermic animals (Kilduff et al. 1982). Some structures did stand out from the homogeneous background in the autoradiographs of hibernating brains. Many of these structures were sensory nuclei which receive primary afferent projections, e.g., the cochlear nucleus, superior colliculus, and the dorsal horn of the spinal cord. Our interpretation of this result is that the first synapse in any sensory pathway remains metabolically active during hibernation and this may account for the sensitivity of the hibernator to environmental stimuli mentioned above. However, since the 2DG method does not distinguish between inhibitory versus excitatory synaptic activity, we cannot eliminate the alternative explanation that the relatively high metabolic activity is due to descending inhibition onto these lower-order sensory nuclei.

Two categories of structures were identified as possibly being important during hibernation: (1) structures that had greatest R2DGU during hibernation and (2) structures which underwent the smallest reduction in R2DGU between euthermia and hibernation. Structures which fell into the first category included sensory nuclei (the cochlear nucleus, inferior and superior colliculus), brainstem and hypothalamic structures (dorsal tegmental nucleus, locus coeruleus, the suprachiasmatic nucleus) and the paratrigeminal nucleus. Structures which fell into the second category and had very small reductions in R2DGU during hibernation include the suprachiasmatic nucleus, the lateral septal nucleus, pontine nuclei, several cerebellar structures, the dorsal horn, and the locus coeruleus (Kilduff et al. 1982). The paratrigeminal nucleus was the only structure which significantly increased its R2DGU during hibernation (+50.3%, $p < 0.05$) (Kilduff et al. 1983). Figure 1 shows autoradiographs through the medulla of a hibernating animal (A) and a euthermic animal (B). In the autoradiograph of the hibernating medulla, the paratrigeminal nuclei (PTN) stand out from the rather homogeneous background as two dorsolaterally positioned black dots which are not seen at all in the autoradiograph of the euthermic medulla. The autoradiograph from the euthermic brain shows a greater range of activities, with the lateral cuneate and inferior olivary nuclei being among the most active structures.

The observations on the paratrigeminal nucleus and the suprachiasmatic nucleus (SCN) were of particular interest. The PTN, located in the dorsolateral medulla, was

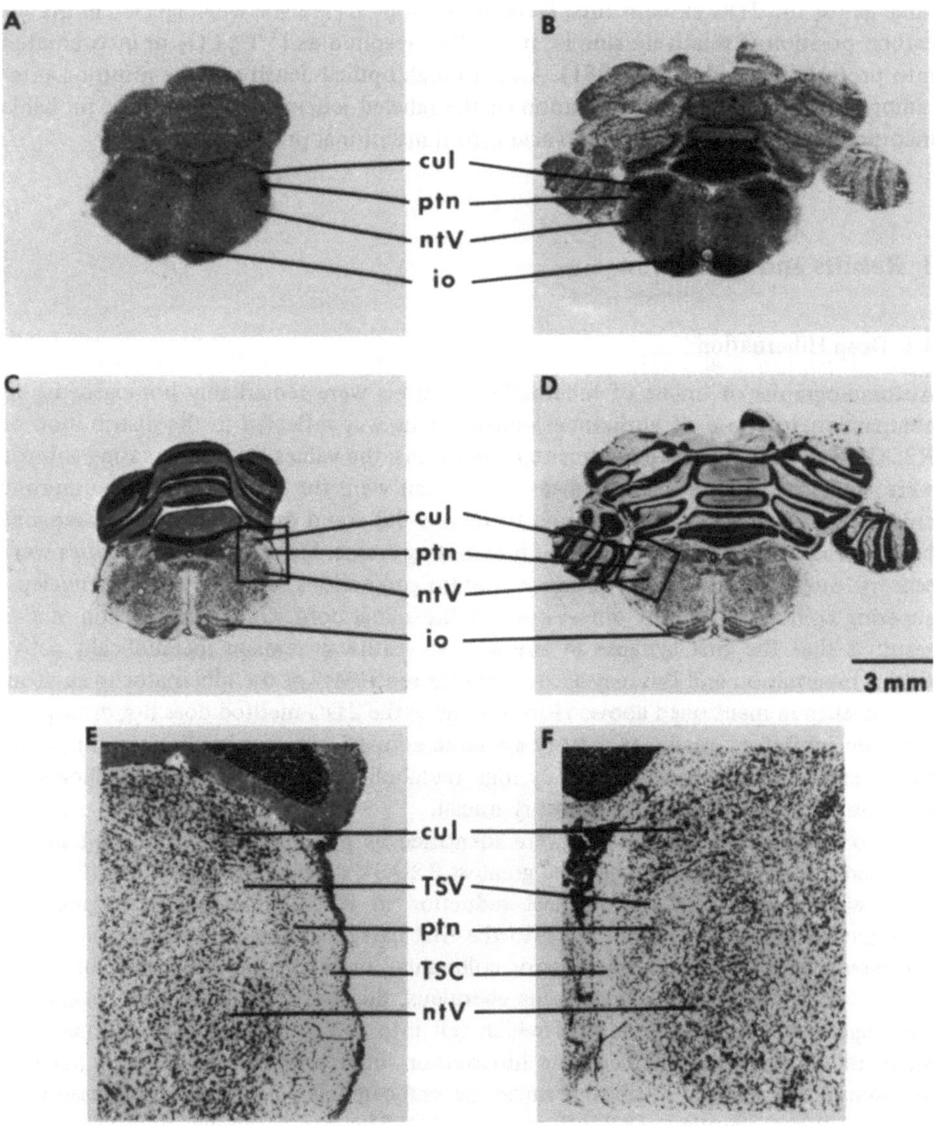

Fig. 1A–F. Autoradiographs of sections through the medullas of a hibernating (A) and an euthermic (B) brain. C and D are the corresponding stained sections. E and F are enlargements of boxed regions in C and D, respectively. *Abbreviations: cul,* lateral cuneate nuc.; *ptn,* paratrigeminal nuc.; *ntV,* nuc. of the tract of the trigeminal; *io,* inferior olive; *TSV,* spinal trigeminal tract; *TSC,* spinal cerebellar tract

first described by Ramon y Cajal in 1909 (Ramon y Cajal 1952) and has been ignored in the neurobiological literature since that time. Its function is unknown, and little is known of its anatomical connections (Chan-Palay 1978a; Somana and Walberg 1979). Chan-Palay (1978b) has noted that this structure has high levels of serotonin and stains immunocytochemically for beta-endorphin and substance P, a peptide characteristic of primary sensory nuclei. Given this information and our observation that the band of 2DG uptake overlying the PTN is continuous caudally with a band overlying the marginal zone of the spinal trigeminal nucleus, we have suggested (Kilduff et al. 1983) that the PTN may have a function similar to that of the marginal zone; namely, it may relay thermal information from the face. We have subsequently observed that the PTN is also the only structure which significantly increases its R2DGU during induced hypothermia (Kilduff et al. 1984). Thus, the function of the PTN may be to monitor environmental temperature, especially under conditions of depressed body temperatures which occur during hibernation and hypothermia. Therefore, the high R2DGU of the PTN may be another example of a lower-order sensory nucleus having maintained activity during hibernation.

The function of the SCN is better understood. Several lines of observation have suggested that the SCN is a circadian pacemaker in mammals (Rusak and Zucker 1979). The SCN of the rat has been shown to undergo a circadian rhythm in its metabolism, as demonstrated by the 2DG method (Schwartz and Gainer 1977; Schwartz et al. 1980). Such a rhythm has also been observed in the SCN of the hamster (Flood and Gibbs 1982), cat, and squirrel monkey (Schwartz et al. 1983), but not in the SCN of the 13-lined (Flood and Gibbs 1982) or golden-hamster ground squirrels (R.Y. Moore, personal communication). Results from our lab also indicate that the R2DGU of the SCN of the golden-mantled ground squirrel remains low throughout the day during euthermia (Fig. 2). Thus, the high R2DGU of the SCN during hibernation in the apparent absence of a circadian rhythm during euthermia in this species is intriguing.

3.2 Entrance into Hibernation

The next step of this research was to measure regional R2DGU during entrance into and arousal from hibernation. If any neural structure is specifically involved in induction and/or termination of hibernation, it would most likely be revealed by changes in metabolic activity during the transitions between euthermia and hibernation. The metabolic activity of the SCN in these experiments showed major state-dependent changes. The SCN is readily apparent on autoradiographs from animals during entrance and deep hibernation, but not from the euthermic or arousing animals (Fig. 3). Experiments on successive stages of entrance showed progressive activation of the SCN. In the T_b range between 30°–25 °C, the SCN is not readily apparent. In the T_b range between 25°–20 °C, the medial portion of the SCN is labeled. In the T_b range between 20°–15 °C, the entire SCN is labeled and a band of high R2DGU extends dorsally into the periventricular nuclei (PVN) of the hypothalamus. These observations suggest that the activity of the SCN is being modulated as the entrance to hibernation proceeds (Fig. 4). The fact that the SCN appears on the autoradiographs at all during entrance and deep hibernation may reflect greater temperature compensation of its activity relative to surrounding hypothalamic nuclei, strongly indicating that this structure is playing an important role during hibernation.

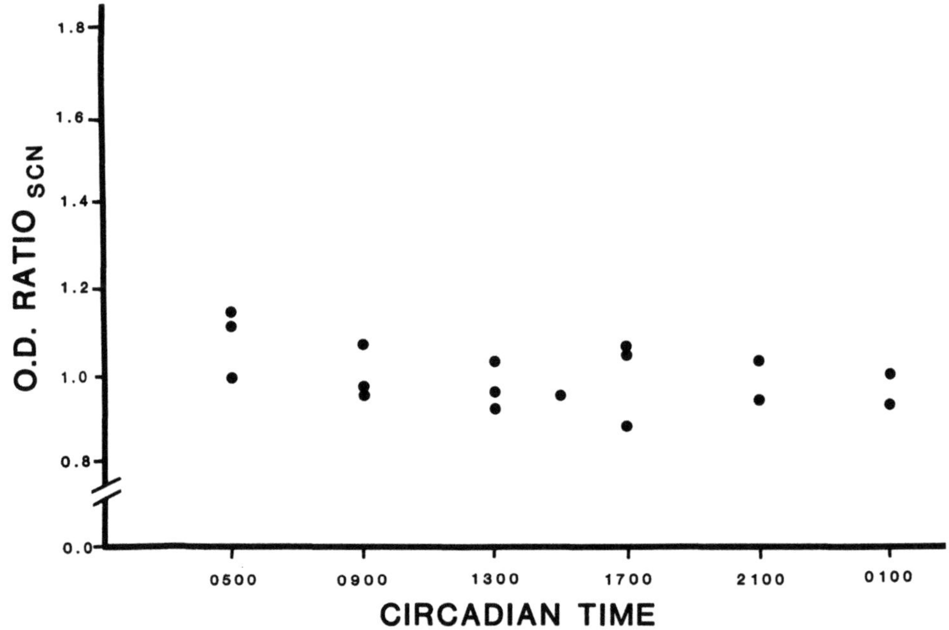

Fig. 2. Distribution of optical density ratios for the suprachiasmatic nucleus (*SCN*) obtained from 17 euthermic animals injected with ^{14}C 2DG at the indicated times during the circadian cycle. Prior to day of experiment, the lights were on at 0700 and off at 1900

The metabolic activation of the periventricular nuclei during the entrance process is of great interest. These nuclei stain immunocytochemically for several peptides and opiates and have axonal processes oriented toward the walls of the third ventricle. The high levels of 2DG uptake in these nuclei during entrance may indicate synthesis and release of a substance, perhaps a peptide, into the cerebrospinal fluid (CSF). Such a peptide could be a neuromodulatory substance involved in the transition to the hibernating state. A blood-borne polypeptide has been implicated in hibernation (Dawe 1978), and substances originally secreted into the CSF eventually appear in the plasma (Wood 1980). Furthermore, there is evidence for CSF-borne substances inducing sleep (Pappenheimer et al. 1975).

In addition to the SCN and the periventricular nuclei, several other structures were selectively labeled during entrance into hibernation. They include the median preoptic nucleus, an area in the lateral hypothalamus which probably corresponds to the neuropil of the median forebrain bundle, and the reticular nucleus of the thalamus.

Fig. 3A–D. [^{14}C] 2DG autoradiographs through the diencephalon of squirrels during (**A**) euthermia, (**B**) entrance, (**C**) deep hibernation, and (**D**) arousal from hibernation. *Panels A1–D1* are enlargements of the areas enclosed by the rectangles in A–D. *Panel A2* presents the cresyl violet-stained section corresponding to *Panel A1*. Note the high 2DG uptake of the SCN and the PVN during entrance (**B**) and deep hibernation (**C**)

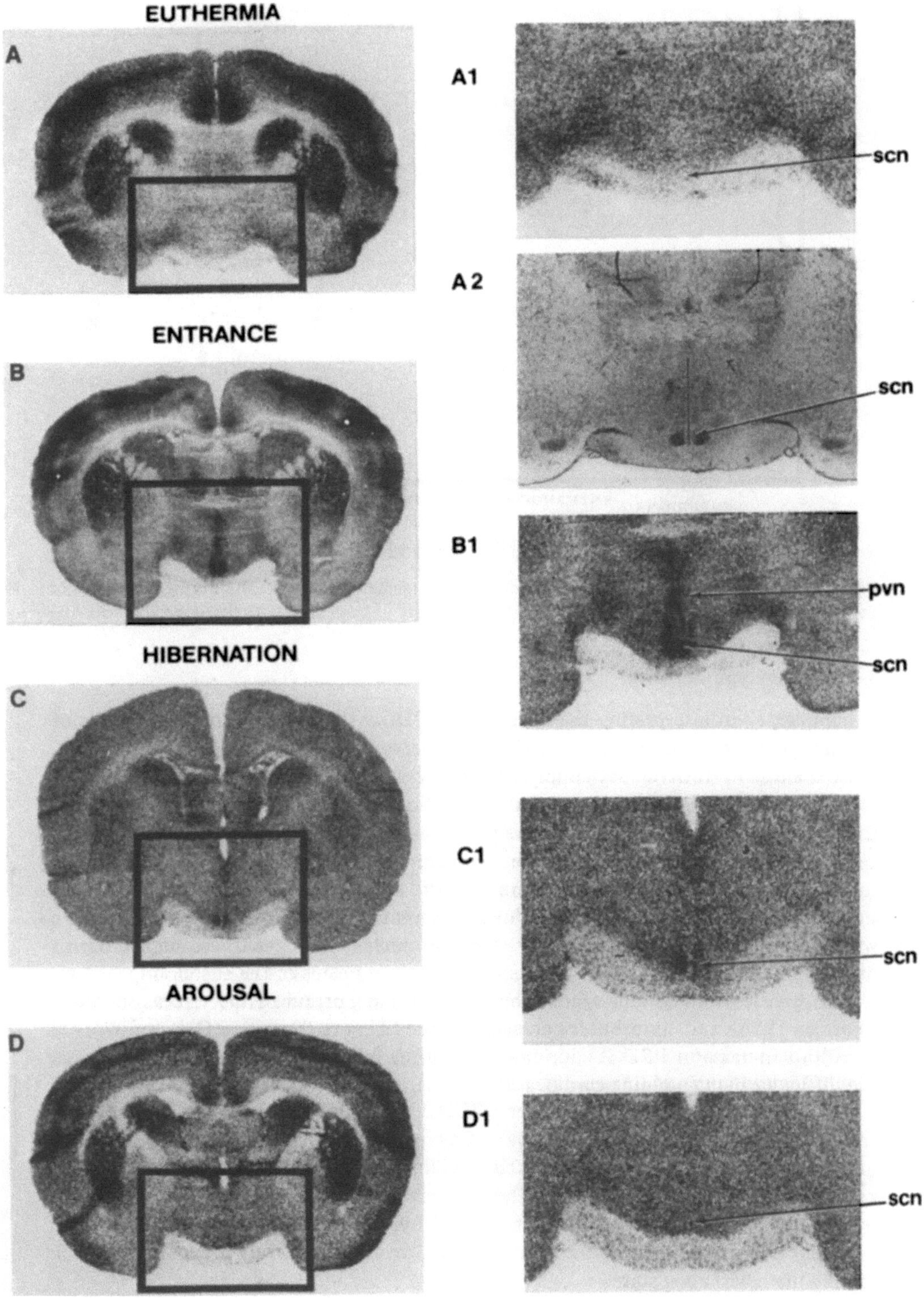
EUTHERMIA
A
A1
scn
A 2
scn
ENTRANCE
B
B1
pvn
scn
HIBERNATION
C
C1
scn
AROUSAL
D
D1
scn

Fig. 3

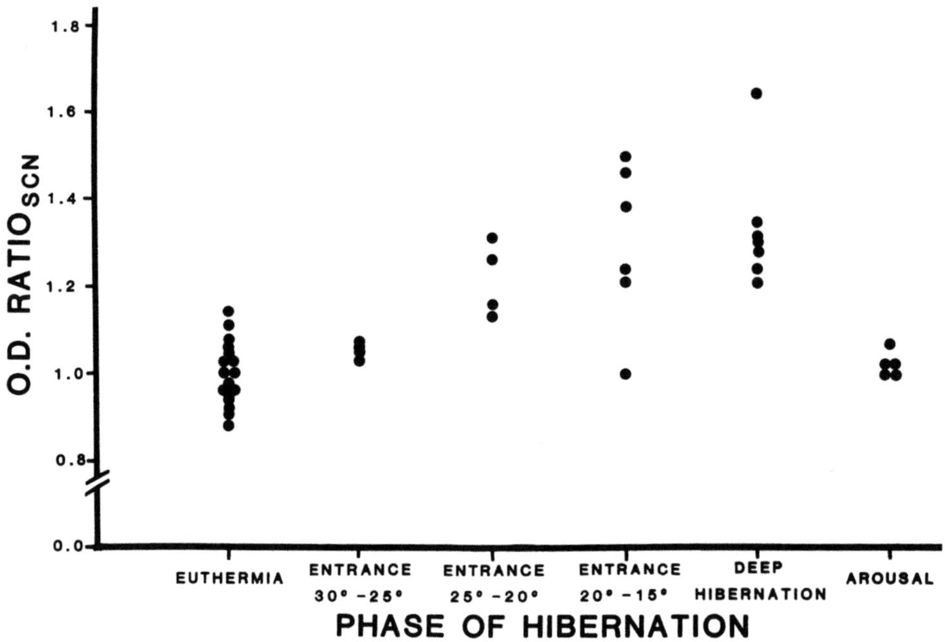

Fig. 4. Optical density ratios of the SCN for animals injected with ^{14}C 2DG during six phases of the hibernation cycle

3.3 Forced Hypothermia

A complexity in interpreting the changes in R2DGU that occur during the events of the hibernation cycle is that temperature of the nervous system is changing dramatically. Most changes in R2DGU may simply be consequences of temperature change and have no functional significance in terms of hibernation control. For this reason we undertook a series of experiments on animals forced into pathological hypothermia. R2DGU values for 85 neural structures were compared from hypothermic and hibernating squirrels at the same body temperature (Kilduff et al. 1984). Only ten structures had significantly different R2DGU values during hibernation than during hypothermia. Nine were higher in hibernation and only one, the nucleus ambiguus, was higher during hypothermia. Since the nucleus ambiguus is a motor nucleus of the autonomic nervous system, receives projections from the cardiovascular portion of the tractus solitarius (Norgren 1978), and contributes parasympathetic fibers to the vagus (Geis and Wurster 1980), its increase in R2DGU during hypothermia relative to hibernation may reflect the difficulty in maintaining cardiovascular integrity.

Of the nine structures with higher R2DGU during hibernation than during hypothermia, three have been noted above as having high R2DGU values during entrance into hibernation. They are the SCN, the medial preoptic nucleus, and the lateral septal nucleus. In addition, high activity of the periventricular nuclei which was so prominent during entrance into hibernation is not seen in hypothermia. These results support the idea that hypothalamic structures are crucially involved in actively inducing and maintaining the hibernating state.

Four other structures with significantly higher R2DGU during hibernation than during hypothermia probably represent secondary adaptations for hibernation rather than primary mechanisms involved in controlling the induction and maintenance of the hibernating state. One of these, the cochlear nucleus, is sensory. Hibernating animals respond to auditory cues. Severely hypothermic animals do not, but we do not know if this is due to a sensory or to a motor deficit. The other three structures are cerebellar (nucleus interpositus, nodulus, and uvula) and their low activity in hypothermia may be related to the fact that the hypothermic animal is flaccid, whereas the hibernator retains postural control and responds to vestibular information.

3.4 ^{14}C Leucine Autoradiography

When animals were injected with ^{14}C leucine during hibernation and the brains subsequently treated as in the 2DG experiments, the resulting autoradiographs reveal possible sites of synthesis and transport of translational products which have incorporated the ^{14}C leucine. The results of these experiments have not been completely analyzed, but one very prominent finding is a high uptake of label by the habenulae and a movement of the label down the fasciculus retroflexus (Fig. 5) which connects the habenulae to the interpeduncular nuclei and to the midbrain reticular formation including the raphe nuclei (Aghajanian and Wang 1977).

4 Conclusions

The conservative general conclusions we can offer at this point in our research is that a number of basal forebrain structures are implicated in the induction and maintenance of the hibernating state. They include the median preoptic nucleus, the lateral septum, the suprachiasmatic nucleus, and the periventricular nuclei. The periventricular nuclei may have a neurosecretory function in this process. The interconnections of these areas with each other, with other forebrain structures, and with the midbrain via the medial forebrain bundle (MFB) may be reflected in the high R2DGU in the lateral hypothalamic area corresponding to the neuropil of the MFB.

Another important route of communication between many of the basal forebrain structures mentioned above and the midbrain is via the stria medullaris, a fiber system interconnecting septal and hypothalamic structures with the habenulae. The habenulae give rise to projections to the interpenduncular nuclei, the midbrain reticular formation, and to the raphe nuclei. This fact makes the observation shown in Fig. 5 from the ^{14}C leucine autoradiographic experiments extremely interesting.

There is good evidence that the habenulae and the fasciculus retroflexus are involved in arousal state control. Arginine vasopressin is synthesized in the pineal and is a potent sleep inducing peptide when injected into the pineal recess near the habenulae (Pavel 1979; Goldstein 1983a). Electrical stimulation of the lateral habenulae can also induce nonREM sleep, but lesions of the lateral habenulae block the AVP effect on sleep (Goldstein 1983b). Just as the pineal can apparently influence arousal states via the

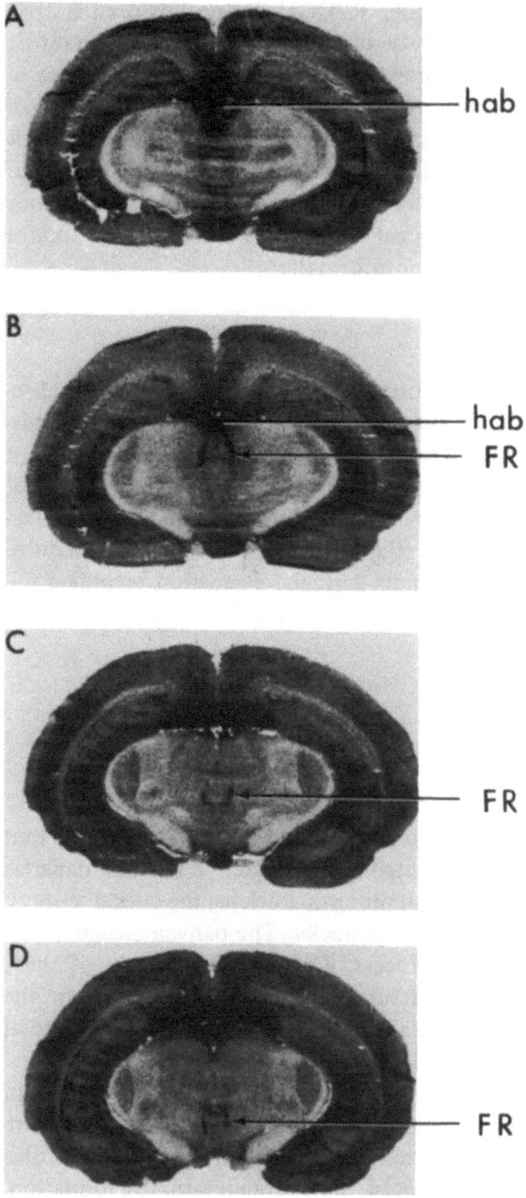

Fig. 5. ^{14}C-leucine autoradiographs through a progressively caudal section in the diencephalon of a hibernating ground squirrel. Note the high level of protein synthesis in the habenula (*hab*) and in the fasciculus retroflexus (*FR*)

habenulae, perhaps basal forebrain nuclei responsible for induction of hibernation do so via the effects of habenular projections on hypnogenic structures in the midbrain.

Use of the 2DG technique has given us a global view of neural activity during hibernation and has suggested some testable hypotheses. Having identified particular structures of interest, such as the PTN, the SCN, the PVN, and the habenulae, the possible roles of these structures in hibernation can be explored with other anatomical and physiological techniques, such as local measures of protein synthesis, receptor localization, electrophysiological and lesion studies. Such combined approaches will deepen our understanding of the remarkable phenomenon of hibernation and shed light upon the neuromodulatory mechanisms underlying arousal state control in general.

References

Aghajanian GK, Wang RY (1977) Habenula and other midbrain raphe afferents demonstrated by a modified retrograde tracing technique. Brain Res 122:229–242

Chan-Palay V (1978a) The paratrigeminal nucleus. I. Neurons and synaptic organization. J Neurocytol 7:405–418

Chan-Palay V (1978b) The paratrigeminal nucleus. II. Identification and inter-relations of catecholamine axons, indoleamine axons, and substance P immunoreactive cells in the neuropil. J Neurocytol 7:419–442

Dawe AR (1978) Hibernation trigger research updated. In: Wang LCH, Hudson JW (eds) Strategies in cold: Natural torpidity and thermogenesis. Academic, New York, p 541

Flood DG, Gibbs FP (1982) Species difference in circadian [^{14}C] 2-deoxyglucose uptake by suprachiasmatic nuclei. Brain Res 232:200–205

Geis GS, Wurster RD (1980) Horseradish peroxidase localization of cardiac vagal preganglionic somata. Brain Res 182:19–30

Glotzbach SF, Heller HC (1976) Central nervous regulation of body temperature during sleep. Science 194:537–539

Goldstein R (1983a) The administration of synthetic vasotocin into various spaces of the ventricular system of the brain and the sleep-wake cycle of the cat. Rev Roum Neurol 21:111–113

Goldstein R (1983b) A gabaergic habenulo-raphe pathway mediation of the hypnogenic effects of vasotocin in cat. Neuroscience 10:941–945

Heller HC, Walker JM, Florant GL, Glotzbach SF, Berger RJ (1978) Sleep and hibernation: Electrophysiological and thermoregulatory homologies. In: Wang LCH, Hudson JW (eds) Strategies in cold: Natural torpidity and thermogenesis. Academic, New York, p 225

Kennedy C, Duda S, Smith CB, Miyaoka M, Ito M, Sokoloff L (1981) Changes in protein synthesis underlying functional plasticity in immature monkey visual system. Proc Natl Acad Sci USA 78:3950–3953

Kilduff TS, Sharp FR, Heller HC (1982) [^{14}C] 2-deoxyglucose uptake in ground squirrel brain during hibernation. J Neurosci 2:143–157

Kilduff TS, Sharp FR, Heller HC (1983) Relative 2-deoxyglucose uptake of the paratrigeminal nucleus increases during hibernation. Brain Res 262:117–123

Kilduff TS, Heller HC, Sharp FR (1984) Paratrigeminal nucleus: a previously unrecognized thermal relay? In: Hales JRS (ed) Thermal physiology. Raven, New York, p 101

Lyman CP, Willis JS, Malan A, Wang LCH (1982) Hibernation and torpor in mammals and birds. Academic, New York

Norgren R (1978) Projections from the nucleus of the solitary tract in the rat. Neuroscience 3: 207–218

Pappenheimer JR, Koski G, Fencl V, Karnovsky ML, Krueger J (1975) Extraction of sleep-promoting factor S from cerebrospinal fluid and from brains of sleep-deprived animals. J Neurophys 38: 1299–1311

Pavel S (1979) Pineal vasotoxin and sleep: involvement of serotonin-containing neurons. Brain Res Bull 41:731–734

Ramon y Cajal S (1952) Histologie du system Nerveux de l'Homme et des Vertebres, vol 1, translated by Azoulay L. Consejo Superior de Investigaciones Cientificas, Madrid

Rusak B, Zucker I (1979) Neural regulation of circadian rhythms. Physiol Rev 59:449–526

Schwartz WJ, Gainer H (1977) Suprachiasmatic nucleus: use of ^{14}C-labeled deoxyglucose uptake as a functional marker. Science 197:1089–1091

Schwartz WJ, Davidsen LC, Smith CB (1980) In vivo metabolic activity of a putative circadian oscillator, the rat suprachiasmatic nucleus. J Comp Neurol 189:157–167

Schwartz WJ, Reppert SM, Egan SM, Moore-Ede MC (1983) In vivo metabolic activity of the suprachiasmatic nuclei: A comparative study. Brain Res 274:184–187

Sokoloff L, Reivich M, Kennedy C, DesRosiers MH, Patlak CS, Pettigrew PD, Sakurada O, Shinohara M (1977) The [^{14}C] deoxyglucose method for the measurement of local cerebral glucose utilization: theory, procedure, and normal values in the conscious and anesthetized albino rat. J Neurochem 28:13–26

Somana R, Walberg F (1979) The cerebellar projection from the paratrigeminal nucleus in the cat. Neurosci Lett 15:49–54

Walker JM, Glotzbach SF, Berger RJ, Heller HC (1977) Sleep and hibernation in ground squirrels (*Citellus* spp.): electrophysiological observations. Am J Physiol 233:R213–R221

Wood JH (1980) Neurobiology of cerebrospinal fluid, vol 1. Plenum, New York

Physiological and Biochemical Aspects of Mammalian Hibernation

L.C.H. WANG and D.J. PEHOWICH[1]

1 Introduction

Hibernation in mammals in response to cold and food shortage in winter is characterized by a profound reduction in metabolism (to 1% or less of normal) and body temperature (to near 0 °C) which may last from a few days to a few weeks. Unlike hibernation in the ectotherms (e.g. frogs and snakes), hibernating mammals are capable of exiting from the depressed metabolic state at any time, rewarming to normal body temperature using exclusively, endogenously produced heat by shivering and non-shivering thermogenesis. Extensive comparative studies on the evolutionary, physiological, biochemical, and neuroendocrinologic aspects have firmly established that mammalian hibernation represents an advanced form of thermoregulation rather than a reversion to primitive poikilothermy, and that it is polyphyletic in origin, acquired independently by members of at least six mammalian orders (see Hudson 1973 and Lyman et al. 1982, for reviews). Because of this latter aspect, divergence in regulatory mechanisms for hibernation is common amongst different species; for instance, Syrian hamsters *(Mesocricetus auratus)* and chipmunks store food whereas ground squirrels and marmots store fat. This leads to differences in at least two aspects in energy metabolism: (1) the endocrine strategy for food digestion and assimilation vs chronic fasting; and (2) the reliance on carbohydrates as the energy source for hibernation and the biochemical emphasis on glycogenolysis vs gluconeogenesis. Thus, one must be cautious when extrapolating mechanistic findings from one species to another because of the heterogeneity in niche specialization and the species-specific solutions (behavioral and physiological) in achieving the depressed metabolic state exemplified by hibernation.

Several recent reviews are available dealing with the general (Hudson 1973; Wang 1984a), neural (Beckman and Stanton 1982; Heller 1979), endocrinologic (Hudson and Wang 1979; Wang 1984b), ionic (Willis 1979; Ellory and Willis 1983), and membrane aspects (Willis et al. 1981) of mammalian hibernation. In addition, proceedings from the recent international symposia on hibernation and related topics (Wang and Hudson 1978; Musacchia and Jansky 1981) and a monograph (Lyman et al. 1982) have also provided extensive coverages on specific topics. The major thrust of this review is to provide selected coverages on recent advances in ecological, physiological, bio-

1 Department of Zoology, University of Alberta, Edmonton, Alberta, T6G 2E9, Canada

Circulation, Respiration, and Metabolism
(ed. by R. Gilles)

chemical, and neuroendocrinologic aspects of mammalian hibernation to acquaint the non-specialists with this fascinating physiological adaptation.

2 Time Pattern and Energetics of Natural Hibernation

To quantify the energy savings derived from natural hibernation, it is necessary to conduct both field and laboratory experiments documenting the timing, duration, and depth of the hibernation bouts in field animals and the metabolic costs associated with entry into, maintenance of and arousal from hibernation at different body and ambient (burrow) temperatures and the energy cost of euthermia between successive hibernation bouts in laboratory animals. Using this approach and employing temperature-sensitive radiotransmitters, Wang (1979) has provided estimates of the above aspects in the Richardson's ground squirrel *(Spermophilus richardsonii)*. In this species, adults commence estivation/hibernation from early to mid-July and young from mid-September but all emerge from hibernation in mid-March. The average duration of torpor ranges from 3–4 days in July and August and gradually lengthens to 15–19 days in December-January and shortens again to 14 days in February and 6 days in March. The body temperature during torpor is within 1°–4 °C of burrow temperature and ranges between 16.4° in July to 2.1 °C in January-February. The duration of euthermia between successive hibernation bouts ranges from 5–25 h depending on the season, longer in February-March than in November-December. Metabolic measurements simulating the body and ambient temperatures observed in the field animals have shown that the proportional costs for different stages of a hibernation bout are as follows: entry into hibernation 12.8%, maintenance of hibernation 16.6%, arousal from hibernation 19.0%, and euthermia between successive hibernation bouts, 51.6%. The energy savings in utilizing hibernation as compared to remaining euthermic and resting for the same duration (up to 8 mol in the adult Richardson's group squirrel) amounts to 87.8% (Wang 1979).

3 Intermediary Metabolism

In species which use fat as the overwintering energy source, enzyme activities associated with lipogenesis increase markedly during pre-hibernation fattening. For instance, in the dormouse (Castex and Sutter 1981), the activity of the adipocyte glucose-6-phosphate dehydrogenase, a key enzyme in the pentose phosphate shunt which generates NADPH for lipogenesis, is six times greater in the fall than in the spring and the insulin-stimulated glucose transport, oxidation, and lipogenesis in the adipocytes are highest when the dormouse is in its hyperphagic, weight-gain phase (Melnyk et al. 1983). The incorporation of ^{14}C-glucose into adipose tissue lipids increases some 88- to 108-fold between June and August in the juvenile Richardson's ground squirrels in preparation for hibernation under field conditions (Bintz and Strand 1983). Lipogenesis during the hibernation season is generally reduced as judged by the absence of insulin-stimulated glucose transport, oxidation and lipogenesis in adipocytes of dormice during its hypophagic, weight-loss phase (Melnyk and Martin 1983), the greatly decreased glucose-6-phosphate dehydrogenase activity in the liver of the hibernating 13-lined ground squirrel and the adipose tissue of the hibernating hedgehog and the reduced incorporation of ^{14}C-glucose into total lipids in the adipose tissue of the hibernating golden-mantled ground squirrel (see Willis 1982).

With respect to energy utilization during hibernation, in species which do feed between successive hibernation bouts (e.g. Syrian hamsters), glucose is utilized to a large extent especially during arousal from hibernation when profound depletion of liver and muscle glycogen occurs (Musacchia and Deavers 1981). In species which don't feed between hibernation bouts, fat is the primary fuel (Willis 1982). In the hibernating big brown bat, *Eptesicus fuscus* (Yacoe 1983a), fat metabolism supports nearly all the energy needs and glucose oxidation is significantly inhibited; muscle mitochondria preferentially oxidize fatty acids over pyruvate and physiological concentrations of palmitoyl-carnitine inhibits pyruvate oxidation. The inhibition of glucose oxidation during hibernation could be due to the inactivation of phosphofructokinase, a key regulatory enzyme controlling glycolytic flux; this enzyme may be reversibly converted from its catalytically active tetrameric to its catalytically inactive dimeric form by the combination of low intracellular pH and low temperature which prevail during the hibernating state (Hand and Somero 1983). This serves to conserve muscle glycogen during hibernation but allows it to be used for shivering thermogenesis during arousal when acidosis is reversed (Malan 1982). In addition to fat, glucose utilization by the central nervous system continues during hibernation especially in areas receiving thermal afferents (e.g., the paratrigeminal nucleus; Kilduff et al. 1983). Without feeding, gluconeogenesis becomes the only means for carbohydrate replenishment. A greater gluconeogenic capacity in liver and kidney cortex (see Willis 1982) and elevated activity of phosphoenolpyruvate carboxylase, a key enzyme in hepatic gluconeogenesis (Behrisch et al. 1981), have all been observed in the hibernating ground squirrels as compared to their summer-active counterparts. The gluconeogenic precursors may be mainly from glycerol released from triglycerides as in the arctic ground squirrel (Galster and Morrison 1975) or amino acids from protein catabolism as in the big brown bat (Yacoe 1983b) and the European hedgehog (Hoo-Paris et al. 1984).

4 Ionic Regulation

For cells to survive at low temperature, high intracellular K^+ and low intracellular Na^+ must be maintained. Tissues (kidney cortex, liver, skeletal muscle, aortic smooth muscle) and cells (kidney cortex, red blood cells) from the hibernators (13-lined ground squirrel, hamster) retain intracellular K^+ much more effectively than those from the non-hibernators (guinea pig, rat) at 5 °C, indicating superior cold tolerance for ionic transport in the hibernators (Willis 1979). In the red blood cell, the greater ability of the hibernators to retain intracellular K^+ at 5 °C is due to both a greater capacity for (Na^+ + K^+) pump-mediated K^+ influx and a much reduced passive K^+ leak than the non-hibernators. The (Na^+ + K^+) pump activity ratio at 5°/37 °C measured as ouabain-sensitive K^+ influx, lies between 1.9%–3.5% in six of seven hibernators and 0.18%–0.78% in eight of nine non-hibernators (Willis et al. 1980), suggesting a generally greater Na^+ pump activity at low temperature in the hibernators although with exceptions. Further studies (Ellory and Willis 1982) in the cold-sensitive (guinea pig) and cold-tolerant (13-lined and Columbian ground squirrels) species indicate that little difference exists in Na^+ pump affinity, number of ouabain binding sites per cell, and the amount of ouabain bound to the cell at 37° and 5 °C between the two groups. The only difference between the two groups appears to be the turnover number for Na^+ pumps at 5 °C; the turnover number is three- to fivefold higher in the hibernators. The reason for this difference is presently unknown but it is not related to the blockage of converting E_2 (K^+ binding) to E_1 (Na^+ binding) form at low temperature but may be related to alteration in partial fluxes between K^+:K^+ and Na^+:Na^+ exchange (Ellory and Willis 1982).

In addition to Na^+ pump, two other transport systems also govern the electrodiffusive (leak) pathways for K^+ (Ellory and Willis 1983): the Cl^--dependent $Na^+ + K^+$ co-transport (furosemide- or bumetanide-sensitive) and the intracellular Ca^{2+}-dependent K^+ efflux (Gardos channel; quinine-sensitive). The Cl^--dependent $Na^+ + K^+$ co-transport is absent at 5 °C in the guinea pig red blood cell but present in the 13-lined ground squirrel (Hall and Willis 1984). However, the existence of this system is highly variable amongst different species and is unlikely to be a universal mechanism for cellular cold tolerance. The existence of the Gardos channel in red blood cells has been verified in both the guinea pig and the 13-lined ground squirrel (Hall and Willis 1984). At 5 °C,

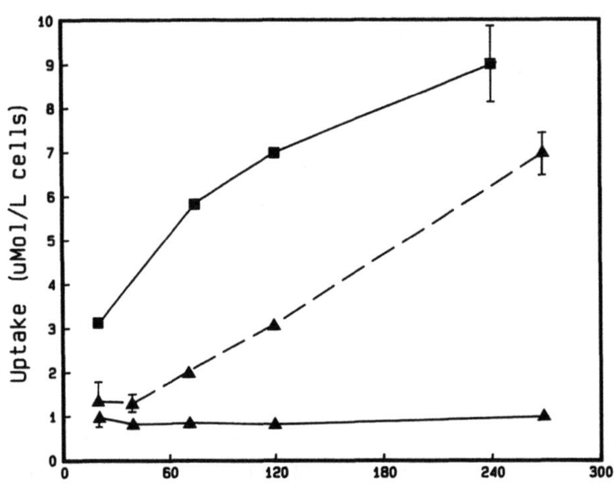

Fig. 1. Time course of $^{45}Ca^{2+}$ uptake at 5 °C in the fresh (–▲–) and ATP depleted (–■–) ground squirrel erythrocytes and in fresh (---▲---) guinea pig erythrocytes. All values are mean ± SE (n=3). *Point without error bar* had errors less than the size of symbol. Uptake was measured in calcium medium of the following composition (m*M*): NaCl (145), KCl (5), $CaCl_2$ (1.5), MOPS (15) buffered to pH 7.6 at 5 °C. Cells were ATP depleted by a 90 min preincubation in calcium medium also containing inosine (10) and iodoacetamide (5). Uptake was terminated by washing cells in calcium-free medium containing: NaCl (150), EGTA (0.5), and MOPS (15)

the quinine-sensitive, Ca^{2+}-dependent K^+ efflux increases significantly in the guinea pig red blood cell but not in the 13-lined ground squirrel, indicating an activation of the Gardos channel in the guinea pig but not in the ground squirrel (Hall and Willis 1984). The lack of activation of the Gardos channel at 5 °C in the 13-lined ground squirrel indicates that either external Ca^{2+} does not enter the red blood cells or that once entered, the excess Ca^{2+} is being pumped outward by the plasma membrane Ca^{2+} pump without increasing the intracellular Ca^{2+} concentration. Initial studies have indicated (Ellory and Hall 1983) that at 5 °C, the passive permeability to Ca^{2+} entry into the red blood cells (measured in ATP starved cells) is similar between guinea pigs and hedgehogs, but the active Ca^{2+} pump for Ca^{2+} efflux in the guinea pig is inhibited. A similar finding has also been confirmed recently in our laboratories (Wolowyk, Wang, Ellory, and Hall, unpublished) in red blood cells from a non-hibernator (guinea pig) and a hibernator (Richardson's ground squirrel; Fig. 1). Two points seem clear from these studies: (1) the inability of the non-hibernators to retain intracellular K^+ in cold may be due to activation of the Gardos channel consequent to failure in preventing a rise in intracellular Ca^{2+} concentration; and (2) in addition to activating the Gardos channel, an undue rise in intracellular Ca^{2+} is also detrimental to cell survival due to alterations in membrane structure and enzymic functions. Thus, the ability of the hibernators to retain intracellular K^+ and cell integrity at low temperature may reside in their superior ability in regulating intracellular Ca^{2+} concentration as compared to their non-hibernating counterparts.

5 Membrane Aspects

The temperature sensitivity of membrane enzyme functions has often been assessed with an Arrhenius plot relating the logarithm of maximum rate against the reciprocal of absolute temperature. Functionally, a continuous (linear) plot within the physiological temperature range for hibernation (2°–37 °C) is interpreted as being "cold resistant" whereas a discontinuous (non-linear) plot, "cold sensitive". The discontinuity at the critical temperature (Tc, where two slopes intercept) is due either to a conformational change of the enzyme protein or a transition of the order-disorder in mobility of membrane lipids surrounding the enzyme, or both (see Charnock 1978). The earlier studies on membrane adaptations at low temperature have been extensively reviewed by Charnock (1978) and Willis et al. (1981). Only a brief summary and some recent studies will be presented here.

The Arrhenius plots of plasma membrane and endoplasmic reticulum membrane ($Na^+ + K^+$)-ATPase from kidney, brain, and heart are discontinuous in both hibernators and non-hibernators and in hibernators between the hibernating and non-hibernating state (see Charnock 1978), suggesting no special change in this enzyme or its lipid environment in conjunction with hibernation. The sarcoplasmic reticulum Ca^{2+}-ATPase from the leg muscles of rats and awake 13-lined ground squirrels also show discontinuous Arrhenius plots; whether the Tc changes in hibernation in the ground squirrel is unknown (see Charnock 1978). In the Syrian hamster (Houslay and Palmer 1978),

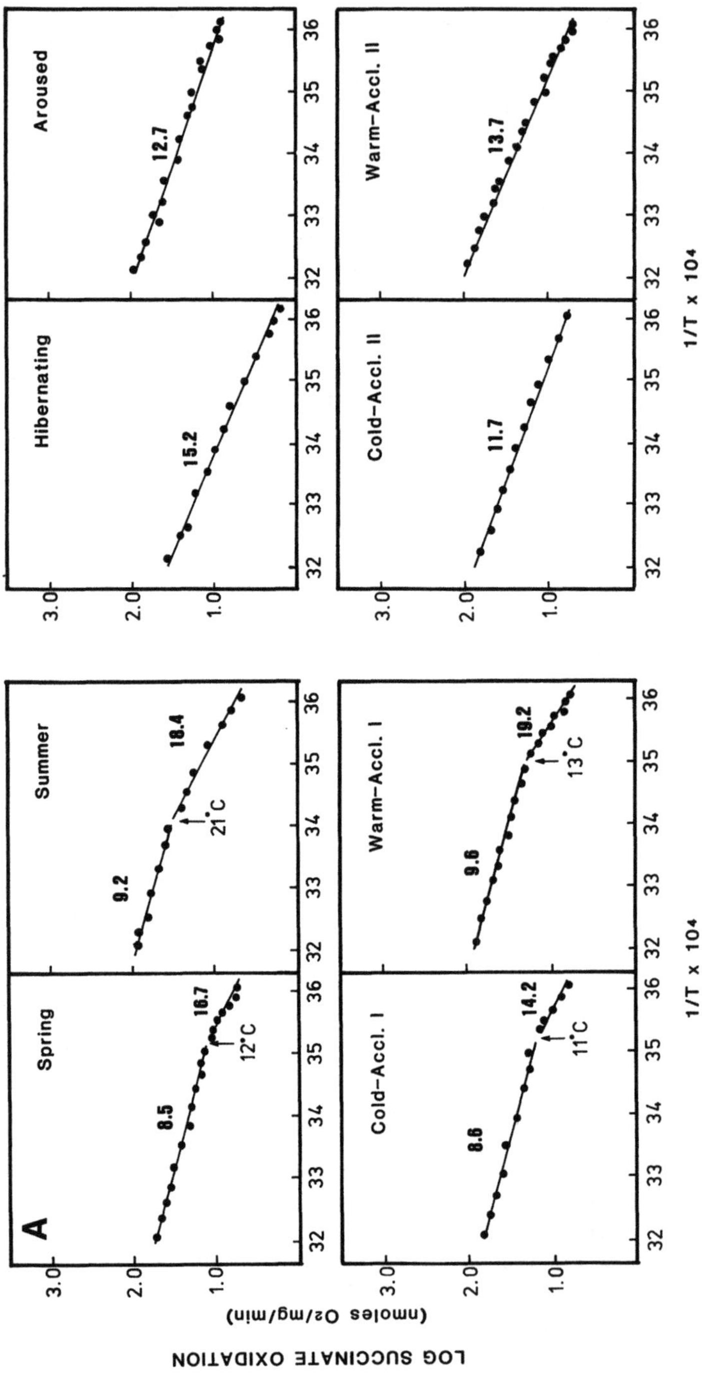
A
Spring
Summer
Hibernating
Aroused
Cold-Accl. I
Warm-Accl. I
Cold-Accl. II
Warm-Accl. II
8.5
16.7
12°C
9.2
18.4
21°C
15.2
12.7
8.6
14.2
11°C
9.6
19.2
13°C
11.7
13.7
LOG SUCCINATE OXIDATION
(nmoles O2/mg/min)
1/T x 10⁴
3.0
2.0
1.0
32
33
34
35
36

Fig. 2A

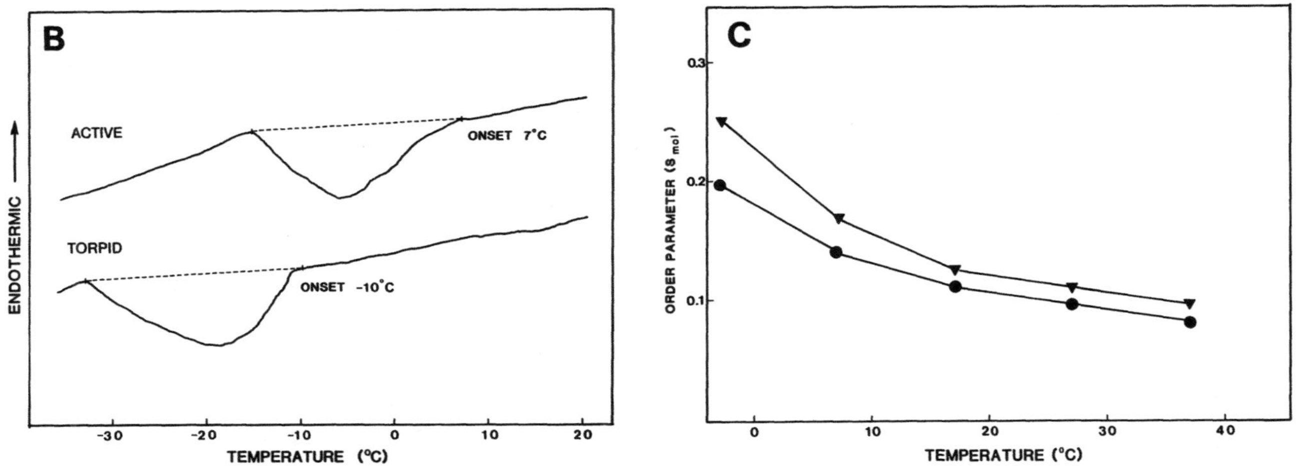

Fig. 2. A Arrhenius plots of succinate oxidation in liver mitochondria of *S. richardsonii* sacrificed in April (Spring, n = 5), July (Summer, N=6), October-November (Warm = 20 °C-acclimated and Cold = 4 °C-acclimated), and in hibernation (body temperature = 5 °C, n=8) and 12–14 h after arousal (body temperature = 37 °C, n=9). *Cald-Accl. I* (n=4) and *Warm-Accl. I* (n=7) are squirrels which showed no weight gain (presumably in their non-hibernating phase) during the 4 week acclimation period whereas *Cold-Accl. II* (n=4) and *Warm-Accl. II* (n=4) are those which did (presumably in their hibernating phase). None of these animals experienced hibernation prior to sacrifice. *Numbers below the arrows* are the critical transition temperature. *Numbers above the lines* are Arrhenius activation energy (Kcal mol^{-1}). Standard errors of means are too small to be represented (from Pehowich and Wang 1984). **B** Typical differential scanning calorimetry cooling curve of inner mitochondrial membrane of summer-active and hibernating *S. richardsonii*. Cooling rate was 10 °C min^{-1}. **C** ^{19}F NMR (254.025 MHz) order parameter profiles between −3° and 37 °C of inner mitochondrial membrane of active (▲) and hibernating (●) *S. richardsonii* (n=3, for each)

Arrhenius characteristics of eight liver plasma membrane enzymes are different between the hibernating and non-hibernating state and the differences appear to be related to the location of the enzyme in the lipid bilayer. Of the three enzymes located on the cytoplasmic side of the bilayer, the Tc remains constant near 25 °C regardless of the hibernating state. Of the single enzyme on the outer bilayer, the Tc decreases from 13° to 4 °C in hibernating hamsters. Of the remaining four enzymes which span both bilayers, their Tc's decrease from 25° and 13 °C in the non-hibernating state to 25° and 4 °C, respectively, in the hibernating state. Measurement with a fluorescent probe (DPH) also indicates an increase in membrane fluidity associated with hibernation as the Tc's for fluorescence polarization decrease from 25° and 13 °C in the active, to 25° and 4 °C, respectively, in the hibernating hamsters. Taken together, these observations indicate a lipid phase transition occurs around 25 °C in the inner membrane and 13 °C in the outer membrane in active hamsters; when hibernating, only the outer membrane shows a decrease in lipid phase transition temperature from 13° to 4 °C. Thus, depending on the physical location of the particular enzyme in the lipid bilayer, its Tc or Tc's is either unchanged or lowered with the onset of hibernation (Houslay and Palmer 1978). In the cardiac membrane, analysis by differential scanning calorimetry and electron spin labelling indicate increased membrane fluidity during hibernation as compared to the active state in the Richardson's ground squirrel (Charnock et al. 1980; Raison et al. 1981). A small increase in fluidity as detected by fluorescence probe (DPH) has also been observed in the brain microsomal membrane of hibernating Syrian hamsters (Goldman and Albers 1979) and European hamsters (Montaudon et al. 1984). However, using the same technique, such an increase in membrane fluidity has not been consistently observed in the brain synaptosomes and kidney microsomes of the hibernating Syrian hamster (Cossins and Wilkinson 1982).

In the mitochondrial membrane, Arrhenius plots of succinate-stimulated oxygen consumption in summer-active ground squirrels are typically discontinuous whereas in the hibernating animals they are continuous (Raison and Lyons 1971; Pehowich and Wang 1981). In a recent study comparing the succinate-stimulated mitochondrial proton ejection, calcium uptake and oxygen consumption in the Richardson's ground squirrel under spring, summer, winter (20 °C- and 4 °C-acclimated), hibernating and aroused conditions (Pehowich and Wang 1984), the discontinuous Arrhenius plots of these rate processes typical of spring and summer animals become continuous in the hibernating and aroused animals (Fig. 2A). More interestingly, winter animals (either 20 °C- or 4 °C-acclimated) which do not show weight gain (presumably in their non-hibernating phase) exhibit discontinuous Arrhenius plots whereas those which have shown significant weight gain (presumably in their hibernating phase) exhibit continuous Arrhenius plots (Fig. 2A) even though these animals have not yet experienced any decrease in body temperature. It is indicative that these changes are not brought out by differences of ambient or body temperature but are related to the endogenous circannual rhythm for hibernation. Differential scanning calorimetric analysis shows the thermal transition temperature in the inner mitochondrial membrane is 10 °C less in the hibernating squirrels than it is in the summer-active squirrels, indicating a decrease in the membrane bulk lipid liquid crystalline – gel phase transition temperature during hibernation (Fig. 2B). Further, NMR analysis using a fluorinated palmitic acid as the probe also indicates that the orientational order parameter of the inner mitochondrial membrane is consistently less (i.e. more fluid) in the hibernating squirrels, especially at low temperature than in the active squirrels (Fig. 2C, Pehowich, Macdonald, McElhaney, and Wang, unpublished). These combined physical and biochemical assessments on seasonal changes of mitochondrial membrane structures and functions indicate that (1) an increase in inner mitochondrial membrane fluidity is associated with hibernation; (2) the timing of membrane lipid changes is dictated by an endogenous rhythm for hibernation and is complete prior to the onset of hibernation; and (3) the membrane fluidity remains constant during the hibernation season and is unaffected by changes of body temperature during periodic arousals.

Based on compositional analysis of lipids from the brain and kidney of hamsters, liver and heart mitochondria of ground squirrels (see Aloia and Pengelley 1979; Willis et al. 1981), and liver mitochondria of European hamsters (Cremel et al. 1979), there does not appear to be any consistent relationship between the unsaturation index of fatty acids and the onset of hibernation. It is possible that the unsaturation index may not be the best indicator to predict the fluidity change of the membrane phospholipids (Willis et al. 1981). It has been shown recently that the change in Arrhenius characteristic of liver and heart mitochondrial membrane lipids induced by dietary lipids is not related to the unsaturation index but to the $\omega6$ to $\omega3$ unsaturated fatty acid ratio (McMurchie et al. 1983). Thus, future studies relating changes in Tc and fluidity to hibernation should perhaps include this particular measurement.

6 The "Hibernation Induction Trigger" and Endogenous Opioids

The "hibernation induction trigger" is a mysterious substance(s) contained in the serum (plasma) of hibernating rodents such as woodchucks and ground squirrels (Dawe and Spurrier 1969). Upon transfusion into the 13-lined ground squirrel, hibernation is induced during the summer, at a time when hibernation is normally absent. Attempts using a similar approach but with other species of ground squirrels (Galster 1978; Abbotts et al. 1979) or hamsters (Minor et al. 1978) as the recipient have not been met with equal success. These results seem to suggest that a species specificity might exist in the recipient but not in the donor. Biochemical efforts (Oeltgen and Spurrier 1981) have not yet produced purified preparations but have shown that the "hibernation induction trigger" is a thermolabile peptide tightly bound to serum albumin. Intravenous injection of this lyophilized albumin fraction to the 13-lined ground squirrel induces summer hibernation in 2 days to 5 weeks thereafter (Oeltgen and Spurrier 1981). When administered via the cerebral ventricle to non-hibernating species such as the macaque monkey, this lyophilized albumin fraction causes bradycardia, hypothermia, behavioral depression and aphagia (Myers et al. 1981). Interestingly, these effects may be retarded or reversed by the opiate antagonist, naloxone or naltrexone, suggesting the possibility that the hibernation induction trigger may be an endogenous opioid (Oeltgen et al. 1982).

There is increasing evidence that the endogenous opioids might be involved in the regulation of hibernation. For instance, increased brain levels of Met- and Leu-enkaphalins (Kramarova et al. 1983) and increased Met-enkaphalin immunoreactivity in specific hypothalamic areas have been observed during hibernation (Nürnberger 1983). By blocking the opiate receptors with naloxone, both the incidence (Kromer 1980) and the duration of hibernation (Llados-Eckman and Beckman 1983) are reduced and premature arousal is triggered (Margules et al. 1979). Further, in the golden-mantled ground squirrel, physical dependence on morphine fails to develop during hibernation but does so in euthermia (Beckman et al. 1981), suggesting greater occupancy of opiate receptors by endogenous opioids during hibernation. Taken together, it is possible that the maintenance of hibernation may be associated with an increased brain opioids activity. Whether this increase may be manifested in the peripheral circulation and appears as the hibernation induction trigger is presently unknown.

Acknowledgments. Research supports for work described herein are from the Natural Sciences and Engineering Research Council of Canada (A6455) to L. Wang and the Alberta Heritage Foundation for Medical Research (1119) to D. Pehowich.

References

Abbotts B, Wang LCH, Glass JD (1979) Absence of evidence for a hibernation "trigger" in blood dialyzate of Richardson's ground squirrel. Cryobiology 16:179–183

Aloia RC, Pengelley ET (1979) Lipid composition of cellular membranes of hibernating mammals. In: Florkin M, Sheer BT (eds) Chem Zool Vol 11. Academic, New York, pp 1–47

Beckman AL, Stanton TL (1982) Properties of the CNS during the state of hibernation. In: Beckman AL (ed) The natural basis of behavior. Spectrum, New York, pp 19–45

Beckman AL, Llados-Eckman C, Stanton TL, Adler MW (1981) Physical dependence on morphine fails to develop during the hibernating state. Science 212:1527–1529

Behrisch HW, Smullin DH, Morse GA (1981) Life at low and changing temperatures: molecular aspects. In: Musacchia XJ, Jansky L (eds) Survival in the cold. Elsevier/North Holland, Amsterdam, pp 191–205

Bintz GL, Strand CE (1983) Radioglucose metabolism by Richardson's ground squirrels in the natural environment. Physiol Zool 56:639–647

Castex Ch, Sutter BChJ (1981) Insulin binding and glucose oxidation in edible dormouse *(Glis glis)* adipose tissue: seasonal variations. Gen Comp Endocrinol 45:273–278

Charnock JS (1978) Membrane lipid phase-transitions: a possible biological response to hibernation? In: Wang LCH, Hudson JW (eds) Strategies in cold: Natural torpidity and thermogenesis. Academic, New York, pp 417–460

Charnock JS, Gibson RA, McMurchie EJ, Raison JK (1980) Changes in the fluidity of myocardial membranes during hibernation: relationship to myocardial adenosinetriphosphatase activity. Mol Pharmacol 18:476–482

Cossins AR, Wilkinson HL (1982) The role of homeoviscous adaptation in mammalian hibernation. J Therm Biol 7:107–110

Cremel G, Robel G, Canguilhem B, Rendon A, Waksman A (1979) Seasonal variation of the composition of membrane lipids in liver mitochondria of the hibernator *Cricetus cricetus*. Relation to intra-mitochondrial intermembranal protein movement. Comp Biochem Physiol 63A:159–167

Dawe AR, Spurrier WA (1969) Hibernation induced in ground squirrels by blood transfusion. Science 163:298–299

Ellory JC, Hall AC (1983) Ca^{2+} transport in hibernators and non-hibernator species' red cells at low temperature. J Physiol (Lond) 344:148P

Ellory JC, Willis JS (1982) Kinetics of the sodium pump in red cells of different temperature sensitivity. J Gen Physiol 79:1115–1130

Ellory JC, Willis JS (1983) Adaptive changes in membrane-transport systems of hibernators. Biochem Soc Trans 11:330–332

Galster WA (1978) Failure to initiate hibernation with blood from the hibernating arctic ground squirrel, *Citellus undulatus*, and eastern woodchuck, *Marmota monax*. J Therm Biol 3:93

Galster WA, Morrison PR (1975) Gluconeogenesis in arctic ground squirrels between periods of hibernation. Am J Physiol 228:325–330

Goldman SS, Albers RW (1979) Cold resistance of the brain during hibernation: changes in the microviscosity of the membrane and associated lipids. J Neurochem 32:1139–1142

Hall AC, Willis JS (1984) Differential effects of temperature on three components of passive permeability to potassium in rodent red cells. J Physiol (Lond) 348:629–643

Hand SC, Somero GN (1983) Phosphofructokinase of the hibernator *Citellus beecheyi*: temperature and pH regulation of activity via influences on the tetramer-dimer equilibrium. Physiol Zool 56:380–388

Heller HC (1979) Hibernation: neural aspects. Annu Rev Physiol 41:305–321

Hoo-Paris R, Castex CH, Sutter BChJ (1984) Alanine-turnover and coversion to glucose of alanine in the hibernating, arousing and active hedgehog *(Erinaceus europaeus)*. Comp Biochem Physiol 78A:159–161

Houslay MD, Palmer RW (1978) Changes in the form of Arrhenius plots of the activity of glucagon-stimulated adenylate cyclase and other hamster liver plasma-membrane enzymes occurring on hibernation. Biochem J 174:909–919

Hudson JW (1973) Torpidity in mammals. In: Whittow GC (ed) Comparative physiology of thermoregulation, vol III. Academic, New York, pp 97–165

Hudson JW, Wang LCH (1979) Hibernation: endocrinologic aspects. Annu Rev Physiol 41:287–303

Kilduff TS, Sharp FR, Heller HC (1983) Relative 2-deoxyglucose uptake of the paratrigeminal nucleus increases during hibernation. Brain Res 262:117–123

Kramarova LI, Kolaeva SH, Yukhananov Ryu, Rozhanets VV (1983) Content of DSIO, enkephalins and ACTH in some tissues of active and hibernating ground squirrels *(Citellus suslicus)*. Comp Biochem Physiol 74C:31–33

Kromer W (1980) Naltrexone influence on hibernation. Experientia 36:581–582

Llados-Eckman C, Beckman AL (1983) Reduction of hibernation bout duration by icv infusion of naloxone in *Citellus lateralis*. Proc Soc Neurosci 9:796

Lyman CP, Willis JS, Malan A, Wang LCH (eds) (1982) Hibernation and torpor in mammals and birds. Academic, New York

Malan A (1982) Respiration and acid-base state in hibernation. In: Lyman CP, Willis JS, Malan A, Wang LCHW (eds) Hibernation and torpor in mammals and birds. Academic, New York, pp 237–282

Margules DL, Goldman B, Finck A (1979) Hibernation: an opioid-dependent state? Brain Res Bull 4:721–724

McMurchie EJ, Gibson RA, Abeywardena MY, Charnock JS (1983) Dietary lipid modulation of rat liver mitochondrial succinate: cytochrome c reductase. Biochim Biophys Acta 727:163–169

Melnyk RB, Martin JM (1983) Lipolytic activity in brown adipocytes during spontaneous weight cycles in dormice. Physiol Behav 31:303–306

Melnyk RB, Mrosovsky N, Martin JM (1983) Spontaneous obesity and weight loss: insulin action in the dormouse. Am J Physiol 245:R396–R402

Minor JD, Bishop DA, Badger CR (1978) The golden hamster and the blood-borne hibernation trigger. Cryobiology 15:557–562

Montaudon D, Robert J, Canguilhem B (1984) Fluorescence polarization study of lipids and membranes prepared from brain hemispheres of a hibernating mammal. Biochem Biophys Res Commun 119:396–400

Musacchia XJ, Deavers DR (1981) The regulation of carbohydrate metabolism in hibernators. In: Musacchia XJ, Jansky L (eds) Survival in the cold. Elsevier/North Holland, Amsterdam, pp 55–75

Musacchia XJ, Jansky L (eds) (1981) Survival in the cold. Elsevier/North Holland, Amsterdam

Myers RD, Oeltgen PR, Spurrier WA (1981) Hibernation "trigger" injected in brain induces hypothermia and hypophagia in the monkey. Brain Res Bull 7:691–695

Nürnberger F (1983) Der Hypothalamus des Igels (*Erinaceus europaeus* L.) unter besonderer Berücksichtigung des Winterschlafes. Cytoarchitektonische und immuncytochemische Studien. Ph.D. Dissertation, University of Marburg, Federal Republic of Germany

Oeltgen PR, Spurrier WA (1981) Characterization of a hibernation trigger. In: Musacchia XJ, Jansky L (eds) Survival in the cold. Elsevier/North Holland, Amsterdam, pp 139–157

Oeltgen PR, Walsh JW, Hamann SR, Randall DC, Spurrier WA, Myers RD (1982) Hibernation "trigger": opioid-like inhibitory action on brain function of the monkey. Pharmacol Biochem Behav 17:1271–1274

Pehowich DJ, Wang LCH (1981) Temperature dependence of mitochondrial Ca^{2+} transport in a hibernating and nonhibernating ground squirrel. Acta Univ Carol Biol 1979:291–293

Pehowich DJ, Wang LCH (1984) Seasonal changes in mitochondrial succinate dehydrogenase activity in a hibernator, *Spermophilus richardsonii*. J Comp Physiol 154:495–501

Raison JK, Lyons JM (1971) Hibernation: alteration of mitochondrial membranes as a requisite for metabolism at low temperature. Proc Natl Acad Sci USA 68:2092–2094

Raison JK, McMurchie EJ, Charnock JS, Gibson RA (1981) Differences in the thermal behaviour of myocardial membranes relative to hibernation. Comp Biochem Physiol 69B:169–174

Wang LCH (1979) Time patterns and metabolic rates of natural torpor in the Richardson's ground squirrel. Can J Zool 57:149–155

Wang LCH (1984a) Mammalian hibernation. In: Morris GJ, Grout B (eds) The effects of low temperatures on biological systems. Edward Arnold, London (in press)

Wang LCH (1984b) Temperature regulation. In: Brush FR, Levine S (eds) Psychoendocrinology, vol II: Behaviors of regulation and species adaptation. Academic, New York (in press)

Wang LCH, Hudson JW (eds) (1978) Strategies in cold: Natural torpiditiy and thermogenesis. Academic, New York

Willis JS (1979) Hibernation: Cellular aspects. Annu Rev Physiol 41:275–286

Willis JS (1982) Intermediary metabolism in hibernation. In: Lyman CP, Willis JS, Malan A, Wang LCH (eds) Hibernation and torpor in mammals and birds. Academic, New York, pp 124–139

Willis JS, Ellory JC, Wolowyk MW (1980) Temperature sensitivity of the sodium pump in red cells from various hibernators and nonhibernator species. J Comp Physiol 138:43–47

Willis JS, Ellory JC, Cossins AR (1981) Membranes of mammalian hibernators at low temperatures. In: Morris GJ, Clarke A (eds) Effects of low temperatures on biological membranes. Academic, New York, pp 121–142

Yacoe ME (1983a) Adjustments of metabolic pathways in the pectoralis muscles of the bat, *Eptesicus fuscus*, related to carbohydrate sparing during hibernation. Physiol Zool 56:648–658

Yacoe ME (1983b) Protein metabolism in the pectoralis muscle and liver of hibernating bats, *Eptesicus fuscus*. J Comp Physiol 152B:137–144

The Adaptation of Membrane Structure and Lipid Composition to Cold

A.R. COSSINS and J.A.C. LEE[1]

1 Introduction

An important adaptive strategy during cold acclimation is the preservation of cellular performance despite the direct effects of cooling (Hazel and Prosser 1974). Membranes play an important part in this adaptive process. Passive permeability, the specific activity of membrane-bound enzymes and the transport activity of epithelia may all become enhanced during cold acclimation to compensate for the effects of cooling (reviewed in Cossins 1983). These different functional adaptations may result from several distinct mechanisms of change. For example, the concentration of membrane-bound proteins and enzymes may be altered, the amount and disposition of subcellular membrane systems may vary and, finally, the chemical composition of the bilayer may change.

It is a long-standing observation that the level of unsaturation of membrane lipids invariably increases during adaptation to cold, and there is now much evidence to show that this applies to microorganisms, and plants as well as animals. Based upon early work with phospholipid monolayers and with liposomes of defined chemical composition, these compositional adjustments have invariably been interpreted in an adaptive framework. Thus, the incorporation of additional cis-olefinic bonds into hydrocarbon chains introduces a kink into the chain which disrupts or offsets the closer packing of adjacent hydrocarbon chains that is induced by cooling. Thus, membranes of cold-acclimated organisms should have a greater cross-sectional area per molecule (i.e. a more expanded bilayer), a lower phase transition temperature and a less constrained molecular mobility than corresponding membranes of warm-acclimated organisms, predictions that were largely borne out by the work of Cullen et al. (1971). So pervasive are these ideas, that an adaptive interpretation is not infrequently made on the basis of changes of lipid saturation.

Since this hypothesis was originally formulated, concepts of membrane structure and of the functional significance of the relative mobility of membrane molecules have been considerably refined, as exemplified by the fluid-mosaic model of Singer and Nicholson (1972). Since many membrane processes and functions are affected by the fluidity of the bilayer (Quinn 1981) it follows that they will also be affected in an adaptive manner by compensatory adjustments of membrane lipid composition and membrane fluidity. The response is thus seen as resulting in the manipulation of the

1 Department of Zoology, University of Liverpool, P.O. Box 147, Liverpool L69 3BX, Great Britain

Circulation, Respiration, and Metabolism
(ed. by R. Gilles)

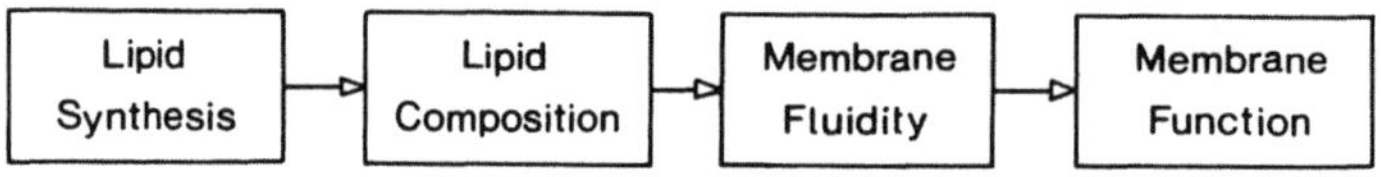

Fig. 1. The relationship between the elements involved in the hypothesis of homeoviscous adaptation

effective solvent viscosity of crucial cellular processes and as such should be regarded as a manifestation of a more general characteristic of cells, namely their ability to create and to maintain a complex set of membranes with distinct chemical compositions, and functional properties. The restoration of membrane fluidity when perturbed by temperature is termed homeoviscous adaptation (Sinensky 1974) but is a process that is probably not limited to temperature. Adaptive responses have been observed to dietary treatments (Cossins and Sinensky 1984), to chronic drug treatment (Littleton 1983), to hydrostatic pressure (Cossins and Macdonald 1984) and to dietary manipulation (Cossins and Sinensky 1984).

The linkage between membrane composition, fluidity and function as visualised by the hypothesis of homeoviscous adaptation is schematically illustrated in Fig. 1. Until recently, the vast majority of studies were concerned with analyses of lipid composition of differently-acclimated organisms and in some cases with demonstrations of functional differences between them. It is only more recently that techniques for measuring membrane fluidity have become available so that the scheme depicted in Fig. 1 can be more rigorously examined.

In this paper we briefly review the evidence for adaptations of membrane fluidity during temperature acclimation of fish and then consider in some detail the quantitative relationship between this phenomenon and the accompanying changes in lipid composition.

2 Fluidity Measurements: Technical Limitations

The term 'fluidity' as applied to biological membranes refers to the relative motional freedom of membrane (usually lipid) molecules. It is worth emphasising at the outset that this term and the term 'microviscosity' are somewhat imprecise in that the organisation and motional characteristics of membrane molecules differ fundamentally from that of bulk hydrocarbon fluids, for which the terms have a specific meaning. Molecules may be subject to different types of motion ranging from isomerisations of hydrocarbon chains (i.e. flexing), to wobbling motion of the whole molecule and finally to lateral diffusion of molecules in the plane of the bilayer. Each type of motion occurs over a rather different time domain and changes in one need not correlate with changes in another (Kleinfeld et al. 1981). No one biophysical technique is capable of providing information on all aspects of motion and consequently the estimate of fluidity provided by each technique is limited to the type(s) of motion sensed by that technique.

At present, spectroscopic probes provide the most accessible information on membrane fluidity. However, our understanding of the values obtained from these techniques depends critically upon the rigor of the theories used in their interpretation. This is nowhere more true than in the case of the fluorescence probe, diphenylhexatriene, (DPH), where polarisation of fluorescence was originally interpreted using assumptions which were subsequently revealed as invalid. It is now recognised that DPH polarisation (or anisotropy) more closely reflects the degree of hindrance to free probe rotation, and is thus a measure of order within the parallel array of hydrocarbon chains (i.e. a static concept) rather than a measure of the rate of wobbling motion (a dynamic concept). Time-resolved fluorescence anisotropy measurements now provide estimates of static order (r_∞) and of rate of motion (D_W, wobbling diffusion coefficient) which relate to the two physical characteristics just described (Cossins 1981).

Finally, it is necessary to recognise that probe studies provide average information on the membrane preparation under investigation. The distribution of probe binding sites within the preparation is largely unknown and hence the degree of heterogeneity that exists is not readily appreciated. The membrane is, therefore, treated as though it was a single hydrophobic phase with discrete and definable properties, rather than as a collection of distinct microdomains with rather different physical properties. This leads to difficulties of interpretation since it is not possible to distinguish a change in membrane fluidity that is caused by changes in *all* binding sites, from those caused by an alteration in only *some* of the binding sites, or from changes in the distribution of probes *between* different binding sites.

3 Adaptations of Membrane Order

Despite these problems there is a clear-cut difference in the DPH polarisation and r_∞ of membranes isolated from cold and warm-acclimated fish. Figure 2 illustrates the general observation of reduced polarisation in membranes of cold-acclimated fish compared to corresponding membranes of warm-acclimated fish. This indicates a reduced hindrance to probe wobbling motion and by inference a reduced membrane order in the former. In fish, this observation is typical of a variety of membrane types ranging from brain myelin to liver mitochondria (reviewed in Cossins 1983) and to carp erythrocytes (Cossins, unpublished observations).

The magnitude of the adaptive response has been estimated from comparison shown in Fig. 2 by measuring the shift of the curve along the temperature axis (in degrees centigrade) as a fraction of the difference in acclimation temperature. This quantity has been termed 'homeoviscous efficacy' (Cossins 1983); a value of one provides for identical polarisation at each acclimation temperature, and a value of zero indicates no adaptive response. Homeoviscous efficacy in fish membranes varies between 0.2–0.5, with higher values being observed in mitochondrial membranes of brain and liver and lowest values in brain myelin. In fish, only one case in which no response has been recorded, namely goldfish epaxial muscle sarcoplasmic reticulum (Cossins et al. 1978).

The time-course of the homeoviscous response has been studied in only a few instances. In goldfish brain synaptosomes, the transition from 5 °C to 25 °C required approx. 10–14 days, whilst the 25 °C–5 °C transition required 30–40 days (Cossins et al. 1977). Comparable values for microorganisms and protozoa are in the order of

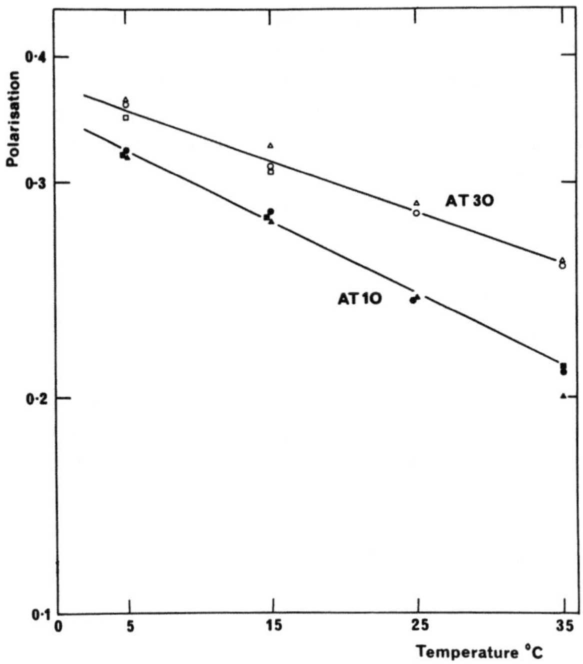

Fig. 2. The effect of thermal acclimation upon the DPH polarisation (on the *ordinate*) of membrane vesicles prepared from carp erythrocyte ghosts. Each *symbol* represents a separate preparation. The labels adjacent to each graph indicate 10 °C-acclimated (AT10) or 30 °C-acclimated carp (AT30). (Cossins, unpublished observations)

a few hours (Martin and Thompson 1978). More recently, Wodtke and Cossins (unpublished observations) have found a more rapid homeoviscous response in endoplasmic reticulum of carp liver. Polarisation changed over a period of 1–2 days following transfer of 25 °C acclimated fish to 5 °C. Interestingly, there are two peaks of fatty acid desaturase activity in these membranes during the transition from warm to cold (E. Wodtke, personal communication), only one of which corresponds to changes in polarisation. Evidently, the second burst of desaturase activity is concerned with compositional modifications which are not detected by DPH polarisation.

4 Lipid Composition and Membrane Fluidity

A great many workers use changes in lipid composition during temperature acclimation as the principal and usually only evidence of adaptive change. In most cases these studies focused on changes in phospholipid fatty acid composition, partly because the observed changes were generally clear-cut and relatively consistent between different studies, but also because of the widely perceived and apparently straightforward relationship between lipid unsaturation and membrane fluidity.

Table 1 illustrates a typical comparison in which saturated fatty acids show a modest reduction, and unsaturated fatty acids a corresponding increase, in membranes of cold-acclimated fish relative to warm-acclimated fish. The crucial questions are whether shifts of this magnitude lead to a significant change in membrane physical properties, and

Table 1. A typical comparison of the fatty acid composition of ethanolamine phosphoglycerides from liver microsomes of cold- and warm-acclimated green sunfish (after Cossins et al. 1980)

Fatty acid	5 °C Acclimation temperature (n = 3)	25 °C Acclimation temperature (n = 4)
16:0	7.5 ± 1.6	13.9 ± 0.8
16:1	3.0 ± 1.1	3.5 ± 0.6
18:0	8.9 ± 0.5	13.3 ± 1.5
18:1	22.4 ± 0.7	13.4 ± 0.7
18:2ω6	5.9 ± 0.7	10.0 ± 1.4
20:4ω6	3.0 ± 0.6	5.0 ± 0.4
22:6ω3	32.4 ± 1.7	27.6 ± 1.6
Others[a]	16.9	13.3
Saturated	16.8 ± 1.8	28.0 ± 2.3
Monosaturated	27.5 ± 1.8	17.6 ± 1.0
Polyunsaturated	55.7 ± 3.6	54.3 ± 2.8
Unsaturation index	294.6 ± 12.5	265.0 ± 10.4
Saturation ratio	0.203 ± 0.03	0.395 ± 0.05

[a] Other components present in trace quantities incluced 17:0, 17:1ω9, 18:3ω3, 20:4ω3, 20:5ω3, 22:0, 22:2ω6, 22:3ω6, 22:4ω6, 22:5ω3, 22:5ω6. Values represent means ± SEM of n separate preparations

what specific aspects of the change in composition are important to adaptations of membrane fluidity? It is worth pointing out that, at present, there is no satisfactory means of predicting the physical consequences of a complex shift in lipid saturation of the type shown in Table 1. This is due mainly to the fact that, until recently, studies with artificial bilayers used lecithins which contained only saturated and monounsaturated fatty acids but not polyunsaturated fatty acids in the combinations in which they occur in nature. In addition, gross hydrocarbon unsaturation as shown in Table 1 is only one factor amongst many which significantly influence bilayer properties and which are only considered in a few highly detailed studies. These factors include double bond position (Barton and Gunstone 1975), positional distribution of fatty acids on phospholipids, phospholipid headgroup composition and cholesterol content (Stubbs and Smith 1984).

In searching for the quantitative relationship between lipid composition and membrane fluidity it is necessary to express the fatty acid composition in some numerical way. A particularly common index is the average number of unsaturated bonds per fatty acid molecule or per 100 molecules; the so-called unsaturation index (UI). This index is based upon the premise that each unsaturated bond has an equivalent effect upon bilayer properties irrespective of whether it is the first unsaturation bond in a hydrocarbon chain, or the fourth or fifth. An alternative index is simply to group the fatty acids into saturated or unsaturated classes, or into n−3, n−6, and n−9 classes according to their biosynthetic origin. The former scheme treats all unsaturated fatty acids as similar in their effects upon membrane fluidity, whilst the latter places more importance on the position of the terminal olefinic bond and length of the terminal uninterrupted segment of the hydrocarbon chain. None of these indices address the problem of how these different fatty acids interact with cholesterol or with intrinsic proteins.

4.1 Physical Studies

A good indication of the relative importance of these various indices is provided by studies of the physical properties of artificial bilayers composed of mixed-acid lecithins containing polyunsaturated fatty acids. Table 2 shows the gel ⇌ liquid-crystalline transition temperature (T_c) observed by differential scanning calorimetry for phosphatidylcholines containing a saturated fatty acid in the sn−1 position and an unsaturated fatty acid in the sn−2 position (Coolbear et al. 1983). The most important point to be gleaned from this data is that each successive double bond did *not* elicit an equivalent reduction in T_c. Introduction of the first double bond caused a reduction in T_c of approx. 50 °C, the second caused a further reduction of 20 °C whilst the third and fourth caused no further reduction but a small increase. According to Coolbear et al. (1983) this is because although each successive double bond causes an increase in the cross-sectional area per molecule, it also reduces the rotational freedom of other carbon-carbon bonds in the chain. These two effects roughly counterbalance each other for highly polyunsaturated fatty acids.

This important conclusion has been substantiated by monolayer studies (Ghosh et al. 1972; Evans and Tinoco 1978) and by a study of time-resolved fluorescence polarisation of DPH in bilayers containing polyunsaturated fatty acids (Stubbs et al. 1981). Very small or no differences were observed between bilayers with different polyunsaturation either in the rate of probe motion (D_w) or extent of motion (r_∞) over a wide range of temperatures. These diverse studies amply demonstrate that there is no systematic or simple relationship between bilayer physical properties and their overall degree of unsaturation (i.e. UI). Based on the foregoing consideration a more appropriate index would comprise the percentage of fatty acids that are saturated or perhaps the ratio of saturated to unsaturated fatty acids (saturation ratio, SR). This could be elaborated perhaps by weighting monounsaturated and polyunsaturated fatty acids differently to take account of their difference in phase transition temperatures.

Table 2. The gel ⇌ liquid crystalline transition temperature (T_c) for artificial bilayers composed of lecithins with increasing numbers of olefinic bonds (after Coolbear et al. 1983)

Lecithin sn-1 Position	Lecithin sn-2 Position	T_c (°C)
18:0	18:0	54–58
18:0	18:1	6.3
18:0	18:2	−16.2
18:0	18:3	−13
18:0	20:4	−12.6
18:1	18:1	−17.6

However, this is not to say that variations in polyunsaturated fatty acids have no significance, but that they do not appear to affect bulk physical properties as measured by several independent techniques. It is clear from the accumulation of polyunsaturated fatty acids in certain tissues and their preservation when deprived of dietary sources (Stubbs and Smith 1984) that they serve an essential though, as yet, undefined function. Moreover, there are numerous observations of a specificity of changes in lipid composition during thermal acclimation (reviewed in Hazel 1980), either between phosphatide classes or between (n–6) or (n–3) families of fatty acids. It is likely that this specificity serves some functional purpose though its precise nature is unknown.

4.2 Correlations During Temperature Acclimation

In the comparison shown in Table 1 there is distinct change in the fatty acid profile in the expected way during thermal acclimation. Thus, the proportions of 16:0 and 18:0 decrease and 18:1 and 22:6 increase on cold acclimation. Note, however, that the proportion of 18:2 decreases. However, in this particular case there is no significant difference in UI whilst SR is significantly different. In other cases UI does show significant changes with acclimation temperature, but clearly, on the basis of this and other examples, UI is not a reliable indicator of fluidity adaptations.

Measurements during the transfer of temperature-acclimated animals to either lower or higher temperatures permits a more extensive study of the relationship between fluidity and saturation than the simple comparison of cold- and warm-acclimated animals. Martin and Thompson (1978) followed DPH polarisation and lipid composition of 39.5 °C acclimated *Tetrahymena* following a shift to 15 °C. Figure 3a shows the relationship in graphical form as calculated from their data. Both UI and SR display the expected systematic changes with polarisation. Cossins et al. (1977) have performed a similar correlation for the brain synaptosome fraction during transfer of temperature-acclimated goldfish to lower or higher temperatures. Although the rates of compositional and fluidity adaptations were appreciably faster during warm acclimation than during cold acclimation, the underlying relationships between polarisation and unsaturation were similar. As a general trend, the expected increases in unsaturation were observed on cooling, and vice versa. However, the UI did not, in general, provide a statistically significant correlation with polarization, whereas SR did (Fig. 3b). The compositional adjustments in goldfish membranes were more pronounced in the ethanolamine and serine/inositol phosphoglycerides compared to the choline phosphoglycerides. In these classes the most clearcut changes were between 18:0 and 18:1, which obviously would have virtually no effect upon UI. An important difference between *Tetrahymena* and fish membranes is the large proportion of highly polyunsaturated fatty acids (4–6 olefinic bonds) found in the latter. The UI is dominated in particular by changes in proportion of 22:6 or 20:5 and small variations in these fatty acids produce a disproportionate effect upon UI. Moreover, it is not uncommon for 22:6 to show paradoxical changes during thermal acclimation (Cossins 1976).

Cossins and Prosser (1978) have examined the relationship between DPH polarisation and lipid composition in a comparative study of fish species from diverse thermal environments and from small mammals. In general, there was increased unsaturation of synaptosomal lipids and decreased values of DPH polarisations as the habitat or body

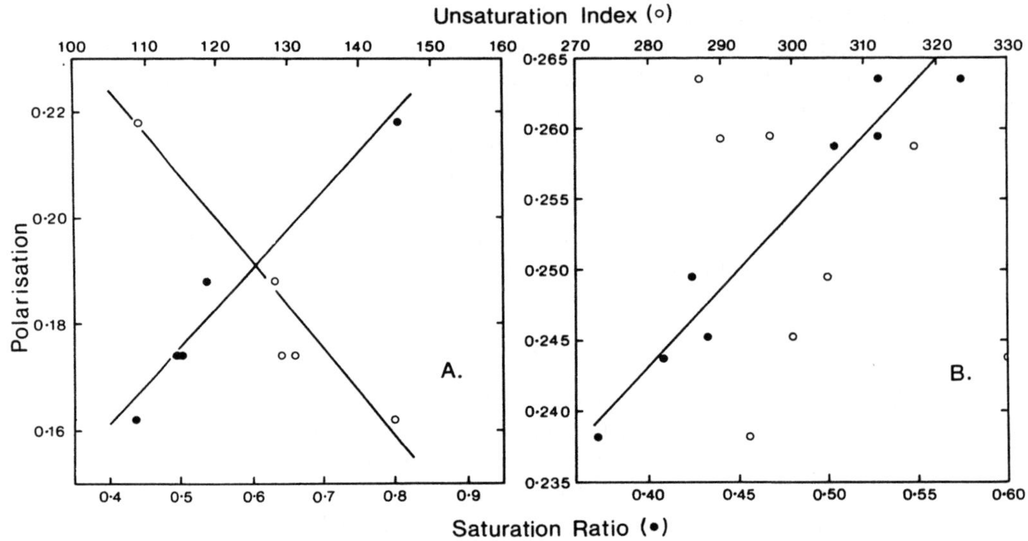

Fig. 3A,B. The relationship between two indices of membrane fatty acid composition and membrane fluidity (as expressed by DPH polarisation) during cold acclimation. The *left-hand graph* is for the microsomal membranes of *Tetrahymena* during transfer of 39.5 °C-acclimated cells to 15 °C (data from Martin and Thompson 1978). The *right-hand graph* is for phosphatidyl-ethanolamine of brain synaptosomes of 5 °C-acclimated goldfish during transfer to 25 °C (after Cossins et al. 1977). The *solid symbols* are for saturation ratio (SR) whilst the *open symbols* are for unsaturation index (UI). The accompanying regression lines were determined by the least squares procedure. The correlation coefficients for SR and UI of *Tetrahymena* were 0.984 ($P = 0.002$) and −0.958 ($P = 0.009$), respectively. The corresponding correlation coefficient for the goldfish study were 0.969 ($P = 0.001$) and 0.33 ($P = 0.05$)

temperature was reduced. The SR consistently provided a more significant correlation than UI, though for the predominant phosphoglycerides both were statistically significant.

4.3 Dietary Studies

Membrane lipid composition may be greatly influenced by dietary lipids and in many experimental studies these variations produce noticeable effects upon fluidity and function. However, there are an increasing number of other studies in which dietary-induced compositional variations do not produce equivalent changes in fluidity (reviewed by Cossins and Sinensky 1984). Di Costanzo et al. (1983) produced dietary-induced modifications of the fatty acid composition, the cholesterol content and the phospholipid/protein ratio of an intestinal brush border preparation of trout. However, the DPH polarisation was not affected. Essentially, similar conclusions have been made by Stubbs et al. (1980) in lymphocytes, by Herring et al. (1980) in *Dictylostelium*, and by Cossins and Wilkinson in an unpublished study of hamster kidney microsomes. In their timely review of dietary manipulation of membrane lipid composition in mammalian cells, Stubbs and Smith (1984) found that in many dietary studies the proportion of saturated fatty acids remained remarkably constant. Modifications to the composition of fatty acids occurred largely by substitution within the unsaturated fatty acids. They suggest that the constancy of saturated fatty acids may relate to the segregation of saturated and unsaturated fatty acids between the sn-1 and sn-2 positions, respectively.

5 Conclusions

The quantitative link between adaptations of membrane fluidity and changes in membrane lipid saturation during temperature acclimation is now firmly established, though it is not possible to conclude that other compositional changes are without effect. Improved correlations will follow from yet more detailed studies of chemically-defined artificial membrane systems containing polyunsaturated fatty acids together with cholesterol and/or proteins. Fatty acid composition in higher animals is highly complex and only part of the important adaptive changes are revealed by a gross comparison of membranes from differently acclimated animals. Virtually no information is revealed by a comparison of cells, tissue or whole animals because the membrane-type composition is heterogeneous and undefined. At present, it is not certain precisely which features of lipid composition are important to adaptations of fluidity or function beyond the gross correlations discussed here.

Physical studies of lecithins containing different levels of unsaturation suggest that the most effective means of adaptively altering membrane fluidity would be to alter the proportion of fatty acids that are saturated. This is demonstrably the case in several studies where DPH polarisation is statistically correlated with changes in the total % of saturated fatty acids or with indices based on this. Correlations with UI are not necessarily observed, particularly in cases where large proportions of highly polyunsaturated fatty acids exist. Thus, in studies where lipid composition only is investigated, more attention should be paid to the characteristics of the observed change. It is not valid to use altered lipid composition as the principal or only evidence of adaptive changes in bulk membrane fluidity. Confirmation with physical techniques is necessary.

References

Barton PG, Gunstone FD (1975) Hydrocarbon chain packing and molecular motion in phospholipid bilayers formed from unsaturated lecithins. Synthesis and properties of sixteen positional isomers of 1,2 dioctadecenoyl-sn-glycero-3-phosphorylcholine. J Biol Chem 250:4470–4476

Coolbear KP, Berde CB, Keough KMW (1983) Gel to liquid-crystalline phase transitions of aqueous dispersions of polyunsaturated mixed-acid phosphatidylcholines. Biochemistry 22:1466–1473

Cossins AR (1976) Changes in muscle lipid composition and resistance adaptation to temperature in the freshwater crayfish *Austropotamobius pallipes*. Lipids 11:307–316

Cossins AR (1981) Steady state and dynamic fluorescence studies of the adaptation of cellular membranes to temperature. In: Beddard GS, West MA (eds) Fluorescent probes. Academic, London, pp 39–80

Cossins AR (1983) The adaptation of membrane structure and function to changes in temperature. In: Cossins AR, Sheterline P (eds) Cellular acclimation to environmental change. Cambridge University Press, Cambridge, UK, pp 3–32

Cossins AR, Macdonald AG (1984) Homeoviscous adaptation under pressure: II. The molecular order of membranes from deep-sea fish. Biochim Biophys Acta 776:144–150
Cossins AR, Prosser CL (1978) Evolutionary adaptation of membranes to temperature. Proc Natl Acad Sci USA 75:2040–2043
Cossins AR, Sinensky M (1984) Adaptation of membranes to temperature, pressure and exogenous lipids. In: Shinitzky M (ed) Physiology of membrane fluidity, vol II. CRC Press, Boca Raton, Florida, p 1–20
Cossins AR, Friedlander MJ, Prosser CL (1977) Correlations between behavioural temperature adaptations of goldfish and the viscosity and fatty acid composition of their synaptic membranes. J Comp Physiol 120:109–121
Cossins AR, Christiansen J, Prosser CL (1978) The adaptation of biological membranes to temperature. The lack of homeoviscous adaptation in the sarcoplasmic reticulum. Biochim Biophys Acta 511:442–454
Cossins AR, Kent J, Prosser CL (1980) A steady state and differential polarised phase fluorimetric study of the liver microsomal and mitochondrial membranes of thermally acclimated green sunfish *(Lepomis cyanellus)*. Biochim Biophys Acta 599:341–358
Cullen J, Phillips MC, Shipley GG (1971) The effect of temperature on the composition and physical properties of the lipids of *Pseudomonas fluorescens*. Biochem J 125:733–742
Di Costanzo G, Duportail G, Florentz A, Leray C (1983) The brush border membrane of trout intestine: Influence of its lipid composition on ion permeability, enzyme activity and membrane fluidity. Mol Physiol 4:279–290
Evans RW, Tinoco J (1978) Monolayers of sterols and phosphatidylcholines containing a 20-carbon chain. Chem Phys Lipids 22:207–220
Ghosh D, Lyman RL, Tinoco J (1972) Behaviour of specific natural lecithins and cholesterol at the air-water interface. Chem Phys Lipids 7:173–184
Hazel JR (1980) The regulation of membrane lipid composition the thermally-acclimated poikilotherms. In: Gilles R (ed) Animals and environmental fitness. Pergamon, Oxford, UK
Hazel JR, Prosser CL (1974) Molecular mechanisms of temperature compensation in poikilotherms. Physiol Rev 54:620–677
Herring FG, Tatischeff I, Weeks G (1980) The fluidity of plasma membranes of *Dictyostelium discoideum*. The effects of polyunsaturated acid incorporation assessed by fluorescence depolarisation and electron paramegnetic resonance. Biochim Biophys Acta 602:1–9
Kleinfeld AM, Dragsten P, Klausner RD, Pjura WJ, Matayoshi ED (1981) The lack of relationship between fluorescence polarisation and lateral diffusion in biological membranes. Biochim Biophys Acta 649:471–480
Littleton JM (1983) Membrane reorganisation and adaptation during chronic drug exposure. In: Cossins AR, Sheterline P (eds) Cellular acclimatisation to environmental change. Cambridge University Press, Cambridge, UK, pp 145–160
Martin CE, Thompson GA Jr (1978) Use of fluorescence polarisation to monitor intracellular membrane changes during temperature acclimation. Correlation with lipid compositional and ultrastructural changes. Biochemistry 17:3581–3586
Quinn PJ (1981) The fluidity of cell membranes and its regulation. Prog Biophys Mol Biol 38:1–104
Sinensky M (1974) Homeoviscous adaptation – a homeostatic process that regulates the viscosity of membrane lipids in *Escherichia coli*. Proc Natl Acad Sci USA 71:522–525
Singer SJ, Nicholson GL (1972) The fluid mosaic model of the structure of cell membranes. Science 175:720–731
Stubbs CD, Smith AD (1984) The modification of mammalian membrane polyunsaturated fatty acid composition in relation to membrane fluidity and function. Biochim Biophys Acta 779: 89–137
Stubbs CD, Tsang WM, Belin J, Smith AD, Johnson SM (1980) Incubation of exogenous fatty acids with lymphocytes. Changes in fatty acid composition and effects on the rotational relaxation time of 1,6-diphenyl-1,3,5-hexatriene. Biochemistry 19:2756–2762
Stubbs CD, Kouyama T, Kinosita K Jr, Ikegami A (1981) Effect of double bonds on the dynamic properties of the hydrocarbon region of lecithin bilayers. Biochemistry 20:4257–4262

Biochemical Adaptation to the Freezing Environment – Structure, Biosynthesis and Regulation of Fish Antifreeze Polypeptides

C.L. HEW[1] and G.L. FLETCHER[2]

1 Introduction

The ability of many organisms to survive in cold environments is of considerable scientific interest and importance. During the past 2 decades, two entirely different animal groups have been investigated for their ability to survive at subzero temperatures. There are the marine fishes inhabiting the ice-laden seawater of the polar regions (Feeney and Yeh 1978; Ananthanarayanan and Hew 1978; DeVries 1982) and the overwintering terrestrial arthropods (Duman 1982). In order to survive, these organism have evolved mechanisms which include seasonal migrations to warmer and/or deeper waters, increased concentrations of glucose, glycerol, sorbitol, and other small molecules. Some organisms also appear to survive in a super-cooled state (Scholander et al. 1957). However, the most intriguing mechanism is the occurrence of an unique class of serum proteins which specifically lowers the freezing temperature below that of the surrounding environment. These proteins, commonly known as antifreeze proteins, have many interesting features. However, in the present communication, only the structural and biosynthetic aspects of these macromolecules from the marine fishes will be discussed.

2 Presence of Antifreeze Glycoproteins and Antifreeze Polypeptides

Seawater temperatures around the world vary tremendously, depending on latitude and season. In the polar and subpolar regions, the water temperatures could be as low as -1.8 °C. However, the serum of most fishes freezes at -0.6 °C and cannot tolerate lower temperatures (Holmes and Donaldson 1969). Scholander et al. (1957) were the first to make observations on freezing resistance in fishes. They reported that the blood sera of Arctic fish had a lower freezing temperature than did the blood sera of fish not adapted to the cold. The lowering of the serum freezing temperature was due to macromolecular materials which were soluble in trichloroacetic acid. DeVries et al. (1970) isolated and characterized these materials from Antarctic fish and found them

1 Research Institute, Hospital for Sick Children, Toronto, Ontario and Department of Clinical Biochemistry, University of Toronto, Toronto, Ontario M5G 1L5, Canada

2 Marine Sciences Research Laboratory, Memorial University of Newfoundland, St. John's, Newfoundland A1C 5S7, Canada

Circulation, Respiration, and Metabolism
(ed. by R. Gilles)

to be glycoproteins of unusual chemical structure. The glycoprotein antifreeze (AFGP) from the two species of Antarctic fish studied, *T. borchgrevinki* and *D. mawsoni* are identical and contain eight polypeptides made up of a repeating tripeptide unit of alanine-alanine-threonine linked to galactose-N-acetylgalactosamine. This unit is repeated up to 50 times to give eight components of 2,600 to 33,000 daltons. Since then, AFGP have been isolated from the Northern cods, *B. saida, E. gracilis, G. ogac* and more recently the Atlantic cod, *G. morhua* and Atlantic tomcod, *M. tomcod* (Fletcher et al. 1982 and references within). All of these AFGP have similar, if not identical structure. The smaller glycopeptides contain proline which replace alanine at certain positions. Amino acid sequence determinations show that the positions of the proline vary from species to species. Physical studies on some of these AFGP suggest either an extended conformation or a completely flexible random coil. Chemical and enzymatic modifications of the carbohydrate side-chain such as periodate oxidation and acetylation show that the hydroxy groups of the carbohydrate moiety are important for the glycoprotein antifreeze function (Feeney and Yeh 1978).

Investigations on the antifreezes of fish species from the North Atlantic coast have shown that there are macromolecular antifreezes which lack the carbohydrate moiety that is characteristic of the AFGP. These antifreeze polypeptides (AFP) have been isolated from the winter flounder, shorthorn sculpin, sea raven, and ocean pout (Duman and DeVries 1976; Hew and Yip 1976; Hew et al. 1980; Slaughter et al. 1981; Hew et al. 1984b,c; Fourney et al. 1984d).

3 Comparative Studies on Antifreeze Polypeptides

Unlike AFGP, AFP are diverse in their chemical structure. The identification, distribution, and characterization of these AFP are facilitated by the combination of two methods, namely, the freezing point osmometer and reverse phase HPLC (Hew and Yip 1976; Hew et al. 1980; Slaughter et al. 1981; Hew et al. 1984a; Fourney et al. 1984d). The osmometer provides a simple and fast monitoring procedure for antifreeze activity in crude serum preparations while reverse phase HPLC has the high resolving power required to fractionate the AFP.

Based on the information available on their amino acid composition, protein sequence, and secondary structure, at least three distinct types of AFP can be recognized. The first is the alanine-rich AFP from the winter flounder (Hew and Yip 1976; Duman and DeVries 1976) and shorthorn sculpin (Hew et al. 1980; Hew et al. 1984b). Both flounder and sculpin AFP share homology in primary and secondary structure. The second group is the cystine-rich AFP from sea raven (Slaughter et al. 1981) and some insect AFP (Schneppenheim and Theede 1980; Duman et al. 1982; Hew et al. 1983). This family is sensitive to reducing agents and possesses a β-turn secondary structure (Slaughter et al. 1981). The third type, represented by ocean pout AFP is neither alanine-rich nor cystine-rich (Hew et al. 1984c). Table 1 shows the amino acid compositions of these three types of AFP and some of their structural characteristics.

Another interesting feature of these AFP is the presence of multiple polypeptides within each species. There are at least 7, 5, 11, and 2 active components in flounder,

Table 1. Amino acid composition of antifreeze polypeptides[a]

	Winter flounder Component No. 6	Shorthorn sculpin SS-8	Ocean pout HPLC-5	Sea raven SR-2
Asx	4	2	4	7
Thr	4	2	7	6
Ser	1	1	2	6
Pro	0	1	6	6
Glx	1	2	7	10
Gly	0–1	1	5	7
Ala	23	27	6	13
Half Cys	0	0	0	8
Val	0	0	8	3
Met	0	1	5	8
Ile	0	1	5	3
Leu	2	2	3	6
Tyr	0	0	1	1
Phe	0	0	1	2
Lys	1	4	5	2
His	0	0	0	3
Arg	1	1	0	2
Trp	0	0	0	+
Molecular weight	3,300	4,000	6,800	9,700
Secondary structure	α-Helix	α-Helix	Expand	β-Turn
Activity inhibited by	Carboxyl group modification	Carboxyl group modification	N.D.	Reducing agents

[a] The number of residues for SR-2, based on amino acid analysis, can only be taken as approximate values. The values for other AFP were based on amino acid and sequence analysis

sculpin, ocean pout, and sea raven AFP respectively as revealed by reverse phase HPLC (Fig. 1). Since individual animals exhibit the same elution profiles as pooled samples, and the serum preparations contain protease inhibitors, it is unlikely that the microheterogeneity is due to the pooling of samples or to proteolytic digestion. This evidence indicated the presence of multiple gene families and de novo post-translational modifications of the nascent chains (Davies et al. 1984; Pickett et al. 1984).

The variation of AFP from different geographical locations is minimal. HPLC analysis of winter flounder AFP from Newfoundland, Nova Scotia, New Brunswick, and Long Island, New York, does not reveal any significant differences in their profiles (Fourney et al. 1984c). This is also true for Newfoundland and New Brunswick ocean pout AFP (Fletcher et al. 1984a).

The mechanisms by which antifreezes inhibit freezing is not fully understood (Feeney and Yeh 1978; DeVries 1982). These comparative studies on their structure are useful in that they may elucidate common features which are essential to their mechanisms of action. On the other hand, the structural diversity of the AFP and AFGP may indicate a diversity of mechanisms.

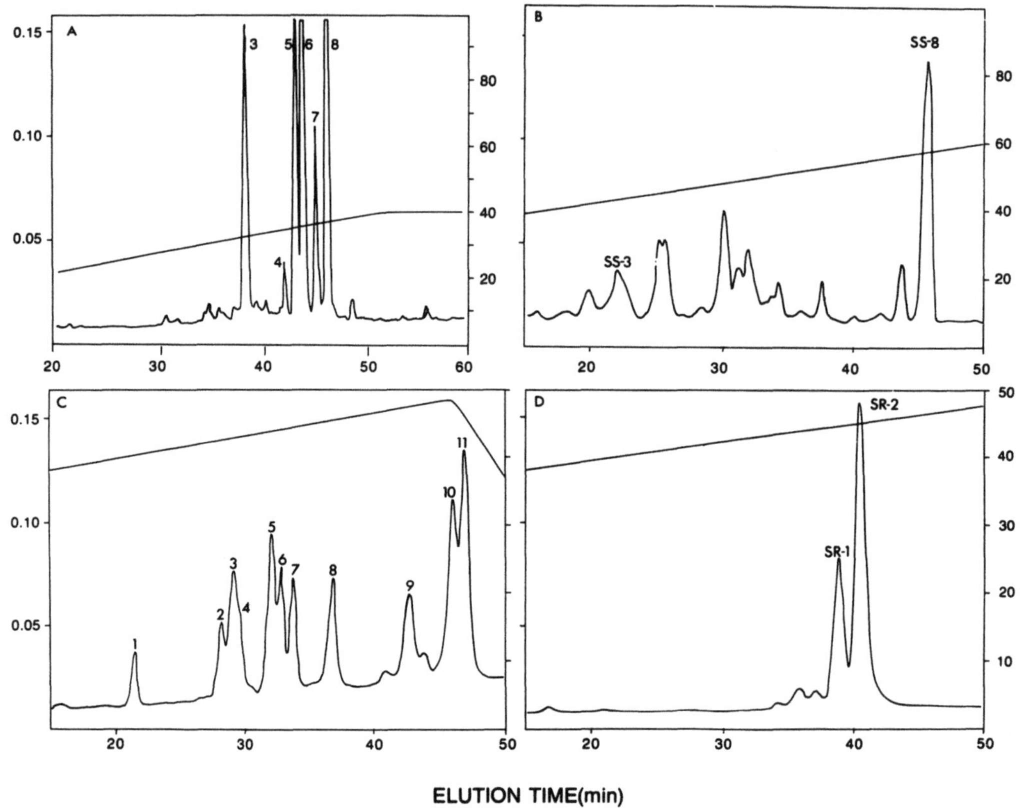

Fig. 1A–D. Analysis of fish antifreeze polypeptides by reverse phase HPLC (**A**) flounder AFP, Altex Ultrasphere ODS column (4.6 mm × 25 cm), flow rate 1 ml min^{-1} with triethylamine phosphate buffer 0.01 *M*, pH 3.0-acetonitrile gradient. AFP are designated from *3 to 9. Component 9* is eluted later in the gradient (~10 min, not shown in figure). The conditions for (**B**) shorthorn AFP, (**C**) ocean pout AFP, (**D**) sea raven AFP were: Water's μBondapak C18 column (7.1 mm × 30 cm), flow rate 1 ml min^{-1}, buffer used was 0.05% trifluoroacetic acid-acetonitrile gradient

4 Primary Structure of Polypeptide Antifreeze

For the past 3 years, significant progress has been made on the structural characterization of these AFP. There are now three known polypeptide sequences for flounder AFP (DeVries and Lin 1977; Davies et al. 1982; Picket et al. 1984), two sequences for shorthorn sculpin (Hew et al. 1984b) and one for ocean pout (Hew et al., unpublished results). Figure 2 shows the amino acid sequences of these AFP.

We have determined the structure of the two major components (6 and 8) of the flounder AFP using both protein and DNA sequencing procedures (Davies et al. 1982; Pickett et al. 1984). The structural variation between these two components consists of three substitution at Lys_{18} → Ala, Glu_{22} → Lys, and Ala_{26} → Asp from component 6 to 8. The amino acid sequence for component 8 represents an extension

Flounder AFP

\#6 (10: Ala; 20: Ala; 30: Ala)

Asp Thr Ala Ser Asp Ala Ala Ala Ala Ala Ala Leu Thr Ala Ala Asn Ala Lys Ala Ala Ala Glu Leu Thr Ala Ala Asn Ala Ala Ala Ala Ala Ala Ala Thr Ala Arg

\#8

Asp Thr Ala Ser Asp Ala Ala Ala Ala Ala Ala Leu Thr Ala Ala Asn Ala Ala Ala Ala Ala Lys Leu Thr Ala Asp Asn Ala Ala Ala Ala Ala Ala Ala Thr Ala Arg

AFP-3

Asp Thr Ala Ser Asp Ala Ala Ala Ala Ala Ala Leu Thr Ala Ala Asx Ala Ala Ala Ala Ala Lys Leu Thr Ala Asx Asx Ala Ala Ala Ala Ala Ala Ala Thr Ala Ala

Sculpin AFP

SS-8 (10: Ala)

Met Asn Gly Glu Thr Pro Ala Gln Lys Ala Ala Arg Leu Ala Ala Ala Ala Ala Leu Ala Ala Lys Thr Ala Ala Asp Ala Ala Ala Lys Ala Ala Ala Lys Ala Ala
Ala Ile Ala Ala Ala Ala Ala Ser Ala

SS-8 (3: Ala)

Met Asp - Ala Pro Ala Ala Arg Ala Ala Ala Lys Thr Ala Ala Asp Ala Leu Ala Ala Ala Lys Lys Thr Ala Ala Asp Ala
Ala Ala Ala Ala Ala Ala Ala

Ocean Pout AFP

Clone 77 (1: Gln; 10: Pro; 20: Met; 30: Ile)

Gln Ser Val Val Ala Thr Gln Leu Ile Pro Ile Asn Thr Ala Leu Thr Pro Ala Met Met Gln Gly Lys Val Thr Asn Pro Ile Gly Ile Pro Ple Ala Glu Met Ser

(40: Gly; 41: Lys; 60: Thr)

Gln Ile Val Gly Lys Gln Val Asn Thr Pro Val Ala Lys Gly Gln Thr Leu Met Pro Asn Met Val Lys Thr Tyr Val Ala Gly Lys

Fig. 2. Amino acid sequences of some antifreeze polypeptides. The C-terminal glycine is absent in mature flounder AFP Nos. 6 and 8. Its presence has been derived from nucleotide sequences. Sculpin AFP SS-3 is aligned with a gap between residue 2 and 3 in order to obtain maximum homology with SS-8

and correction of AFP-3 published earlier by DeVries and Lin (1977). Aside from delineating the position of Asn and Asp, the sequence in component 8 has arginine in the C-terminal in place of alanine. One explanation for this discrepancy may lie in the post-translational modification that the AFP undergo. The cDNAs of both components 6 and 8 contain a glycine codon proceding the termination codon, though glycine is not present in the mature peptides. If glycine is removed from the C-terminal position after synthesis, it is conceivable that in some samples, an additional residue (i.e. arginine) is also removed. Indeed, the amino acid composition of one of the minor component 5 is consistent with the loss of arginine from the C-terminal position of component 6 (Fourney et al. 1984d). The presence of alanine at residue 37 in AFP-3 could be due to a carry-over from residue 36 during automated Edman degradation.

While the structural variation between the two major flounder AFP is due to three separate point mutations as discussed earlier, the difference between the two sculpin AFP (SS-8 and SS-3) is due instead to a deletion of 11 amino acid residues at the N-terminal of SS-3 (Hew et al. 1984b). The activity of SS-3 is much lower probably as a consequence of this deletion. Consistent with our earlier experiments on enzymic digestion and peptide mapping (Hew et al. 1980), amino acid sequences comparison of flounder and sculpin AFP has confirmed that these two AFP are structurally homologous. When aligned for maximum homology, component 6 and SS-8 are 60% identical. Furthermore, both components contain three structural repeats of an 11-amino acid fragment of Thr-x-x-polar aa-xxxxxxx where x is alanyl residues. This repeat forms an amphiphilic helical structure where all the hydrophilic amino acid side chains are clustered together. These structural features will be useful in understanding the mechanism of action of AFP.

In contrast, the amino acid sequence of an ocean pout AFP deduced from a cDNA clone has no structural similarities whatsoever with the flounder and sculpin AFP (Hew et al., unpublished results). Besides being a larger polypeptide of 65 amino acid residues, ocean pout AFP has no obvious repeating structure. Both point mutation and post-translational modification contribute to the heterogeneity observed in ocean pout AFP as shown by reverse phase HPLC (Hew et al., unpublished results).

The structure of sea raven AFP is also being investigated in our laboratories. Due to the vast differences in its amino acid composition, it is unlikely that the sea raven AFP will share any common features with the other AFP. Similarly, except for amino acid composition, little structural information is available for insect antifreezes.

5 Biosynthesis of Antifreeze

The seasonal cycle of AFP production and degradation in the winter flounder proved invaluable in the establishment of the biosynthetic events controlling AFP synthesis. AFP are synthesized in the liver during the fall and winter resulting in winter plasma AFP level of 5 to 10 mg ml^{-1}, accounting for approx. 10%–20% of the total plasma proteins (Slaughter and Hew 1982). The two major components, namely 6 and 8, are encoded by two closely related yet distinct liver poly A^+ mRNA of approx. 7.5 S

(Hew and Yip 1976; Lin 1979; Davies and Hew 1980). Each mRNA synthesizes a 82 amino acid residue precursor protein, the prepro AFP (Davies et al. 1982; Pickett et al. 1984). As demonstrated by automated Edman degradation as well as by cDNA cloning and nucleotide sequencing, both the pre and pro sequences in component 6 and 8 are identical. The difference in components 6 and 8 therefore occurs only in the mature polypeptides. The C-terminal glycine in both polypeptides is removed by post-translational modification.

The pre sequence in the precursor protein consists of 21 amino acid residues with a cluster of hydrophobic amino acids which is typical for signal peptides of most secretory proteins. Its function, apparently is to facilitate the vectorial discharge of the nascent polypeptide chains across the cisternal membranes (Blobel and Dobberstein 1975). Unlike the pre sequence, the function of the 23 amino acid residues of pro segment in the flounder AFP is unknown. This pro sequence is rich in alanine and proline. It shares a structural feature with honey-bee promelittin where every other amino acid is either alanine or proline. Kreil et al. (1980) have isolated an aminodipeptidase which has a substrate specificity for these two amino acids. However, the presence of such an enzyme in the winter flounder and its possible role in processing of the proAFP has yet to be established.

The pre sequence in the flounder AFP is tightly coupled with the synthesis and secretion of the nascent chain and hence has a transient half-life. The presence of the proAFP, on the other hand, can be readily demonstrated using both in vitro and in vivo method during the winter months (Hew et al. 1978). It is apparent that the timing of the ability to synthesize the precursor correlates closely with the seasonal appearance of the serum AFP. Similarly, using liquid hybridization techniques to measure the AFP mRNA. Pickett et al. (1983a) demonstrated that the levels of AFP mRNA fluctuate seasonally. From November to mid-winter, 0.5% of the total liver RNA is AFP mRNA, and by late July, the level of this mRNA declines to 0.0007% of the total RNA (Table 2). While there is a detectable but minute amount of AFP mRNA in the summer, the abundance and the 700-fold increase of AFP mRNA in the winter would suggest that transcriptional control mechanisms play an important role in regulating AFP synthesis.

Table 2. Effect of hypophysectomy on the level of antifreeze polypeptide and RNA in the winter flounder[a]

	Freezing temperature in °C	Liver weight % Body weight	AFP determined by RIA	PolyA$^+$ RNA % Total RNA	AFP mRNA % Total RNA
Summer control (July)	−0.62°	0.90	0.07 mg ml^{-1}	0.50	0.0007
Hypophysectomized (July)	−1.22°	1.67	7.8 mg ml^{-1}	1.36–2.30	0.5
Winter control (November)	−1.21°	1.65	8.0 mg ml^{-1}	1.01–1.50	0.5

[a] The animals were hypophysectomized in April and the measurements were performed in July with seawater temperature 8°–10 °C

In order to accommodate the synthesis of these alanine-rich polypeptides, Picket et al. (1983b) have shown that in the liver of the winter fish, there is a 40% increase of alanine tRNA acceptor capacity over tRNA in the summer months. In contrast, the acceptor capacities for other amino acids show no seasonal differences. Furthermore, the alanyl tRNA synthetase functions best between 0° and 5 °C, which is the seawater temperature when antifreeze synthesis occurs, while prolyl and valyl tRNA synthetases are most active between 20° and 30 °C. These experiments demonstrate that in addition to the transcriptional regulation of the AFP mRNA level, the difference in temperature optima and the seasonal variation in tRNA levels and isoaccepting species may both serve to optimize AFP synthesis by increasing the translational efficiency of its mRNA.

6 Regulation of AFP Synthesis

All of our studies to date indicate that the annual cycle of AFP levels in winter flounder may be endogenous and regulated by photoperiod, temperature and the pituitary gland (Fletcher 1977; Petzel et al. 1980; Fletcher 1981).

The major environmental factor controlling the timing of the plasma AFP cycle appears to be photoperiod. Long day lengths (14 h or greater) or short nights in the fall delay the appearance (by 1 to 2 months) and result in a decreased concentration of AFP in the plasma (Fletcher 1981). In addition, the level of AFP mRNA in fish exposed to long day lengths (15 h light) is equally delayed and never reaches half of the mean value attained by control fish (Fourney et al. 1984b).

Water temperature does not appear to play any role in initiating or preventing the initiation of AFP synthesis in the fall, nor does it play a role in terminating AFP synthesis in the spring. It only appears to affect the rate of AFP degradation and clearance from the plasma (Fletcher 1981). Thus, flounder exposed to a normal photoperiod will synthesize AFP on schedule regardless of water temperature (0°–10 °C). In the spring (March), when AFP synthesis has ceased, the time at which plasma AFP disappear depends upon the water temperature. Unusually cold water temperature during the spring will result in a low AFP clearance rate and thus extend the time the fish are protected from freezing.

Hypophysectomy activates AFP mRNA production and AFP biosynthesis regardless of photoperiod and temperature (Table 2). The AFP which appear in the plasma of these hypophysectomized animals are identical to those observed in normal flounder during the winter (Fourney et al. 1984a). AFP synthesis can be prevented or stopped by intraperitoneal injections of pituitary extracts or by implanting pituitary glands under the skin (Fletcher et al. 1984b). The chemical nature of the pituitary inhibitor of AFP is unknown, although preliminary experiments indicate that it is a polypeptide. Our current hypothesis is that this pituitary AFP inhibitor is normally secreted during the summer months. With the approach of winter, the release of this inhibitor is suppressed, apparently via the central nervous system (Fletcher et al. 1984b) and consequently AFP mRNA transcription resumes. Consistent with this hypothesis, hypophysectomy mimics the suppression of the pituitary AFP inhibitor and the AFP genes are reactivated. The release of the pituitary factor is tuned to environmental conditions.

Long day lengths appear to allow the continued release of the AFP inhibitor into winter.

The seasonal appearance of insect antifreezes like those in the winter flounder is also influenced by environmental factors and endocrines. Insects, such as the beetle exposed to a 10 h photoperiod or less have significantly elevated antifreeze level over those maintained on a 11 h photoperiod or more at 20 °C (Horwath and Duman 1983a). Temperature appears to play a more subordinate role to that of the photoperiod response in controlling antifreeze production. On the other hand, both photoperiod and temperature are important in the spring clearance of antifreeze. Unlike the flounder AFP where the pituitary plays an inhibitory role in suppressing AFP synthesis, the juvenile hormone induces antifreeze production in the larvae of the beetle, *Dendroides canadensis* (Horwath and Duman 1983b). It will be interesting to compare these two systems in terms of mechanisms controlling AFP mRNA production.

7 Conclusion

Antifreeze glycoproteins and antifreeze polypeptides present an intriguing and important area for the study of protein structure, function and evolution. The structural information accumulated during the past few years should enable one to carry out further experimentation such as chemical modifications and to examine their mode of action in further detail. Biosynthetic studies provide a model for the study of environmental and endocrinological control of gene expression. Moreover, studies of these proteins and their genes may be of considerable practical value in such areas as salmon sea ranching and sea-pen culture and in the biological control over forestry pests such as the spruce budworm. Experiments are now in progress in our laboratories in attempting to improve the freezing tolerance of Atlantic salmon by gene transfer techniques.

Acknowledgment. The Research cited in this review was supported in part by operating grants from MRC and NSERC of Canada and by a strategic grant from NSERC, Canada.

References

Ananthanarayanan VS, Hew CL (1978) Structural studies on the freezing-point depressing protein and its synthetic analog. In: Srinavasan B (ed) Biomolecular structure, conformation and evolution. Pergamon, New York, 39:191–198

Blobel G, Dobberstein B (1975) Transfer of proteins across membranes II. Reconstitution of functional rough microsomes from heterologous components. J Cell Biol 67:852–862

Davies PL, Hew CL (1980) Isolation and characterization of the antifreeze protein messenger RNA from winter flounder. J Biol Chem 255:8729–8734

Davies PL, Roach AH, Hew CL (1982) DNA sequence coding for an antifreeze protein precursor from winter flounder. Proc Natl Acad Sci USA 79:335–339

Davies PL, Hough C, Scott GK, Ng N, White BN, Hew CL (1984) Antifreeze protein genes of the winter flounder. J Biol Chem 259:9241–9247

DeVries A (1982) Biological antifreeze agents in cold water fishes. Comp Biochem Physiol 73A: 627–640

DeVries AL, Lin Y (1977) Structure of a peptide antifreeze and mechanism of absorption to ice. Biochem Biophys Acta 495:380–392

DeVries A, Komatsu SK, Feeney RE (1970) Chemical and physical properties of freezing-point depressing glycoproteins from Arctic fishes. J Biol Chem 245:2901–2913

Duman JG (1982) Insect antifreeze and ice nucleating agents. Cryobiology 19:613–627

Duman JG, DeVries AL (1976) Isolation, characterization and physical properties of protein antifreezes from the winter flounder, *Pseudopleuronectes americanus*. Comp Biochem Physiol 53B: 375–380

Duman JG, Horwarth KL, Chancey A, Patterson JL (1982) Antifreeze agents of terrestrial anthropods. Comp Biochem Physiol 73A:545–555

Feeney RE, Yeh Y (1978) Antifreeze proteins from fish bloods. Adv Protein Chem 32:191–282

Fletcher GL (1977) Circannual cycles of blood plasma freezing point and Na^+ and Cl^- concentrations in Newfoundland winter flounder *Pseudopleuronectes americanus* correlated with water temperature and photoperiod. Can J Zool 55:789–795

Fletcher GL (1981) Effects of temperature and photoperiod on the plasma freezing point depression. Cl^- concentration and protein "antifreeze" in winter flounder. Can J Zool 59:193–201

Fletcher GL, Hew CL, Joshi SB (1982) Isolation and characterization of antifreeze glycoproteins from the frost fish, *Microgadus tomcod*. Can J Zool 60:348–355

Fletcher GL, Hew CL, Li XM, Haya K, Kao MH (1984a) Year-round presence of high levels of plasma antifreeze peptides in a temperate fish, Ocean Pout. *Macrozoarces americanus*. Can J Zool (in press)

Fletcher GL, King MT, Hew CL (1984b) How does the brain control the pituitary's release of antifreeze synthesis inhibition? Can J Zool 61:839–844

Fourney RM, Fletcher GL, Hew CL (1984a) Accumulation of winter flounder antifreeze messenger RNA after hypophysectomy. Gen Comp Endocrinol 54:392–401

Fourney RM, Fletcher GL, Hew CL (1984b) The effect of long day length on liver antifreeze mRNA in the winter flounder. Can J Zool 62:1456–1460

Fourney RM, Hew CL, Joshi SB, Fletcher GL (1984c) Comparison of antifreeze polypeptides from Newfoundland, Nova Scotia, New Brunswick and Long Island winter flounder. Comp Biochem Physiol 180:191–196

Fourney RM, Hew CL, Joshi SB, Kao MS (1984d) The heterogeneity of antifreeze polypeptides from the winter flounder, *Pseudopleuronectes americanus*. Can J Zool 62:28–23

Hew CL, Yip C (1976) The synthesis of freezing piont-depression protein of the winter flounder *Pseudopleuronectes americanus* in *Xenopus laevis* oocytes. Biochem Biophys Res Commun 71: 845–850

Hew CL, Liunardo N, Fletcher GL (1978) In vivo biosynthesis of the antifreeze protein in the winter flounder – evidence for a larger precursor. Biochem Biophys Res Commun 85:421–427

Hew CL, Fletcher GL, Ananthanarayanan VS (1980) Antifreeze protein from the shorthorn sculpin, *Myoxocephalus scorpius*: Isolation and characterization. Can J Biochem 58:377–383

Hew CL, Kao MH, So YP, Lim KP (1983) Presence of cystinecontaining antifreeze proteins in the spruce budworm. *Chorisoneura fumiferana*. Can J Zool 61:2324–2328

Hew CL, Joshi S, Wang NC (1984a) Analysis of fish antifreeze polypeptides and proteins by reverse phase HPLC. J Chromatogr 296:213–219

Hew CL, Joshi S, Wang NC, Kao MH, Ananthanarayanan VS (1984b) Primary and secondary structures of antifreeze polypeptides in the shorthorn sculpin. *Myoxocephalus scorpius*. Eur J Biochem (accepted)

Hew CL, Slaughter D, Joshi SB, Fletcher GL, Ananthanarayanan VS (1984c) Antifreeze polypeptides from the Newfoundland ocean pout. *Macrozoaces americanus*: presence of multiple and compositionally diverse components. J Comp Physiol 155:81–88

Holmes WN, Donaldson EM (1969) The body compartments and distribution of electrolytes. In: Hoar WS, Randall DJ (eds) Fish physiology, vol 1. Academic, New York, pp 1–83

Horwarth KL, Duman JG (1983a) Photoperiodic and thermal regulation of antifreeze protein levels in the beetle *Dendroides canadensis*. J Insect Physiol 29:907–917

Horwath KL, Duman JG (1983b) Induction of antifreeze protein production by Juvenile hormone in Larvae of the beetle, *Dendroides canadensis*. J Comp Physiol 151:233–240

Kreil G, Haime L, Sucharek G (1980) Stepwise cleavage of the pro part of promelittin by dipeptidylpeptidase IV. Eur J Biochem 111:49–58

Lin Y (1979) Environmental regulation of gene expression. J Biol Chem 254:1422–1426

Petzel D, Reixman H, DeVries AL (1980) Seasonal variation of antifreeze peptide in the winter flounder, *Pseudopleuronectes americanus*. J Exp Zool 211:63–69

Pickett MH, Hew CL, Davies PL (1983a) Seasonal variation in the levels of antifreeze protein mRNA from the winter flounder. Biochem Biophys Acta 739:97–104

Pickett MH, White BN, Davies P (1983b) Evidence that translational control mechanisms operate to optimize antifreeze protein production in the winter flounder. J Biol Chem 258:14762–14765

Picket MH, Scott G, Davies P, Wang N, Joshi S, Hew CL (1984) Sequence of an antifreeze protein precursor. Eur J Biochem 143:35–38

Schneppenheim R, Theede H (1980) Isolation and characterization of freezing point depressing peptides from larvae of *Tenebrio molitor*. Comp Biochem Physiol 67B:561–568

Scholander PF, VanDane L, Kanwisher JW, Hammel HT, Gordon MS (1957) Supercooling and osmoregulation in arctic fish. J Cell Comp Physiol 49:5–24

Slaughter D, Hew CL (1982) Radioimmunoassay for the antifreeze proteins of the winter flounder: Seasonal profile and immunological cross-reactivity with other fish antifreeze proteins. Can J Biochem 60:824–829

Slaughter D, Fletcher GL, Ananthanarayanan VS, Hew CL (1981) Antifreeze proteins from the sea raven, *Hemitripterus americanus*: Further evidence for diversity among fish polypeptide antifreeze. J Biol Chem 256:2022–2026

Subject Index

MIX
Papier aus verantwortungsvollen Quellen
Paper from responsible sources
FSC® C105338

If you have any concerns about our products,
you can contact us on
ProductSafety@springernature.com

In case Publisher is established outside the EU,
the EU authorized representative is:
Springer Nature Customer Service Center GmbH
Europaplatz 3, 69115 Heidelberg, Germany

Printed by Libri Plureos GmbH
in Hamburg, Germany